D1608928

Encyclopedia of
TOXICOLOGY

Volume 1   **A–E**

Encyclopedia of
TOXICOLOGY

Volume 1   **A—E**

Editor-in-Chief

Philip Wexler

Toxicology and Environmental Health Information Program
National Library of Medicine
National Institutes of Health
Bethesda, Maryland

Academic Press

San Diego   London   Boston   New York   Sydney   Tokyo   Toronto

The following articles are U. S. government works in the public domain:

Carbamate Pesticides
Combustion Toxicology
Medical Surveillance
Nerve Gases
Nitrogen Mustards
Organotins
Risk Assessment, Ecological
Sarin
Soman
Sulfur Mustard
Tabun
VX

This book is printed on acid-free paper. ∞

Academic Press
a division of Harcourt Brace & Company
525 B Street, Suite 1900, San Diego, California 92101-4495, USA
http://www.apnet.com

Academic Press Limited
24-28 Oval Road, London NW1 7DX, UK
http://www.hbuk.co.uk/ap/

Library of Congress Card Catalog Number: 98-84114

International Standard Book Number: 0-12-227220-X (set)
International Standard Book Number: 0-12-227221-8 (volume 1)
International Standard Book Number: 0-12-227222-6 (volume 2)
International Standard Book Number: 0-12-227223-4 (volume 3)

PRINTED IN THE UNITED STATES OF AMERICA
98 99 00 01 02 03 MM 9 8 7 6 5 4 3 2 1

For my parents, Yetty and Will,
My wife, Susan,
and my son, Jacob
—with love.

Contents

Volume 1

Guide to Encyclopedia   xix
Foreword   xxiii
Preface   xxxv

A

Absorption   1
Academy of Toxicological Sciences   7
ACE Inhibitors   7
Acenaphthene   9
Acetaldehyde   10
Acetaminophen   11
Acetic Acid   13
Acetone   14
Acetonitrile   16
Acetylaminofluorene   17
Acetylcholine   18
Acetylene   20
Acetylsalicylic Acid   21
Acids   22
Aconitum Species   23
Acrolein   24
Acrylamide   25
Acrylic Acid   26
Acrylonitrile   27
Adiponitrile   28

Aerosols   30
A-esterase   30
Aflatoxin   33
Agency for Toxic Substances and
    Disease Registry   34
Alachlor   35
Alar   36
Albuterol   37
Aldicarb   38
Aldrin   39
Algae   40
Alkalies   41
Alkyl Halides   42
α-Methylfentanyl   44
α-Naphthyl Thiourea   45
Aluminum (Al)   46
Aluminum Phosphide   48
Amdro   49
American Academy of
    Clinical Toxicology   50
American Association of Poison
    Control Centers   50
American Board of Toxicology   51
American College of
    Medical Toxicology   53
American College of Toxicology   53
American Conference of Governmental
    Industrial Hygienists   55

Ames Test   56   Benadryl   130
Aminoglycosides   58   Benomyl   131
Amiodarone   59   Benz[a]anthracene   132
Ammonia   60   Benzene   133
Ammonium Nitrate   61   Benzene Hexachloride   134
Ammonium Perchlorate   62   Benzidine   136
Amphetamine   63   Benzo[a]pyrene   137
Amphibians   64   Benzodiazepines   139
Amyl Nitrate   65   Benzyl Alcohol   141
Analytical Toxicology   66   Benzyl Benzoate   143
Androgens   69   Beryllium (Be)   144
Animal Models   70   Beta Blockers   145
Anthracene   73   Biguanides   146
Anticholinergics   74   Bioaccumulation   148
Antimony (Sb)   76   Bioconcentration   151
Anxiolytics   77   Biomagnification   152
Aramite   79   Biomarkers, Environmental   153
Arsenic (As)   80   Biomarkers, Human Health   154
Arum   82   Biotransformation   159
Asbestos   82   Bismuth (Bi)   174
Ascorbic Acid   84   Bleach   174
Aspartame   85   Blood   176
Astemizole   86   Boric Acid   182
Atrazine   87   Boron (B)   183
Atropine   87   Botulinum Toxin   184
Avermectin   89   Brodifacoum   185
Azinphos-Methyl   90   Bromine   187
                                                  Bromobenzene   188
                                                  Bromoform   189
                  B   Buckthorn   190
                                                  Butadiene, 1,3-   191
Bacillus cereus   93   Butane   192
BAL   93   Butter Yellow   193
Baneberry   94   Butyl Ether   194
Barbiturates, Long-Acting   95   Butyl Ethyl Ether   195
Barbiturates, Short-Acting   97   Butyl Methyl Ether   196
Barium (Ba)   99   Butyl Nitrite   196
Baygon   100   Butylamines   197
BCNU   101   Butylated Hydroxyanisole   198
Behavioral Toxicology   102   Butylated Hydroxytoluene   199
Belladonna Alkaloids   128

Butyraldehyde, *n*-   201
Butyric Acid   202
Butyronitrile   203
Butyrophenones   203

C

Cadmium (Cd)   207
Caffeine   209
Calcium Channel Blockers   211
Calomel   212
Camphor   213
Captafol   214
Captan   215
Carbamate Pesticides   216
Carbamazepine   218
Carbaryl   219
Carbofuran   221
Carbon Dioxide   222
Carbon Disulfide   223
Carbon Monoxide   224
Carbon Tetrabromide   226
Carbon Tetrachloride   227
Carbonyl Sulfide   228
Carboxylesterases   229
Carboxylic Acids   232
Carcinogen Classification Schemes   232
Carcinogen–DNA Adduct Formation and DNA Repair   237
Carcinogenesis   246
Cardiovascular System   268
Castor Bean   288
Catecholamines   288
Cell Proliferation   290
Centipedes   295
Cephalosporins   295
Charcoal   296
Chemical Industry Institute of Toxicology   297
Chloral Hydrate   298

Chloramphenicol   299
Chlorbenzilate   300
Chlordane   301
Chlordecone   303
Chlordimeform   304
Chlorine   306
Chlorine Dioxide   307
Chlorobenzene   308
Chloroform   310
Chloromethyl Ether, bis-   311
Chlorophenols   312
Chlorophenoxy Herbicides   313
Chloropicrin   315
Chloroquine   316
Chlorothalonil   317
Chlorpheniramine   318
Chlorpromazine   319
Chlorpyrifos   322
Chlorzoxazone   324
Cholesterol   325
Choline   325
Cholinesterase Inhibition   326
Chromium (Cr)   340
Chromosome Aberrations   342
Chrysene   344
Ciguatoxin   346
Cimetidine   346
Cisplatin   347
Clean Air Act   349
Clean Water Act   350
Clofibrate   351
Clonidine   352
Clostridium perfringens   353
Cobalt (Co)   353
Cocaine   355
Codeine   356
Coke Oven Emissions   357
Colchicine   358
Combustion Toxicology   360
Comprehensive Environmental Response, Compensation, and Liability Act   374

Coniine	375	Dibromochloropropane	458
Consumer Product Safety Commission	376	Dibutyl Phthalate	459
Copper (Cu)	376	Dicamba	460
Corrosives	378	Dichlone	461
Corticosteroids	379	Dicholorobenzene	461
Cosmetics	380	Dichloroethanes	462
Cotinine	382	Dichloroethylene, 1,1-	463
Coumarins	383	Dichloroethylene, 1,2-	465
Creosote	384	Dichloropropene, 1,3-	467
Cromolyn	385	Dichlorvos	468
Cyanamide	386	Dieldrin	469
Cyanide	387	Diesel Exhaust	470
Cyclodienes	389	Diesel Fuel	471
Cyclohexane	390	Diethyl Ether	474
Cyclohexene	391	Diethylamine	475
Cyclophosphamide	392	Diethylene Glycol	475
Cypermethrin	393	Diethylstilbestrol	477
Cysteine	395	Diflubenzuron	478
		Difluoroethylene, 1,1-	478
		Digitalis Glycosides	479
		Dimethoate	481

D

		Dimethyl Sulfoxide	483
2,4-D	397	Dinitroanilines	484
Dalapon	398	Dinitrophenols	485
DDT	399	Dinitrotoluene	487
Decane	400	Dinoseb	489
DEET	401	Dioctylphthalate	490
DEF	402	Dioxane, 1,4-	492
Deferoxamine	403	Dioxins	494
DEHP	404	Diphenhydramine	496
Delaney Clause	405	Diphenoxylate	497
Deltamethrin	405	Diphenylhydrazine	498
Deodorants	406	Disc Batteries	499
Detergent	407	Distribution	500
Developmental Toxicology	408	Disulfiram	503
Dextromethorphan	453	Dithiocarbamates	504
Diazinon	454	Diuron	506
Diazoxide	455	Dominant Lethal Tests	506
Dibenz[a,h]anthracene	456	Dose–Response Relationship	508
Dibenzofuran	457		

E

E. coli 515

Ecological Toxicology 515

Ecological Toxicology, Experimental Methods 516

EDTA 520

Effluent Biomonitoring 521

Emergency Response 527

Endocrine System 531

Endrin 535

Environmental Hormone Disruptors 536

Environmental Processes 538

Environmental Protection Agency, National Health and Environmental Effects Research Laboratory 540

Environmental Toxicology 541

Eosinophilia–Myalgia Syndrome 541

Epichlorohydrin 554

Epidemiology 555

Ergot 560

Erythromycin 561

Ethane 562

Ethanol 563

Ethanolamine 565

Ethchlorvynol 566

Ethene 567

Ethyl Acetate 568

Ethyl Acrylate 569

Ethylamine 570

Ethylbenzene 571

Ethyl Bromide 572

Ethylene Glycol 573

Ethylene Glycol Monoethyl Ether 575

Ethylene Glycol Mono-*n*-butyl Ether 576

Ethylene Imine 577

Ethylene Oxide 578

European Commission 579

European Society of Toxicology 584

Excretion 585

Exposure 588

Exposure Assessment 592

Exposure Criteria 594

Eye Irritancy Testing 598

Volume 2

F

Federal Insecticide, Fungicide and Rodenticide Act 1

Fentanyl 2

Fentanyl Derivatives, Illicit 3

Fenthion 4

Fenvalerate 5

Fluorine 6

Fluoxetine 7

Folic Acid 8

Folpet 9

Food Additives 10

Food and Agriculture Organization of the United Nations 20

Food and Drug Administration 25

Food, Drug, and Cosmetic Act 27

Forensic Toxicology 28

Formaldehyde 33

Formamide 34

Formic Acid 35

Foxglove 36

Freons 37

Furan 39

Furfural 40

G

Gallium (Ga) 43

Gastrointestinal System 44

Generally Recognized As Safe (GRAS) 51

Ginger Jake 52

Ginseng 52
Glutathione 53
Glutethimide 54
Glyceraldehyde 55
Glycerol 56
Glycol Ethers 57
Glyphosate 60
Gold (Au) 61
Good Laboratory Practices 61
Guaifenesin 65

H

Hazard Identification 67
Hazardous Waste 70
Helium 73
Hemlock, Poison 73
Hemlock, Water 74
Heparin 74
Heptachlor 76
Heptane 77
Heptanone 78
Heroin 79
Hexachlorobutadiene 81
Hexachlorocyclopentadiene 82
Hexachlorophene 83
Hexane 85
Holly 88
Hormesis 88
Host-Mediated Assay 92
Hydrangea 94
Hydrazine 95
Hydrobromic Acid 96
Hydrochloric Acid 97
Hydrocodone 98
Hydrofluoric Acid 100
Hydrogen Peroxide 101
Hydrogen Sulfide 102
Hydroiodic Acid 105
Hydromorphone 105

Hydroperoxides, Organic 107
Hydroxylamine 107
Hymenoptera 108
Hypersensitivity, Delayed-Type 109
Hypoglycemics, Oral 110

I

Ibuprofen 113
Immune System 114
Indole 139
Indoor Air Pollution 140
Information Resources in Toxicology 148
International Agency for Research on Cancer 170
International Life Sciences Institute—North America 173
International Programme on Chemical Safety 173
International Register of Potentially Toxic Chemicals 177
International Union of Toxicology 180
Investigative New Drug Application 182
In Vitro Test 183
In Vivo Test 184
Iodine 186
Iron (Fe) 187
Isocyanates 189
Isoniazid 189
Isoprene 190
Isopropanol 191

J

Jequirity Bean 195
Jimsonweed 196

K

Kerosene 197
Kidney 198

L

LD$_{50}$/LC$_{50}$	227
Lead (Pb)	230
Levels of Effect in Toxicological Assessment	233
Levothyroxine	234
Lidocaine	235
Life Cycle Assessment	237
Lily of the Valley	244
Limonene	245
Lindane	245
Liothyronine	246
Lipid Peroxidation	247
Lithium (Li)	253
Liver	253
Loperamide	262
Loxapine	262
LSD	264
Lye	265
Lyme Disease	266

M

Malathion	269
Mancozeb	270
Manganese (Mn)	271
Marijuana	272
Maximum Allowable Concentration	274
Maximum Tolerated Dose	274
Mechanisms of Toxicity	275
Medical Surveillance	279
Meperidine	283
Meprobamate	285
Mercaptoethanol, 2-	286
Mercapturic Acid	287
Mercury (Hg)	288
Mescaline	289
Metallothionein	290
Metals	291
Methadone	292

Methamidophos	293
Methane	294
Methanol	295
Methaqualone	297
Methomyl	298
Methoprene	299
Methoxychlor	299
Methoxyethanol	301
Methoxypsoralen	302
Methyl Acrylate	302
Methylamine	303
Methyl Bromide	304
Methylcholanthrene, 3-	305
Methyl Disulfide	306
Methyldopa	307
Methylene Chloride	308
Methyl Ethyl Ketone	310
Methylenedioxymethamphetamine	311
Methyl Isobutyl Ketone	312
Methylmercury	314
Methylnitrosourea	316
Methyl Parathion	317
Methyl *t*-Butyl Ether	318
Methyprylon	319
Metronidazole	320
Mevinphos	321
Microtox	322
Minoxidil	324
Mistletoe	325
Mitomycin C	326
Mixtures	326
Modifying Factors of Toxicity	327
Molecular Toxicology—Recombinant DNA Technology	335
Molybdenum (Mo)	343
Monoamine Oxidase Inhibitors	344
Monosodium Glutamate	345
Morning Glory	346
Morphine	347
Mouse Lymphoma Assay	348

Mouthwash 350
Multiple Chemical Sensitivities 351
Mushrooms, Coprine 354
Mushrooms, Cyclopeptide 355
Mushrooms, Ibotenic Acid 357
Mushrooms, Monomethylhydrazine 358
Mushrooms, Muscarine 359
Mushrooms, Psilocybin 360
Mustard Gas 362
Mutagenesis 363
Mycotoxins 369

N

Naphthalene 371
Naphthylisothiocyanate, 1- 373
National Center for
   Toxicological Research 373
National Environmental Policy Act 374
National Institute for Occupational
   Safety and Health 375
National Institute of Environmental
   Health Sciences 376
National Institutes of Health 377
National Library of Medicine/TEHIP 378
National Toxicology Program 378
Nematocides 380
Nerve Gases 380
Neurotoxicity, Delayed 385
Neurotoxicology: Central and Peripheral 389
Niacin 414
Nickel (Ni) 415
Nicotine 417
Nitric Oxide 418
Nitrite Inhalants 419
Nitrites 420
Nitrobenzene 421
Nitroethane 423
Nitrogen Mustards 424
Nitrogen Tetraoxide 426
Nitromethane 427

Nitrosamines 428
N-Nitrosodimethylamine 429
Nitrous Oxide 430
Noise: Ototraumatic Effects 431
Norbormide 438
Nutmeg 439

O

Occupational Safety and Health Act 441
Occupational Safety and
   Health Administration 442
Occupational Toxicology 442
Octane 450
Oil, Crude 451
Oil, Lubricating 453
Oleander 454
Opium 455
Organisation for Economic
   Co-operation and Development 456
Organochlorine Insecticides 464
Organophosphate Poisoning,
   Intermediate Syndrome 465
Organophosphates 467
Organotins 471
Oxygen 472
Ozone 473

P

Paraquat 475
Parathion 476
Paregoric 478
Penicillin 478
Pentachloronitrobenzene 479
Pentachlorophenol 480
Pentane 481
Pentazocine 482
Perchloric Acid 483
Periodic Acid 484
Permethrin 484

Permissible Exposure Limit	485	Potassium (K)	584
Peroxisome Proliferators	486	Primidone	585
Pesticides	494	Procainamide	586
Petroleum Distillates	495	Prometryn	588
Petroleum Ether	496	Propachlor	588
Petroleum Hydrocarbons	497	Propane	589
Peyote	500	Propanil	590
Pharmacokinetic Models	501	Propargite	591
Pharmacokinetics/Toxicokinetics	503	Propazine	591
Phenanthrene	512	Propene	592
Phencyclidine	512	Propionic Acid	593
Phenol	514	Proposition 65	594
Phenothiazines	515	Propoxur	594
Phenylmercuric Acetate	516	Propoxyphene	595
Phenylpropanolamine	518	Prostaglandins	596
Phenytoin	519	Prunus Species	597
Phorbol Esters	520	Pseudophedrine	598
Phosgene	522	Psychological Indices of Toxicity	599
Phosphoric Acid	523	Pyrethrin/Pyrethroids	610
Photoallergens	523	Pyridine	611
Photochemical Oxidants	528	Pyridoxine	612
Phthalate Esters Plasticizers	529	Pyriminil	613
Picloram	530		
Picric Acid	531		
Piperonyl Butoxide	532		
Platinum (Pt)	533		
Poinsettia	534		
Poisoning Emergencies in Humans	534		
Pokeweed	550		
Pollution, Air	551		
Pollution Prevention Act	559		
Pollution, Soil	560		
Pollution, Water	566		
Polybrominated Biphenyls	573		
Polychlorinated Biphenyls	574		
Polycyclic Aromatic Amines	576		
Polycyclic Aromatic Hydrocarbons	577		
Polyethylene Glycol	577		
Polymers	578		
Population Density	580		

V o l u m e   3

Q

Quartz	1
Quinidine	2
Quinine	3

R

Radiation Toxicology	5
Radon	19
Ranitidine	20
Red Dye No. 2	21
Red Tide	22
Reproductive System, Female	23

Reproductive System, Male   38
Reserpine   49
Resistance to Toxicants   50
Resource Conservation and
    Recovery Act   51
Respiratory Tract   52
Rhododendron Genus   86
Rhubarb   87
Riboflavin   87
Rifampin   88
Risk Assessment, Ecological   89
Risk Assessment, Human Health   103
Risk Characterization   109
Risk Communication   110
Risk Management   113
Rotenone   114

S

Saccharin   117
Safe Drinking Water Act   118
Salicylates   118
Salmonella   120
Sarin   121
Saxitoxin   124
Scombroid   125
Scorpions   126
Selenium (Se)   127
Sensitivity Analysis   128
Sensory Organs   130
Sertraline Hydrochloride   135
Shampoo   136
Shellfish Poisoning, Paralytic   137
Shigella   138
Short-Term Exposure Limit   138
Sick Building Syndrome   139
Silver (Ag)   144
Sister Chromatid Exchanges   144
Skeletal System   146
Skin   151

Snake, Crotalidae   179
Snake, Elapidae   180
Society of Environmental Toxicology
    and Chemistry   182
Society of Toxicology   183
Sodium (Na)   185
Sodium Fluoroacetate   186
Sodium Sulfite   187
Solanum Genus   188
Soman   189
Speed   192
Spider, Black Widow   193
Spider, Brown Recluse   194
Staphylococcus aureus   196
State Regulation of Consumer Products   197
Structure–Activity Relationships   198
Strychnine   201
Styrene   202
Sudan Grass   203
Sulfites   205
Sulfur Dioxide   205
Sulfur Mustard   206
Surfactants, Anionic and Nonionic   209

T

2,4,5-T   211
Tabun   213
TCDD   215
Tellurium (Te)   216
Terbutaline   217
Terfenadine   218
Tetrachloroethane   219
Tetrachloroethylene   220
Tetrachlorophenoxyacetic Acid   222
Tetrahydrofuran   222
Tetranitromethane   224
Tetrodotoxin   225
Thalidomide   226
Thallium (Tl)   227

Theophylline 228

Thiamine 230

Thiazide Diuretics 230

Thioacetamide 232

Thioxanthenes 232

Thiram 234

Threshold Limit Value 235

Thyroid Extract 236

Tin (Sn) 237

Tissue Repair 238

Titanium (Ti) 244

Tobacco 245

Tobacco Smoke 247

Toluene 248

Toluene Diisocyanate 250

Toluidine 251

Toxicity, Acute 252

Toxicity, Chronic 259

Toxicity, Subchronic 264

Toxicity Testing, Alternatives 269

Toxicity Testing, Aquatic 273

Toxicity Testing, Behavioral 282

Toxicity Testing, Carcinogenesis 289

Toxicity Testing, Dermal 293

Toxicity Testing, Developmental 305

Toxicity Testing, Inhalation 319

Toxicity Testing, Reproductive 328

Toxicology 338

Toxicology, Education and Careers 341

Toxicology, History of 347

Toxic Substances Control Act 354

Toxic Torts 356

Triadimefon 371

Trichloroethanes 372

Trichloroethylene 372

Tricyclic Antidepressants 375

Trihalomethanes 376

Trinitrotoluene 377

U

Uncertainty Analysis 379

Uranium (U) 381

Urethane 382

V

Valproic Acid 385

Vanadium 386

Vanillin 387

Veterinary Toxicology 388

Vinyl Chloride 403

Vitamin A 404

Vitamin D 405

Vitamin E 406

VX 407

W

Warfarin 411

Wisteria 412

X

Xylene 415

Y

Yew 419

Yohimbine 420

Z

Zinc (Zn) 423

Glossary of Key Terms 425

Contributors 433

Index 441

Guide to the Encyclopedia

The *Encyclopedia of Toxicology* is a comprehensive reference work providing information on the adverse effects of chemicals on biological systems. It offers a cogent overview of each selected topic to inform a broad spectrum of readers, ranging from research professionals and medical practitioners to students to the interested general public.

In order that you, the reader, will derive maximum benefit from your use of the Encyclopedia, we have provided this Guide. It explains how the Encyclopedia is organized and how the information within it can be located.

Content

Within the covers of a single self-contained work, the *Encyclopedia of Toxicology* gives complete scientific descriptions of substances that are harmful, or potentially harmful, to living organisms. It also describes the actions, effects, detection, and treatment of these substances. The substances discussed here run the gamut from poisons that were well known to the ancients (e.g., Hemlock) to products that are of concern to modern consumers (e.g., Red Dye No. 2).

As a crucial meeting point of chemistry and biology, the discipline of toxicology encompasses not only the chemical substances themselves (e.g., Ammonia, Asbestos, Diazinon, Mustard Gas) but also the biological processes and mechanisms of toxicity (e.g., Absorption, Bioconcentration, Excretion). Beyond this, the Encyclopedia also includes entries on a wide variety of other topics, including body organs and systems (e.g., Liver, Respiratory Tract), toxic organisms (Mushrooms, Spiders), environmental issues (Diesel Exhaust, Red Tide), concepts (Median Lethal Dose), procedures (Ames test), and organizations (Society of Toxicology).

Organization

The *Encyclopedia of Toxicology* is organized to provide the maximum ease of use for its readers. It contains more than 750 separate articles, all arranged in a single alphabetical sequence according to title. Thus the Encyclopedia begins with the entry "Absorption" and ends with "Zinc."

The A-to-Z text of the *Encyclopedia of Toxicology* is presented in three separate volumes. In addition, Volume 3 also includes the Subject Index for the entire work, as well as a Glossary of Key Terms.

So that they can be easily located, article titles generally begin with the key word or phrase indicating the topic, with any descriptive terms following. For example, "Barbiturates, Short-Acting" is the article title rather than "Short-Acting Barbiturates," because the technical term *Barbiturates* is the key word rather than the adjective *Short-Acting*. Similarly, "Toxicity Testing, Aquatic" is the article title, not "Aquatic Toxicity Testing." This technique also allows the grouping of related articles that can be usefully consulted together; in this case, a series of articles on toxicity testing.

Table of Contents

A complete table of contents for the *Encyclopedia of Toxicology* appears at the front of each volume. Thus it is possible for a reader to consult the entire article list for the Encyclopedia within any of the volumes. This list of article titles represents topics that have been carefully selected by the Editor-in-Chief, Philip Wexler, and the Associate Editors (see p. ii for a list of these editors).

Article Format

An important feature of the *Encyclopedia of Toxicology* is the structure in which chemical entries are presented. The text of an entry, regardless of the particular topic being addressed, will follow a standard sequence, with uniform headings. The purpose of this consistent format is to enable a reader to locate information quickly and easily within the article.

For example, any article on a substance will present the same categories of information about it, such as its uses and its toxic effects on humans. Each category will appear in the same place in the article, under the same prominent heading. The following is the standard sequence for an article on a toxic substance:

- CAS identifying number
- other names (e.g., acetaminophen/Tylenol)
- pharmaceutical class
- chemical structure
- uses
- exposure pathways
- toxicokinetics
- mechanism of toxicity
- human toxicity
- clinical management
- animal toxicity

Index

The Subject Index in Volume 3 contains more than 6,000 entries. The entries are listed alphabetically and

indicate the volume and page number where the information on this topic will be found. Within the alphabetical entry for a given topic, references to its subtopics also appear alphabetically.

Cross-References

Most articles in the *Encyclopedia of Toxicology* have cross-references to other articles. These references appear as a list below the end of the text for the article, under the heading "Related Topics." The references indicate related entries that can be consulted for further understanding of the given topic or for more information about a similar topic.

For example, the article "Bioaccumulation" (the accumulation in the body of industrial chemicals from the environment) has references to Bioconcentration, Biomagnification, and Biotransformation, which are articles describing similar processes of environmental absorption. It also has cross-references to Ecological Toxicology, Effluent Monitoring, Environmental Toxicology, Pesticides, and Water Pollution, articles that discuss the sources and effects of bioaccumulation.

Bibliography

In addition to the cross-references, many articles in the *Encyclopedia of Toxicology* contain references to other sources outside the Encyclopedia itself. These appear as a Further Reading section at the end of the article. This section lists recent sources that will aid the reader in locating more detailed information about the topic.

The Further Reading feature is for the benefit of the reader. It is not intended to represent a complete listing of all materials consulted by the author or authors in preparing the article.

Glossary

So that this Encyclopedia may be useful to as wide a readership as possible, a Glossary of Key Terms has been provided in Volume 3, preceding the Index. For the purpose of the article text, it is important to use the technical vocabulary of the science of toxicology, in the interest of accuracy, brevity, and consistency. However, some of these technical terms may not be entirely familiar to the nonprofessional reader. Therefore they are defined in the Glossary in relatively simple terminology. The Glossary section includes approximately 300 entries.

Companion Works

The *Encyclopedia of Toxicology* is one component of an extensive, ongoing series of multivolume reference works in the life sciences published by Academic Press. Other such works include the *Encyclopedia of Human Biology, Encyclopedia of Cancer, Encyclopedia of Immunology, Encyclopedia of Virology,* and *Encyclopedia of Microbiology,* as well as the forthcoming *Encyclopedia of Reproduction.*

Foreword

I am delighted to have the opportunity to introduce the *Encyclopedia of Toxicology* to its readers. It is an excellent work that deserves to be found on many desks and bookshelves.

During the past four decades, toxicology has developed more rapidly than any other science in a comparable period of time. From a discipline largely descriptive of deleterious effects on intact animals, toxicology now embraces all aspects of modern biology from molecular biology to highly sophisticated instrumental analysis. The philosophical basis has shifted from routine risk analysis based primarily on mortality or pathological endpoints to an analysis of mechanisms of toxic action and the development of new paradigms of risk assessment that include mechanistic considerations. All these changes have not been brought about without considerable increases in the numbers of practicing toxicologists and training programs and in research funding. More important in the current context, however, is the explosive growth in toxicological literature.

Although specialists must still utilize the original, peer-reviewed, research literature to keep abreast of developments in their own specialized fields, other means of communication must be utilized in keeping up with general trends. Public interest in toxic chemicals and concern about their toxic effects has generated much controversy and much regulation at the national, state, and local levels. Although this interest has been instrumental in provoking much of the growth in toxicology as a science, it has also created the need for many more generalists than are to be found in other sciences.

The toxicological literature, therefore, must serve many masters. General works are needed at several levels of detail, from the dictionary with short entries to multivolume monographs. Works such as the *Encyclopedia of Toxicology* play a critical role at the intermediate level of this spectrum, offering more detail than dictionaries while remaining accessible to the generalist in risk assessment, regulation, teaching, consulting, and other activities. The encyclopedia is equally valuable to specialists seeking information outside the confines of their own narrow areas of specialization. The selection of topics is excellent and the contributors are highly qualified. This is an outstanding contribution that no serious toxicologist nor library serving toxicologists can afford to be without.

Ernest Hodgson
William Neal Reynolds Professor of Toxicology
North Carolina State University

Preface

There are many fine general and specialized monographs on toxicology, most of which are addressed to toxicologists and students in the field and a few to laypeople. This encyclopedia of toxicology does not presume to replace any of them but rather is intended to fulfill the toxicology information needs of new audiences by taking a different organizational approach and assuming a middle ground in the level of presentation by borrowing elements of both primer and treatise.

The encyclopedia is broad ranging in scope, although it does not aspire to be exhaustive. The idea was to look at basic, critical, and controversial elements in toxicology, which are those elements that are essential to an understanding of the subject's scientific underpinnings and societal ramifications. As such, the encyclopedia had to cover not only key concepts, such as dose response, mechanism of action, testing procedures, endpoint responses, and target sites, but also individual chemicals and classes of chemicals. Despite the strong chemical emphasis of the book, we had to look at concepts such as radiation and noise, and beyond the emphasis on the science of toxicology, we had to look at history, laws, regulation, education, organizations, and databases. The encyclopedia also needed to consider environmental and ecological toxicology to somewhat counterbalance the acknowledged emphasis on laboratory animals and humans because, in the end, all our connections run deep.

In terms of the chemicals, we the editors of this book made a personal selection based on our own knowledge of those with relatively high toxicity, exposure, production, controversy, newsworthiness, or other interest. The chemicals do not represent a merger of regulatory lists or databases of chemicals: They are what we consider to be, for one reason or another, chemicals of concern to toxicology. The book was not intended as a large-scale compendium of toxic chemicals, several of which already exist.

In the tradition of many standard encyclopedias, scientific and otherwise, the encyclopedia is organized entirely alphabetically. Other than in a few useful but smaller scale dictionaries, this style of arrangement has not been done before for toxicology. This organization, along with a detailed index and extensive cross-references, should help the reader quickly arrive at the needed information.

Next, although this book should be of use to the practicing toxicologist, it is geared more to others who, in the course of their work, study, or for general interest, need to know about toxicology. This would include the scientific community in general, physicians, legal and regulatory professionals, and laypeople with some scientific background. Toxicologists needing to brush up on or get a quick review of a subject other than their own specialty would also benefit from it, but toxicologists seeking an in-depth treatment should instead consult a specialized monograph or the journal literature.

The encyclopedia is meant to give relatively succinct overviews of sometimes very complex subjects. Formal references and footnotes were dispensed with because these seemed less relevant to the encyclopedia's goals than a simple list of recommended readings designed to lead the reader to more detailed information on a particular subject entry. The entry on Information Resources leads readers to print and electronic sources of information in toxicology.

First and foremost, thanks go to the Associate Editors and contributors, whose efforts are here in print. Yale Altman and Linda Marshall, earlier Aquisitions Editors for the books, were of great assistance in getting the project off the ground. Tari Paschall, the current Aquisitions Editor, and Monique Larson, Senior Production Editor, both of Academic Press, have with great expertise and efficiency brought it to fruition. Organization and formatting of the original entry manuscripts were handled with skill, patience, and poise by Mary Hall with the help of Christen Bosh and Jennifer Brewster.

My work on the *Encyclopedia of Toxicology* was undertaken as a private citizen, not as a government employee. The views expressed are strictly my own. No official support or endorsement by the U.S. National Library of Medicine or any other agency of the U.S. Federal Government was provided or should be inferred.

—*Philip Wexler*

Absorption

Introduction

Absorption is the process by which a chemical crosses the various membrane barriers of the body before it enters the bloodstream. The main sites of entry are the gastrointestinal tract, the lungs, and the skin. In drug therapy, other convenient, but more rarely used, portals of entry are the intravenous, subcutaneous, and intramuscular routes.

The absorption of a chemical from the site of exposure is regulated by the biologic membrane surrounding the various cells that line the tissue compartments of the body. The membrane is composed principally of phospholipids forming an oriented bilayer, 7–9 nm thick. The more polar hydrophilic (attracted to water) ends of the phospholipids project into the aqueous media on each side of the membrane, and the hydrophobic (repelled by water) fatty acid tails form a barrier to water in the inner space of the membrane. Proteins are embedded throughout the lipid bilayer and have various functions. One of these is to act as active carriers for certain molecules across the membrane. Proteins can also form pathways or small pores through the membrane, serving as aqueous channels and allowing passage of water across them.

Before discussing absorption in more detail, it is important to consider mechanisms by which chemicals cross membranes. These mechanisms are of interest not only for absorption but also for all other processes (distribution, biotransformation, and excretion) involved in the disposition of chemicals because they also require passage through membranes.

Chemicals can cross membranes by one or more of the following mechanisms: passive diffusion, facilitated diffusion, active transport, filtration, and endocytosis.

Passive Diffusion

This is the mechanism by which lipophilic (hydrophobic) uncharged molecules find a passage across the membrane by solubilizing within the lipids of the membrane. The driving force for this process is the concentration gradient of the chemical between each side of the membrane, allowing molecules to be transported from the side with higher concentration to the side with lower concentration. Passive diffusion, therefore, requires no energy expenditure by the cell; it is not saturable or subject to competition between molecules.

Factors that govern passive diffusion are:

(1) The lipid solubility of a chemical: This is a characteristic that is usually expressed in terms of the ability of the chemical to distribute between separate oil and water phases. The more a chemical dissolves in oil, or its substitute octanol, the more lipid-soluble it is and the more easily it will cross membranes.

(2) The electrical charge (degree of ionization) of a chemical: As a rule, chemicals that are electrically

neutral permeate more easily through the lipid phase of a membrane by virtue of their higher degree of lipid solubility. For several therapeutic agents that are weakly charged molecules, the pH of the aqueous environment will have considerable influence on the degree of ionization of the chemicals and hence on their lipid solubility and membrane permeation.

(3) The molecular size of a chemical: Passive diffusion is normally limited to molecules whose molecular weight does not exceed 500 Da. However, a small molecule will cross membranes more rapidly than a larger one of equal lipophilicity.

Facilitated Diffusion

Facilitated diffusion is very similar to passive diffusion with the difference that transfer across membranes is assisted by the participation of carrier proteins embedded in the membrane bilayer. Again, the direction of passage will be from the side of the membrane with high concentration of a chemical to the side with low concentration; this also occurs without energy expenditure by the cell. Such a process is somewhat specific in the sense that it applies to molecules that are able to bind to a carrier protein. Absorption of nutrients such as glucose and amino acids across the epithelial membrane of the gastrointestinal tract occurs by facilitated diffusion. Since a finite number of carriers are available for transport, the process is saturable at high concentrations of the transported molecules and competition for transport may occur between molecules of similar structure.

Active Transport

Active transport requires a specialized carrier molecule, a protein, and the expenditure of cellular energy. Transfer across membranes can therefore occur against a concentration gradient. The carrier system is saturable. In addition, molecules with similar structural features may compete for transport by the carrier.

Active transport is of limited importance for absorption of chemicals; it plays an important role, however, in the elimination of chemicals by the liver and the kidneys.

Filtration

Small water-soluble and small charged molecules, such as methanol and salts, respectively, may cross the gastrointestinal epithelial membrane through minute pores or water channels (less than 4 nm) in the membrane. Filtration is also an important function for urinary excretion: Renal glomeruli possess rather large pores that allow passage into the urine of various solutes contained in blood, including small proteins.

Endocytosis

Endocytosis is a specialized form of transport by which very large molecules and insoluble materials are engulfed by invagination of the absorptive cell membrane, forming intracellular vesicles. This process is responsible for the absorption of certain dyes by mucosal cells of the duodenum (pinocytosis). In the lung, alveolar macrophages scavenge insoluble particles, such as asbestos fibers, and may transport them into the lymphatic circulation (phagocytosis).

Absorption by the Gastrointestinal Tract

The major role of the gastrointestinal tract is to provide for efficient absorption of essential nutrients contained in ingested foods and liquids. It is also an important route for absorption of drugs and toxicants. The entire surface of the gastrointestinal tract is very large, being 200 times that of the body surface; the barrier between the contents of the tract and the bloodstream is easily crossed, consisting essentially of an epithelium only one cell thick. The anatomy of the gastrointestinal tract is illustrated in Fig. A-1. Absorption occurs mostly by passive diffusion of lipid-soluble, electrically neutral (non-ionized) molecules.

The degree of ionization of many therapeutic drugs, which are usually weak electrolytes, is directly dependent upon the pH of the gastrointestinal content. The pH will therefore have considerable influence on the absorption of such chemicals; absorption will tend to occur at sites where the drugs are present as neutral molecules. At the low acidic pH of the stomach (1–3), most weak organic acids such as acetylsalicylic acid will be nonionized and will diffuse passively across the

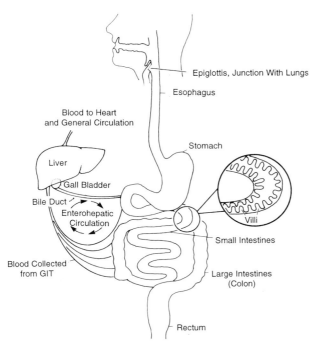

FIGURE A-1.   *The anatomy of the gastrointestinal tract (reproduced with permission from Smith, R. P., A Primer of Environmental Toxicology, p. 70. Lea & Febiger; Philadelphia, 1992).*

gastric mucosa at a rate that will be proportional to the concentration gradient of the nonionized form. On the other hand, weak organic bases will diffuse more easily through the mucosa of the small intestine in which pH is higher (5–8). However, the bulk of absorption does not necessarily occur at the site where pH is optimal for electrical neutrality of the molecules. The very large surface area of the small intestine, due to the presence of finger-like projections, namely the villi and the microvilli, favors the diffusion of substances even at pH values for which the degree of ionization is not maximal; as a consequence, the small intestine is the region of the gastrointestinal tract that is most effective in the absorption of chemicals.

A small number of chemicals may be absorbed using facilitated diffusion (antimetabolic nucleotides), active transport (lead and 5-fluouracil), or pinocytosis (dyes and bacterial endotoxins).

Chemicals that reach the bloodstream by absorption through the gastrointestinal tract will move directly to the liver, where they will normally undergo metabolic biotransformation to more or less active chemical forms, even before they gain access to the various tis-

sues of the body; this phenomenon is known as the first-past effect.

Among factors that may modify gastrointestinal absorption of ingested chemicals, the presence of food in the tract is one of the most important. The presence of food in the stomach will delay the absorption of weak organic acids at that site. The presence of lipid-rich food will delay the emptying of the gastric content into the intestine and thus also delay the absorption of chemicals. Conversely, an empty stomach facilitates absorption, a situation that is almost always beneficial in drug therapy.

Chemical interactions in the gastrointestinal tract between nutrients and drugs may considerably reduce the absorption of some drugs: Calcium ions from dairy products form insoluble and therefore nonabsorbable complexes with the antibiotic tetracycline. On the other hand, certain drugs are irritants to the gastrointestinal tract (nonsteroidal anti-inflammatory drugs and potassium chloride tablets) and must be ingested with food.

Enterohepatic circulation provides an example of a special case of intestinal absorption. Certain chemicals, like methyl mercury, after undergoing biotransformation in the liver, are excreted into the intestine via the bile. They then can be reabsorbed in the intestine, sometimes after enzymatic modification by intestinal bacteria. This process can markedly prolong the stay of chemicals in the body. It can be interrupted by antibiotics that destroy the intestinal bacterial flora.

Absorption through the Skin

Normal skin represents an effective, but not perfect, barrier against the entry of chemicals present in the environment. There are two major structural components to the skin: the epidermis and the dermis (Fig. A-2).

The epidermis is formed of several layers of cells, with the outermost layers, approximately 10 μm thick, consisting of dried dead cells forming the *stratum corneum*. The latter, whose cells are rich in a filament-forming protein called keratin, represents the major structural component of the barrier to passage of chemicals through the skin. Chemicals may move through the various cell layers of the epidermis by passive diffusion, more slowly through the *stratum corneum*, but more rapidly through the inner layers of live epidermal

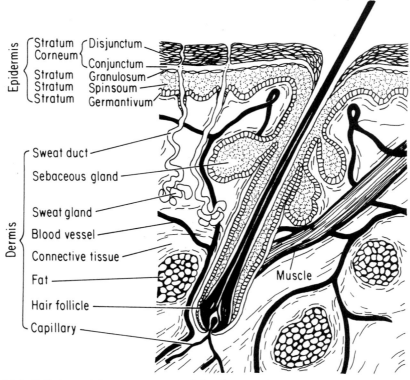

FIGURE A-2.   The organization of the skin as a biologic barrier (reproduced with permission from Smith, R. P., A Primer of Environmental Toxicology, p. 73, Lea & Febiger; Philadelphia, 1992).

cells (*stratum granulosum, stratum spinosum,* and *stratum germinativum*).

The epidermis rests upon and is anchored onto a much thicker base of connective and fatty tissues, the dermis, whose major structural components are proteins called collagen and elastin; these proteins provide the skin with tensile strength and elasticity. The dermis also contains small blood vessels (capillaries), nerve endings, sebaceous glands, sweat glands, and hair follicles. Small pores in the epidermis that allow passage for sweat and sebum glands, as well as hair shafts, are not an important route of entry for chemicals. Once a chemical has crossed the epidermis by passive diffusion and gained access to the dermis, diffusion into the bloodstream occurs rapidly.

The *stratum corneum* is much thicker in areas where considerable pressure and repeated friction occur, like palms and soles; absorption is therefore much slower in these areas. Conversely, the *stratum corneum* is ex-

tremely thin on the skin of the scrotum. In general, skin surfaces of the ventral aspect of the body represent barriers that are easier to cross than those of the dorsal aspect.

Mechanical damage to the *stratum corneum* by cuts or abrasions of the skin or chemical injury by local irritation with acids or alkalis, for example, are likely to facilitate the entry of chemicals through the skin. This may also be the case in subjects suffering from certain skin diseases.

Lipid-soluble chemicals like organophosphate insecticides, tetraethyl lead, certain organic solvents, and certain dyes like aniline are relatively well absorbed through the skin. Percutaneous absorption is facilitated by increasing peripheral dermal blood flow, as might occur when the ambient temperature is elevated. Under the same conditions, and in the presence of elevated sweating, the degree of hydration of the skin will increase considerably, increasing by the same occasion the permeability of the *stratum corneum* to foreign chemicals; this observation is of special interest to workers in occupational settings.

Absorption by the Lung

The fundamental physiological role of the lung is to allow gas exchange, extracting oxygen from the ambient air and eliminating carbon dioxide as a catabolic waste. When performing this function, the human adult lung is exposed each day to approximately 10,000 liters of more or less contaminated air. The lung can therefore become an important portal of entry for airborne chemicals present in the environment.

Extraneous substances are presented to the lung as gases or vapors or as liquid or solid particles; they may reach various regions of the respiratory tract, where they will undergo absorption into the bloodstream or elimination even before they can reach the latter.

In terms of its anatomical and functional relationship with the contaminated atmospheric environment, the respiratory tract can be divided into three regions: the nasopharyngeal, the tracheobronchiolar, and the alveolar regions (Fig. A-3). The major part of the absorptive process takes place in the alveolar region, due principally to its large surface area (80 m² in an adult human) and the extreme thinness of the cellular barrier (less than 1 μm) between the air-side of the alveolar sac (lined with epithelial cells) and the lumen of the lung capillaries (lined with endothelial cells).

When discussing absorption of chemicals through the respiratory tract, it is practical to consider separately gases and vapors, on the one hand, and particles on the other hand.

Gases and Vapors

How much and at what location a contaminant gas or vapor will be absorbed in the respiratory tract is determined primarily by the solubility of the contaminant. The more water-soluble agents (sulfur dioxide and ketonic solvents) may dissolve in the aqueous fluid lining the cells of the more proximal region of the respiratory tree, even before they reach the alveolar region. They may then undergo absorption by passive diffusion or passage through membrane pores. When,

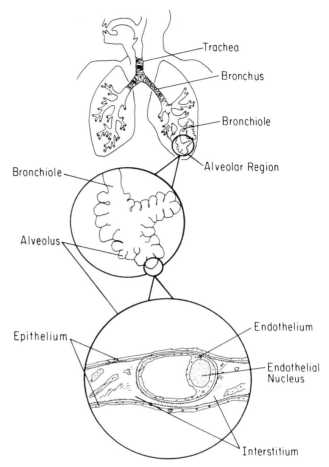

FIGURE A-3. *The anatomy of the respiratory tract from trachea to alveolus. (reproduced with permission from Smith, R. P., A Primer of Environmental Toxicology, p. 67, Lea & Febiger; Philadelphia, 1992).*

in addition, water-soluble contaminants are very reactive substances, like formaldehyde, they may form stable molecular complexes with cell components as proximally as the nasopharyngeal region. By virtue of these mechanisms, the alveolar region of the lung is partially protected against potential injury by certain gases and vapors.

Lipid-soluble contaminants diffuse passively through the thin alveolar–vascular cell barrier of the alveolar sac and then dissolve into the blood according to the ability of the contaminant to partition between alveolar air and circulating blood. Substances that are very soluble in blood are rapidly transported into the bloodstream. For these substances, like styrene and xylene, the amount absorbed will be greatly enhanced by increasing the rate and the depth of respiration, as it is likely to happen when doing strenuous physical work. On the other hand, substances that are poorly soluble in blood have limited capacity for absorption due to rapid saturation of blood. For these substances, like the solvents cyclohexane and methyl chloroform, the amount absorbed may be increased only by increasing the blood perfusion rate in the lung; that is, by enhancing the replacement of saturated blood circulating in the lung capillaries. This can be achieved, for example, when doing work requiring heavy muscular activity.

Particles

Liquid (sulfuric acid and cutting fluids) and solid (silica dusts, asbestos fibers, and microorganisms) particles may become airborne and form respirable aerosols. According to their size and diameter, inhaled particles may be deposited in different anatomical regions of the respiratory system. Once deposited, particles may dissolve locally or may undergo removal to other regions of the respiratory tree.

The surface of the cells lining the tracheobronchial tree and the surface of most of the cells lining the nasopharyngeal region are covered with a layer of relatively thick mucous material; in the alveolar region, cells are lined with a thin film of fluid. The aqueous environment provided by these surface liquids favors at least partial dissolution and eventually absorption of water-soluble particles, especially those present as liquid droplets. Various defense mechanisms may help to remove less soluble particles from their site of deposition.

Particles larger than 5 μm in diameter are usually deposited by inertial impaction on the surface of the nasopharyngeal airways. They may be removed by coughing, sneezing, or nose wiping.

Particles with diameters between 1 and 5 μm are deposited in the tracheobronchial region as a result of either inertial impaction at airway bifurcations or gravitational sedimentation onto other airway surfaces. Undissolved particles may then be removed by the action of the mucociliary defense system working as an escalator; particles trapped in the mucus are propelled toward the pharynx by the action of thin cilia located on the surface membrane of specialized cells. Once in the pharynx, the particles may be swallowed. The

efficiency of the escalator defense system may be greatly impaired by various environmental contaminants, like sulfur dioxide, ozone, and cigarette smoke, that are known to paralyze the activity of the ciliated cells and consequently the upward movement of the mucus.

Particles ranging between 0.1 and 1.0 μm in diameter reach the alveolar region, where they finally hit cellular walls as a result of their random movement within minute air sacs. Removal of particles in this region of the lung is much less efficient. Some of the particles may eventually reach the tracheobronchiolar escalator system, either as engulfed material within alveolar macrophages or as naked particles transported by the slow movement of the fluid lining the alveoli. Other possible mechanisms involve transport of the particles into the lymphatic system, either within macrophages or by direct diffusion through the intercellular space of the alveolar wall.

Particles smaller than 0.1 μm are not usually deposited in the lung, entering and exiting the airways together with inhaled and exhaled air.

Often, particulate matter acts as a carrier for gases, vapors, and fumes adsorbed onto their surface (solid particles) or dissolved within them (liquid particles); this increases the residence time of such pollutants in specific areas of the lung and imposes an additional task on the pulmonary defense mechanisms.

The most striking example of this synergistic effect is the one observed between sulfur dioxide, a respiratory tract irritant, and suspended particles, both being typical components of urban air pollution. This explains why nowadays guideline values for exposure to sulfur dioxide in the presence of particulate matter are lower than those for exposure to sulfur dioxide alone. Similar concerns can be expressed for combinations comprising exhaust particles from diesel engines and certain carcinogens like polycyclic aromatic hydrocarbons, as well as cigarette smoke and certain other carcinogens like aromatic amines.

Chemicals absorbed by the lung reach the systemic circulation directly and are therefore immediately available for distribution to the various tissues of the body: brain, kidneys, liver, muscles, skin, bones, and others.

Further Reading

Rozman, K., and Klaassen, C. D. (1996). Absorption, distribution, and excretion of toxicants. In *Casarett and Doull's Toxicology. The Basic Science of Poisons*

(C. D. Klaassen, Ed.), 5th ed., pp. 91–112. McGraw-Hill, New York.

Roebuck, B. D. (1992). Absorption, distribution, and excretion of chemicals. In *A Primer of Environmental Toxicology* (R. P. Smith, Ed.), pp. 59–76. Lea & Febiger, Philadelphia.

—*Jules Brodeur and Robert Tardif*

Related Topics

Biotransformation
Distribution
Excretion
Exposure
Gastrointestinal System
Modifying Factors of Toxicity
Pharmacokinetics/Toxicokinetics
Respiratory Tract
Skin
Toxicity Testing, Dermal
Toxicity Testing, Inhalation

Academy of Toxicological Sciences

The Academy of Toxicological Sciences was established in 1981 for the purpose of recognizing and certifying toxicologists in order to ensure the competence and experience of professional practitioners whose work affects the public welfare. Recognition and certification are accomplished by the peer review process, a time-honored mechanism for scientists to evaluate one another. The academy bases certification on formal training, proven ability, and experience. Demonstrated achievement, rather than the potential for achievement, is the substance of the academy's evaluation process. Thus, an individual certified as a fellow in toxicology by the Academy of Toxicological Sciences is a qualified person who actively practices toxicology and who has been evaluated by the peer review process by the academy according to its bylaws.

Candidates for certification must have broad knowledge of toxicology and demonstrate substantive involvement in toxicological activities. To apply, an applicant submits an application form and supporting documentation to the secretary–treasurer of the academy. The board of directors reviews the credentials of applicants twice a year in the spring and fall. The criteria for certification in toxicology by the academy are divided into three sections: (1) education and training, (2) professional experience, and (3) demonstration of scientific judgment and recognition. Following review by the board of directors, candidates are notified in writing of the board's decision.

Successful candidates are certified as fellows of the academy for a period of 5 years. Every 5 years, each fellow is recertified by submitting a current *curriculum vitae* for the board's review and vote.

In 1995, there were 125 fellows of the Academy of Toxicological Sciences.

For additional information, contact the Academy of Toxicological Sciences, Secretariat, 9200 Leesburg Pike, Vienna, Virginia 22182. Telephone: 703-893-5400; fax: 703-759-6947.

—*David M. Krentz and Harihara M. Mehendale*

(Adapted from information supplied by the Academy of Toxicological Sciences.)

Related Topics

American Board of Toxicology
American Academy of Clinical Toxicology
American College of Medical Toxicology
American College of Toxicology
Society of Environmental Toxicology and Chemistry
Society of Toxicology

ACE Inhibitors

◆ PREFERRED NAME: Angiotensin-converting enzyme inhibitors

- REPRESENTATIVE COMPOUNDS: Benazepril; captopril; enalapril; enalaprilat; fosinopril; lisinopril; quinapril; ramipril
- SYNONYMS: Lotensin (CAS: 86541-75-5); Capoten (CAS: 62571-86-2); Vasotec (CAS: 75847-73-3); Vasotec IV (CAS: 84680-54-6); Monopril (CAS: 88889-14-9); Prinivil; Zestril (CAS: 76547-98-3); Accupril (CAS: 85441-61-8); Altace (CAS: 87333-19-5)
- PHARMACEUTICAL CLASS: Angiotensin-converting enzyme inhibitors

Uses

ACE inhibitors are used in the management of hypertension and congestive heart failure.

Exposure Pathways

Ingestion is the most common route for both accidental and intentional exposures. Enalaprilat is available for parenteral administration and toxicity could occur via this route.

Toxicokinetics

The extent of oral absorption varies from 25% (lisinopril) to 75% (captopril). The rate of absorption also varies from 0.5 hr (captopril and enalopril) to 7 hr (lisinopril). Reported volumes of distribution range from 0.7 liters/kg (captopril) to 1.8 liters/kg (lisinopril). All of the ACE inhibitors, except for captopril and lisinopril, are metabolized in the liver to active metabolites. Excretion is via both the urine and the feces. The half-life ranges from 13 hr (enalapril) to 17 hr (ramipril).

Mechanism of Toxicity

The ACE inhibitors affect the renin–angiotensin system. This system has effects on blood pressure as well as fluids and electrolyte balance. Renin modulates the formation of angiotensin I from angiotensinogen. Angiotensin I is then converted via angiotensin-converting enzyme to angiotensin II. Angiotensin II is a potent vasoconstrictor that also causes increased aldosterone secretion. Aldosterone is responsible for sodium and water retention. The ACE inhibitors interfere with the conversion of angiotensin I to angiotensin II and, therefore, cause vasodilation as well as sodium and water loss. Literature exists which supports a relationship between angiotensin and the beta endorphins. Angiotensin II is thought to be inhibited by endogenous beta endorphin. *In vitro* studies have demonstrated that captopril can inhibit encephalinase, the enzyme which degrades endorphins. Interference with endorphin metabolism should result in prolonged effects from these opiate-like neurotransmitters. Also, the opiate antagonist naloxone is thought to interfere with beta-endorphin's inhibition of angiotensin II. An interaction between angiotensin and bradykinin may also exist. ACE is identical to kinase II, which is responsible for inactivation of bradykinins. Accumulation of bradykinins may cause a decrease in blood pressure by a direct vasodilatory mechanism or through stimulation of prostaglandin release and/or synthesis.

Human Toxicity

The literature contains limited information about ACE inhibitor toxicity. Adverse effects observed at therapeutic doses include cough, dermal reactions, blood dyscrasias, and hypogeusia. The clinical effects observed following ACE inhibitor poisoning or overdose are a direct extension of their therapeutic effects. Ingestions involving small amounts of ACE inhibitors may result in limited or no toxic effects. Clinical effects that may occur include hypotension with or without a reflex tachycardia, changes in level of consciousness that are directly related to vascular changes, and hyperkalemia. Hyperkalemia can occur as a response to sodium loss. Only a few cases of profound hypotension have been reported. In each of these cases, blood pressure returned to normal within 24 hr of ingestion. Delayed hypotension, at 19 and 25 hr, has been observed following ingestion of captopril. One death has been attributed to an ACE inhibitor. This was in a 75-year-old male who ingested captopril and the calcium channel blocker diltiazem. Because this was a coingestion, it is not certain that captopril was the primary cause of death.

Clinical Management

Supportive care, including airway management as well as cardiac and blood pressure monitoring, should be provided to unstable patients. Witnessed ingestion of small amounts of an ACE inhibitor in children can be managed with observation at home. Ingestion of unknown amounts, exposure to coingestants, and intentional ingestions require monitoring and management in the emergency department. Some form of gastric decontamination should be considered following ingestion of a toxic amount of these agents. Syrup of ipecac can be administered to induce emesis in recent pediatric ingestions in which no change in level of consciousness has occurred. In recent ingestions involving coingestants and/or in which a change in level of consciousness has occurred, gastric lavage can be utilized to decontaminate the stomach. Because the ACE inhibitors are adsorbed by activated charcoal, it can be utilized to prevent their absorption in the gastrointestinal tract. Hypotension following ACE inhibitor ingestion has been managed with fluids alone or in combination with vasopressors such as dopamine. A limited number of case reports exist which describe a need for dopamine to treat hypotension. If profound hypotension resistant to dopamine were to occur, other vasopressors, such as epinephrine and norepinephrine, can be used. Laboratory analysis should be used to monitor electrolytes, especially sodium and potassium. ACE inhibitor serum concentrations are not readily available and have little if any clinical utility. Because ACE inhibitors may potentiate the effects of the opiate-like beta-endorphins, some authors have suggested the use of naloxone to reverse their toxicities. Successes and failures with naloxone have been described in case reports. Because naloxone has limited adverse effects, its use could be considered in the management of serious ACE inhibitor toxicity. One case report describes the use of the experimental exogenous angiotensin II to counter severe ACE inhibitor toxicity. The pharmacokinetic characteristics of the ACE inhibitors, limited protein binding, and small volume of distribution make them amenable to hemodialysis. Because major morbidity is uncommon with these agents, the need for dialysis is limited.

—*Daniel J. Cobaugh*

Related Topics

Beta Blockers
Calcium Channel Blockers
Cardiovascular System
Gastrointestinal System

Acenaphthene

- CAS: 83-32-9
- SYNONYMS: 1,2-Dihydroacenaphthylene; 1,8-dihydroacenaphthalene; 1,8-ethylenenaphthalene; acenaphthylene; naphthyleneethylene; periethylenenaphthalene
- CHEMICAL CLASS: Arene belonging to the class of polycyclic aromatic hydrocarbons
- CHEMICAL STRUCTURE:

Uses

Acenapthene is a chemical intermediate used to produce naphthalimide dyes, which are used as fluorescent whitening agents, and used in manufacturing plastics, insecticides, and fungicides.

Exposure Pathways

Skin contact is the most common accidental exposure pathway. Acenapthene may irritate or burn skin. Its vapor can be poisonous if inhaled.

Toxicokinetics

Little information is available regarding acute exposure to acenaphthene. It is biotransformed in the liver. On the basis of a mouse oral subchronic study in which hepatotoxicity was seen as the major effect, the NOAEL and LOAEL were 175 and 350 mg/kg/day, respectively. Acenaphthene is devoid of any mutagenic activity in *Salmonella typhimurium* assay. The nitro-derivatives of acenaphthene have tumorigenic potential.

Mechanism of Toxicity

5-Nitroacenaphthene causes toxicity by the reduction of the nitro function to corresponding hydroxylamine. These arylhydroxylamines may be either direct-acting mutagens or may become so following nonenzymic conversion to aryl nitronium ions or they may be esterified to the corresponding electrophilic hydroxamic acid esters. Acenaphthene can bind to hemoglobin to cause methemoglobinemia.

Human Toxicity

Acenaphthene may be poisonous if inhaled or absorbed through skin. The vapor may cause dizziness or suffocation. Acenaphthene may cause vomiting if swallowed in large quantity. It can cause methemoglobinemia.

Human exposure data are not available. Currently, acenaphthene is under review by U.S. EPA for evidence of human carcinogenic potential. This does not imply that this agent is necessarily a carcinogen. The nitroderivative of acenaphthene (5-nitroacenaphthene) is a possible carcinogen to humans.

Clinical Management

The victim should be moved to fresh air and emergency medical care should be provided. If the victim is not breathing, artificial respiration should be provided; if breathing is difficult, oxygen should be administered. In case of contact with the eyes, the eyes should be flushed immediately with running water for at least 15 min. Affected skin should be washed with soap and water. Contaminated clothing and shoes should be removed and isolated at the site. If methemoglobinemia occurs and is severe, treatment with methylene blue and oxygen is recommended.

Animal Toxicity

Acenaphthene can cause hepatotoxicity in rats and mice. No information is available regarding the toxicity of acenaphthene in cats and dogs.

—*Sanjay Chanda*

Related Topic

Polycyclic Aromatic Hydrocarbons

Acetaldehyde

- ◆ CAS: 75-07-0
- ◆ SYNONYMS: Acetic aldehyde; acetylaldehyde; ethylaldehyde

- ◆ CHEMICAL CLASS: Aldehydes
- ◆ CHEMICAL STRUCTURE:

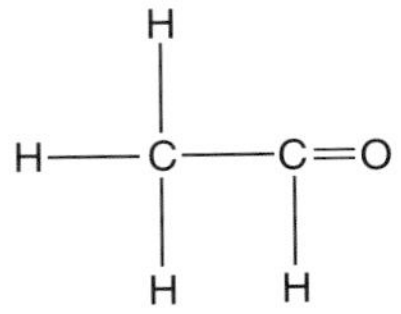

Uses

Acetaldehyde is used in the manufacturing of various chemicals such as acetic acid, pyridine, peracetic acid, pentaerythritol, 1,3-butylene glycol, and chloral. It is also used in the silvering of mirrors, leather tanning, fuel compositions, preservatives, paper processing, glues, cosmetics, dyes, plastics, and rubber.

Exposure Pathways

Industrial exposures to acetaldehyde are most likely to occur by inhalation with potential for skin and eye contact. Accidental ingestion is also possible.

Toxicokinetics

Following inhalation exposure, acetaldehyde is deposited in the nasal cavity and upper respiratory tract, and eventually some traces can be absorbed into the blood and be distributed throughout the body. The uptake of acetaldehyde in the nasal cavity is influenced by its solubility and inspiratory flow rate. Perhaps acetaldehyde uptake in the nasal tissue is dependent on its reaction with tissue substrates that become depleted at high exposure concentrations. The acetaldehyde vapor can be metabolized in the nasal cavity by the mixed-function oxidase and carboxylesterase systems. Further metabolism takes place in the liver to a number of metabolites and some unchanged acetaldehyde that can be excreted in the urine. Most of the free acetaldehyde is excreted in the exhaled breath.

Mechanism of Toxicity

Acetaldehyde is soluble in the mucous membranes of the upper respiratory tract causing irritation of the sensory nerve endings. There is also depression of the mucociliary defense system. The direct action of acetaldehyde in the skin and eyes is the result of irritation to these tissues.

Human Toxicity

Inhalation exposures to acetaldehyde can result in irritation of the upper respiratory tract. Inhalation at concentrations ranging from 100 to 200 ppm can cause irritation to the mucous membranes.

Skin and eye contact with liquid acetaldehyde can produce a burning sensation, lacrimation, and blurred vision. Unacclimated subjects experienced eye irritation at 50 ppm after a 15-min exposure. Some more sensitive persons exhibited eye irritation at 25 ppm for a 15-min exposure.

A STEL ceiling of 25 ppm for acetaldehyde was recommended to prevent excessive eye irritation and lacrimation and potential injury to the respiratory tract. Data from some animal studies suggest that acetaldehyde is teratogenic. According to ACGIH, the recent identification of nasal and laryngeal carcinomas indicate that acetaldehyde should be considered an A3 animal carcinogen.

Clinical Management

Exposures by inhalation should be monitored for respiratory tract irritation, bronchitis, or pneumonitis. Humidified supplemental 100% oxygen should be administered.

Gastric lavage may be indicated soon after ingestion of acetaldehyde followed by administration of activated charcoal slurry mixed with a saline cathartic or sorbitol. Exposed eyes should be irrigated with copious amounts of tepid water for at least 15 min. If eye irritation, pain, swelling, lacrimation, or photophobia persist, the patient should be seen in a health care facility.

Animal Toxicity: Acute

The oral LD_{50} for acetaldehyde in rats has been reported to be 1930 mg/kg and the 4-hr LC_{50} is approximately 13,300 ppm. Acetaldehyde is a severe eye irritant to rabbits at 40 mg and mildly irritating to rabbit skin at 500 mg. Rats were exposed to acetaldehyde concentrations ranging from 400 to 5000 ppm in a 4-week subchronic inhalation study at 6 hr per day, 5 days per week. At 1000 and 2200 ppm, the rats exhibited growth retardation, polyuria, and nasal epithelial degeneration. At 400 ppm, there was slight degeneration of the olfactory epithelium. A 52-week chronic inhalation study in hamsters exposed to 1500 ppm acetaldehyde produced growth retardation, slight anemia, increased enzyme and protein content in the urine, and increased kidney weight. There were distinct histopathological changes in the nasal mucosa and trachea, including hyperplasia, squamous cell metaplasia, and inflammation.

Inhalation exposure to acetaldehyde has produced nasal tumors in rats and laryngeal tumors in hamsters. Male and female rats were exposed to acetaldehyde 6 hr per day, 5 days per week for 28 months at concentrations of 0, 750, 1500, or 3000 ppm. A concentration-related incidence of squamous cell carcinomas of the respiratory epithelium was observed in both male and female rats. A statistically significant number of adenocarcinomas occurred in the olfactory epithelium of both sexes of rats exposed at all three acetaldehyde concentrations. Male and female hamsters were exposed to acetaldehyde 7 hr per day, 5 days per week at concentrations gradually reduced from 2500 to 1650 ppm for 52 weeks. Both sexes of acetaldehyde-exposed hamsters developed laryngeal tumors consisting of squamous cell carcinomas and adenosquamous cell carcinomas.

Data from studies with rats suggest that acetaldehyde is teratogenic. Fetuses from dams injected intraperitoneally with acetaldehyde concentrations ranging from 50 to 100 mg/kg on Day 10, 11, or 12 of gestation produced a significant increase in fetal resorptions, growth retardation, and an increase in malformations, including digital anomalies, cranial and facial malformations, and delayed skeletogenesis. It was concluded that acetaldehyde interfered with placental function via the maternal–placental nutrient exchange, resulting in retarded growth.

Acetaldehyde has been shown to induce mutagenic changes in many assays. In mammalian *in vitro* assays, acetaldehyde produced sister chromatid exchanges and chromosomal breaks and aberrations in mammalian *in vitro* assays.

—*Edward Kerfoot*

Related Topic

Respiratory Tract

Acetaminophen

- ◆ CAS: 103-90-2

- ◆ SYNONYMS: APAP; 4′-hydroxyacetanilide; *N*-acetyl-*p*-aminophenol; paracetamol; Tylenol

♦ PHARMACEUTICAL CLASS: Acetaminophen is a synthetic nonopioid congener of acetanilid in the para-aminophenol class.

♦ CHEMICAL STRUCTURE:

$$HO - \bigcirc - NHCOCH_3$$

Uses

Acetaminophen is used as an analgesic and as an antipyretic.

Exposure Pathways

Ingestion is the most common route of both accidental and intentional exposure to acetaminophen. It is available in oral and rectal dosage forms.

Toxicokinetics

Acetaminophen is absorbed rapidly during therapeutic use but absorption may be delayed after large ingestions, producing peak blood concentrations at approximately 4 hr postingestion. The absorption phase is highly dependent on the presence of food and other chemicals which may influence the rate of acetaminophen absorption. In adults, the majority of acetaminophen is metabolized in the liver to inactive glucuronide and sulfate conjugates. In young children, sulfate conjugation predominates. A very small amount undergoes biotransformation via the P450-mixed-function oxidase pathway to *n*-acetyl-*p*-benzoquinoneimine (NAPQI). Protein binding is 5–50%. The volume of distribution approximates 1 liter/kg. The water-soluble glucuronide and sulfate conjugates are eliminated via the kidneys. Approximately 2–5% is eliminated in the urine as unchanged acetaminophen. The half-life of therapeutic dose is 1–3 hr. In overdose patients, this may be increased to more than 4 hr and may even exceed 12 hr in patients with severe acetaminophen-induced liver toxicity.

Mechanism of Toxicity

The toxic metabolite NAPQI is normally detoxified by endogenous glutathione to cysteine and mercapturic acid conjugates. In acetaminophen overdosage, the amount of NAPQI increases and depletes endogenous glutathione stores. The NAPQI is then thought to bind to hepatocytes and cause cellular death. Hepatic necrosis, as a consequence of hepatocellular death, develops and then results in the development of clinical and laboratory findings that are consistent with liver damage. A similar mechanism is postulated for the renal damage that occurs in some patients who suffer from acetaminophen toxicity.

Human Toxicity: Acute

Hepatotoxicity is the primary toxic insult from acetaminophen overdosage. Most of the severe clinical manifestations are secondary to hepatic damage, which may occur following acute doses in excess of 150 mg/kg or a total of 7.5 g. Plasma acetaminophen concentrations should be obtained to determine the probability of acetaminophen-induced hepatotoxicity: The Rumack–Matthews Nomogram should be used to assess the risk of hepatotoxicity. Levels in excess of 200 μg/ml at 4 hr postingestion are associated with a high probability that some degree of hepatotoxicity will develop. The clinical presentation follows four distinct stages. During the first 24 hr postingestion, gastrointestinal irritation manifest as nausea and/or vomiting is common. It is uncommon to see the overt signs and symptoms of hepatotoxicity in the first stage. The second stage (24–48 hr postingestion) is characterized by resolution of the initial symptoms, and the patient may be relatively asymptomatic. However, it is during this period of time when the laboratory tests will reveal evidence of hepatoxicity via elevations in hepatic transaminases. Most cases do not progress to stage three, but in those which proceed, the consequences of hepatotoxicity develop. The patient may develop hypoglycemia, coagulopathies, jaundice, and other problems consistent with hepatic encephalopathy. Hepatorenal syndrome may also develop. Patients who proceed to and survive stage three progress to the fourth stage of recovery, which commences 5 or 6 days after the initial ingestion. After several months, the normal liver histology will be restored with little or no fibrotic evidence of the hepatotoxicity.

Chronic Ingestion

The chronic ingestion of excessive amounts of acetaminophen may produce similar toxicity but in a more insidious fashion. Age, chronic ethanol abuse, and pre-existing disease may be contributing factors.

Clinical Management

Basic and advanced life-support measures should be utilized as necessary. Gastrointestinal decontamination procedures should be used as deemed appropriate to the patient's level of consciousness and the history of the ingestion. Activated charcoal may be used to adsorb acetaminophen or concomitant ingestants. Patients who have toxic blood concentrations as determined from the Rumack–Matthew Nomogram should receive treatment with *N*-acetylcysteine (NAC). NAC is hepatoprotective by acting as a glutathione surrogate or precursor which restores endogenous glutathione. A loading dose of 140 mg/kg is administered orally, followed by 70 mg/kg administered orally every 4 hr for an additional 17 doses. A single dose of activated charcoal will not significantly interfere with the absorption and antidotal effects of NAC. Intravenous NAC administration is investigational in the United States.

Animal Toxicity

Cats and dogs are exquisitely sensitive to acetaminophen (see Species Differences under Veterinary Toxicology). Toxicity may include methemoglobinemia. NAC can be used in animals.

—*Edward P. Krenzelok*

Related Topics

Glutathione
Kidney
Lipid Peroxidation
Liver
Mechanisms of Toxicity
Poisoning Emergencies in Humans
Veterinary Toxicology

Acetic Acid

- ◆ CAS: 64-19-7
- ◆ PREFERRED NAME: Glacial acetic acid

- ◆ SYNONYMS: Acido acetico; vinegar acid; methanecarboxylic acid; ethanoic acid; pyroligneous acid; UN2789 (DOT); UN2790 (DOT)
- ◆ PHARMACEUTICAL CLASS: Pharmaceutical aid (acidifier)
- ◆ CHEMICAL STRUCTURE:

$$CH_3COOH$$

Uses

Acetic acid is used in the manufacturing of various acetates, acetyl compounds, cellulose acetate, acetate rayons, plastics, and rubber. It is also used in tanning, as laundry sour, in printing calico, and in dying silk. It is an acidulant and preservative in food. It is a solvent for gums, resins, volatile oils, and many other substances. Acetic acid is widely used in commercial organic synthesis.

Exposure Pathways

Contact with skin or ingestion are the most common exposure pathways.

Mechanism of Toxicity

Acetic acid causes toxicity by coagulative necrosis, meaning that the acid denatures all tissue protein to form an acid proteinate. As a result both structural and enzymatic proteins are denatured, and cell lysis is blocked. Therefore, cell morphology is not greatly interrupted. In addition, an eschar is formed which delays further corrosive damage and helps reduce systemic absorption. Thus, damage, especially with small quantities of acid, is frequently limited to local sites of injury to the skin or the gastrointestinal tract rather than the systemic response.

Human Toxicity

Acetic acid is corrosive to skin and gastric mucosa. Repetitive exposure to acetic acid may cause erosion of dental enamel, bronchitis, and eye irritation.

Clinical Management

Exposure should be terminated as soon as possible by removal of the patient to fresh air. The skin, eyes, and mouth should be washed with copious water. A 15- to

20-min wash may be necessary to neutralize and remove all residual traces of the contaminant. Contaminated clothing and jewelry should be removed and isolated. Contact lenses should be removed from the eye to avoid prolonged contact of the acid with the area. A mild soap solution may be used for washing the skin and as an aid to neutralize the acid but should not be placed into the eye. No cream, ointment, or dressing should be applied to the affected area. Emesis should be avoided in case of ingestion. If a large quantity has been swallowed, gastric lavage should be considered. Dilution with water may be the solution for small quantities swallowed. Under no circumstances should carbonated beverages ever be used because of large quantities of carbon dioxide gas released that distend the stomach.

Animal Toxicity

The toxicity of acetic acid in animals is the same as that found in humans.

—Sanjay Chanda

Acetone

- CAS: 67-64-1
- SYNONYMS: 2-Propanone; β-ketopropane; dimethyl ketone; ketone propane; methyl ketone; propanone
- CHEMICAL CLASS: Hydrocarbon ketone
- MOLECULAR FORMULA: C_3H_6O

Uses

Acetone is a common solvent used for fats, oils, waxes, resins, rubber, plastics, lacquers, varnishes, and rubber cements. It is used in the manufacturing of a number of chemicals, including methyl oxide, acetic acid, acetic anhydride, diacetone alcohol, chloroform, iodoform, and bromoform. It is also used in the extraction of various principles from animal and plant substances, in purifying paraffin, and in processes for hardening and dehydrating tissues in the laboratory. Acetone is found in paint and varnish removers and in many consumer products, such as nail polish remover.

Pharmaceutically, acetone is used in the preparation of vitamin intermediates and antiseptic solutions. The rapid evaporation rate of acetone makes it a useful solvent for cleaning and drying precision parts. About one-fourth of the acetone produced is used in the manufacture of methacrylates and one-third is used as solvent.

Exposure Pathways

Exposures to acetone may occur via inhalation, ingestion, or dermal contact. Acetone is well absorbed through inhalation and to a lesser extent through the skin.

Toxicokinetics

The pulmonary retention of acetone has been estimated to be about 45%. Only a small amount is reduced. Acetone absorbed during 8 hr at 200 ppm will be completely metabolized or excreted within 16 hr. Acetone is readily absorbed into the bloodstream because of its solubility in water and thus is transported rapidly throughout the body. A man breathing an estimated concentration of 22 mg/liter (9300 ppm) for 5 min absorbed 71% of inhaled acetone; two men breathing 11 mg/liter (4650 ppm) for 15 min each absorbed 76 or 77% (mainly via lungs and urine.). Excretion in humans is rapid. The ratio of the three primary routes for excretion are approximately 40–70% through breathing, 15–30% through the urine, and 10% through skin.

Acetone may be converted to 1,2-propanediol, which enters the glycolytic pathway and possibly the one carbon pool. Acetone has been shown to be converted to lactate in mice. The rate-limiting step appears to be the conversion of acetone to a hydroxylated intermediate.

Mechanism of Toxicity

Acetone destroys biomembranes, probably because it is very lipid soluble. Continued exposure will defat skin, producing irritation, rash, and redness.

The mechanism involved with the potentiation is not completely known; however, it has been reported that the potentiation phenomenon observed with various haloalkane substances is closely related to the presence of exogenous or endogenous ketones. Acute administration of acetone to rats resulted in measurable levels of isopropanol in blood. Acetone can potentiate haloalkane's hepatotoxicity.

Human Toxicity

The primary human health effects after acute exposure to acetone include the following:

- Early emotional lability (exhilaration, boastfulness, talkativeness, remorse, and belligerency)
- Impaired motor coordination (slowed reaction time, slurred speech, and ataxia)
- Sensory disturbances (diplopia and vertigo)
- Flushing of face, rapid pulse, and sweating
- Nausea and vomiting
- Eventual incontinence of urine and feces
- Drowsiness, stupor, and finally coma, with impaired or absent tendon reflexes: Convulsive episodes may indicate hypoglycemia
- Pupils dilated or normal
- Peripheral vascular collapse (shock; hypotension, tachycardia, cold pale skin, and hypothermia)
- Slow stertorous respiration
- Death from respiratory or circulatory failure or from aspiration pneumonitis
- Effects during convalescence: postalcoholic headache and gastritis

Acute exposure may cause respiratory tract irritation, sore nose and throat, and cough. Effects are similar to those of ethyl alcohol exposure with greater anesthetic potency. In acute cases, a latent period may be followed by restlessness and vomiting leading to hematemesis and progressive collapse with stupor. Workers who were exposed to 1000 ppm, 3 hr per day for 7–15 years, complained of chronic inflammation of airways, stomach, and duodenum. Prolonged or repeated skin contact may defat the skin and produce dermatitis. Toxic and lethal concentrations in human blood are reported to be 200–300 and 550 μg/ml, respectively.

A total of 659 males occupationally exposed to acetone and other solvents were divided into nine unrelated groups working in plastic boat, chemical, plastic button, paint, and shoe factories. Urine samples were collected at the beginning of the work shift and at the end of the first half of the shift. A close relationship between the average environmental solvent concentration (mg/m^3) measured in the breathing zone and the urinary concentration of unchanged solvent (μg/liter) was observed. A biological equivalent exposure limit (56 mg/liter) corresponding to the environmental ACGIH TLV (58 mg/liter) was recommended for acetone. The biological exposure data for urine collected over 4 hr during random sampling for at least 1 year could be used to evaluate individuals or groups of workers for long-term exposure and the probability of noncompliance.

Ingestion of 200 ml of acetone produced severe coma, hyperglycemia, and acetonuria in an adult. This dose is approximately 2 or 3 ml/kg. About 2 or 3 ml/kg acetone for children may be considered a toxic oral dose. U.S. EPA states that acetone is "not classifiable as to human carcinogenicity" (Class D). Mutagenicity and genotoxicity studies reviewed in the IRIS database were negative except for one study that reported chromosomal aberrations.

Clinical Management

The symptomatic patient may need medical supervision (monitoring) for up to 30 hr because of the prolonged elimination half-life of acetone. In case of accidental ingestion, emesis is not recommended because of the potential for central nervous system (CNS) depression and subsequent aspiration. Gastric lavage should be considered in recent substantial ingestions. Activated charcoal is recommended. The patient should be treated symptomatically and blood glucose should be monitored.

For inhalation exposures, the victim should be removed from the source to fresh air. Supportive care should be provided. If cough or difficulty in breathing develops, the patient should be evaluated for respiratory tract irritation, bronchitis, or pneumonitis. Supplemental oxygen (100% humidified) should be administered with assisted ventilation as required.

Exposed eyes should be irrigated with copious amounts of tepid water for at least 15 min. If irritation, pain, swelling, tearing, or sensitivity to light persist, the victim should be seen in a health care facility. In case of dermal exposure, the exposed area should be washed thoroughly with soap and water. The victim should see a physician if irritation or pain persists.

Animal Toxicity

In a 90-day rat toxicity study (dose administered via gavage) a NOEL of 100 mg/kg/day and a LOAEL of 500 mg/kg/day were identified. The study showed increased liver and kidney weights and signs of nephrotoxicity. Acetone can cause moderate corneal injury in rabbits.

Male mice and rats were exposed for varying time periods to vapor levels of 12,600–50,600 ppm acetone. Unconditioned performance and reflex tests were used to measure CNS depression. Animals breathing acetone took 9 hr to recover from a 5-min exposure (blood levels were a reliable index for depression).

Female Sprague–Dawley rats were given 0.5, 1, or 2.5 ml/kg of acetone once by gavage. Sodium phenobarbital (100 mg/kg) was administered once a day for 3 days. The animals were killed 24 hr after the last dose. Livers were homogenized and microsomes were prepared by differential centrifugation. Microsomal lipids were extracted with a 2 to 1 chloroform methanol mixture. The extracted samples were assayed for total phosphate or resuspended in saline and assayed for cholesterol. Treatment with acetone did not cause alterations in the concentrations of total phospholipid and total cholesterol in microsomal membranes. Acetone had no effect on microsomal N-demethylation of aminopyrine; however, at the high dose, it significantly increased the metabolism of acetonitrile to cyanide.

Acetone was evaluated by the standard plate incorporation method in the Ames Salmonella reverse-mutation assay with strains TA98, TA100, TA1535, TA1537, and TA1538. Experiments were conducted in triplicate with and without metabolic activation (S9 fraction from Aroclor-treated Sprague–Dawley rats). Results were negative in these strains.

In studies of acetone-potentiated liver injury induced by haloalkanes, acetone is usually given by gavage, whereas industrial exposure to acetone normally occurs by inhalation. It was of interest to verify if the route of administration influences the potentiation. Male Sprague–Dawley rats were exposed for 4 hr to acetone vapors or treated orally with acetone; the minimal effective dose levels for potentiating CCl_4-induced liver injury were estimated to be 2500 ppm and 0.25 mg/kg, respectively. Groups were treated with acetone using 0.4, 1, 2, 4, or 6 times the minimal effective dose. Half

of each group was killed at various time intervals after treatment for blood acetone measurements by gas chromatography; the other half was challenged with CCl_4 (0.1 and 1 kg, intraperitoneal) 18 hr after acetone and killed 24 hr later. Plasma alanine aminotransferase (ALT) activity and bilirubin concentrations were measured. Inhalation and oral administration of acetone both potentiated CCl_4 toxicity. Rats exposed respectively to acetone vapors (10 daily exposures) and subsequently challenged with CCl_4 exhibited liver toxicity that was not significantly different from that of rats subjected to a single exposure. Correlations between ALT activities and maximal blood acetone concentrations were found to be linear and significant for both routes. For a given blood acetone concentration, however, toxicity was least severe following acetone exposure by inhalation.

—*Leyna Mulholland*

Related Topic

Skin

Acetonitrile

- ◆ CAS: 75-05-8
- ◆ SYNONYMS: Ethyl nitrile; methyl cyanide; ethanenitrile; cyanomethane; methanecarbonitrile
- ◆ CHEMICAL CLASS: Aliphatic nitriles
- ◆ MOLECULAR FORMULA: CH_3CN

Uses

Acetonitrile is used in the chemical industry as an intermediary in the synthesis of several chemicals and products such as acetophene, thiamine, acetamidine, α-naphthaleneacetic acid, nitrogen-containing compounds, acrylic fibers, nitrile rubber, pesticides, pharmaceuticals, and perfumes. It is also used as a polar solvent for both organic and inorganic compounds and in nonaqueous titrations.

Exposure Pathways

Exposure to acetonitrile can occur through the oral, dermal, and inhalation routes. Symptoms of poisoning have been observed in persons exposed through these three routes.

Toxicokinetics

Acetonitrile can be acutely lethal when absorbed in high doses. Acetonitrile is metabolized in the organism to cyanide. Once metabolized, the mechanism of action is the same as expected for cyanide poisoning (see entry for Cyanide). Onset of cyanide poisoning may be delayed 8 or more hours as metabolism is required to produce the cyanide metabolite. Toxicity may be prolonged for up to 3 days in some cases.

Mechanism of Toxicity

Acetonitrile is slowly metabolized in the liver to hydrogen cyanide. Toxicity is produced by the combined effect of circulating acetonitrile and cyanide.

Human Toxicity

Signs and symptoms of exposure will be determined by the dose of acetonitrile. Exposure to low doses will produce nausea, salivation, vomiting, headache, and lethargy. Exposure to higher doses may produce cyanide intoxication characterized by extreme weakness, lethargy, respiratory depression, metabolic acidosis, tachycardia, shock, coma, seizures, and possibly death.

Clinical Management

The major goal of treatment is to maintain respiration, blood circulation, and vital signs and to prevent further absorption of acrylonitrile into the systemic circulation. If ingested, absorption can be prevented or minimized by instituting gastric lavage or by giving activated charcoal and a cathartic. Gastric lavage is effective only if performed soon after ingestion.

Treatment of acetonitrile poisoning is similar to that of cyanide poisoning (see Cyanide). This includes immediate therapy with 100% oxygen and assisted ventilation, if necessary. Seizures can be controlled by giving diazepam, phenobarbital, or phenytoin intravenously at appropriate doses. Therapy should also include correction of the metabolic acidosis to combat cyanide poisoning. Cyanide poisoning is treated by the intravenous administration of sodium nitrite and sodium thio-sulfate. Care should be taken to maintain treatment for as long as acetonitrile is being metabolized to cyanide.

Animal Toxicity

The cyanogenic mode of action of acetonitrile has been observed in animals. Atmospheres containing up to 16,000–32,000 ppm acetonitrile are lethal to dogs. In rats, the oral LD_{50} has been measured to range from 200 mg/kg (in young rats) to 3800 mg/kg (age unspecified), whereas the inhalation LC_{50} has been determined to be 7500 ppm following an 8-hr exposure. The acute dermal lethal dose has been investigated in rabbits. The LD_{50} through the dermal route has been determined to be 980 mg/kg. Subchronic exposures to low acetonitrile concentrations in the air (665 ppm or less) produced pulmonary inflammation and minor changes in body weights, hematocrit, hemoglobin, and liver and kidney function.

—Heriberto Robles

Acetylaminofluorene

- ♦ CAS: 28322-02-3
- ♦ SYNONYMS (4-ISOMERS): *N*-Fluorene-4-yl-acetamide; *N*-4-fluorenylacetamide; *N*-fluoren-4-yl-acetamide; 4-acetylaminofluorene
- ♦ CHEMICAL CLASS: Aromatic amine
- ♦ CHEMICAL STRUCTURE:

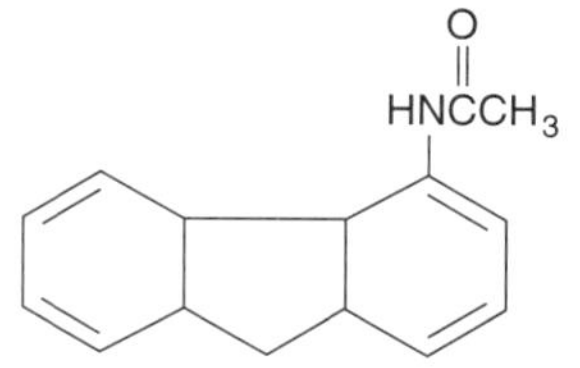

Uses

Acetylaminofluorene is found as a contaminant in coal gasification processes. It has no known use.

Exposure Pathways

Skin contact is the most common accidental exposure pathway. Acetylaminofluorene emits toxic fumes of nitrous oxides when heated to decompose and can be toxic when inhaled.

Toxicokinetics

Acetylaminofluorene is biotransformed in the liver. 2-Acetylaminofluorene can stimulate cytochrome P-450 1A1 isozyme (CYP1A1) activity, inducing both CYP1A1 and CYP1A2 proteins, whereas 4-acetylaminofluorene modestly increases CYP1A2 but does not influence CYP1A1.

Mechanism of Toxicity

4-Acetylaminofluorene is not carcinogenic; 2-acetylaminofluorene is carcinogenic. 2-Acetylaminofluorene can be metabolized to form *N*-hydroxy-acetylaminofluorene and 2-aminofluorene, which may covalently bind to the DNA and macromolecules. Ring hydroxylation, however, leads to the formation and excretion of water-soluble conjugates (e.g., glucuronides) of the respective hydroxylated metabolites and detoxification.

Human Toxicity

Human exposure data are not available. 2-Acetylaminofluorene is thought to be carcinogenic to humans.

Clinical Management

In case of contact with the eyes, the eyes should be immediately flushed with running water for at least 15 min. Affected skin should be washed with soap and water. If vapor is inhaled, the victim should be moved to fresh air and emergency medical care provided. If the victim is not breathing, artificial respiration should be provided; if breathing is difficult, oxygen should be administered. Contaminated clothing and shoes should be removed and isolated at the site.

Animal Toxicity

No information is available on the toxicity of acetylaminofluorene in cats and dogs. 2-Acetylaminofluorene is carcinogenic to the mouse and rat. The LD_{50} of 4-acetylaminofluorene in mice is 364 mg/kg by the intraperitoneal route. When fed to male rats (0.05% of diet) for 3 or 4 weeks, 4-acetylaminofluorene caused proliferation of agranular endoplasmic reticulum and glycogen depletion in hepatocytes. The same treatment when continued for 10 months produces conspicuous morphological alterations in pancreatic granular endoplasmic reticulum together with mitochondrial damage and focal cytoplasmic degradation.

—*Sanjay Chanda*

Related Topic

Liver

Acetylcholine

- ◆ CAS: 51-84-3
- ◆ Synonyms: Acecoline; choline acetate; arterocholine; 2-(acetoxy)-*N, N, N*-trimethylethanaminium; ethanaminium; 2-(acetyloxy)-*N, N, N*-trimethyl
- ◆ Pharmaceutical Class: Neurohumoral transmitter
- ◆ Chemical Structure:

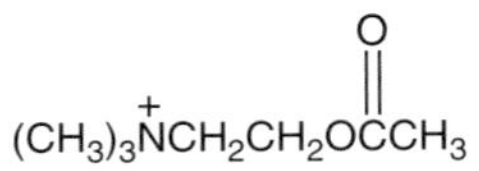

$$(CH_3)_3 \overset{+}{N} CH_2CH_2O \overset{\overset{O}{\|}}{C} CH_3$$

Uses

Acetylcholine is present naturally in the body. Commercial drugs used as cholinergic agonists mimic the action of acetylcholine.

Exposure Pathways

Acetylcholine is present in the body as a neurohumoral transmitter. Acetylcholinesterase is the enzyme responsible for breakdown of acetylcholine in the synapse. Any drug classified as cholinergic agonist (that mimics the action of acetylcholine) or anticholinesterase agent (e.g., organophosphorus pesticides, which block the action of acetylcholinesterase and hence stop the breakdown of acetylcholine in the synapse) can increase the level of acetylcholine in the body.

The most common exposure pathways for the cholinergic agonists are ingestion or contact to the eye. Acetylcholine chloride is available as an intraocular solution, methacholine chloride is available as a powder, bethnacol chloride is available as tablets, and carbachol is available as an ophthalmic solution. Common exposure pathways to anticholinesterase agents are ingestion, dermal or ocular contact, or inhalation.

Toxicokinetics

Acetylcholine is broken down by the acetylcholinesterase enzyme to choline and acetate. The time required for hydrolysis of acetylcholine is less than a millisecond. If the enzyme is depleted or inhibited, then excessive acetylcholine accumulation in the body can cause toxicity. Symptoms are salivation, lacrimation, urination, diarrhea, muscle tremor, and fasciculation.

Mechanism of Toxicity

Cholinergic agents can increase the acetylcholine level at the synaptic junction and cause rapid firing of the postsynaptic membrane. Anti-acetylcholinesterase agents block the acetylcholinesterase enzyme and thus increase the acetylcholine level in the synapse causing rapid firing of the postsynaptic membrane.

Human Toxicity

Acetycholine agents are contraindicated in persons with asthma, hyperthyroidism, coronary insufficiency, and peptic ulcer. The bronchoconstrictor effect may precipitate asthma, hyperthyroid patients may develop atrial fibrillation, hypotension induced by these agents can reduce coronary blood flow, and gastric acid secretion caused by these agents can aggravate the symptoms of peptic ulcer. Excessive acetylcholine can also cause flushing, sweating, bradycardia, hypotension, abdominal cramps, belching, diarrhea, sensation of tightness in the urinary bladder, involuntary defecation and urination, penile erection, difficulty in visual accommodation, headache, salivation, and lacrimation. It can also cause paralysis of the respiratory muscles. Central nervous system effects include ataxia, confusion, slurred speech, loss of reflexes, Cheyne–Stokes respiration, and

finally coma. The time of death after a single acute exposure ranges from 5 min to 24 hr depending on the route, dose, and agent of exposure (among other factors).

Chronic toxicity can cause polyneuritis, which starts with mild sensory disturbances, ataxia, weakness, and ready fatigability of legs, accompanied by fasciculation, muscle twitching, and tenderness to palpitation. In severe cases, the weakness may progress eventually to complete flaccid paralysis that, over the course of weeks or months, is often succeeded by a spastic paralysis with a concomitant exaggeration of reflexes. During these phases, muscles show marked wasting.

Clinical Management

Exposure should be terminated as soon as possible either by removal of the patient or by fitting the patient with a gas mask if the atmosphere remains contaminated. Contaminated clothing should be removed immediately; the skin and mouth should be washed with copious amounts of water. Gastric lavage should be conducted if necessary. Artificial respiration should be administered if required, and administration of oxygen may be necessary. If the convulsion persists, diazepam (5–10 mg intravenously) or sodium thiopental (2.5% intravenously) should be administered, and the patient should be treated for shock. Atropine should be administered in sufficiently large doses, but atropine is without any effect against peripheral neuromuscular activation and subsequent paralysis. Pralidoxime (1 or 2 g infused intravenously) should be administered for all the peripheral effects.

Animal Toxicity

The toxic effects seen in animals are similar to those noted in humans.

—*Sanjay Chanda*

Related Topics

A-esterase
Anticholinergics
Cholinesterase Inhibition
Neurotoxicology: Central and Peripheral
Respiratory Tract
Organophosphates

Acetylene

- CAS: 74-86-2
- SYNONYMS: Acetylene; ethine; ethyne; narcylen; welding gas
- CHEMICAL CLASS: Aliphatic hydrocarbon (C_nH_{2n-2})
- CHEMICAL STRUCTURE:

$$H—C \equiv C—H$$

Uses

Acetylene is used primarily for welding; it is also a source of manufacturing carbon black, acetaldehyde, and acetic anhydride.

Exposure Pathways

Industrial exposures to acetylene are most likely to occur by inhalation, with potential for skin and eye contact.

Toxicokinetics

The toxicokinetics for acetylene are similar to other simple asphyxiants causing tissue hypoxia. A high concentration of acetylene can deplete the atmosphere of available oxygen. This lack of oxygen does not afford adequate oxygen transport by hemoglobin of the red blood cells. Unoxygenated blood is then distributed throughout the systemic circulation causing tissue hypoxia. Following inhalation, acetylene is rapidly absorbed in the blood, distributed throughout the body, and deposited in the brain, producing central nervous system effects. Profuse salivation occurs at 700,000 ppm, with narcosis and full anesthesia at 800,000 ppm, followed by stimulated respiration and increased blood pressure leading to unconsciousness and death. Elimination can occur mostly through exhalation of unchanged acetylene from the lungs and some elimination by excretion in the urine.

Mechanism of Toxicity

Acetylene is classified as a simple asphyxiant because it is physiologically inert but can deplete the atmosphere of available oxygen when present in high con-

centrations and deprive the tissues of necessary oxygen. Signs of asphyxia will be noted when atmospheric oxygen is reduced to concentrations of 15 or 16% or less due to the high levels of acetylene. The tissues that are most sensitive to hypoxia are the brain and heart.

There are also potential effects of the known impurities (phosphine, arsine, hydrogen sulfide, carbon disulfide, and carbon monoxide) when the concentration of acetylene increases above 30,000 ppm.

Human Toxicity

The major concern with acetylene exposure is its action as a simple asphyxiant. Therefore, in high concentrations (60,000 ppm and higher), acetylene can deplete the atmosphere of available oxygen and cause asphyxiation due to lack of oxygen in the body resulting in tissue hypoxia. At 100,000 ppm, there is a slight intoxication which leads to a staggering gait and general incoordination as the acetylene concentration increases to 300,000 ppm. Unconsciousness occurs in 5–7 min at around 350,000 ppm. Profuse salivation occurs at 700,000 ppm with narcosis and full anesthesia at 800,000 ppm, followed by stimulated respiration and increased blood pressure leading to unconsciousness and death.

At acetylene concentrations above 30,000 ppm, there is also a concern for possible toxicity due to the major impurities [(phosphine ([less than 95 ppm]), arsine ([less than 3 ppm]), hydrogen sulfide, carbon disulfide, and carbon monoxide)]. The toxicity of these contaminants can pose an additional threat to human health. ACGIH classifies acetylene as a simple asphyxiant; a TLV is not recommended since the available oxygen is the limiting factor. The minimal oxygen content should be 18% by volume under normal atmospheric pressure (OSHA standard is 19.5%). Special care should be taken for the potential explosion hazard when the lower explosive limit (2.5%) is reached.

Clinical Management

The exposed individual should be removed from the toxic environment and given 100% humidified supplemental oxygen with assisted ventilation as required. If hypoxia has been prolonged, the patient should be evaluated for neurologic sequelae and supportive treatment provided. Exposed eyes should be irrigated and treated in a health care facility if irritation and pain persist. Dermal exposure should be treated as indicated for frostbite injury.

Animal Toxicity

Because of its history as an anesthetic, there is much information regarding the acute exposure of animals to acetylene, but there is little information regarding chronic exposure. NIOSH has summarized the acute studies of acetylene on dogs, cats, and rabbits. In dog studies, there was a rapid anesthetic effect at 500,000 ppm, which was pronounced at 750,000 ppm. In the cat, reduced respiration and increased blood pressure was demonstrated at 400,000–800,000 ppm. In the rabbit, reduced respiration and increased blood pressure was experienced at 600,000–800,000 ppm.

—Edward Kerfoot

Acetylsalicylic Acid

♦ CAS: 50-78-2

♦ PREFERRED NAME: Aspirin

♦ SYNONYMS: 2-Acetoxybenzoic acid; *o*-carboxyphenyl acetate; acetisal; acetonyl; acetophen; acetosal; acetylin; 2-(acetyloxy) benzoic acid

♦ PHARMACEUTICAL CLASS: Synthetic derivative of salicylic acid

♦ CHEMICAL STRUCTURE:

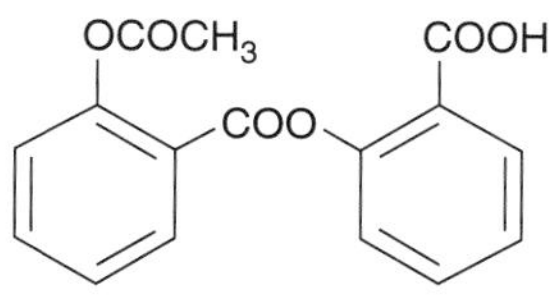

Uses

Aspirin is an analgesic, antipyretic, and antiinflammatory agent.

Exposure Pathways

Ingestion is the most common route of both accidental and intentional exposure. Aspirin is available in oral and rectal dosage forms.

Toxicokinetics

Absorption of aspirin occurs by passive diffusion across the gastrointestinal membrane and is influenced by gastric pH. The presence of food delays the absorption of aspirin. Unhydrolyzed aspirin does not undergo capacity-limited metabolism and does not accumulate in plasma following large doses. The elimination half-life of aspirin in plasma is approximately 15–20 min. Aspirin is excreted in the urine as glycine, or glucuronide conjugate, or as free salicylic acid.

Mechanism of Toxicity

Since some percentage of aspirin is excreted as such, at high doses it causes severe acidosis. When nonionized salicylic acid in the gastric mucosa enters mucosal cells, large amounts of salicylate can accumulate because of the dissociation of the ionic species at the intracellular pH. As a result, gastric mucosal damage may occur.

Human Toxicity

Hypersensitivity to aspirin occurs in 0.2% of the general population. Aspirin is an acute irritant to the skin and eyes. Direct contact with the eyes is painful. Aspirin can cause severe ulceration of the conjunctiva and superficial injury to the eye. Aspirin is a known respiratory allergen. It can cause acidosis if a large volume is ingested. A large oral dose can also cause a burning pain in the throat, deep and rapid breathing, apathy, anorexia, irritability, restlessness, disorientation, delirium, convulsions, coma, and death due to cardiovascular collapse or respiratory failure. Renal or hepatic insufficiencies increase the chance of toxicity. Aspirin also has anticoagulant action. In addition, the use of aspirin is contraindicated in children and adolescents with febrile viral illnesses because of the risk of Reye's syndrome.

There is no evidence that long-term use in therapeutic doses causes fetal damage. Long-term use may cause damage to the gastric mucosa. Aspirin may cause iron-deficiency anemia during long-term therapy. Chronic administration can decrease plasma protein-bound iodine and thyroidal uptake.

Clinical Management

Absorption of salicylate from the gastrointestinal tract can be reduced by emesis or gastric lavage. Syrup of ipecac can be used to induce emesis. After gastric lavage, activated charcoal can be instilled to retard further absorption of remaining aspirin. Treatment is largely symptomatic. In patients presented with acidosis, low blood pH can be corrected by the use of bicarbonate solution administered intravenously. If potassium deficiency occurs, it should be treated by adding cationic fluid once it has been determined that the urine formation is adequate.

Animal Toxicity

Cats are more susceptible to aspirin than other species of animals because cats do not metabolize the drug rapidly. Parenteral administration in cats causes gastric ulceration because of cyclooxygenase inhibition. In days, aspirin may cause irritant action on gastric mucosa, nausea, and vomiting. Aspirin can also cause gastric hemorrhage in dogs if used for a long time (4 weeks).

—*Sanjay Chanda*

Related Topic

Salicylates

Acids

♦ CHEMICAL CLASS: Acids

♦ CHEMICAL STRUCTURE: Varies

Uses

Acids have a wide range of uses. The specific use depends on the specific acid.

Exposure Pathways

Dermal contact, inhalation, or ingestion are the most common exposure pathways.

Toxicokinetics

The toxicokinetics depends on the specific type of acid.

Mechanism of Toxicity

Acids cause toxicity by coagulative necrosis, meaning that the acid denatures all tissue protein to form an acid proteinate. As a result both structural and enzymatic proteins are denatured, and cell lysis is blocked. Therefore, cell morphology is not greatly interrupted. In addition, an eschar is formed which delays further corrosive damage and helps reduce systemic absorption. Thus, damage, especially with small quantities of acid, is frequently limited to local sites of injury to the skin or the gastrointestinal tract, rather than the systemic response.

Human Toxicity

Acids are corrosive to skin and gastric mucosa. Repetitive exposure to acids may induce mucosal forestomach hyperplasia.

Clinical Management

Exposure should be terminated as soon as possible by removal of the patient to fresh air. The skin, eyes, and mouth should be washed with copious amounts of water. A 15- to 20-min wash may be necessary to neutralize and remove all residual traces of the contaminant. Contaminated clothing and jewelry should be removed and isolated. Contact lenses should be removed from the eye to avoid prolonged contact of the acid with the area. A mild soap solution may be used for washing the skin and as an aid to neutralize the acid, but it should not be placed into the eye. No cream, ointment, or dressing should be applied to the affected area. Emesis should be avoided in case of ingestion. If a large quantity has been swallowed, then gastric lavage should be considered. Dilution with water may be effective for small quantities swallowed. Under no circumstances should carbonated beverages ever be used because of large quantities of carbon dioxide gas released that distends the stomach.

Animal Toxicity

The toxic effects seen in animals are the same as those seen in humans.

—*Sanjay Chanda*

Related Topics

Alkalies
Corrosives
Gastrointestinal System
Skin

Aconitum Species

♦ SYNONYMS: *Aconitum napellus—Ranunculaceae* (buttercup) family; monkshood; wolfsbane; helmet flower; friar's cap; soldier's cap; aconite

Description

Aconitum are perennial herbs with a blackish tuberous rootstock that gives rise to several palmate or cleft leaves. Wild plants often have blue-mauve flowers. Cultivated flowers range in color from rich blue to dark purple, purple, white, or yellow. They are bilaterally symmetrical with five membered flowers; the uppermost is shaped like a large, downward-opening hood. This feature gives the genus its name and distinguishes it from the larkspur. It grows from 1 to 5 ft high. The ripe follicles contain many seeds. Aconitum occur naturally in the northern temperature zones of North America, Great Britain, Europe, and Asia. It usually prefers shady, moist places. Many cultivated forms and species are grown widely outdoors and in gardens.

Exposure Pathways

The most common route of exposure is ingestion of any parts of the plant, root, or powders and paste derived from it. Dermal exposure is also possible. Aconitine can be rapidly absorbed through membranes and intact skin. Therefore, handling the plants and tubers can produce toxicity.

Toxicokinetics

Aconitine is very rapidly absorbed after ingestion, usually within a few minutes. Absorption also occurs with dermal contact.

Mechanism of Toxicity

Aconitum species contains potent diperpenoid ester alkaloids including aconitine, mesaconitine, and jesaconitine, the three major toxins. The nitrogen of these compounds is usually ethylated or methylated to make them alkamines. The diterpenes are of relatively low toxicity, but the esterified norditerpene bases have higher toxicity. If the ester functions are hydrolized, toxicity lessens to that of ordinary diterpenes. The pow-erful cardiac agent, aconite, has a vagal action that causes slowing of the heart. There is a more direct effect on cardiac muscle, which causes a spontaneous auricular rate increase and spontaneous ventricular beating. Atrial fibrillation is uncommon. Aconitum roots contain physiologically active catecholamine analogs and can also possess hypoglycemic actions.

The toxicity of any particular aconite plant varies depending on the amount of diterpenes versus the number of norditerpenes in relation to the amount of esterification of the norditerpenes. All parts of the plant contain toxic alkaloids, with the content and composition of these varying throughout the year. It is most toxic in its preflowering stage.

Human Toxicity: Acute

Acontin's minimal lethal dose is 3–6 mg. One gram of fresh *Aconitum napellus* may contain 2–20 mg of aconitine. Therefore, small amounts of this plant can be lethal. The effects produced by aconite poisoning are similar to that of veratrum alkaloids (veratrine) with the exception of the parasthesias being more prominent and persistent. A burning sensation and tingling of the mouth, lips, tongue, and throat occur almost instantly, within 10–20 min. This is usually followed within 2–6 hr by nausea, salivation, violent emesis, generalized paresthesias, weakness, and extreme pain. Colickly diarrhea, skeletal muscle paralysis, cardiac rhythm disturbances, convulsions, and death may follow in up to 8 hr. Cardiac toxicity often complicates serious aconitine poisoning with hypotension, conduction delays, and dysrhythmias within 6 hr. Respiratory paralysis is often the cause of death.

Human Toxicitiy: Chronic

Chronic toxicity is not expected.

Clinical Management

Basic and advanced life-support measures should be utilized as necessary. Decontamination with syrup of ipecac is essentially contraindicated due to extensive vomiting, rapid onset of symptoms, and possible respiratory paralysis. Gastric lavage may be indicated if performed soon after ingestion, followed by activated charcoal. Fluid and electrolytes need to be frequently monitored and replaced as necessary with extensive vomiting and diarrhea.

Most arrhythmias are refractory to drug management. Atropine, quinidine, isoproterenol, digitalis, parasympathomimetics, and inorganic ions have been tried. For hypotension, intravenous fluids should be administered and Trendelenburg positioning used. If this is unsuccessful, dopamine and norepinephrine should be tried. There is no specific antidote. No specific laboratory tests are available.

Treatment is symptomatic and supportive after decontamination. Since toxicity is unpredictable due to alkaloid variability, observation for 2–4 hr is recommended. Symptomatic patients should be hospitalized for 24 hr with cardiac monitoring.

Animal Toxicity

General symptoms in animals primarily include vomiting, colic, bloating, bradycardia, bradypnea, muscle weakness, paralysis, and dilated pupils. Death is usually due to cardiopulmonary failure. Again, no specific antidote or treatment is available. The estimated lethal dose of aconitine is 2 or 3 mg in a dog and 10–12 mg in a horse.

—*Lanita B. Myers*

Related Topic

Cardiovascular System

Acrolein

- ◆ CAS: 107-02-8
- ◆ SYNONYMS: Acrylaldehyde; allyl aldehyde; ethylene aldehyde; 2-propenal
- ◆ CHEMICAL CLASS: Aldehydes
- ◆ CHEMICAL STRUCTURE:

$$\begin{array}{c} H \\ \diagdown \\ C = C - C = O \\ \diagup \quad | \quad | \\ H \quad \ H \quad H \end{array}$$

Uses

Acrolein is used as an intermediate in the manufacture of glycerol, polyurethane, and polyester resins, methionine, pharmaceuticals, and herbicides.

Exposure Pathways

Industrial exposures to acrolein are most likely to occur by inhalation with potential for skin and eye contact. Accidental ingestion is also possible.

Toxicokinetics

Following inhalation exposure, acrolein can be deposited in the nasal cavity and respiratory tract and eventually some traces can be absorbed into the blood and be distributed throughout the body. The uptake of acrolein in the nasal cavity is influenced by its solubility and inspiratory flow rate. Perhaps acrolein uptake in the nasal tissue is dependent on its reaction with tissue substrates that become depleted at high exposure concentrations. The acrolein vapor can be metabolized in the nasal cavity by the mixed-function oxidase and carboxylesterase systems. Further metabolism takes place in the liver to glycidaldehyde and a number of metabolites that can be excreted in the urine as well as some unchanged acrolein. Most of the free acrolein is excreted in the exhaled breath.

Mechanism of Toxicity

Acrolein is soluble in the mucous membranes of the upper respiratory tract causing irritation of the sensory nerve endings. There is also depression of the mucociliary defense system. The direct action of acrolein on the skin and eyes is the result of irritation to these tissues.

Human Toxicity

Acrolein is an acute hazard exhibiting irritation to the upper respiratory tract, skin, and eyes. Irritation to the mucous membranes occurs as low as 0.25 ppm within 5 min and marked irritation of the eyes and nose at 1 ppm for 5 min. Fatalities have occurred at exposures to concentrations of 150 ppm for 10 min, resulting in pulmonary edema and tracheobronchitis. The lowest lethal concentration reported is 10 ppm.

Skin irritation with erythema, edema, and sensitization can occur from prolonged or repeated contact with acrolein. Liquid splashes to the eye can cause corneal damage and exposures to concentrations of 0.25 ppm may cause eye irritation, lacrimation, conjunctivitis, lid edema, fibrinous or purulent discharge, and corneal injury. There is inadequate evidence in humans for chronic toxicity or carcinogenicity.

The TLV-TWA of 0.1 ppm is sufficiently low to minimize irritation to most exposed individuals and a 15-min STEL of 0.3 ppm is also recommended.

Clinical Management

Exposures by inhalation should be monitored for nasal and respiratory tract irritation, bronchitis, or pneumonitis. Humidified supplemental 100% oxygen should be administered.

Gastric lavage may be indicated soon after ingestion of acrolein followed by administration of activated charcoal slurry mixed with saline cathartic or sorbitol. Oxygen, in combination with intubation and mechanical ventilation, may be required in severe cases. Exposed eyes should be irrigated with copious amounts of tepid water for at least 15 min. If eye irritation, pain, swelling, lacrimation, or photophobia persist, the patient should be seen in a health care facility.

Animal Toxicity

The oral LD_{50} for acrolein in rats is 46 mg/kg. An animal inhalation exposure study at 10 ppm for 3.5 hr resulted in respiratory irritation in cats. Another inhalation study on rats exposed to 8 ppm for 4 hr resulted in the death of one animal while all the animals died at 16 ppm. A subchronic inhalation study with rats exposed to 4 ppm, 6 hr per day, for 60 days resulted in 32/57 animal deaths due to bronchiolar necrosis and focal emphysema.

The carcinogenicity potential of acrolein has not been adequately determined in chronic studies, but glycidaldehyde, a potential metabolite of acrolein, is considered to be carcinogenic. When tested in the *Salmonella* assay, acrolein was weakly positive. It was negative in *Drosophila* sex-linked recessive lethal test and negative for chromosome aberrations when tested in cultured Chinese hamster ovary cells; however, there was an increase in the frequency of sister-chromatid exchanges.

—*Edward Kerfoot*

Related Topics

Combustion Toxicology
Respiratory Tract

Acrylamide

- ◆ CAS: 79-06-01
- ◆ SYNONYMS: Acrylic amide; propenamide
- ◆ CHEMICAL CLASS: Amide
- ◆ MOLECULAR FORMULA: C_3H_5NO
- ◆ CHEMICAL STRUCTURE:

$$H_2C{=}CH{-}CO{-}NH_2$$

Uses

The primary use of acrylamide is in the production of polymers and copolymers. Polyacrylamide is used as a flocculant in wastewater treatment plants, as a coagulant in the treatment of potable water, and as a sizing agent in the paper and permanent press fabric industries. Acrylamide monomer is used to produce grouts and soil stabilizers for the construction of tunnels, dams, foundations, and roadways.

Exposure Pathways

Exposures to acrylamide monomer are most likely to occur in an occupational environment. The low vapor pressure of acrylamide (7×10^{-3} torr at 20°C) suggests that dermal contact (e.g., during grout applications) will be the most significant route of exposure in an occupational setting. However, inhalation exposures may result from pyrolysis during fires at tunneling and mining sites where acrylamide is used as a soil stabilizer.

Toxicokinetics

Acrylamide is well absorbed via the gastrointestinal and respiratory tracts. It is also well absorbed through the skin but less rapidly than observed through the gastrointestinal tract; a significant portion of the dermally applied dose remains in the skin. After absorption, acrylamide is rapidly metabolized. The major route of acrylamide biotransformation is conjugation with glutathione in the liver. Upon absorption into the blood, acrylamide is rapidly distributed throughout the body with an apparent volume of distribution equal to total body water. With the exception of erythrocytes, acrylamide and/or its metabolites do not exhibit preferential bioconcentration in any body tissue. Although equally distributed, these compounds appear to persist in the skin and testes.

Acrylamide is excreted primarily via the kidneys. About 60% of the administered dose appears in the urine within the first 24 hr of exposure. Metabolites of acrylamide constitute the majority of the dose excreted in the urine; only about 2% of the dose is excreted as the parent compound. Acrylamide and/or its metabolites are subject to enterohepatic circulation; about 6% of the applied dose is eliminated in the feces. About 5% of the dose is expired as CO_2.

Mechanism of Toxicity

Although the mechanism of acrylamide toxicity is unknown, it is generally thought that acrylamide must be transformed to an as yet unidentified intermediate before toxicity can become manifest.

Human Toxicity

Upon chronic exposure, acrylamide produces a motor and sensory polyneuropathy in which the distal regions of the longest and largest axons appear to be preferentially affected. These effects may be manifested by weakness, parethesias, fatigue, as well as decreased pinprick sensation and reflexes. Recovery generally occurs within a year following cessation of exposure although severe exposures may result in permanent peripheral nerve damage. Acrylamide may also cause allergic contact dermatitis.

Clinical Management

Clinical management involves removal from exposure and treatment of symptoms.

Animal Toxicity

Studies in several animal species indicate that acrylamide causes dose-related neurotoxic effects. Acrylamide has been observed to produce testicular lesions at high dose levels that also result in neurotoxicity. In a lifetime study, rats exposed to acrylamide in their drinking water developed neoplasms in multiple organ systems. The acute oral LD_{50} for acrlyamide in mice, rats, rabbits, and guinea pigs was determined to range between 107 and 180 mg/kg.

—Ralph Parod

Related Topics

Neurotoxicology: Central and Peripheral
Pollution, Water
Polymers

Acrylic Acid

♦ CAS: 79-10-7

♦ SYNONYMS: Acroleic acid; ethylenecarboxylic acid; propene acid; propenoic acid; vinylformic acid; 2-propenoic acid; RCRA waste number U008; UN 2218 (DOT)

♦ CHEMICAL CLASS: Copolymer

♦ CHEMICAL STRUCTURE:

$$CH_2=CHCOOH$$

Uses

Acrylic acid derivatives treated with heparin are used to coat surfaces of clinical equipment. Acrylic acid is also used as a copolymer component in aerosol hair spray, in plastics, in molding powder for signs, in paint formulations, in leather finishing, in paper coatings, and in latex applications to prevent premature coagulation. It is also used in the production of hydrogels used for contact lenses.

Exposure Pathways

Inhalation or skin contact are the most common exposure pathways. Acrylic acid is available as a colorless liquid.

Toxicokinetics

The excretion half-life of acrylic acid has been found to be 20–40 min. Both *in vivo* and *in vitro* studies of acrylic acid metabolism have produced strong evidence that the metabolism proceeds by a mitochondrial biochemical pathway for proprionic acid metabolism that normally functions in the body at the final stages of breakdown of fatty acids and the production of intermediates for the tricarboxylic acid cycle. It is primarily excreted as carbon dioxide through the lungs. 3-Hydroxypropionate has been found to be a major metabolite. Part of acrylic acid also binds to glutathione and excreted as the cysteine conjugate in the urine.

Some part of acrylic acid can also be converted to acrylyl-CoA and reacts with glutathione to be excreted as cysteine conjugate.

Mechanism of Toxicity

Acrylic acid causes toxicity by rapid polymerization in the presence of light, heat, and oxygen and thereby interfering with incorporation of thymidine into DNA and uracil into RNA.

Human Toxicity: Acute

Acrylic acid is corrosive to skin. Acrylic acid vapor can cause light to moderate skin and eye irritation. It can also cause forestomach edema.

Human Toxicity: Chronic

Repetitive exposure to acrylic acid may induce mucosal forestomach hyperplasia.

Clinical Management

Exposure should be terminated as soon as possible by moving the victim to fresh air. The skin, eyes, and mouth should be washed with copious amounts of water. Contaminated clothing should be removed and isolated. The victim should be kept calm and normal body temperature should be maintained. Artificial respiration should be provided if the breathing has stopped. Treatment is usually symptomatic.

Animal Toxicity

Acrylic acid has been tested on mice, rats, and rabbits. The toxicity of acrylic acid in animals is similar to that found in humans.

—Sanjay Chanda

Related Topic

Polymers

Acrylonitrile

- ◆ CAS: 107-13-1
- ◆ SYNONYMS: Acritet; carbacryl; propenenitrile; ventox; vinyl cyanide; TL 314
- ◆ CHEMICAL CLASS: Industrial chemical, solvent
- ◆ MOLECULAR FORMULA: C_3H_3N

Uses

Acrylonitrile is used in the manufacture of acrylic fibers and in the plastic surface coatings and adhesive industries. It is also used as a pesticide/fumigant. It is a chemical intermediate in the synthesis of antioxidants, pharmaceutical dyes, surface-active agents, and reactions requiring the cyanoethyl group.

Exposure Pathways

Accidental exposure can occur via dermal contact, ingestion, or inhalation. Acrylonitrile is found in cigarette smoke. It does not occur naturally.

Toxicokinetics

Acrylonitrile is absorbed by way of inhalation, ingestion, and percutaneously. Rats treated with [^{14}C] acrylonitrile via oral or intravenous routes produced radioactivity in the blood, liver, kidneys, lungs, adrenal cortex, and stomach mucosa. Significant amounts are retained in the plasma. Acrylonitrile is metabolized to a lesser extent in humans than in rodents. Acrylonitrile metabolism in humans follows first order kinetics and half-life of approximately 8 hr. The elimination of acrylonitrile from the plasma of rats is biphasic, with a half-life of 3.5–5.8 and 50–77 hr in the a and b phases, respectively.

There are four major pathways of metabolism for acrylonitrile: formation of glucuronides, direct reaction with glutathione to form cyanoethyl mercapturic acid, direct reaction with the thiol groups of proteins, and epoxidation to 2-cyanoethylene oxide. N-acetyl-S-(2-cyanoethyl)-L-cysteine is a major urinary metabolite in human volunteers exposed to 5–10 mg.

Mechanism of Toxicity

Acrylonitrile owes some of its toxicity to cyanide generation, which inhibits cellular respiration (see Cyanide). Preinduction of microsomal MFO with Arochlor 1254 greatly enhanced the toxicity of acrylonitrile and caused a threefold increase in cyanide levels in rats. Therefore, metabolic activation appears to be necessary in the toxicity of acrylonitrile. The direct reaction of acrylonitrile with the SH groups of proteins and its epoxide metabolite are also thought to be responsible for its effects.

Human Toxicity: Acute

Mild exposures usually involve eye irritation, headache, nausea, and weakness. Serious exposures may result in asphyxia and death. The order of onset of symptoms is weakness in the limbs, dyspnea, burning sensation in the throat, dizziness and impaired judgment, cyanosis and nausea, irregular breathing, seizures, and death. Seizures and cardiac arrest may occur without warning.

Human Toxicity: Chronic

Chronic exposures have been associated with liver damage. Acrylonitrile is a suspected human carcinogen.

Clinical Management

In oral exposure, gastric lavage may be performed soon after ingestion or in patients who are comatose or at risk of convulsing. The volume of lavage return should approximate the volume given. Charcoal slurry, aqueous or mixed with saline cathartic or sorbitol, may be administered. The usual charcoal dose is 30–100 g in adults and 15–30 g in children (1 or 2 g/kg in infants). In case of inhalation exposure, the patient must be moved to fresh air for respiratory distress. If cough or difficulty in breathing develops, evaluation for respiratory tract irritation, bronchitis, or pneumonitis must be performed. For eye exposure, eyes must be washed with copious amounts of tepid water for at least 15 min. If irritation, pain, lacrimation, or photophobia

persist, the patient should be removed to a health care facility. For dermal exposure, the exposed area must be washed thoroughly with soap and water.

Animal Toxicity

Sufficient evidence has been collected to suggest that acrylonitrile is an animal carcinogen. It was found to be both embryotoxic and teratogenic in experimental animals.

—Raja S. Mangipudy and Harihara M. Mehendale

Related Topics

Combustion Toxicology
Polymers
Respiratory Tract

Adiponitrile

- ◆ CAS: 111-69-3
- ◆ SYNONYMS: Adipic acid dinitrile; adipic acid nitrile; 1,4-dicyanobutane; hexanedinitrile; tetramethylene cyanide
- ◆ MOLECULAR FORMULA: $C_6H_8N_2$

Uses

Adiponitrile is a starting chemical intended for synthesis of hexamethylenediamine to make nylon, corrosion inhibitors, and rubber accelerators. It is also used for synthesis of adipoguanamine used as an extractant for aromatic hydrocarbons.

Exposure Pathways

Adiponitrile may be inhaled, swallowed, or absorbed through skin.

Adiponitrile could potentially be released to the environment in the effluent or emissions from plants manufacturing adiponitrile, hexamethyelenediamine, or nylon-66. If released to soil, aerobic biodegradation may be an important removal mechanism. Although adiponitrile has the potential to undergo extensive leaching, biodegradation should limit movement through soil. Volatilization from soil surfaces is not expected to be significant. If released to water, aerobic biodegradation may again be an important removal mechanism.

Toxicokinetics

Seventy percent of the dose (about 50 mg/kg) administered subcutaneously to guinea pigs was eliminated as thiocyanate in urine. After application of adiponitrile to depilated skin, skin penetration was suggested by increased thiocyanate in urine. Greater quantities were absorbed when skin was abraded. Based on the ratio between administered adiponitrile dose and quantity of cyanide detected, it was shown that a greater part of the dose was metabolized to cyanide. Cyanide thus released is the principle cause of toxicity (see Cyanide).

Mechanism of Toxicity

Adiponitrile's mechanism of toxicity is similar to cyanide because it can potentially liberate cyanide in the body spontaneously. It forms a stable complex with ferric iron in the cytochrome oxidase enzymes, thereby inhibiting cellular respiration. Cyanide affects primarily the central nervous system (CNS), producing early stimulation followed by depression. It initially stimulates the peripheral chemoreceptors (causing increased respiration) and the carotid bodies (thereby slowing the heart). Early CNS, respiratory, and myocardial depression result in decreased oxygenation of the blood and decreased cardiac output. These effects produce both stagnation and hypoxemic hypoxia in addition to cytotoxic hypoxia from inhibition of mitochondrial cytochrome oxidase.

Human Toxicity

Vapors are irritating to the eyes and respiratory system at higher concentrations. Humans may experience tightness in the chest, headache, weakness with difficulty in standing, and vertigo, cyanotic, rapid respirations, low blood pressures, and tachycardia. Mental confusion and tonic clonic contractions of limbs and facial muscles may occur after exposure to higher concentrations. Contact with skin and eyes may cause burns. Adiponitrile may be fatal if absorbed through skin, inhaled, or swallowed.

The ACGIH TLV is 2 ppm (with a skin designation indicating the potential significant contribution to the overall exposure by the cutaneous route). Short-term inhalation limits are not available. Adiponitrile is not classifiable as a human carcinogen because no human or animal cancer data are available.

Environmental Regulation

Adiponitrile is produced as an intermediate or final product by a process covered under regulatory performance standards that have been promulgated to protect the atmosphere from equipment leaks of volatile organic compounds (VOCs) in the synthetic organic chemical manufacturing industry (SOCMI). The intended effect of these standards is to require all newly constructed, modified, and reconstructed SOCMI process units to use the best demonstrated system of continuous emission reduction for equipment leaks of VOCs, considering costs, non-air quality health and environment impact and energy requirements.

Clinical Management: Emergency Treatment

The affected person should be removed from exposure to adiponitrile immediately. Contaminated clothes should be removed and the patient sponged to avoid any absorption through skin. Immediate cardiopulmonary resuscitation should be administered. If the victim breathes with difficulty, oxygen should be given. In case of ocular contact, the eyes should be flushed with copious amounts of water for at least 20 min. In cases of ingestion, vomiting should be induced. Mouth-to-mouth resuscitation should be avoided in order to prevent self-poisoning.

Clinical Management: Medical Treatment

The goal of medical treatment is to eliminate the cyanide formed in the body. Sodium nitrate, amylnitrate, and thiosulfate should be administered. Sodium nitrite should be administered intravenously very slowly. Amylnitrite can also be inhaled from ampules. Later, sodium thiosulfate should be administered.

Sodium nitrate reacts with hemoglobin in the red blood cells forming methemoglobin, which in turn can react with the free cyanide ion forming cyanmethemoglobin, thereby binding free cyanide and preventing its reaction with cytochrome oxidase enzymes in the cells. Cyanmethemoglobin dissociates slowly into free cyanide plus methemoglobin. The cyanide released by dissociation of cyanmethemoglobin then reacts with the thiosulfate ion forming thiocyanate, a relatively nontoxic compound that is excreted in the urine.

Animal Toxicity

The oral LD_{50} is 155 mg/kg in rats, 172 mg/kg in mice, and 22 mg/kg in rabbits. The subcutaneous LD_{50} in the guinea pig is 50 mg/kg. Adiponitrile was negative for mutagenicity in salmonella with or without bioactivation.

—Shashi Kumar Ramaiah and
Harihara M. Mehendale

Aerosols

◆ DESCRIPTION: Aerosols consist of very finely subdivided liquid or solid particles dispersed in and surrounded by a gas.

◆ PHARMACEUTICAL CLASS: Aerosols are systems ranging from those of colloidal nature to systems consisting of "therapeutic packages." Aerosols are classified as follows:

(A) Liquified-gas systems
    Two-phase:
        Space-spray
        Surface-coating
        Dispersion or suspension
    Three-phase:
        Two-layer
        Foam
            Stabilized
            Quick-breaking
(B) Compressed-gas systems
    Solid-stream dispensing
    Foam dispensing
    Spray dispensing
(C) Separation of propellant from concentrate systems
    Piston type
    Flexible type
    Atomizer type
    Mechanical systems
    Latex diaphragm

Uses

Many therapeutically active ingredients are administered or applied to the body by means of the aerosol dosage form, including agents such as epinephrine, isoproterenol, antibiotics, antiseptics, steroids, and ergotamine. Oral aerosols have been used for the symptomatic treatment of asthma as well as for the treatment of migraine headaches, whereas topical aerosols find use in numerous dermatological manifestations.

Human Toxicity

The inflammability and toxicity of the propellant needs to be considered. Additionally, the topical effects of the propellants must be determined.

Clinical Management

Supportive care must be instituted for patients accidentally exposed to aerosol contents via topical, inhalation, or oral routes.

—Raja S. Mangipudy and
Harihara M. Mehendale

A-esterase

The phosphorothioate insecticides, such as chlorpyrifos and methyl parathion (Fig. A-4), are some of the most commonly used organophosphorus insecticides in the United States. Interestingly, these compounds have little capacity to inhibit the enzyme

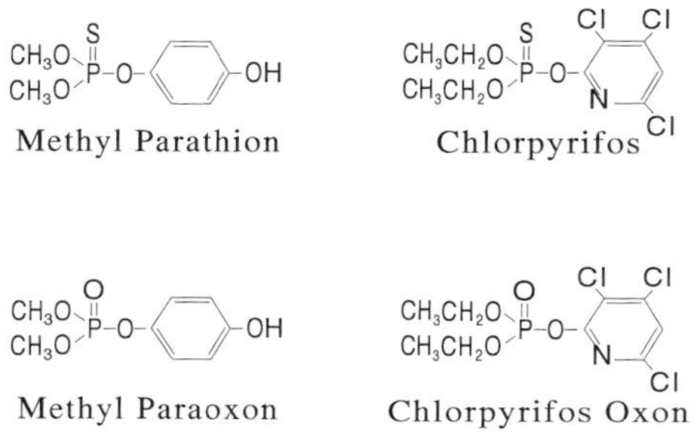

CH₃O—P(=S)(—O—⟨benzene⟩—OH) structure
Methyl Parathion

CH₃CH₂O—P(=S)(—O—⟨pyridine with Cl,Cl,Cl⟩) structure
Chlorpyrifos

CH₃O—P(=O)(—O—⟨benzene⟩—OH) structure
Methyl Paraoxon

CH₃CH₂O—P(=O)(—O—⟨pyridine with Cl,Cl,Cl⟩) structure
Chlorpyrifos Oxon

FIGURE A-4. *Structures of the phosphorothioate insecticides chlorpyrifos and methyl parathion, and their corresponding oxygen analogs.*

acetylcholinesterase and are, therefore, not highly toxic themselves. However, they are converted by the liver to potent acetylcholinesterase inhibitors termed oxons (such as chlorpyrifos oxon and methyl paraoxon; Fig. A-4), which are responsible for the toxicities observed following exposure to phosphorothioate insecticides. Once the oxons have been produced from the parent insecticides, one of the ways in which these highly toxic compounds can be metabolized by a variety of species is through their hydrolysis by an enzyme(s) termed A-esterase(s) (Fig. A-5). Since the products of these hydrolysis reactions are usually of low toxicity, A-esterase(s) catalyzes the detoxification of these oxons. Consequently, A-esterase(s) likely plays an important role in the protection of mammals against phosphorothioate insecticide toxicity. For example, although paraoxon and chlorpyrifos oxon have about the same capacity

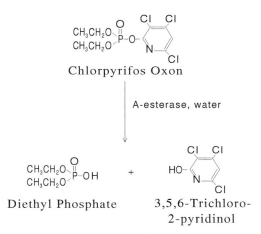

FIGURE A-5. *Hydrolysis of chlorpyrifos oxon by A-esterase. Although water is also involved in the reaction, it is usually ignored because it is present at an extremely high and constant concentration.*

to inhibit acetylcholinesterase, the insecticide chlorpyrifos is about 10 times less toxic to laboratory mice and rats than is parathion, probably because chlorpyrifos oxon is detoxified much more avidly by A-esterase(s).

The term A-esterase originally referred to an enzyme(s) in the serum that metabolized carboxylic esters and was insensitive to inhibition by organophosphates (in contrast to the B-esterases, which are inhibited by organophosphates). Later this activity was shown to be associated with the detoxification of paraoxon, leading to the use of the term A-esterase to refer to enzymes that hydrolyze organophosphates. Since the original discovery of A-esterase, several enzymes have been identified which can hydrolyze certain organophosphates, and hence have been referred to as A-esterases, even though it is currently not known if they also hydrolyze carboxylic esters.

Currently, there is much confusion regarding the nomenclature of A-esterase(s), and the term A-esterase is by no means universally endorsed. This enzyme(s) has been referred to by many different names, including paraoxonase, aryl-ester hydrolase, arylesterase, organophosphate hydrolase, organophosphorus compound hydrolase, and organophosphorus acid anhydrolase. Part of the confusion in the classification of this enzyme(s) appears to result from the presence of different forms (which may or may not be related) within an organism, as well as different forms within different species. For example, an enzyme that can detoxify paraoxon has been isolated from the bacteria *Pseudomonas diminuta*. This enzyme requires zinc for activity and can also detoxify the parent phosphorothioate insecticide parathion. In contrast, mammals seem to have at least two kinds of enzymes that could be called A-esterase, neither of which can detoxify parathion. The first detoxifies the compound diisopropylfluorophosphate and requires magnesium, manganese, or cobalt for activity. The second detoxifies oxons, such as chlorpyrifos oxon and paraoxon, and requires calcium for activity. Although the term A-esterase has been sometimes applied to all these enzymes, it more often is used to refer to the enzyme(s) that requires calcium and detoxifies the oxygen analogs of phosphorothioate insecticides.

Considerable species differences in A-esterase activity exist, ranging from very low or nonexistent in certain birds and fish, to very high in rabbits. Species differences in A-esterase activity could account, at least in part, for species differences in the relative sensitivity to certain phosphorothioate insecticides. For example, birds are much more susceptible to the toxicity of pirimiphos methyl than are mammals.

In mammals, A-esterase(s) has been identified in several tissues, with the highest activity usually found in the blood and liver. It is now known that the liver synthesizes A-esterase(s) and secretes it into the blood. In Caucasians, serum A-esterase, as determined by the metabolism of paraoxon, displays a genetic polymorphism that results in two phenotypes: type A (low activity) and type B (high activity) with a frequency of about 0.7 and 0.3, respectively. In contrast, this bimodal distribution of activity could not be demonstrated in sample populations from African and Asian populations. Since type A and type B A-esterase activities seem to be differentially affected by sodium chloride, their differences are likely qualitative in nature and not just quantitative. However, it must be emphasized that the toxicological significance of this genetic polymorphism is currently unclear. It is not known if individuals with low A-esterase activity are more susceptible to phosphorothioate insecticide poisoning than are individuals with high A-esterase activity. Moreover, further confusing the issue of the toxicological significance of this genetic polymorphism is the observation that the detoxification of chlorpyrifos oxon by serum A-esterase in Caucasians does not display multimodality, even though a bimodal distribution of A-esterase activity toward paraoxon can be demonstrated in those same individuals.

As outlined in Fig. A-6, the hydrolysis of paraoxon by human serum A-esterase(s) is very similar to the phosphorylation of B-esterases, such as acetylcholinesterase, by paraoxon. Both reactions involve an initial binding of paraoxon to the enzyme, followed by a rapid conformational change that produces diethyl phosphate and p-nitrophenol from paraoxon. P-nitrophenol is quickly released from the enzyme, leaving diethyl phosphate covalently bound to enzyme. At this point, A-esterase quickly releases diethyl phosphate as a result of interacting with a water molecule. However, B-esterases, such as acetylcholinesterase, retain the diethyl phosphate for a much longer period of time, thereby resulting in inhibition of the enzyme.

While A-esterase(s) and B-esterases interact kinetically with paraoxon in a similar fashion (Fig. A-6), the molecular events occurring at their active sites during catalysis are probably very different. The active site of B-esterases such as acetylcholinesterase has been well characterized and contains a serine residue which is phosphorylated by paraoxon at the hydroxyl group. In contrast, the active site of A-esterase(s) has not been studied as extensively but likely does not contain a serine residue that participates in the hydrolysis of paraoxon. Furthermore, a free sulfhydryl group is required for A-esterase activity, although it is not known if this sulfhydryl group is located at the active site. Additionally, A-esterase(s) requires a divalent cation like calcium for activity, whereas B-esterases do not.

The function of mammalian A-esterase(s) (when it is not detoxifying organophosphates) is currently unknown. A naturally occurring substrate for this enzyme has not yet been identified. However, in blood this enzyme seems to be associated with the high-density lipoprotein fraction, and therefore could play some role in lipoprotein metabolism. Interestingly, serum A-esterase activity, as measured by hydrolysis of paraoxon, is thought to be lower in patients after myocardial infarction, although it is not known if the lower activity is a direct result of the myocardial infarction or if it indicates an increased risk to coronary heart disease.

Further Reading

Aldridge, W. N., and Reiner, E. (1972). *Enzyme Inhibitors as Substrates. Interactions of Esterases of Organophosphorus and Carbamic Acids.* Elsevier, New York.

Chambers, J. E., and Levi, P. E. (Eds.) (1992). *Organophosphates: Chemistry, Fate, and Effects.* Academic Press, New York.

Reiner, E., Aldridge, W. N., and Hoskin, F. C. G. (Eds.) (1989). *Enzymes Hydrolyzing Organophosphorus Compounds.* Ellis Horwood, West Sussex, UK.

—*Lester G. Sultatos*

$$E + S \underset{k_{-1}}{\overset{k_1}{\rightleftarrows}} \left(E\text{-}S \rightleftarrows E \overset{P_2}{\underset{P_1}{\diagdown}} \right) \underset{k_{-2}}{\overset{k_2}{\rightleftarrows}} \underset{\substack{+ \\ P_1}}{E\text{-}P_2} \underset{k_{-3}}{\overset{k_3}{\rightleftarrows}} E + P_2$$

FIGURE A-6. *Kinetic mechanism for the interaction of paraoxon (S) with A-esterase (E) or acetylcholinesterase (E). P-nitrophenol (P_1) is the first product released, whereas diethyl phosphate (P_2) is the second.*

Related Topics

Cholinesterase Inhibition
Delayed Neurotoxicity

Neurotoxicology: Central and Peripheral
Organophosphate Poisoning, Intermediate
   Syndrome
Organophosphates

Aflatoxin

- SYNONYMS: Aflatoxins B1; B2; B3; B4; G1; G2; M1; M2

- DESCRIPTION: Aflatoxins are naturally occurring bisfuranocoumarin compounds produced from the molds *Aspergillus flavus* and *Aspergillus parasiticus*. The aflatoxins are highly fluorescent. The "B" refers to blue, the "G" signifies green fluorescence. "M" aflatoxins are fungal metabolites present in milk. Aflatoxin B1 is the most potent. Aflatoxins are contaminants in corn, peanuts, tree nuts, cotton seed, and certain meats. They have also been found in hypoallergenic milk.

- CHEMICAL STRUCTURE:

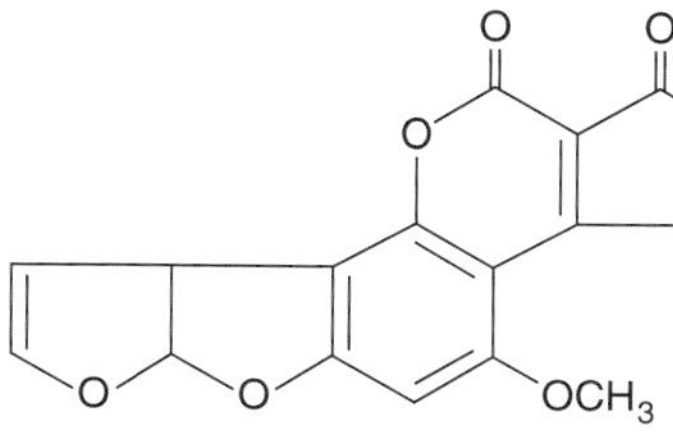

Exposure Pathways

Ingestion and dermal contact are possible routes of exposure.

Toxicokinetics

Aflatoxins are well absorbed orally. Exposure to human skin results in slow absorption. Aflatoxins are rapidly cleared from blood. Sixty-five percent of an initial dose of aflatoxin B1 is removed from the blood within 90 min and excreted primarily in the bile. The plasma half-life of aflatoxin is short, and it is excreted slowly as multiple moieties as a result of extensive metabolism. When estimated in human liver homogenates, the parent compound had an estimated half-life of 13 min.

Aflatoxins are metabolized by the NADPH-dependent enzyme system using cytochrome P450. *In vitro* liver metabolism studies have shown five different types of metabolic pathways for aflatoxin B1: reduction, hydroxylation, hydration, O-demethylation, and epoxidation. All of these products contain hydroxide groups which allow them to be conjugated with glucuronic acid and sulfate, thus becoming detoxified.

Mechanism of Toxicity

Aflatoxins combine with DNA, suppressing DNA and RNA synthesis. This leads to structural changes in cell nucleoli and reduction of protein synthesis. Formation of reactive DNA adducts causes cancer.

Human Toxicity

Aflatoxin poisoning is difficult to diagnose early in humans. The first clinical symptoms are anorexia and weight loss. Aflatoxins are associated with hepatocellular damage and necrosis, cholestasis, hepatomas, acute hepatitis, periportal fibrosis, hemorrhage, jaundice, fatty liver changes, cirrhosis in malnourished children, and Kwashiorkor. There is evidence of transplacental transport of aflatoxin by the fetoplacental unit. Aflatoxins are proven human carcinogens.

Clinical Management

Acute aflatoxin toxicity should be treated with decontamination procedures and good supportive care. With chronic ingestions, the primary treatment remains supportive in nature. Elevation of serum alkaline phosphatase is a good indicator of aflatoxin toxicity.

Animal Toxicity

Aflatoxins are carcinogenic in animals. The carcinogenic potential seems to be increased in malnutrition, especially pyridoxine deficiency. It has been proposed that aflatoxin B1-2,3-oxide (metabolite of aflatoxin B1) is the actual carcinogen.

—*Raja S. Mangipudy and
Harihara M. Mehendale*

Related Topics

Carcinogen–DNA Adduct Formation and DNA
Repair
Immune System
Mutagenesis
Mycotoxin
Veterinary Toxicology

Agency for Toxic Substances and Disease Registry

The agency for toxic Substances and Disease Registry (ATSDR), located in Atlanta, Georgia, is a federal agency created in 1980 by the Comprehensive Environmental Response, Compensation, and Liability Act (CERCLA), or what is more commonly known as Superfund legislation. Congress enacted Superfund as part of its response to two highly publicized and catastrophic events: discovery of the Love Canal hazardous waste site in Niagara Falls, New York, and an industrial fire in Elizabethtown, New Jersey, which set off the release of highly toxic fumes into the air in a densely populated area. Congress also created ATSDR to implement the health-related sections of laws that protect the public from hazardous wastes and environmental spills of hazardous substances.

In 1983, the secretary of the Department of Health and Human Services by administrative order established ATSDR as a separate agency of the Public Health Service. In 1984, amendments to the Resource Conservation and Recovery Act (RCRA) authorized ATSDR to conduct public health assessments at RCRA sites when requested by the U.S. EPA, states, or individuals, and to help EPA decide which substances should be regulated and at what levels those substances threaten human health.

In June 1983, ATSDR was formally organized to begin in concert with U.S. EPA, the Centers for Disease Control (now the Centers for Disease Control and Prevention), and the National Institute of Environmental Health Sciences to address CERCLA, one of the most challenging and innovative environmental laws relating to public health.

Following the reauthorization of Superfund in 1986 under the Superfund Amendments and Reauthorization Act (SARA), the agency received major new mandates. SARA broadened ATSDR's responsibilities in the areas of public health assessments, establishment and maintenance of toxicological databases, information dissemination, and medical education; new groups within ATSDR were organized to carry out the new tasks. By August 1989, the agency had assumed its current structure.

Agency Mission

The mission of ATSDR is to prevent exposure and adverse human health effects and diminished quality of life associated with exposure to hazardous substances from waste sites, unplanned releases, and other sources of pollution in the environment. ATSDR works closely with state, local, and other federal agencies to reduce or eliminate illness, disability, and death that result from exposure of the public and workers to toxic substances at waste disposal and spill sites.

As the lead agency within the Public Health Service responsible for implementing the health-related provisions of CERCLA, ATSDR was charged with assessing the presence and nature of health hazards at specific Superfund sites, helping to prevent or reduce further exposure and the illnesses that result and expanding what is known about the health effects of exposure to hazardous substances.

Range of agency activities

The following is a summary of the activities assigned to ATSDR in 1980 under the original Superfund statute:

- Determine the extent of danger to public health from a release or threatened release of a hazardous substance. (This mandate covers the range of public health assessment and other support activities provided to U.S. EPA, states, and other federal agencies at emergency, immediate-removal, and remedial Superfund sites.)

- Conduct periodic surveys and screening programs to determine the relationships between exposure to hazardous substances and illness. (This mandate includes *in vivo* and *in vitro* toxicologic testing, human epidemiologic studies, and establishment of surveillance systems.)

- Establish and maintain a registry of serious diseases and illnesses and registries of all persons environmentally exposed to hazardous substances whenever inclusion of such persons in registries would be scientifically appropriate or valuable for long-term follow-up or specific scientific studies.

- Establish and maintain a comprehensive and publicly accessible inventory of literature on the health effects of hazardous substances.

- When public health emergencies are caused or are believed to be caused by exposure to hazardous substances, assist, consult, and coordinate with private or public health care providers in providing medical care and testing exposed individuals, including collecting and analyzing laboratory specimens as may be indicated by specific exposures.

- Establish and maintain a complete list of areas closed to the public or otherwise restricted in use because of hazardous substance contamination.

For additional information, contact ATSDR at 1600 Clifton Road NE (E-60), Atlanta, GA 30333. Telephone: 404-639-000; fax: 404-639-0522.

—*David M. Krentz and Harihara M. Mehendale*

(Adapted from information supplied by ATSDR.)

Related Topics

Comprehensive Environmental Response, Compensation and Liability Act
Emergency Response
Hazardous Waste

Information Resources in Toxicology
National Institute of Environmental Health Sciences
Resource Conservation and Recovery Act

Alachlor

- CAS: 15972-60-8
- SYNONYMS: Alachlore; Alanex; Alanox; Alatox 480; Lasso; Lasagrin
- CHEMICAL CLASS: Herbicide
- CHEMICAL STRUCTURE:

Uses
Alachlor is a preemergence herbicide registered by Monsanto in 1969. It is used as an herbicide for grasses, broadleaf seeds, corn, sorghum, soybeans, peanuts, cotton, vegetables, and forage crops.

Exposure Pathways
Dermal exposure is most common, although exposure via oral/parenteral route and ocular contact are also possible.

Toxicokinetics
Alachlor is absorbed orally. Dermal absorption may be linear over time for the duration of exposure. Excretion via kidneys is the major route of elimination.

Mechanism of Toxicity
This agent is a mucous membrane irritant. The exact mechanism of potential teratogenic changes is still being investigated.

Human Toxicity

Alachlor has little acute toxicity in mammals. It is a mild eye and mucous membrane irritant. Hepatic damage was seen in mice taking large doses in chronic studies. Acute exposure resulted in a decrease in erythrocyte number and hemoglobin in rodents. Dermal exposure may cause allergic reaction. The primary concern with alachlor is its carcinogenic and mutagenic potential. This agent has shown positive mutagenic response to several bacteria and *in vitro* models. U.S. EPA listed alachlor as a probable human carcinogen in 1984.

Clinical Management

There are few acute symptoms. Treatment is symptomatic and supportive. There are no specific antidotes. In cases of oral exposure, measures to decrease absorption may be useful. Emesis may be induced after careful consideration. For dermal exposure, decontamination by washing the exposed area thoroughly with soap and water is recommended. In cases of inhalation exposure, the victim must be moved to fresh air and monitored for respiratory distress. In cases of eye exposure, the eyes should be irrigated with copious amounts of tepid water for at least 15 min. If irritation, pain, swelling, lacrimation, or photophobia persist, the person should be seen in a health care facility.

—*Raja S. Mangipudy and Harihara M. Mehendale*

Related Topic

Pesticides

Alar

- ◆ SYNONYMS: Aminozide; daminozide; DMSA; B-995; kylar; aminocide
- ◆ CHEMICAL CLASS: Organic acid

- ◆ CHEMICAL STRUCTURE:

$$HOOCCH_2CH_2CONHN(CH_3)_2$$

Uses

Alar is used as a translocated plant growth regulator. It reduces internode elongation; induces heat, drought, and frost resistance; and produces darker foliage and stronger stems. It also produces earlier and multiple flowers and fruits. A spray is often applied at the rate of 1500–10,000 ppm. It is systemic (i.e., it is taken up by the fruit). Its residues cannot be washed off or removed by peeling. Use of alar in apples caused environmental concern a few years ago; it has now been banned in the United States.

Exposure Pathways

Dermal contact and ingestion are routes of exposure.

Toxicokinetics

A breakdown product of alar is an asymmetrical 1,1-dimethylhydrazine and is excreted renally.

Mechanism of Toxicity

The growth retardant action of alar has been attributed to the formation of dimethylhydrazine, which inhibits tryptamine oxidation.

Human Toxicity

There are little data on mammalian toxicity. No human case reports are available. Based on animal data, alar should be low in toxicity. U.S. EPA has determined that alar does not represent an imminent health hazard.

Clinical Management

No human cases have been reported so treatment recommendations are speculative. Dermal contamination probably requires no treatment other than decontamination. For gastric contamination caused by swallowing, treatment by emesis, gastric lavage, and/or activated charcoal may be indicated. Patients should be monitored for central nervous system (CNS) depression, ptosis (drooping eyelid), and liver functional abnormalities if significant amounts (greater than 8 g) have been ingested.

Animal Toxicity

The primary toxic effects seen in animals include ptosis, CNS depression, gastrointestinal irritation, and possibly liver functional abnormalities.

*—Raja S. Mangipudy and
Harihara M. Mehendale*

Albuterol

♦ CAS: 18559-94-9

♦ Synonyms: Salbutamol; Ventolin; Proventil

♦ Pharmaceutical Class: Selective β-2 adrenergic agonist

♦ Chemical Structure:

HOCH₂

HO— (ring) —CHCH₂NHC(CH₃)₃
                |
               OH

Uses

Albuterol is used as a bronchodilator and in the prevention of premature labor. Unlabeled use includes treatment of hyperkalemia.

Exposure Pathways

Ingestion is the most common route of accidental and intentional exposure to albuterol. Inappropriate overuse of the inhalation solution may also be seen. It is available in tablets and a syrup for oral use, as a solution for inhalation (base) or as a sulfate for inhalation (rotocaps), as a solution for injection, and as a solution for intravenous infusion for parenteral use.

Toxicokinetics

Oral albuterol is readily absorbed from the gut. There is significant first-pass conjugation with 50% bioavailability of an ingested dose. From 21 to 30% of an inhaled dose is available for absorption. Parenteral absorption is 100%. Sulfate conjugation is the primary metabolic pathway. There appears to be no direct biotransformation of albuterol in the lungs. The volume of distribution is 156 ± 38 liters. The percentage of protein binding is unknown. Albuterol, as both the sulfate and sulfate conjugates, is eliminated via the kidneys. With oral dosing, 28% of albuterol is excreted unchanged in the urine and 64% unchanged with intravenous dosing. Albuterol follows first-order kinetics. The half-life is 3–5 hr with oral dosing, 2–7 hr with inhalation, and 5.5–6.9 hr with intravenous dosing.

Mechanism of Toxicity

The primary mechanism of albuterol is the stimulation of adenyl cyclase, which catalyzes cyclic adenosine monophosphate (AMP) from adenosine triphosphate (ATP). In the liver, buildup of cyclic AMP stimulates glycogenolysis and an increase in serum glucose. In skeletal muscle, this process results in increased lactate production. Direct stimulus of sodium/potassium ATPase in skeletal muscle produces a shift of potassium from the extracellular space to the intracellular space. Relaxation of smooth muscle produces a dilation of the vasculature supplying skeletal muscle, which results in a drop in diastolic and mean arterial pressure (MAP). Tachycardia occurs as a reflex to the drop in MAP or as a result of β-1 stimulus. β-adrenergic receptors in the locus ceruleus also regulate norepinephrine-induced inhibitory effects, resulting in agitation, restlessness, and tremor.

Human Toxicity

The toxic events of albuterol overdose follow its β-adrenergic agonist activity. The effects of albuterol overdose are usually mild and benign, although they can be prolonged. Maximal metabolic and cardiovascular changes in overdose will be seen in the first 4–8 hr. Cardiovascular effects are usually limited to a sinus tachycardia and widened pulse pressure. Although there may be a drop in diastolic pressure, the systolic pressure is maintained by increased cardiac output from the tachycardia. Transient hypokalemia can be seen, caused by a shift of extracellular potassium to the intracellular space. A transient metabolic acidosis can be seen due to increased lactate production. Restlessness, agitation, and tremors are common in albuterol overdose.

Clinical Management

Basic and advanced life-support measures should be utilized as necessary. Albuterol overdoses rarely require treatment beyond gastrointestinal decontamination. Activated charcoal effectively adsorbs albuterol. The hypokalemia produced reflects a transient shift in potassium location rather than a true deficit of potassium; therefore, only rarely is there a need for external replacement therapy. Since arrhythmias beyond an increase in rate have not occurred with albuterol overdose, a conservative approach to the tachycardia is recommended. In the rare event of complications, intravenous propranolol rapidly and effectively reverses the symptoms of albuterol poisoning.

Animal Toxicity

Apparently albuterol is relatively benign in animals as well. Agitation, vomiting, and lethargy may be seen. In rats, the oral LD_{50} was greater than 2000 mg/kg; the inhalation LC_{50} could not be determined.

—Henry A. Spiller

Aldicarb

♦ CAS: 116-06-3

♦ SYNONYMS: 2-Methyl-2(methylthio)-propionaldehyde O-(methylcarbamoyl)oxime; Aldecarb; Aldicarbe; Temik ; AI3-27093, ENT 27093, OMS 771, NCI 08640, UC 21149, RCRA Waste Number P070

♦ CHEMICAL CLASS: Carbamate pesticide

♦ CHEMICAL STRUCTURE:

$$CH_3-S-\underset{\underset{CH_3}{|}}{\overset{\overset{CH_3}{|}}{C}}-CH=N-O-\underset{\underset{CH_3}{|}}{\overset{\overset{O}{||}}{C}}-NH$$

Uses

Aldicarb is a systemic insecticide used to control a wide variety of insects, mites, and nematodes. Major sites of application include ornamentals, cotton, and some fruit and vegetable crops.

Exposure Pathways

The most common exposure routes are dermal (during processing, packaging, or application) and oral (through consumption of products containing aldicarb residues). Exposure may also occur through unprotected handling of treated plants or soil. Although inhalation of fine particles and dusts of aldicarb has been reduced through improvements in applicator design, inhalation is still a significant route of exposure.

Toxicokinetics

Aldicarb is readily absorbed from all routes of exposure. Oxidation reactions rapidly convert aldicarb to aldicarb sulfoxide of which a small portion may then be slowly oxidized to aldicarb sulfone. Both the parent compound and its oxidized metabolites can be converted to their respective oximes and nitriles, which may ultimately be converted to aldehydes, acids, and alcohols. Animal studies have indicated aldicarb and its metabolites are distributed to many different tissues but no evidence of accumulation has been found. In the various tissues examined, aldicarb residues were not detected more than 5 days after exposure. The presence of aldicarb in fetal tissue indicates placental transfer in pregnant rats. Various aldicarb metabolites have been found in the milk of cows acutely treated with aldicarb.

Animal studies have indicated the major route of excretion to be urinary. Most studies have shown at least 80% of the original dose to be eliminated within 24 hr. Aldicarb is excreted primarily as aldicarb sulfoxide and sulfoxide oxime; the parent compound is excreted only in trace amounts. Biliary metabolites have been shown to undergo reabsorption and urinary excretion.

Mechanism of Toxicity

Aldicarb basically acts as a poor substrate for acetylcholinesterase, the enzyme responsible for degrading the neurotransmitter acetylcholine. Aldicarb reversibly binds to and inhibits the enzyme, resulting in accumulation of acetylcholine in the synapse, hyperstimulation of cholinergic receptors, and "cholinergic crisis." The enzyme is fairly rapidly reactivated through spontaneous decarbamoylation or via hydrolysis.

Human Toxicity: Acute

The acute effects of aldicarb exposure are due to acetyl-cholinesterase inhibition and may include the typical symptoms of cholinergic overstimulation such as the SLUDGE syndrome (salivation, lacrimation, urination, diarrhea, gastrointestinal cramping, and emesis), respiratory depression, bronchospasms, increased bronchial secretions, pulmonary edema, blurred vision, miosis, headache, tremors, muscle fasiculations, convulsions, mental confusion, and coma. Respiratory failure is the cause of death.

Human Toxicity: Chronic

Researchers have examined the possible effects of aldicarb on the induction of peripheral neuropathies. Currently, insufficient evidence exists to indicate any significant long-term health risk associated with aldicarb exposure.

Clinical Management

Activated charcoal may be used to reduce absorption from the gastrointestinal tract. Syrup of ipecac-induced emesis may be used if the patient is conscious and alert. Emesis is contraindicated if the patient is, or may become, convulsive. For inhalation exposure, the patient should be removed from the exposure area and observed for signs of breathing difficulty. In cases of dermal exposure, contaminated clothing should be removed and disposed of. Any exposed areas of skin should be repeatedly washed with soap and water. For eye contact, flush the eyes with generous amounts of lukewarm water for a minimum of 15 min.

Clinical management is basically symptomatic and supportive. Artificial ventilation with 100% humidified oxygen is necessary in cases of respiratory paralysis. An endotracheal tube may be required to maintain the airway. Erythrocytic acetylcholinesterase (AChE) levels, arterial blood gases, and cardiac function should be monitored. Because of the rapid reactivation of AChE, significant enzyme inhibition may not be seen unless blood samples are drawn and assayed within approximately 2 hr of the exposure. Atropine should be given intravenously to relieve the muscarinic (i.e., SLUDGE) symptoms. Convulsions may be treated with intravenous diazepam. Phenytoin may be used if the convulsions are recurrent. Pralidoxime (2-PAM) is indicated in cases of mixed exposure to both carbamates and organophosphorus compounds but is contraindicated in cases of carbamate-only exposure.

Animal Toxicity

In rats, the oral LD_{50} is 0.6–0.8 mg/kg, the dermal LD_{50} is 2.5 mg/kg. The inhalation LC_{50} in rats is 200 mg/m^3/5 hr. The oral LD_{50} in mice is 0.3 mg/kg.

—*Paul R. Harp*

Related Topics

A-esterase
Carbamate Pesticides
Cholinesterase Inhibition
Neurotoxicology: Central and Peripheral
Pesticides
Pollution, Water

Aldrin

- ◆ CAS: 309-00-2
- ◆ PREFERRED NAME: 1,2,3,4,10,10-Hexachloro-1,4,4a,5,8,8a-hexahydro-endo-1,4-exo-5,8-dimethanonaphtSynonyms: Aldrex; Altox; Drinox; Octalene; Toxadrin
- ◆ CHEMICAL CLASS: Synthetic organochlorine insecticide
- ◆ CHEMICAL STRUCTURE:

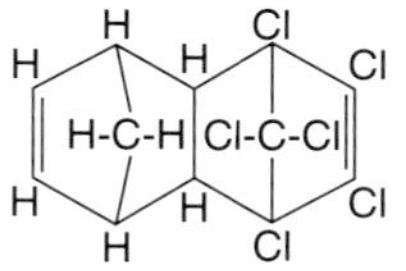

Uses

Aldrin is used as an insecticide.

Exposure Pathways

The most important exposure routes for aldrin are oral and dermal.

Toxicokinetics

Aldrin is readily absorbed through the gastrointestinal tract via the hepatic portal vein. The toxicity of aldrin is almost as high via the dermal route indicating extensive dermal absorption.

Epoxidation by cytochromes P450 of aldrin to dieldrin occurs both in the liver and in the lungs. Dieldrin is metabolized by liver microsomal enzymes to less toxic metabolites, including *cis*-aldrinol, 9-hydroxy dieldrin, and photoconverted to the more toxic 2-ketodieldrin. *cis*-Aldrinol is epimerized to *trans*-aldrindiol, which is further metabolized to aldrin diacid. Like other organochlorine insecticides, adipose tissue is the major storage tissue followed by liver, brain, and blood. The water-soluble metabolites of aldrin detoxification are excreted to a large extent (90%) in feces and in urine. Dieldrin is also found in mothers' milk.

Mechanism of Toxicity

Dieldrin binds to the γ-aminobutyric acid receptor and inhibits chloride ion flux. The result is similar to that described for chlordane or other cyclodiene compounds.

Human Toxicity

The toxicity of aldrin is similar to that of dieldrin and other organochlorine cyclodiene insecticides (see Dieldrin). The central nervous system (CNS) is the primary target. Convulsion is the major symptom. Patients may also experience nausea, vomiting, hyperexcitability, and coma.

Clinical Management

Treatment is symptomatic. Activated charcoal has been reported to increase the rate of excretion after oral exposure has ceased. Phenobarbital or diazepam are used when anticonvulsant therapy is necessary. In severe cases of aldrin-induced convulsions, muscle paralysis may be necessary in addition to anticonvulsant therapy.

Animal Toxicity

Animal toxicity with aldrin is similar to that of dieldrin. In addition to its effects on the CNS, dieldrin increases hepatocarcinogenesis. Dieldrin also has been reported to be immunotoxic suppressing macrophage function and T cell-dependent humoral immune functions.

—Benny L. Blaylock

Related Topics

Organochlorine Insecticides
Pesticides

Algae

♦ SYNONYMS: Blue-green (cyanobacteria and cyanophytes); anatoxin A (2-acetyl-9-azabicyclo(4.2.1) non-2, 3-ene); gastrointestinal illness (barcoo sickness, barcoo fever, and barcoo spews)

Exposure Pathways

Algae are found in stagnant water or slow moving streams. Human exposures are common but rarely result in toxicity due to the distasteful appearance of the water, but exposures have occurred as a result of consumption of contaminated water while swimming.

Toxicokinetics

Absorption of toxins occurs rapidly from the gastrointestinal tract. The greatest concentrations are found in the liver. Lesser amounts are found in the kidneys and remain detectable for up to 24 hr.

Mechanism of Toxicity

Toxins are released when the algae cell dies and produces strong neuromuscular blocking agents. There are two main types of toxins in these cyanobacteria: neurotoxic alkaloids (anatoxins) and hepatotoxic peptides. These are also referred to as fast death factor, microcystine, cyanoginosin, cyanoviridin, and cyanogenosin. The toxins include alkaloids, polypeptides, pteridines, and lipopolysaccharides (endotoxins).

Anatoxin A is a potent depolarizing neuromuscular blocking agent that is active at both the nicotinic and muscarinic receptors. Anatoxin-A(S) is a peripheral acting organophosphorous anticholinesterase agent. Inhibition of cholinesterase in the blood, lungs, and muscle occurs. It is also known to contain hepatotoxic peptides. Aphantoxin is a potent and specific inhibitor of the voltage-dependent sodium ion channel. Microcystin is a hepatotoxic algaltoxin. It causes cell injury due to interference with intracellular calcium flux, not extracellular.

Human Toxicity: Acute

Local irritant effects can be seen. These include swelling of the lips, conjunctivitis, and earaches. Allergic reactions (rhinitis and pruritus maculopapular rash) can also develop. Gastroenteritis (appearing as enteritis or amebic dysentery) including nausea, vomiting, abdominal cramps, and diarrhea can begin within 3–5 hr after ingestion. This resolves within 1 or 2 days. Transient liver enzyme elevation is common. Muscle weakness and pain in joints and limbs can occur. Although not commonly seen, respiratory paralysis can occur. Muscle weakness and joint pain have also been reported.

Human Toxicity: Chronic

Chronic exposure to these toxins may cause continued effects with likely exacerbations.

Clinical Management

Treatment is supportive. Emesis is not likely to be effective since the toxins are absorbed so quickly. Activated charcoal may be used but its efficacy has not been established. Liver enzymes should be monitored. Fluid and electrolyte balance should be closely monitored since vomiting and diarrhea may be severe. Local irritation and allergic reactions would be treated as indicated by symptoms.

Animal Toxicity

In severe cases, toxins act so rapidly that animals die before leaving the water in which they were drinking or they die nearby. Death may be due to respiratory arrest caused by paralysis, liver damage, or gastrointestinal tract hemorrhage. Symptoms include profuse salivation, gaping mouth, irregular respiration, ataxia, spastic twitching, violent tremors, and bloody diarrhea. Petechial hemorrhage of the heart is consistently found on autopsy. If death does not occur, recovery is uneventful without treatment.

—*Regina M. Rogowski*

Related Topics

Red Tide Toxicity
Veterinary Toxicology

Alkalies

- ♦ PREFERRED NAME: Bases
- ♦ SYNONYM: Strong alkalies
- ♦ CHEMICAL CLASS: Inorganic alkalies (e.g., sodium hydroxide, potassium hydroxide, and sodium hypochlorite)

Uses

Depends on the specific alkali. Alkalies are primarily used as cleaning agents, bleaches, and unslaked lime.

Exposure Pathways

Skin contact, ingestion, and inhalation are the most common exposure pathways.

Toxicokinetics

The toxicokinetics varies depending on the type of alkali.

Mechanism of Toxicity

Alkalies cause toxicity by liquefaction necrosis, meaning that the alkali destroys the cell membrane and cell integrity and thereby causes cell lysis.

Human Toxicity: Acute

Alkalies can burn skin, mucous membrane, and eyes almost immediately on contact. However, the absence of burns, irritation, erythema, or other such signs in the oral or circumoral area does not necessarily indicate that esophageal injury does not exist. Inhalation of the fumes may cause pulmonary edema or pneumonitis.

Human Toxicity: Chronic

Burns that at the time of injury appear to be mild can sometimes go on to cause opacification, vascularization, ulceration, or perforation.

Clinical Management

Exposure should be terminated as soon as possible by removing the victim to fresh air. The skin, eyes, and mouth should be washed with copious amounts of water. A 20- to 30-min wash may be necessary to neutralize and remove all residual traces of the contaminant. Contaminated clothing and jewelry should be removed and isolated.

Treatment may require instillation of a local anesthetic to treat the blepharospasm (spasmodic winking from involuntary contraction of the orbicular muscle of the eyelids). Oral ingestion requires immediate dilution therapy with water or milk. Antidotes such as vinegar or lemon juice are absolutely contraindicated. Emesis should be avoided in case of ingestion.

Animal Toxicity

The toxicity of alkalies in animals is the same as that in humans.

—Sanjay Chanda

Related Topics

Acids
Corrosives
Gastrointestinal System
Skin

Alkyl Halides

- ♦ SYNONYMS: Halogenated hydrocarbons; haloalkanes
- ♦ REPRESENTATIVE COMPOUNDS: Methyl bromide; methyl chloride; methyl iodide; dichloromethane; tetrachloroethane; carbon tetrachloride; trichlorethene; trichloroethylene; a number of fluorinated hydrocarbons (e.g., freons)
- ♦ DESCRIPTION: Alkyl halides are practically insoluble in water. They are miscible in all proportions with liquid hydrocarbons and are, in general, good solvents for many organic substances. Most of the common organic halides are liquids. Like alkanes, halogen compounds are insoluble in and inert to cold concentrated sulfuric acid. In a series of alkyl halides, the boiling point rises with an increase in molecular weight due to the presence of either a heavier halogen atom or a larger alkyl group. Bromides boil at temperatures distinctly higher than the corresponding chlorides, and iodides are higher boiling than the bromides. Increase in the halogen content decreases their flammability. In contact with an open flame or a very hot surface fluorocarbons may decompose into highly irritant and toxic gases such as chlorine, hydrogen fluoride, or chloride and even phosgene.

- ♦ MOLECULAR FORMULA: $R(X)_n$, where R is a hydrocarbon alkyl group and X is a halogen. One or more halogens may be present in one compound.

Uses

Many halogenated hydrocarbons have important commercial applications. For example, trichloroethene is a common dry cleaning solvent. The fluorinated hydrocarbons (freons) are used as refrigerants, industrial solvents, fire extinguishers, local anesthetics, and glass chillers, but mainly as propellants in aerosol products. Methyl bromide, methyl chloride, and methyl iodide are used as refrigerants in chemical synthesis and as fumigants. Methyl bromide is used with carbon tetrachloride in fire extinguishers. Methylchloroform is used as a solvent for cleaning, degreasing, and in paint removers. Dichloromethane is used in paint removers and as an industrial solvent. Tetrachloroethane is used as a solvent in industry and occurs as a contaminant in other chlorinated hydrocarbons. It is occasionally present in household cleaners. Carbon tetrachloride is used as a solvent and intermediate in many industrial processes.

Method of Preparation

Alkyl halides can be prepared by addition of the halogen or hydrogen halides to alkenes, as well as by substitution of a halogen for a hydrogen in an alkane. The most important method of preparing alkyl halides is by reaction between an alcohol and a hydrogen halide.

Exposure Pathways

Inhalation and dermal and ocular contact are common routes of exposure.

Toxicokinetics

Fluorocarbon compounds are lipid-soluble and, thus, generally well absorbed through the lung. Absorption after ingestion is much lower than after inhalation. Most of the fluorinated hydrocarbons are immediately absorbed.

There is a significant accumulation of fluorocarbons in the brain, liver, and lungs compared to blood levels, signifying a tissue distribution of fluorocarbons similar to that of chloroform. Fluorocarbons are concentrated in body fat where they are slowly released into blood at a concentration that should not cause any risk of cardiac sensitization.

Fluorocarbons are excreted by the lungs and the parent compound is eliminated in about 15 min.

Human Toxicity: Acute

Freons are very toxic when inhaled in high concentrations and/or for extended periods. Inhalation of fluorinated hydrocarbons such as those caused by leaking air conditioners or refrigerators usually results in transient eye, nose, and throat irritation. Palpitations and light headedness are also seen. Headache was a common complaint, reported in 71% of 31 workers exposed to bromotrifluoromethane in one incident. Inhalation of halides at sufficient concentrations associated with deliberate abuse, or spills or industrial use occurring in poorly ventilated areas, has been associated with ventricular arrhythmias, pulmonary edema, and sudden death. Fluorinated hydrocarbons are believed to cause arrhythmias by sensitizing the myocardium to endogenous catecholamines. Freon solvents are degreasers. Dermal contact with fluorinated hydrocarbons may result in defatting, irritation, or contact dermatitis. Severe frostbite was reported as a rare effect of severe freon exposure. Mucosal necrosis and perforation of the stomach developed in one patient after ingesting a small amount of trichlorofluoromethane. Fluorocarbons containing bromine are more toxic than the corresponding chlorine compounds. There is a significant interpatient variation following exposure to fluorocarbons and it is difficult to predict symptoms following exposure. Compounds like dibromochloropropane, in which occupational exposure has affected

male fertility, have now been removed from the market. Following acute exposure to methyl bromide, chloride, or iodide, nausea and vomiting, blurred vision, vertigo, weakness or paralysis, oliguria or anuria, drowsiness, confusion, hyperactivity, coma, convulsions, and pulmonary edema are noted. Pulmonary edema and bronchial pneumonia are most often the cause of death. Skin contact causes irritation and vesiculation.

Methyl chloroform and dichloromethane are central nervous system (CNS) depressants. Methyl chloroform sensitizes the myocardium to catecholamine-induced arrhythmias. Following exposure to tetrachloroethane, irritation of the eyes and nose, followed by headache and nausea, is observed. Cyanosis and CNS depression progressing to coma may appear after 1–4 hr.

Human Toxicity: Chronic

A syndrome of impaired psychomotor speed, impaired memory and learning, has been described in workers with chronic occupational exposure to fluorinated hydrocarbons. Skin irritation and defatting dermatitis upon prolonged or repeated contact with the skin to trichloromonofluoromethane have been reported. An excess of CNS symptoms was seen in a group of workers chronically exposed to trichloromonofluromethane. Repeated exposure to methyl bromide, methyl chloride, and methyl iodide will cause blurring of vision, numbness of the extremities, confusion, hallucinations, somnolence, fainting attacks, and bronchospasm. Methyl iodide is a potential carcinogen. Chronic toxicity has not been reported with dichloromethane. Headache, tremor, dizziness, peripheral paresthesia, or anesthesia have been reported after chronic inhalation or skin exposure to tetrachlorethane.

Clinical Management

This management is intended for use in the absence of a specific treatment protocol for a product or a chemical. Symptomatic and supportive care is the primary therapy. The general approach to a poisoned patient is to first assess the vital signs of the patient followed by assessing the route of administration for potential toxicity. Measures to prevent further absorption of the compound may be useful. Move victims of inhalation exposure from the toxic environment and administer 100% humidified supplemental oxygen with assisted ventilation as required. Exposed individuals should have a careful and thorough medical examination performed to look for abnormalities. Patients with fluorohydrocarbon poisoning should not be given epinephrine or similar drugs because of the tendency of fluorohydrocarbon to induce cardiac arrhythmia, including ventricular fibrillation. Monitoring including complete blood count, urine analysis, and liver and kidney function tests is suggested for patients with significant exposure. Activated charcoal or gastric lavage may be indicated to prevent further absorption. Exposed eyes should be irrigated with copious amounts of tepid water for at least 15 min. If irritation, pain, swelling, lacrimation, or photophobia persist after 15 min of irrigation, an ophthalmologic examination should be performed.

Animal Toxicity

Deliberate ocular exposure in rabbits to liquid Freon 12 produced effects related to the duration of exposure. Severe corneal damage with opacity occurred following exposure for 30 sec. In dogs, inhalation of fluorinated hydrocarbon vapors causes bradycardia followed by deterioration to ventricular fibrillation in some animals. Some of the chlorinated hydrocarbon solvents, such as methylene chloride and chloroform, have caused cancer in several species of experimental animals and are suspect human carcinogens.

—*Swarupa G. Kulkarni and
Harihara M. Mehendale*

α-Methylfentanyl

- SYNONYMS: *N*-[1-(1-methyl-2-phenylethyl)-4-piperidinyl]-*N*-phenylpropana China white
- PHARMACEUTICAL CLASS: α-Methylfentanyl is a narcotic analgesic, a designer drug derived from fentanyl (see Fentanyl and Fentanyl Derivatives, Illicit).
- CHEMICAL STRUCTURE:

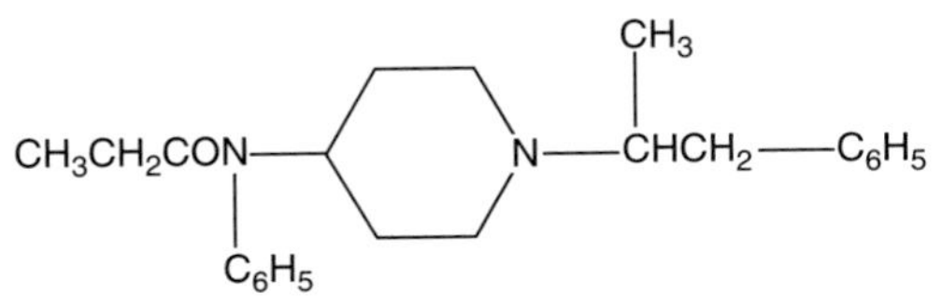

Uses

α-Methylfentanyl is not medically used per se, although it is a derivative of fentanyl with higher analgesic effects. α-Methylfentanyl is a designer drug that has been synthesized for its analgesic and euphoric effects. Due to its high potency (1000–2000 times more potent than heroin) and fast-acting narcotic analgesia, it has high abuse potential and is sold on the street as synthetic heroin. α-Methylfentanyl also has a high abuse potential in racing horses for its analgesic and stimulant actions. Therefore, α-methylfentanyl is a controlled substance listed in the U.S. *Code of Federal Regulations*, Title 21, Part 1308.11 (1987).

Exposure Pathways

Most common exposure pathways to α-methylfentanyl are via intramuscular and intravenous injection.

Toxicokinetics

Limited information indicates that α-methylfentanyl is rapidly absorbed and distributed to the central nervous system (CNS). Elimination of this drug is primarily via the kidneys.

Mechanism of Toxicity

α-Methylfentanyl is believed to exert its toxic effects by binding to opiate receptors (μ-agonist) at many sites in the CNS.

Human Toxicity

α-Methylfentanyl is fast acting. At high doses it causes euphoria, marked muscular rigidity, and respiratory depression. The literature indicates that α-methylfentanyl overdose deaths are primarily due to respiratory paralysis. In addition to its high potency and fast action, further danger is due to its poor ability to mix with the cutting agents used with illicit drugs. Information on chronic toxicity is not available.

Clinical Management

Since α-methylfentanyl exerts its toxicity as μ-agonist, its toxicity can be managed with narcotic μ-antagonists such as naloxone and respiratory support with resuscitative equipment.

Animal Toxicity

In experimental horses, α-methylfentanyl induces locomotive responses (as quantified by counting the number of footsteps taken per unit of time), indicating that it is a morphine-like narcotic agonist in horses. The maximum effect can be seen about 10 min after treatment of horses with greater than 4 μg/kg body weight. Information on chronic toxicity is not available.

—Abraham Dalu

α-Naphthyl Thiourea

- ◆ CAS: 86-88-4
- ◆ PREFERRED NAME: ANTU
- ◆ SYNONYMS: α-Naphthyl urea; 1-(1-naphthyl)-2-thiourea
- ◆ MOLECULAR FORMULA: $C_{11}H_{10}N_2S$
- ◆ CHEMICAL STRUCTURE:

$$\underset{\text{NH}}{} \overset{\overset{\text{S}}{\underset{||}{}}}{\underset{}{}} \text{C} - \text{NH}_2$$

- ◆ DESCRIPTION: ANTU is a gray crystalline odorless powder with a bitter taste. It has a molecular weight of 202.7. It has a melting point of 198°C and does not ignite readily. On contact with strong oxidizers it may cause fire and explosions. Fire may produce irritating or poisonous gas. Hazardous decomposition products include sulfur dioxide, oxides of nitrogen, and carbon monoxide.

Uses

ANTU is a single-dose rodenticide used as a bait and tracking powder and is specifically used against Norway rats. It is ineffective against all species of field rodents. It is used in baits in concentrations of 1–3%. Because of its specificity for Norway rats and its ten-

dency to cause resistance, this poison rapidly lost popularity and is no longer manufactured.

Exposure Pathways

Inhalation, ingestion, and dermal contact are possible routes of exposure.

Toxicokinetics

Limited data on the toxicokinetics of ANTU are available. However, absorption does occur following oral administration. ANTU toxicity in the rat is thought to depend on metabolic activation via the hepatic and lung microsomal enzymes to form a hydrosulfide and α-naphthyl urea.

Mechanism of Toxicity

ANTU toxicity in the rat is thought to depend on metabolic activation via the hepatic and lung microsomal enzymes to form a hydrosulfide and α-naphthyl urea. The metabolites are covalently bound to lung macromolecules. However, it is not known if these metabolites are produced in humans. ANTU presumably acts on some enzyme system involving the sulfhydryl group. Analogous pulmonary edema is produced by sulfhydryl inhibitors, such as alloxan, iodoacetamide, and oxophenarsine. Production of oxygen free radicals via the cyclooxygenase pathway has been implicated in mediating ANTU-induced lung damage. Following exposure to ANTU, there are a number of biochemical events, such as alteration in carbohydrate metabolism, adrenal stimulation, and interaction of the chemical with sulfhydryl groups, but none of these appear to bear any relationship to the observed signs of toxicity.

Human Toxicity: Acute

The estimated mean lethal dose in humans is 25 g/70 kg. ANTU is classified as being moderately toxic. No human fatalities have been reported. It is probably not toxic to humans except in large amounts. Although it appears that humans are resistant to ANTU intoxication, probably because insufficient quantities are ingested, poisonings have occurred, with tracheobronchial hypersecretion of a white, nonmucous froth containing little protein, pulmonary edema, and respiratory difficulty. Inhalation of ANTU powder may result in dyspnea, rales, cyanosis, and pulmonary edema or effusion. A case of contact eczema due to handling a rat poison containing ANTU as a base has been reported.

Human Toxicity: Chronic

Chronic exposure to ANTU led to the investigation of two cases of bladder cancer in two rodent operators. Therefore, the use of ANTU was restricted to professional operators. Available data were inadequate to evaluate carcinogenicity in humans. Chronic sublethal exposure may result in antithyroid activity and hyperglycemia.

The ACGIH TLV-TWA and the OSHA PEL-TWA are both 0.3 mg/m^3.

Clinical Management

For ingestion, emesis is indicated unless the patient becomes comatose or shows convulsions. Emesis is most effective if initiated within 30 min of ingestion. Syrup of ipecac can be used for inducing emesis. Charcoal slurry, aqueous or mixed with saline cathartic, or sorbitol may be used. Treatment would be by liberal gastric lavage, the substance being only slightly soluble. Ventilation and oxygenation with close arterial blood gas monitoring should be maintained in case of pulmonary edema. In case of an inhalation exposure, the patient should be moved to fresh air and monitored for respiratory distress. The person should also be evaluated for respiratory tract irritation, bronchitis, or pneumonitis. Humidified supplemental oxygen (100%) with assisted ventilation may be administered as required. Exposed eyes should be irrigated with copious amounts of tepid water. For dermal exposure, the affected skin should be washed with soap and water. No antidotes are established.

Since ANTU is a sulfhydryl blocking agent, cysteine has been tried in rats and was effective in some cases. There is no human experience with cysteine and its use is not recommended.

Animal Toxicity

Toxicity of ANTU is strikingly higher for wild Norwegian rats than for other species. Mice and dogs rank next in susceptibility. Young animals are less resistant. Rats that survive sublethal doses develop a high degree of tolerance owing partly to refusal to eat freely. Symptoms in rats appear within 12–25 min with a sharp fall of body temperature, huge pleural and intraalveolar edema, anuria, dyspnea, and death. Blood sugar rises to nearly three times the normal level within 2.5 hr with a severe fall of liver glycogen and failure to deposit liver glycogen. Observation in experimental animals indicate that the principal organ affected is the lung; pulmonary edema and pleural effusion develop due to the action of ANTU on pulmonary capillaries causing marked edema of the subepithelial spaces of the alveolar walls; pericardial effusion is less marked.

Dogs are quite susceptible to toxicity but may be protected by prompt vomiting. Pulmonary effusion in dogs showed an increase in albumin globulin ratio. Hemorrhagic glomerular nephritis has been seen after acute exposure in rats. Hyperglycemia has been reported in experimental animals.

Animal Toxicity: Chronic

Chronic exposure in rats results in stunted growth, thinning and coarsening of hair, deformities of the legs and feet, hyperplasia of the thyroid and splenic pulp, hyaline changes in the hepatic cells, decreased thickness of the adrenal cortex, and calcified tubular casts. Continued administration to cats produces fatal intrahepatic obstructive jaundice without pulmonary lesion. Available data were inadequate to evaluate the carcinogenicity of ANTU in experimental animals.

—Swarupa G. Kulkarni and
Harihara M. Mehendale

Related Topic

Pesticides

Aluminum (Al)

- CAS: 7429-90-5
- SELECTED COMPOUNDS: Aluminum oxide, Al$_2$O$_3$ (CAS: 1344-28-1); aluminum chloride, anhydrous, AlCl$_3$ (CAS: 7446-70-0); aluminum hydroxide, Al(OH)$_3$ (CAS: 21645-51-2)
- CHEMICAL CLASS: Metals

Uses

Aluminum is used alone or in alloys for aircraft, utensils, electrical conductors, and various apparatus. Fine or coarse aluminum powder is used in manufacturing paint and explosives and in the thermite process. Anhydrous aluminum chloride is used in the cracking of petroleum, in the manufacturing of rubbers and lubricants, and as a catalyst. Hexahydrate aluminum chloride is used in antiperspirants and in various processes including area disinfection, wood preservation, fabric dyeing, parchment paper manufacturing, and crude oil refinement. Aluminum hydroxide is used as an absorbent, emulsifier, and ion exchanger. It is also used in chromatography, in glass manufacturing, and as an ingredient in detergents and printing inks. Many antacids have been mainly composed of aluminum compounds and calcium carbonate.

Exposure Pathways

Aluminum is the most plentiful metallic element in the earth's crust. The primary exposure pathway for aluminum is ingestion (e.g., food and water); however, inhalation (of aluminum oxide) is a significant exposure pathway in industrial settings. For the general population, daily intake of aluminum averages approximately 20 mg (from 9 to 36 mg/day). Certain patients on renal dialysis may have additional exposure from aluminum contained in the dialysate. Currently, most dialysis water is pretreated with ion-exchange resins to remove aluminum.

Toxicokinetics

The human body attempts to maintain a balance between aluminum exposure and aluminum content of body tissues so that little is absorbed. Approximately 1 or 2% of ingested aluminum is absorbed through the gastrointestinal tract; however, many anions (e.g., citrate, lactate, malate, and ascorbate) can increase oral absorption of aluminum. For example, citrate and ascorbate may be present in the gastrointestinal tract after ingestion of citrus fruits (e.g., oranges and grapefruits). Approximately 50% of serum aluminum is bound to albumin and transferrin. Although it is not an essential element, aluminum can inhibit absorption of other elements such as fluoride, calcium, iron, and even phosphates. Little information is available regarding the effect of aluminum on enzyme reactions. Aluminum is mainly excreted through the liver in the bile. The bones and lungs have the largest concentrations of aluminum, indicating that bone may be a "sink" for aluminum in the body.

Mechanism of Toxicity

Aluminum binds diatomic phosphate and possibly depletes phosphate, which can lead to osteomalacia. High aluminum serum values and high aluminum concentrations in the bone interfere with the function of vitamin D. The incorporation of aluminum in the bone may interfere with deposition of calcium; the subsequent increase of calcium in the blood may inhibit release of parathyroid hormones by the parathyroid gland. The mechanism by which aluminum concentrates in the brain is not known; it may interfere with the blood–brain barrier.

Human Toxicity

Inhalation of fine aluminum dusts or bauxite dusts (which are mainly aluminum oxide) may produce fibrosis of the lungs. Fibrosis grossly limits the areas in the lungs in which oxygen and carbon dioxide are exchanged. This leads to restrictive and sometimes obstructive pulmonary disease.

Absorption of aluminum has been associated with renal failure, intestinal contraction interference, and adverse neurological effects. Patients on hemodialysis with ordinary tap waters (containing aluminum) have experienced temporary dementia (dialysis encephalopathy syndrome); even fatal neurologic syndromes have been recorded following long-term dialysis treatment (3–7 years). In these patients, increased aluminum concentrations are noted in brain, muscle, and bone tissues. This effect has been attributed to the aluminum salt content in the dialysis water. When the aluminum is removed from the water (using ion-exchange resins), adverse neurological symptoms are not observed.

Other neurological abnormalities have been attributed to absorbed aluminum. Alzheimer's patients show certain distinct abnormalities such as a structure called the neuritic plaque, which is made up of deteriorating nerve endings, and a tangle of nerve fibers not observed in the brains of normal, healthy adults. Aluminum is found in these neurofibrillary tangles; however, the association of the presence of aluminum in these tangles has yet to be elucidated in the etiology of Alzheimer's disease.

The ACGIH TLV-TWA is 10 mg/m^3 for aluminum metal dust, 5 mg/m^3 for aluminum welding fumes, and 10 mg/m^3 for aluminum oxide.

Clinical Management

In recent years, the chelating agent deferoxamine has been used to treat aluminum overload (see Deferoxamine). This drug is not absorbed from the gut and must be administered subcutaneously or intravenously.

Animal Toxicity

Aluminum exhibits marked differences in its effects on animals at different points in their life span and in different species. This is particularly important as it relates to developing an animal model for human toxicity. Cats and rabbits are aluminum sensitive, monkeys display delayed effects, and rats appear immune to aluminum's effect on the brain. Increasing brain concentrations of aluminum in cats and rabbits result in subtle behavioral changes, including learning and memory deficits and poor motor function. These changes progress to tremor, incoordination, weakness, and ataxia. Seizures and death follow within 3 or 4 weeks of initial exposure. This effect may result from aluminum competing with or somehow altering calcium metabolism in several organ systems including the brain; brain calcium rises with aluminum exposures.

Aluminum is not carcinogenic in experimental animals. There are not any reports on the mutagenicity or teratogenicity of this element.

—Arthur Furst, Shirley B. Radding,
and Kathryn A. Wurzel

Related Topics

Metallothionein
Metals

Aluminum Phosphide

- ◆ CAS: 20859-73-8
- ◆ SYNONYMS: Celphos; aluminum monophosphide; Cleophos; Detia-GasEx-T, Ex-B; Delica Gastoxin; Phostoxin; aluminum fosfide; ALP
- ◆ CHEMICAL CLASS: Phosphide fumigant
- ◆ MOLECULAR FORMULA: AlP

Uses

Aluminum phosphide is used as a fumigant, especially in graineries.

Exposure Pathways

Ingestion is the primary route of exposure. Aluminum phosphide is available in tablet form, which is mixed with water to produce phosphine gas. Inhalation of phosphine gas is a possible route of exposure.

Toxicokinetics

Aluminum phosphide tablets, if swallowed, release phosphine gas, which is readily absorbed by the intestinal mucosa. Inhaled phosphine is readily absorbed by the lungs. There is no literature on the quantitative or qualitative distribution and metabolism of phosphine. It is effective on all major organs suggesting that it is widely distributed in the body. Regardless of route of exposure, some phosphine is exhaled by the lungs.

Mechanism of Toxicity

Aluminum phosphide in contact with water releases phosphine gas. Phosphine is a highly toxic gas especially to organs of high oxygen flow or demand. In mammals, phosphine blocks the cytochrome c oxidase, the terminal electron acceptor in the mitochondrial electron transport chain. Surprisingly, the activity of this enzyme in insects is inhibited very little. In beetles there was inhibition of catalase activity, which was proportional to the phosphine exposure.

Human Toxicity: Acute

The LCL_0 (inhalation) in humans is 1000 ppm/5 min. Symptoms are rapid in onset. Exposure to high concentrations of phosphine leads to a profound drop in blood pressure. Lower concentrations cause pulmonary edema. Initial symptoms include respiratory, cardiac, circulatory, and cerebral difficulties with extreme gastrointestinal irritation followed later by renal and hepatic toxicity. Acute symptoms include severe pulmonary irritation, cough, headache, chest tightness, dizziness, lethargy, and stupor. The onset of symptoms is usually within a few hours of exposure.

Human Toxicity: Chronic

Exposure to sublethal doses over long periods may produce toxic symptoms. Chronic exposure to phosphine may be identified through evaluation of blood and urine phosphorus levels.

Clinical Management

Life-support measures for respiratory and cardiovascular functions should be provided. In cases of inhalation exposure, the victim should be moved to fresh air. Patients should be monitored for respiratory distress. Emergency airway support and 100% humidified supplemental oxygen with assisted ventilation may be needed. Circulatory support should be provided to avoid possible pulmonary edema, which may be anticipated. Cardiac, hepatic, and renal functions should be monitored closely. Fluid and electrolytes should be administered as needed. In case of seizures, diazepam or phenytoin can be administered. Dialysis should be considered as symptomatic care of severe renal damage. In case of ingestion, aluminum phosphide will release phosphine in the stomach when it comes in contact with water. If ipecac is administered, it should not be followed with fluids. Gastric lavage is also dangerous. If activated charcoal is administered, it should be mixed with sorbitol and not water.

Animal Toxicity

The inhalation LC_{50} is 380 mg/m^3/2 hr in mice and 70 mg/m^3/2 hr in cats. The LC_{50} in rats is 11 ppm/4 hr. Animals exposed to high concentrations of phosphine show immobility, restlessness, urge to escape, ataxia, pallor, epileptiform seizures, and death, sometimes within 30 min. Exposure to a lower concentration causes similar symptoms but the onset and progression of action is slower. Vomiting may occur in some species.

—Sushmita M. Chanda

Related Topic

Pesticides

Amdro

- CAS: 67485-29-4
- SYNONYMS: Combat; Maxforce; pyramdron; aminohydrazone; hydramethylnon

- CHEMICAL CLASS: Trifluoromethyl amidinohydrazone
- CHEMICAL STRUCTURE:

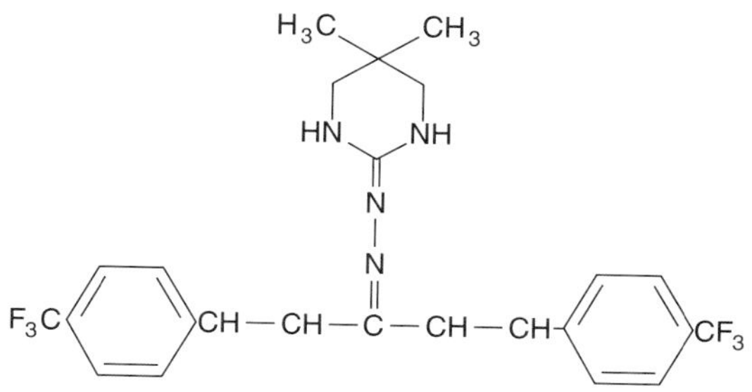

Uses

Amdro is used as an insecticide mainly for ants and cockroaches.

Exposure Pathways

Ingestion is the primary route of exposure. Dermal exposure is also possible.

Toxicokinetics

Amdro is poorly absorbed by the oral route but is absorbed through the skin. Amdro is poorly metabolized in the body with greater than 95% being excreted in the feces in the unchanged form. Elimination in rats dosed orally with hydramethylnon is 72% of the dose in 24 hr and 92% of the dose in 9 days.

Amdro is rapidly degraded in the environment by photolysis or more slowly by hydrolysis. The approximate half-life is 1 hr in direct sunlight.

Mechanism of Toxicity

Amdro is a slowly activating stomach poison. The exact mechanism of toxicity is unclear.

Human Toxicity

Children who ingest small amounts of amdro have symptoms of diarrhea. A case study showed that one-half pound of amdro ingested by an adult diabetic patient produced no specific symptoms except diarrhea. Information on the chronic toxicity of amdro is unavailable.

Clinical Management

The exposed area should be thoroughly washed with soap and water. If pain or irritation continues, a physician should be consulted. Eyes should be washed with copious amounts of room-temperature water for 15 min in cases of eye contamination. If irritation, pain, swelling, lacrimation, or photophobia persists after 15 min of irrigation, medical attention is necessary. Emesis is necessary only when large amounts (greater than 1 ounce of bait/kg) are ingested. In such cases, ipecac may be used for inducing emesis. Activated charcoal slurry with or without saline cathartic or sorbitol may also be administered. Basic respiratory and cardiovascular function support should be utilized.

Animal Toxicity

The oral LD_{50} in rats is 1131–1300 mg/kg. In rabbits, the dermal LD_{50} is greater than 5 g/kg. According to the manufacturer of Combat, a medium-sized dog (20 kg) would experience adverse effects from bait itself only after eating 250 trays.

—*Sushmita M. Chanda*

Related Topic

Pesticides

- Foster a better understanding of the principles and practice of clinical toxicology.
- Encourage development of new therapies and treatment in clinical toxicology.
- Facilitate information exchange among individual members and organizations interested in clinical toxicology.

For membership or other information, please contact the American Academy of Clinical Toxicology at P.O. Box 8820, 777 East Park Drive, Harrisburg, PA 17105-8820. Telephone: (717) 558-7847; fax: (717) 558-7841.

—*Harihara M. Mehendale*

Related Topics

Academy of Toxicological Sciences
American College of Medical Toxicology
American College of Toxicology
International Union of Toxicology
Society of Toxicology

American Academy of Clinical Toxicology

The American Academy of Clinical Toxicology (AACT) was established in 1968 as a not-for-profit, multidisciplinary organization uniting scientists and clinicians in the advancement of research, education, and prevention of diseases caused by chemicals, drugs, and other toxins.

The founders of AACT established the academy to

- Promote the study of health effects of toxins on humans and animals.
- Unite into one group scientists and clinicians whose research, clinical, and academic experience focuses on clinical toxicology.

American Association of Poison Control Centers

The American Association of Poison Control Centers, Inc. (AAPCC) is a nationwide organization of poison centers and interested individuals. (For a complete listing of poison centers in the United States, refer to Poisoning Emergencies in Humans.)

Objectives

- To provide a forum for poison centers and interested individuals to promote the reduction of morbidity and mortality from unintentional poisonings through public and professional education and scientific research

- To set voluntary standards for poison center operations

Activities

- Certification of regional poison centers and poison center personnel
- Interaction with private and governmental agencies whose activities influence poisoning and poison centers
- Development of public and professional education programs and materials
- Collection and analysis of national poisoning data

Publications and Awards

The AAPCC's bulletin, *The Poison Line*, is published six times each year in *Veterinary and Human Toxicology*. Annual recognition awards are presented to individuals who have made a significant contribution to poison control. Research fellowship and annual meeting awards are also provided.

Meetings

Each year, the association holds an annual meeting, which includes scientific presentations, staff sessions, business meetings, and committee meetings.

Membership Categories

- Institutional members: United States poison centers
- Associate institutional members: Hospitals and industrial participants
- Canadian associates: Members of the Canadian Association of Poison Control Centers
- Other foreign institutions
- Individual members, including emeritus, Canadian, and non-Canadian

For additional information, contact the American Association of Poison Control Centers in Washington, DC. Telephone: (202) 362-7217. fax: (202) 362-8377.

—David M. Krentz and Harihara M. Mehendale

(Adapted from information supplied by AAPCC.)

Related Topics

Agency for Toxic Substances and Disease Registry
Food and Drug Administration
National Institutes of Health
Poisoning Emergencies in Humans
Veterinary Toxicology

American Board of Toxicology

Introduction

The American Board of Toxicology, Inc. (ABT) certifies individuals in general toxicology through a process that evaluates expert knowledge as demonstrated by education, experience, and passage of a comprehensive written examination. Each certified individual is initially recognized by being designated as a Diplomate of the American Board of Toxicology (D.A.B.T.) for a period of 5 years.

Other ABT objectives are to encourage the study of the science of toxicology and to stimulate its advancement by promulgation of standards for professional practice. It is ABT policy that diplomates demonstrate a continual commitment to excellence in the science of toxicology. Successful achievement of these goals as outlined by the board will result in an individual maintaining recognition as a diplomate by the ABT.

Performance Criteria

The ABT has identified three performance criteria by which a diplomate will be evaluated pursuant to recertification. These criteria are (1) active practice of toxicology, (2) continuing education, and (3) maintaining expert knowledge in general toxicology. Each diplomate, at the beginning of the fourth year of their current certification, will be required to apply for recertification. ABT will review activities in each of the three performance areas and notify the diplomate of acceptable progress or deficiencies that need to be addressed. If, in the opinion of the board, a diplomate is not compliant with each of the three criteria at the end of the fifth certification year, that diplomate may be required to successfully pass the formal certification examination. Diplomates who are compliant with each of the three performance criteria will be certified for an additional 5 years.

Active Practice of Toxicology

Active practice is defined as performing, directing, or managing toxicology activities such as research, testing, teaching, clinical practice, or regulation.

Continuing Education

A successful program of continuing education may encompass a myriad of diverse activities. The study of published texts, periodicals, or scientific journals germane to toxicology are means by which diplomates routinely maintain or expand their knowledge of toxicology. Other evidence of a commitment to continued education is attendance at specific programs where toxicology themes are presented in a comprehensive or in-depth manner. Such programs are often held during general or annual meetings of the Society of Toxicology, American College of Toxicology, FASEB, Environmental Mutagen Society, Teratology Society, American Association for Cancer Research, or chapter meetings of the Society of Toxicology. Attendance Forum or Target Organ conferences also provide opportunities to maintain or expand a diplomate's knowledge of toxicology.

Maintaining Expert Knowledge of General Toxicology

It is held that an objective mechanism is required for the diplomate and ABT to gauge the success of their efforts to maintain expert knowledge in general toxicology. A recertification examination prepared by the ABT is to serve in this evaluation process. Diplomates will have the opportunity to privately complete the recertification examination during the fourth year of their certification period using their own reference material as needed. The completed examination will be graded by ABT and returned to the diplomate. The diplomates will be furnished a comparison of their results with the performance of peers for each subject area. Stimulated by these results, the diplomate would be expected to tailor a continuing education program that addresses those subject areas in which his or her knowledge appears to have diminished. The ABT may ask a diplomate to complete specific portions of the recertification exam to assess the success of his or her focused continuing education program.

Summary of Recertification Process

Diplomates maintain personal files of activities germane to the Active Practice and Continuing Education criteria for certification (e.g., name of meeting attended, number of hours, title, topics, and faculty). Each diplomate is required to be recertified every 5 years in order to maintain the diplomate status. In addition to maintaining active practice of toxicology during the first 3 years, this procedure involves submission of credentials during the fourth year and fulfilling other requirements during the fifth year.

For additional information, contact the American Board of Toxicology, P.O. Box 30054, Raleigh, NC, 27622-0054. Telephone: 919-782-0036; fax: 919-782-3851.

—David M. Krentz and
Harihara M. Mehendale

(Adapted from information supplied by the American Board of Toxicology.)

Related Topics

Academy of Toxicological Sciences
American Academy of Clinical Toxicology
American College of Medical Toxicology

American College of Toxicology
Society of Environmental Toxicology and
 Chemistry
Society of Toxicology

American College of Medical Toxicology

History

The college was formerly known as the American Board of Medical Toxicology (ABMT). The ABMT offered specialty certification in medical toxicology at a time when the American Board of Medical Specialties (ABMS) did not recognize subspecialty certification in toxicology. When the ABMS approved formal recognition of medical toxicology as a subspecialty in September 1992, the ABMT discontinued its function as a certifying body. It was reincorporated in September 1993 as the American College of Medical Toxicology (ACMT), a specialty society providing support and representation for medical toxicologists.

Purpose

The purpose of ACMT is to operate as a nonprofit organization to advance the science, study, and practice of medical toxicology by fostering the development of medical toxicology in its provision of emergency, consultation, forensic, legal, community, and industrial services; and by otherwise striving to advance and elevate the science, study, and practice of medical toxicology.

The college is a nonprofit organization which is not involved in authorizing or designating any political lobby action.

Membership

Active members of ACMT are physicians who have been certified by the American Board of Medical Toxi-

cology and/or by the Sub-Board in Medical Toxicology of the American Board of Medical Specialties.

In addition to active members, the college accepts applications for international and associate membership. International members are physicians licensed to practice medicine in countries outside the United States, who practice medical toxicology as a substantial portion of their professional activities. Associate members are physicians licensed to practice medicine who have completed a residency training program in a primary medical specialty and who are enrolled in or have completed a training program in medical toxicology.

Fellow Designation

Members, international members, and members emeritus of ACMT who have met additional criteria as established by ACMT shall be designated as "Fellow of the American College of Medical Toxicology" and shall be entitled to use the title "FACMT."

For more information contact the American College of Medical Toxicology, 777 E. Park Drive, P.O. Box 8820, Harrisburg, PA 17105-8820. Telephone: 717-558-7846; fax: 717-558-7841

—Harihara M. Mehendale

Related Topics

Academy of Toxicological Sciences
American Academy of Clinical Toxicology
American Board of Toxicology
American College of Toxicology
Society of Environmental Toxicology and
 Chemistry
Society of Toxicology

American College of Toxicology

Introduction

The American College of Toxicology (ACT) is dedicated to providing an interactive forum for the advancement and exchange of toxicologic information between industry, government, and academia. The goal of the college is to bring together people having common interests in the broad field of toxicology. This includes not only those individuals involved in toxicology but also those from related disciplines: analytical chemistry, biology, biological statistics, computer science, physiology, toxicokinetics, pathology, teratology, genetic toxicology, molecular biology, experimental psychology, immunology, cancer biology, and animal husbandry. The college recognizes that application of the science of toxicology is multifaceted encompassing many disciplines. All of these relate to modern toxicology and attempt to address present and future problems. Toxicology can be defined in the classical sense as the scientific study of the effects of toxicants on biological systems. However, the explosion of scientific knowledge has made infinitely more complex the statement of the toxicological problem and understanding the solution. Modern toxicology offers a unique challenge since it involves quantitative interpolation of data from high to low dose and its eventual extrapolation from simple life forms and animals to humans and the environment. The prediction of toxic effects upon all stages of development and the use of computer models represent frontiers of knowledge in toxicology that in some cases are in their infancy.

ACT also recognizes that the interests and problems of its members are disparate and not only stem from the performance of their responsibilities but also are significantly impacted by government regulations, industrial practices, and societal perception.

ACT is committed to addressing the toxicological issues of the day and those it anticipates will arise in the future. Its interests lie in disseminating information to and among its members so that their combined talents and creative insights may further the practice of their science. To do so, the college brings together the necessary experts to debate and discuss unique and creative approaches to problems which hopefully will better serve the needs and interests of the communities in which we live and the society we serve.

Mission Statement

Mission: The mission of the American College of Toxicology is to educate and lead professionals in industry,

government and related areas of toxicology by actively promoting the exchange of information and perspective on the current status of safety assessment and the application of new developments in toxicology.

Strategic Objectives

- Focus on interdisciplinary exchange of scientific information, especially as scientific information is used in regulation
- Sponsor scientific and educational programs in toxicology.
- Present the ideals and opinions to its membership
- Disseminate information of the results of toxicological research, standards and practices through the College journal and newsletter
- Serve in other capacities in which the College can function more efficiently as a group than as individuals

Activities

Activities include annual meetings and workshops. ACT publishes a newsletter (quarterly), and the *International Journal of Toxicology* (formerley *Journal of the American College of Toxicology,* (bimonthly). The *International Journal of Toxicology* publishes fully refereed papers covering the entire field of toxicology, including research in risk assessment, general toxicology, carcinogenicity, safety evaluation, reproductive and genetic toxicology, epidemiology and clinical toxicology, mechanisms of toxicity, new approaches to toxicological testing, and alternatives to animal testing. Reviews and major symposia in the field are included.

Membership

ACT membership is by election after submission of an application and supporting documentation. There are three types of individual membership: full, associate, and student. Full membership is for qualified individuals who have conducted and published original research in toxicology. Associate membership is for individuals with critical interests in toxicology who have not reached full membership status. Student membership is for qualified predoctoral students.

Corporate membership is available for corporations, associations, and other organizations that support the activities of the college.

For additional information, contact the American College of Toxicology at 9650 Rockville Pike, Bethesda, MD 20814. Telephone: 301-571-1840; Fax: 301-571-1852. e-mail: eKagan@act.faseb.org Home Page: http://Landaus.com/toxicology//.

—*Harihara M. Mehendale*

Related Topics

Academy of Toxicological Sciences
American Academy of Clinical Toxicology
American Board of Toxicology
American College of Medical Toxicology
Society of Environmental Toxicology and
    Chemistry
Society of Toxicology

American Conference of Governmental Industrial Hygienists

The American Conference of Governmental Industrial Hygienists (ACGIH) is an organization devoted to the administrative and technical aspects of worker health and safety. The ACGIH serves as a medium of exchange of ideas and experiences to facilitate the promotion of standards, recommendations, and techniques in occupational and environmental hygiene.

Once each year, ACGIH publishes the *Threshold Limit Values for Chemical Substances and Physical Agents and Biological Exposure Indices*. ACGIH Threshold Limit Values (TLVs) are widely used and cited guidelines for occupational health; they cover more than 750 chemical compounds. The values refer to "airborne concentrations of substances and represent conditions under which it is believed that nearly all workers may be repeatedly exposed day after day without adverse health effects." The ACGIH Chemical Substances TLV committee also categorizes the carcinogenicity of chemicals in the same publication. Data on exposures to physical agents (e.g., sound, cold and heat stress, and radiation) are presented as well.

According to its bylaws, ACGIH objectives are

- To promote and encourage the coordination of industrial hygiene, occupational and environmental health, and safety in the prevention of death, illness, and injury in all aspects and phases of federal, state, local, territorial, and international agencies;

- To encourage the interchange of experiences and knowledge among industrial hygiene, occupational and environmental health, and safety professionals and in the occupational/ environmental health community at large;

- To collect and make accessible to all those engaged in industrial hygiene, occupational and environmental health, and safety such information and data as may be of assistance to them in the proper fulfillment of their duties;

- To collect and make available information, data, and reports to federal, state, and local agencies, international organizations, and the general public which would assist in providing more adequate services;

- To engage in such activities and to hold annual and such other meetings as may be necessary and proper to carry out the objectives of the conference.

Throughout its more than 50 years of service to the profession, ACGIH has endeavored to provide the opportunities, information, and other resources needed by those who protect worker health and safety. ACGIH technical committees, journal publications, symposia and conferences, and other programs all play major roles in achieving the conference's primary goal— worker health protection.

Applied Occupational and Environmental Hygiene is a monthly journal that includes association news, book reviews, a calendar of events, employment listings, information on new products and literature, and professional course listings.

Other ACGIH publications include the annual *ACGIH Transactions* and proceedings from the ACGIH annual meeting in May.

For more information, contact the American Conference of Governmental Industrial Hygienists (ACGIH), Building D-7, 6500 Gateway Avenue, Cincinnati, OH 45211-4438. Telephone: (513) 742-2020; fax: (513) 661-7195.

—David M. Krentz and
Harihara M. Mehendale

(Adapted from information supplied by the American Conference of Governmental Industrial Hygienists.)

Related Topics

Agency for Toxic Substances and Disease Registry
Chemical Industry Institute of Toxicology
Information Resources in Toxicology
Medical Surveillance
National Center for Toxicological Research
National Institute for Occupational Safety and Health
Occupational Safety and Health Association
Occupational Toxicology
Permissible Exposure Limit
Short-Term Exposure Limit
Threshold Limit Values

Ames Test

The Ames test, named for Bruce Ames of the University of California at Berkeley, evaluates the capability of chemicals to induce base pair and frameshift histidine (his) reversion mutations in tester strains that are histidine auxotrophs of the bacterium *Salmonella typhimurium* in the absence or presence of exogenous liver homogenates (e.g., the 9000g supernatant fraction, termed S9). Because evidence suggests that a high percentage of chemicals that elicit a mutagenic response in this assay are potential animal and human mutagens and carcinogens, with few positive results for noncarcinogens, and because the test is efficient, relatively economical, and can indicate mechanisms of chemical interaction with DNA, it is required for essentially all regulatory submissions and is considered by many to be the cornerstone of evaluations for genotoxicity.

It was not until the late 1950s and the 1960s that, as a result of genetic research, scientists expressed concerns that chemicals might be hazardous to the germline of humans and suggested that routine toxicity testing of chemicals should include assays for mutagenicity. The scientific foundations of mutagenicity tests in *Salmonella* also date from this time, but the selected tester strains could have come from any number of bacterial species, particularly *Escherichia coli*, that have been extensively used for genetic research. Reasons for development of the test in *Salmonella* include the ready availability of a large collection of histidine auxotrophs, the experience of Ames and colleagues in testing for reversion of the histidine auxotrophs to prototrophy (histidine independence), and the relative ease with which S9, developed by Heinrich Malling in 1971, could be utilized in the test system. The latter was necessary because, as shown by James and Elizabeth Miller, due the limited metabolic capacity of the bacterial and animal cells a number of chemicals found mutagenic and carcinogenic in whole animals are without genetic effects *in vitro* in the absence of a provision for the metabolic conversion of the chemicals to electrophiles capable of interacting with nucleic acids and proteins.

The test initially defined by Ames in 1971 was progressively revised with more sensitive *Salmonella* strains until publication of the definitive version of the Ames test in 1975, which was further defined in 1983. Although additional *Salmonella* strains have been developed since that time, the major contributions of Ames and colleagues in the early to mid-1970s that have justifiably led to designation of bacterial tests for mutagenicity as ''Ames tests'' include the provision of a good set of tester strains, definition of a protocol that works with most mutagens and carcinogens, and an initial demonstration of a 90% correlation between mutagenicity and carcinogenicity for about 300 chemicals, which led them to assert that ''mutagens are carcinogens.''

Although, today one of the *Salmonella* strains is often replaced with an *E. coli* strain that detects A-T base pair mutations by reverting to tryptophan prototrophy, the most extensively used Ames test features five unique strains of *Salmonella* that are histidine auxotrophs by virtue of G-C mutations in the histidine operon. They were designed for sensitivity in detecting point mutations, specifically frameshift mutations and base pair substitutions, which result in the bacteria's reversion to histidine prototrophy ($his^- \rightarrow his^+$). Frameshift mutagens cause the addition or deletion of single or multiple base pairs in the DNA molecule. Base pair mutagens cause a base change in the DNA; in a reversion assay, this change may occur at the site of the original mutation or at a second site in the bacterial chromosome. When histidine-dependent cells are grown on minimal medium agar plates containing a trace of histidine, only those cells that revert to histidine independence are able to form colonies. The small amount of histidine allows all the plated bacteria to undergo a few divisions; in many cases, this growth is essential for mutagenesis to occur, and the revertant colonies are easily visible against the slight background lawn.

Basic microbial mutagenesis protocols use the plate incorporation approach or a preincubation modification of this approach. In the standard plate incorporation protocol, the test material, bacteria, and either S9 or buffer are added to liquid top agar which is then mixed, and the mixture is immediately poured on a plate of bottom agar. After the agar gels, the bacteria are incubated, at 37°C, for 48–72 hr before the resulting colonies are counted. The preincubation modification, in which the test material, bacteria, and S9 mixture (if used) are incubated for 20–30 min at 37°C before top agar is added, mixed, and the mixture poured on a plate of bottom agar, is used for materials that may be poorly detected in the plate incorporation

assay. Increased activity with preincubation in comparison to plate incorporation is attributed to the fact that the chemical, bacteria, and S9 are incubated at higher concentrations (without agar present) than in the standard plate incorporation test. Other modifications of the assays provide for exposing the bacteria to measured concentrations of gases in closed containers, the use of metabolic activation systems from a variety of species, and protocols for testing urine extracts, extracts of biomedical materials, and ''micro'' assays for test materials of limited availability.

The Ames test has now been used to evaluate several thousand chemicals and, although the correlations of results with rodent carcinogenicity are no longer as high as 90%, few rodent noncarcinogens have been found positive in the test. This has, unfortunately, led many to believe that only an Ames test is required to assess potential genetic hazards, to define only those chemicals that are Ames-positive as mutagenic, and, conversely, to prematurely abandon otherwise beneficial chemicals that yield a positive result in this test. However, tests in bacteria are limited in that they are incapable of assessing some important genetic and cellular processes that are present in mammalian cells and, even with S9, *in vitro* tests cannot completely mimic *in vivo* metabolic fate processes. Hence, although undoubtedly the Ames test will continue to be the first test utilized and will continue to be considered as the cornerstone of genotoxicity, it must be used in conjunction with other *in vitro* and *in vivo* assays to fully assess the potential genetic hazards of chemicals.

—*Ann D. Mitchell*

Related Topics

Analytical Toxicology
Carcinogen–DNA Adduct Formation and DNA
  Repair
Chromosome Aberrations
Developmental Toxicology
Dominant Lethal Tests
Host-Mediated Assay
Molecular Toxicology
Mouse Lymphoma Assay
Mutagenesis
Sister Chromatid Exchange
Toxicity Testing

Aminoglycosides

- REPRESENTATIVE COMPOUNDS: Amikacin; gentamicin; kanamycin; neomycin; netilmicin; paromomycin; streptomycin; tobramycin
- PHARMACEUTICAL CLASS: Antimicrobial agents. These drugs contain amino sugars in glycoside linkage.
- CHEMICAL STRUCTURE: Aminoglycosides are antimicrobial agents with dissimilar structures; it is impossible to represent aminoglycosides with a single general structure. The following is the structure of streptomycin.

$$R = CH_3NH$$

Uses

Aminoglycosides are antimicrobial agents used to treat infections caused by gram-negative (G⁻) organisms. Some of the aminoglycosides have been widely used for preparation of the bowel for surgery and as adjunct to the therapy of hepatic coma.

Exposure Pathways

Ingestion is the most common route for both accidental and intentional exposure to aminoglycosides. Dermal route of exposure is also possible specially with some of chronic topical application of aminoglycosides such as 1% neomycin.

Toxicokinetics

Generally speaking, aminoglycosides are poorly absorbed from the stomach or respiratory tract. The extent of absorption varies with a specific agent, ranging from as low as 0.2% to as high as 9%. Protein binding of aminoglycoside is from as low as 0–3% to as high as 11% depending on the agents. The volume of distribution for aminoglycosides ranges from 0.16 to 0.34 liters/kg. Greater than 90% of aminoglycosides are excreted unchanged through the kidney. The half-life of the therapeutic doses range from 1.5 to 3.2 hr.

Mechanism of Toxicity

The mechanism of toxicity for aminoglycosides is unclear. Pharmacologically, aminoglycosides act directly on the ribosome to inhibit protein biosynthesis and induce specific misreading of the genetic code. Secondary effects upon a variety of polymerization of amino acids may account for their bactericidal activity.

Human Toxicity: Acute

Overdoses may result in renal damage or ototoxicity (deafness and vertigo) depending on the dose and duration. Retinopathy, visual loss, and conjunctival necrosis have been also associated with this class of antibacterial agent. Irreversible damage of the auditory and vestibular functions of the eighth cranial nerve can occur but this is thought to be related to dose and duration of treatment.

Human Toxicity: Chronic

Chronic topical application of 1% neomycin to a large wound precipitated severe hearing loss in an adult within 3 weeks following application. Serious toxicity is a major limitation to the usefulness of the aminoglycosides, and the same spectrum of toxicity is shared by all drugs in this class.

Clinical Management

With overdose of aminoglycosides, the first effort is mobilized in supporting respiratory and cardiovascular functions. For oral ingestion, treatment focuses on preventing absorption with emesis and/or activated charcoal if appropriate. Since overdoses of aminoglycosides are also associated with renal damage, maintaining urine output with intravenous fluids is recommended if necessary.

Animal Toxicity

Several investigators have assessed the toxic effects of high doses of aminoglycosides in animals. Studies with dogs, rabbits, rats, and guinea pigs treated with doses ranging from 7.5 to 120 mg/kg/day in single and divided doses for 10–29 days suggest that less frequent glycoside administration is associated with less nephrotoxicity as assessed by serum creatinine levels, the glomerular filtration rate, and histopathology. A single study in rats assessing ototoxicity based on cochlear histology reported a lack of toxicity regardless of administration frequency.

—Abraham Dalu

Related Topic

Sensory Organs

Amiodarone

- ◆ CAS: 1951-25-3; 19774-82-4
- ◆ SYNONYMS: Amiodarone hydrochloride; Cordarone
- ◆ PHARMACEUTICAL CLASS: Class III antiarrhythmic agent; an iodinated benzofuran derivative
- ◆ CHEMICAL STRUCTURE:

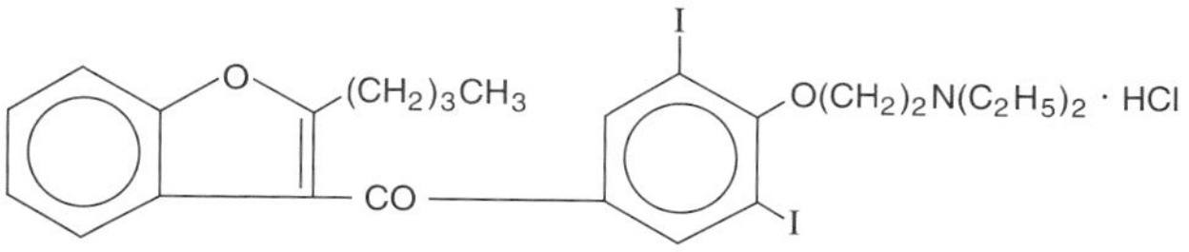

Uses

Amiodarone is used in the suppression and prevention of documented life-threatening ventricular arrhythmias, nonsustained ventricular tachycardia, frequent ventricular premature contractions, supraventricular arrhythmias, and paroxysmal reentrant supraventricular tachycardias that are not responsive to other agents. Amiodarone is not a first-line antiarrhythmic.

Exposure Pathways

Ingestion is the most common route of both accidental and intentional exposure to amiodarone. It is available in an oral dosage form.

Toxicokinetics

Amiodarone is slowly absorbed with an absolute bioavailabilty of 22–86%. Peak plasma concentrations occur within 3–7 hr (range, 2–12 hr). In the absence of a loading dose, steady-state plasma concentrations are not attained for 1–5 months or longer. The onset of antiarrhythmic activity is variable; a response may occur in 2 or 3 days but is generally delayed by 1–3 weeks with maximal effect within 1–5 months. Amiodaron is extensively metabolized. It appears to undergo first-pass metabolism in the gut wall or liver and may undergo enterohepatic circulation. The major metabolite, desethylamiodarone, may possess some antiarrhythmic activity; concentrations are 0.5–2 times that of the unchanged drug during chronic dosing. The average volume of distribution is 65.8 liters/kg (range, 0.9–148 liters/kg). The drug is concentrated in the skin, skeletal muscle, adipose tissue, lung, liver, myocardium, and other body tissues. Desethylamiodarone appears to accumulate in the same body tissues as the parent drug; however, adipose tissue mainly contains amiodarone. Approximately 96% is protein bound. The drug is distributed into breast milk in concentrations exceeding that of maternal plasma.

Less than 1% is excreted in the urine unchanged. Elimination is at least biphasic. The initial phase half-life following a single dose is 3.2–79.7 hr with a terminal half-life of 25 days (range, 9–44 days). One overdose case displayed a terminal half-life of 31 hr. The terminal half-life of the parent drug following chronic dosing is 14–100 days. The terminal half-life of desethylamiodarone is 60–92 days (range, 20–118 days).

Mechanism of Toxicity

Amiodarone prolongs the action potential duration in all cardiac cells with resultant prolongation of refractoriness. Consequently, repolarization is altered without affecting spontaneous depolarization. Sinus node function and automaticity are depressed. Amiodarone is a noncompetitive inhibitor of α and β receptors. This effect is not mediated through receptor blockade; the mechanism is unknown. The mechanism by which thyroid metabolism is altered is also unclear but most likely involves an intracellular effect rather than a central or peripheral effect of the drug. The production of amiodarone–phospholipid complexes within some organs may be the cause of some of this drug's adverse effects. Pulmonary toxicity seen following chronic use is probably the result of a hypersensitivity reaction or other immune-mediated response.

Human Toxicity: Acute

The minimum toxic or lethal dose has not been defined as few cases of amiodarone overdose have been reported. Ingestion of 8 g of amiodarone in a 20-year-old healthy adult resulted in diaphoresis, without a change in blood pressure, 12 hr postingestion. A slight bradycardia and QT prolongation developed 2 or 3 days postingestion. Symptoms associated with chronic therapy did not occur.

Human Toxicity: Chronic

Monitoring of amiodarone or desethylamiodarone levels is thought to be of little clinical value in predicting efficacy or toxicity; the therapeutic range is stated to be 1–2.5 μg/ml. Chronic use of amiodarone has been associated with photosensitivity, hypothyroidism, hyperthyroidism, gynecomastia, hepatotoxicity, pulmonary fibrosis, and asymptomatic corneal deposits. The skin may develop a slate-blue color, especially in areas exposed to the sun.

Clinical Management

Because of delayed absorption, gastric decontamination procedures may be useful beyond several hours postingestion. Basic and advanced life-support measures should be utilized as necessary. Hemodialysis is not of value. Guidelines for specific management are lacking as too few reports of acute amiodarone overdose have been reported.

Animal Toxicity

The oral LD_{50} in dogs is 3 g/kg. Single large doses, up to 3 g/kg, in dogs have caused emesis, tremors, and hindlimb paresis.

—Elizabeth J. Scharman

Related Topic

Cardiovascular System

Ammonia

◆ CAS: 7664-41-7
◆ CHEMICAL CLASS: Nitrogen family
◆ CHEMICAL STRUCTURE:

$$H-\underset{\underset{H}{|}}{N}-H$$

Uses

Ammonia is used as a fertilizer and a refrigerant and in steel treatment. It is also used in the manufacture of nitric acid, hydrazine hydrate, hydrogen cyanide, acrylonitrile, and in many other applications.

Exposure Pathways

Industrial exposures to ammonia are most likely to occur by inhalation with potential for skin and eye contact. Accidental ingestion is also possible.

Toxicokinetics

Following inhalation exposure, ammonia is deposited in the nasal cavity and respiratory tract, and eventually, some traces can be absorbed into the blood and be distributed throughout the body. The average nasal retention of ammonia by human subjects was found to be 83%. The uptake of ammonia in the nasal cavity is influenced by its solubility and inspiratory flow rate. Perhaps ammonia uptake in the nasal tissue is dependent on its reaction with tissue substrates that become deleted at high exposure concentrations. The ammonia vapor can be metabolized in the nasal cavity by the mixed-function oxidase and carboxylesterase systems. Further metabolism takes place in the liver to a number of metabolites and some unchanged ammonia can be excreted in the urine. Most of the free ammonia is excreted in the exhaled breath.

Mechanism of Toxicity

Ammonia is soluble in the mucous membranes of the upper respiratory tract causing irritation of the sensory nerve endings. There is also depression of the mucociliary defense system. The direct action of ammonia on the skin and eyes is the result of irritation to these tissues.

Human Toxicity

Ammonia can be detected by odor in concentrations ranging from 1 to 5 ppm. Exposures between 20 and 25 ppm have caused complaints and discomfort in uninjured workers, while concentrations of 100 ppm caused definite irritation of the respiratory tract and eyes. Severe irritation of the respiratory tract, skin, and eyes have been observed following massive ammonia exposures ranging from 400 to 700 ppm. Concentrations ranging from 2500 to 4500 ppm are dangerous for even short exposures. NIOSH cited a number of references indicating that ammonia concentrations as low as 50 ppm are moderately irritating. NIOSH has published a REL-TWA of 25 ppm and recommended a 15-min STEL of 35 ppm. NIOSH also established an IDLH value of 500 ppm. ACGIH established a TLV TWA of 25 ppm for ammonia, whereas OSHA set a 15-min STEL of 35 ppm for ammonia.

Clinical Management

Exposures by inhalation should be monitored for respiratory tract irritation, bronchitis, or pneumonitis. Humidified supplemental 100% oxygen should be administered.

Gastric lavage may be indicated soon after ingestion of ammonia followed by administration of an activated charcoal slurry mixed with a saline cathartic or sorbitol. Oxygen, in combination with intubation and mechanical ventilation, may be required in severe cases. Exposed eyes should be irrigated with copious amounts of tepid water for at least 15 min. If eye irritation, pain, swelling, lacrimation, or photophobia persist, the patient should be seen in a health care facility.

Animal Toxicity

Animal studies have demonstrated that ammonia concentrations as low as 3 ppm for 7 or 8 min have stopped rat respiratory tract cilia from beating, whereas decreased ciliary activity in rabbits was observed at approximately 100 ppm. Pigs were exposed to 12, 61,

103, or 145 ppm of ammonia in a 5-week subchronic study. At the higher concentrations, there was a reduction in feed consumption and lost body weight. Although these effects may have been a secondary effect of irritation, possible systemic effects cannot be ruled out. The NTP has not conducted genotoxic, chronic, or carcinogenic studies on ammonia.

—*Edward Kerfoot*

Related Topics

Respiratory Tract
Sensory Organs
Veterinary Toxicology

Ammonium Nitrate

- ♦ CAS: 6484-52-2

- ♦ SYNONYMS: Ammonium nitrat; ammonium saltpeter; German saltpeter; Norway saltpeter; nitrate d'ammonium; nitrate of ammonium; Herco prills; Merco prills; Varioform I; AN

- ♦ DESCRIPTION: Ammonium nitrate is found as colorless or white to gray crystals or odorless beads with a molecular weight of 80.06 and specific gravity of 1.725 g/cm. It has a melting point of 169.5°C and boils at 210°C with evolution of nitrous oxide. It forms chloramines on chlorination and is incompatible with acetic acid; acetic anhydride, hexamethylene tetramine acetate, and nitric acid mixture; ammonia; aluminum, calcium nitrate, and formamide mixture; metals; alkali metals; and combustible agents.

- ♦ CHEMICAL STRUCTURE:

$$NH_4NO_3$$

Uses

Ammonium nitrate is used commonly in fertilizers; in pyrotechnics, herbicides, and insecticides, as well in the manufacture of nitrous oxide. It is used as an absorbent for nitrogen oxides, an ingredient of freezing mixtures, an oxidizer in rocket propellants, and a nutrient for yeast and antibiotics. It is also used in explosives (especially as an oil mixture) for blasting rocks and in types of mining.

Nitrates and nitrites are used to cure meats and to develop the characteristic flavor and pink color, to prevent rancidity, and to prevent growth of *Clostridium botulinum* spores in or on meats.

Exposure Pathways

Common exposure pathways are via products in which ammonium nitrate is used. Nitrates are also found in water from soils, rocks, decomposing organic matter; vegetables like beets, radish, lettuce, celery, and spinach. It is also found in secretions like saliva and formed in the mouth and gut due to bacterial action.

Toxicokinetics

Nitrates are well absorbed from the gastrointestinal tract producing peak blood levels only 40 min after ingestion. They may also be absorbed through abraded or damaged skin. Nitrates are converted to nitrites by various bacteria in the stomach and intestines of animals and humans. Approximately 14–31% of nitrate is excreted via the kidneys. The mean renal clearance for nitrates is approximately 26 ml/min. About 40% is excreted as nitrites in the urine. It is also recycled through the saliva.

Mechanism of Toxicity

Nitrate and nitrites can combine with secondary amines to form dimethylnitrosamines, which are acutely toxic and cause centrilobular necrosis, fibrous occlusion of central veins, and pleural and peritoneal hemorrhages in animals. In the body nitrates are converted to nitrites, which can oxidize hemoglobin to methemoglobin and lead to cyanosis.

Human Toxicity: Acute

Ammonium nitrate is irritating to the eyes, nose, throat, and mucous membranes. Inhalation of this compound can cause severe lung congestion, coughing, difficulty in breathing, and increased acid urine. Exposure to large amounts can cause systemic acidosis and abnormal hemoglobin. It is considered to have low toxicity since it causes readily reversible tissue changes which disappear when exposure stops.

In the body nitrates are converted to nitrites, which can oxidize hemoglobin to methemoglobin and lead to cyanosis. They also cause unconsciousness, dizziness, fatigue, shortness of breath, nausea, and vomiting. The skin is warm and sweaty and later becomes cold due to vasodilation. It causes coronary blood vessel contraction, bradycardia, atrial fibrillation, cardiac ischemia, headache, convulsions, and diarrhea.

Nitrate transferred through breast milk causes methemoglobinemia in the infant. Infants are more predisposed to nitrate-related toxicity than adults due to decreased ability to secrete gastric acid, higher levels of fetal hemoglobin, and diminished enzymatic capability to reduce methemoglobin to hemoglobin.

No data are available on the teratogenicity or mutagenicity of ammonium nitrate.

Human Toxicity: Chronic

Chronic ingestion of 5 mg/kg/day is considered unacceptable. Common findings associated with nitrate poisoning include unconsciousness, dizziness, fatigue, shortness of breath, nausea, vomiting, coma, cyanosis, dyspnea, and pallor.

Clinical Management

Absorption should be prevented by dilution with 4–8 ounces of milk or water or by gastric lavage in patients who are comatose or at a risk of convulsing. Charcoal or saline cathartic may also be given. Emesis may be induced if initiated within 30 min of ingestion. Methylene blue is used to treat methemoglobinemia. Diazepam is administered (maximum rate 5 mg/min) to control seizures. Recurrent seizures are controlled by phenytoin or phenobarbital. An EKG should be monitored while administering phenytoin. Dopamine or norepinephrine are administered to control hypotension.

—Prathibha S. Rao and Harihara M. Mehendale

Ammonium Perchlorate

- ◆ CAS: 7790-98-9
- ◆ SYNONYMS: Ammonium perchlorate; perchloric acid; ammonium salt; UN 0402; UN 1442

◆ DESCRIPTION: Ammonium perchlorate is an odorless, water-soluble, white crystalline solid with a molecular weight of 117.5, specific gravity of 1.95, and density of 1.95. It has a decomposition temperature of 130°C and explodes at 380°C. Ammonium perchlorate decomposes at 130°C. It is a strong oxidizing agent. It reacts with combustible materials, heat, moisture, zinc, ammonium sulfate, ammonium nitrate, and metallic powders to cause explosion and ignition. It is incompatible with nitryl perchlorate, potassium permanganate, potassium iodide tetroxide, and metals. It also causes explosion when mixed with sugar and carbon and on contact with copper pipes.

◆ CHEMICAL STRUCTURE:

$$ClO_4 — H_4N$$

Uses

Ammonium perchlorate is an explosive agent used as a component of fireworks, flash powders, explosives, smokeless jet, and rocket propellants. It is also used in oxidizing, engraving, or etching compounds, and as a reagent in analytical chemistry.

Exposure Pathways

Inhalation and dermal and ocular contact are possible routes of exposure.

Mechanism of Toxicity

Ammonium perchlorate is a direct irritant of the eyes, skin, and mucous membranes. It may also have an unspecified antithyroid activity.

Human Toxicity

Ammonium perchlorate is a direct irritant of the eyes, skin, and mucous membranes. Exposure to irritating and toxic fumes of ammonia can cause severe respiratory tract irritation or pulmonary edema. It presents a significant explosion and fire hazard. It also has an unspecified antithyroid activity.

Clinical Management

The victim should be removed to fresh air and monitored for respiratory distress. Early intravenous administration of corticosteroids is recommended to prevent or treat noncardiogenic pulmonary edema. Inhalation of sympathomimetic agents is used to treat bronchospasm and wheezing. Absorption can be prevented by dilution with 4–8 ounces of milk or water. Absorption can also be prevented by gastric lavage in patients who are comatose or at the risk of convulsing. Charcoal, saline, or other cathartics can also be used. Cathartics should be avoided in patients with ileus or impaired renal function.

—*Prathibha S. Rao and Harihara M. Mehendale*

Amphetamine

◆ CAS: 300-62-9

◆ SYNONYMS: 1-Phenyl-2-aminopropane; dl-α-methylphenethylamine; phenylisopropylamine. In drug abuse, the word "amphetamine" can also refer to a number of related compounds such as methamphetamine and other analogs with similar activity (e.g., meth, speed, wire, cross-tops, ice, dexies, black beauties, and hearts).

◆ PHARMACEUTICAL CLASS: Central nervous system stimulant

◆ CHEMICAL STRUCTURE:

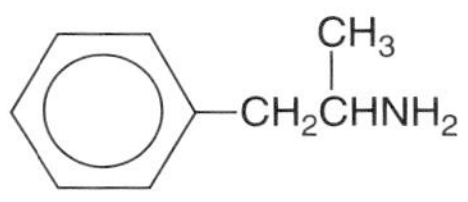

Uses

Amphetamine is used in the treatment of attention-deficit hyperactivity disorder and narcolepsy. It is also a drug of abuse.

Exposure Pathways

Oral and intravenous use are probably the most common routes of exposure. Amphetamines can also be used via nasal insufflation and smoking.

Toxicokinetics

Peak concentrations after oral ingestion range from 1 to 4 hr, depending on the specific amphetamine. Some pharmaceutical preparations are sustained release, with slower absorption rates. More than 50% of a dose undergoes hepatic metabolism, and approximately 30% is excreted unchanged in urine. The amount of unmetabolized drug recovered in urine is greater with acidic urine pH. The apparent volume of distribution is 3–5 liters/kg and also varies by specific drug. The half-life ranges from 8 to 15 hr. Analytical methods used should distinguish the specific compound present since other compounds are structurally similar and may cross-react with anti-amphetamine antibodies.

Mechanism of Toxicity

Amphetamine blocks reuptake of dopamine and norepinephrine at the synapse, increases catecholamine release, and may inhibit monoamine oxidase. These mechanisms increase the activity of catecholamines, which produce the central nervous system (CNS) and sympathetic activity.

Human Toxicity: Acute

Toxicity primarily involves the CNS and cardiovascular system. CNS effects include increased alertness, restlessness, decreased appetite, irritability, stereotyped repetitive behavior, and insomnia with low doses. With increased doses confusion, panic reactions, aggressive behavior, hallucinations, seizures, delirium, coma, and death can occur. Intracranial bleeding can result. Trauma is common secondary to the changes in behavior and decreased judgment. Frequent use results in fatigue, paranoia, and depression. Cardiovascular effects include increased heart rate and blood pressure, chest pain, myocardial ischemia or infarction, dysrhythmias, cardiovascular collapse, and death. Other effects include increased temperature, rhabdomyolysis, increased respiratory rate, flushing, sweating, and dilated pupils. Spasm of peripheral blood vessels can result in cold, numb, and painful extremities. Peak plasma concentrations range from 30 to 110 μg/ml, depending on dose and drug. Average blood concentrations in fatalities are approximately 9 mg/liter, with a wide range of reported values.

Human Toxicity: Chronic

Chronic use can result in psychosis and cardiomyopathy.

Clinical Management

After assessment of airway, breathing, and circulation with necessary supportive care, decontamination of the gastrointestinal tract should be undertaken if it is possible that an oral ingestion occurred. Determination of toxic doses is difficult in chronic users of amphetamines due to the development of tolerance. Oxygen and benzodiazepines should be administered as needed for agitation, shortness of breath, or chest pain. Increased blood pressure can be managed with vasodilators or phenothiazines. Phenothiazines may be necessary for psychosis and cooling and rehydration as standard treatment for increased temperature and rhabdomyolysis.

Animal Toxicity

Animal models describe changes in behavior with toxicity and withdrawal.

—*William A. Watson*

Further Reading

Baselt, R. C., and Cravey, R. H. (1989). *Disposition of Toxic Drugs and Chemicals in Man*, 3d ed., pp. 49–52. Year Book Medical Publishers, Chicago.

Amphibians

Description

The class of amphibia contains approximately 2600 species and is divided into anura, frogs, and toads; and urodela, salamander, and newts. The most toxic members of amphibia are toads (anura) belonging to the family of *Bufonidae*; frogs of the families *Atelopodidae*, *Dendrobatidae*, *Discoglossidae*, *Hylidae*, *Phylomedusae*, *Pipidae*, and *Ranidae*; newts of the genera *Taricha* and *Triturus* and salamanders of the genus *Salamandra*.

Amphibian Toxins

The chemical compositions of amphibian toxins are highly diversified. Amphibians secrete substances to prevent desiccation, control the growth of microorganisms on skin, and discourage predators. These secretions have cytotoxic and/or hemolyzing effects.

Toxicity: Toads

The toxins from *Bufo* species of toads are venom complexes that have a distinct cardioactive digitalis-like action. Toxic signs include profuse salivation with pulmonary edema, cardiac arrhythmia, hypertension, and prostration. Convulsions and death due to cardiac arrest may occur as early as 15 min after exposure to the toxin. Susceptible populations include children and pet dogs or cats playing with toads.

Toxicity: Salamanders

Tetrodotoxin and additional toxic components are found associated with this group (see Tetrodotoxin). Toxic effects are noted at 10 mg/kg body weight. Toxic signs include tingling of the oral cavity with salivation, muscle weakness, motor incoordination, skin numbness, vomiting, diarrhea, and generalized paralysis with convulsions and death in severe cases.

Clinical Management

The victim's mouth should be washed out with copious amounts of water, and atropine should be administered to control salivation. Barbiturates are used to control convulsions, and calcium gluconate may be used to control some physiologic effects. Phenoxybenzamine and propranolol have been used experimentally to block α- and β-adrenergic receptors. Life-support therapy may be used to maintain respiration and other vital functions.

—Prathibha S. Rao and
Harihara M. Mehendale

Amyl Nitrate

- ◆ CAS: 463-04-7
- ◆ SYNONYMS: 1-Nitropentane; *n*-amyl nitrite; nitrous acid pentyl ester; pentyl nitrite

- ◆ PHARMACEUTICAL CLASS: Antidote for cyanide poisoning; vasodilator
- ◆ CHEMICAL STRUCTURE:

$$CH_3—CH—(CH)_3—CH_2—CH_2—O—NO$$

Uses

Amyl nitrate is a vasodilator that acts by relaxing vascular smooth muscle.

Exposure Pathways

Inhalation is the most common route of exposure.

Toxicokinetics

Amyl nitrate is a volatile liquid and is rapidly absorbed from the lungs. It is rapidly hydrolyzed to nitrite ion and the corresponding alcohol. About 60% of the ion is metabolized by the body. About 60% of the nitrites are excreted unchanged in the urine.

Mechanism of Toxicity

Nitrites bind to hemoglobin causing oxidation of hemoglobin to methemoglobin, which is unable to transport oxygen. When methemoglobinemia exceeds 10–15%, cyanosis may be detected.

Human Toxicity: Acute

Amyl nitrate causes methemoglobinemia, unconsciousness, dizziness, fatigue, shortness of breath, nausea, and vomiting. The skin is initially warm and sweaty and later becomes cold due to vasodilation. It causes coronary blood vessel contraction, bradycardia, atrial fibrillation, cardiac ischemia, headache, convulsions, and diarrhea.

Nitrate transferred through breast milk can cause methemoglobinemia in infants. Infants are more predisposed to nitrate-related toxicity than adults due to their decreased ability to secrete gastric acid, higher levels of fetal hemoglobin, and diminished enzymatic capability to reduce methemoglobin to hemoglobin. Nitrites can combine with secondary amines to form dimethylnitrosamines, which are acutely toxic to the liver and cause centrilobular necrosis, fibrous occlusion of central veins, and pleural and peritoneal hemorrhages in animals.

Human Toxicity: Chronic

Chronic ingestion of 5 mg/kg/day is considered unacceptable. Common findings associated with nitrate poisoning include unconsciousness, dizziness, fatigue, shortness of breath, nausea, vomiting, coma, cyanosis, dyspnea, and pallor.

Clinical Management

Absorption should be prevented by dilution with 4–8 oz. of milk or water or by gastric lavage in patients who are comatose or at a risk of convulsing. Charcoal or saline cathartic may also be given. Emesis may be induced if initiated within 30 min of ingestion. Methylene blue is used to treat methemoglobinemia. Diazepam is administered (maximum rate 5 mg/min) to control seizures. Recurrent seizures are controlled by phenytoin or phenobarbital. The EKG should be monitored while administering phenytoin. Dopamine or norepinephrine can be administered to control hypotension.

—Prathibha S. Rao and
Harihara M. Mehendale

Analytical Toxicology

Introduction

Analytical toxicology is a term covering the vast range of analytical techniques (from chemistry and immunology) utilized to identify and quantitate chemical entities as part of toxicological research and testing. Examples of the objectives of such analysis include the following:

- Determining levels of exposure to potential toxicants via air, water, or food

- Verifying exposure levels to doses for animals in experimental studies

- Determining levels of xenobiotics and their metabolites in animal studies

- Screening blood and urine for the presence of illicit drugs or their metabolites

- Measuring levels of endogenous compounds and molecules to evaluate organ function and damage (clinical chemistry)

- Identifying metabolites and macromolecular adjuncts to identify mechanisms of action

The diagnosis and treatment of health problems induced by chemical substances and the closely allied field of therapeutic drug monitoring rely on analytic toxicology. Although the analytes are present in matrices similar to those seen in forensic toxicology, the results must be reported rapidly to be of use to clinicians in treating patients. This requirement of a rapid turnaround time limits the number of chemicals that can be measured because methods, equipment, and personnel must all be available for an instant response to toxicological emergencies.

Occupational and regulatory toxicology require analytic procedures for their implementation or monitoring. In occupational toxicology, the analytic methods used to monitor TLVs and other means of estimating the exposure of workers to toxic hazards may utilize simple, nonspecific, but economical screening devices. However, to determine the actual exposure of a worker, it is necessary to analyze blood, urine, breath, or another specimen by employing methods similar to those used in clinical or forensic toxicology. For regulatory purposes, a variety of matrices (e.g., food, water, and air) must be examined for extremely small quantities of analytes. Frequently, this requires the use of sophisticated methodology with extreme sensitivity. Both of these applications of analytic toxicology impinge on forensic toxicology because an injury or occupational disease in a worker can result in a legal proceeding, just as a violation of a regulatory law may.

Other applications of analytic toxicology occur frequently during the course of experimental studies. Confirmation of the concentration of dosing solutions and monitoring of their stability often can be accomplished with the use of simple analytic techniques. The bioavailability of a dose may vary with the route of administration and the vehicle used. Blood concentrations can be monitored as a means of establishing this important parameter. In addition, an important feature in the study of any toxic substance is the characterization of its metabolites as well as the distribution of the parent drug, together with its metabolites, to various tissues. This requires sensitive, specific, and valid analytic procedures. Similar analytic studies can be conducted within a temporal framework to gain an understanding of the dynamics of the absorption, distribution, metabolism, and excretion of toxic chemicals.

Analysis of Common Toxic Substances

It is evident that analytic toxicology is intimately involved in many aspects of experimental and applied toxicology. Because toxic substances include all chemical types and because the measurements of toxic chemicals may require the examination of biological or nonbiological matrices, the scope of analytic toxicology is broad. Nevertheless, a systematic approach and a reliance on the practical experience of generations of forensic toxicologists can be used in conjunction with the sophisticated tools of analytic chemistry to provide the data needed to understand the hazards of toxic substances more completely. "All substances are poisons: There is none which is not a poison." Analytic toxicology potentially encompasses all chemical substances. Forensic toxicologists learned long ago that when the nature of a suspected poison is unknown, a systematic, standardized approach must be used to identify the presence of most common toxic substances. An approach which has stood the test of time was first suggested by Chapuis in 1873 in *Elements de Toxicologie*. It is based on the origin or nature of the toxic agent. Such a system can be characterized as follows:

1. Gases
2. Volatile substances
3. Corrosives
4. Metals
5. Anions and nonmetals
6. Nonvolatile organic substances
7. Miscellaneous

Closely related to this descriptive classification is the method for separating a toxic agent from the matrix in which it is embedded. The matrix is generally a biological specimen such as a body fluid or a solid tissue. The agent of interest may exist in the matrix in a simple solution or may be bound to protein and other cellular constituents. The challenge is to separate the toxic agent in sufficient purity and quantity to permit it to be characterized and quantified. At times, the parent compound is no longer present in large enough amounts to be separated. In such cases, known metabolites may indirectly provide a measure of the parent substance. With other substances, interaction of the poison with tissue components may require the isolation or characterization of a protein adduct. Methods for separation have long provided a great challenge to analytic toxicologists. Only recently have methods become available which permit direct measurement of some analytes without prior separation from the matrix.

Gases

Gases are most simply measured by means of gas chromatography. Some gases are extremely labile, and the specimen must be collected and preserved at temperatures as low as that of liquid nitrogen. Generally, the gas is carefully liberated by incubating the specimen at a predetermined temperature in a closed container. The gas, freed from the matrix, collects over the specimen's "headspace," where it can be sampled and injected into the gas chromatograph. Other gases, such as carbon monoxide, interact with proteins. These gases can be carefully released from the protein, or the adduct can be measured independently, as in the case of carboxyhemoglobin.

Volatile Substances

Volatile substances are generally liquids of a variety of chemical types. The temperature at which they boil is sufficiently low that older methods of separation utilized microdistillation or diffusion techniques cannot be utilized. Gas–liquid chromatography is the simplest approach for simultaneous separation and quantitation in favorable cases. The simple alcohols can be measured by injecting a diluted body fluid directly onto the column of the chromatography. A more common approach is to use the headspace technique, as is done for gases, after incubating the specimen at an elevated temperature.

Corrosives

Corrosives include mineral acids and bases. Many corrosives consist of ions which are normal tissue constituents. Clinical chemical techniques can be applied to detect these ions when they are in great excess over normal concentrations. Because these ions are normal constituents, the corrosive effects at the site of contact of the chemical, together with other changes in blood chemistry values, can confirm the ingestion of a corrosive substance.

Metals

Metals are encountered frequently as occupational and environmental hazards. Elegant analytic methods are available for most metals even when they are present at extremely low concentrations. Classical separation procedures involve destruction of the organic matrix by chemical or thermal oxidation. This leaves the metal to be identified and quantified in the inorganic residue. Unfortunately, this prevents a determination of the metal in the oxidation state or in combination with other elements, as it existed when the metal compound was absorbed. For example, the toxic effects of metallic mercury, mercurous ion, mercuric ion, and dimethyl mercury are all different. Analytic methods must be selected which determine the relative amount of each form present to yield optimal analytic results.

Toxic Anions and Nonmetals

Toxic anions and nonmetals are a difficult group for analysis. Some anions can be trapped in combination with a stable cation, after which the organic matrix can be destroyed, as with metals. Others can be separated from the bulk of the matrix by dialysis, after they are detected by colorimetric or chromatopathic procedures. Still others are detected and measured by ion-specific electrodes. There are no standard approaches for this group, and other than phosphorus, they are rarely encountered in an uncombined form.

Nonvolatile Organic Substances

Nonvolatile organic substances constitute the largest group of substances which must be considered by analytic toxicologists. This group includes drugs, both prescribed and illegal, pesticides, natural products, pollutants, and industrial compounds. These substances are solids of liquids with high boiling points. Thus, separation procedures generally rely on differential extractions, either liquid–liquid or solid–liquid in nature. These extractions often are not efficient, and recovery of the toxic substance from the sample matrix may be poor. When the nature of the toxic substance is known, immunoassay procedures are useful because they allow a toxicologist to avoid using separation procedures.

Such compounds can be classified as

Organic strong acids

Organic weak acids

Organic bases

Organic neutral compounds

Organic amphoteric compounds

Miscellaneous

Finally, a miscellaneous category must be included to cover the large number of toxic agents that cannot be detected by the routine application of the methods described previously. Venoms and other toxic mixtures of proteins or uncharacterized constituents fall into this class. Frequently, if antibodies can be grown against the active constituent, immunoassay may be the most practical means of detecting and measuring these highly potent and difficult to isolate substances. Unfortunately, unless highly specific monoclonal antibodies are used, the analytic procedure may not be acceptable for forensic purposes. Frequently, specific analytic procedures must be developed for each analyte of this type. At times, biological endpoints are utilized to semiquantify the concentration of the isolated product.

Analytical Techniques

Due to the increased levels of sensitivity of analytical techniques and a range of legal requirements (including Good Laboratory Practices and issues in potential litigation), particular care must be taken in collecting and handling samples to both avoid contamination and maintain a chain of custody of samples and sample records. There are a vast variety of techniques now employed in analysis, as outlined briefly below.

Chromatography

    Thin layer

    Gas

    High-performance liquid chromatography (HPLC)

    Mass spectrophotomology

Photometry/spectroscopy

Spectrophotomology (ultraviolet, IR, and visible light)

    Flame photometry

    Atomic absorption

    Nuclear magnetic resonance (NMR) spectroscopy

    Electron spin resonance (ESR) spectrophotometry

    Raman spectroscopy

Immunoassays

    Radioimmunoassay (RIA)

    Enzyme immunoassay (EIZ)

    Fluorescent immunoassay (FIA)

Isotopic labeling

    Positron emission tomography (PET)

Magnetic resonance imaging (MRI)

Further Reading

Arnold, D. L., Grice, H. C., and Krewski, D. R. (1990). *Handbook of in Vivo Toxicity Testing.* Academic Press, San Diego.

Blanke, R. V., and Decker, W. J. (1986). Analysis of toxic substances. In *Textbook of Clinical Chemistry* (N. W. Tietz, Ed.), pp. 1670–1744. Saunders, Philadelphia.

Caldwell, W. S., Byrd, G. D., deEthizy, J. D., and Crooks, P. A. (1994). Modern instrumental methods for studying mechanisms of toxicity. In *Principles and Methods of Toxicology* (A. W. Hayes, Ed.), 3d ed., pp. 1335–1390. Raven Press, New York.

Cravey, R. H., and Baselt, R. C. (Eds.) (1981). *Introduction to Forensic Toxicology*, pp. 3–6. Biomedical Publications, Davis, CA.

Popov, A. I., and Hallenga, K. (Eds.) (1991). *Modern NMR Techniques and Their Application in Chemistry.* Dekker, New York.

Rose, M. E., and Johnstone, R. A. W. (1982). *Mass Spectrometry for Chemists and Biochemists.* Cambridge Univ. Press, Cambridge, UK.

Sanders, J. K. M., and Hunter, B. K. (1987). *Modern NMR Spectroscopy: A Guide for Chemists.* Oxford Univ. Press, Oxford, UK.

Whateley, T. L. (1988). Radiochemistry and radiopharmaceuticals. In *Practical Pharmaceutical Chemistry* (A. H. Beckett and J. B. Stanlake, Eds.), 4th ed., Part 2, pp. 501–534. Athlone Press, London.

Wolfe, R. R. (1992). *Radioactive and Stable Isotope Tracers in Biomedicine: Principles and Practice of Kinetic Analysis.* Wiley-Liss, New York.

Yergey, A. L., Edmonds, C. G., Lewis, I. A. S., and Vestal, M. L. (1990). *Liquid Chromatography/Mass Spectrometry, Techniques and Applications.*, Plenum, New York.

—*Shayne C. Gad*

Related Topics

Ames Test

Animal Models

Carcinogen–DNA Adduct Formation and DNA Repair

Corrosives

Forensic Toxicology

Good Laboratory Practices

In Vitro

In Vivo

Metals

Occupational Toxicology

Toxicity Testing

Androgens

- SYNONYM: Male sex hormones

- PHARMACEUTICAL CLASS: Androgens are natural and synthetic congeners of the steroid class of compounds.

- CHEMICAL STRUCTURE:

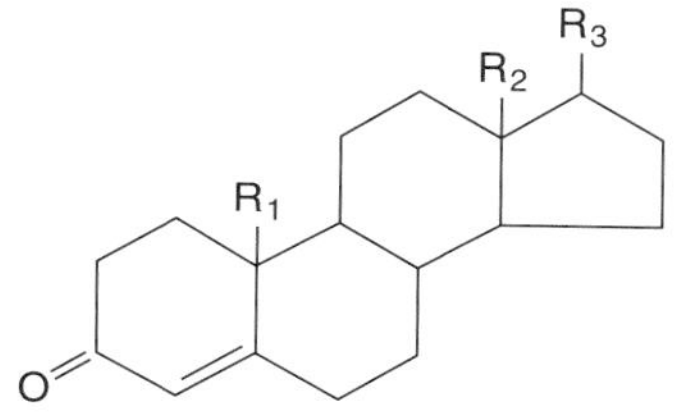

	R_1	R_2	R_3
1. Testosterone	H	H	OH
2. Nandrolone decanoate	H	CH_3	$O\text{-}\overset{\displaystyle O}{\overset{\|}{C}}\text{-}C_9H_{10}$
3. Testosterone propionate	CH_3	CH_3	$O\text{-}\overset{\displaystyle O}{\overset{\|}{C}}\text{-}CH_2CH_3$
4. Ethylestrenol	H	CH_3	$\overset{\displaystyle OH}{\underset{}{|}}\text{-}CH_2CH_3$

Uses

Therapeutic indications for androgens are deficient endocrine functions of the testes, like hypogonadism, treatment of refractory anemias in men and women, and hereditary angioneurotic edema. Testosterone has been known to have a palliative effect in some cases of breast cancer and in osteoporosis.

Toxicokinetics

Injected as an oil, androgens are so quickly absorbed, metabolized, and excreted that the effect is very small. Esters of testosterone are more slowly absorbed and are more effective. The majority of the androgens are inactivated primarily in the liver and involve oxidation of the hydroxy groups and reduction of the steroid ring. Alkylation at the 17 position retards hepatic metabolism and hence is effective orally.

About 90% of the androgens are excreted in the urine; 6% appear in the feces after undergoing enterohepatic circulation. Small amounts are also excreted as soluble glucuronide and sulfate conjugates. Many of the synthetic androgens have a longer half-life. Unaltered compounds are excreted in the urine and feces.

Human Toxicity

Water retention due to sodium chloride (salt) is a common manifestation which leads to weight gain. Edema is also found in patients with cardiac heart failure, renal insufficiency, liver cirrhosis, and hypoproteinemia. When large doses are used to treat neoplastic diseases, compounds with 17-alkyl substitutions can cause cholestatic hepatitis; at high doses jaundice is the most common clinical feature with accumulation of bile in the bile capillaries. Jaundice usually develops after 2–5 months of therapy. It can be detected by increases in plasma aspartate aminotransferase and alkaline phosphatase. Development of hepatic carcinoma has also been reported.

Untoward effect: Androgens may have a virilizing effect with development of male characteristics in women and feminizing effects in men.

Clinical Management

Edema due to salt retention is generally treated with diuretics targeted at increased sodium excretion.

—*Prathibha S. Rao and*
Harihara M. Mehendale

Related Topics

Endocrine System
Reproductive Toxicology, Female
Reproductive Toxicology, Male
Toxicity Testing, Reproductive

Animal Models

Introduction

The use of animals in experimental medicine, pharmacological study, and toxicological assessment is a well-established and essential practice. Whether serving as a source of isolated cells or tissues, a disease model, or as a prediction for drug or other xenobiotic action in humans, experiments in animals have provided the necessary building blocks that permitted the explosive growth of medical and biological knowledge in the latter half of the twentieth century. Animal experiments also have served rather successfully as identifiers of potential hazards to and toxicity in humans for synthetic chemicals with many intended uses.

Animals have been used as models for centuries to predict what chemicals and environmental factors would do to humans. The earliest uses of experimental animals are lost in prehistory, and much of what is recorded in early history about toxicology testing indicates that humans were the test subjects. The earliest clear description of the use of animals in the scientific study of the effects of environmental agents appears to be by Priestley (1792) in his study of gases. The first systematic use of animals for the screening of a wide variety of agents was published by Orfila (1814) and was described by Dubois and Geiling (1959) in their historical review. This work consisted of dosing test animals with known quantities of agents (poisons or drugs) and included the careful recording of the resulting clinical signs and gross necropsy observations. The use of animals as predictors of potential ill effects has grown since that time.

Current Animal Studies

The current regulatorily required use of animal models in acute testing began by using them as a form of instrument to detect undesired contaminants. For example, canaries were used by miners to detect the presence of carbon monoxide—a case in which an animal model is more sensitive than humans. By 1907, the U.S. FDA started to protect the public by the use of a voluntary testing program for new coal tar colors in foods. This was replaced by a mandatory program of testing in 1938, and such regulatorily required animal testing programs have continued to expand until recently.

The knowledge gained by experimentation on animals has undoubtedly increased the quality of our lives, an observation that most reasonable people would find difficult to dispute, and has also benefited animals. As is the case with many tools, animals have sometimes been used inappropriately. These unfortunate instances have helped fuel an increasingly vituperative animal rights movement. This movement has encouraged a measure of critical self-appraisal on the part of scientists concerning the issues of the care and usage of animals. The Society of Toxicology, for example, has established Animals in Research Committees, and has published guidelines for the use of animals in research and testing. In general, the purpose of these committees is to foster thinking on the four "Rs" of animal-based research: reduction, refinement, research into replacements, and responsible use.

The media commonly carry reports that state that most (if not all) animal testing and research is not predictive of what will happen in humans and, therefore, such testing is unwarranted. Many of the animal rights groups also present this argument at every opportunity and reinforce it with examples that entail seemingly great suffering in animals but which add nothing to the health, safety, and welfare of society. This is held to be especially the case for safety testing and research in toxicology. Animal rights activists try to "prove" this point by presenting examples of failure; for example, thalidomide may be presented as an example without pointing out that, in the case of thalidomide, there was lack of adequate testing (or of interpretation of existing test results) prior to marketing. In light of the essential nature of animal research and testing in toxicology, this is equivalent to seeking to functionally disarm us as scientists. Our primary responsibility (the fourth "R") is to provide the information to protect people and the environment, and without animal models we cannot discharge this responsibility.

When confronted with this argument, all too many toxicologists cannot respond with examples to the contrary. Indeed, many may not even fully understand the argument at all. Also, very few are familiar enough with some of the history of toxicity testing to be able to counter with examples where it has not only accurately predicted a potential hazard to humans but also where research has directly benefited both humans and animals. There are, however, many such examples. Demonstrating the actual benefit of toxicology testing and research with examples that directly relate to the everyday lives of most people and not esoteric, basic research findings (which are the most exciting and interesting products to most scientists) is not an easy task. Examples that can be seen to affect neighbors, relatives, and selves on a daily basis would be the most effective. The problem is that toxicology is, in a sense, a negative science. The things we find and discover are usually adverse. Also, if the applied end of our science works correctly, then the results are things that do not happen (and therefore are not seen).

If we correctly identify toxic agents (using animals and other predictive model systems) in advance of a product or agent being introduced into the marketplace or environment, then generally it will not be introduced (or it will be removed) and society will not see death, rashes, renal and hepatic diseases, cancer, or birth defects (for example). Also, as these things already occur at some level in the population, it would seem that seeing less of them would be hard to firmly tie to the results of toxicity testing that rely on animals. In addition, the fact that animals are predictive models for humans is controversial.

Origins of Predictive Animal Testing

The actual record of evidence for the predictive value of animal studies and how they have benefited man and domestic animals will be reviewed in the following two sections. However, the negative image needs to be rebutted. First, it must be remembered that predictive animal testing in toxicology, as we now know it, arose largely out of three historical events.

The "Lash Lure" Case

Early in the 1930s, an untested eyelash dye containing *p*-pheylenediamine (Lash Lure) was brought onto the market in the United States. This product (as well as a number of similar products) rapidly demonstrated that it could sensitize the external ocular structures, leading to corneal ulceration with loss of vision and at least one fatality.

The Elixir of Sulfanilamide Case

In 1937, an elixir of sulfanilamide dissolved in ethylene glycol was introduced into the marketplace. One hundred and seven people died as a result of ethylene glycol toxicity. The public response to these two tragedies helped prompt U.S. Congress to pass the Federal Food, Drug, and Cosmetic Act of 1938 (FD&C Act). This law mandated the premarket testing of drugs for safety in experimental animals. The most compelling evidence that should be considered is "negative." It is a fact that since the imposition of animal testing as a result of these two cases, no similar occurrence has happened—even though society uses many more consumer products and pharmaceuticals today than during the 1930s.

Thalidomide

The use of thalidomide, a sedative–hypnotic agent, led to some 10,000 deformed children being born in Eu-

rope (see Thalidomide). This in turn led directly to the 1962 revision of the FD&C Act—requiring more stringent testing. Current testing procedures (or even those at the time in the United States, where the drug was never approved for human use) would have identified the hazard and prevented this tragedy. In fact, tragedies like this have not occurred in Europe or the United States except when the results of animal tests have been ignored. Table A-1 presents an overview of cases in which animal data predicted adverse effects in humans.

Birth defects, for example, have occurred with isotretinoin (Accutane) where developmental toxicity had been clearly established in animals and presented on labeling, but the drug has continued to be used by potentially pregnant women.

Research into replacements for test animals, such as cellular cultures, organs harvested from slaughter houses, computer modeling, and physical/chemical systems, has been extensive. While each of these have their utility, they will not replace animals for the foreseeable future. Some degree of animal use will continue.

Choosing an Animal Model

Choosing the appropriate animal model for a given problem is sometimes guesswork and often a matter of

TABLE A-1
Animal Models That Predicted Adverse Effects of Xenobiotics in Humans

Agent	Effect	Animal species	In human
Thalidomide	Phocomelia	Rat	N/Y
Accutane	Developmental toxicity of CNS (neural tube defects)	Rat, rabbit, dog, primate	Y
AZT	Bone marrow depression	Dog, rat, monkey	Y
Valproic acid	Cleft palate	Rat, mouse, rabbit	Y
Cyclosporine	Nephropathy, reversible immune response suppression (essential aid to organ transplantation)	Rat, monkey	Y
Benoxaprofen (Oraflex)	Hepatotoxicity, photosensitivity	No Guinea pig	Y Y
Zomepirac (Zomax)	Anaphylactic shock	No	Y
MPTP	Parkinsonism	Monkey	Y
Cyclophosphamide	Hemorrhagic cystitis	Rat, dog	Y
Mercury	Encephalopathy	Rat, monkey	Y
Diethylene glycol	Nephropathy	Rat, dog	Y
Razoxin	Myelomonocytic leukemia	Mouse	Y

convenience. One often uses a species with which one is most familiar, with little consideration as to whether the chosen species is actually the most appropriate for the problem at hand. For example, the rat is probably a poor model for studying the chronic toxicity of any new nonsteroidal anti-inflammatory drug (NSAID) because the acute gastrointestinal toxicity will probably mask any other toxic effects. The guinea pig is less sensitive to most NSAIDs than the rat and closer in sensitivity to humans and would, therefore, be a more appropriate species for investigating the chronic (non-gastrointestinal) toxicity of an NSAID. This practice of not rationally choosing an appropriate species for an experiment undoubtedly results in questionable science. This alone should be considered a waste of animals and resources. It results also in additional, and sometimes duplicative, experiments.

Further Reading

Gad, S. C., and Chengelis, C. P. (Eds.) (1992). *Animal Models in Toxicology*. Dekker, New York.

—Shayne C. Gad

Related Topics

Analytical Toxicology
In Vitro Test
In Vivo Test
Respiratory Tract
Society of Toxicology
Toxicity Testing, Alternatives
Toxicity Testing, Aquatic
Toxicity Testing, Behavioral
Toxicity Testing, Carcinogenesis
Toxicity Testing, Dermal
Toxicity Testing, Developmental
Toxicity Testing, Inhalation

Anthracene

- CAS: 120-12-7
- SYNONYMS: Anthracin; paranaphthalene; green oil; tetra oil N2G

- DESCRIPTION: Solid white to yellow crystals, weak aromatic odor, sinks in water; B.P., 342°C; M.P., 218°C; molecular weight, 178.22; density/specific gravity, 1.25 at 27 and 4°C; octanol–water coefficient = 4.45; soluble in absolute alcohol and organic solvents. Maximum absorption occurs at 218 nm.
- CHEMICAL CLASS: Polycyclic aromatic hydrocarbon
- CHEMICAL STRUCTURE:

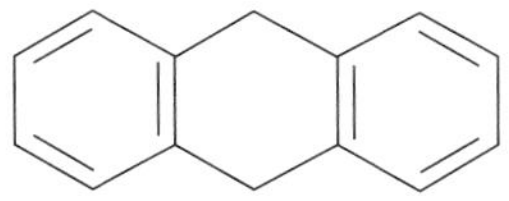

Exposure Pathways

Inhalation is the primary exposure pathway. Natural occurring sources include a high boiling fraction of coal tar, consisting of anthracene, phenanthrene, and other solid hydrocarbons as well as acridine. Other sources include volcanoes and forest fires. Artificial sources include exhaust from motor vehicles and other gasoline and diesel engines; cigarette, marijuana, and cigar smoke; emissions from coal-, oil-, and wood-burning stoves, furnaces and power plants; smoke and soot. Air pollution sources include coke oven emissions, space heating installation burning, emissions from typical European gasoline engines, dielectric in the manufacture of battery electrodes, and electric arc furnace electrodes; felt, roof, and paper manufacturing; and alumina reduction.

Toxicokinetics

Polycyclic aromatic hydrocarbons were detected in human fat and liver and their average concentrations were 1100 and 380 ppt, respectively. Anthracene was found at high levels in the liver and fat. When administered orally to animals 70–80% of the dose is excreted unchanged in the feces but metabolites present in rat urine include *N*-acetyl-*S*-(1,2-dihydro-2-hydroxy-1-anthryl)-cysteine and conjugates of *trans*-1,2-dihydroanthracene-1,2-diol, and 1,2-dihydroxyanthracene. The cysteine conjugate is decomposed by mineral acids to yield 1-anthrylmercapturic acid, 1- and 2-anthrols, and anthracene. Rats metabolize anthracene into *trans*-9,10-dihydroanthracene-9,10-diol, which gives rise to anthrone and several hydroxylated metabolites.

Human Toxicity: Acute

Anthracene is photosensitizing. It can cause acute dermatitis with symptoms of burning, itching, and edema, which are more pronounced in the exposed bare skin regions. Other symptoms are lacrimation, photophobia, edema of the eyelids, and conjunctival hyperemia. The acute symptoms disappear within several days after cessation of contact. Systemic effects of industrial anthracene manifest themselves by headache, nausea, loss of appetite, slow reactions, and adynamia.

Human Toxicity: Chronic

Chronic exposure may lead to inflammation of the gastrointestinal tract, patchy areas of increased yellow-brown pigment changes, loss of skin pigment, thinning or patchy thickening of skin, skin warts, skin cancer, and pimples. Repeated breathing of "fumes" especially from heated anthracene may cause a chronic bronchitis with cough and phlegm. Repeated exposure of male scrotum can cause skin thinning and increased skin pigmentation.

No occupational exposure limits have been established for anthracene. However, safe work practices should always be followed.

Clinical Management

No specific treatments have been prescribed. The patient should be moved to fresh air in case of respiratory distress.

Animal Toxicity

Anthracene showed no mutagenic activity in *Salmonella thyphimurium* TA100 and TA98 with and without addition of rat liver microsomes (S9) and no carcinogenic activity in Swiss albino mice. A significant increase in the formation of nonneoplastic melanotic tumors was observed among first and second generation progeny of *Drosophila melanogaster* that had been exposed chronically as larvae to low concentrations of anthracene.

—Prathibha S. Rao and
Harihara M. Mehendale

Related Topics

Coke Oven Emissions
Polycyclic Aromatic Hydrocarbons

Anticholinergics

◆ SYNONYMS: Parasympatholytics; cholinergic blockers; sympatholytics; antispasmodics

◆ CHEMICAL CLASS: As a class, anticholinergics include the antihistamines, atropine and homatropine; anti-Parkinsonian agents like benzotropine, procyclidine, and trihexyphenidyl; the antimuscarinics of which atropine is the prototype; and antispasmodics like dicyclomine and oxybutymin. Most antimuscarinics are aminoalcohols or their derivatives (usually esters or ethers), aminoamides, or other amines. Antimuscarinics can be divided into two groups. These are the naturally occurring alkaloids and their semisynthetic derivatives like atropine, homatropine, scopolamine, and hyoscyamine and the synthetic amine compounds such as anisotropine, dicyclomine, and ipratropium.

Uses

Anticholinergics have a wide range of therapeutic uses: prior to anesthesia; as a prophylactic for preventing motion sickness; in symptomatic control of Parkinson's disease; in abnormal slowing of the heart in poisoning with organophosphates and other cholinergic drugs; and in the treatment of peptic ulcer and irritable bowel syndrome.

Since these agents act as smooth muscle relaxants they are used as antispasmodics and may be used to reduce spasms of the stomach. Antimuscarinics appear to be useful in the treatment of gastrointestinal hypersecretory states (e.g., Zollinger Ellison syndrome) and may be used in conjunction with a H_2 receptor antagonist. In addition to atropine, belladonna, and other semisynthetic derivatives, a number of other synthetic compounds are also used in gastrointestinal disorders. These compounds consist of a large blocking group linked by a short chain to a strongly basic tertiary or quaternary group. Synthetic drugs used as antispasmodics or antisecretory agents in gastrointestinal disorders include oxyphenonium bromide, isopropamide iodide, mepenzolate bromide, and dicyclomine.

Antimuscarinics are potent bronchodilators and are used in the treatment of chronic bronchitis and asthma.

Toxicokinetics

Antimuscarinics having a quaternary ammonium group are incompletely absorbed from the gut since these are completely ionized. The tertiary amine antimuscarinics are readily absorbed from the gut. The presence of food may reduce absorption. Quaternary ammonium antimuscarinics exhibit poor lipid solubility, do not cross the blood–brain barrier, and thus exhibit minimal central nervous system (CNS) effects. Also due to their poor lipid solubility they do not penetrate the eye and are unlikely to appear in the milk. Atropine and other tertiary amines are capable of crossing the CNS. Atropine is capable of crossing the placenta and has been stated to distribute into milk in small quantities. It is oxidized primarily in the liver. Atropine is apparently metabolized in the liver to tropic acid, tropine, and possibly esters of tropic acid and glucuronide conjugate.

Antimuscarinics are mainly eliminated in urine as unchanged drug and its metabolites. Following oral administration substantial amounts of antimuscarinics may be eliminated in feces as unabsorbed drug.

Mechanism of Toxicity

Antimuscarinics competitively inhibit the action of acetylcholine or other cholinergic stimuli at the muscarinic receptor. At usual doses these have little or no effect on the cholinergic stimuli at nicotinic receptors. Autonomic ganglia, where cholinergic transmission involves nicotinic receptors, produce a partial cholinergic block at relatively high doses. Receptors at various sites are not equally sensitive to inhibitory effects of antimuscarinics. Atropine acts by competitive antagonism at the receptor sites of the effector organs. It may also inhibit responses to histamine, serotonin, and norepinephrine and may block transmission at the autonomic ganglia and the skeletal neuroeffector junction.

Human Toxicity

Single, 10-mg oral doses of atropine have produced signs of acute toxicity in adults. Children are more susceptible than adults to the toxic effects of atropine. Deaths have been reported in children following ingestion of 10 mg of atropine.

Acute overdosage with antimuscarinics produces both peripheral and CNS symptomatology. The quaternary ammonium compounds do not readily penetrate the CNS and thus exhibit minimal central effects even at toxic doses. Patients with anticholinergic toxicity will typically show peripheral symptoms including dry mouth, thirst, fixed dilated pupils, flushed face, fever, hot, dry, red skin, urinary retention, hyperthermia, hypotension, tachycardia, and increased respiratory rate. In addition to tachycardia, cardiac manifestations may include EKG abnormalities similar to those produced by quinidine. Speech and swallowing may be impaired in association with blurred vision. Other peripheral signs and symptoms may include nausea and vomiting.

In large doses, atropine induces stimulation of the CNS, which in humans is characterized by overactive coordinated movements, hallucinations, and delirium. After the stimulation has lasted for some time, depression sets in and may proceed to complete paralysis of the CNS, which is fatal through cessation of respiration. In infants, particularly those ingesting antihistamines, paradoxical excitement may occur subsequently followed by a more characteristic CNS depression. CNS manifestations may resemble acute psychosis characterized by incoherence, confusion, hallucinations, delusions, paranoia, and abnormal motor behavior.

In severe overdosage, CNS depression, circulatory collapse, and hypotension may occur. Coma and skeletal muscle paralysis may also occur followed by death due to respiratory failure. Acute overdosage with quaternary ammonium antimuscarinics may produce a curariform neuromuscular block and ganglionic blockade manifested as respiratory paralysis.

Clinical Management

Immediate treatment should include instituting emesis, with syrup of ipecac or gastric lavage, followed by administration of activated charcoal and saline cathartics if the patient is not comatose. Induced emesis may be ineffective in ingestion of antihistaminics related to phenothiazines or in massive ingestion. The use of physostigmine should generally be reserved for treatment of patients with extreme delirium or agitation. Physostigmine in a dose of 0.5–2 mg administered intravenously, which can be repeated every 30 min as needed, may be used to alleviate symptoms like confusion, agitation, or coma. Other cholinergic antagonists have not been useful since they do not cross the blood–brain barrier. Measures such as forced diuresis and dialysis have not yet been shown to be effective. Fluid therapy and other standard treatments of shock should be administered as needed.

—*Swarupa G. Kulkarni and Harihara M. Mehendale*

Related Topics

Cholinesterase Inhibition
Gastrointestinal System
Neurotoxicology: Central and Peripheral
Organophosphate Poisoning, Intermediate Syndrome
Organophosphates

Antimony (Sb)

- ◆ CAS: 7440-36-0
- ◆ SELECTED COMPOUNDS: Antimony hydride (stibine), SbH_3 (CAS: 7803-52-3); antimony pentachloride, $SbCl_5$ (CAS: 7647-18-9); antimony pentoxide, Sb_2O_5 (CAS: 1314-60-9); antimony trichloride, $SbCl_3$ (CAS: 10025-91-9); antimony trioxide, Sb_2O_3 (CAS: 1309-64-4); antimony trisulfide, Sb_2S_3 (CAS: 1345-04-6)
- ◆ CHEMICAL CLASS: Metals

Uses

Antimony is used in white metal, which is any of a group of alloys having relatively low melting points. White metal usually contains tin, lead, or antimony as the chief component (e.g., Britannia and Babbitt). Antimony is used as a hardening alloy for lead, especially in storage batteries and cables, bearing metal, type metal, solder, collapsible tubes and foil, sheet and pipe, semiconductor technology, and pyrotechnics. It is also used in thermoelectric piles and for blackening iron or coatings. Antimony-containing compounds are employed in producing materials used in refrigerators, air conditioners, aerosol sprays, paints, and flame-proofing agents. Approximately half of the antimony used in the United States is recovered from lead-based battery scrap.

Antimony is also used medicinally (e.g., antimony potassium tartrate as an emetic and antimony as an antiparasitic agent).

Exposure Pathways

In industrial settings, inhalation is the primary exposure pathway for antimony compounds; these compounds are found in air pollution near factories. For the general population, ingestion is a common exposure pathway because antimony is found in certain foods and in some drinking water. In addition, medicines containing antimony are administered orally

Toxicokinetics

Normally, antimony is absorbed slowly when ingested or administered orally. Many antimony compounds are gastrointestinal irritants. The emetic antimony potassium tartrate is easily absorbed and, within 24 hr, 50% is excreted in the urine (hamsters). Antimony can concentrate in lung tissue, the thyroid gland, the adrenal glands, the kidneys, and the liver. The trivalent compounds concentrate in the red blood cells and liver and the pentavalent compounds in the blood plasma. Both forms are excreted in feces and urine, but generally, more trivalent compounds are excreted in urine and more pentavalent compounds in feces.

Presumably by reacting with the sulfhydryl groups, antimony can inhibit oxidative and phosphorylating enzymes like monoamine oxidase, succinoxidase, pyruvate brain oxidase, and phosphofructokinase. Inhibition of these enzymes can alter activities such as glucose metabolism and nerve transmission.

Mechanism of Toxicity

Antimony toxicity often parallels that of arsenic, although antimony salts are less readily absorbed than arsenic. It is presumed that antimony, like arsenic, complexes with sulfhydryl groups of essential enzymes and other proteins. By analog, antimony can uncouple oxidative phosphorylation, which would inhibit the production of energy necessary for cellular functions. Antimony's trivalent compounds are more toxic than its pentavalent compounds.

Human Toxicity

Accidental poisonings can result in acute toxicity, which produces vomiting and diarrhea. Most information regarding antimony toxicity has been obtained from industrial exposure experiences. Occupational exposures usually occur through inhalation of dusts containing antimony compounds. Inhalation of antimony compounds produces different effects at different concentrations. Chronic inhalation of low concentrations causes rhinitis and irritation of the trachea. At high concentrations, acute pulmonary edema occurs, and bronchitis may occur (the bronchitis may lead to emphysema). Inhaled antimony concentrates in lung tissue; as a result, pneumoconiosis with obstructive lung disease has been recorded. Workers exposed to antimony trisulfide (used as a pigment and in match production) at concentrations greater than 3.0 mg/m³ experienced heart complications and died. In addition, a temporary skin rash, called "antimony spots," can occur in persons chronically exposed to antimony in the workplace. Inhalation of antimony hydride (stibine gas) can lead to hemolytic anemia, renal failure, and hematuria. Stibine gas is produced when antimony alloys are treated with acids.

The ACGIH TLV-TWA for antimony is 0.5 mg/m³. ACGIH classifies antimony as a suspected human carcinogen.

Clinical Management
The oil-soluble BAL (British anti-Lewisite; 2,3-dimercaptopropanol) administered intramuscularly appears to be the antidote of choice for antimony poisoning.

Animal Toxicity
Rats exposed to a dose level of 4.2 mg/m³ airborne antimony trioxide dust for 1 year were reported to develop lung tumors; at a dose level of 1.6 mg/m³, lung tumors were not found. Guinea pigs exposed to airborne antimony trioxide developed interstitial pneumonia. Antimony administered intravenously to experimental animals resulted in abnormal electrocardiograms. Oral feeding of antimony to rats does not induce an excess of tumors or teratogenesis; however, specific compounds appear to be mutagenic in some strains of bacillus.

—Arthur Furst, Shirley B. Radding,
and Kathryn A. Wurzel

Related Topic
Metals

Anxiolytics

- ◆ Synonyms: Minor tranquilizers; antianxiety drugs; sedative-hypnotics; benzodiazepines (BZDs)
- ◆ Pharmaceutical Class: This class of compounds includes the BZDs like diazepam (Valium) and oxazepam (Serax), chlordiazepoxide (Librium), meprobamate (carbamate derivative), and related compounds, and buspirone (aryl piperazine derivative), which is an anxioselective drug. A miscellaneous group of drugs includes certain antihistaminic and anticholinergic drugs that are difficult to classify (e.g., hydroxyzine and buclizine).

Uses
Anxiolytics are used for preoperative relief of anxiety, for conscious sedation, as hypnotics in the treatment of insomnia, for short-term relief of symptoms of anxiety, or for the management of anxiety disorders. BZDs are also used for the management of agitation associated with alcohol withdrawal, for their anticonvulsant properties, and as skeletal muscle relaxants. BZDs are preferred over barbiturates since these are less likely to produce tolerance and physical dependence and are remarkably safe in large suicidal doses.

Toxicokinetics
The BZDs meprobamate and buspirone are well absorbed from the gut. Plasma concentration of the BZDs and their metabolites exhibit considerable interpatient variation. Onset and duration of action varies depending on the BZD and the route of administration. BZDs are widely distributed into body tissues and cross the blood–brain barrier. Generally, BZDs and their metabolites cross the placenta. The concentration of diaze-

pam in fetal circulation has been reported to be equal to or greater than the maternal plasma concentration. The drugs and their metabolites are distributed into milk. BZDs and their metabolites are highly bound to plasma proteins. Meprobamate is uniformly distributed throughout the body and is 20% bound to plasma proteins. It is capable of crossing the placenta and is distributed in milk. Buspirone is extensively distributed and is 95% bound to plasma proteins, mainly albumin. These are metabolized in the liver and may undergo conjugation. Meprobamate is metabolized to form the 2β-hydroxymeprobamate and glucosyluronide and glucuronide conjugate of meprobamate. Buspirone is metabolized in the liver, mainly via oxidation, to form the hydroxylated metabolites which may further undergo conjugation. BZD metabolites are excreted principally in urine. Meprobamate is excreted mainly via the urine. Buspirone is excreted principally in the urine and to a lesser extent in feces.

Mechanism of Toxicity

The advantage of using BZDs is that they have a larger therapeutic index. The exact sites and mode of action of the BZDs have not been elucidated. However, their effects seem to be mediated through the inhibitory neurotransmitter γ-aminobutyric acid (GABA). Allosteric interaction of central BZD receptors with GABA$_A$ receptors and subsequent opening of chloride channels are involved in eliciting the central nervous system (CNS) effects of the drugs. These drugs appear to act at the limbic, thalamic, and hypothalamic levels of the CNS. In usual doses BZDs appear to have very little effect on the autonomic nervous system, respiration, or the cardiovascular system. These do not produce extrapyramidal side effects or interfere with the autonomic nervous system function. The mechanism of action of meprobamate is unknown. The mechanism of action of buspirone probably involves several neurotransmitter systems.

Human Toxicity: Acute

The BZDs have a low order of toxicity unless ingested with other CNS depressants. Deep coma is rare. The BZDs have been known to cause dose-dependent adverse CNS effects. BZD overdosage may result in somnolence, impaired coordination, slurred speech, confusion, coma, and diminished reflexes. Hypotension, seizures, respiratory depression, and apnea may also occur. Although cardiac arrest has been reported, death from overdosage of BZDs in the absence of concurrent ingestion of alcohol and other CNS depressants is rare.

BZDs should be avoided during the first trimester and at delivery. Malformation and CNS dysfunction have been described in infants born of mothers using BZDs during pregnancy. Both animal data and human epidemiological studies suggest that BZDs are teratogens.

Severe anaphylactic reactions following intravenous administration of diazepam have been reported. Meprobamate causes toxicity similar to that of a barbiturate overdosage. Death may result from respiratory failure or hypotension. Limited information is available about the acute toxicity of buspirone. Effects are merely extensions of pharmacological effects. Nausea, vomiting, dizziness, drowsiness, miosis, and gastric distention may be seen.

Human Toxicity: Chronic

Tolerance and psychologic and physical dependence may occur following prolonged use of BZDs. Such effects may occur following short-term use of BZDs particularly at high doses. Drowsiness, ataxia, slurred speech, and vertigo may be seen on dependence. Withdrawal symptoms, including anxiety, agitation, tension, dysphoria, anorexia, insomnia, sweating, blurred vision, irritability, tremors, and hallucinations, may be seen. Milder withdrawal symptoms such as insomnia have also been reported. Since some BZDs and their metabolites have long elimination half-lives, withdrawal symptoms may not occur until several days after the drug has been discontinued. Meprobamate causes physical dependence similar to that seen with barbiturate dependence. No physical dependence on buspirone administration has been seen.

Clinical Management

Emesis is not recommended following an overdosage of BZD because of the potential of CNS depression. Gastric lavage soon after ingestion and activated charcoal/cathartic may be administered. Pulse, respiration, and blood pressure should be monitored and the patient should be closely observed. Intravenous fluids should be administered and adequate airway maintained. Hypotension may be controlled, if necessary, by intravenous administration of norepinephrine or metaraminol. Although some manufacturers recommend use of caffeine and sodium benzoate to combat CNS depression, most authorities believe that caffeine and other analeptic agents should not be used. Flumazenil (BZD antagonist) may be used in treatment. Flumazenil is an adjunct to and not a substitute for appropriate supportive and symptomatic therapy. Flumazenil (0.2–3 mg) intravenously in 0.2- to 0.3-mg increments for BZD overdose in adults and 0.2–1 mg intravenously in 0.2- to 0.3-mg increments for reversal of BZD sedation in adults may be used. Gradual dosage tapering is required. Occasionally temporary reinstitution of BZD therapy to suppress withdrawal symptoms may be necessary. Initial withdrawal symptoms may be managed with phenobarbitone or diazepam, followed by decreasing the dose by about 10% per day of the initial dose required to control symptoms. Treatment of BZD physical dependence consists of cautious and gradual withdrawal of the drug using a dosage tapering schedule. In the case of meprobamate toxicity general supportive therapy should be maintained. Forced diuresis may be beneficial. In the case of withdrawal symptoms, the patient may be stabilized on phenobarbitone, which is then withdrawn over 10–14 days. No specific antidote is available for the treatment of an overdosage of buspirone and treatment involves symptomatic and supportive care.

—*Swarupa G. Kulkarni and Harihara M. Mehendale*

Related Topics

Barbiturates, Long-Acting
Barbiturates, Short-Acting
Neurotoxicology: Central and Peripheral

Aramite

♦ CAS: 140-57-8
♦ SYNONYMS: 2-(*p*-butyl phenoxy)-1-methylethyl-2-chloroethylsulfite; 2-(*p*-butyl phenoxy) isopropyl-2-chloroethyl sulfite; Aracide; Aramit; Aratron; Ortho-mite

♦ DESCRIPTION: Aramite is a clear light-colored oil with a melting point of −31.7°C and a boiling point of 175°C at 0.1 mm of Hg. It is noncorrosive and has a specific gravity of 1.145 at 20°C. It is practically insoluble in water and is miscible with many organic solvents. When heated to decomposition, it emits highly toxic fumes of chlorides and oxides of sulfur [SO(X)].

♦ CHEMICAL STRUCTURE:

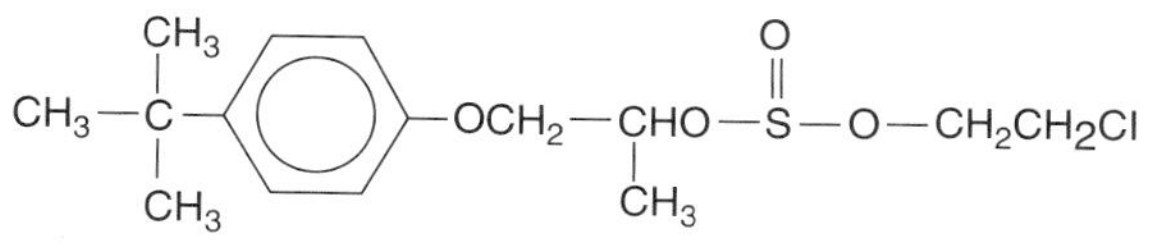

Uses

Aramite was formerly used as an antimicrobicide agent and as a miticide.

Exposure Pathways

Although exposure through oral consumption of contaminated fruits is possible, it should no longer be occurring since the use of aramite has been discontinued voluntarily on the basis of oncogenicity according to the U.S. EPA. Occupational exposure through dermal contact and inhalation of aerosols and dusts is possible.

Human Toxicity: Acute

Acute exposure to aramite in an undiluted form may cause skin irritation.

Human Toxicity: Chronic

This compound is classified as a human carcinogen based on sufficient data from clinical bioassays and animal data. No data are available on the number of workers who were actually or potentially exposed to aramite during its manufacture and formulation. The lowest published lethal dose/concentration in humans is 429 mg/kg.

Animal Toxicity

A large oral dose causes central nervous system depression of long duration in laboratory mammals. The principal autopsy finding was hemorrhagic syndrome involving particularly the lung. Undiluted aramite and its concentrated solution are irritating to the skin and conjunctiva of experimental animals. Aramite has been found to give electroretinographic indications of intoxication of retinal photoreceptors when injected into mice and when applied to the eyeball. Increased incidence of liver tumors and/or neoplastic nodules in three strains of male and female rats and males of one strain of mice and extrahepatic biliary system tumors were noted in dogs following chronic oral exposure.

—*Swarupa G. Kulkarni and Harihara M. Mehendale*

Related Topic

Pesticides

Arsenic (As)

- ◆ CAS: 7400-38-2
- ◆ SELECTED COMPOUNDS: Arsenic trichloride, $AsCl_3$ (CAS: 7784-34-1); arsenic trioxide, As_2O_3 (CAS: 1327-53-3); arsenic pentoxide, As_2O_5 (CAS: 1303-28-2); sodium arsenite, $NaAsO_2$ (CAS: 7784-46-5); sodium dibasic arsenate, Na_2HAsO_4 (CAS: 7778-43-0); arsine, AsH_3 (CAS: 7784-42-1)
- ◆ CHEMICAL CLASS: Metals

Uses

Odorless and tasteless, arsenic was one of the earliest poisons. For decades, many forms of arsenic were used medicinally (e.g., to treat syphilis). Although arsenic use has decreased dramatically, certain forms are still used in sheep dips, rat poison, wood preservatives, weed killers, and pesticides. An organic arsenic compound, arsanilic acid, is used to promote growth in poultry and swine. Recently, arsenic has been used in electronics (e.g., gallium arsenide). In addition, certain alloys of arsenic are used in special glasses.

Exposure Pathways

The primary exposure pathway for arsenic is ingestion (e.g., water and food); inhalation is a minor exposure pathway; and dermal absorption is negligible. The main source appears to be drinking water. The major information on arsenic toxicity symptoms comes from regions where the natural arsenic content of the drinking water is high. Many foods contain arsenic but at relatively low levels [vegetables, meat, or specialty seafood (e.g., crustacea)]. Wines may also contain arsenic from the use of pesticides on the grapes.

Although industrial pollution is a source of airborne arsenic (mainly in the form of arsenic trioxide), the arsenic concentration in urban air is low, ranging from 1 ng to tenths of a microgram per cubic meter of industrial air. Occupational exposure to arsenic is primarily associated with smelting industries (roasting of ore) or manufacturing of arsenic-containing pesticides.

Toxicokinetics

Arsenic is readily absorbed from the gastrointestinal tract and lungs. Recent studies show that trivalent arsenic is oxidized to pentavalent (and vice versa) in the human body. Metabolism of arsenic compounds results in excretion of methylated arsenic, mainly dimethylarsenic acid.

Arsenic is widely distributed in most tissues of the body. It concentrates mainly in the liver, kidneys, lungs, skin, and less in bone and muscle. Chronic exposure may result in hair, nail, and skin accumulation as well. Mee's lines (transverse white bands across fingernails) may be used to help identify the onset of arsenic exposure. Arsenic can also cross the placenta. Blood levels are not a good indicator of exposure levels due to the short half-life of arsenic compounds.

There is large individual variation in the elimination of arsenic via the urine. Organic arsenic compounds are eliminated faster than inorganic (both forms have short half-lives, in the range of hours). Less than 10% of organic arsenic is excreted in the feces; in approximately 3 days, up to 80% is excreted in the urine.

Mechanism of Toxicity

Arsenic affects mitochondrial enzymes and impairs tissue respiration, which seems to be related to the cellular toxicity of arsenic. Arsenic reacts with thiol groups, especially enzymes or cofactors with two thiols. A variety of enzymes (including those related to respiration) are altered when arsenic binds with these thiol groups. If a cofactor like dihydrolipoic acid is essential for enzyme activity, arsenic binding to the two thiol groups will prevent the enzyme reaction.

Arsenic inhibits succinic dehydrogenase and stimulates mitochondrial adenosine triphosphatase (ATPase) activity by uncoupling oxidative phosphorylation. Certain hypotheses suggest that arsenic can replace phosphorus in many biochemical reactions; it can inhibit ATP formation during glycolysis.

Selenium is somewhat of an antagonist of arsenic (i.e., it reduces the effect of arsenic). Accumulating evidence seems to suggest that arsenic is an essential element for some animals (rats, chicks, minipigs, and goats) but this has not been established in humans.

Human Toxicity

Pentavalent arsenics are less toxic than trivalent arsenics. Soluble arsenic compounds are the most toxic.

Acute

Ingestion of arsenic may cause extreme irritation to the mucous membranes. Acute ingestion of greater than 70 mg of arsenic leads to a variety of symptoms, including gastrointestinal distress, difficulty in swallowing, fever, nausea, vomiting, and diarrhea. Acute arsenic exposure can affect the heart, the peripheral nervous system, and the blood-forming elements. Cardiac arrhythmias and other heart effects can be fatal. Effects to the peripheral nervous system include sensory loss, which can be reversed when arsenic is no longer ingested. Symptoms of anemia and leukopenia are also reversible.

Some arsenic compounds have specific toxic manifestations. Arsine, a colorless gas with an unpleasant garlic-like odor, induces hemolysis, which in turn leads to pigmented urine, jaundice, anemia, and heart failure. Some inorganic arsenic compounds (especially potential war gases) are vesicants (agents that cause burning and destruction of tissue both internally and externally).

Chronic

Chronic exposure (even continuous low doses) leads to jaundice, which may lead to cirrhosis of the liver and, in turn, elevation of liver enzymes. In Taiwan and Chile, peripheral vascular diseases have been observed in individuals exposed to high arsenic concentrations in drinking water. These pathologies have not been

reported in many other countries that also have high arsenic concentrations in drinking water. These peripheral vascular effects have led to Raynard's phenomenon and gangrene of the lower extremities (black foot disease).

The skin appears to be a major target of arsenic toxicity. Dermatitis is noted at first with erythema, itching, and then swelling and a mottled appearance. After longer exposure, melanosis (abnormal pigmentation) appears on different parts of the body; hyperkeratosis (thickening of the skin) and warts may appear.

IARC considers all arsenic compounds carcinogenic to humans. Angiosarcoma has been reported in vineyard workers chronically exposed to arsenic-containing water, Fowler's solution, and arsenic-containing pesticides. Lung cancer among workers at copper smelters has been associated with arsenic. In Taiwan, various types of skin cancer have been reported among the population drinking water with high arsenic concentrations.

ACGIH lists arsenic as a confirmed human carcinogen. The ACGIH TLV-TWA is 0.01 mg/m^3 for elemental arsenic and inorganic compounds (except arsine). The ACGIH TLV-TWA for arsine is 0.16 mg/m^3.

Clinical Management

Within 4–6 hr of oral exposure, purging (using syrup of ipecac) and gastric lavage is essential. Common antidotes include from 3 to 5 mg/kg BAL (British anti-Lewisite; 2,3-dimercaptopropanol) administered intramuscularly over a specified period of time. Penicillamine administered orally has also been used; however, serious side effects (e.g., optic neuritis) have been associated with this antidote.

Recently, certain synthetic, water-soluble dimercapto compounds have been found especially effective and may become the oral therapy of choice for arsenic intoxication. These compounds include DMSA (meso-2,3-dimercaptosuccinic acid) and the sodium salt of 2,3-dimercaptopropane-1-sulfonate.

Animal Toxicity

Attempts to induce cancer using arsenic in experimental animals have either been negative or, at best, equivocal. One hypothesis is that arsenic is a powerful promoter, if not an initiator, of the cancer process. Although not mutagenic in many tests, arsenic can produce chromosome aberrations. It causes malformations in offspring of pregnant rats; however, there are no epidemiology studies to suggest teratogenic effects in humans.

A problem with many studies on arsenic toxicity concerns the use of rats. Rodents differ completely from humans in how arsenic affects the blood, is biotransformed, and is excreted.

*—Arthur Furst, Shirley B. Radding
and Kathryn A. Wurzel*

Related Topics

Metals
Mutagenesis
Pollution, Soil
Pollution, Water
Sensory Organs
Skin
Veterinary Toxicology

Arum

- ◆ SYNONYM: Araceae
- ◆ DESCRIPTION: Common plants included in the family Araceae are the philodendron, dieffenbachia, pothos, syngonium, caladium, aglaonema, and monstera.

Exposure Pathways
Routes of exposure include accidental ingestion and dermal contact.

Mechanism of Toxicity
Calcium oxalate crystals are commonly found in all members of this family. It is postulated that needle-like calcium oxalate monohydrate crystals with grooved ends are located within specialized cells known as idioblasts. Inside each idioblast is a bundle of raphides, which are ejected from the plant when pressure such as squeezing is applied. These raphides are then imbed-

ded in the skin or mucous membranes. There are also other proteolytic enzymes found in various species.

Human Toxicity
Symptoms include skin irritation consisting of erythema and vesiculation. Ingestion of plant material may result in local mouth or throat irritation and resultant swelling. Salivation and dysphagia may also be present. Systemic effects are extremely rare. Ocular exposures will result in pain and photophobia, followed by eyelid edema and corneal disturbances such as abrasions and chemosis.

Clinical Management
Treatment for dermal exposure should include irrigation of the contaminated area followed by cool compresses. Treatment for oral exposures should consist of removing any plant material from the oral cavity and administering cool liquids. Significant toxicity is rare. Irrigation should be performed in instances of ocular contamination. Further care should be symptomatic and supportive.

Animal Toxicity
Feline ingestions of philodendron result in central nervous system excitability, seizures, renal failure, and encephalitis.

—Rita Mrvos

Asbestos

- ◆ CAS: 1332-21-4
- ◆ SYNONYMS: Asbestos fiber; actinolite; amianthus; amosite; amphibole; anthophylite; ascarite; chrysotile; crocidolite; fibrous grunerite; serpentine; tremolite
- ◆ DESCRIPTION: Asbestos is a generic name that refers to a group of naturally occurring hydrated mineral silicates. They are characterized by fibers

or bundles of fine single crystal fibrils. Asbestos is a fibrous material, insensitive to heat and chemical attack, widely used in textiles, insulation, ceiling and floor tile, reinforced cement, industrial water filters, gaskets, and brake linings.

Exposure Pathways

The primary exposure pathways are inhalation, accidental ingestion, or dermal exposure.

Fine dust particles that are able to reach the alveolar area of the lung include crocidolite at a concentration of 0.05×10^6 fibers/m^3 (0.025 mg/m^3), and chrysotile, amosite, anthophylite, tremolite, actinolite at a concentration of 1×10^6 fibers/m^3 (0.05 mg/m^3) when there is more than 2.5% asbestos in the dust and at a concentration of 2 mg/m^3 when there is less than or equal to 2.5 weight% asbestos in fine dust.

A fiber is defined as having a length greater than 5 μm, a diameter less than 3 μm, and length/diameter greater than 3:1, equivalent to 1 fiber/ml.

Toxicokinetics

Asbestos fibers may be ingested through food or drink or by swallowing of inhaled asbestos. Asbestos fibers may be deposited in the gastrointestinal tract. There is no systemic absorption of asbestos fibers and they do not appear to stimulate an inflammatory reaction or any other adverse effect in the gastrointestinal tract. Asbestos fibers may penetrate the skin but are not absorbed.

The most common route of entry into the body is by inhalation. When asbestos is inhaled, larger fibers (10–20 μm long) tend to be filtered out in the upper airways or collide with the walls of the conducting airway walls in the lungs where they are captured in the respiratory mucous. These fibers are then removed by cilia of the tracheobroncheal tree and are swallowed. Asbestos fibers less than 10 μm long may eventually reach the alveoli. In autopsy lung specimens, asbestos fibers of 5–200 μm have been found in alveoli. Very small fibers may be engulfed by alveolar macrophages and transported to lymph nodes. In human autopsies, asbestos fibers have also been found in the thoracic diaphragm and chest wall. Once deposited in alveoli, asbestos fibers remain permanently embedded as asbestos bodies (ferruginous bodies) and are not excreted. Asbestos is not absorbed and is not metabolized in the body.

Mechanism of Toxicity

Asbestos produces its toxic effects by direct contact with lung tissue or by stimulating an acute or chronic inflammatory reaction in the tissue.

Human Toxicity

Inhalation of asbestos produces a disease called asbestosis, which is characterized by interstitial fibrosis of lung parenchyma. All types of asbestos fibers can cause asbestosis but crocidolite is most potent. The first symptoms of asbestosis are dyspnea with exertion and reduced exercise tolerance. Lung function abnormalities can include decreases in vital capacity, residual volume, functional residual capacity, and lung compliance. The disease can progress to massive pulmonary fibrosis. In these cases, the diffuse fibrosis and contraction of lung tissue causes constriction of the pulmonary vasculature, leading to pulmonary hypertension, which may lead to death. Asbestos causes a fibrous pleuritis in which the pleural membrane thickens to encase the lung in a rigid fibrous capsule. There is formation of pleural plaques. Radiologic evidence of asbestos-induced lung damage is not present at least until 5 years after exposure. The most important physical sign is the presence of high-pitched fine crepitations (crackles) at full inspiration, which persist after coughing. Lung function is decreased. The total lung volume is decreased especially the forced vital capacity.

Handling asbestos without gloves can cause corns (asbestos warts), which are areas of thickened skin surrounding implanted fibers.

Chronic exposure to asbestos can cause lung cancer, bronchogenic carcinoma. The latency period for lung cancer is 20–30 years. Symptoms of lung cancer may include chest pain, chronic cough, hemoptysis, and decreased exercise tolerance. Mesothelioma is another malignant disease associated with asbestos. The latency period is 35–40 years. Asbestos is the only known cause of this tumor. The first symptoms of mesothelioma are those associated with pleural irritation such as cough and chest pain.

The ACGIH TLV-TWAs (8-hr exposure) ranges from 0.2 fibers/cc to 2 fibers/cc, depending on the form of asbestos (0.2 fibers/cc for crocidolite; 0.5 fibers/cc for amosite; 2 fibers/cc for chrysotile and other forms).

Clinical Management

There is no effective treatment for asbestosis. The only preventive measure is to keep asbestos fibers out of the lungs. Once asbestosis has started, further inhalation of asbestos fibers causes acute inflammatory reactions, which can worsen the disease. At the time of diagnosis, lung cancer is usually too advanced for successful treatment. Long-term survival rates are low following surgery and treatment with chemotherapeutic agents or radiation. There is no effective treatment for mesothelioma, and death usually occurs within 1 year after diagnosis. Radiation therapy and chemotherapy may prolong survival.

Animal Toxicity

Animals exposed to asbestos over a period of time can develop mesothelioma.

—*Arvind K. Agarwal*

Related Topics

Carcinogenesis
Immune System
Indoor Air Pollution
Pollution, Water
Respiratory Tract
Toxic Torts

Ascorbic Acid

- ◆ CAS: 50-81-7
- ◆ SYNONYMS: Vitamin C; acidum; antisorbutic vitamin; ascurbicum; cevitamic acid; 2,3-didehydro-L-threo-hexono-1,4-lactone; E300; L-ascorbic acid; L-xyloascorbic acid; L-3-ketothreohexuronic acid lactone
- ◆ PHARMACEUTICAL CLASS: Water-soluble vitamin
- ◆ MOLECULAR FORMULA: $C_6H_8O_6$
- ◆ CHEMICAL STRUCTURE:

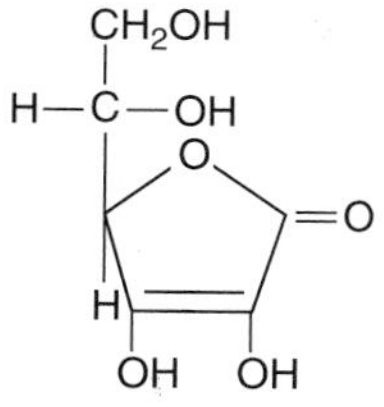

Uses

Ascorbic acid is used as a nutritional supplement during deficiency states (scurvy). Ascorbic acid needs may increase during chronic illness, infection, trauma, pregnancy, and lactation. It has also been used as a urinary tract acidifier. It purportedly is a cure for the common cold.

Exposure Pathways

Routes of exposure are oral, intravenous, intramuscular, and subcutaneous. Dietary sources of ascorbic acid include citrus fruits, tomatoes, cantaloupe, raw peppers, and green leafy vegetables.

Toxicokinetics

Ascorbic acid is readily absorbed from the gastrointestinal tract; however, absorption may be delayed with large doses. It is metabolized hepatically, reversibly oxidized to dehydroascorbic acid, and metabolized to inactive ascorbate-2-sulfate and oxalic acid. Protein binding is 25%, and ascorbic acid is widely distributed into body tissue. It is renally excreted. Elimination increases with higher doses.

Mechanism of Toxicity

Metabolism of ascorbic acid can lead to deposition of oxalate crystals in kidney tissue.

Human Toxicity: Acute

Toxicity is unlikely following acute ingestions of even 100 times the recommended daily allowance. Side effects include nausea, vomiting, diarrhea, abdominal pain, headache, dizziness, and flushing of the skin.

Human Toxicity: Chronic

Chronic megadoses of vitamin C may precipitate formation of calcium oxalate renal stones, oxalate nephropathy, and renal failure. The amount required to cause this is variable from 2 to 8 g per day. Bone oxalate deposits have also been reported. Esophageal and dental erosion are possible with tablet ingestion. Heinz body hemolytic anemia has been seen in premature infants.

Clinical Management

Acute ingestions seldom require treatment. Dilution is recommended to reduce the risk of esophageal and gastrointestinal irritation. During chronic excessive use, patients should be instructed to discontinue the supplement and observe for signs of rebound scurvy. Any toxic symptom should be treated symptomatically.

Animal Toxicity

Acute toxicity is not expected, and it would be unlikely for animals to be given chronic vitamin C overdoses.

—*Denise L. Kurta*

Aspartame

- SYNONYMS: Nutrasweet; 3-amino-*N*-(-carboxyphenethyl)succinamic acid *N*-methyl ester; aspartylphenylalanine methyl ester; *N*-1-aspartyl-1-phenylalanine 1-methyl ester; canderel; dipeptide sweetener; Equal; methyl aspartylphenylalanate; 1-methyl-*N*-1--aspartyl-1-phenylalanine; sweet dipeptide
- CHEMICAL CLASS: Esterified dipeptide sweetener
- CHEMICAL STRUCTURE:

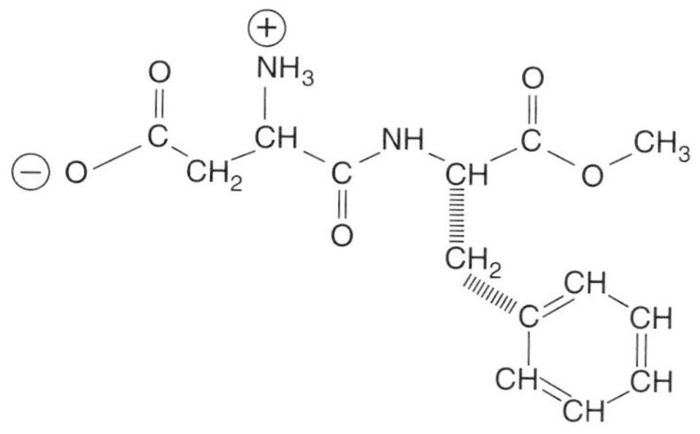

Uses

Aspartame is a synthetic nonnutritive sweetener used as a food additive and tabletop sweetener.

Exposure Pathways

The primary exposure pathway is intentional or accidental ingestion.

Toxicokinetics

Aspartame is rapidly absorbed from the intestines. In the intestines, it is completely metabolized to its constituent amino acids (aspartic acid and phenylalanine) and methanol. This methanol is rapidly absorbed and distributed uniformly to body water. In the liver, the methanol is metabolized to formaldehyde and formic acid and finally to carbon dioxide. Aspartic acid is either converted to alanine and carbon dioxide or to alanine and oxaloacetate. Aspartic acid can also be incorporated to body proteins as an amino acid. Phenylalanine can be incorporated into proteins as an amino acid or metabolized to tyrosine.

Mechanism of Toxicity

The actual mechanism of toxicity is not known but the toxicity may be directly due to elevated blood levels of the individual components of aspartame, namely, aspartic acid and phenylalanine.

Human Toxicity

No acute toxicity by aspartame has been reported. The potential of acute toxicity due to aspartame by ingesting consumer products seems highly unlikely. Aspartame alone or in conjunction with glutamate might contribute to mental retardation, undesirable behavioral effects, and endocrine dysfunction. The most common complaint is headache. It is possible that certain individuals are unusually sensitive to aspartame, which may be responsible for symptoms such as mood alterations, anxiety, irritability, insomnia, and fatigue.

Although methanol is a known toxic chemical, the amount of methanol produced by aspartame biotransformation even in large doses is not enough to cause toxic effects. Because aspartame is completely degraded (10% of aspartame by weight is methanol), ingestion of 50 mg aspartame/kg body weight (approximately 30 cans of diet soda) will result in the ingestion of 5 mg methanol/kg body weight, which is less than the amount of methanol formed during consumption of many foods including fruits and vegetables.

Abuse doses of aspartame (200 mg/kg body weight; approximately 24 liters of diet soda) can increase plasma phenylalanine levels in normal persons but can exceed the toxic threshold in individuals with phenylketonuria disease. Children (especially infants) can be vulnerable to brain damage caused by excessive stimulation of the brain by aspartate. These findings are not confirmed.

Clinical Management

No specific aspartame-related clinical toxicology cases have been reported. In sensitive individuals, basic life-support measures should be utilized and symptomatic therapy should be instituted.

Animal Toxicity

No information on aspartame-induced toxic effects in animals has been reported.

—Arvind K. Agarwal

Astemizole

◆ CAS: 688-44-77-9

◆ SYNONYM: Hismanal

◆ PHARMACEUTICAL CLASS: A benimidazole derivative H-1 receptor antagonist

◆ CHEMICAL STRUCTURE:

Uses

Astemizole is indicated for the symptomatic relief of seasonal allergic rhinitis and chronic idiopathic urticaria.

Exposure Pathways

Ingestion is the route of both accidental and intentional exposures to astemizole.

Toxicokinetics

Astemizole is rapidly absorbed and reaches peak plasma concentrations in 1 hr. When taken with food, the drug's absorption is reduced by 60%. Astemizole has a slow onset of action and its effects last up to 24 hr. It is extensively metabolized in the liver. The principle metabolite, desmethylastemizole, may have some antihistaminic activity.

The drug is 96.7% protein bound and has a very large volume of distribution (250 liters/kg). Approximately 40–50% of the dose is excreted in the urine by 4 days, with 50–70% eliminated in the feces by 14 days. All of the dose is eliminated as metabolites. Astemizole's half-life is biphasic: 20 hr for the distribution phase and 7–11 days for the elimination phase.

Mechanism of Toxicity

The mechanism of the cardiotoxic effects of astemizole is not well understood. Evidence in animal models suggests the cardiotoxic effects may result at least in part from blockade of the potassium channel involved in repolarization of cardiac cells. It is also postulated that H-3 receptors (mediating a regulatory feedback mechanism) may be involved in the development of cardiovascular toxicity. Anticholinergic effects appear to be unlikely causes of the cardiac effects.

Human Toxicity

Unlike other antihistamines, astemizole is considered "nonsedating" and lacks the anticholinergic properties. Serious cardiac effects, including prolongation of the QT interval, arrhythmias (i.e., ventricular tachycardia, Torsades de Pointes, ventricular fibrillation, and heart block), arrest, hypotension, palpitations, syncope, dizziness, and death have been reported in patients receiving astemizole. These cardiotoxic effects are usually associated with higher than recommended doses and/or increased plasma concentrations of the drug and its active metabolite. Although rarely reported, cardiotoxic effects may occur at the recommended dose and at doses two or three times the recommended dose (10 mg daily). Patients with impaired liver function and geriatric patients may be at risk for the development of cardiovascular toxicity. Concomitant use of azole-derivatives (i.e.,. itraconazole and ketoconazole) and macrolide antiinfectives (i.e., erythromycin) with astemizole may also increase the risk of toxicity. These antiinfectives interfere with the metabolism of astemizole resulting in an increased serum concentration.

Clinical Management

Basic and advanced life-support measures should be utilized as necessary. Appropriate gastrointestinal decontamination procedures should be administered based on the history of the ingestion and the patient's level of consciousness. Close EKG monitoring should be instituted for a minimum of 24 hr.

—*Carla M. Goetz*

Related Topic

Cardiovascular System

Atrazine

- ◆ CAS: 1912-24-9
- ◆ SYNONYMS: Atrasol; Atranex; Atratol; Gesaprim; Primatol; Crisazine; 2-chloro-4-ethylamino-6-isopopylamine-s-triazine
- ◆ CHEMICAL CLASS: Triazine herbicide
- ◆ CHEMICAL STRUCTURE:

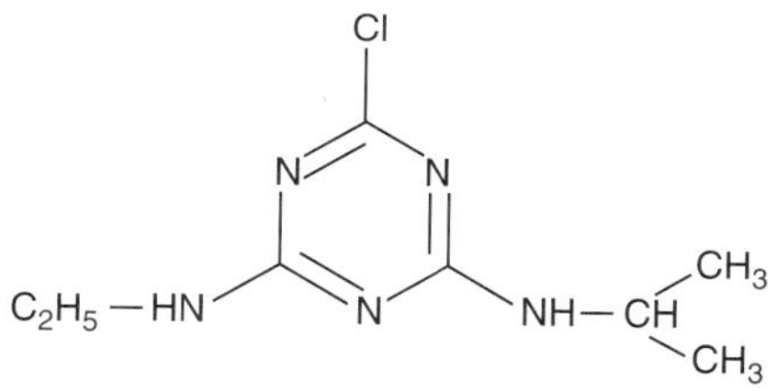

$$C_2H_5-HN \quad \text{(triazine ring with Cl)} \quad NH-CH(CH_3)_2$$

Uses

For decades, atrazine has been the most heavily used herbicide in the United States. Atrazine is used for selective and nonselective weed control in various field crop and industrial applications.

Exposure Pathways

The ocular and dermal routes are the primary exposure pathways. Ingestion of atrazine appears to present little hazard.

Toxicokinetics

Atrazine is readily absorbed (80%) from the gastrointestinal tract and undergoes dealkylation and dechlori-

nation. After oral administration, higher levels are noted in the lungs, liver, and kidneys. With dermal exposures, the highest levels were measured in liver and muscle. Urinary excretion is the major route of elimination in mammals. A small amount is also excreted in the feces.

Mechanism of Toxicity

The triazine herbicides are selective inhibitors of the Hill reaction in photosynthesis.

Human Toxicity

Dermal exposure to atrazine can cause skin rash, erythema, blisters, and edema. The ACGIH TLV-TWA for atrazine is 5 mg/m^3.

Clinical Management

Treatment is symptomatic.

Animal Toxicity

Atrazine has low acute toxicity in mammals. The oral LD_{50} in rats is approximately 2 g/kg. The dermal LD_{50} and inhalation LC_{50} (1 hr) values in rats are approximately 3 g/kg and 700 mg/m^3, respectively. The oral LD_{50} values in mice and rabbits are about 1.5 g/kg and 750 mg/kg, respectively. Rats exposed to high dosages of atrazine showed changes in arousal and motor function, dyspnea, hypothermia, and spasms. With lethal oral dosages, death occurred rapidly (within 12–24 hr). Atrazine alone or in the presence of metabolic activating enzymes is not mutagenic in standard assays.

—*Jing Liu*

Related Topics

Pesticides
Pollution, Water

Atropine

- ◆ CAS: 51-55-8
- ◆ SYNONYMS: AtroPen Auto Injector; Antrocol; Atropine Sulfate Injection

- PHARMACEUTICAL CLASS: Antimuscarinic agent; anticholinergic agent
- CHEMICAL STRUCTURE:

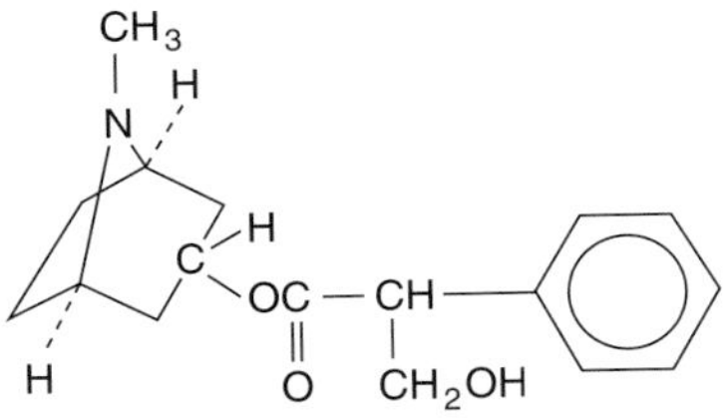

Uses

Atropine is used in the treatment of sinus bradycardia with hemodynamic instability, preoperatively to decrease secretions, and in the treatment of peptic ulcer disease, irritable bowel syndrome, and carbamate and organophosphate poisoning. It is also used in ophthalmic preparations to induce mydriasis and cyclopegia.

Exposure Pathways

Ingestion is the most frequent route of exposure. Exposure can also occur following application of eye solutions and via subcutaneous, intramuscular, intravenous, and inhalation routes. Accidental overdosage may occur when atropine is administered in the management of organophosphate or carbamate insecticide poisoning.

Toxicokinetics

In therapeutic doses, atropine is well absorbed. In toxic doses, absorption may be prolonged due to a decrease in gastric motility. Fifty percent is protein bound. The volume of distribution is 2–4 liters/kg. Atropine is metabolized in the liver to tropic acid, tropine, esters of tropic acid, and glucuronide conjugates. Elimination follows first-order kinetics. Up to 60% of atropine is excreted unchanged in the urine. The elimination half-life is 2–4 hr in adults but may be longer in children.

Mechanism of Toxicity

Atropine antagonizes acetylcholine competitively at the neuroreceptor site. Atropine inhibits acetylcholine from exhibiting its usual action but does not decrease production. Cardiac muscle, smooth muscle, and the central nervous system are most affected by the antagonism of acetylcholine.

Human Toxicity: Acute

Overdosage of atropine results in signs and symptoms of anticholinergic syndrome. Patients will exhibit warm flushed skin as a result of peripheral vasodilitation. Mydriasis may occur due to antagonism of acetylcholine in the muscles of the iris. Urinary retention, thirst, hallucinations, and decreased bowel sounds may occur. Tachycardia can occur as a result of vagal blockade. The anticholinergic syndrome may be delayed and can occur in cycles.

Human Toxicity: Chronic

Chronic ingestion of greater than therapeutic amounts may produce symptoms of the anticholinergic syndrome.

Clinical Management

Basic and advanced life-support measures should be utilized as necessary. Gastric decontamination procedures should be used based on the patient's history and current symptomatology. Gastric emptying may be beneficial, even if delayed due to a decrease in gastrointestinal motility. Activated charcoal can be given to adsorb atropine. The mainstay of treatment is supportive care. Physostigmine, a cholinesterase inhibitor, can be given in patients exhibiting severe anticholinergic effects such as seizures, hypertension, and arrhythmias. Extracorporeal elimination means are ineffective.

Animal Toxicity

Animals are at risk for anticholinergic poisoning from atropine. Toxicity is similar to that in humans. Gastrointestinal decontamination and supportive care should be utilized.

—*Bridget Flaherty*

Related Topics

Anticholinergics
Carbamate Pesticides
Organophosphate Poisoning, Intermediate Syndrome
Organophosphates
Poisoning Emergencies in Humans (Physostigmine)

Avermectin

- CAS: 73989-17-0
- CHEMICAL CLASS: Avermectins are a group of eight closely related natural products with anthelmintic potency. They are macrolide type antibiotics isolated from soil bacteria *Streptomyces avermitilis*. The avermectins are a family of macrocyclic lactones that consist of four major (A1a, A2a, B1a, and B2a) components and four homologous minor (A1b, A2b, B1b, and B2b) components. Avermectins are designated as A1, A2, B1, and B2, referring to mixtures of the homologous pairs containing at least 80% of the *a* component and 20% of the *b* component.
- CHEMICAL STRUCTURE:

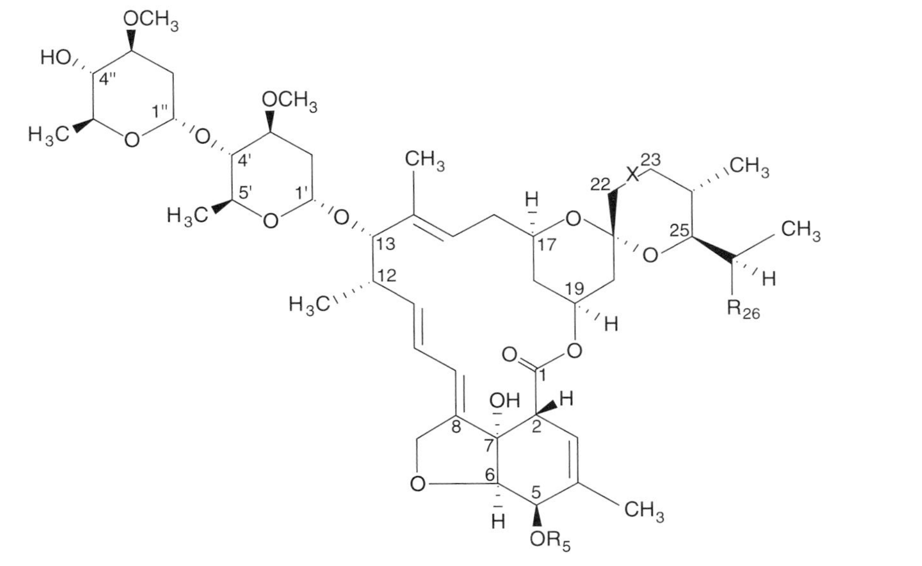

Uses

The most commonly used avermectins are abamectin (avermectin B1) and ivermectin (22,23-dihydroavermectin B1). Avermectin B1 is the most effective of the avermectin family of natural products against agriculturally important insects and mites. It is used in agriculture as a foliar spray under the name abamectin. It is sold under the trade names Affirm, Avid, Agrimec, Agri-mek, Vertimec, and Zephyr for crop protection (e.g., citrus, ornamentals, cotton, vegetables, pears, strawberries, and tree nuts) and for the control of agricultural and household arthropod pests.

Avermectin is also used to control ectoparasites mostly in cattle and horses. A synthetic derivative of avermectin B1, ivermectin, is used as an animal antiparasiticide, primarily to control dog heart worms. Ivermectin is also used in human beings to prevent the serious effects of filarial nematode *Onchocerca volvulus*, the causative agent of the tropical disease onchocerciasis, also known as river blindness. It is marketed

Components A: $R_5 = CH_3$ Components a: $R_{26} = C_2H_5$ Components 1: $X = -CH = CH-$

Components B: $R_5 = H$ Components b: $R_{26} = CH_3$ Components 2: $X = -CH_2 = \overset{OH}{\underset{\equiv}{CH}}-$

under the trade names Cardomec, Equvalan, Ivomec, Heartgard 30, Mectizan, and Zimectrin for veterinary and human purposes.

Exposure Pathways

Routes of exposure include accidental ingestion, accidental injection stab, dermal contact, and inhalation.

Toxicokinetics

In healthy subjects, peak plasma ivermectin concentrations of approximately 50 μg/liter are recorded after the administration of a 12-mg dose in either tablet or capsule form. Information on the distribution and elimination of ivermectin in humans is not available. In healthy subjects, the reported half-life is 28 hr, the volume of distribution is 46.9 liters, plasma protein binding is 93%, and oral clearance is 1.2 liters/hr. The metabolite 3'-O-dimethyl-22,23-dihydroavermectin B1a is detected in the urine and 22,23-dihydroavermectin B1a monosaccharide is detected in the feces. Ivermectin and its metabolites are excreted in bile and eventually into feces with minimal excretion (less than 1.0%) in urine.

Mechanism of Toxicity

The actual mechanism in human beings is not fully understood but avermectin may have a direct effect, stimulating the release of neurotransmitter γ-amino butyric acid.

Human Toxicity

The toxic effects due to ivermectin are dose dependent and most commonly include myalgia; rash; node tenderness; swelling of nodes, joints, limbs, or face; itching; fever; and chills. Ivermectin at doses of 50–200 μg/kg causes transient hematological changes. Prolongation of prothrombin time due to ivermectin has been reported. Moderate to severe toxic effects are manifested within 48 hr and may include pruritus, painful skin edema (especially of limbs and face), arthralgia (mostly in knees, ankles, and elbows), bone pain, thoracic pain, malaise, headache, and fever. No deaths due to ivermectin have been reported. Accidental stabbing with veterinary injection of ivermectin has produced irritation at the injection site, nausea, vomiting, abdominal pain, tachycardia, hypotension, hypothermia, urticaria, and a stinging sensation in the eye. No human

toxicity due to abamectin has been reported, but, being an analog of ivermectin, similar toxic effects are expected.

Clinical Management

No specific antidote is available. Analgesics or antihistamines may be used to control the symptoms. Supportive and symptomatic therapy is suggested.

Animal Toxicity

Ivermectin is used in dogs to control heart worms and for topical application for control of mites. The toxic effects are dose related and include symptoms such as mydriasis, depression, tremors, ataxia, emesis, drooling, coma, and death. The LD_{50} in dogs is 80 mg/kg body weight. Death is possible above 40 mg/kg. Daily oral administration of 1 or 2 mg/kg for 14 weeks in dogs causes mydriasis and weight loss. Initial signs of toxicity are hyperexcitation, enhanced responses to stimuli, followed by coarse motor tremors. Physostigmine, 1 mg twice a day, has a limited benefit in treating these dogs, and then only in more severely depressed and/or comatose stages of ivermectin toxicity. Its use as the sole therapeutic agent or complete antidote for ivermectin toxicity is not recommended. The unapproved use of products designed for large animals result in toxic deaths in dogs. Ivermectin is not indicated for cats. Abamectin causes mydriasis in dogs.

—Arvind K. Agarwal

Related Topic

Pesticides

Azinphos-Methyl

◆ CAS: 86-50-0

◆ SYNONYMS: Cothion-methyl; Gusathion; Methyl Guthion

- CHEMICAL CLASS: Synthetic organophosphorous insecticide in the phosphorothionate class
- CHEMICAL STRUCTURE:

Uses

Azinphos-methyl is a nonsystemic insecticide and acaricide.

Exposure Pathways

The dermal and inhalation routes are the primary exposure pathways for azinphos-methyl. It is available as emulsifiable concentrates, wettable powders, and dusts.

Toxicokinetics

The parent compound is well absorbed by all routes of exposure. As with other organophosphorothionate compounds, oxidative desulfuration of the parent compound by the mixed-function oxidase system produces the active oxon. Numerous other reactions (e.g., oxidative demethylation/dearylation and glutathione-mediated demethylation/dearlylation) constitute detoxification processes. As with other organophosphorothionate compounds, azinphos-methyl distributes readily throughout the body. Approximately 70% of the radioactivity from an intravenous dose of azinphos-methyl was recovered in the urine of human volunteers within 5 days of exposure.

Mechanism of Toxicity

The toxicity of azinphos-methyl is due to indirect (i.e., it requires metabolic activation to the oxygen analog) inhibition of acetylcholinesterase, resulting in excessive stimulation of cholinergic synapses within the central and peripheral nervous systems.

Human Toxicity: Acute

Inhalation, dermal absorption, or ingestion of azinphos-methyl may result in systemic intoxication due to inhibition of the enzyme acetylcholinesterase. The onset and systemic development of symptoms vary with the route of entry and may be delayed up to 12 hr. First symptoms may be nausea, increased salivation, lacrimation, blurred vision, and constricted pupils. Other symptoms include vomiting, diarrhea, abdominal cramping, dizziness, and sweating. After inhalation, respiratory symptoms may be pronounced at first (e.g., tightness of chest, wheezing, and laryngeal spasms). If the poisoning is severe, then weakness, muscle twitching, confusion, ataxia, slurred speech, convulsions, low blood pressure, cardiac irregularities, loss of reflexes, and coma may occur. In extreme cases, death may occur due to a combination of factors including respiratory arrest, paralysis of respiratory muscles, or intense bronchoconstriction. Complete symptomatic recovery from sublethal poisoning usually occurs within 1 week of complete removal of the source of exposure. Azinphos-methyl is a weak skin irritant. It can cause mild irritation to the cornea, iris, and/or conjunctiva with irritation typically resolving within 4 days.

Human Toxicity: Chronic

Acetylcholinesterase inhibition sometimes persists for weeks, thus repeated exposure to small amounts of this material may result in unexpected signs of acetylcholinesterase depression such as malaise, weakness, and anorexia that resemble other illnesses (e.g., influenza).

Clinical Management

For exposure to eyes, eyelids should be held open and the eyes flushed with copious amounts of water for 15 min. For exposure to skin, affected areas should be washed immediately with soap and water. The victim should receive medical attention if irritation develops and persists.

For exposure through inhalation, the victim should be removed to fresh air and, if not breathing, given artificial ventilation. The victim should receive medical attention as soon as possible.

First aid for ingestion victims would be to induce vomiting, keeping in mind the possibility of aspiration of solvents. Gastric decontamination should be performed within 30 min of ingestion to be most effective. Initial management of acute toxicity is establishment and maintenance of adequate airway and ventilation. Atropine sulfate in conjunction with 2-PAM can be administered as an antidote. Atropine by intravenous injection is the primary antidote in severe cases. Test injections of atropine (1 mg in adults, 0.15 mg/kg in children) are initially administered, followed by 2–4 mg (in adults) or 0.015–0.05 mg/kg (in children) every 10 to 15 min until cholinergic signs (e.g., diarrhea, salivation, and bronchial secretions) decrease. High doses of atropine over several injections may be necessary for effective control of cholinergic signs. If lavage is performed, endotracheal and/or esophageal control is suggested. At first signs of pulmonary edema, the patient should be placed in an oxygen tent and treated symptomatically.

Animal Toxicity

In contrast to most organophosphate pesticides, azinphos-methyl can cause toxicity in fish when runoff from agriculture following heavy rainfall occurs shortly after field application. The toxicity potential of azinphos-methyl in fish is considerably higher than with most other organophosphates (e.g., 96-hr TLM in bluegills: azinphos-methyl = 5.2 μg/liter, methyl parathion = 1900 μg/liter). Azinphos-methyl has similar oral acute toxicity potential in rats (LD$_{50}$ values about 10–25 mg/kg) compared to parathion and methyl parathion but is markedly less potent by the dermal route. Chromosomal abnormalities were reported in cultured Chinese hamster ovary and human cells at high concentrations, but carcinogenicity was not reported in either rats or mice following bioassay.

—*Thuc Pham*

Related Topics

A-esterase
Acetylcholine
Anticholinergics
Carboxylesterases
Cholinesterase Inhibition
Delayed Neurotoxicity
Neurotoxicology: Central and Peripheral
Organophosphate Poisoning, Intermediate
   Syndrome
Organophosphates
Pesticides

Bacillus cereus

◆ DESCRIPTION: *Bacillus cereus* is a spore-forming, motile, gram-positive, large, long, and rod-shaped bacterium.

Exposure Pathways

Ingestion of a contaminated food product is the most common route of exposure.

Mechanism of Toxicity

Bacillus cereus food poisoning symptoms are caused by an enterotoxin. It is not an invasive process. *Bacillus cereus* spores are found in grains and spices. The spores can germinate if food is not refrigerated after cooking. They germinate and release gastrointestinal toxins. The spores are heat stable and are not destroyed when the food is reheated prior to eating it.

Human Toxicity

Bacillus cereus can produce two different types of presentation: vomiting and diarrhea. One type of *B. cereus* toxin produces vomiting within 2–4 hr after the exposure and symptoms resolve within 10 hr. This enterotoxin is heat-stable and affects the upper gastrointestinal tract; therefore, this emetic form of *B. cereus* has a rapid onset and shorter clinical course.

The second type of toxin produces diarrhea. The incubation period is 6–16 hr and may resolve within 36 hr. This enterotoxin is heat labile and affects the lower gastrointestinal tract. This form mimics staphylococcal food poisoning.

Clinical Management

Bacillus cereus-induced illness is self-limited. Most cases only require symptomatic and supportive care and fluid replacement. Culturing stool is not generally useful in the diagnosis of *B. cereus*. The symptoms are similar but milder than those of cholera.

—*Vittoria Werth*

BAL

◆ CAS: 59-52-9
◆ PREFERRED NAME: 2,3-Dimercaptopropanol
◆ SYNONYMS: British anti-Lewisite (BAL); dimercaprol; dimercaptol; dimercaptopropanol; 2,3-dimercapto-1-propanol; 2,3-dimercaptopropan-1-ol; 2,3-dimercaptal-1-propanol; 1,2-dithioglycerol; 2,3-dithiopropanol; USAF ME-1

◆ CHEMICAL STRUCTURE:

$$CH_2SH$$
$$|$$
$$CHSH$$
$$|$$
$$CH_2OH$$

Uses

BAL is a chelating agent used as an antidote for treatment of metal poisonings especially arsenic, mercury, and gold (see Arsenic and Mercury). It was developed when an antidote to arsenic in the chemical warfare agent "Lewisite" was needed during the World War II.

Exposure Pathways

BAL is administered intramuscularly in oil.

Toxicokinetics

BAL cannot be administered orally. It is given as a deep intramuscular injection as a 10% solution in oil. The therapeutic dose of BAL in acute arsenic poisoning is 3–5 mg/kg body weight intramuscularly every 4 hr for the first 2 days. Thereafter the dose is decreased and the interval between injections prolonged. Peak concentrations of BAL occur within 30–60 min after administration. The half-life is short and metabolic degradation and excretion are completed in about 4 hr. BAL is excreted as unidentified thiols in urine. A large amount of BAL is also excreted in bile.

Mechanism of Toxicity

BAL is believed to compete with tissue sulfhydryl groups and interferes with cellular respiration. It also competes with metallic cofactors of metabolic enzyme systems.

Human Toxicity

Acute toxic effects due to BAL can occur during normal therapy or in overdoses. BAL causes pain and sterile abscesses at the injection site. It causes a rise in systolic and diastolic arterial pressures accompanied by tachycardia. The rise in pressure is proportional to the dose administered. Accompanying signs and symptoms include nausea and vomiting; headache; burning sensation in the lips, mouth, and throat; tingling in the hands; rash; a feeling of constriction, even pain, in the throat, chest, or hands; conjunctivitis, lacrimation; blepharal spasm, rhinorrhea, and salivation; burning sensation in the penis; sweating of the forehead, hand, and other areas; abdominal pain; and occasional appearance of painful sterile abscesses. Many of these symptoms are accompanied by a feeling of anxiety, weakness, and unrest. There is a transient reduction in the percentage of polymorphonuclear lymphocytes. Children develop fever, which persists until BAL is withdrawn. BAL may cause hemolytic anemia in patients deficient in glucose-6-phosphate dehydrogenase. It is contraindicated in patients with hepatic insufficiency. BAL can potentially cause kidney damage. When applied locally to skin, it produces redness and swelling. It is an irritant to eyes and mucous membranes.

Long-term administration of BAL is unnecessary. There are no reports on the long-term toxic effects of BAL.

Clinical Management

There is no specific antidote. Normal supportive measures are recommended. Some symptoms can be relieved by administration of an antihistamine. Alkalinization of urine may prevent kidney damage.

Animal Toxicity

BAL can be administered to animals. It produces toxic effects on heart function.

—*Arvind K. Agarwal*

Related Topic

Metals

Baneberry

◆ SYNONYMS: *Actaea pachypoda* (white baneberry); *Actaea rubra* (red baneberry); doll's eyes; cohosh; snakeberry; coral berry

- ◆ DESCRIPTION: Baneberry is a tall perennial herb that grows in most woodlands throughout the United States and Canada. It has large compound leaves, small white flowers, and either red or white berries (depending on the region).

Exposure Pathways

The leaves, berries, and small white flowers may be ingested. Dermal exposure to the plant parts may also occur.

Toxicokinetics

The plant is poorly absorbed due to low solubility. Large amounts may cause systemic toxicity. With the ingestion of six or more berries, gastrointestinal symptoms may occur within 30 min. The symptoms usually disappear in 3 hr. The toxic principle, protoanemonin, is excreted by the kidneys.

Mechanism of Toxicity

The toxic effects of baneberry result from the irritant and vesicant effect of protoanemonin on mucous membranes. Protoanemonin is released through enzymatic cleavage.

Human Toxicity

The ingestion of the berries of the plant results in an initial burning sensation, increased salivation, and mucosal irritation, resulting in oral ulcerations. This is followed by acute stomach cramps and vomiting within 30 min. Dizziness, headache, and delirium were noted 1 hr after ingestion. The symptoms usually disappear 3 hr after ingestion. Prolonged dermal contact with the juice of berries or leaves may result in burning and skin irritation. As the toxic principle protoanemonin is excreted, inflammation of the urinary tract results in hematuria and dysuria.

Clinical Management

Treatment is supportive. Careful evaluation and management of fluid and electrolyte imbalance is required. Gastric decontamination is effective if performed within the first hours after exposure. Tissue damage in the oral area from the effect of protoanemonin should be assessed before any gastrointestinal decontamination. Monitoring renal output is necessary.

Animal Toxicity

The toxic effects in animals are similar to those in humans after ingestion of baneberry. Added symptoms include seizure activity and paralysis in livestock that ingest large amounts of the plant.

—*Denise A. Kuspis*

Barbiturates, Long-Acting

- ◆ REPRESENTATIVE COMPOUNDS: Barbital (CAS: 57-44-3); mephobarbital (CAS: 115-38-8); phenobarbital (CAS: 50-06-6)
- ◆ SYNONYMS:

  Barbital—diethylbarbituric acid; diethylmalonylurea; barbitone; DEBA

  Mephobarbital—methylphenobarbital; Mebaral

  Phenobarbital—phenylethylmalonylurea; barbenyl; barbiphenyl; dormiral; phenylbarbital; 5-ethyl-5-phenylbarbituric acid; Solfoton

- ◆ PHARMACEUTICAL CLASS: Barbituric acid derivative
- ◆ CHEMICAL STRUCTURE: (general structure)

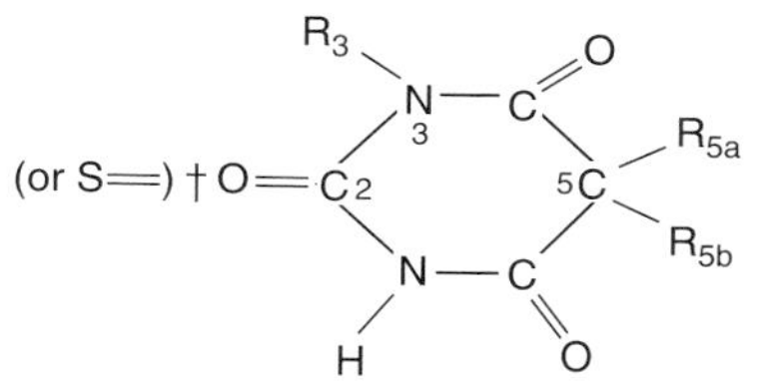

Uses

The long-acting barbiturates are used primarily for short-term (up to 2 weeks) treatment of insomnia. Other uses include treatment of anxiety or psychosis, preoperative sedation, and control of seizures. Long-acting barbiturates are drugs of abuse.

Exposure Pathways

The most common route of exposure to the long-acting barbiturates is ingestion of oral dosage forms. Phenobarbital is also available for parenteral administration.

Toxicokinetics

Approximately 50–90% of the long-acting barbiturates are slowly absorbed from the gastrointestinal tract. Absorption is more rapid when ingested on an empty stomach and in the presence of alcohol. The onset of action varies from 30 to 60 min for mephobarbital. The peak serum concentration following an oral dose of phenobarbital is not reached for 8–12 hr. Mephobarbital is primarily metabolized by N-demethylation to form phenobarbital. Phenobarbital is metabolized by the hepatic microsomal enzyme system to an inactive metabolite.

The long-acting barbiturates are extensively distributed to all body tissues and fluids with highest concentrations achieved in the brain, liver, and kidneys. The apparent volume of distribution for phenobarbital is 0.5–1.0 liters/kg. Approximately 20–45% is bound to plasma proteins. A minimal amount of mephobarbital is eliminated unchanged in the urine. Phenobarbital has a long elimination half-life of approximately 2–6 days. Approximately 25% of a dose is eliminated unchanged in the urine with the remainder eliminated as inactive metabolites. The pKa of phenobarbital (7.24) is similar to physiologic pH; as a result, the elimination of unchanged drug is significantly influenced by changes in the urine pH. Alkalinization of the urine can therefore enhance the elimination of unchanged drug. This is referred to as ion trapping.

Mechanism of Toxicity

The mechanism of long-acting barbiturate toxicity is not completely understood. Barbiturates are known to decrease the excitability of presynaptic and postsynaptic membranes where neurotransmission is mediated by γ-aminobutyric acid (GABA). However, it is unclear to what extent GABA is involved. The central nervous system (CNS) is particularly sensitive to these effects; however, with intoxication, the cardiovascular system and other peripheral functions are also depressed.

Human Toxicity: Acute

Doses of 8 mg/kg of phenobarbital will likely cause signs and symptoms of toxicity. The estimated potentially fatal dose in nondependent adults is 6–10 g. Overdose will produce CNS depression ranging from drowsiness to profound coma. Patients who overdose with long-acting barbiturates may be comatose for several days. Severe intoxication may result in cardiovascular depression leading to hypotension, cardiovascular collapse, and cardiac arrest. Apnea and respiratory arrest also may occur. Depression of the gastrointestinal tract may cause an ileus. Comatose patients may develop bullous skin lesions primarily over areas of pressure (e.g., elbows and knees).

Human Toxicity: Chronic

Chronic use of high doses of the long-acting barbiturates may produce psychologic and physical dependence. Abrupt discontinuation of therapy may result in withdrawal signs and symptoms. Mild withdrawal may include weakness, anxiety, muscle twitching, insomnia, nausea, and vomiting. Severe withdrawal may consist of hallucinations, delirium, and seizures. Unlike opioid withdrawal, long-acting barbiturate withdrawal may be life-threatening.

Clinical Management

Basic and advanced life-support measures should be implemented as necessary. Gastrointestinal decontamination procedures should be used as appropriate based on the patient's level of consciousness and history of ingestion. Activated charcoal can be used to adsorb the long-acting barbiturates. Multiple-dose activated charcoal therapy (every 2–6 hr for 24–48 hr) effectively enhances the nonrenal elimination of phenobarbital. It may be effective with other long-acting barbiturates. The patient's level of consciousness and vital signs should be monitored closely. Obtunded patients with reduced gag reflex should be intubated. Respiratory support including oxygen and ventilation should be provided as needed. There is no antidote for the long-acting barbiturates. If hypotension occurs it should be treated with standard measures including intravenous fluids, Trendelenburg positioning, and dopamine hydrochloride by intravenous infusion. Forced alkaline diuresis may enhance the elimination of the long-acting barbiturates. Hemodialysis is effective for removing the long-acting barbiturates but should be reserved for severe cases when standard supportive measures are inadequate. The occurrence of withdrawal signs and symptoms indicates the need to reinstitute barbiturate therapy and gradually reduce the dose until discontinued.

Animal Toxicity

Animals may be affected by the long-acting barbiturates much in the same way as humans. Lethargy, coma, shallow respirations, incoordination, and depressed reflexes may occur. Standard supportive measures should be employed.

—*Gregory P. Wedin*

Related Topics

Anxiolytics
Neurotoxicology: Central and Peripheral
Poisoning Emergencies in Humans

Barbiturates, Short-Acting

♦ REPRESENTATIVE COMPOUNDS: Amobarbital; aprobarbital; butabarbital sodium; butethal; cyclobarbital; heptabarbital; hexobarbital; methohexital sodium (CAS: 309-36-4); pentobarbital; secobarbital sodium; talbutal; thiamylal sodium; thiopental sodium

♦ SYNONYMS:

Amobarbital—5-ethyl-5-isoamylbarbituric acid; Amytal

Aprobarbital—5-allyl-5-isopropylbarbituric acid; allylisopropylmalonylurea; Alurate

Butabarbital sodium—sodium 5-Sec-butyl-5-ethylbarbiturate; secbutobarbitone sodium; Butisol sodium; Butalan; Sarisol

Butethal—5-ethyl-5-butylbarbituric acid

Cyclobarbital—5-(1-cyclohexenyl)-5-ethylbarbituric acid; tetrahydrophenobarbital; $C_{12}H_{16}N_2O_3$

Heptabarbital—5-(1-cyclohepten-1-yl)-5-ethylbarbituric acid; 5-ethyl-cycloheptenylbarbituric acid; $C_{13}H_{18}N_2O_3$

Hexobarbital—*N*-methyl-5-cyclohexenyl-5-methylbarbituric acid; $C_{12}H_{16}N_2O_3$

Methohexital sodium—Compound 25398; enallynymalnatrium; sodium methohexital; methohexitone sodium; 1-methyl-5-(1-methyl-2-pentynyl)-5-(2-propenyl)-2,4,6-(1H,3H,5H)-pyrimidinetrion sodium salt; 5-allyl-1-methyl-5-(1-methyl-2-pentynyl) barbituric acid sodium salt; α-dl-1-methyl-5-allyl-5-(1-methyl-2-pentynyl) barbituric acid sodium salt; Brevital sodium

Pentobarbital—5-ethyl-5-(1-methylbutyl)-barbituric acid; Nembutal

Secobarbital sodium—quinalbarbitone sodium; 5-allyl-5(1-methylbutyl)-barbituric acid; Seconal

Talbutal—Lotusate

Thiamylal sodium—Surital

Thiopental sodium—thiopentone sodium; Pentothal sodium

♦ PHARMACEUTICAL CLASS: Barbituric acid derivative
♦ CHEMICAL STRUCTURE: (general structure)

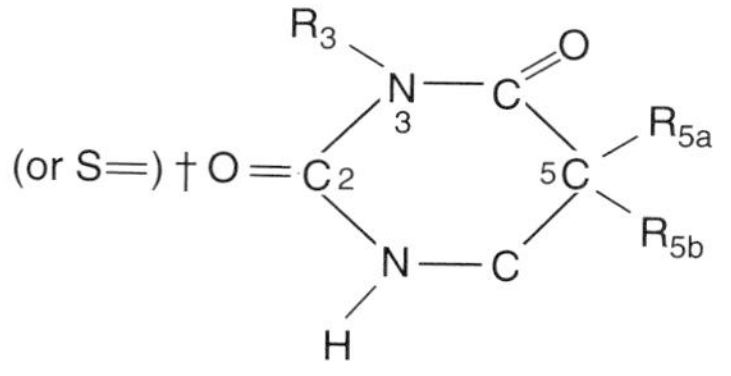

Uses

The short-acting barbiturates are used primarily for short-term (up to 2 weeks) treatment of insomnia. They are also used in the treatment of anxiety or psychosis, for preoperative sedation, for control of seizures, and as basal or general anesthetics. Short-acting barbiturates are drugs of abuse.

Exposure Pathways

The most common route of exposure to the short-acting barbiturates is ingestion of oral dosage forms. Several of these agents are also available for parenteral administration (intramuscular or intravenous).

Toxicokinetics

The short-acting barbiturates are rapidly and completely absorbed from the gastrointestinal tract. The sodium salts are absorbed more rapidly than the acids by all routes. Absorption is more rapid when ingested on an empty stomach and also in the presence of alcohol. The onset of action varies from 10 to 30 min. The short-acting barbiturates are all extensively metabolized by the hepatic microsomal enzyme system.

The short-acting barbiturates are rapidly distributed to all body tissues and fluids with the highest concentrations achieved in the brain, liver, and kidneys. The apparent volume of distribution ranges from 0.6 to 1.9 liters/kg. Inactive metabolites of the short-acting barbiturates are eliminated in the urine. Only aprobarbital, which is less lipid soluble, has a significant fraction (13–24%) that is eliminated unchanged in the urine. The elimination half-life ranges from 1 to 48 hr.

Mechanism of Toxicity

The mechanism of short-acting barbiturate toxicity is not completely understood. These barbiturates are known to decrease the excitability of presynaptic and postsynaptic membranes where neurotransmission is mediated by gamma-aminobutyric acid (GABA). However, it is unclear to what extent GABA is involved. The central nervous system (CNS) is particularly sensitive to these effects, but with intoxication the cardiovascular system and other peripheral functions are also depressed.

Human Toxicity: Acute

Doses of 3–5 mg/kg of most short-acting barbiturates will cause toxicity in children. The estimated potentially fatal dose in nondependent adults is 3 to 6 g. Overdose will produce CNS depression ranging from drowsiness to profound coma. Severe intoxication will result in cardiovascular depression, which may lead to hypotension, cardiovascular collapse, and cardiac arrest. Apnea and respiratory arrest also may occur. Depression of the gastrointestinal tract may cause an ileus. Comatose patients may develop bullous skin lesions primarily over areas of pressure.

Human Toxicity: Chronic

Chronic use of high doses of the short-acting barbiturates may produce psychologic and physical dependence. Abrupt discontinuation of therapy may result in withdrawal signs and symptoms. Mild withdrawal may include weakness, anxiety, muscle twitching, insomnia, nausea, and vomiting. Severe withdrawal may consist of hallucinations, delirium, and seizures. Unlike opioid withdrawal, short-acting barbiturate withdrawal may be life-threatening.

Clinical Management

Basic and advanced life-support measures should be implemented as necessary. Gastrointestinal decontamination procedures should be used as appropriate based on the patient's level of consciousness and history of ingestion. Activated charcoal can be used to adsorb the short-acting barbiturates. The patient's level of consciousness and vital signs should be monitored closely. Obtunded patients with reduced gag reflex should be intubated. Respiratory support including oxygen and ventilation should be provided as needed. There is no antidote for the short-acting barbiturates. If hypotension occurs, it should be treated with standard measures including intravenous fluids, Trendelenburg positioning, and dopamine hydrochloride by intravenous infusion. Forced alkaline diuresis is of no value for the short-acting barbiturates, with the possible exception of aprobarbital. Charcoal hemoperfusion can be used to remove the short-acting barbiturates from serum in severe cases when standard supportive measures are inadequate. The occurrence of withdrawal signs and symptoms indicates the need to reinstitute barbiturate or substitute alternative benzodiazepine therapy and gradually reduce the dose until discontinued.

Animal Toxicity

Animals may be affected by the short-acting barbiturates much in the same way as humans. Lethargy, coma, shallow respirations, incoordination, and depressed reflexes may occur. Standard supportive measures should be employed.

—*Gregory P. Wedin*

Related Topics

Anxiolytics
Neurotoxicology: Central and Peripheral
Poisoning Emergencies in Humans

Barium (Ba)

- CAS: 7440-39-3
- SELECTED COMPOUNDS: Barium chloride, $BaCl_2$ (CAS: 10361-37-2); barium sulfate, $BaSO_4$ (CAS: 7727-43-7)
- CHEMICAL CLASS: Metals

Uses

Barium is found in various alloys, paints, soap, paper, photographic chemicals, explosives, and rubber, and is used in the manufacture of ceramics and glass. Some of its compounds are used as mordants in fabric dying and in the preparation of phosphors. One major use is in a slurry of ground barite ($ZnS + BaSO_4$) for gas and oil drilling. Barium fluorosilicate has been used as an insecticide and some barium compounds are used as rodenticides. Medicinally, barium sulfate, being very sparingly soluble, is used as a radiopaque contrast material for X-ray diagnostic purposes.

Exposure Pathways

Exposure pathways for barium primarily consist of ingestion (e.g., food and water) and inhalation. Barium is relatively abundant in nature; hence, most food contains small amounts of barium. Brazil nuts have very high barium concentrations (from 3 to 4000 ppm). It is also found in drinking water from natural deposits in certain regions. Barium is also detected in the air of most cities.

Toxicokinetics

Soluble barium compounds are absorbed by the lungs and gastrointestinal tract and small amounts are accumulated in the skeleton. The highest concentration of barium in the body is present in the lungs. Although some barium is excreted in the urine, it is reabsorbed by the renal tubules. It is primarily excreted in feces.

Mechanism of Toxicity

Ingestion of toxic doses of barium affects the muscles; especially the heart. Barium has a digitalis-type effect on the heart. Ventricular fibrillation and slowed pulse rate are noted. This may be related to barium's tendency to displace potassium; the resulting potassium deficiency causes muscle weakness.

Human Toxicity

The toxicity of barium is related to the solubility of the compound. Barium sulfate, being very sparingly soluble, is relatively nontoxic. Soluble barium salts are toxic by ingestion (e.g., acetate, chloride, nitrate, sulfide, as well as carbonate and hydroxide compounds). Ingestion results in nausea, vomiting, stomach pains, and diarrhea. Severe gastrointestinal irritation is followed by muscle twitching and then a flaccid muscular paralysis. Barium can activate catecholamines, resulting in muscle twitching and other nervous system effects. Gastroenteritis, decreased pulse rate, and ventricular fibrillation may also occur.

Soluble compounds also irritate skin, eyes, and mucous membranes and can be absorbed following inhalation. Inhalation of insoluble sulfate and oxide, as dusts, produces a pneumoconiosis called baritosis, which is a relatively benign condition that is usually reversible with cessation of exposure.

The ACGIH TLV-TWA is 0.5 mg/m^3 for soluble barium compounds and 10 mg/m^3 for barium sulfate.

Clinical Management

Addition of sodium sulfate as a lavage solution may precipitate the very insoluble barium sulfate. As potassium deficiency occurs in acute poisoning, serum potassium and cardiac rhythm must be monitored closely. Administration of intravenous potassium appears beneficial. As renal failure is also a concern, urinary output also must be monitored closely.

Animal Toxicity

In guinea pigs, barium has been noted to cause various changes in the blood and pathological changes in the bone marrow, spleen, and liver.

*—Arthur Furst, Shirley B. Radding
and Kathryn A. Wurzel*

Related Topic

Metals

Baygon

- ◆ CAS: 114-26-1
- ◆ SYNONYMS: Propoxur; Aprocarb; Unden; 2-isopropoxyphenyl *N*-methylcarbamate
- ◆ CHEMICAL CLASS: Carbamate insecticide
- ◆ CHEMICAL STRUCTURE:

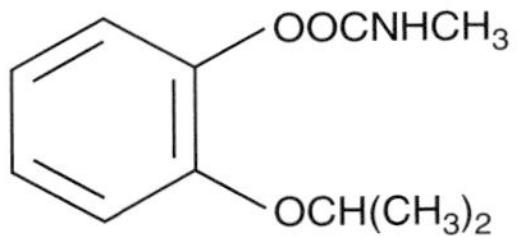

Uses

Baygon is an insecticide.

Exposure Pathways

Ingestion and inhalation are common routes of exposure in humans.

Toxicokinetics

Baygon is rapidly and almost completely absorbed during its normal transit through gastrointestinal tract. In addition, baygon can penetrate into skin, mucus membranes, and the respiratory tract. Baygon is metabolized to 10 different organosoluble and 6 water-soluble products in insects, whereas incubation with rat liver microsomes produced 11 metabolites. Hydroxylation of the *N*-methyl moiety is the major metabolic reaction with mammalian enzymes compared to ring hydroxylation in the 5-position with insect enzyme. Little information is available on the distribution of baygon. The sites in which residues have been reported are the liver, kidneys, brain, fat, and muscle. The half-life in the rat is of the order of 3 or 4 hr. The major portion of the administered dose in human was eliminated in urine as a conjugate of 2-isopropoxyphenol, presumably as the glucuronide. In rats approximately 25% of the administered dose was obtained as volatile compounds, whereas 70% was found in the urine.

Mechanism of Toxicity

Baygon is an effective insecticide by virtue of its ability to inhibit acetylcholinesterase in the nervous system. As a cholinesterase inhibitor, baygon behaves as a synthetic neurohormone that produces its toxic action by interrupting the normal action of the acetylcholinesterase so that the substrate acetylcholine accumulates at synaptic junction. It can also inhibit other esterases. The carbomylation of the enzyme is unstable and the regeneration of acetylcholinesterase is relatively rapid. A pro-oxidation mode of action has been proposed for certain toxic manifestation of baygon.

Human Toxicity

Health hazards for humans occur mainly from occupational overexposure to baygon, resulting in poisoning characterized by cholinergic symptoms caused by the inhibition of the enzyme acetylcholinesterase. A single relatively small oral dose of baygon (0.36 mg/kg) in volunteers may produce symptoms of short duration, like blurred vision, nausea, pallor, sweating, tachycardia, and vomiting. Higher doses if divided into portions and taken within a relatively brief period may be tolerated without symptoms. Pronounced and symptomless daily depression and reactivation of cholinesterase was observed in persons who are occupationally exposed to baygon.

The acceptable daily intake (FAO/WHO) is 0.02 mg/kg body weight.

Clinical Management

Atropine should be administered to the patient immediately. In an adult, the usual dose of atropine is 1–4 mg as sulfate (or other salt) given intramuscularly or intravenously (if there is no cyanosis). Repeated doses may be required depending on the recurrence of toxic symptoms. Mucus and other respiratory secretions may have to be aspirated continuously. Artificial respiration may be needed. Dehydration and electrolyte imbalance can be overcome by intravenous infusion of isotonic saline.

Animal Toxicity

Depression of acetylcholinesterase activity from plasma, erythrocyte, and brain was observed during baygon exposure to rats and mice. The acute oral LD_{50} of the pure compound for rats is 95–104 mg/kg body weight. Baygon feeding through diet up to 4500 ppm for 30 days did not show any delayed neurotoxic signs of poisoning either during the period of feeding or in the posttreatment observation period of 4 weeks. Baygon feeding at different levels for 2 years in dogs resulted in mortality at 2000 ppm. Dietary consumption below 750 ppm did not affect appearance, behavior, food consumption, or growth of animals. Baygon administration to rats during gestation can cause embryotoxicity.

—*Madhusudan G. Soni*

Related Topics

A-esterase
Acetylcholine
Carbamate Pesticides
Cholinesterase Inhibition
Neurotoxicology: Central and Peripheral
Pesticides

BCNU

- CAS: 154-93-8
- SYNONYMS: Carmustine; *N,N-bis* (2-chloroethyl)-*N-nitrosourea*; BiCNU; *carmubris; nitrumon*
- PHARMACEUTICAL CLASS: Alkylating agent

- CHEMICAL STRUCTURE:

$$ClCH_2CH_2N \overset{NO}{\underset{|}{N}} - \overset{O}{\underset{\|}{C}} - NHCH_2CH_2Cl$$

Uses

BCNU has been used in human medicine as an antineoplastic agent (alone or in combination with other agents) in the treatment of Hodgkin's lymphoma, multiple myeloma, and in primary or metastatic brain tumors.

Exposure Pathways

Intravenous injection is the most common route of exposure. Doses range from 100 to 250 mg/m² body surface for courses of 2 or 3 days.

Toxicokinetics

In animal experiments, BCNU is rapidly absorbed, following different routes of ingestion. A few minutes after administration, no unchanged BCNU can be detected in plasma. BCNU undergoes spontaneous decomposition under physiological conditions to release both alkylating and carbamoylating entities. In addition to chemical decomposition, BCNU may be denitrosated enzymatically to its corresponding urea. BCNU is rapidly distributed to most tissues including brain and cerebrospinal fluid. The volume of distribution approximates 0.18 liter/kg. Approximately 80% of the drug appears in the urine within 24 hr as degradation products. BCNU is reported to have a biological half-life of less than 20 min.

Mechanism of Toxicity

It is generally assumed that BCNU exerts its cytotoxicity through the liberation of alkylating and carbamoylating moieties. Alkylating entity particularly chloroethyl carbonium ion is strongly electrophilic and can alkylate a variety of biomolecules, including the purine and pyrimidine bases of DNA. The interstrand cross-linking is generally associated with the cytotoxicity of BCNU. The carbomylation of lysine residues of protein can inactivate certain DNA repair enzymes thus interfering with repair processes.

Human Toxicity

Various cytotoxic effects of BCNU in humans are reported. The drug is not a vesicant, but local burning pain has been reported after intravenous administration. Nausea and vomiting occur approximately 2 hr after injection. Flushing of the skin and conjunctiva, central nervous system toxicity, esophagitis, diarrhea, interstitial pulmonary fibrosis, and renal and hepatic toxicity have been reported. Although bone marrow suppression is observed, this drug characteristically causes an unusually delayed onset of leukopenia and thrombocytopenia. The nadir of the leukocyte and platelet counts may not reach normal levels until 6 weeks after treatment.

Clinical Management

Most of the adverse reactions of BCNU are reversible if detected early. When such effects or reactions do occur, the drug should be reduced in dosage or discontinued and appropriate corrective measures should be taken according to the clinical judgment of the physician. Blood counts should be monitored weekly for at least 6 weeks after the dose. Baseline pulmonary function studies, hepatic functional tests, and periodic renal functional tests should be monitored. No proven antidotes have been established for BCNU overdosage.

Animal Toxicity

In dogs, high doses of BCNU resulted in severe bone marrow hypoplasia with delayed, reversible thrombocytopenia. The other major toxicities observed were cardiopulmonary (pulmonary edema, myocardial infarction, and pericardial hemorrhage), intestinal mucosal damage with hemorrhage, renal toxicity, and delayed hepatotoxicity. Similar toxicity was seen in monkeys except that cardiopulmonary toxicity did not occur. In rats initially well tolerated doses may cause death later. There is sufficient evidence for the carcinogenicity of BCNU in rats. BCNU is embryo and feto-lethal in rats and rabbits at doses nontoxic to the mother and can induce a variety of teratogenic effects in rats.

—*Madhusudan G. Soni*

Related Topics

(see discussions of alkylating agents in entries listed below)
Carcinogen—DNA Adduct Formation and DNA Repair
Mutagenesis

Behavioral Toxicology

Introduction

Behavioral toxicology is that scientific discipline which studies the effects of nontherapeutic chemicals and toxicants on human behavior, the ultimate output of the nervous system, and also seeks to determine how such effects are caused. The impetus for such studies has come from multiple sources (see Psychological Indices of Toxicity). Both human and experimental animal studies have been carried out to assess the behavioral consequences arising from exposures to chemicals used in the workplace as well as those dispersed in the environment. These efforts have been important in determining safe exposure and risk levels, as well as in furthering our understanding of these chemicals. A second force behind many such studies has been the need to screen newly synthesized chemicals for any potential adverse behavioral effects before their introduction into use, efforts which are obviously carried out only in experimental laboratory contexts.

Human behavior is, of course, extremely diverse and complex, comprised numerous different functions, any or all of which might be perturbed by exposure to a toxicant. Thus, understanding how a chemical affects human behavior may require a determination of its effects across these different behavioral functions. Furthermore, some human behaviors require an integration of several different behavioral domains. If we think about learning in a classroom, for example, in addition to cognitive functions, sensory functions are needed to process the information presented, and motor functions are required for executing the correct response. Thus, in the event that a chemical is suspected to produce effects on cognitive functions, the possibility that such effects, instead, result indirectly from changes in sensory or motor functions must always be considered.

The entire range of behavioral functions and the tests designed to evaluate them cannot be presented here. This entry first presents the types of methods that comprise the test batteries used in screening newly developed chemicals for behavioral toxicity. While screening batteries are extremely useful in providing a preliminary assessment of adverse behavioral effects, they are less useful for elaborating the actual nature of the behavioral deficits or for yielding an understanding of their underlying behavioral and neurobiological mechanisms. For such purposes, more specific tests of various behavioral functions are utilized. Such higher-order tests, in particular those related to sensory, motor, and cognitive functions, are subsequently presented in this entry and are followed by some discussion of the testing methods utilized in experimental animals to determine adverse behavioral effects of chemicals during the course of development as well as some of the test methods unique to human populations.

Screening Batteries

Because a newly developed chemical may have effects on any of the numerous behavioral functions that comprise a behavioral repertoire, screening batteries must necessarily assess a wide variety of functions with sufficient sensitivity to suggest potential behavioral toxicity even in a single behavioral domain. These screening batteries are typically executed in studies using rats and mice and generally consist of two components: a functional observational battery (FOB) and a measure of motor activity (see below). FOBs include an array of measures, generally of unlearned or instinctive behaviors, designed to detect any indications of gross changes in reflexes and in gross motor or sensory function. Most FOBs are relatively easy to implement since there is typically no behavioral training or special equipment required for any of the behavioral measures utilized as they are carried out and scored by an experienced observer. An FOB may include measures of general integrity, such as any signs of convulsions, palpebral closure, lacrimation, piloerection, salivation, and vocalizations. In addition, assessments of sensory capability, based on measures such as response to a finger snap or a tail-pinch, righting reflex, and assessments of motor function, as evaluated by the posture or gait of the animal, catalepsy, hindlimb foot splay, forelimb and hindlimb grip strength, and the time to begin ambulating, may be included. Finally, any signs of arousal or stress can be measured, such as ease of removal and handling, the animal's response to touch or approach, and urination and defecation. In addition, certain physiological responses, including body temperature and body weight, are measured. These evaluations are often carried out in two different environments: a familiar one, such as the animal's home cage, and an unfamiliar flat surface of some type. This series of measures can be made relatively rapidly on each animal, consistent with the goal of screening of new compounds across a wide range of doses. In the event that behavioral activity of the chemical is indicated in such a screening test, more advanced and specific behavioral procedures would be required to delineate the precise nature of the behavioral impairment.

One question that has arisen with respect to the use of FOBs is whether changes in only one or two of the numerous measures made are really indicative of neurotoxicity. For example, how is a change in two seemingly unrelated measures interpreted (e.g., vocalizations and hindlimb grip strength). One answer that has been suggested is that neurotoxicity would be indicated by similar changes within a behavioral domain. Thus, changes in both forelimb and hindlimb grip strength would be indicative of altered motor function. Some have contended that if the toxicant under test produces body weight changes, then any changes also observed in the FOB may simply be due to "sickness syndrome" or general malaise of the animal, not neurotoxicity. This is not necessarily a valid conclusion, however, since body weight changes may occur totally independently of any observed FOB effects. In fact, FOB changes are often reported in the absence of any body weight changes.

Motor Function

Motor function is a critical component of human behavior because it embodies the ultimate execution of a behavioral response. The feats of highly skilled athletes provide one example of incredibly refined motor performance, but even everyday functions such as walking or driving to work depend on adequate motor capabilities. Motor behavior is not a unitary behavioral function, but rather one with many different components. Various motor responses entail such aspects as strength, coordination and endurance, precision and duration, frequency of occurrence, and for ambulation, gait and balance as well. Measurement of these different aspects of motor function obviously requires different procedures. As is the case for measurement of virtually all behavioral functions, the paradigms for assessing motor capabilities range from simple assessments to more complex technologies. The former provide easily implemented but generally less specific and selective measures of function and the more advanced procedures provide specific measures of motor function as distinct from changes in sensory or motivational processes but may require some training of the experimental subject.

Motor Activity Levels

Motor activity measures the frequency of occurrence of integrated movements and/or ambulation of the organism over some designated period of time, a behavior that generally occurs at some baseline level in mammalian species and which may be altered by exposure to a toxicant. A measure of motor activity is typically one component of a screening battery, and most studies of motor activity are carried out using rodents. Generally, the animals are placed on a horizontal surface, which could be square, rectangular, or even a maze such as a T-shaped apparatus, and the number of defined movements over a specified time period are recorded. In nonautomated versions, movements are typically recorded by an observer who, it is hoped, has no information with respect to the treatment condition (toxicant-exposed or control) of the subject which might bias the recording of data. In most automated versions, the movements of the animal either interrupt a light beam or trip a switch which then records a count.

Measurements of motor activity have been used to evaluate the potential central nervous system (CNS) effects of a wide variety of drugs and toxicants. One of the advantages of such measures of motor function is that no training is required of the subject. In addition, measures of motor activity can be made repeatedly across time so that the time course, including the onset and reversibility of toxicant effects, can be determined. In these types of repeated measurement experiments, moreover, an animal can serve as its own control, meaning that the experimenter looks for a change in the animal's normal pattern of motor activity after receiving the toxicant compared to the pattern observed before the treatment.

The experimenter must be cognizant of the fact that different devices for measuring motor activity may measure markedly different aspects of motor function. For example, in devices such as a figure-eight maze, an animal may rear up on its hindlimbs in front of a light beam. This response will break the beam of light and be counted as a response. In contrast, in devices like the open field shown in Fig. B-1, the investigator tallies the number of squares in a rectangular field entered by the animal with all four paws, and a rearing response would not be counted. Such differences often negate the direct comparisons of various studies of the effects of a toxicant on motor activity and also underscore the importance of precise specifications of the response(s) being recorded in any given study with a particular device. Failure to do so may result in seemingly inconsistent results. Other influences must also be considered in the interpretation of changes in motor activity. For example, motor activity levels are known to be influenced by a variety of nonmotoric variables, such as time of day at which testing is carried out (rodents are nocturnal and show greater activity levels during dark hours), room lighting, and odors. As this list indicates, changes in sensory capabilities (perceived difference in the room odors or lighting) or circadian (nocturnal) rhythms could influence measures of motor activity independently of any direct toxicant-induced changes.

Motor activity may be a relatively insensitive measure of toxicant-induced changes when it relies on relatively gross measures such as total counts per unit time. For example, in the open field test mentioned previously and shown in Fig. B-1, the total number of squares entered into by the animal over a period of time is measured. However, the same total number may be achieved through very different patterns of behavior. For example, the organism might show an initial period of rapid movement followed by immobility or, alternatively, cycles of high activity followed by low activity or, finally, even a continuous but moderate rate of ambulation. All three could lead to the same total number of squares entered, but the disparate patterns suggest differences in behavior that are not being captured.

Strength, Coordination, and Endurance

Weakness and fatigue are common complaints resulting from exposures to a number of different chemicals. Both simple and more complex approaches to measuring these facets of motor function are available. A simple and commonly used procedure that has the advantage of not requiring any specific training of the animal

FIGURE B-1.   *Open field apparatus in which the number of squares entered by the mouse or rat is counted over a fixed period of time (reproduced with permission from Cory-Slechta, D. A., Behavioral measures of neurotoxicity, Neurotoxicology 10, 271–296, 1989).*

is the rotarod device shown in Fig. B-2. A rat or mouse is placed on a rotating cylinder, the speed of which can be manipulated, and the time the animal remains on the rotating device before falling onto the plate below is recorded. Falling off more quickly may be an indication of changes in coordination and/or endurance. As with motor activity, time spent on the rotarod can be measured repeatedly, and a stable baseline performance can be generated across experimental sessions against which the impact of toxicants may be compared.

The difficulties with such an approach are also evident in Fig. B-2. Mice frequently attempt to scramble up the dividers; some attempt to run backwards. Others begin to jump off the device and will not remain on the device no matter how frequently they are replaced on the rotarod. As these examples indicate, the rotarod device thus measures aspects of behavior in addition to coordination and endurance which must obviously

be considered in interpreting such data. In other words, one cannot necessarily be certain that decreased time spent on a rotarod after toxicant treatment necessarily reflects changes in endurance and coordination or, for example, reflects increased distraction.

More advanced techniques that rely on learned behavior of animals (i.e., operant behavior) provide better controls for such nonmotoric behavioral factors and thus provide a more specific indication of changes in endurance and coordination. For example, rats can be trained to depress a lever with a specified amount of force in order to obtain food delivery. The amount of force required to depress the lever can then be successively increased until the maximal force that can be exerted is reached. In addition, the force that the animal can sustain over time can also be measured as an indication of endurance. Thus, the true force capabilities of each individual rat can be determined. Decreases in the

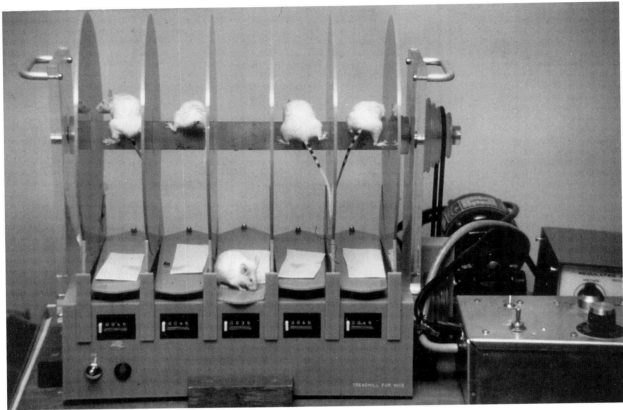

FIGURE B-2.   *Illustration of the rotarod apparatus for mice. Each mouse is placed on the rotating cylinder; speed of revolution can be manipulated and time on rotarod typically constitutes the dependent variable of interest (reproduced with permission from Cory-Slechta, D. A., Behavioral measures of neurotoxicity, Neurotoxicology 10, 271–296, 1989).*

speed and uniformity (rate) at which these responses occur would suggest motivational impairments as distinct from true force capabilities decrements.

Gait and Balance

Walking, running, and many other motor responses depend on intact gait and balance, and such functions may be particularly vulnerable to chemicals that affect the peripheral nervous system. One simple procedure that has been devised to assess postural dysfunction is known as hindlimb splay. In this procedure, the hindpaws of a mouse or rat are dipped in ink and the animal is then dropped from a fixed height onto a piece of paper below as can be seen in Fig. B-3. An increase in the distance between the hindlimbs upon landing is indicative of damage to the peripheral nervous system

with consequent effects on gait and ambulation. This approach is simple in that the rodent does not have to be specifically trained for the task, and this measurement can be made repeatedly across time without extensive equipment requirements so that time to onset and recovery of a toxicant's effects can be followed. However, hindlimb splay may not be a totally specific measure of altered motor function. Sensory disturbances, for example, might alter landing foot distance as well.

A more advanced type of approach, an automated hindlimb movement detection apparatus, is shown in Fig. B-4. In this scheme, a TV—microprocessor system is utilized to record the placement of a rat's hindpaws as it traverses from one rung to the next in a running wheel analogous to those offered in pet stores for rodents. Computer analysis of the recording provides a measurement of both quantitative and temporal characteristics of stepping, such as correct small steps and

FIGURE B-3. *Illustration of the hindlimb splay procedure: The hind paws of the mouse are dipped in ink, the animal is then dropped from a fixed height onto a piece of paper, and the distance between some parameter of the hindlimbs is measured (reproduced with permission from Cory-Slechta, D. A., Behavioral measures of neurotoxicity, Neurotoxicology 10, 271–296, 1989).*

large steps, missteps, and the temporal parameters of these movements. Thus, an experimenter can measure with great precision how different parameters of gait differ before and after exposure to a toxicant. The animal need not be explicitly trained, and this approach provides a relatively specific measure of motor function per se. Procedures for measuring bodily sway in children have also been used in behavioral toxicology studies. In these procedures, the child stands on foam or on a hard surface under conditions of either eyes opened or

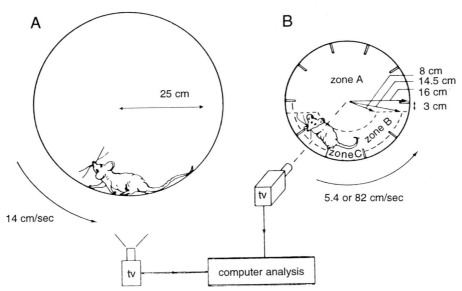

FIGURE B-4.   *Automated hindlimb movement apparatus. (A) The camera can register the movements of the dyed soles of the paws of the rat from below. (B) The wheel with a transparent front facing the axially mounted color TV camera (reproduced with permission from Tanger H. J., Vanwersch, R. A. P., and Wolthius, O. L., Automated quantitative analysis of coordinated locomotor behaviour in rats,* J. Neurosc. *Methods 10, 237–245, 1984).*

closed, and the extent of the sway of his or her body is measured utilizing strain gauges.

Sensory Function

A wide range of sensory functions provide us with information about the environment. These functions include our abilities to hear, see, smell, and detect movement, vibration, and pain. Deficits in sensory function often constitute some of the earliest or even the most pronounced manifestations of chemical exposures that affect the nervous system. As with the measurement of motor function, both simple and more advanced procedures are available to measure sensory function. Since almost all sensory procedures require the subject to make motor responses to indicate whether it has detected some sensory stimulus, changes in motor capabilities could conceivably be measured instead of sensory changes. Thus, while the simple procedures do not require any training of the subject, the experimenter must recognize that any changes measured may be due to changes in sensory function or motor function or both. The more complex techniques

do require extensive training of the subject before it is possible to measure sensory function, but they also offer the possibility of differentiating any contribution of motor abnormalities from sensory changes. The more advanced procedures can also be used across species, including rats, nonhuman primates, and humans, thus alleviating some of the questions that arise with respect to the extrapolation of findings across species.

In most procedures used to evaluate sensory function, a sensory stimulus is presented to the subject, and a response by the subject, either learned or unlearned depending on the specific procedure, then indicates whether the subject has detected that stimulus. The stimulus may vary from one presentation to the next in its important dimensions such as frequency and intensity, yielding a complete profile of sensory capabilities for that specific sensory modality. For example, in measuring hearing, tones differing in their loudness and pitch are used so that hearing along the entire spectrum is measured.

A clinical neurological examination often includes components designed to measure sensory function, but they can be relatively insensitive; thus, subtle changes in sensory function might not be detected. One of the simpler experimental procedures used to test sensory function is referred to as reflex modification and is based on unlearned reflexes, in particular the startle reflex. A stimulus such as a loud noise can elicit a startle response (i.e., a startle reflex). It is also known that a stimulus presented prior to the presentation of that loud noise (prestimulus) can measurably decrease the magnitude of the startle response. Thus, a prestimulus is detectable if the magnitude of the subsequent startle response is decreased. The prestimulus can be varied in intensity and frequency dimensions during a testing session to produce a complete profile of sensory changes in a particular modality. For example, across the trials of a test session, the intensity and frequency of an auditory stimulus can be modified, and a threshold (e.g., the intensity for a given tone frequency that inhibits startle on 50% of its presentations) can be determined for each tone frequency, generating a classical audiogram. The advantages of reflex modification include its utility across different stimulus modalities (e.g., visual auditory and proprioceptive), its potential utilization across species, and the absence of any requirement for training subjects. It has already been utilized to demonstrate auditory impairments in response to exposures to neurotoxic compounds such as trimethyltin.

One factor that must be considered in reflex modification procedures is that a less pronounced startle response could result from alterations in motor function per se rather than deficits in the subject's ability to detect the prestimulus. For this reason, it is imperative that some trials be interspersed throughout each test session in which no prestimulus is presented, only the startle stimulus. This allows the experimenter to determine whether the magnitude of the startle response remains constant after a chemical has been administered. If so, then any changes in the amplitude of the startle response during prestimulus trials necessarily reflect altered sensory function. Another caution regarding this procedure is that the startle reflex itself may diminish over time. Thus, the number of trials in an experimental session must be carefully controlled.

The more advanced methods for the measurement of chemical-induced changes in sensory function are deemed operant psychophysical procedures. These methods have been used in almost identical forms across a range of sensory modalities and in numerous species, including rodents (rats and guinea pigs), chinchillas, pigeons, nonhuman primates, and humans. Figure B-5 depicts an example of both a human and a nonhuman primate being tested for sensitivity to a vibratory stimulus presented to the hand using operant psychophysical methods. Here the subject is typically required to make a specified response within some particular period of time to signify that a stimulus presentation was detected. Experimental training of the subject is required before any sensory capabilities can be precisely gauged. In Fig. B-5, the subjects were required to hold down a key when a tone sounded. If they detected a vibratory stimulus delivered to the fingertips during the tone delivery, they released the key and received a reward. To determine how much subjects were simply guessing as to whether a vibratory stimulus was presented, some trials involved no vibratory stimulus presentation. On those trials, the subjects were rewarded for releasing the key only after the tone ended, to indicate that they had detected no vibratory stimulus. As with measures of sensory function such as reflex modification, the various parameters of the sensory modality being evaluated are varied from trial to trial (e.g., intensity, magnitude, and frequency) allowing a determination of that specific sensory function along its significant dimensions. Since changes in sensory function may sometimes be quite selective (e.g., hearing loss for high frequency tones but not low frequency tones), the ability to map sensory changes along the entire spectrum of its significant dimensions is an important component of these methods.

There are several different variations of the methods by which stimuli are presented in the operant psychophysical procedures. In the method of constant stimuli, the subject is presented with several different values (e.g., intensities) of the stimulus in a random sequence or order across trials. The proportion of stimulus presentations detected at each intensity is then calculated, and the value yielding a 50% detection response is deemed the threshold. The method of limits presents a series of stimulus intensities which begin either well above or well below the presumed threshold value. The stimulus value is then either progressively decreased or increased, respectively, until a change in the subject's ability to report the stimulus presentation occurs. The intensity of the stimulus at which this change in detectability occurs is designated as the threshold. In the up-and-down, staircase, or titration method of stimulus presentation, the threshold is continuously tracked by

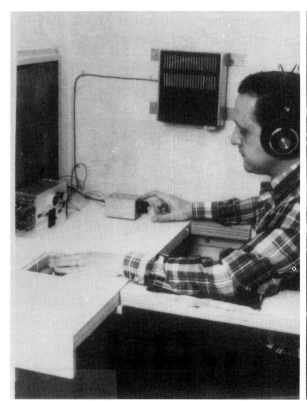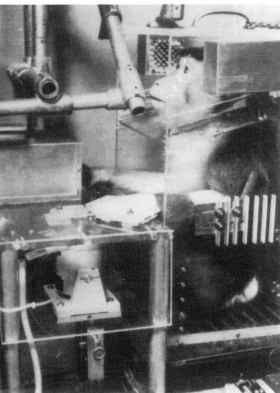

FIGURE B-5. *Photograph of a human and nonhuman primate working on a vibration sensitivity paradigm. In each case, the left hand of the subject is placed atop a device that delivers the vibratory stimulus, while the right hand holds down a telegraph key to be released when the subject detects the vibratory stimulus. A spout at the level of the monkey's mouth delivers a squirt of fruit juice for a correct response (reproduced with permission from Maurissen, J. P. J., Effects of toxicants on the somatosensory system, Neurotoxicol. Teratol. 1 (Suppl.), 23–31, 1979).*

raising or lowering the stimulus intensity depending on whether the subject correctly detected the stimulus. If the subject fails to detect the stimulus, presumably because it is below the threshold for detection, the intensity is raised on the next trial; if the stimulus was detected (i.e., was above threshold), the intensity is then lowered on the next trial. In this fashion, the stimulus intensity can be titrated around the threshold value of the subject.

One of the advantages of operant psychophysical procedures over methods such as reflex modification is that stimulus presentation and subsequent responses occur on a continuous or response-dependent basis. In other words, a response of the subject is recorded, and the next stimulus is presented. In the reflex modification procedure, stimuli are presented during trials which

are experimenter initiated and which are separated by specified time interval. The continuous procedures permit the experimenter to measure the rate of responding over time and the time required to respond following stimulus presentations (latency). These measures provide the experimenter with information as to any possible motor dysfunction or motivational problems that the subject may experience as a result of chemical administration which could contribute to behavioral changes in operant psychophysical procedures. Motor dysfunction might increase the latency to respond following stimulus presentations, while an unmotivated subject might be expected to show periods of nonresponding. Armed with this information, the experimenter can proceed to determine which behavioral changes result from true sensory loss.

Cognitive Function

One of the major concerns aroused by exposures to chemicals that affect the nervous system is their potential to adversely impact cognitive functions such as learning and memory. Such a concern certainly has precedent. Lead exposure at high levels can leave children with permanent mental retardation. Recently, it has been demonstrated that even very low levels of lead exposure (i.e., environmental exposures) can produce subtle changes in cognitive processes. Pesticides are known to exert pronounced effects on cholinergic neurotransmitter systems, the very system that has been repeatedly implicated as a causative factor in Alzheimer's disease.

Learning

Learning might be defined simply as an enduring change in behavior that results from experience with changes in environmental events. As a topic of long historical interest in psychology and neuroscience, there are numerous different methods that have been applied to the study of learning ranging from the relatively simple to the more complex and advanced paradigms. Methods for assessing learning involve the processing of sensory stimuli, the execution of motor responses, and a motivated subject. Difficulties in distinguishing the contributions of sensory, motor, and motivational deficits from learning deficits are frequently encountered when using relatively simple learning paradigms. Some of the more complex procedures are designed to specifically differentiate such functions from learning and thus allow the experimenter to determine whether the chemical has specific effects on learning per se as distinct from changes in sensory or motor function, motivational levels, or other nonspecific behavioral alterations.

Many of the earliest studies of learning utilized rats as experimental subjects and required them to run mazes of various shapes and sorts, generally from a start box to a goal box where some type of food reward (reinforcement) was available. For example, in a T-maze, named because of its shape, the subject is reinforced for running from the start box at the base of the T to that arm of the T which has been designated as the correct arm and contains the goal box where food is located. The designation of which arm (stimulus) is correct may be based on side (the right side is correct), color (the arm painted black), or some other stimulus feature. Choosing the wrong arm at the choice point means no reinforcement. After entering an arm and either being reinforced or not, the animal is removed, and after a period of time (the intertrial interval) the animal is placed back in the start box and another trial initiated. Learning under such conditions is typically measured as the number of trials required to reach a specified accuracy level or until behavior reaches a stable level of accuracy, at which point it is stated that the subject has learned to "discriminate" between the correct and incorrect arm. The experimenter may compare two groups of rats in such an experiment: one treated with a chemical and one not treated, with the latter serving as a control group indicating "normal" performance under the particular experimental conditions.

More complex versions of mazes soon emerged in response to the need for more difficult tasks because the T-maze was a relatively simple problem for a rodent to solve and thus not always adequately sensitive to effects of drugs or chemicals. Moreover, once the animal learned which was the correct arm, learning is no longer being measured, only the performance of an already learned response. Two different approaches were offered to circumvent these limitations. One was the construction of more complicated mazes, such as the Hebbs–William maze, which is actually a series of mazes. The correct route to the goal box in this device can be modified as needed by moving the various arms and boundaries into new configurations and thus requiring the subject to learn a new problem, allowing a repeated assessment of learning.

A second approach is embodied in reversal learning. Using this approach, the correct and incorrect arms (stimuli) in the maze are reversed after the subject initially learns which is correct. For example, after the rat learns to run to the right arm of the maze with 90% accuracy, the discrimination is reversed, such that the left arm of the maze is now the rewarded arm. After criterion accuracy is achieved following this reversal, the designation of correct and incorrect stimuli may be reversed again, allowing the repeated measurement of learning over time. Eventually, however, this scheme is also learned by the subject, a phenomenon known as "learning to learn," such that it comes to learn each successive reversal problem with maximal efficiency (i.e., after only one or two trials).

All maze procedures have limitations that must be considered when interpreting data obtained with these methods. One related particularly to their use with rodents is that subjects leave an odor trail in the maze that can influence the behavior of rodents subsequently tested in the maze. While the experimenter can clean the maze between subjects, it must also be noted that the rodent's sense of smell is much more sensitive than humans, making it difficult to be certain that indeed no odors are still present. Another potential problem is that these procedures obviously require interactions between the experimenter and the subject during the course of testing because the rat must be constantly retrieved from the arms and replaced in the start box. This raises the distinct possibility of both subject and experimenter bias, unless the experiment can be carried out by an individual with no knowledge of any treatment (e.g., exposure to drug or chemical) of any subjects.

Another limitation of simple maze methods for assessing learning is that they do not selectively measure learning. While a decrease in the speed of reaching a 90% accuracy level may be observed in response to a chemical treatment, it may not necessarily be due to alterations in cognitive processing since changes in either motor performance or sensory capabilities may impact performance in the maze, altering learning independently of any real cognitive changes. Impaired motor function may increase the time taken to reach the goal box; and sensory deficits may cause the subject to be unable to utilize the environmental stimuli that normally guide its path to the goal box. Motivational changes (e.g., if the reward becomes less appealing) may clearly retard the rate at which learning occurs. Changes in motor, sensory, and motivational functions as potential contributors to the observed effect may then have to be ruled out in separate additional experiments.

The Morris water maze is a method increasingly being used in behavioral studies for evaluating learning. A rodent, usually a rat, is placed in a large tub of water that has been made opaque by the addition of a substance such as nonfat milk powder. Since most laboratory rodents do not prefer water, the reward is escaping from the water by locating a platform submerged just below the surface of the water which is not visible to the rat. Learning is measured as a decrease in the time to locate the hidden platform across successive trials; a learning deficit is suggested by a slower decrease in that time requirement or a greater number of trials to reliably locate the platform. The procedure is relatively simple and imposes no requirement to train the animal or to impose food restriction on the subject. However, despite its ostensible simplicity, it suffers from many of the same limitations as non-water-based mazes. First, the procedure is nonautomated and thus requires subject–experimenter interaction which can introduce bias into the results. Furthermore, since it is a relatively simple problem, the maze may be learned rapidly and, thus, it is of little utility for experiments aimed at understanding the time course of a chemical's effects on learning. This problem can be alleviated to some extent by moving the platform to a new location each time the subject has mastered the previous location. Although one might suspect that odor trails would not be a factor in a water maze, it has indeed been shown that odors are present and can be utilized by other subjects later placed in the maze. In addition, water temperature plays an important role in this task since age-related deficits in learning in the maze can be alleviated by warming up old rats between trials.

Finally, a rat that requires a greater number of trials or exhibits a slower decrease in the time to locate the platform following chemical treatment is not necessarily exhibiting a learning deficit. Changes in motor capabilities may affect swimming performance and thus lengthen the time it takes the subject to swim to the submerged platform, even if it knows the location of the platform. It is known that subjects rely on environmental cues to find the platform; thus, changes in sensory capabilities could mean an inability of the subject to detect the necessary environmental cues, a deficit which would also increase the length of time the subject required to reach the platform. These alternative explanations of any deficits in swimming time must be ruled out by additional experiments before one can reasonably conclude that a cognitive impairment is present. One way to achieve this is to have the same subjects, in other circumstances, simply be required to swim to a platform, the location of which never changed. This response would also require intact motor and sensory capabilities in performing the same response but no learning since the platform location remains constant. The observation of treatment effects when the platform is moved around, in the absence of any treatment effects when the platform remains fixed, would provide support for an interpretation that the chemical induced learning deficits.

Another maze procedure frequently utilized to evaluate learning (and memory) with rodents is the radial-arm maze. The device itself consists of a central circular area from which eight arms radiate outward like the spokes of a wheel. At the end of each of the eight arms is some type of reinforcer, usually a food pellet. In essence, the subject has access to eight reinforcement deliveries, one in each of the eight arms, and the accuracy and speed (efficiency) with which the subject learns to retrieve all eight reinforcer deliveries is measured. Obviously, under these conditions, the most efficient performance is to obtain all eight reinforcements without revisiting an arm from which the food has already been obtained. The measure of learning is the number of trials required for the subject to reach some specified level of efficiency in the maze. One way of further increasing task difficulty is to provide reinforcement only in a specified number of the arms (e.g., four of the eight) and to change which of the arms provides reinforcement over time or trials. The radial arm maze obviously presents a more difficult problem to the subject than the T-maze or other simple mazes, but the possibility of interference from motor or sensory deficits produced by chemical treatment still remains. Thus, an increase in the number of trials to reach efficient performance is not necessarily indicative of a learning deficit with this method. By measuring the time of entry into each arm, investigators can begin to get some indication of whether changes in overall activity levels are affecting performance.

In addition to mazes, learning can be measured in Skinner boxes, also known as operant chambers, and these types of approaches have been used across a variety of species, notably rodents, pigeons, nonhuman primates, and humans. In such chambers are some type of response device, speakers, and/or lights for presentation of auditory or visual stimuli, respectively, as well as some type of reinforcement delivery device. An operant chamber configured for a rat is shown in Fig. B-6. Discrimination paradigms are among the simplest measures of learning in operant chambers. In such procedures, a response is reinforced in the presence of one stimulus, the "correct" stimulus, but not in the presence of another stimulus. Accuracy is defined as the percentage of the total responses that occur in association with the correct stimulus, and learning can be assessed as the number of experimental sessions required by the subject to achieve a criterion level of accuracy.

One distinct advantage of discrimination paradigms in the operant chamber is that behavior can occur at any time (i.e., the frequency with which it occurs is not constrained by trials as necessitated by the requirement of moving the animal from the goal box back to the start box in most maze-based methods). When responding can occur at any time, the rate or frequency of responding over time can be measured and used to gauge the possibility that motor deficits or motivational insufficiencies may contribute to any observed changes in learning accuracy. A decrease or slow-down in rate of responding would suggest such possibilities.

Another advantage of operant chamber-based procedures is the enormous flexibility they provide for behavioral assessment and the ease with which they can be carried out in these devices. For example, conditional discrimination problems, which are more difficult discrimination tasks, can be easily implemented in operant chambers. Matching to sample is one such method. In this task, the subject first makes a designated response to indicate that it is attending to the task. Subsequently, a sample stimulus is presented for a short period of time. This is followed, after an interval of time, by the presentation of two or more stimuli, and reinforcement is contingent upon a response to the stimulus that matches the sample stimulus. The accuracy and speed with which the subject learns to match the sample stimulus is, of course, the measure of learning. Such tasks can be used with different species by simply increasing or decreasing the number of choice stimuli or the similarity of the stimuli appropriately. Because the procedure includes an initiating response on the part of the subject, a measure of rate of responding is possible, again providing information on potential motoric or motivational contributions to any deficits observed in matching accuracy.

Even the more complex matching to sample discrimination problems are eventually mastered by the subject, in which case discrimination reversals may be implemented in which the stimuli associated with reinforcement and with nonreinforcement are repeatedly switched. Eventually, however, the subjects will learn the reversal concept as well, such that they come to solve each reversal problem with maximal efficiency; that is, on the basis of only one or two responses (e.g., which stimulus is correct today?).

FIGURE B-6.   *An operant chamber or Skinner box for a rodent. The front wall (right) shows two response levers, below which is situated a food pellet trough. Food pellets are dispensed from a feeder located behind the front wall into the pellet trough. Above and to the left of the levers are two keys which can be illuminated with various colors and used for external or environmental stimuli. Above and to the right of the rightmost lever is a speaker through which auditory stimuli can be projected and used as external (environmental) stimuli (reproduced with permission from Reynolds, G. S., A Primer of Operant Conditioning, Scott Foresman, Glenview, IL, 1968).*

The most advanced method for the assessment of learning is the repeated learning paradigm, also called repeated acquisition, sequence acquisition, or response sequence learning. It specifically addresses the limitations discussed previously. This method actually originated for the measurement of learning in human subjects and has since been adapted for a variety of species. In repeated learning, the subject must make a sequence of responses for reinforcement, and the correct sequence changes with each successive experimental test session. Because the procedure thus requires subjects to learn a new sequence of responses each day, learning can be measured repeatedly across time. A high rate of errors is typically evident during the early part of each test session, as the subject begins to learn the correct sequence for the specific session. The error rate gradually declines as the session progresses, and reinforcers for completing the correct sequence of responses occur at an increasing rate. The ability to measure learning repeatedly across time with this task provides the basis for the measurement of the time course of a chemical's effects (i.e., the time to onset of any learning disabilities and their potential reversibility). This is a particular advantage given that many of the chemicals being developed are intended for chronic use.

The control for changes in sensory, motor, motivational, or other nonspecific behavioral changes as potential contributors to apparent chemical-induced learning impairments comes when the repeated acquisition task is run in conjunction with a "performance" task in what is known as a multiple schedule format. The performance component also requires the subject to emit a sequence of responses for reward, but in this case, the sequence of responses stays constant across time. Thus, the subject simply performs an already learned response. In the multiple schedule format, the repeated learning and performance components are presented alternately during the course of the experimental session, with a transition between them occurring either on the basis of time or on the number of reinforcers the subject has earned (e.g., after 15 min or 30 food deliveries switch from repeated learning to performance). Thus, during some portions of the test session, the subject is responding on the repeated acquisition task, while at other times during the session, the performance baseline is operative. Typically, different environmental stimuli, such as different colored lights, are used to indicate to the subject whether the performance or the repeated learning component is in effect.

Both the repeated learning and the performance tasks require intact motor and sensory capabilities, as well as appropriately motivated subjects. Learning per se is only required during the repeated acquisition task; the performance task simply requires completion of an already learned response sequence. Thus, if a toxicant or treatment has selective effects on learning per se, impairments in accuracy should only be evident during the repeated learning components of the session. If these changes arise, however, as a result of nonspecific behavioral changes (i.e., from sensory, motor, or motivational impairments), then accuracy impairments would be expected in both the repeated learning and performance components of the session since both require these behavioral capabilities. The elegance of this technique derives not only from its ability to distinguish learning effects from other types of behavioral changes but also from its ability to do so in the same subject during the same test session.

Behavior of a normal rat under these conditions is depicted in the top of Fig. B-7. In this diagram, the top tracing shows correct responses, which cumulate vertically; time is represented horizontally. The *P* indicates the performance components of the session, whereas the *A* indicates the repeated learning components. This 1-hr behavioral test session began with a performance component and was followed by the repeated learning component, once again by the perfor-

mance component, and finally by the repeated learning component. Illumination of lights in the operant chamber signaled to the subject that the performance component was operative, while turning out the lights signaled the repeated learning component was in effect. Each short pip mark in the top tracing indicates where the rat earned a food delivery for correctly completing the sequence of three responses required by the schedule. The bottom tracing shows the concurrent errors that occurred.

As Fig. B-7 shows, this well-trained rat exhibited a relatively high level of accuracy during the first performance component, earning a steady rate of food rewards and making few errors. The switch to the repeated learning component is accompanied by a dramatic increase in errors and a decline in the number of food rewards earned, as the subject begins to learn the correct sequence of three responses for this specific session. Behavior during the second performance component is again composed of a steady rate of food rewards and the occurrence of relatively few errors. The second presentation of the repeated learning component is marked by both a gradual increase in the rate at which food rewards were earned and a decrease in the number of errors relative to levels occurring in the first presentation of the repeated learning component, consistent with a gradual learning of the correct sequence for this session.

The bottom set of tracings shows behavior under the same conditions for a rat that has been exposed from weaning to a relatively low level of lead in drinking water. It shows, in a rather dramatic fashion, a selective effect of lead on learning processes per se, as distinct from nonspecific behavioral changes. Specifically, behavior during both presentations of the performance component is unimpaired in that a substantial rate of food deliveries and a minimal rate of errors is evidenced. In contrast, there is no evidence of learning during either presentation of the repeated learning component of the schedule in that virtually no food deliveries were obtained and a very high rate of errors was sustained. In fact, the rat continued to make errors throughout the entire time in the repeated learning components. Thus, in this case, the effects of lead on accuracy were restricted to the repeated learning component of the schedule. These impairments could not have been due to deficits in motor or sensory function, or inappropriate motivation, since behavior in the

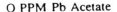

O PPM Pb Acetate

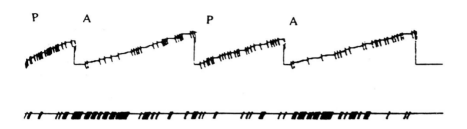

250 PPM Pb Acetate

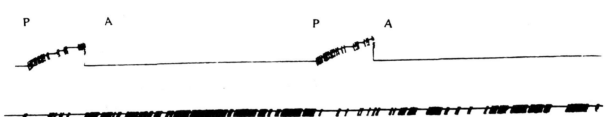

FIGURE B-7. *Behavior of a control rat (top) and a rat exposed to 250 ppm lead acetate in drinking water from weaning (bottom) on a multiple repeated acquisition (A; repeated learning) and performance (P) schedule of reinforcement. The top line of each record shows correct responses cumulating vertically with pips indicating food delivery for the completion of the correct sequence of responses; the bottom line shows errors. Time is represented horizontally (reproduced with permission from Cohn and Cory-Slechta, unpublished data).*

performance components, which also required such functions, was perfectly normal.

Although an effect of a toxicant on behavior during the repeated learning but not during the performance component of a multiple schedule is strong evidence of a chemical's selective effects on cognitive functions, there are other factors that should be taken into consideration. Some investigators subscribe to the idea that a selective effect of a chemical on learning means that its effects should be evident across a variety of learning tasks. While this notion has some validity, it should not be considered a necessary condition since, as has already been described, all learning paradigms are not equal. The extent to which different learning tasks selectively measure learning per se, as distinct from sensory, motor, or motivational influences, clearly differs, as does the possible "contamination" of the learning measure by changes in other behavioral properties. This is not to diminish the importance of these other types of behavioral effects, be they sensory or motor, for example, since such processes are clearly essential for integrated behavioral function, including cognitive functioning. Another important consideration is that the ability to detect effects of a chemical upon learning may depend to a large extent on the degree of task difficulty. It is well established that learning tasks that are relatively easy (i.e., those resulting in relatively high levels of accuracy) will be less sensitive to disruption either by drugs or by toxicants than are tasks of greater difficulty.

Memory

Memory, or remembering, is behavioral recall (i.e., the preservation of learned behavior over time). A distinction is often made between what is referred to as short-term or working memory, occurring over relatively short delay periods, and long-term or reference memory, considered more permanent memory. Obviously, the temporal parameters associated with what is designated as short- and long-term memory are species dependent.

The measurement of memory is typically based on the persistence of a previously learned response following some time delay; differences in recall accuracy are compared before and after delay intervals. Typically, the longer the delay, the greater the decrease in accuracy. An impairment of memory by a chemical accelerates the rate at which accuracy decreases with increasing delay values.

Both simple and more advanced techniques are available to evaluate memory. Again, however, many of the ostensibly simple tasks cannot differentiate memory deficits per se from deficits produced by changes in other behavioral functions, be they motor or sensory functions, or in level of motivation. For example, an inability to execute the response as efficiently (motor impairment) may in essence mean that the delay interval for the subject is functionally longer, thus indirectly impairing accuracy. Alternatively, a treatment which somehow increased the speed of responding could cause the subject to respond before adequately evaluating stimulus options, and thus decrease accuracy independently of a real change in remembering. Here, again, the more advanced methods include the capabilities for differentiating real effects of a chemical upon remembering from those caused by other behavioral consequences of the exposure.

One widely used simple measure of memory is passive avoidance. In this task, the subject, most often a rodent, is placed in a chamber that has two quite distinct compartments. The subject receives a shock in the compartment it prefers (spends most time in), engendering an association between the shock and the distinctive characteristics of that compartment. At some later time (i.e., after some delay interval), the subject is placed back into this two-compartment chamber, and memory is evaluated on the basis of the time (referred to as latency) that elapses before the subject steps back into the side of the chamber in which it previously received shock. The contention is that the longer the subject waits to enter that compartment, the better it remembers the shock it received there.

While changes in latency on this task are produced by a variety of drugs and chemical treatments, the interpretation of these changes can be problematic. If, for example, the chemical causes hyperactivity, the subject might reenter the shocked compartment sooner even if it does remember its association with shock. If the treatment disrupts sensory capabilities, altering perceived distinctions between the compartments, this too may result in a more rapid reentry into the compartment in which the subject had previously been shocked. If the administration of a chemical causes a sedative effect in the subject, rendering it less mobile, the time to reenter the shocked compartment may be increased relative to that seen in nontreated controls, but this would not be considered facilitation of memory. Again, such possible alternative interpretations must necessarily be worked out in additional experiments or with additional manipulations.

The more advanced procedures for memory evaluation do require more extensive training of the subject, but they also control for some of the possible confounds mentioned previously. There are two general types of more advanced procedures for the assessment of memory. One uses the previous responding of the subject as the event to be remembered, such as in the delayed alternation paradigm. In this procedure, the subject has access to at least two response manipulanda and is required to alternate its responses on the two for reward after some delay interval ends. That is, a response on manipulanda A initiates a delay interval, after which a response on B produces reward. This event initiates another delay interval, after which a return to manipulanda A produces reward. Typically a series of delay intervals are tested in each session, with the length of the delay interval varied randomly across the trials of a session and the specific lengths of the delays appropriate to species. Responses during the delay start the delay over again, thus increasing the time to reward. On this task, then, the subject has to remember which response manipulanda it responded on before the delay interval started in order to respond correctly after the delay. Typical behavior observed under these conditions is a decrease in remembering (accuracy) as the length of the delay interval increases. A chemically induced impairment of memory would then be manifest as a more pronounced decrease in accuracy as delay length increases than is observed under nontreatment (control) conditions.

Critical to the interpretation of any memory-related deficits with the delayed alternation task is the inclusion of a zero-second or no-delay condition. The no-delay condition requires no memory, as there is no delay. Therefore, if a treatment is impairing accuracy under the no-delay condition as well as at the various delay intervals, it is likely that the effects are due to changes in behavioral processes other than remembering. The pattern of change consistent with a selective memory impairment of a toxicant, then, is one composed of no change in accuracy at the zero-second delay but a more pronounced decrease in accuracy with increasing delay values relative to nontreated control subjects.

An example of an apparently selective impairment of memory independently of changes in other behavioral processes is shown in Fig. B-8. As can be seen, the accuracy levels of a group of nontreated normal rats (solid circles) declines as the delay value increases, as expected. Corresponding data for a group of rats treated with the organic metal, trimethyltin, are shown in the open circles. In this group, accuracy was unaffected at the zero-second delay but decreased more rapidly than did that of normal rats as delay value lengthened.

Other methods for measurement of memory function rely on explicit discrimination tasks. The matching to sample task described earlier is one example. In this paradigm, a sample stimulus is presented briefly to the subject. The subject must then pick the sample stimulus when subsequently presented with multiple stimulus options (i.e., the subject must match the sample). When delay intervals are imposed between the presentation of the sample stimulus and the subsequent presentation of multiple stimuli, the task becomes a memory task. In this case, the subject must remember the sample stimulus in order to perform correctly. As in the delayed alternation procedure, delay intervals of various lengths are used, including the no-delay or zero-delay condition, and a delay function similar to that shown in Fig. B-8 is expected. Many of the caveats mentioned with respect to interpreting memory effects in the delayed alternation task likewise apply to the delayed matching to sample paradigm. Separation of a chemical effect arising directly from changes in memory processes rather than from changes in motor, sensory, or motivational functions depends on the inclusion of a no-delay condition. Furthermore, as with learning paradigms, the contention that if a chemical produces a true memory deficit it will be observed across different memory tasks must be tempered by the fact that not all memory paradigms produce an equally selective measure of memory.

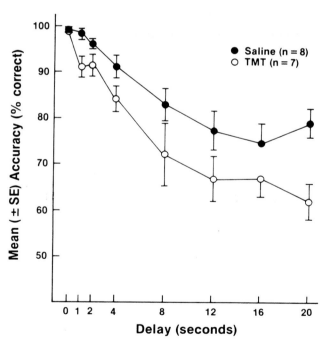

FIGURE B-8. *Effects of 7 mg/kg trimethyltin on delayed alternation performance. Lower accuracy values were evident in TMT-treated rats at all delay values, but no impairment was seen in the zero-second delay condition, consistent with a specific effect on memory function (reproduced with permission from Bushnell, P. J., Effects of delay, intertrial interval, delay behavior and trimethyltin on spatial delayed response in rats, Neurotoxicol. Teratol. 10, 237–244, 1988).*

Schedule-Controlled Operant Behavior

Learned voluntary behavior is a function of the consequences that follow it. If a response is followed by a reinforcing stimulus, the rate of that response subsequently increases; if followed by a punishing stimulus, or by the absence of a reinforcing stimulus, the rate of responding subsequently decreases. In addition to determining the subsequent frequency of that response, these consequence stimuli will also determine the intensity and temporal pattern with which that response will be emitted in the future.

In the real world, consequence stimuli do not necessarily follow every occurrence of the response. In fact, typically, consequences follow the response on an intermittent basis. Paychecks, for example, are typically distributed on a weekly, biweekly, or even monthly basis, not after each instance of work-related activity that occurs. The pianist plays the entire piece of music before the audience applauds. This strategy of intermittent reinforcement of responding actually provides greater behavioral efficiency and economy as well as greater response strength and persistence than does continuous reinforcement. A response that has been reinforced after every occurrence declines much more rapidly when reinforcement is withheld (extinction) than does one that has been reinforced on an intermittent schedule.

The term schedule of reinforcement refers to the nature of the rules governing the allocation of consequences for a particular response. Behavioral performance controlled by a schedule of reinforcement is referred to as schedule-controlled operant behavior. These schedules of reinforcement are critical because they govern the rate and pattern of responding in time which underlie other behavioral functions. For example, the rate of learning may well be influenced by the underlying schedule of reinforcement. If reinforcement of the correct response during a learning task is too infrequent, the task may not be adequately learned or not learned at all. Likewise, remembering that response, as in a memory task, may depend on the extent to which it was sufficiently reinforced to begin with.

Consequence stimuli can occur on the basis of time elapsing or on the basis of the number of responses that have occurred or both. In the human environment, schedules of reinforcement exhibit a remarkable complexity. For the purposes of understanding how these various reinforcement schedules or payoff schemes control the frequency and the pattern of behavior in time, simpler versions were initially studied in a laboratory context. As the understanding of simple reinforcement schedules evolved, increasingly complex schedules that more closely mimicked the human environment were elaborated and examined in laboratory experiments.

One of the important aspects of schedule-controlled behavior that deserves note is the remarkable similarity of behavior patterns generated by these schedules across a wide variety of species, even when type of response and type of consequence stimuli differ—a phenomenon of obvious importance for the issue of cross-species extrapolation because it shows the similarity and contiguity of such behavioral processes across species.

Simple Schedules of Reinforcement

There are four simple schedules of reinforcement: the fixed interval (FI) and the variable interval (VI), both of which are temporally based reinforcement schedules, and the fixed ratio (FR) and the variable ratio (VR) schedules, both of which are response-based schedules. The FI and the VI schedules both stipulate that a certain amount of time must elapse from the occurrence of a previously reinforced response before a response will again produce reinforcement. On the FI schedule, that time interval remains constant and the parameter value of the schedule indicates the length of that temporal interval (e.g., FI 1 min means that the first response occurring at least 1 min after the preceding reinforced response will result in reinforcement). On the VI schedule, the length of the interval varies from one interval to the next and the parameter value of the schedule indicates the average of the different interval lengths. For example, on a VI 1 min schedule, the average time between reinforcement opportunities is 1 min, but each interval may be either longer or shorter. Responses during the interval have no specific consequence attached to them on either the FI or VI schedules.

Because of the differences in the way in which they schedule reinforcement, the FI and VI control quite different rates of responding (responses per unit time) and patterns of responding, as can be seen in Fig. B-9. The FI schedule generates a characteristic "scallop" pattern of responding, which includes little or no responding (i.e., pausing) immediately after reinforcement (indicated by the short pip marks), followed by a gradual increase in the rate of responding as the time of reinforcement availability again approaches. In the human environment, studying for an exam has features that are characteristic of FI performance: Little or no responding early in the semester but a gradual increase as the time of the exam approaches. While one might expect that the performance under such conditions would be characterized by a single response as soon as the interval ends, such a pattern would require the subject to have perfect timing capabilities. Responding

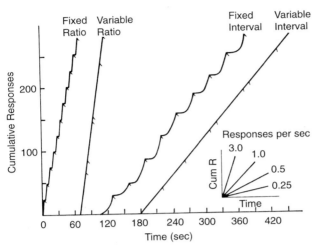

FIGURE B-9. Schematic cumulative records of performance on the fixed ratio (FR), variable ratio (VR), fixed interval (FI), and variable interval (VI) schedules of reinforcement. Responses are cumulated vertically over time. Each downward deflection of the pen represents reinforcement delivery; horizontal lines indicate pausing (reproduced with permission from Seiden, L. S., and Dykstra, L. A., Psychopharmacology: A Biochemical and Behavioral Approach, Van Nostrand Reinhold, New York, 1977).

at a very rapid rate as the end of the interval approaches ensures that reinforcement delivery will occur with minimal delay as soon as it is available.

The pattern of responding on the VI schedule differs from that on the FI (Fig. B-9) in that no pausing occurs after reinforcement delivery. Instead, the subject continues to respond at a steady and relatively uniform rate over time. The absence of pausing on the VI schedule is thought to reflect the lack of predictability of reinforcement. On the VI schedule, reinforcement may be available immediately after a previous reinforcement delivery since the interval length varies. Thus, pausing after reinforcement could result in a reduction in the rate or number of reinforcement deliveries. One example of VI-maintained behavior sometimes cited is that of getting a busy signal when calling someone on the telephone. The caller continues to redial and is eventually reinforced by a ringing sound on the other end. The persistent redialing reflects the variable length or interval of telephone conversations and, therefore, the unpredictability of when the line will no longer be busy.

In the other two simple reinforcement schedules, reinforcement availability is based on the number of occurrences of the designated response. On a FR schedule, the completion of the number of responses specified by the schedule parameter value is required for each reinforcement delivery. An FR 100 schedule, then, requires 100 occurrences of the designated response for reinforcement delivery. The classic examples of FR schedules are the piecework systems that operated in factories early in U.S. history, where workers were paid for each piece or unit they produced. The FR schedule generates its own characteristic behavior pattern which consists of a pause or period of no responding after each reinforcement delivery, followed by an abrupt transition to a very rapid rate of responding—a pattern known as "break and run" and shown in Fig. B-9.

A VR schedule also requires the occurrence of a designated number of responses for reinforcement delivery, but the response requirement varies from one reinforcement delivery to the next in an unpredictable fashion. The parameter value of the schedule indicates the average response requirement. Thus, on a VR 100 schedule, the average number of responses required for reinforcement is 100, but the actual number varies from one reinforcement delivery to the next. Perhaps the most obvious example of behavior maintained by a VR schedule is that of gambling. A slot machine may pay off on the average once every 100 plays, but the number of plays between payoffs varies in an unpredictably way; thus, one play that results in a payoff may follow immediately after a preceding payoff or may follow only after a large number of subsequent plays. The VR schedule maintains the highest rates of responding of the four simple schedules (Fig. B-9). In essence, it is characterized by a continuous high rate of responding without pausing after reinforcement deliveries.

Like the VI, the pattern of responding on the VR schedule reflects the lack of predictability of reinforcement availability. Since reinforcement availability may always be imminent, pausing would delay reinforcement. The high rates characteristic of VR and FR schedules are thought to be due to the ratio basis of reinforcement in that the faster the response requirement is completed, the faster reinforcement is available. Increases in rates of responding on interval-based schedules such as the FI and VI cannot accelerate the availability of reinforcement; one must still wait for the time interval to end.

Complex Schedules of Reinforcement

As mentioned previously, the complexity of reinforcement schedules encountered in the human environment is much greater than those embodied in the simple schedules studied in the laboratory. Combinations and variants of the simple schedules of reinforcement produce greater approximations of this complexity. One such example is a multiple schedule of reinforcement, in which component schedules alternate over the course of a behavioral test session. On a multiple FI–FR schedule, for example, the session would begin with an FI schedule in effect and would be indicated to the subject by some explicit stimulus (e.g., illumination of a red light). After some specified period of time elapsed or after the delivery of designated number of reinforcers on the FI schedule, the red light would change to a green light, and the schedule would switch to an FR. The FR schedule component would then remain in effect until a designated time had elapsed or a specified number of reinforcers had been delivered, and would be followed by a switch back to the FI component, and so on. After training on this schedule, patterns of behavior characteristic of each schedule component emerge; thus, during the FI component, a scalloped pattern of responding is maintained, whereas during the FR component break-and-run performance is exhibited. In addition, after experience on the schedule, the colored light stimuli associated with each schedule component come to exert strong control over behavior, such that performance appropriate to the schedule occurs immediately upon switching the color of the light. That is, these stimuli serve as discriminative stimuli signaling the schedule in effect.

This arrangement allows the experimenter to measure two very different types of schedule-controlled performances in the same subject during the same test session, making it a highly efficient experimental paradigm. This permits a determination as to whether a chemical may have selective effects on certain schedules (e.g., change FI performance without affecting the FR). If the compound being evaluated affects the control of the stimulus lights over responding, it might be manifest as a delay in transition to schedule-appropriate behavior whenever the light colors switched.

A mixed schedule of reinforcement is identical to a multiple schedule of reinforcement, except that there are no external stimuli provided to the subject to indicate that the operative reinforcement schedule has switched. Thus, the only indication to the subject as to "what pays off" is the feedback it receives from its own behavior. This minimizes the extent of stimulus control over behavior relative to that of a comparable multiple schedule of reinforcement.

A chained schedule of reinforcement, like a multiple schedule of reinforcement, also has different external stimuli associated with each component of the schedule, but it requires the completion of a sequence of components for reinforcement delivery. Thus, on a chained FI–FR schedule, a red light may signal that the FI component is in effect. Completion of the FI with the first response after the interval ends produces the external stimulus (e.g., green light) associated with the FR component. Completion of the response requirement during the FR component then produces reinforcement, and the chain subsequently begins over—a course which continues throughout the behavioral session. A tandem schedule is identical to a chained schedule, but like the mixed schedule, it provides no external stimuli to signal which component schedule is in effect.

The schedule which probably most resembles those operative in the human environment is known as a concurrent schedule of reinforcement. In the real world, we are routinely faced with a multitude of simultaneously operative schedule options with various schedule conditions and consequences, and we must make choices among them. The foraging (food seeking) environment of many species likewise provides such concurrent options with differential probabilities of reinforcement among which species must make choices. Concurrent schedules provide an experimental analog of this facet of the environment and require the subject to make choices among component reinforcement schedules and reinforcers. For example, in an operant chamber such as shown in Fig. B-6, different response manipulanda might be associated with different but simultaneously available reinforcement schedule options, perhaps associated with different reinforcing events as well. In some cases, once the subject chooses one option, the alternative schedule options are no longer available for some period of time. Others allow subjects to switch back and forth between schedule options. These types of schedules allow experimenters to ask questions about how much behavior the subject is willing to emit for specified reinforcers, preferences for reinforcers and response patterns, relative magnitude of reinforcement and allocation of behavior depending on effort and reinforcement availability.

Measurement of Schedule-Controlled Behavior

The universal measure of schedule-controlled behavior is the rate of responding, which is simply the total number of responses divided by total time. While this is a useful measure of behavior, it provides no indication of other aspects of schedule-controlled behavior, such as the extent of pausing or the patterns of behavior over time. For such purposes, a more fine-grained analysis or microanalysis of performance must be undertaken.

One such measure, applicable to both FR and FI schedules, is postreinforcement pause (PRP) time, which is simply measured as the time from reinforcement delivery until the first response occurs in the next interval (FI) or ratio (FR). For the FI schedule, one may be interested in the extent to which the scalloped pattern of performance occurs as an indication of the extent to which responding is controlled by the contingencies operative on the schedule. For this, one of two measures is utilized: the index of curvature or the quarter life. Index of curvature simply utilizes a mathematical formula to indicate how the observed scallop deviates from a straight line that would be generated by a constant rate of responding throughout the interval. Quarter life measures the time it takes for the first 25% of responses in the interval to occur.

Another measure of schedule-controlled behavior is that provided by the distribution of the times between successive responses or interresponse times (IRTs). These can be generated as a frequency distribution and have been shown to be important targets of chemical exposure. For example, lead exposure appears to affect primarily the very short IRTs on FI schedules. Many different drugs from a variety of different classes have been shown to increase the frequency of long IRTs and to decrease the frequency of short IRTs on an FI schedule—a phenomenon known as "rate-dependency" and which results in a more uniform and less scalloped pattern of responding. Rates of responding can also be calculated on schedules of reinforcement after the PRP or the IRTs longer than some designated

time (pauses) have been subtracted out. This results in a "truer" rate of responding and is known as running rate.

Behavioral Teratology

Behavioral teratology, or neurobehavioral teratology, is often referred to separately from behavioral toxicology. Behavioral teratology focuses on the behavioral impact of toxic exposures occurring prenatally or during early development. In some cases, these studies may only track the consequences of chemical exposures into early postnatal life, but in others effects may be studied well into the juvenile and even adult stages of the life cycle. Because the possibility has been raised that such exposures may accelerate the processes of aging, some studies are now beginning to follow subjects throughout the life span. Behavioral teratology studies typically include a series of tasks designed to evaluate multiple behavioral functions. Consequently, such experiments may include assessment of the development of various reflexes and developmental landmarks (e.g., eye opening), performance on a functional battery (FOB), motor activity, sensory capabilities, learning, and even schedule-controlled operant behavior. In addition, some such experiments may include evaluation of behaviors deemed "species specific" (i.e., behaviors that are innate and unique to that species), such as the ontogeny of aggression, play, or vocalization in rodents.

In cases in which outcome is followed through maturity, many of the behavioral paradigms that have already been described are utilized. Assessment of behavioral changes early in life, however, may require modification of such procedures and even the development of specialized behavioral preparations. One example of such a specialized preparation which concurrently measures sensory and motor capabilities (though not independently) is that known as "homing behavior," a behavior utilized by rodent pups to locate the nest should they wander. In this procedure, a rat pup is placed in the center of a rectangular apparatus, one side of which contains clean bedding material, whereas the other side contains bedding material from the home nest with its scent familiar to the pup. The time taken for the pup to orient to or to reach the home cage bedding is then measured. Such a test is deemed apical because it requires the integration of both motor and sensory capabilities.

There are certain issues uniquely related to behavioral teratology studies that require special consideration. One is that of toxicant effects on the dam (mother). Since the behavior of the dam may ultimately influence behavior and development of the offspring, great care must be taken to determine whether any observed effects of a chemical in the offspring are direct effects of the toxicant itself or whether they arise indirectly as a result of the effect of the compound on the dam's behavior. This is typically done by using a variety of fostering procedures. A cross-fostering procedure distributes the pups of treated dams to dams who are treatment free, in which case there should be no chemical-induced changes in maternal behavior.

There are also issues related to statistical analyses of the data that are unique to behavioral teratology. The offspring of a given litter are not considered as individual subjects since, as members of a litter, they have all experienced factors of the fetal environment which may be unique to their dam. This means that the total number of subjects in a treatment group is really equivalent to the total number of litters represented in that group, a factor which can change the degrees of freedom in the statistical analyses.

Human Testing

Behavioral toxicological studies in humans have focused primarily on adults occupationally exposed to chemicals and children exposed to toxicants environmentally. There is frequently a good deal of overlap in the specific behavioral functions evaluated in each case, although the tests utilized must be age appropriate. However, many of the studies of children include measurements of developmental profiles and landmarks which are not relevant to studies of occupationally exposed adults. In adults, in contrast, assessment of exposure-related symptomatology is possible. Both types of studies also generally assess a broad variety of behavioral functions and may include tests of motor function, sensory capabilities, complex or cognitive behaviors, attentional processes, and vigilance, usually in the context of a standardized test or test battery. In the past, many such functions would be evaluated as part of a neurological or clinical examination. However, it has become increasingly clear that such examinations,

meant to diagnose disease or brain damage, are neither sufficiently sensitive nor quantitative for purposes of detecting subtle effects of toxicants and ultimately for setting standards of exposure.

The test batteries commonly used in human studies have come primarily from the field of clinical neuropsychology, in which human testing has predominated. Behavioral measures such as are utilized in experimental animal studies are rarely included in human studies, although this approach is likely to increase in the future given the relatively broad scope of functions tapped by test batteries and the difficulties encountered in subsequently defining the specific behavioral process affected by a chemical. In part this overall emphasis on broad testing of behavioral functions has been driven by the lack of any information on the behavioral properties of many of these chemicals as well as by the need to establish dose–effect and dose–response relationships.

Many of the same issues raised with respect to experimental animal studies also apply to human testing and to the choices of particular tests to be utilized. There are numerous tests that can be utilized for measurement of behavioral functions in humans, and questions remain as to the correct choice. One consideration related to the various tests is deemed validity and refers to the degree to which the test actually measures the behavioral function that it was designed to measure. For example, does a test of memory really evaluate memory function? In addition, how specifically does the test measure that function? The related issue was raised in experimental animal studies in which the possibility that changes in motor, sensory, motivational processes, etc. might contaminate a measure of memory function, and appropriate controls were included in the more advanced procedures to evaluate those possibilities.

Another important issue relates to the reliability of the test. That is, how reproducible or consistent are the test results across multiple administrations? Inadequate reliability almost guarantees that a subtle toxicant effect will not be detected against a background of scores of broad individual variability that will be present in any normal population. An issue that has not received adequate attention is the sensitivity of these tests to detect toxicant effects, a factor that is of particular importance if the test results are used in the context of setting exposure standards. If a particular test indicates effects of lead, for example, at a blood lead concentration of 40 μg/dl, one may wonder whether this represents the bottom limits of sensitivity of the test or the actual blood lead value at which such effects occur. In other words, could the test have detected effects at even lower levels of exposure if it had been more sensitive? A deficiency in test sensitivity could mean that exposure standards will be set at levels that are too high and will not protect the exposed populations.

A related question of relevance, particularly to tests of achievement such as the so-called intelligence tests, is standardization. This refers to the population from which the normative scores for the test were collected. This issue is often raised in the interpretation of intelligence tests for populations that are culturally and socially distinct from the populations of white middle-class English-speaking children from which normal scores for such tests have typically been derived.

Developmental Assessments

As mentioned previously, several unique considerations affect the assessment of toxicant-induced behavioral changes in children. One such consideration is the rapid development that children undergo from birth through even the preschool and early school stages. Moreover, this development is marked by wide individual differences in the rate at which it occurs and, for some facets of behavior, gender-related differences as well. An additional difficulty is that many of the behavioral processes that are of particular interest, such as complex cognitive behavior, are more difficult to evaluate at a young age. While it seems clear that children certainly have both learning and memory capabilities even from birth, assessment of such changes has typically relied on tests which may require language or motor skills well beyond the capabilities represented by these early stages of development.

Because of this rapid change in the behavioral repertoire over the course of early development, the tests that are utilized in studies of children tend to differ at different ages. One test frequently utilized in the first few days after birth is the Brazelton Neonatal Behavioral Assessment Scale, which is composed of two subscales. The first taps a range of behavioral items such as habituation and responsiveness to environmental stimuli. The second primarily measures a variety of unconditioned reflexes. While the Brazelton scale is obviously limited in the extent to which it can tap

cognitive functions, or define specific behavioral deficits, its utility in detecting drug-induced changes has been established.

A recently developed technique for infant assessment is embodied in the Fagan Test of Infant Intelligence, which assesses visual recognition memory. In this test, an infant faces a display with two screens. On one screen, a visual stimulus is presented for a specified period of time. Subsequently, that visual stimulus is projected on one screen and, at the same time, another visual stimulus is projected onto another screen. An observer records the amount of time the infant spends gazing at each screen. Normal infants look away from the visual stimulus which they have already seen and spend more time gazing at the novel stimulus, a trait which has been shown to correlate with higher scores later in development on the Stanford–Binet intelligence test.

A widely used test at a slightly later stage of development is the Bayley Scales of Infant Development, appropriate to children from 2 to 30 months of age. The test is composed of three subscales: motor, mental, and behavioral. Each is arranged with respect to chronological development. One of the advantages of this test is the ability to carry out repeated testing over the normed age range.

As children reach preschool and school age, the number of test choices available increases. For example, the McCarthy Scales of Children's Abilities provides an analog of an intelligence test score by combining the scores from its five subscales into a general cognitive index score. Its applicability extends from children aged 2.5 to 8.5 years. Like the Bayley Scales, it too allows for repeated measurement over time, which is a particular advantage for longitudinal studies; utilization of the same test instrument over time, given appropriate reliability of the instrument, provides greater assurance of the continuity and of the onset or disappearance of an effect than does the use of different instruments at different ages.

Various intelligence tests are available for preschool age children, such as the Weschler Preschool and Primary Scale of Intelligence (WPPSI). The advantage of this particular instrument is that it represents an extension of the well-standardized and widely used Weschler Intelligence Scale for Children (revised; WISC-R). The WISC-R is an intelligence test for children of 6 years of age or older; the WPPSI extends this age range to include children of ages 4–6.5 years. In addition, both rely on the same two subscales, verbal and performance, to measure a variety of behavioral functions, thus providing a type of continuity from the preschool to the school-aged child for repeated assessment of behavioral function.

One of the major concerns with developmental and intelligence tests such as the WISC-R and others is to be able to rule out contributions from the numerous sociodemographic and other variables known to covary with intelligence test score. Variables which may potentially modulate intelligence include birth weight, length of gestation, maternal age, birth order, parental education, parental IQ, socioeconomic status, and quality of the home environment. Appropriate statistical controls or subject matching must be undertaken to evaluate the contributions of these variables to outcome measures.

While these developmental and intelligence tests may clearly be important to the determination of the levels and conditions of exposures to a toxicant associated with adverse behavioral function, they are less useful, as noted previously, in providing a precise delineation of the behavioral functions actually affected by a chemical. Measures such as intelligence test scores are global measures in that they rely on the integration of all behavioral functions. Even performances on subscales of these tests are jointly dependent on integrative motor, sensory, and cognitive functions. Thus, even a preferential deficit on a verbal scale, which is clearly geared toward cognitive function, may not provide a precise understanding of the nature of the behavioral deficit.

To achieve a true understanding of the behavioral processes affected by a chemical will necessarily require direct measurement of those specific functions, much as is done in the experimental animal studies described previously. Some neuropsychologists have recognized this problem and have begun to employ measures of specific behavioral functions such as learning, memory, sustained attention, and abstract thinking in an attempt to determine the source of the deficits in global intelligence test scores produced by lead exposure. An alternative approach is to utilize many of the behavioral tasks already employed in experimental animal studies—tasks which are designed to evaluate specific functions and which have already been widely used across species, including humans, in other research contexts. The repeated learning paradigm actually originated in studies using human subjects and was later adapted for

nonhuman primates and rodents. Procedures such as delayed matching to sample and operant psychophysical procedures have also been used across species with appropriate parametric modifications. These types of paradigms may play a more significant role in future developmental studies of children because they provide direct and specific measures of behavioral functions that are more difficult to differentiate in standardized tests.

Adult Assessments

Assessments of behavioral toxicity in adults frequently occur in the context of occupational exposures to chemicals. Like studies carried out in school-aged children, these evaluations rely almost exclusively on standardized tests, including intelligence tests. They also tend to employ a broad variety of tests so that numerous behavioral functions can be tested, particularly when the effects of a toxicant are ill-defined. As such, the same considerations must be taken into account with respect to the choices of tests utilized. These include validity, reliability, and sensitivity, as well as standardization issues related to the population from which test norms were derived.

One distinction between many of the studies of behavioral toxicity in children and adults is that while the former have tended to be primarily longitudinal in nature, following the effects of toxicant exposures to children across the course of development, most of the occupational exposure studies are cross-sectional studies that encompass only a single time point of measurement of behavioral function. This no doubt reflects the added difficulties of carrying out studies in the workplace, where it may be more difficult to obtain appropriate amounts of the subjects' time for behavioral evaluation and where resistance to such experiments may be encountered from either the employer or the employee.

One of the most common inclusions in these test batteries that have been utilized in studies of occupational behavioral toxicity is the Weschler Adult Intelligence Scale (WAIS), which is actually a battery of tests subsumed under verbal and performance subscales which, combined, provide a full-scale intelligence test score. The series of verbal tests includes information (general information questions), comprehension (interpretation test), arithmetic, similarities (between nouns), digit span (repeating sequences of digits), and vocabulary. The performance tests include digit symbol (associating digits with symbols), picture completion, block design (duplicating block patterns), picture arrangement, and object assembly. Because of the obvious overlap of behavioral functions in some of these subtests, and the consequent global and nonspecific nature of any change detected in full-scale intelligence test scores, some investigators have opted to use only selected tests from the battery to provide a shortened version of the WAIS for occupational behavioral toxicity studies.

Two different test batteries, the World Health Organization (WHO) Neurobehavioral Core Test Battery and the Neurobehavioral Evaluation System (NES), are currently the most widely used test batteries in occupational behavioral toxicology studies. Both include components of the WAIS described previously in addition to other psychometric tests of behavioral function. The WHO Neurobehavioral Core Test Battery is a pencil and paper-administered test battery, whereas the NES is a computerized test battery that has been translated into several languages and in fact presents a more extensive set of tests than does the WHO in that it includes tests of psychomotor performance, cognition, memory and learning, and perceptual ability and affect.

Memory dysfunction has been a frequent complaint in populations of workers exposed to various neurotoxicants and is tapped by several different tests used in a human testing context. One of the most widely used for this purpose is the digit span that constitutes one of the WAIS subtests. As indicated earlier, this test requires the subject to recall a series of digits, and the length of the list is successively increased contingent upon the subject correctly recalling the members of the list. In some cases, words or letters are utilized instead of digits. As is the case with experimental animal studies, more complex versions of these tests have been devised and implemented. In procedures such as continuous recognition memory or memory scanning, subjects may be shown a list of digits or letters and then shown, after a delay interval, a longer list of various digits or letters and asked to recall those that were on the original list. Analogies to such tests are embodied in procedures such as the Benton Visual Recognition Test, which requires a subject to reproduce a drawing or geometrical design.

Paired-associates learning is also frequently used in a memory context in occupational exposure test batteries as well as in studies with children. In these paradigms, a list of paired words is read to the subject, who must then recall the second member of the pair when the first is read after a delay. The task can be made relatively simple by using pairs which have some type of obvious relationship or made more difficult by having pairs with no apparent relationship. The test can be used in a memory context by including a delay between the experimenter's reading of the list and the subject recalling the second member of each pair. In addition, the task can be used in a repeated learning context, much as the repeated acquisition paradigm described earlier, by using new lists of paired-associates after the subject masters the initial list. This particular approach has a long history of use with human subjects and has been found to be sensitive to toxicants such as lead.

Measures of vigilance, attention, or distractibility are also frequently included in assessments of occupational behavioral toxicology. These range from very simple procedures, such as reaction time, to more complex tasks, such as simulated cockpit or tracking tasks. Even reaction time can be varied from a very simple to a highly complex procedure. In a simple reaction time task, the subject is typically presented with some type of screen on which a single visual stimulus will appear at intermittent and unpredictable intervals. The subject must respond on the single response manipulanda as soon as the stimulus appears. Complex reaction time presents the subject with multiple stimuli as well as multiple response options. For example, there may be four different stimuli, each of which is presented at random and unpredictable intervals. The appropriate response depends on which of the stimuli is presented, and the subject is asked to respond on the appropriate manipulanda as quickly as possible after detecting the stimulus. Thus, the more complex reaction time task involves not only attending to the screen to detect the stimulus presentation but also making a decision as to the correct manipulanda and then executing the response. Obviously, the number of options can be modified to fit the experimental situation.

One of the important parameters of the reaction time task is the rate at which the attention of the subject deteriorates, such that reaction time is slowed down or even that the subject misses stimulus presentations entirely. This rate of deterioration of performance will depend on many factors, one of which is the rate at which stimuli are presented to the subject and another being the length of the session during which reaction time is measured. While one might intuitively think that the slower the rate of presentation, the more rapid the rate of deterioration of performance, in fact sometimes the opposite is true. A very rapid rate of presentation of stimuli can render the subject exhausted and less alert or less motivated. With respect to session length, one typically expects to see a gradual decrement in performance as the session progresses, such that an adequately long session must be implemented to catch this function. Finally, a critical variable in reaction time studies is the prominence of the stimuli used. In fact, this parameter can be manipulated to change the sensitivity and difficulty of the task.

Reaction time tasks are by no means limited to the presentation of discrete visual stimuli. Other variants have included those in which the subject must respond to a stimulus that is different from a continuously presented array of stimuli. The so-called clock test is one example. In this procedure, the subject is instructed to respond when the hand of a clock ticks off 2 seconds at once rather than the typical 1-second tick; the 2-second tick is an infrequent and unpredictable occurrence. In other situations, a continuous presentation of letters or numbers may be presented and the subject instructed to respond to one particular letter or number whenever it appears.

Pursuit and tracking tests represent even more complex versions of vigilance assays. In these kinds of tasks, subjects must continuously monitor a stimulus which drifts off a home position on the dial. The situation can be made quite complex, as in flight simulators in which there may be multiple dials which must be continuously monitored and returned to the home position, with drift occurring at varying rates on each dial across time. The various vigilance tasks described previously have a long experimental history and have been shown to be sensitive to a wide variety of influences, including fatigue and various drugs and chemical exposures.

The kinds of vigilance procedures described obviously require reasonably intact motor function and are often interpreted with that in mind. However, these techniques also depend on sensory processes. In fact, assuming intact motor functions, vigilance tasks such as those described can be adapted to provide some indication of sensory function changes by modifying the saliency (intensity) of the sensory stimuli used in the paradigm. However, more direct approaches to the

evaluation of sensory function following occupational or environmental exposures to toxicants are provided by the types of operant psychophysical procedures elaborated previously. In fact, psychophysical procedures were developed using human subjects and only later adapted for various species of experimental animals. The psychophysical procedures clearly provide more direct and straightforward assessments of sensory detection capabilities in the absence of confounding changes in a subject's motor capabilities or motivation to respond.

Assessments of motor function are often included in the neuropsychological test batteries utilized in occupational exposure studies. Typically, these tend to be relatively simple measures of motor capabilities, probably for two reasons. The first is that the inclusion of vigilance tasks such as those described previously depends on motor coordination in addition to sensory capabilities; therefore, toxicant-induced changes in such performances may already be indicative of motor impairment. This can then be pursued by inclusion of some additional and more direct assessments of motor function in the battery. The second reason relates to logistical reasons and practicalities. Test batteries such as the WHO Neurobehavioral Core Test Battery and the Neurobehavioral Evaluation System are typically taken to the site where measurements of subjects are to be made. Thus, portability is a major consideration, and more complex assessments of motor function would incur greater equipment needs. Since the purpose of these batteries is to screen for adverse effects, studies providing more precise delineations of affected functions can be pursued at a later time.

Simple tests of motor function utilized are generally those such as finger tapping in which subjects are asked to tap a key or a button at as rapid a rate as possible for a designated period of time. The subjects may be asked to carry out this task with the preferred hand as well as with the alternate hand. In some cases, toe tapping has been used in addition to finger tapping. Other batteries have relied on the tests of manual dexterity that are frequently used in screening prospective applicants for some types of factory work jobs. One of the most frequently used of such tests is the Santa Ana test, which requires the subject to remove pegs from a hole and to reinsert them into the hole after turning them 180°. The measurement of interest in this case is the number of pegs that are successfully rotated within the specified time interval. The Purdue Pegboard test is likewise used in this capacity. It requires the proper orientation and placement of pins in a series of holes. Such tests have indeed successfully defined subjects occupationally exposed to chemicals from those non-exposed.

One final common inclusion in many studies of occupational behavioral toxicology and in some test batteries is assessments of symptoms experienced by those exposed to chemicals. While this might be perceived as an ostensibly simple procedure, it entails numerous potential confounds. These evaluations are typically administered via questionnaires. Items for the questionnaire must be carefully constructed with respect to not only the choices of items but also the wording of the text and the manner in which the response is recorded. Clearly, the motivation of the subject in answering the questions must be considered. One problem can arise when the list of symptoms are only those that are associated with the toxicant of concern. It is necessary to include symptoms that are not associated with the particular toxicant under evaluation so that some assessment of the tendency of the subject to respond positively to all symptoms can be evaluated. Several such evaluations of subjective and mood states are available. The most widely used is the Profile of Mood States (POMS), which consists of 65 adjectives of various moods that the subject answers according to a 5-point rating scale. The POMS has been used extensively in the evaluation of the acute effects of central nervous system drugs and toxicants.

Further Reading

Annau, Z., and Eccles, C. U. (Eds.) (1986). *Neurobehavioral Toxicology.* Johns Hopkins Univ. Press, Baltimore.

Cory-Slechta, D. A. (1989). Behavioral measures of neurotoxicity. *Neurotoxicology* 10, 271–296.

Cory-Slechta, D. A. (1994). Neurotoxicant-induced changes in schedule-controlled behavior. In *Principles of Neurotoxicology* (L. W. Chang, Ed.), pp. 313–344. Dekker, New York.

Maurissen, J. P. J. (1979). Effects of toxicants on the somatosensory system. *Neurotoxicol. Teratol.* 1(Suppl.), 23–31.

Reynolds, G. S. (1968). *A Primer of Operant Conditioning.* Scott Foresman, Glenview, IL.

Russell, R. W., Flattau, P. E., and Pope, A. M. (Eds.) (1990). *Behavioral Measures of Neurotoxicity.* National Academy Press, Washington, DC.

Seiden, L. S., and Dykstra, L. A. (1977). *Psychopharmacology: A Biochemical and Behavioral Approach.* Van Nostrand Reinhold, New York.

Tanger, H. J., Vanwersch, R. A. P., and Wolthius, O. L. (1984). Automated quantitative analysis of coordinated locomotor behaviour in rats. *J. Neurosci. Methods* 10, 237–245.

Tilson, H. A., and Mitchell, C. L. (Eds.) (1992). *Neurotoxicology.* Raven Press, New York.

Weiss, B., and Cory-Slechta, D. A. (1994). Assessment of behavioral toxicity. In *Principles and Methods of Toxicology* (A. W. Hayes, Ed.), 3d ed., pp. 1091–1156. Raven Press, New York.

Weiss, B., and O'Donoghue, J. L. (Eds.) (1994). *Neurobehavioral Toxicity, Analysis and Interpretation.* Raven Press, New York.

Yanai, J. (Ed.) (1984). *Neurobehavioral Teratology.* Elsevier, Amsterdam.

Acknowledgments

Supported by NIEHS Grants ES05903, ES05017, and ES01247.

—*Deborah A. Cory-Slechta*

Related Topics

Indoor Air Pollution
Multiple Chemical Sensitivities
Neurotoxicology: Central and Peripheral
Occupational Toxicology
Pesticides
Psychological Indices of Toxicity
Sensory Organs
Sick Building Syndrome

Belladonna Alkaloids

- ◆ REPRESENTATIVE COMPOUNDS: Atropine (see separate entry); scopolamine; homatropine
- ◆ PHARMACEUTICAL CLASS: Naturally occurring antimuscarinic drugs
- ◆ CHEMICAL STRUCTURE:

Atropine   Scopolamine   Homatropine

Uses

Belladonna alkaloids are used in clinical medicine for their ability to block the effects of parasympathetic nerve stimulation.

Exposure Pathways

Ingestion is the most common route of both accidental and intentional exposure to belladonna alkaloids. They are available in oral and tincture forms.

Toxicokinetics

The belladonna alkaloids are absorbed rapidly from the gastrointestinal tract. They also enter the circulation when applied locally to the mucosal surfaces of the body. Absorption from intact skin is limited. Belladonna alkaloids are metabolized mainly in liver to glucuronide conjugates after metabolic hydroxylation of the aromatic ring. The volume of distribution for important belladonna alkaloids, atropine and scopolamine, is approximately 1.7 and 1.4 liter/kg, respectively. Elimination of belladonna alkaloids is rapid. About half of the atropine is excreted unchanged in urine. Traces of atropine are found in various secretions including breast milk. The elimination half-life for intravenously injected atropine ranges from 1.9 to 4.3 hr.

Mechanism of Toxicity

Toxic doses of belladonna alkaloids prominently lead to central excitement. The earliest symptoms of atropine toxicity are due to blocking of acetylcholine receptor sites on cells of organs innervated by the craniosacral division of the visceral effector nervous system. Scopalamine differs from atropine intoxication in that the cardiac rate is rarely increased and cerebral excitement is of short duration. Although the mechanism of action of these compounds is well studied the mechanism of toxicity is not entirely understood.

Human Toxicity

The deliberate or accidental ingestion of belladonna alkaloids is a major cause of toxicity in humans. The most dangerous and spectacular manifestation of poisoning arises from the intense excitation of the central nervous system (CNS). Infants and young children are especially susceptible to the toxic effects of atropinic drugs. In adults, delirium or toxic psychoses without undue peripheral manifestations have been reported after instillation of atropine eye drops. Transdermal preparation of scopolamine has been reported to cause toxic psychoses especially in children and in the elderly. Serious intoxication may occur in children who ingest berries or seeds containing belladonna alkaloids. In case of full-blown poisoning, the syndrome may last 48 hr or longer. Depression and circulatory collapse are evident only in cases of severe intoxication; the blood pressure declines, respiration becomes inadequate, and death due to respiratory failure may follow after a period of paralysis and coma.

Clinical Management

The diagnosis is suggested by the widespread paralysis of organs innervated by parasympathetic nerves. Intramuscular injection of physostigmine may be used for confirmation. If the typical salivation, sweating, and intestinal hyperactivity do not occur after physostigmine injection, intoxication with atropine or a related agent is almost certain. Measures to limit intestinal absorption should be initiated without delay if the poison has been taken orally. The most effective antagonist to the CNS manifestation is physostigmine salicylate in doses of 0.5–2.5 mg by any parenteral route. Since physostigmine is metabolized fast, repeated doses may be needed. If marked excitement is present and more specific treatment is not available, diazepam is the most suitable agent for sedation and for the control of convulsions. Large doses should be avoided. Artificial respiration may be necessary. Ice bags and alcohol sponges help to reduce fever, especially in children.

Animal Toxicity

Cats, dogs, and birds are sensitive to belladonna alkaloid toxicity; horses and oxen less so; and pigs, goats, and sheep are comparatively resistant to the alkaloids. Parenteral administration of lethal doses of atropine to young rabbits produces two distinctly different types of deaths. About half of the animals died promptly in a convulsive state perhaps comparable to the commonly encountered clinical syndrome of central excitement, but a smaller group suffered delayed deaths in about 2 weeks with endarteritis obliterans in the distal portion of the injected limb. Chronic belladonna alkaloid or atropine poisoning has evidently not been encountered as a clinical entity, but the parenteral administration of large doses of atropine (16 mg/kg daily) for periods of 1–3 weeks produces in young puppies a syndrome clinically similar to advanced fibrocystic disease of the pancreas.

—*Madhusudan G. Soni*

Related Topics

Cholinesterase Inhibition
Neurotoxicology: Central and Peripheral
Poisoning Emergencies in Humans (Under
    Physostigmine)

Benadryl

- ◆ CAS: 147-24-0
- ◆ SYNONYMS: Diphenhydramine; 2 diphenylmethoxy-*N*; *N*-dimethyl-ethanamine; BAX; benocten; dolestan; wehydryl
- ◆ PHARMACEUTICAL CLASS: Benadryl is a congener of H1-receptor antagonist type.

- ◆ CHEMICAL STRUCTURE:

$$C_6H_5\text{—}CHOCH_2CH_2N\underset{CH_3}{\overset{CH_3}{<}}$$
$$C_6H_5$$

Uses

Benadryl is used as an antihistaminic and motion sickness suppressor.

Exposure Pathways

Ingestion is the most common route of both accidental and intentional exposure to benadryl. It is available in oral, injectable, and topical forms.

Toxicokinetics

Benadryl is well absorbed from the gastrointestinal tract. Following oral administration, peak plasma concentrations are achieved in about 2 hr, with a fairly rapid drop toward normal levels in 6 hr. Benadryl is extensively metabolized in humans by two successive N-demethylations. The resulting primary amine is then further oxidized to a carboxylic acid. The metabolism of benadryl occurs principally in the liver although many other organs, including the lung and kidney, metabolize significant quantities. The drug is widely distributed throughout the body including the central nervous system (CNS). The highest concentrations were found in the lungs. The apparent volume of distribution has been found to range from 3.3 to 6.8 liter/kg with some ethnic variations. The major route of elimination is through urine. The metabolic products of benadryl are excreted in far greater concentration than unaltered benadryl. The plasma elimination half-life of benadryl is about 8 hr.

Mechanism of Toxicity

The majority of the toxic effects in human beings due to benadryl are probably unrelated to a specific histamine blocking action. The underlying mechanisms of toxic reactions of benadryl remains obscure.

Human Toxicity

The most common side effect in humans is sedation in varying degrees. Other untoward reactions referable to the CNS are dizziness, tinitus, fatigue, ataxia, and blurred vision. Anorexia and other gastrointestinal symptoms are not uncommon. These manifestations are generally noted when the plasma concentration reaches 30–40 ng/ml. When benadryl is ingested in an amount distinctly greater than the customary dose, any one of many toxic syndromes may result. In young children central stimulation, convulsions, and high fever appear to be the dominant signs, followed by profound cardiorespiratory depression and finally resulting in death due to vascular collapse. There is no evidence of cumulative toxic effects.

Clinical Management

In acute oral overdosage, gastric lavage is indicated immediately. Cautious use of short-acting barbiturates in repeated doses for the control of central nervous stimulation may be helpful. Stimulants, such as caffeine or amphetamine, may occasionally be useful in supportive treatment of CNS depression. Artificial respiration and oxygen therapy as necessary. A specific antidote is lacking; therefore, the therapy must currently be entirely symptomatic.

Animal Toxicity

Benadryl can produce convulsive or postconvulsive depressant deaths in laboratory animals like mice, rats, rabbits, guinea pigs, cats, and dogs. Chronic administration is usually well tolerated although some visceral changes are occasionally noted.

—*Madhusudan G. Soni*

Benomyl

- ◆ CAS: 17804-35-2
- ◆ SYNONYMS: Agrocit; Fundazole; benomyl 50W; Benlate; Benlate T-; methyl 1-(butylcarbamoyl)-2-benzimidazolylcarbamate
- ◆ CHEMICAL CLASS: Benzimidazole fungicide

◆ CHEMICAL STRUCTURE:

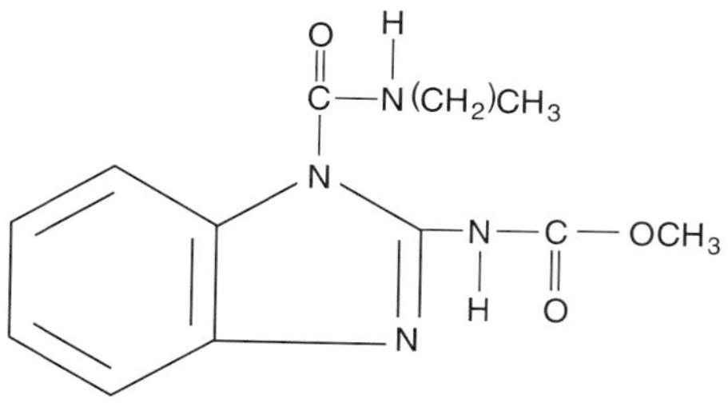

Uses

Benomyl is a protective and eradicant fungicide, effective against a wide range of fungi affecting fruits, nuts, vegetables, turf, and field crops.

Exposure Pathways

Benomyl is available as a powder (Benlate and Benlate T). Dermal and oral routes are the most common exposure pathways.

Toxicokinetics

Benomyl is poorly absorbed as it is degraded in the gastrointestinal tract. Blood levels after oral administration are only 1/10 of those found after intraperitoneal injection in rats. Benomyl is hydroxylated and/or methylated, then conjugated and promptly excreted in the urine. There is minimal or no tissue storage of benomyl or metabolites based on a 2-year feeding study in rats and dogs. In rats, the major metabolites of benomyl are converted to sulfate and/or glucuronide conjugates and excreted in urine (78%) and feces (8.7%). In mice, rabbits, and sheep, 44–71% of benomyl metabolites are found in urine and about 21–46% in feces.

Mechanism of Toxicity

Benomyl is a microtubule-disrupting agent in fungi. This agent may cause chromosomal aberrations (e.g., aneuploidy). There is very little evidence of benomyl toxicity in mammals, however. Benomyl itself does not have any direct effect on acetylcholinesterase. Under certain conditions, however, benomyl breaks down to produce carbendazim and butyl isocyanate, of which the isocyanate is an irreversible inhibitor of acetylcholinesterase with comparable potency to some active organophosphorous compounds.

Human Toxicity: Acute

Benomyl is a potential mild skin, eye, and respiratory tract irritant. Systemic poisoning is rare. Contact dermatitis has been reported in occupationally exposed workers. The skin lesions, which consisted of redness and edema, occurred on the back of hands, forearms, and in other places not covered by clothing. These lesions generally clear within 3 weeks. Hyperpigmentation and photosensitization have also been reported.

Human Toxicity: Chronic

No study has reported chronic toxicity of benomyl.

Clinical Management

Dermal decontamination should be accomplished by repeated washing with soap. Leather clothing can absorb benomyl; any contaminated leather clothing should therefore be discarded. Exposed eyes should be irrigated with copious amounts of room-temperature water for at least 15 min. Emesis can be induced in cases of recent ingestion. In such cases, ipecac can be used to induce emesis. Emesis is not encouraged if the patient is comatose or convulsing. Activated charcoal slurry with or without saline cathartic and sorbitol may be used.

Animal Toxicity

Benomyl has very low acute toxicity in laboratory animals. The oral LD_{50} value in rats is greater than 10 g/kg.

—*Sushmita M. Chanda*

Related Topics

Chromosomal Aberrations
Organophosphates
Pesticides

Benz[a]anthracene

- CAS: 56-55-3
- SYNONYMS: 1,2-Benzanthracene; 2,3-benzphenanthrene; 2,3-benzophenanthrene; tetraphene; naphthanthracene; benzanthrene; BA

- CHEMICAL CLASS: Polycyclic aromatic hydrocarbon
- CHEMICAL STRUCTURE:

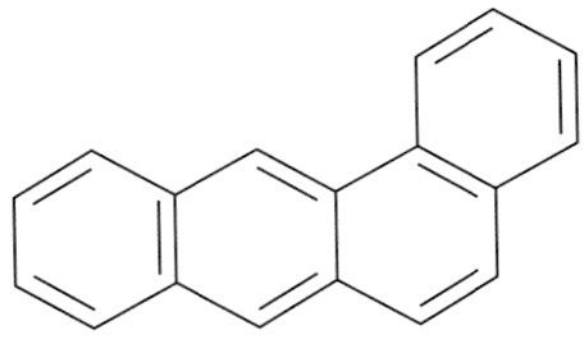

Uses

There is no commercial production or known use of this compound.

Exposure Pathways

Human exposure to benz[a]anthracene occurs primarily through smoking of tobacco, inhalation of polluted air, and by ingestion of food and water contaminated by combustion effluents.

Toxicokinetics

Like benzo[*a*]pyrene, benz[a]anthracene may cross the gastrointestinal lining, pulmonary endothelium, or percutaneous barriers. Benz[a]anthracene is biotransformed to five dihydrodiols and a number of phenolic metabolites by P450 mixed-function oxidases. Detectable levels of benz[a]anthracene can be observed in most internal organs from minutes to hours after administration. Regardless of route of administration, once metabolized, hepatobiliary excretion and elimination through feces is the major route.

Mechanism of Toxicity

Benz[*a*]anthracene has been proposed to exert toxic effects through irreversible (covalent) binding of its electrophilic metabolites to nucleophilic sites within biological molecules. The species thought to be responsible for the genotoxic effects of benz[a]anthracene are diol epoxides. The genotoxic effects are due to the reactions of hard electrophiles derived from the diol epoxides with DNA. The ease of production of these hard electrophiles is related to the extent of delocalization of positive charge formed during the formation of benzyl carbonium ion intermediate.

Human Toxicity

No case reports or epidemiological studies on the significance of benz[a]anthracene exposure to humans are available. However, coal-tar and other materials which are known to be carcinogenic in humans may contain benz[a]anthracene.

Benz[a]anthracene alone is not regulated; however, all polycyclic aromatic hydrocarbons or volatile coal tar products together are regulated. The WHO has established 0.2 µg/l as the limit for aromatic hydrocarbons in a domestic water supply. The OSHA limit in workplace air (coal tar volatiles) is 0.2 mg/m³.

Animal Toxicity

Benz[a]anthracene has been shown to be carcinogenic to experimental animals. Benz[a]anthracene given by several routes of administration has proven to be carcinogenic in mouse. Subcutaneous injection of 5 mg benz[a]anthracene daily to rats from the first day of pregnancy resulted in fetal death and resorption. Benz[a]anthrancene has been shown to induce benzo[a]-pyrene hydroxylase activity in the rat placenta.

—*Madhusudan G. Soni*

Related Topic

Polycyclic Aromatic Hydrocarbons

Benzene

- ◆ CAS: 71-43-2
- ◆ SYNONYMS: Cyclohexatriene; benzol; coal naphtha; benzole; phenyl hydride
- ◆ CHEMICAL CLASS: Aromatic hydrocarbon
- ◆ MOLECULAR FORMULA: C_6H_6
- ◆ CHEMICAL STRUCTURE:

Uses

The toxicological properties of benzene have substantially reduced its industrial use. However, benzene is found in petroleum products such as gasoline and diesel fuel. In the past, benzene was used as a solvent for oils, resins, rubber, varnishes, lacquers, and waxes; as a chemical intermediate in the manufacture of pharmaceuticals, adhesives, and coatings; and as a solvent for dyes and inks. Its current use may be limited to the production of synthetic organic chemicals and plastics. Products in which benzene is used as a raw material include polystyrene plastics, polyester resins, synthetic rubber, phenol, nylon, aniline, detergents, and chlorobenzenes.

Exposure Pathways

Human exposure to benzene occurs as a result of exposure to petroleum products. Benzene is volatile at room temperatures. Exposure in the workplace can occur mainly via inhalation and dermal contact. Oral exposure can occur from accidental or intentional consumption of benzene-containing products.

Toxicokinetics

Benzene is lipid soluble and volatile at room temperature. As such, benzene readily crosses the alveolar membranes and is taken up by circulating blood in pulmonary vessels. The lung also serves as an excretion pathway for unmetabolized benzene. Benzene can also be readily absorbed from the gastrointestinal tract and from intact skin. Circulating benzene is preferentially taken up by lipid-rich tissues such as adipose and nervous tissue. Benzene has also been detected in the bone marrow, liver, kidneys, lungs, and spleen.

The human liver can metabolize benzene through a number of metabolic pathways. The major end products of benzene metabolism include phenol, catechol, and quinol. These metabolic products are subsequently conjugated with inorganic sulfate and glucuronic acid in various degrees before being excreted in the urine. A small fraction of the catechol derived from benzene metabolism is oxidized to hydroxyhydroquinol or transformed to mucuronic acids.

Mechanism of Toxicity

Benzene can be irritating to mucus membranes. Acute exposures to high concentrations can produce pulmonary irritation and edema, dermatitis, and gastrointestinal irritation (if consumed). Chronic exposure to benzene produces bone marrow depression. Experimental evidence indicates that benzene's bone marrow toxicity is mediated by one or more of its metabolites. For example, inhibition of benzene metabolism by administration of toluene or partial hepatectomy protects bone marrow against benzene damage, and benzene metabolites, such as 1,2-dihydroxybenzene (catechol), 1,4-dihydroxybenzene (quinol), and 1,2,4-trihydroxybenzene(hydroxyhydroquinol), have been shown to inhibit cell mitosis.

Human Toxicity

Acute exposure to high doses of benzene in air (at concentrations in excess of 3000 ppm) causes symptoms typical of organic solvent intoxication. Symptoms may progress from excitation, euphoria, headache, and vertigo, in mild cases, to central nervous system (CNS) depression, confusion, seizures, coma, and death from respiratory failure in severe exposures.

The major toxicological manifestation of chronic benzene exposure in humans is bone marrow depression. Clinical manifestations include anemia, leucopenia, and thrombocytopenia. In severe cases, bone marrow aplasia develops. Later stages of toxicity are manifested by pancytopenia and aplastic anemia. Death may result from aplastic anemia or from leukemia. The U.S. EPA and IARC classify benzene as a known human carcinogen. This classification was given to benzene in view of strong epidemiological and experimental evidence.

Clinical Management

The victim should be removed from the contaminated atmosphere. Contaminated clothing should be removed and the affected area should be washed with soap and water. Supportive treatment should be provided. In cases of ingestion, vomiting should not be induced. Benzene or organic solvents containing benzene can cause acute hemorrhagic pneumonitis if aspirated into the lungs. Activated charcoal can be given to minimize absorption from the gastrointestinal tract. Charcoal can be given in a slurry or mixed with sorbitol or a saline cathartic. The recommended doses of activated charcoal are 30–100 g for adults, 15–30 g for children, and 1 or 2 g per kilogram for infants. The indicated doses can be prepared in a slurry by mixing charcoal in a diluent at a rate of 10 g charcoal per 80 ml diluent.

Animal Toxicity

The literature on the toxicological properties of benzene in laboratory animals is extensive. Benzene can cause severe eye irritation and moderate skin irritation. When given orally, benzene is moderately toxic. The oral LD_{50} in rats and mice is 3400 and 4700 mg/kg, respectively. The median lethal dose through inhalation has been evaluated in rats, mice, dogs, and cats. In these laboratory species, the LC_{50} ranges from 31,887 mg/m^3 in mice to 170,000 mg/m^3 in cats. The effects of lifetime exposure to benzene have also been evaluated in laboratory animals. Chromosomal abnormalities in bone marrow cells have been reported to appear in rats, rabbits, mice, and amphibians as a consequence of experimental benzene exposure.

—*Heriberto Robles*

Related Topics

Blood
Carcinogenesis
Indoor Air Pollution
Neurotoxicology: Central and Peripheral
Pollution, Water
Respiratory Tract
Skin

Benzene Hexachloride

- ◆ CAS: 608-73-1
- ◆ PREFERRED NAME: BHC (technical-grade BHC is a mixture of eight isomers)
- ◆ SYNONYMS: HCCH; HCH; 1,2,3,4,5,6-hexachlorocyclohexane; hexachlor; hexachloran; benzahex; benzex; hexator; kotol
- ◆ CHEMICAL CLASS: Halogenated hydrocarbon. Technical-grade BHC consists of 65–70% α-BHC, 6–8% β-BHC, 12–15% γ-BHC, and approximately 10% of other isomers and compounds.
- ◆ CHEMICAL STRUCTURE: Isomers differ on the spatial positions of the chlorine atoms on the boat and chair forms.

CL
CL CL
CL CL
CL

Uses

The only known use of BHC is as an insecticide.

Exposure Pathways

More than 90% of BHC intake in humans originates from food.

Toxicokinetics

BHC is absorbed through all portals including the intact skin. Different isomers of BHC are reported to be absorbed rapidly from the gastrointestinal tract and transferred exclusively to blood. Metabolism of BHC mainly takes place in the liver by four enzymatic reactions. Dehydrogenation, dechlorination, and hydroxylation is via P450 mixed-function oxidases, whereas dehydrochlorination is carried out by cytosolic enzymes. The endproducts of biotransformation are di-, tri-, tetra-, penta-, and hexachloro compounds. Within a few hours of uptake, BHC is distributed to all organs and tissues. The highest concentrations are found in adipose tissues and skin. In a long-term high-level BHC feeding study, it was shown that adipose tissue retains more α isomer than β and γ isomers. BHC is excreted rapidly in urine and feces after metabolic degradation. The excreted metabolites are either free or conjugated forms of glucuronic or sulfuric acids of *N*-acetyl cysteine.

Mechanism of Toxicity

BHC produces a variety of neurological effects in insects and mammals. However, at both levels of the nervous system (peripheral and central), the mechanism of toxic action of BHC is poorly understood. Central nervous system stimulation appears to be due to blockade of the effects of γ-aminobutyric acid. *In vitro* BHC isomers are reported to increase the calcium uptake of isolated rat brain synaptosomes. In addition, γ isomer

of BHC has been shown to inhibit the uptake of chloride ions at inhibitory synapses in the brain, and it is this mode of action that is now widely considered to account primarily for the convulsant activity of this insecticide. The results of studies on initiation–promotion, on mode of action, and on mutagenicity indicate that tumorigenic effects of BHC in mice results from nongenetic mechanisms.

Human Toxicity: Acute

Clinical signs of intoxication can appear from a few minutes to some hours after BHC ingestion. Ingestion of large (unspecified) doses of BHC has led to muscle and kidney necrosis and in one case to pancreatitis. The symptoms of poisoning include nausea, restlessness, headache, vomiting, tremor, ataxia, and tonic–clonic convulsion. Digestive tract inflammation, hemorrhage, coma, and death have also been reported after poisoning.

Human Toxicity: Chronic

Chronic liver damage (cirrhosis and chronic hepatitis) were observed in liver biopsies from eight workers heavily exposed to BHC or DDT or both for periods ranging from 5 to 13 years. Several case reports indicate a relationship between exposure to BHC and the occurrence of aplastic anemia.

The exposure limit for γ-BHC in most countries is 0.5 mg/m^3. The FAO/WHO acceptable daily intake of γ-BHC is 0.008 mg/kg body weight.

Clinical Management

Gastric decontamination by lavage and saline cathartics should be carried out. Oil laxatives should not be used because they promote BHC absorption. Pentabarbital or phenobarbital in adequate amounts or calcium gluconate intravenously in conjunction with anticonvulsants may be used in the control of convulsions.

Animal Toxicity

Acute toxicity of BHC has been investigated in numerous studies in a variety of species and strains via different routes. Signs of acute poisoning in rats include diarrhoea, hypothermia, epistaxis, and convulsions; death is due to respiratory failure. Feeding of BHC (10–1600 mg/kg diet) for life span to rats resulted in decreased body weight and an increase in mortality at 800 mg/kg and above. Fatty degeneration and focal necrosis of the liver were observed at higher doses. Chronic nephritis with glomerular fibrosis and hyaline deposits was seen in rats fed 800 mg/kg diet BHC.

—*Madhusudan G. Soni*

Related Topic

Pesticides

Benzidine

- CAS: 92-87-5
- SYNONYMS: *P*-Diaminodiphenyl; 4,4'-diaminobiphenyl; 4,4'-diaminodiphenyl; [1,1'-biphenyl]-4,4'diamine
- CHEMICAL CLASS: Synthetic aromatic hydrocarbon with two benzene rings covalently bonded to one another (1,1), substituted by amino group at 4,4'
- CHEMICAL STRUCTURE:

$$H_2N \text{—} \bigcirc \text{—} \bigcirc \text{—} NH_2$$

Uses

Benzidine is used as an intermediate in the production of azo dyes, sulfur dyes, fast color salts, naphthol, and other dye compounds. To date, more than 250 benzidine-based dyes have been reported. These dyes are primarily used for dyeing textiles, paper, and leather products.

Exposure Pathways

The primary routes of potential human exposure to benzidine are inhalation, ingestion, and dermal contact. Benzidine may get into the respiratory tract from accidental release into the air; into the gastrointestinal tract from contaminated fingers, cigarettes, or food; and onto the skin directly from contaminated clothing and gloves.

Toxicokinetics

Benzidine is rapidly absorbed through skin in solid and vapor form. It is also quickly absorbed through lung on inhalation and from the gastrointestinal tract on consuming contaminated water and food substances. It is a lipophilic substance, hence easily stored in fat tissues, and it firmly binds to cell membrane receptors. Absorption by the body greatly depends on occupational or accidental exposure, as well as through the colored foodstuffs. Benzidine is metabolized to aromatic amine by intestinal microflora or liver azoreductase. Liver is the chief organ of metabolism where benzidine is converted to more reactive, toxic, and mutagenic (carcinogenic) *N*-hydroxyarylamides and *N*-hydroxyarylamine, considered to be proximate carcinogens. *N*-hydroxyarylamides are converted to the ultimate carcinogens through conjugation with sulfuric, acetic, or glucuronic acids. *N*-acetoxyarylamines are also produced as metabolites and are highly reactive mutagens and carcinogens. Benzidine is slightly soluble in water but highly soluble in organic solvents. It strongly binds to organic matters. To a great extent it is stored in fatty tissues because of its high lipophilicity. Glutathione transferases play an important role in elimination of reactive metabolites of benzidine. Sulfonation, carboxylation, deamination, or substitution of an ethyl alcohol or an acetyl group for the hydrogen in the amino groups leads to a decrease in mutagenic activity of benzidine metabolites and it also leads to easy elimination, primarily through urine and feces.

Mechanism of Toxicity

Benzidine is metabolized to highly toxic, reactive metabolites, like *N*-hydroxy arylamides and *N*-hydroxy arylamines, which act as procarcinogens and are more mutagenic than parent compounds. They strongly bind to DNA and cell receptors. On conjugation with sulfuric, acetic, and glucuronic acids they form ultimate carcinogens. Benzidine metabolites on acetylation produce more reactive N-acetoxyarylamines known to be the cause of bladder cancer in dye industry workers.

Human Toxicity

Benzidine is carcinogenic in humans. An IARC study on dye industry workers reported that there is a direct correlation between the incidence of bladder cancer in the occupationally benzidine exposed workers and the incidence of this cancer in workers decreasing after a reduction in industrial exposure. Benzidine causes vomiting, nausea, damage to blood, liver, and kidneys, and hematuria (bloody urine) on poisoning.

Clinical Management

There is no antidote for benzidine poisoning. Since it produces reactive metabolites, administration of free radical scavengers should alleviate the toxicity. Complexation of ions with copper and in the form of hydrochloride is known to decrease its mutagenic effects.

Animal Toxicity

There is sufficient evidence for the carcinogenicity of benzidine in experimental animals. When administered in the diet, benzidine induced bladder carcinomas in dogs, multiple mammary carcinomas in rats, and liver cell tumors in hamsters of both sexes. When administered by subcutaneous injection, benzidine induced malignant tumors of the Zymbal gland (ear) and liver hepatocellular carcinomas in mice of both sexes; hepatomas, malignant tumors of Zymbal gland, and local sarcomas in male rats; and malignant tumors of the Zymbal gland, mammary adenocarcinomas, and amyloid leukemia in female rats. When administered by intraperitoneal injection, benzidine induced Zymbal gland adenomas and carcinomas and malignant mammary tumors in female rats. The lethal dose in dogs is 400 mg/kg by the subcutaneous route and 200 mg/kg by the oral route.

—Vaman C. Rao

Related Topics

Carcinogenesis
Mutagenesis
Occupational Toxicology

Benzo[a]pyrene

- CAS: 50-32-8
- SYNONYMS: BAP; B[a]P; BP; 3,4-benzopyrene; 6,7-benzopyrene; 3,4-benzpyrene; 3,4-benz[a]pyrene
- DESCRIPTION: Benzo[a]pyrene (BP) has a faint aromatic odor. It has a boiling point of greater than 360°C at 760 mm Hg and a melting point of 179–179.3°C. It has a specific gravity of 1.351. Crystals of BP may be monoclinic or orthorhombic.

  On contact with strong oxidizers, BP may cause fire or explosion. BP is light labile and is oxidized by chromic acid and by ozone.

  BP is found in fossil fuels and occurs in products of incomplete combustion. It is present in charcoal, chimney sweepings, and coal tar.
- CHEMICAL CLASS: Polycyclic aromatic hydrocarbon
- MOLECULAR FORMULA: $C_{20}H_{12}$
- CHEMICAL STRUCTURE:

Uses

In research, BP is used extensively as a positive control in a variety of laboratory mutagenicity and carcinogenicity short-term tests. It is not produced commercially in the United States.

Exposure Pathways

The primary routes of exposure to BP are inhalation and ingestion.

Toxicokinetics

Polycyclic aromatic hydrocarbons (PAHs) are absorbed following ingestion, inhalation, and dermal exposure. Following absorption, PAHs enter the lymph and then the bloodstream. BP is readily absorbed from the intestinal tract and tends to localize primarily in body fat and fatty tissues such as the breast. Disappearance of BP from blood and liver of rats following a single intravenous injection is very rapid, having a half-life in blood of less than 5 min and a half-life in liver of 10 min. In blood and liver, the initial rapid elimination phase is followed by a slower disappearance phase, lasting 6 hr or more. A rapid equilibrium is established between BP in blood and that in liver. The fast disappearance of the compound from blood is due to metabolism and distribution in tissues. BP is known to cross the placenta in mice and rats. ^{14}C metabolites were secreted into the bile of rats within 7 min of receiving an intravenous dose of ^{14}C BP. Pretreatment of animals with this carcinogen enhanced biliary secretion of ^{14}C radiolabel. PAHs are primarily metabolized enzymatically in the liver and kidneys. Additional sites of PAH metabolism include the adrenal glands, testes, thyroid, lungs, skin, and sebaceous glands. PAHs are metabolized by aryl hydrocarbon hydroxylase. The ultimate carcinogen of CYP450 metabolism of BP is 7,8-dihydro-7,8-diol-9,10-epoxide. The predominant metabolites of BP in mammals are 3- and 9-hydroxy BP, BP-1,6-quinone and BP-3,6-quinone, BP-4,5-dihydrodiol, BP-7,8-dihydrodiol, and BP-9,10-dihydrodiol. Human liver microsomal fractions were characterized for differences in the metabolism of BP. Pronounced interindividual differences in the composition of microsomal proteins in the molecular weight range of 49,000–60,000 were found. Large variation among human liver microsomal samples was also seen in BP metabolism. The results indicate the presence of seven or eight different forms of CYP450 in human liver microsomes and interindividual variations seen in metabolism may partly be explained by variations in the distribution of these isozymes.

Human Toxicity: Acute

In general, PAHs have a low order of acute toxicity in humans. BP may cause skin irritation with rash, redness, and/or a burning sensation. Exposure to sunlight and the chemical together can increase these effects. BP can irritate and/or burn the eyes on contact.

Human Toxicity: Chronic

Long-term health effects can occur at some time after exposure to BP and can last for months or years. BP is a probable carcinogen in humans. There is some evidence that it causes skin, lung, and bladder cancer

in humans and animals. BP has caused cancer in the offspring of animals exposed to the substance during pregnancy. Many scientists believe that there is no safe level of exposure to a carcinogen. Cancer is the most significant toxicity associated with PAHs. The first occupational cancer described was that of scrotal cancer in chimney sweeps exposed to PAHs in soot and ash. Studies have noted increased lung cancer and a suggestion of increased gastrointestinal cancer incidence in the coal carbonization and coal gasification industries. BP has been observed to produce epithelial hyperplasia and inhibition of connective tissue growth on human fetal lung cultures. Since tobacco smoke contains BP, smoking may increase the risk of lung cancer with exposure to BP. BP on the skin in the presence of sunlight and/or ultraviolet light also increases the risk of skin cancer. Persistent nodules diagnosed as squamous epithelioma developed in a man who had been exposed to BP for 3 weeks while carrying out an experiment in mice. BP may damage the developing fetus. There is some evidence that BP may affect the sperm and the testes. BP may be transferred to nursing infants through mother's milk. Repeated exposure to substances that contain BP can cause skin changes such as thickening, darkening, and pimples. Later skin changes include loss of color, reddish areas, thinning of the skin, and warts. Bronchitis may result from repeated exposure to BP-containing mixtures. Coke oven workers exposed to BP had significantly depressed levels of IgG and IgA compared to cold-rolling mill workers

Workplace Controls and Practices

A class I, type B biological safety hood should be used when working with BP in a laboratory. The following work practices are recommended: (1) Contaminated clothing should be removed immediately and laundered by individuals who have been informed of the hazards of exposure to BP; (2) eye wash fountains should be provided for emergency use. Emergency shower facilities should be available if there is a possibility of skin exposure. On skin contact, affected skin should be washed immediately to remove the chemical; (3) eating, smoking, or drinking should be prohibited where BP is handled; and (4) protective clothing (suits, gloves, footware, and headgear) should be donned before work. Workers in industries that produce coal or coal tar products and those who tar road surfaces and roofs are at maximum risk.

Under RCRA, BP must be managed as a hazardous waste according to federal and/or state regulations. The U.S. EPA federal drinking water standard is 0.2 mg/liter. The NIOSH occupational exposure recommendations are 0.1 mg/m^3 for cyclohexane extractable fraction and 0.1 mg/m^3, 10-hr TWA for coal tar products.

Clinical Management

Because of the low acute toxicity associated with PAHs, induced emesis is not recommended. Activated charcoal/cathartic may be used. On inhalation exposure, the patient should be moved to fresh air and monitored for respiratory distress. If cough or difficulty in breathing develop, evaluation for respiratory tract irritation, bronchitis, or pneumonitis should be performed. Humidified supplemental oxygen (100%) should be administered with assisted ventilation as required. On ocular exposure, the eyes should be irrigated for at least 15 min with tepid water. On dermal exposure, the affected area should be washed thoroughly with soap and water. Patients developing dermal hypersensitivity reactions may require treatment with systemic or topical corticosteroids or antihistamines. Treatment of gastric, lung, or skin cancer is no different from that for the same cell type.

Animal Toxicity: Acute

Mild hepatotoxicity and nephrotoxicity have been observed in rats exposed to PAHs. Intraperitoneal administration of BP to rats produced an immediate and sustained reduction in growth rate of young rats. A single topical exposure of BP in acetone increased the mitotic rate of epidermal cells. Single oral administration of 100 mg BP to 50-day-old Sprague–Dawley rats produced mammary tumors. Single intraperitoneal administration of 10 mg BP produced two mammary and two uterine carcinomas among 10 Wistar rats within 1 year.

Animal Toxicity: Chronic

In rats chronically fed PAHs, agranulocytosis, anemia, leukopenia, and pancytopenia have been observed. There is sufficient evidence suggesting that BP is carcinogenic to experimental animals. Exposure to BP caused a dose-dependent increase in the pulmonary tumor burden of mice administered B16F10 melanoma cells intravenously 1 day after the last of a 14-day exposure to BP. Biweekly administration of BP in oil by stomach tube produced papillomas of the stomach in hamsters. Biweekly painting with 0.3% solution of BP in benzene for 400 days produced 1 carcinoma and 10 papillomas among 10 rabbits. A possible causal relation between BP/diribonucleoside adduct formation and papilloma formation in Sencar mice was found. Among rats fed 1 mg BP per gram of diet during pregnancy many resorptions and dead fetuses were observed but only one malformed fetus was noted from seven litters. BP has been shown to be embryotoxic and teratogenic in mice. A reduction in fertility in male and female offspring was observed in mice following exposure *in utero*.

—*Swarupa G. Kulkarni and Harihara M. Mehendale*

Related Topics

Immune System
Occupational Toxicology
Polycyclic Aromatic Hydrocarbons
Pollution, Water

Benzodiazepines

◆ REPRESENTATIVE COMPOUNDS: Alprazolam (CAS: 28981-97-7); chlordiazepoxide; clonazepam; clorazepate dipotassium; clorazepate monopotassium; diazepam (CAS: 439-14-5); estazolam; flunitrazepam; flurazepam; halazepam; lorazepam; midazolam; nitrazepam; oxazepam (CAS: 604-75-1); prazepam; quazepam; temazepam; triazolam

◆ SYNONYMS:

Alprazolam—Xanax; $C_{17}H_{13}ClN_4$; 8-chloro-1-methyl-6-phenyl-4H-1,2,4-triazolo(4,3-A)(1,4) benzodiazepine

Chlordiazepoxide—Librium; 7-chloro-2-methyl-amino-5-phenyl-3H-1,4-benzodiazepine-4-oxide hydrochloride

Clonazepam—Klonopin; 5-(2-chlorophenyl)-1,3-dihydro-7-nitro-2H-1,4-benzodiazepin-2-one

Clorazepate dipotassium—Tranxene; Cloraze-Caps; ClorazeTabs; GenENE; $C_{16}H_{11}ClK_2N_2O_4$

Clorazepate monopotassium—Azene

Diazepam—Valium; $C_{16}H_{13}ClN_2O$; 7-chloro-1-methyl-5-phenyl-1,3-dihydro-2H-1,4-benzodiazepin-2-one

Estazolam—Prosom; 8-chloro-6-phenyl-4H-1,2,4-triazolo(4,3-A)-1,4-benzodiazepine

Flunitrazepam—Rohypnol

Flurazepam—Dalmane; 7-chloro-1-[2-(diethylamino)ethyl]-5-(o-fluorophenyl)-1,3-dihydro-2H-1,4-benzodiazepine-2-one dihydrochloride

Halazepam—Paxipam

Lorazepam—Ativan; 7-chloro-5-(o-chlorophenyl)-1,3-dihydro-3-hydroxy-2H-1,4-benzodiazepin-2-one

Midazolam—Hypnovel; Versed; $C_{18}H_{13}$ $ClFN_3.HCl$; 8-chloro-6-(2-fluorophenyl)-1-methyl-4H-imiddazo[1,5-a][1,4]benzodiazepine hydrochloride

Nitrazepam—Mogadon; dihydro-7-nitro-5-phenyl-1H-1,4-benzodiazepin-2-one

Oxazepam—Serax; 7-chloro-1,3-dihydro-3-hydroxy-5-phenyl-2H-1,4-benzodiazepin-2-one

Prazepam—Verstran; 7-chloro-1-(cyclopropylmethyl)-1,3-dihydro-5-phenyl-2H-1,4-benzodiazepin-2-one

Quazepam—Doral; $C_{17}H_{11}ClF_4N_2S$; 7-chloro-5-(o-fluorophenyl)-1,3-dihydro-1-(2,2,2-trifluoroethyl)-2H-1,4-benzodiazepine-2-thione

Temazepam—Restoril; 3-hydroxydiazepam; $C_{16}H_{13}ClN_2O_2$; 7,-chloro-1,3-dihydro-3-hydroxy-1-methyl-5-phenyl-2H-1,4-benzodiazepin-2-one

Triazolam—Halcion; 8-chloro-6-(o-chlorphenyl)-1-methyl-4H-S-triazolo-(4,3-A)(1,4)benzodiazepine

- ◆ PHARMACEUTICAL CLASS: 5-Aryl-1,4-benzodiazepines
- ◆ CHEMICAL STRUCTURE:

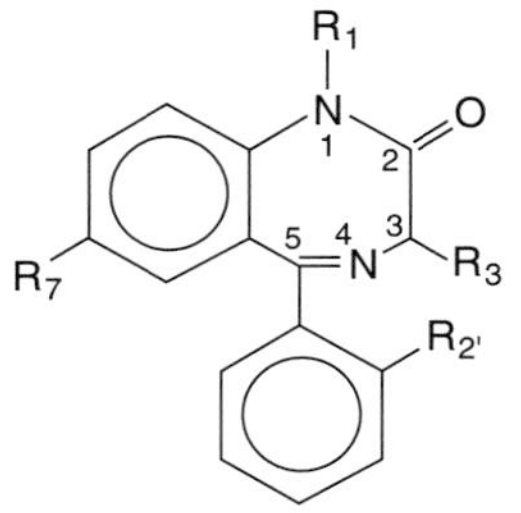

Uses

The benzodiazepines are primarily used as sedative–hypnotic agents and for the treatment of anxiety. They also may be administered preoperatively for anterograde amnesia of perioperative events and light anesthesia. Benzodiazepines are also commonly used as muscle relaxants and anticonvulsants. They are also drugs of abuse.

Exposure Pathways

The most common route of exposure to the benzodiazepines is ingestion of oral dosage forms. Several of these agents are also available for parenteral administration (intramuscular or intravenous).

Toxicokinetics

The benzodiazepines are generally well absorbed from the gastrointestinal tract. Most of clorazepate dipotassium is decarboxylated to desmethyldiazepam (nordiazepam) prior to absorption. The time to peak concentration of the benzodiazepines ranges from 0.5 to 8 hr after ingestion.

Prazepam and flurazepam undergo extensive first-pass metabolism to active metabolites. The benzodiazepines are all extensively metabolized by microsomal enzyme systems in the liver. The metabolites of many benzodiazepines are pharmacologically active and are biotransformed much more slowly than the parent compounds. Benzodiazepines that are metabolized to inactive compounds include lorazepam, midazolam, oxazepam, temazepam, and triazolam. The benzodiazepines and their active metabolites are widely distributed into body tissues and readily cross the blood–brain barrier and placenta. All are highly bound to plasma proteins. The elimination half-lives of the benzodiazepines range from 1 to 50 hr. The half-lives of active metabolites, however, may be as long as 200 hr. These metabolites are ultimately conjugated, largely with glucuronic acid, to inactive compounds which are excreted primarily in the urine.

Mechanism of Toxicity

The benzodiazepines potentiate neuronal inhibition which is mediated by γ-aminobutyric acid (GABA). Specific binding sites for benzodiazepines have been identified which are proximal to GABA receptors and GABA-regulated chloride channels. This action occurs at all levels of the neuroaxis.

Human Toxicity

The acute toxic effects of the benzodiazepines are relatively mild. Lethargy, ataxia, and slurred speech occur commonly. Grade 0 to 1 coma without other organ system involvement is characteristic with even large overdose of these agents. Toxic doses have not been established. Death is extremely uncommon following the ingestion of benzodiazepines alone. If coingested with other central nervous system (CNS) depressant drugs or alcohol, then more severe manifestations and death may result. In such cases profound coma, respiratory depression, hypotension, and hypothermia may develop. Abrupt discontinuation of therapy with the benzodiazepines is likely to result in a withdrawal syndrome consisting of anxiety, agitation, insomnia, tremors, headache, and myalgias. In more severe cases nausea, vomiting, diaphoresis, hyperpyrexia, psychosis, seizures, and death may occur.

Clinical Management

Basic and advanced life-support measures should be implemented as necessary. Gastrointestinal decontamination procedures should be used as appropriate based on the patient's level of consciousness and history of ingestion. Activated charcoal can be used to adsorb the benzodiazepines. The patient's level of consciousness and vital signs should be monitored closely. Obtunded patients with reduced gag reflex should be intubated. Respiratory support, including oxygen and ventilation, should be provided as needed. Flumazenil (Romazicon) is a benzodiazepine antagonist which can reverse the CNS depressant effects of these agents. It is indicated for the treatment of serious benzodiazepine overdose and reversal of conscious benzodiazepine sedation. Because overdose with benzodiazepines generally results in only mild toxicity, it has limited clinical utility in this setting. It must be used with caution in mixed drug overdoses as seizures can develop, particularly if tricyclic antidepressants have been coingested. Also, it can induce benzodiazepine withdrawal in dependent patients. If hypotension occurs it should be treated with standard measures including intravenous fluids, Trendelenburg positioning, and dopamine hydrochloride by intravenous infusion. Forced diuresis, hemoperfusion, and hemodialysis are of no value for the benzodiazepines. If withdrawal develops, it should be managed with phenobarbital or diazepam with a gradual dose reduction.

Animal Toxicity

Animals may be affected by the benzodiazepines much in the same way as humans. Lethargy, coma, shallow respirations, incoordination, and depressed reflexes may occur. Dogs may show a contradictory response (CNS excitement) following exposure. Standard supportive measures should be employed.

—*Gregory P. Wedin*

Related Topics

Anxiolytics
Neurotoxicology: Central and Peripheral
Poisoning Emergencies in Humans (under Flumazenil)

Benzyl Alcohol

◆ CAS: 100-51-6

◆ SYNONYMS: Benzene carbinol; benzene methanol; benzoyl alcohol; phenyl carbinol; phenyl methanol; hydroxymethyl benzene; α-hydroxy toluene

◆ DESCRIPTION: Benzyl alcohol is a water–white liquid with a faint aromatic odor and a sharp burning taste. It has a molecular weight of 108.13 and a specific gravity of 1.045. Aqueous solution of benzyl alcohol is neutral. Benzyl alcohol decomposes to benzaldehyde slowly when exposed to air. Benzyl alcohol is incompatible with oxidizing agents. Problems may occur when polystyrene syringes are used with certain types of drug products containing benzyl alcohol since these agents can extract and dissolve the plastic. At times the rubber tip may release a constituent to the drug product.

- ◆ CHEMICAL CLASS: Alcohol
- ◆ MOLECULAR FORMULA: $C_6H_5CH_2OH$
- ◆ CHEMICAL STRUCTURE:

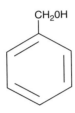

Uses

Benzyl alcohol is primarily used as a solvent and an antimicrobial preservative, but it has also found use as an antiseptic and local anesthetic.

Toxicokinetics

Body tissue possibly takes up benzyl alcohol rapidly and releases it slowly into the blood-stream. Rabbits when given 1 g (subcutaneously) of benzyl alcohol eliminated 300–400 mg of hippuric acid within 24 hr. Rabbits eliminated 65.7% of a dose of 0.4 g of benzyl alcohol as hippuric acid in the urine. The plasma half-life of benzyl alcohol administered as a 2.5% solution in saline was found to be approximately 1.5 h in dogs injected intravenously at doses of 52 and 105 mg/kg.

Benzyl alcohol is oxidized by the liver alcohol dehydrogenase. Humans readily oxidize benzyl alcohol to benzoic acid, which, after conjugating with glycine, is rapidly eliminated as hippuric acid in the urine. Within 6 h after taking 1.5 g of benzyl alcohol orally, human subjects eliminated 75–85% of the dose in urine as hippuric acid. Benzyl alcohol yields benzaldehyde in rabbits and phenol in guinea pigs. If the dose is sufficiently high to allow the rate of formation of benzoic acid to exceed that of hippuric acid some of the benzoic acid is excreted as benzoylglucuronide.

Mechanism of Toxicity

Benzyl alcohol is oxidized by the liver to benzoic acid, and then conjugated with glycine to form hippuric acid. Metabolic acidosis can be explained by a direct effect of benzoic acid and/or secondary lactic acid production through depression of cellular metabolism. Benzyl alcohol is a weak local anesthetic with disinfectant properties.

Human Toxicity: Acute

High doses of benzyl alcohol cause nausea, vomiting, diarrhea, central nervous system (CNS) depression and vertigo. Dilute solutions (1%) produce local anesthesia and slight irritation when instilled into the eye. Pure benzyl alcohol produces corneal necrosis. Following acute exposure lethargy, seizures, intraventricular hemorrhage, and neurological sequelae (cerebral palsy, developmental delay) have been seen in neonates with parenteral benzyl alcohol toxicity. Metabolic acidosis was a common finding with parenteral toxicity in neonates. Thrombocytopenia was a delayed feature of parenteral toxicity in neonates. Deaths associated with intravenous or endotracheal administration of benzyl alcohol-containing solutions in neonates were preceded by symptoms of respiratory distress progressing to gasping respirations, metabolic acidosis, CNS depression, hypotension, renal failure, and occasionally seizures and intracranial hemorrhage. Thrombocytopenia was a delayed feature of parenteral toxicity in neonates. Severe striated keratopathy, progressing to chronic edema of cornea, was noted following intraocular use of a sodium chloride solution containing 2% benzyl alcohol.

Human Toxicity: Chronic

Chronic exposure to benzyl alcohol would presumably produce effects similar to those from acute exposure. No other industrial illness is known from benzyl alcohol. No reproductive effects on humans are known.

The odor threshold for benzyl alcohol is 5.5 ppm.

Clinical Management

Treatment is supportive following exposure. The victim should be monitored for CNS and respiratory depression, metabolic acidosis, and hypotension. Ipecac-induced emesis is not recommended. On ocular exposure, the eyes should be irrigated for at least 15 min with tepid water. On dermal exposure, the exposed area should be washed with soap and water. If irritation, pain, swelling, lacrimation, or photophobia persist, the victim should be seen in a health care facility.

Animal Toxicity

When injected into chickens, benzyl alcohol produced birth defects of the CNS and skeleton. Doses of 0.2 ml/kg or more to dogs by stomach tube induced emesis and defecation. This was apparently due to irritation of the gastric mucosa. Diuresis was more pronounced in the rabbit than in the dog, after administration of benzyl alcohol by various routes. Mice suffered respiratory stimulation, respiratory and muscular paralysis, convulsions, and CNS depression following a subcutaneous injection. A decrease in arterial blood pressure of rabbits, cats, and dogs was seen following intravenous injection of benzyl alcohol. No such decrease in arterial blood pressure was noted following oral administration to dogs. Benzyl alcohol displayed antiarrhythmic, antifibrillatory effects when injected intravenously into dogs and rats with spontaneous and drug-induced arrhythmias. Instillation of pure benzyl alcohol into rabbit conjunctival sac produces corneal necrosis which is resolved after several weeks. The undiluted material when applied to depilated skin of guinea pigs for a period of 24 hr caused moderately strong primary irritation, and there was evidence of systemic symptoms with death from application of less than 5 ml/kg.

Because its primary effects are expected to be irritation and perhaps mild CNS depression, benzyl alcohol is class I for general toxicity (may cause reversible effects which are generally not life-threatening). From the one study in chickens, it is in class B for reproductive hazard (few effects in animals but no human data). The actual human reproductive hazard is known.

—Swarupa G. Kulkarni and Harihara M. Mehendale

Benzyl Benzoate

- ◆ CAS: 120-51-4
- ◆ SYNONYMS: Ascabin-; Ascabiol-; Ascarbin-; Benzylate-; Scabanca-; Tenutex-; Vanzoate-; Venzoate-; benzoic acid phenylmethyl ester; benzy alcohol benzoic ester
- ◆ CHEMICAL CLASS: Benzoic acid phenylmethyl ester
- ◆ CHEMICAL STRUCTURE:

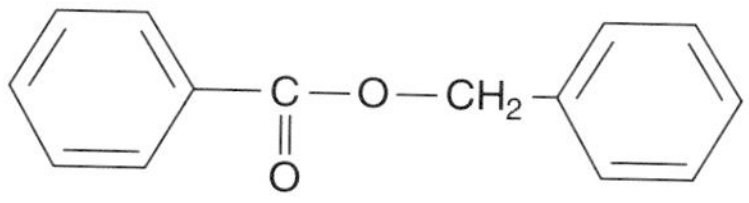

Uses

Benyzl benozate is used as an acaricide, scabicide, and pediculicide in veterinary hospitals and as a repellent for chiggers, ticks, and mosquitos.

Exposure Pathways

Dermal exposure is the most common route of exposure. Benzyl benzoate occurs naturally in balsams of Peru and Tolu and other essential oils. It is also available in liquid, emulsion, and lotion dosage forms.

Toxicokinetics

Benzyl benzoate is rapidly absorbed from the stomach. It is rapidly hydrolyzed to benzoic acid and benzyl alcohol, which is subsequently hydrolyzed to benzoic acid. Benzoic acid is conjugated with glycine to give benzoylglycine or hippuric acid and with glucuronic acid to give benzoylglucuronic acid. The conjugates are rapidly eliminated in urine in varying ratios depending on species and dose.

Mechanism of Toxicity

Benzyl benzoate is a local irritant.

Human Toxicity: Acute

Benzyl benzoate is a moderately toxic compound. It may cause slight allergenic responses, which may disappear after the end of exposure. If used as an acaricide, it may cause peristalsis of the intestine, diarrhea, intestinal colic, enterospasm, pylorospasm, spastic constipation, contraction of the seminal vesicles, hypertension, and bronchospasms.

Human Toxicity: Chronic

Very little is known about the chronic effects of benzyl benzoate.

Clinical Management

Basic life-support measures for respiratory and cardiovascular function should be utilized. Dermal decontamination should be accomplished by repeated washing with soap. Exposed eyes should be irrigated with copious amounts of room-temperature water for at least 15 min.

Animal Toxicity

Benzyl benzoate has low acute toxicity in laboratory animals. The oral LD_{50} value for rats is greater than 1 g/kg. If applied too frequently or to a large area, it can induce systemic signs of toxicity including salivation, piloerection, muscular incoordination, tremors, progressive paralysis of hindlimbs, prostration, violent convulsions, dyspnea, and death. Cats are especially susceptible to such toxic effects of benzyl benzoate. In contrast, dogs can tolerate 10 times the dose that is lethal to rats, rabbits, and cats.

—Sushmita M. Chanda

Related Topic

Pesticides

Beryllium (Be)

- ◆ CAS: 7440-41-7
- ◆ SELECTED COMPOUND: Beryllium oxide, BeO (CAS: 1304-56-9)
- ◆ CHEMICAL CLASS: Metals

Uses

Beryllium is used in nuclear reactors as a neutron reflector or moderator. In the aerospace industry, beryllium is used in inertial guidance systems, and beryllium alloys (consisting of copper or aluminum) are used in structural material. Beryllium oxide is used as an additive in glass, ceramics, and plastics and as a catalyst in organic reactions. In the past, beryllium was widely used in the manufacture of fluorescent lights and neon signs.

Exposure Pathways

The primary exposure pathway for beryllium is inhalation. In the general environment, airborne beryllium is produced from combustion of coal or crude oils used for power generation. Inhalation, ingestion, and dermal contact are possible exposure pathways in certain workplace settings. Exposure to minute amounts of beryllium occurs with ingestion of some foods and drinking water.

Toxicokinetics

Beryllium is not well absorbed by any route; oral absorption of beryllium is less than 0.01% and probably only occurs in the acidic stomach environment. About half of inhaled beryllium is cleared in about 2 weeks; the remainder is cleared slowly and the residual becomes fixed in the lung (granulomata). The half-life of beryllium in rat blood is approximately 3 hr. Beryllium is distributed to all tissues. High doses generally go to the liver and then are gradually transferred to the bone. Most beryllium concentrates in the skeleton. Beryllium is excreted in the urine; however, the fraction of administered dose excreted in urine is variable.

Mechanism of Toxicity

Beryllium compromises the immune system. Enzymes catalyzed by magnesium or calcium can be inhibited by beryllium; succinic dehydrogenase is activated. Beryllium exposure leads to a deficiency in lung carbon monoxide diffusing capacity. Hypercalcemia can occur (excess of calcium in the blood).

Human Toxicity

The major toxicological effects of beryllium are on the lung. Acute exposure to soluble beryllium compounds (e.g., fluoride, an intermediate in the ore extraction process) irritates the entire respiratory tract, may produce acute chemical pneumonitis, and can result in fatal pulmonary edema. Chronic exposure to insoluble beryllium compounds, particularly the oxide, leads to berylliosis (a chronic granulomatous disease), which begins with a cough and chest pains. In most cases, these symptoms soon lead to pulmonary dysfunction.

Hypersensitivity, which appears to be mediated by the immune system, may also occur following exposure. This means that future exposure to beryllium may produce health effects at concentrations lower than those generally associated with the effect (the individual becomes much more sensitive to beryllium).

Effects of beryllium exposure include enlargement of the heart (which can lead to congestive heart failure), enlargement of the liver, and kidney stones. Finger "clubbing" is often seen with berylliosis.

Skin lesions are the most common industrial exposure symptom. Three distinct skin lesions have been noted following exposure to beryllium: dermatitis, ulceration, and granulomas. There appears to be an immunological component to chronic beryllium disease, including the dermal responses.

Although available information from epidemiological studies is insufficient to confirm human carcinogenesis, the data strongly suggest beryllium is associated with cancer in humans. ACGIH classifies beryllium as a suspected human carcinogen. The ACGIH TWA-TLV for beryllium and compounds is 0.002 mg/m^3.

Clinical Management

Chelation has been used to treat beryllium toxicity; however, no one agent is recommended over another. Aurin tricarboxylic acid has been used to protect primates from beryllium overdose, but human trials have not been conducted.

Animal Toxicity

Although beryllium produces cancer in more than one animal species (lung cancer in rats and monkeys; osteogenic sarcoma in rabbits), it does not appear to be teratogenic. *In vitro* studies indicate beryllium will induce morphological transformations in mammalian cells, but beryllium is not mutagenic in bacterial systems.

*—Arthur Furst, Shirley B. Radding
and Kathryn A. Wurzel*

Related Topics

Metals
Respiratory Tract

Beta Blockers

- ♦ PREFERRED NAME: Beta adrenergic blockers
- ♦ REPRESENTATIVE COMPOUNDS: Acebutolol; atenolol; betaxolol; bisoprolol; carteolol; esmolol; labetalol; metoprolol; nadolol; penbutolol; pindolol; propranolol; sotalol; timolol
- ♦ SYNONYMS: Sectral (CAS: 37517-30-9); Tenormin (CAS: 29122-68-7); Kerlone (CAS: 63659-19-8); Zebeta (CAS: 66722-44-9); Cartrol (CAS: 51781-21-6); Brevibloc (CAS: 81161-17-3); Normodyne; Trandate (CAS: 32780-64-6); Lopressor (CAS: 37350-58-6); Corgard (CAS: 42200-33-9); Levatol (CAS: 38363-32-5); Visken (CAS: 13523-86-9); Inderal (CAS: 318-98-9); Betapace (CAS: 959-24-0); Blocadren (CAS: 26921-17-5)
- ♦ PHARMACEUTICAL CLASS: Beta adrenergic blockers

Uses

Beta blockers are used in the treatment of hypertension, angina pectoris, supraventricular arrhythmias, supraventricular tachycardia, sinus tachycardia, ventricular tachycardia, myocardial infarction, pheochromocytoma, migraine headache, and essential tumor.

Exposure pathways: Ingestion is the most common route for both accidental and intentional exposures to the beta blockers. Esmolol, labetalol, metoprolol, and propranolol are all available for parenteral administration; therefore, toxicity can occur via this route. Beta blockers are also administered as ocular medications and systemic toxicity can occur following administration by this route.

Toxicokinetics: The extent of absorption varies widely from 30% (nadolol) to 100% (labetalol, betaxolol). The rate of absorption is rapid for nonsustained-release preparations. Sustained release preparations (i.e., Inderal LA) are more slowly absorbed and can have delayed and prolonged clinical effects following poisoning/overdose. The degree of protein binding has a wide range from 0% (sotalol) to 98% (penbutolol). Most of the beta blockers have significant hepatic metabolism (i.e., at least 50%). Nadolol and sotalol are not metabolized and are excreted unchanged in the urine. Absolute bioavailability is often limited by significant first-pass metabolism. Esmolol is metabolized by esterases in the cytosol of red blood cells. Both renal and fecal elimination occur. Nadolol and sotalol are excreted primarily via the kidneys. Elimination half-life ranges from 0.15 hours (esmolol) to 24 hours (nadolol).

Mechanism of toxicity: The toxicities of the beta blockers are directly related to their pharmacologic effects. These agents block the effects of catecholamines such as epinephrine and norepinephrine on the beta-1 and beta-2 receptors. Beta-1 receptors are located in the heart, kidneys, and eyes. Toxicity is most often due to antagonism of the cardiac beta-1 receptors. Cardiac beta-1 stimulation results in increases in sinoatrial rate; myocardial contractility; and increased atrial, atrioventricular node, and ventricular conduction velocity. The beta blockers decrease heart rate. Contractility, and conduction. Beta-2 receptors are found in the bronchioles, vasculature, intestines, uterus, pancreas, adipose tissue, and the liver. Stimulation of bronchial and vascular beta-2 receptors causes smooth muscle relaxation with resultant bronchial dilation and vasodilation.

Human toxicity: The primary clinical effects observed in beta blocker toxicity are cardiovascular in nature. Direct cardiac effects include bradycardia (sinus, atrioventricular node, and ventricular), all degrees of atrioventricular block, bundle branch blocks, and asystole. Ventricular arrhythmias may occur secondary to bradycardia. Torsades de pointes has been associated with chronic toxicity of sotalol. Hypotension occurs and is most often due to decreased cardiac output. Central nervous system effects of these drugs including lethargy, coma, and seizures are secondary to the cardiovascular toxicities. Although seizures and coma could also be secondary to hypoglycemia. Bronchospasm can occur secondary to beta-2 blockade. Hypoglycemia and hyperkalemia can occur.

Clinical management: Advanced life-support measures should be instituted as deemed appropriate. A baseline 12-lead electrocardiogram should be obtained. Continuous cardiac and blood pressure monitoring should be initiated. Syrup of ipecac induced emesis should be avoided due to the potential for a rapid decrease in level of consciousness. Other gastric decontamination procedures should be initiated based on the history of the ingestion and the patient's neurologic status. Whole bowel irrigation may be useful following ingestions of sustained release preparations (i.e., Inderal LA). Bradyarrhythmias and conduction disturbances should be managed with atropine and a pacemaker. Isoproterenol can be effective in increasing heart rate and contractility, but should be used with caution due to its arrhythmogenic and vasodilatory potential. Ventricular arrhythmias should be managed with class IB antiarrhythmics (e.g., lidocaine) and overdrive pacing. Class IA and IC antiarrhythmics should be avoided due to their potential to interfere with conduction. Hypotension should be managed initially with normal saline solution. If decreased cardiac output is responsible for hypotension, dobutamine, amrinone, or isoproterenol can be used. Glucagon has been effective in increasing myocardial contractility in beta blocker toxicity. Glucagon stimulates production of cyclic adenosine monophosphate (AMP), which enhances contraction. Initial intravenous doses of 50 to 100 ug/kg have been used. These are followed by infusions of 70 ug/kg/hr. If cardiogenic shock is resistant to traditional measures, an intraaortic balloon pump and cardiopulmonary bypass should be considered. If systemic vascular resistance is low, vasopressors such as dopamine and norepinephrine should be administered. Hemodialysis or hemoperfusion may be effective in removing acebutolol, atenolol, nadolol, and sotalol.

—*Daniel J. Cobaugh*

Related Topics

ACE inhibitors
Calcium Channel Blockers
Cardiovascular System

Biguanides

♦ CAS: 5188-42-1 (chromate salt); 6272-66-8 (dinitrate salt)

♦ SYNONYMS: Guanyl guanidine; diguanides; amidinoguanidine; Phenformin; Metformin; Buformin

- ◆ CHEMICAL CLASS: Synthetic aliphatic amino alkanes with a large number of amino substituted groups.
- ◆ CHEMICAL STRUCTURE:

HN, NH
H_2N—C—N(H)—C—NH_2

Biguanide

Modified form of biguanides:

Phenyl—CH_2—CH_2—N(H)—C(=NH)—N(H)—C(=NH)—NH_2

Phenformin:
(β-Phenethyl biguanide)

CH_3—CH_2—CH_2—CH_2—N(H)—C(=NH)—N(H)—C(=NH)—NH_2

Metformin:
(n-butylbiguanide)

H_3C—N(CH_3)—C(=HN)—N(H)—C(=NH)—NH_2

Buformin:
(1,1-dimethyl biguanide)

Uses

Biguanides are employed for oral use in the management of mild to moderately severe, non-insulin-dependent (type II) diabetes mellitus in obese patients who are usually above 40 years of age. The disease should have adult onset for the administration of this drug.

Exposure Pathways

The exposure route for this drug is through gastrointestinal absorption after oral treatment.

Toxicokinetics

Phenformin is approximately 50% absorbed from the gastrointestinal tract. It is not highly protein bound in the bloodstream (less than 20%). Phenformin is distributed throughout the major organs. Phenformin is metabolized in the liver by hydroxylation and produces *N-P*-hydroxy-*B*-phenyl-ethyl biguanide as metabolite. About 66% of the biguanide is excreted unchanged and the remaining 33% as metabolite. Phenformin's half-life in the plasma is 7–15 hr versus metformin's 1.5 hr and buformin's 4–6 hr. Metformin and buformin are excreted largely in unchanged form. The renal clearances of buformin, metformin, and phenformin are 393, 440, and 42–262 ml/min, respectively.

Mechanism of Toxicity

Modification of the basic biguanide structure results in difference in potency, metabolism, excretion, and probably toxicity. Phenformin appears initially to produce a gastric mucosal irritability, which may predispose to a number of gastrointestinal symptoms, including gastric hemorrhage. Biguanides induce an increase in peripheral gluconeogenesis and a decrease in intestinal absorption of glucose, vitamin B_{12}, and bile acids. The biguanides do not usually lower the blood sugar in a normal individual unless ethanol or another hypoglycemic agent is simultaneously ingested or there is severe hepatic insufficiency. Phenformin generally lowers the blood sugar level in diabetic and nutritionally starved patient.

Phenformin may act on the cell membrane to decrease oxidative phosphorylation, produce tissue anoxia, increase peripheral glucose uptake (pasteur effect) and lead to lactic acidosis (accumulation of lactic acid) by inhibition of lactic acid metabolism, and hypoglycemia.

Human Toxicity

Phenformin is the only biguanide to have been marketed in the United States and removed by the U.S. FDA in 1977 from approval for use because of its association with the development of lactic acidosis, a metabolic aberration that results in mortality rate of 50–75% and that is still poorly understood. Ethanol intake prior to administration of phenformin in therapeutic doses or excessive dosage appears to predispose the patient to the development of lactic acidosis with a possibly serious outcome. Phenformin and its other relative biguanides are still sold in European and other countries worldwide.

Biguanides are known to cause vomiting, nausea, abdominal cramps, weight loss, epigastric discomfort, agitation, confusion, lethargy, seizures, pulmonary hypertension, tachycardia, myocardial infarction, and hypothermia.

Daily therapeutic doses recommended for humans for three different biguanides are as follows: buformin, 100 mg; metformin, 500 mg ($\times$3) and 850 mg. ($\times$2); phenformin, 25 mg.

Clinical Management

Management of biguanide toxicity is primarily supportive. Its elimination from the body may be enhanced by dialysis or by enhancing excessive urination.

Animal Toxicity

No detailed systematic animal study has been reported to date.

—Vaman C. Rao

Bioaccumulation

Introduction

Man-made chemicals can accumulate in organisms via transfers from water and food. Residues of many persistent, lipophilic chemicals may be bioaccumulated several orders of magnitude above the concentrations found in the environment. Passive partitioning of chemicals is the major process controlling bioaccumulation. As a result, the ability of a chemical to bioaccumulate correlates well with its octanol–water partition coefficient (log K_{ow}). Chemical characteristics that limit bioaccumulation include ionization, high molecular weight, and susceptibility to biotransformation. The ability to bioaccumulate is characteristic of chemicals that have historically posed significant environmental problems.

Bioaccumulation, bioconcentration, and biomagnification address how plants and animals accumulate man-made chemicals from the physical environment—soil, water, and air. Although they all address the same

endpoint—the degree to which chemicals are accumulated by the biota—they are not the same process (see definitions below). The concept of bioaccumulation was unknown before World War II and did not receive significant attention until Rachael Carson published *Silent Spring* in 1962. Her book stimulated decades of research into the environmental fate and effects of man-made materials and the evolution of these concepts.

Initially, the bioaccumulation of DDT [p,p′-DDT or 2,2-*bis*(*p*-chlorophenyl)-1,1,1-trichloroethane] received the most attention. It is a potent, inexpensive, broad-spectrum insecticide that does not pass through the skin of man but is readily absorbed by insects. It was the first of the chlorinated hydrocarbons, like PCBs and dioxins, that caused substantial environmental damage. Even today, we do not fully comprehend the processes that dictate a chemical's fate (i.e., where they go) and effects (i.e., what they do) over time and space following their release into the environment. However, we do know a great deal about the physical–chemical factors that control the accumulation of synthetic organic chemicals by aquatic organisms. Hence, that will be the primary focus of this entry, but the accumulation of other materials by plants, birds, and worms will also be addressed.

Definitions

Bioaccumulation: Bioaccumulation is the process by which living organisms accumulate chemicals both directly from the water and from dietary sources. The concentration of a chemical attained in the organism may or may not exceed the concentration in the source(s).

Bioaccumulation factors (BAFs): A bioaccumulation factor is an estimate of the tendency of an environmental contaminant to accumulate in living organisms as a result of their combined exposure to water, food, and sediments.

Bioconcentration: Bioconcentration is the process by which organisms accumulate chemicals directly from the water: the passive partitioning of chemicals between organisms and water which is well correlated with their *n*-octanol partition coefficients.

Fugacity: Fugacity is a measure of a molecule's "urge" to flee or escape from a system or the chemical potential of a substance in a particular environmental compartment, like the atmosphere or fish fat. For example, at 14°C the fugacity capacity (Z) for oxygen is about 1.5 gmol/m³ atm (10 mg/liter) in water and about 40 gmol/m³ atm in air. Hence, at equilibrium, the concentration of oxygen in air is 27× that of water.

Biomagnification: Biomagnification is the condition that arises whenever a particular chemical residue is transferred more efficiently than energy transfer in the form of biomass, which causes chemical concentrations to increase with increasing steps of ecological food chains. The fugacity of very hydrophobic (i.e., log K_{ow} greater than 6) persistent chemicals is elevated in the gastrointestinal tract of fish as food is digested, which makes it possible for the fugacity of chemical residues in predators to exceed that in prey species.

Hydrophobic or lipophilic: Hydrophobic or lipophilic compounds are sparingly soluble in water and highly soluble in lipids or body "fats." This tendency is usually quantified by the partition coefficient in water vs *n*-octanol, which is denoted by log P or log K_{ow}. For example, at equilibrium about 1×10^6 times more DDT is found in a volume of octanol than an equivalent volume of water; therefore, it has a K_{ow} of 10⁶ or log K_{ow} = 6.

Uptake rate: The uptake rate is the rate at which a body accumulates a lipophilic substance. Initially, this may be modeled as a zero-order process that is directly proportional to the concentration of chemical dissolved in the water (where k_1 is the uptake rate constant).

Depuration rate: The depuration rate is the rate at which an organism eliminates a foreign chemical (i.e., is purified). Kinetically this can usually be expressed by as a simple first order rate constant (k_2) but biphasic patterns are not uncommon.

Bioconcentration factor (BCF): The bioconcentration factor is the ratio between the truly dissolved concentration in the water and the concentration in the fish at equilibrium. BCF = C_t/C_w,

where C_t is the concentration in tissue (either whole organism or specified tissue) and C_w is the concentration in the water. For a simple model the BCF at steady state is defined as k_1/k_2, which has been shown to be directly correlated with log K_{ow}.

Key Principles

A chemical material is accumulated by an organism whenever the uptake rate exceeds the depuration rate. The uptake rate for organic chemicals increases slightly and the depuration rate decreases dramatically as log K_{ow} increases. Uptake rates also appear to increase and depuration rates decrease as the weight of fish increases. Likewise, depuration rates generally decrease as the size of organisms increases, which may contribute to trophic-level increases in the bioaccumulation of chemicals often observed in the field. Chemical residues are believed to be accumulated by passive diffusion; no energy is expended to enhance their transfer across biological membranes like cell walls or the gastrointestinal tract.

Biomagnification of chemicals essentially requires the chemicals to move against a thermodynamic gradient (i.e., from low fugacity in prey to higher fugacity in the predator), which suggests processes other than passive partitioning are involved. However, digestion and absorption of food (i.e., prey) in the gastrointestinal tract reduces the fugacity capacity of the food and raises fugacity of a chemical in the gastrointestinal tract, which provides the thermodynamic gradient required for passive diffusion from prey to predator. Furthermore, species differences in lipid content may provide an intrinsic thermodynamic gradient for chemicals to partition across. Finally, the utilization of different compartments may contribute to trophic-level differences in bioaccumulation (i.e., surface adsorption for algae, single-compartment partitioning for invertebrates, and multiple-compartment partitioning for fish).

Limitations

The accumulation of stable, nonpolar organic chemicals is largely controlled by their hydrophobicity (which is usually well defined by their log K_{ow}). Many hydrophobic materials may not be accumulated by fish either because their aqueous solubility is truly negligible or because their molecular volume (e.g., molecular width greater than 9.5 Å or total molecular surface area greater than 400 Å²) exceeds that which can cross cell membranes. Likewise, if a hydrophobic material is also poorly soluble in lipids, as characterized by their poor solubility in octanol, the log–log relationship between bioconcentration and hydrophobicity breaks down. Finally, work that was conducted with a wide range of chlorinated hydrocarbons demonstrated that direct uptake from the water of the "super-lipophilic" chemicals (log K_{ow} greater than 6) is considerably reduced but that dietary routes could be significant unless hindered by a particular structural or physico-chemical property that interfered with membrane transport.

Unfortunately, there does not appear to be any simple, over-riding principle or force to describe the accumulation of polar, readily metabolized and/or reactive organics and metals. The bioavailability and bioaccumulation of metals is exceptionally complex because valence state changes resulting from relatively small shifts in pH or oxygen status can have a profound effect on their physical properties.

Organisms are equipped with a wide array of enzymes that function to increase the aqueous solubility (i.e., reduce lipophilicity) of chemicals. Biotransformations can also modify the structure of chemicals to render them susceptible to active elimination processes. Hence, biotransformations usually enhance the depuration of chemicals and thereby limit bioaccumulation.

Aqueous solubility is generally inversely correlated with lipid solubility and it, like log K_{ow}, has been used as a measurable physical property that may be correlated with the ability of an organism to accumulate a particular material from its environment. However, the aqueous solubility of a material may become so reduced that membrane transfer becomes limiting. Thus, although it is generally true that non-ionic, non-degradable, lipophilic chemicals will be accumulated by aquatic organisms in direct proportion to their log K_{ow} or inverse to their aqueous solubility, there are many exceptions. For example, in the laboratory the addition of small amounts of dissolved organic matter to the exposure water can often serve to make materials with a log K_{ow} of greater than 6 largely unavailable; they are not readily taken up directly from the water. However, if they are contained in the sediments they may be readily transferred through the food-chain and be biomagni-

fied. Consequently, how well a worm, plant, bird, insect, or fish may accumulate a particular chemical moiety should never be assumed a priori; it should be determined experimentally.

Further Reading

De Wolf, W., De Bruijn, J. H. M., Seinen, W., and Hermens, J. L. M. (1992). Influence of biotransformation on the relationship between bioconcentration factors and octanol–water partition coefficients. *Environ. Sci. Technol.* **26**, 1197–1201.

Doucette, W. J., and Andren, A. W. (1987). Correlation of octanol/water partition coefficients and total molecular surface area for highly hydrophobic aromatic compounds. *Environ. Sci. Technol.* **21**(8), 821–824.

Gobas, F. P. C., Zhang, X., and Wells, R. (1993). The mechanism of biomagnification and food chain accumulation of organic chemicals. *Environ. Sci. Technol.* **27**(13), 2855–2863.

Hamelink, J. L., and Spacie, A. (1977). Fish and chemicals: The process of accumulation. *Annu. Rev. Pharmacol. Toxicol.* **17**, 167–177.

LeBlanc, G. A. (1995). Trophic-level differences in the bioconcentration of chemicals: Implications in assessing environmental biomagnification. *Environ. Sci. Technol.* **28**(1), 154–160.

Mackay, D. (1979). Finding fugacity feasible. *Environ. Sci. Technol.* **13**(10), 1218–1223.

Spacie, A., and Hamelink, J. L. (1982). Alternative models for describing the bioconcentration of organics in fish. *Environ. Toxicol. Chem.* **1**(4), 309–320.

—*Jerry L. Hamelink and Gerald A. LeBlanc*

Related Topics

Bioconcentration
Biomagnification
Biomarkers, Environmental
Biotransformation
Ecological Toxicology
Effluent Biomonitoring
Environmental Processes
Environmental Toxicology
Pesticides
Pollution, Water
Risk Assessment, Ecological
Toxicity Testing, Aquatic

Bioconcentration

Introduction

The reader is strongly encouraged to read the entry on bioaccumulation before reading this entry as it should help place this material in context.

Soon after Rachael Carson published *Silent Spring* in the early 1960s, George Woodwell published a paper in volume **216**(3) of *Scientific American* that graphically illustrated the basic mechanism by which organisms were believed to accumulate elevated concentrations of contaminants from their environment. He, and others, viewed transfers through the food-chain as the only route by which vertebrates like fish and birds could accumulate these materials. However, several field studies with DDT in the late 1960s demonstrated that it did not "behave" like a nutrient or radioactive fallout, and that algae, invertebrates, and fish could accumulate residues directly from the water. Soon thereafter a number of researchers established the relationship between the *n*-octanol partition coefficient of a chemical material and its accumulation by fish. That is the primary focus of this section—the direct accumulation or sorption of organic materials by biota from environmental media.

Definitions

Bioconcentration: Bioconcentration is the process whereby organisms accumulate chemicals through nondietary routes or as the uptake of a chemical directly from the abiotic environment (i.e., media) resulting in a concentration of chemical in the organism that is higher than the environmental concentration.

Uptake rate: The uptake rate is the rate at which a body accumulates a lipophilic substance. Initially, this may be modeled as a zero-order process that is directly proportional to the concentration of chemical dissolved in the water (where k_1 is the uptake rate constant).

Depuration rate: The depuration rate is the rate at which an organism eliminates a foreign chemical (i.e., is purified). Kinetically this can usually be expressed by as a simple first order rate constant (k_2) but biphasic patterns are not uncommon.

Bioconcentration factor (BCF): The bioconcentration factor is the ratio between the truly dissolved concentration in the water and the concentration in the fish at equilibrium. BCF = C_t/C_w, where C_t is the concentration in tissue (either whole organism or specified tissue) and C_w is the concentration in the water. For a simple model the BCF at steady state is defined as k_1/k_2, which has been shown to be directly correlated with log K_{ow}.

Key Principles

By definition bioconcentration is a physical partitioning process. Residues in the media, water, soil, or air are accumulated directly by the biota. None of the material needs to be transferred through the food chain. Hence, bioconcentration can be viewed as the result of competing rates of uptake and depuration (i.e., BCF = k_1/k_2). However, the exact mechanical principle involved is not necessarily clear. Algae in pools treated with DDT appeared to adsorb, not absorb or partition, residues from the water, while partitioning does appear to be operative in terrestrial vegetation. Hence, the specific mechanics behind the process of bioconcentration in the environment are not explicitly obvious.

Metabolic biotransformations are generally considered to be nonexistent or of a negligible rate. However, in many cases this is not true. Fish are often able to metabolize a chemical which usually renders it more water soluble or lowers the K_{ow}, causing it to display a substantially lower BCF than one would expect based on the initial K_{ow} of the material alone. This kind of deviation is more pronounced and obvious the higher the initial K_{ow} simply because depuration is usually greatly enhanced, and any small increase in depuration can dramatically lower the BCF. Fortunately, it is also easier to see this kind of effect with the higher initial K_{ow} materials because k_2, the depuration rate, is strongly inversely correlated with log K_{ow}. Thus, if k_2 is substantially faster than expected, based on K_{ow}, look for metabolic biotransformation products in the effluent water.

The BCF for a chemical is commonly determined in the laboratory by continuously exposing a number of fish to a uniform concentration of the chemical, under flow-through conditions, for several days or weeks and then transferring the survivors to untreated water. Individual fish are periodically taken out of the system and analyzed for the chemical of interest and, perhaps, fat content. Ideally, the fish would reach steady-state (i.e., a plateau in body concentrations resulting from essentially equivalent amounts being taken-up and depurated over a period of time) during the exposure period because the BCF could then be directly determined as the ratio between the body burden at steady-state versus the average exposure concentration. However, because BCF may also be defined as being equal to k_1/k_2, one really need only to determine these rate constants well in order to derive the BCF for most chemicals. Since this can usually be accomplished with only a few days to a week of exposure (with exposure time increasing as K_{ow} increases) and a similar amount of time for the depuration phase, this information is generally provided about new non-ionic, metabolically stable organic chemicals.

Limitations

The forces, events, and physicochemical properties that may limit bioconcentration are well described by M. G. Barron (see Further Reading) and they will not be elucidated in detail here. Suffice to say that bioconcentration may be profoundly influenced by temperature, water quality, species, the size, weight, and fat content of the organism, and/or growth, as influenced by diet and season. Both lipid content and fatty acid composition appear to play pivotal roles since species differences and individual variability may often be accommodated by normalizing to lipid content, but it is only one determinant of accumulation. A large number of forces operate to determine how well a material is bioconcentrated, including molecular size or shape, true aqueous or lipid solubility (or the lack thereof), and biotransformations. Many of these limitations have been explored well enough for one to anticipate their existence. For example, materials having a cross-sectional surface area greater than 9.5 Å are not likely to be bioconcentrated. Hence, materials like the high-molecular-weight polydimethylsiloxane polymers are not readily accumulated by fish. However, one must recognize that determining a material does not undergo appreciable bioconcentration does not automatically make it acceptable in the environment. There are a score of both direct and indirect ways that man-made materials may have adverse effects on the environment and being bioconcentrated is simply an easily quantified route by which the biota may accumulate them.

Further Reading

Barron, M. G. (1990). Bioconcentration. *Environ. Sci. Technol.* **24**, 1612–1618.

Hamelink, J. L., Waybrant, R. C., and Ball, R. C. (1971). A proposal: Exchange equilibria control the degree chlorinated hydrocarbons are biologically magnified in lentic environments. *Trans. Am. Fish. Soc.* **100**(2), 207–214.

Spacie, A., and Hamelink, J. L. (1982). Alternative models for describing the bioconcentration of organics in fish. *Environ. Toxicol. Chem.* **1**(4), 309–320.

—*Jerry L. Hamelink*

Related Topics

Bioaccumulation
Biomagnification
Biomarkers, Environmental
Ecological Toxicology
Effluent Biomonitoring
Environmental Processes
Environmental Toxicology
Pesticides
Pollution, Water
Risk Assessment, Ecological
Toxicity Testing, Aquatic

Biomagnification

The reader is strongly encouraged to read the section on bioaccumulation first because many of the words used to describe certain key processes, including biomagnification, are defined therein.

Biomagnification is one kind of bioaccumulation. In the literature, one is more likely to see reference to something like food-webs or food-chain transfers than use of the term biomagnification, but they all refer to the same process: how elevated concentrations of contaminants contained in the water or sediments or soils are accumulated by the biota. For all practical purposes, biomagnification of organic chemicals does not become significant until log K_{ow} and log BCF are greater than 5; then it increasingly becomes the driving force and direct uptake, which is defined as bioconcentration, becomes negligible as K_{ow} increases. Conversely, nutrients and the elements contained in radioactive fallout rely entirely on transfers through the food-chain. However, in order to be biologically magnified, food digestion in the gastrointestinal tract must increase the chemical fugacity of the food by altering the fugacity capacity of the food.

The process of biomagnification has been well defined with a relatively simple model for quantifying the uptake and clearance of organic chemicals from food and water by fish. In this model the fugacity of the food (i.e., zooplankton) is set equal to the water, as observed experimentally. The plankton are eaten by small fish (i.e., Vol = 1 cm³), the small fish are eaten by medium-sized fish (i.e., Vol = 100 cm³), and the medium-sized fish are eaten by large fish (i.e., Vol = 10000 cm³). By running the model sequentially, using fish from one level as food for the next, with a log K_{ow} = 6 chemical, the fugacity increases from 2.50×10^{-4} Pa in the water and plankton to 13.8×10^{-4} Pa in the large fish. Consequently, the fish fugacity becomes an increasing multiple of the water fugacity with each step of the food chain. This effect/reality becomes even more pronounced when one goes from fish to birds. Birds are not nearly as energy efficient as fish, so the fugacity of this kind of chemical in a fish-eating bird can easily become a multiple of that found in the fish. That is biomagnification at its best, and that is why the predatory birds have been so heavily impacted by organochlorine residues in the environment.

Further Reading

Clark, K. E., Gobas, F. A. P. C., and Mackay, D. (1990). Model of organic chemical uptake and clearance by fish from food and water. *Environ. Sci. Technol.* **24**(8), 1203–1213.

Gobas, F. A. P. C., Zhang, X., and Wells, R. (1993). Gastrointestinal magnification: The mechanism of biomagnification and food chain accumulation of organic chemicals. *Environ. Sci. Technol.* **27**(13), 2855–2863.

Hamelink, J. L., Waybrant, R. C., and Yant, P. R. (1977). Mechanisms of bioaccumulation of mercury and chlorinated hydrocarbon pesticides by fish in lentic ecosystems. In *Fate of Pollutants in the Air and Water Environments—Part 2* (I. H. Suffet, Ed.), pp. 261–282. Wiley, New York.

—*Jerry L. Hamelink*

Related Topics

Bioaccumulation
Bioconcentration
Biomarkers, Environmental
Ecological Toxicology
Environmental Processes
Environmental Toxicology
Pesticides
Pollution, Water
Risk Assessment, Ecological
Toxicity Testing, Aquatic

Biomarkers, Environmental

The term "biomarker" has been used increasingly by environmental toxicologists to denote early adaptive or aberrant biochemical, physiological, or histopathological changes in organisms in response to exposure to chemical and physical stresses. The concept is similar to the "critical effect," as used by some mammalian toxicologists to highlight the most sensitive effect known, with the assumption that if the critical effect can be avoided then the organism is protected against all other effects. The applications of clinical chemistry, histopathology, and physiological measurements to environmental problems as biomarkers has often required adaptations of well-established methodologies that are used routinely in general toxicology to uncover early and subtle effects.

Biomarkers can be indicators of exposure and/or of early effects. Direct measurements of the chemical can be useful indicators of exposure, but they have limited utility for many substances which are readily metabolized and excreted. Otherwise, substances may be found to accumulate before they elicit effects.

Most early responses in individuals tend to be adaptive and tend to be reversible when the exposure ceases. When these early responses occur in most individuals before adverse effects can be found at the population level, then this biomarker of effects can be used as a warning sign for potential effects at the population level. Ideally, biomarkers can be measured in small samples of individuals and provide adequate warnings against potential effects in large populations.

Examples of phenomena that have been used as biomarkers include the following:

- Stimulation or suppression of the mixed-function oxidase system, often assessed through cytochrome P450 levels

- Assessment of recent exposures to polycyclic aromatic hydrocarbons (PAHs) through measurements of biliary PAH metabolites by their fluorescence

- Evaluation of chromosome damage and DNA adduct formation

- Assessments of early effects on the immune system through the evaluation of mitogenic responses of lymphocytes

Practically all of the responses that have been used as biomarkers also exhibit natural variabilities in relation to various factors including season, age, sex, reproductive status, and nutrition. Therefore, they need to be used with caution and with proper control strategies. The utility of biomarkers as a tool in environmental toxicology depends on their sensitivity, variability, diagnostic specificity, and applicability to field conditions.

Further Reading

Huggett, R. J., *et al.* (1992). *Biomarkers.* Lewis, Boca Raton, FL.
McCarthy, J. F., and Shugart, L. R. (1990). *Biomarkers of Environmental Contamination.* Lewis, Boca Raton, FL.

—*Rolf Hartung*

Related Topics

Biomarkers, Human Health
Risk Assessment, Environmental

Biomarkers, Human Health

Introduction

In the field of environmental health research, one often has a problem determining if exposures to a specific chemical or to a chemical mixture have induced an adverse health effect. In the 1980s, a promising new field of research, called biological markers (or biomarkers), began to develop as an aid for linking toxicant exposures with potential health effects. A biological marker has been defined by the National Research Council (NRC) of the National Academy of Sciences (NAS) as an exogenous substance or its metabolite or the product of an interaction between a xenobiotic agent and some target molecule or cell that is measured in a compartment within an organism. Biomarkers are always found within an organism and can be used to demonstrate the relationship between exposure, internal dose, dose to target organ, biologically effective dose, initial biological effects and induced adverse health effects (Figure B-10). In the absence of biomarkers, epidemiology studies have relied on external indicators of exposure (area or personal monitors, questionnaires, histories) to find associations between past exposures and subsequent disease development. Such tools are not highly accurate, and most only measure exposures to populations, not individuals. New molecular techniques have provided the tools for assessing the dose received by an individual by measuring the chemical adducts formed with macromolecules in the body; the same molecular techniques also provide information on the initial changes induced by the biologically effective dose, changes that may eventually lead to disease. These new biologic markers take into account

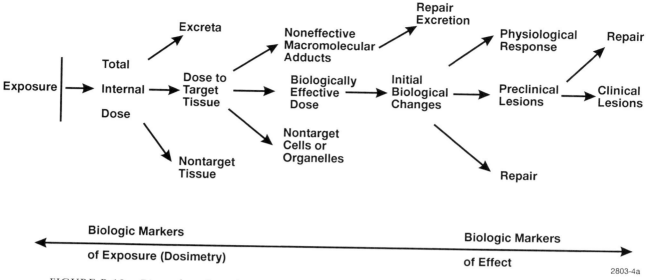

FIGURE B-10. Biomarkers for risk assessment. Toward the left are biomarkers of exposure (dosimetry); most of these markers represent values obtained from toxicokinetic studies. Toward the right are biological markers of effect; many of these markers are standard signs and symptoms familiar to clinicians. The goal of biomarker research is to obtain more information on the link between biologically effective doses and the early, initial biological changes that can lead to disease; such values will come from studies on mechanism of disease induction.

individual variability in processing of potential toxicants and show promise for providing more accurate information for the assessment of the effect of exposures to noxious agents on the induction of adverse health effects in individuals.

The most useful biomarkers are those that are chemical-specific, and are (1) quantitatively relatable to the degree of the prior or ongoing exposure or (2) quantitatively relatable to, or predictive of, later developing disease. Ideally, one would want the biomarker of the extent of exposure to be also predictive of health outcome, but it is rare that sufficient information is available to make such predictions. The emphasis on quantitation is to meet the requirements for setting regulations for allowable exposures that are protective of health. If one only needs to know if an exposure has occurred, the presence of a biological marker specific for the chemical of concern may be all that is needed. However, for the purposes of risk assessment, that is, determining the potential for a given exposure to an exogenous substance to cause adverse health effects, one needs to quantitate the amount of biomarker present.

There are also practical requirements for biomarkers that are to be used in human studies. The biomarker should be in a relatively available tissue or fluid. For example, urine and breath are body fluids that are easily sampled; sampling blood is an invasive process, but can be done; however, liver tissues from humans cannot be sampled for DNA adducts except at the time of autopsy. The assays for the marker of interest should not be so expensive that the cost of a study using the marker is prohibitive. Finally, the marker must be validated for what the marker indicates in terms of either exposure or health outcome. Otherwise, the results of the biomarker assays can not be interpreted.

Biomarkers have been described in terms of markers of exposure, markers of effects, and markers of susceptibility. These types of biomarkers are described below, along with some strategies for their use in risk assessment.

Biomarkers of Exposure

The NRC/NAS has defined biological markers of exposure as exogenous substances or their metabolites or the product of an interaction between a xenobiotic agent and some target molecule or cell that is measured in a compartment within an organism.

Biomarkers of exposure are all measures of internal substances, and are thus biological markers of the dosimetry that results from exposures. Markers of interest include measures of total internal dose, such as blood, urine or breath levels of a chemical; markers of the dose to a target organ, which may be in the form of macromolecular adducts formed between the chemical or its metabolite and the organ tissue; and the biologically effective dose, which can only be measured if the mechanism of disease induction is known in sufficient detail to suggest what entities might represent the biologically effective dose. An example of the latter might be a specific DNA adduct known to lead to a disease.

Some of these biomarkers are the parameters measured in classical pharmacokinetic (or toxicokinetic) studies, such as blood, urine, or breath levels of a substance. Other biomarkers have become available only recently due to the development of new techniques that allow detection, and sometimes quantitation of macromolecular adducts formed from the interaction of the chemical of interest or its metabolite with blood proteins or tissue DNA. Even if the biomarkers of exposure are not thought to lead to disease, such biomarkers can be useful, if one can link them quantitatively to other biomarkers that do lead to disease. For example, if one knows the quantitative relationship between levels of hemoglobin adducts formed during exposure to a specific chemical (example of an adduct formed with a blood protein and an adduct that is not known to lead to disease) and levels of liver DNA adducts (example of an adduct formed in the target tissue and a biomarker of the biologically effective dose) from the same exposure to a liver carcinogen, one could theoretically use the more available blood adducts as a predictor of the biologically effective dose.

Strategies for Use of Biologic Markers of Exposure to Assess Prior Exposures

Many commonly measured pharmacokinetic values, such as parent compound or metabolites in exhaled breath, blood, or urine, macromolecular adducts or degradation products of such adducts that appear in urine, can be used as biomarkers of exposure. To make quantitative assessments of the relationship of such markers to prior exposures, one must determine the rate of formation and removal (clearance) of the marker. From this information one can predict the

steady-state concentrations of the marker following various exposure scenarios. Also, with information on the rate of formation and removal of a marker, and the factors that influence those rates (such as gender, dose, repeated exposures, route of exposure, rate of exposure), one can develop a mathematical model that will describe the concentration of the marker under different exposure conditions. While the concentration of the marker cannot be used to indicate a unique exposure scenario, the marker can indicate the types of exposure regimens that would produce the measured level of the biomarker.

From a practical viewpoint, one cannot use human populations to determine the rate of formation and clearance of markers and the influence of various factors on those rates. Therefore, most toxicokinetic studies are conducted in animal models. From detailed studies in animals, mathematical models are derived based on the animal toxicokinetic data, animal physiological data, and the physical/chemical properties of the compounds of interest (such as partition coefficients). The models, which are often referred to as physiologically based pharmacokinetic models, can then be modified for use in making predictions for humans by substituting human physiological data into the model and using the results of metabolic rate studies conducted with human tissues in vitro. The validity of such modified models must then be verified by limited studies in humans.

A second strategy for the use of biomarkers to help describe prior exposures is to make use of a battery of biomarkers with differing half-lives. Some biomarkers of exposure have half-lives of minutes or hours (volatile parent compound in exhaled breath, some blood or urinary metabolites); other biomarkers may be present for days or weeks (some DNA adducts, blood albumin adducts); while others may accumulate over longer periods of time due to longer half-lives (blood hemoglobin adducts, some DNA adducts, products of DNA repair in the urine). There are also differences in the fraction of the internal dose of a chemical that is converted to each type of biomarker. In general, some of the markers with shorter half-lives, such as urinary metabolites, represent large fractions of the internal dose, while macromolecular adducts, many of which have longer half-lives, represent only a small fraction of the dose. By combining knowledge of the half-lives of markers and the amount of marker formed relative to the total dose, one can obtain more information about a prior exposure using a battery of biomarkers than a single bio-marker. For example, if multiple markers of a single chemical are determined in an individual, one should be able to distinguish between someone who has had a recent exposure, someone who is receiving an ongoing exposure, and someone who was exposed repeatedly in the past but has had no recent exposures. If someone has had only a recent single exposure to a chemical, the shorter half-lived, more abundant biomarkers in the form of urinary metabolites should be readily de-tectable, but there should be very little of the longer half-lived, less abundant DNA adducts present. If the person has had an ongoing exposure for many years to the same chemical, there should be relatively high amounts of both the urinary metabolites and the DNA adducts. If the person was exposed some time ago, but not recently, than only the longer-lived DNA adducts or hemoglobin adducts may be detectable.

Biomarkers of Effect

The NRC/NAS defines a biomarker of effect as any change in a biological system that is qualitatively or quantitatively predictive of health impairment or po-tential impairment resulting from exposure. While a distinction is made between biomarkers of exposure and biomarkers of effects, in practice, the two areas overlap. For example, DNA adducts may be markers of exposure, but if they occur at specific sites known to induce mutations leading to cancer, the adducts may also be biomarkers of effect.

Many types of responses may be made to toxicants. Some of the induced effects may be merely physiological responses that are not deleterious. Other responses may be deleterious, but are quickly repaired. But some re-sponses represent the earliest indicators of a change that, if persistent, can lead to an adverse health effect. The persistence and amplification of such a response leads to a clinical disease state. The most useful bio-marker of an effect is one that is definitive for a specific adverse health effect, is quantitatively predictable for health outcome, and occurs early enough in the process that its detection allows intervention in the disease pro-cess. Signs and symptoms that occur in later stages of a disease process are the tools of clinical medicine; the goal of scientists working on biomarkers is to discover pre-clinical markers of the early stages of a disease process when intervention is still possible. Examples

of early markers of a disease process might be a pre-neoplastic proliferative lesion, or an increase in a cyto-kine that is associated with a fibrotic process or the under-or over-expression of a gene known to be associ-ated with a disease process.

To be able to relate markers of exposure to health outcome, one needs to know which markers can be associated with the disease outcome and the degree of that association. That is, given the presence of a certain level of a biomarker, what is the probability of getting a disease? This query is certain to made by participants in any study in which biomarkers are assayed in work-ers or in the general public. Currently, we have very little information on which to base an answer. The inability to use biological markers of either exposure or effect to predict health outcome represents a major gap in our knowledge and decreases the potential use-fulness of the markers. What is needed are valid mark-ers of risk.

Strategies for Improving the Ability to Link Biomarkers of Effect to Disease Outcome

Perhaps the most fruitful area of research for discover-ing biomarkers of exposure that can be linked to disease outcome is the study of mechanisms of disease induc-tion. One cannot define a marker of a "biologically effective dose" unless one knows the mechanism by which the biological effect is induced. Likewise, one cannot know the earliest biological events that lead to a disease, unless one understands the mechanism of disease induction. Such mechanistic studies should pro-vide the markers for steps that link the biologic markers represented by traditional toxicokinetic measurements and the biologic markers represented by traditional clinical markers of disease.

In addition to knowledge on the mechanism of dis-ease induction, one must define the quantitative rela-tionship between the level of the marker and the proba-bility of progression to an adverse health effect. Pharmaco–or toxicodynamic modeling describing the kinetics of disease development is required in a manner similar to the toxicokinetic modeling used to describe the kinetics of internal dosimetry. For example, if one wants to use chemical-specific DNA adducts to predict cancer induction, the following pieces of information are required. First one must identify the various DNA adducts formed by the chemical. Then one must determine the biological half lives of each adduct (How long will they be present before they are repaired?) and the mutagenic potential of each adduct (How much harm will the adducts cause if they are present?). If adducts are formed that have relatively long half-lives and high mutagenic potential, one can determine if the mutations induced by the adduct in *in vitro* studies are present in tumors induced by the chemical. Once enough is known about the disease induction to form an hypothesis for the process, one can use intervention studies, in which the proposed path to disease is blocked, to validate the path as the active disease generating process. Finally, toxicodynamic models can be generated that describe the quantitative relationship between adduct levels and cancer induction. Such models will require knowledge of cellular dynamics involved in tumor formation.

Biomarkers of Susceptibility

Indicators of individual or population differences that influence the response to environmental agents are called biomarkers of susceptibility. These might include such characteristics as an enhanced metabolic capacity for converting a chemical to its reactive, more toxic metabolite or an enhanced capacity to detoxicate reactive metabolites or differences in number of receptor sites that are critical for a specific response. A specific example is the inherited deficiency in the enzyme, alpha-1-antitrypsin, which is associated with an increased susceptibility to development of emphysema. Such markers can be quite valuable in providing information that will allow protection of susceptible populations. Knowledge of the mechanisms of susceptibility can also be important in designing therapy for a disease. However, the use of such markers is fraught with legal and ethical problems, because identification of persons with enhanced susceptibility to adverse health effects from exposure to chemicals could lead to discrimination against those persons in obtaining certain jobs and insurance.

Uses of Biomarkers

As mentioned in the beginning of this section, a major potential use for biological markers is to link environmental exposures causally and quantitatively to health effects in individuals or populations. If the whole chain of events illustrated in Figure B-10 can be defined in a quantitative fashion for a single toxicant, the exposure to such a toxicant could be regulated to prevent adverse health effects with a reasonable degree of certainty. This would avoid over-or underestimation of the risk from such a toxicant. In practice, this use of biomarkers is still in its infancy because of insufficient data to fill in all the information illustrated in Figure B-10. Strategies for improving our ability to use biomarkers to quantitate the ability of environmental agents to produce adverse health effects have been discussed. Another practical use for biomarkers is to detect and quantitate prior or ongoing exposures to specific chemicals; biomarkers have been successfully used in biological monitoring programs in industry, but have only recently been used to monitor environmental exposures. Medical research is seeking biomarkers to detect early stages of a disease to allow intervention; to determine the effectiveness of intervention strategies; and to detect cells at risk from a toxicant. Finally, research is ongoing, particularly in the field of genetics, to find inherited biomarkers of susceptibility for use in the detection and protection of sensitive populations.

Further Reading

Henderson RF. Strategies for use of biological markers of exposure. *Toxicol. Lett.* 1995;82/83:379–383.

National Research Council, Committee on Biological Markers. Biological markers in environmental health research. *Environ. Health Perspect.* 1987;74:3–9.

National Research Council. Biologic Markers in Pulmonary Toxicology. Washington, DC: National Academy Press; 1989.

National Research Council. Biologic Markers in Reproductive Toxicology. Washington, DC: National Academy Press; 1989.

National Research Council. Biologic Markers in Immunotoxicology. Washington, DC: National Academy Press; 1992.

National Research Council. Environmental Neurotoxicology. Washington, DC: National Academy Press; 1992.

Mendelsohn ML, Peeters JP, Normandy, MJ, eds. Biomarkers and Occupational Health, Progress and Perspectives. Washington, DC: Joseph Henry Press; 1995.

—Rogene F. Henderson

Related Topics

Analytical Toxicology

Biomarkers, Environmental

Carcinogen-DNA Adduct Formation and
  DNA-Repair

Epidemiology

Mechanisms of Toxicity

Medical Surveillance

Molecular Toxicology

Pharmacokinetic Models

Pharmacokinetics/Toxicokinetics

Risk Assessment, Human Health

Biotransformation

Introduction

Biotransformation refers to the process during which lipophilic (fat soluble), xenobiotic (foreign), or endobiotic (endogenous) chemicals are converted in the body by enzymatic reactions to products that are more hydrophilic (water soluble). In this context, metabolism and metabolic transformation are synonymous with biotransformation. A xenobiotic is a relatively small (molecular weight less than 1000), nonnutrient chemical that is foreign to the species where metabolism occurs.

The major purpose of biotransformation is to chemically modify (metabolize) poorly excretable lipophilic compounds to more hydrophilic chemicals that are readily excreted in urine and/or bile. Without metabolism, lipophilic xenobiotics accumulate in biota, increasing the potential for toxicity. Examples of such compounds are highly halogenated polychlorinated biphenyls (PCBs) and polychlorinated dibenzofurans (TCDD and dioxins) that occur as tissue residues in humans. On the other hand, biotransformation is normally not required for xenobiotics with high water solubility because of rapid excretion in urine.

Two or more sequential enzymatic reactions are routinely required to convert lipophilic chemicals to metabolites that are efficiently excreted. R. T. Williams, a pioneer in biotransformation studies, classified these pathways as phase I (oxidation, reduction, and hydrolysis reactions) and phase II (conjugation reactions). Normally, a phase I reaction precedes its phase II counterpart, but some compounds contain functional groups that are sites for direct conjugation (e.g., $-OH$, $-COOH$, and $-NH_2$). Frequently, the biological activity of a chemical decreases (termed "detoxication") during metabolism but this is not always the case. Both phase I and phase II reactions can function in "toxication" or metabolic activation processes as well, and this is a fundamental mechanism for the formation of many chemical toxicants. Multiple classes of toxic compounds, including polycyclic aromatic hydrocarbon-derived carcinogens and mutagens, are formed by cytochrome P450-dependent oxidative metabolism, the most common toxication pathway.

The highest concentration of xenobiotic metabolizing enzymes is routinely found in liver, but epithelial cells of extrahepatic tissues, such as lung, kidney, intestine, placenta, and eye, also have activity. Relative to liver, extrahepatic tissues do not normally play a major quantitative role in the biotransformation of foreign compounds, including drugs. Extrahepatic organs, however, can be extremely important in the metabolic activation of xenobiotics and resultant target organ toxicity because the ratio of activation to detoxication enzyme activity is frequently higher in these cells than in hepatocytes (i.e., bioactivation predominates over detoxication and results in the formation of concentrations of active metabolites that overwhelm the capacity of detoxication pathways). The contribution of intestinal flora to the *in vivo* metabolism of xenobiotics can also be significant, especially for chemicals that require anaerobic (oxygen-deficient) reduction as a quantitatively important pathway.

Oxidation Reactions

Oxidation is the most common metabolic reaction for lipophilic xenobiotic and endobiotic compounds, in part because most mammalian tissues are well oxygenated.

Cytochrome P450 Monooxygenase System

The cytochrome P450 (P450)-dependent monooxygenase system is concentrated in the endoplasmic reticulum of cells and is referred to as a microsomal enzyme system. This P450 system is composed of multiple forms or isozymes of P450 belonging, in mammals, to 10 distinct gene families as well as the flavoprotein, NADPH-P450 reductase. This monooxygenase system is known as a "universal" oxidase because it catalyzes the oxidation of a multitude of lipophilic compounds including both xenobiotics (antioxidants, carcinogens, drugs, environmental pollutants, food additives, hydrocarbons, and pesticides) and endobiotics (bile acids, cholesterol, eicosanoids, fatty acids, lipid hydroperoxides, retinoids, and steroid hormones). Several excellent reviews of the P450 monooxygenase system are available, including those which focus on this system in humans (see Further Reading).

With several classes of xenobiotic substrates, including chemical carcinogens such as benzo(a)pyrene or the mycotoxin, aflatoxin B1, some metabolites are more toxic than the parent chemical, a process termed toxication. Endogenous compounds can also be bioactivated by P450. For example, arachidonic acid is metabolized to four isomeric epoxyeicosatrienoic acids which have potent physiological and/or pathobiological effects in multiple tissues and cell types. Consequently, the P450 system is extremely important in toxicology (toxication and detoxication of both endogenous and exogenous substances), pharmacology (rate-limiting step in the metabolism of many drugs, drug–drug interactions, and individual qualitative and quantitative differences in drug metabolism due to genetic differences), and physiology (formation and metabolism of endobiotics that function as intercellular and/or intracellular messengers).

The multiple forms of P450 vary in their ability to metabolize chemicals. There are also large differences in the P450 isozyme profile of different tissues and cell types. In lung, for example, the highest concentrations of P450 are normally found in (epithelial) Clara and alveolar type II cells but lower amounts occur in ciliated, goblet, and vascular endothelial cells as well as alveolar macrophages. The selective modulation (relative increase or decrease in concentration) of P450 isozymes in a single tissue or cell type can have pronounced effects on the metabolism of both endogenous and ex-

$$RH + O_2 + NADPH + H^+ \longrightarrow ROH + H_2O + NADP^+$$

FIGURE B-11. *Overall reaction that occurs during the cytochrome P450-dependent oxidation of a substrate, RH.*

ogenous substances and on chemical-mediated target organ and/or cell toxicity by altering the balance between toxication and detoxication reactions.

The overall oxidation of a substrate, RH, by P450 is summarized in Fig. B-11, in which reduced nicotinamide-adenine dinucleotide phosphate (NADPH) is shown as the required cofactor.

The sequence of events that occurs during a P450 catalyzed monooxygenase reaction is now well understood (Fig. B-12). P450 contains one molecule of iron in its prosthetic heme group, normally in the oxidized or ferric (Fe^{3+}) state. Oxidized P450 first interacts with one substrate molecule to form an enzyme substrate complex. Next, the first electron is donated to this complex from NADPH via NADPH-P450 reductase, reducing the ferric iron to its ferrous (Fe^{2+}) form. The P450(Fe^{2+})-substrate complex subsequently reacts with molecular oxygen to form an oxycytochrome P450 ternary complex which accepts a second electron from NADPH via P450 reductase to form the equivalent of a two-electron reduced complex of hemoprotein, oxygen, and substrate. This complex dissociates to yield oxidized substrate, P450(Fe^{3+}) and water. The

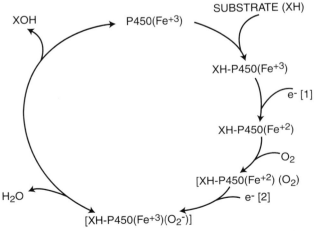

FIGURE B-12. *Simplified scheme for the oxidation of a substrate by the cytochrome P450 monooxygenase system (P450). The first electron [1] is donated from NADPH via NADPH-cytochrome P450 reductase; the second electron [2] is also normally donated from NADPH via this route.*

overall reaction is termed a monooxygenation because one atom of molecular oxygen is transferred to the substrate and the other is incorporated into water.

Some of the important reactions catalyzed by the P450 monooxygenase system include aliphatic hydroxylation, aromatic hydroxylation, epoxidation, heteroatom (N-, O-, and S-)dealkylation, nitrogen oxidation, oxidative deamination, oxidative dehalogenation, oxidative denitrification, and oxidative desulfuration. Most of these reactions result from the initial oxidation of a carbon atom, another reason that P450 is so important in the oxidative biotransformation of lipophilic chemicals. Some P450-catalyzed oxidation reactions are illustrated in Table B-1.

The microsomal P450 system is most highly concentrated in liver, but it is also present in many extrahepatic tissues including lung, kidney, placenta, small intestine, skin, adrenal, testis, ovary, eye, pancreas, mammary gland, aorta walls, brain, nasal epithelial membranes, colon, salivary glands, prostate, heart, lymph nodes, spleen, thymus, and thyroid. A second P450 monooxygenase system is found in the mitochondria of steroid-metabolizing tissues (adrenal, ovary, and testis) and is normally involved with the oxidative biosynthesis of steroids such as cholecalciferol, cortisone, and deoxycorticosterone. The mitochondrial P450 system has a much higher degree of substrate specificity than does the microsomal system; it is the latter that is viewed as a universal oxidase.

Flavin-Containing Monooxygenases

There is also a P450-independent monooxygenase enzyme family, termed the flavin-containing monooxygenases (FMOs), that is localized in the endoplasmic reticulum of virtually all nucleated mammalian cells. Five distinct structural classes of FMOs are known and these enzymes contain the coenzyme flavin adenine dinucleotide (FAD) and also require NADPH as a cofactor. A major difference between the FMOs and P450 is that the former do not oxidize carbon atoms. However, FMOs do oxidize many nitrogen-, sulfur-, selenium-, and phosphorus-containing xenobiotics (Table B-2).

Since there are many drugs and environmental pollutants that contain sulfur, it is of considerable interest that FMO preferentially catalyzes the oxidation of sulfur in compounds containing both nitrogen and sulfur. Thus, FMO is an important enzyme system for the oxidation of selected classes of xenobiotics, and its

spectrum complements that of the P450 system because the latter prefers oxidation of carbon atoms. Other ways in which FMO enzymes differ from many microsomal P450 isozymes include their apparent lack of induction (increased enzyme concentration) or repression (decreased enzyme concentration) by environmental factors and their more limited role in metabolic activation. Consequently, the P450 system is of greater significance in chemical toxicology.

Alcohol and Aldehyde Dehydrogenases

An extremely important metabolic pathway for alcohols and aldehydes is oxidation to aldehydes and ketones and to carboxylic acids, respectively. Mammalian liver alcohol dehydrogenases are a family of zinc-containing, cytosolic NAD^+-dependent enzymes that catalyze the oxidation of primary and secondary aliphatic, arylalkyl, and cyclic alcohols. Aromatic alcohols (phenols), however, are not substrates for these enzymes. Alcohol dehydrogenases are widely distributed in mammalian tissues, with the highest concentrations occurring in liver. As shown in Fig. B-13, alcohol dehydrogenases also catalyze the reverse reaction—reduction of aldehydes to primary alcohols in the presence of reduced nicotinamide adenine dinucleotide (NADH).

However, the *in vivo* reduction of aldehydes by this enzyme is not normally a quantitatively important reaction because aldehydes are rapidly oxidized to their corresponding carboxylic acid derivatives by aldehyde dehydrogenase. Alcohol dehydrogenase is a very important enzyme for the metabolism of ethanol.

Aldehyde dehydrogenases are also widely distributed in mammalian tissues, with the highest concentration in liver. Both aliphatic and aromatic aldehydes are readily oxidized to carboxylic acids by this enzyme in the presence of NAD^+, the required cofactor (Fig. B-14).

Although this is a reversible reaction *in vitro*, the carboxylic acids formed are either converted rapidly to their ester glucuronide derivatives (a phase II reaction catalyzed by UDP-glucuronosyltransferase; see below) or, if polar enough, are excreted unchanged. Consequently, the reverse reaction is generally not of significance *in vivo*.

Monoamine Oxidases

The monoamine oxidases are localized in the outer membrane of the mitochondria of cells and are widely

TABLE B-1
Examples of Important Reactions Catalyzed by the Microsomal P450
Monooxygenase System

Aliphatic Hydroxylation

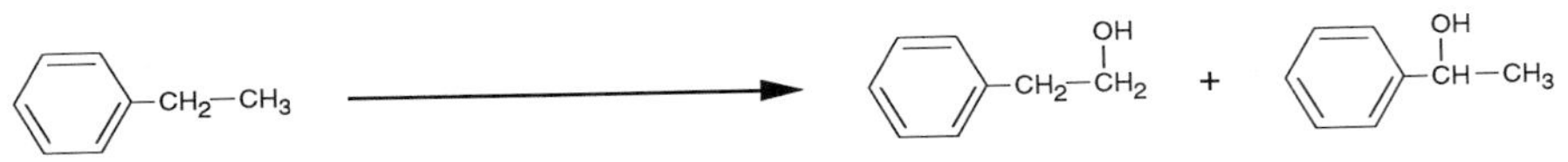

Aromatic Hydroxylation

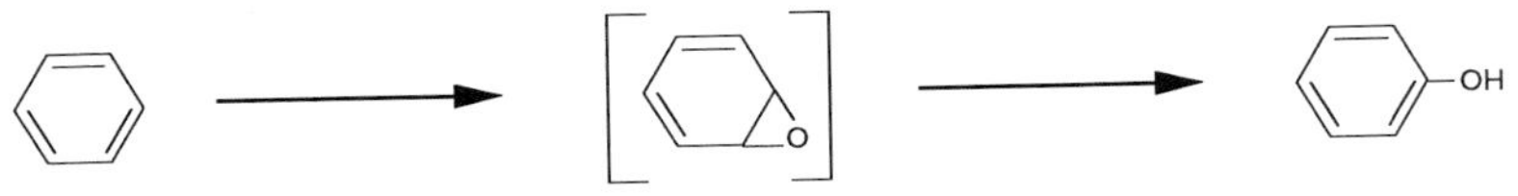

Epoxidation

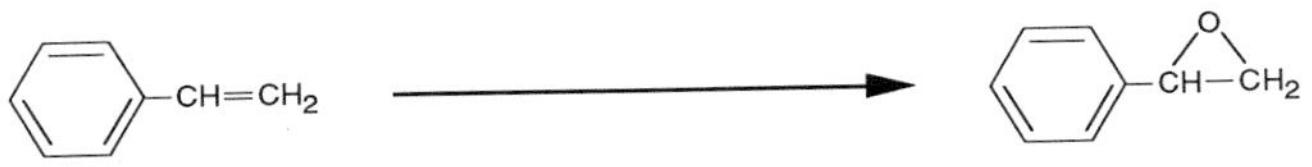

N-Dealkylation

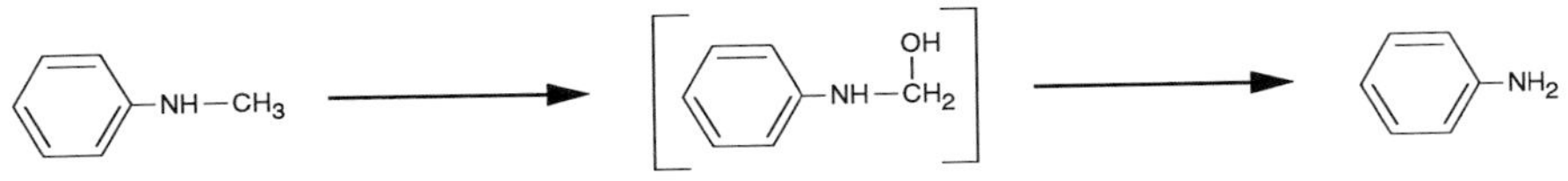

O-Dealkylation

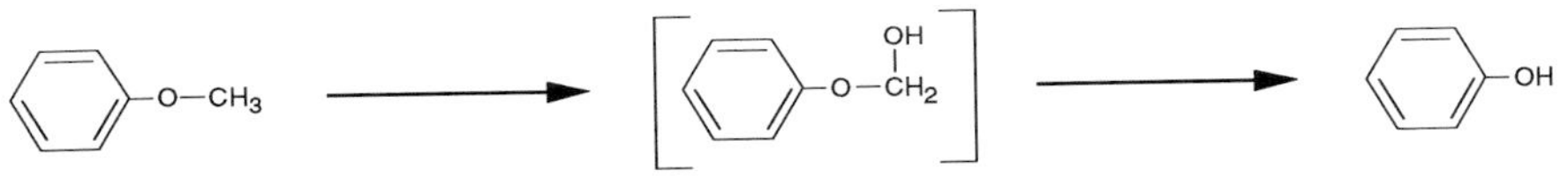

S-Dealkylation

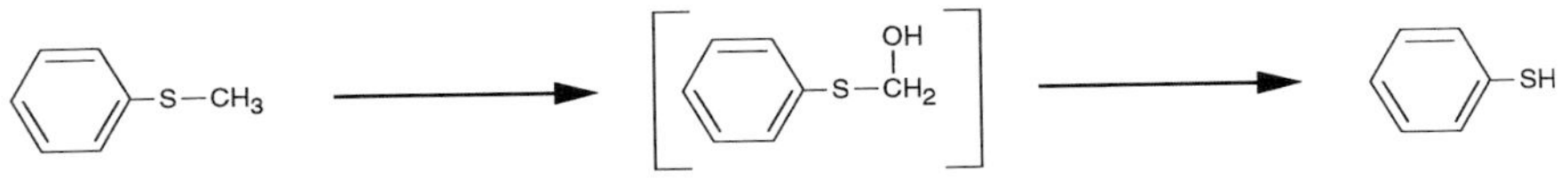

continues

distributed in most mammalian tissues, with exceptions being the erythrocyte and plasma. This enzyme system catalyzes the oxidative deamination of a wide variety of xenobiotic and endobiotic (e.g., neurotransmitter) monoamines (Fig. B-15).

Monoamine oxidases are flavoproteins that contain one molecule of FAD per molecule. There are two ma-jor types of monoamine oxidase (A and B), whose concentration varies in tissues of the same species. In general, the A form of the enzyme is more active with endogenous neurotransmitter amines (serotonin, nor-epinephrine, and epinephrine), whereas the B form is more active toward xenobiotic amines such as 2-phenethylamine.

Continued

Nitrogen Oxidation

Oxidative Deamination

Oxidative Dehalogenation

Oxidative Denitrification

Oxidative Desulfuration

H_2O_2-Dependent Peroxidases

Easily oxidized phenols and arylamines are excellent substrates for peroxidase-catalyzed one-electron oxidation reactions. These reactions are very important in toxicology because of the reactivity and toxicity of the free radicals (molecules with a highly reactive unpaired electron) formed. The best studied example of this type is the cooxidation of xenobiotics catalyzed by the hydroperoxidase activity of prostaglandin H synthase. This enzyme, which converts arachidonic acid to prostaglandin (PG) H_2, has two distinct enzyme sites: cyclooxygenase, which oxidizes arachidonic acid to PGG_2, and hydroperoxidase, which reduces PGG_2 to

PGH_2. PGG_2 reduction requires the donation of single electrons that can come from a xenobiotic and result in its conversion to a free radical. Many chemicals that are oxidized to toxic products, including acetaminophen, 2-aminofluorene, diethylstilbestrol, benzo(a)pyrene 7,8-dihydrodiol, and 4-phenetidine, are bioactivated to free radicals during reduction of PGG_2 to PGH_2 (Fig. B-16).

Prostaglandin H synthase activity is high in several extrahepatic sites that are targets for chemical-mediated toxicity but which contain very low amounts of P450 monooxygenase activity. These include skin, kidney medulla, lung of certain species, and platelets. It is now generally accepted that prostaglandin H syn-

TABLE B-2

Examples of Important Reactions Catalyzed by Microsomal
Flavin-Dependent Monooxygenases

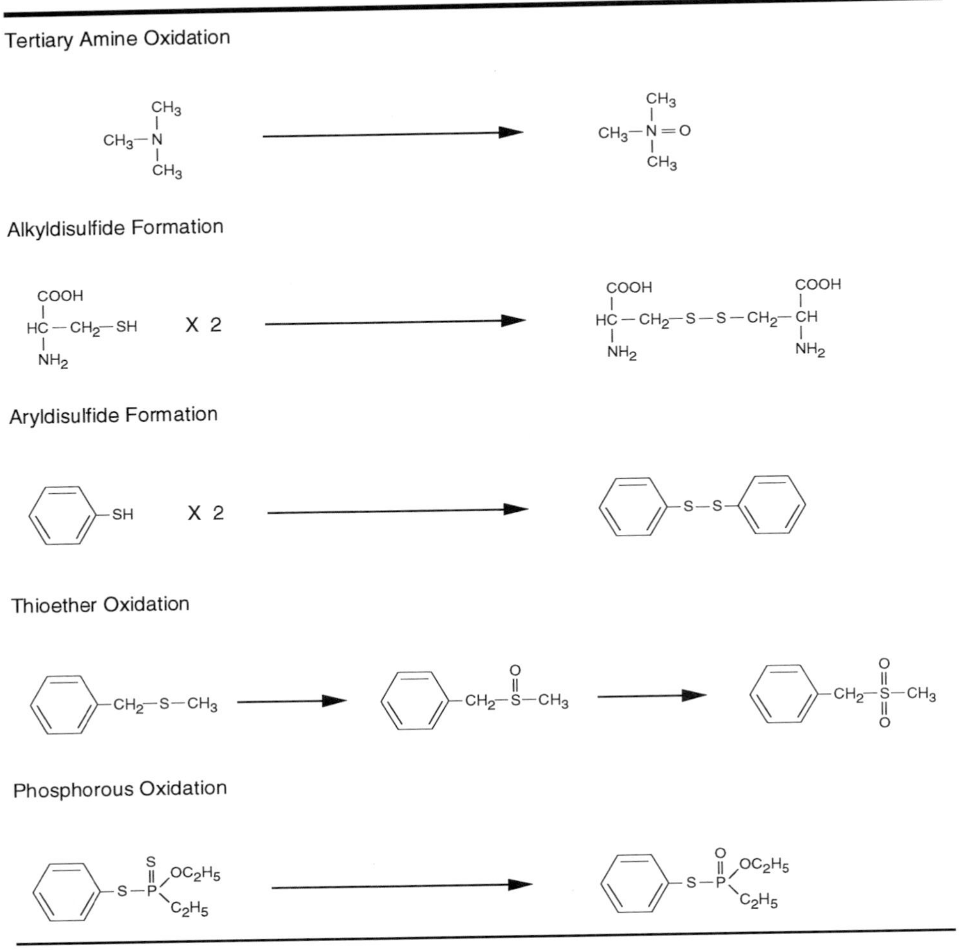

Tertiary Amine Oxidation

Alkyldisulfide Formation

Aryldisulfide Formation

Thioether Oxidation

Phosphorous Oxidation

thase hydroperoxidase activity is important for the metabolic activation of amines and phenols, some of which are converted to potent mutagens and carcinogens, particularly in cells deficient in P450 monooxygenase activity but high in prostaglandin synthesis activity.

Other peroxidases are also involved in bioactivation of easily oxidized compounds. Oxyhemoglobin in erythrocytes can oxidize arylamines to products that cause methemoglobinemia; chloroperoxidase and myeloperoxidase of activated polymorphonuclear leukocytes and macrophages bioactivate certain drugs including various sulfonamides by N-oxidation to reactive nitroso products that contribute to adverse drug reactions; and diethylstilbestrol, a transplacental

$$CH_3CH_2OH + NAD^+ \rightleftharpoons CH_3CHO + H^+ + NADH$$

FIGURE B-13.   *Oxidation of ethanol and reduction of acetaldehyde by alcohol dehydrogenase and the appropriate form of NAD+.*

$$CH_3CHO + NAD^+ \rightleftharpoons CH_3COOH + H^+ + NADH$$

FIGURE B-14.   *Oxidation of acetaldehyde and reduction of acetic acid by aldehyde dehydrogenase and the appropriate form of NAD+.*

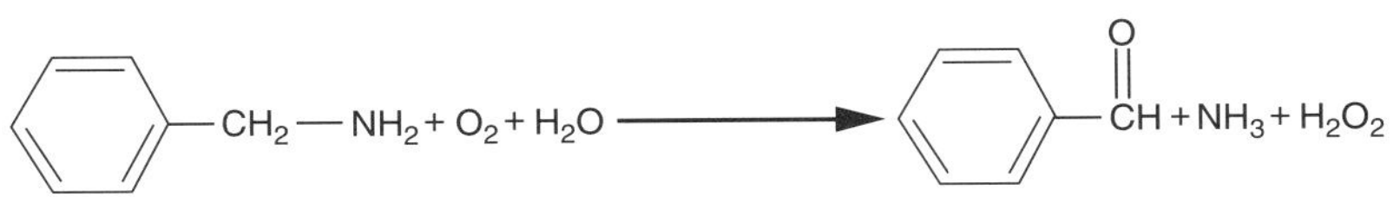

$$RCH_2NH_2 + O_2 + H_2O \longrightarrow RCHO + NH_3 + H_2O_2$$

FIGURE B-15. *Oxidation of a substituted methylamine by monoamine oxidase.* R *can be an alkyl (CH₃-) or aryl (C₆H₅-) substituent.*

carcinogen, is oxidized by estrogen-inducible peroxidases in the reproductive tract.

These few examples emphasize that H_2O_2-dependent peroxidases can activate aromatic alcohols (phenols) and aromatic amines to reactive free radicals, which are often very toxic.

Reduction Reactions

Several functional groups, including nitro, azo, tertiary amine N-oxide, aldehyde, ketone, sulfoxide, and alkyl polyhalide, are reduced by mammals *in vivo*. Toxic free radicals are often formed as intermediates during reduction. Although some of these reactions, or more accurately the initial sequence of the reactions, occur under aerobic conditions *in vitro*, anaerobic conditions are generally required for the complete reduction of xenobiotics. Those reactions that go to completion *in vivo* are either reductions of carbonyl groups or are catalyzed by the intestinal microflora. Reduction that occurs anaerobically is of much less toxicological concern due to the decreased formation of toxic oxygen free radicals.

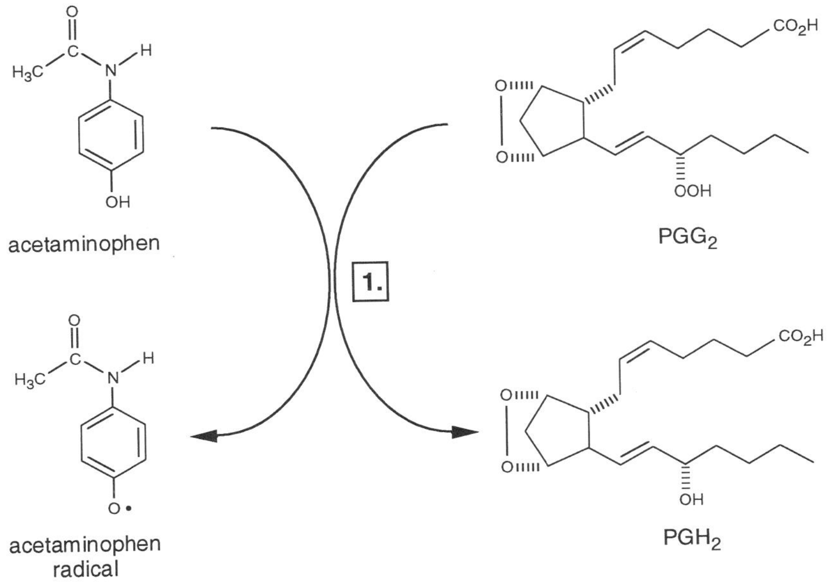

FIGURE B-16. *Conversion of acetaminophen to its reactive free radical by cooxidation mediated by the hydroperoxidase activity of prostaglandin H synthase-catalyzed reduction of prostaglandin G₂ (PGG₂) to prostaglandin H₂ (PGH₂).*

Cytochrome P450-Dependent Reactions

Several reduction reactions occur when subcellular fractions of liver enriched in fragments of the endoplasmic reticulum (microsomes) are incubated with NADPH (or NADH) under anaerobic or aerobic conditions. Such reactions can be catalyzed by the intact P450 monooxygenase system or only by its flavoprotein component, NADPH-P450 reductase.

In addition to being oxidatively metabolized, many polyhalogenated alkanes are converted by a P450-dependent, one-electron reduction pathway to a free radical intermediate and inorganic halide. The best studied example of this reaction is the reduction of carbon tetrachloride (CCl_4) to chloroform ($CHCl_3$), which occurs *in vitro* under aerobic or anaerobic conditions and *in vivo*. The trichloromethyl radical formed (CCl_3) is believed to be a major contributor to CCl_4-mediated hepatotoxicity. Halothane, trichlorofluoromethane, hexachloroethane, pentachloroethane, and DDT are other halogenated compounds that are substrates for this P450-dependent reductive pathway.

Several other classes of xenobiotics are also efficiently reduced by the P450 monooxygenase system under anaerobic conditions. These include tertiary amine N-oxides (converted to tertiary amines), hydroxylamines (primary amines), and hydrazo derivatives (primary amines).

Flavoprotein-Dependent Reactions

The first step of the NADPH-dependent reduction of aromatic nitro and azo compounds by hepatic microsomes is catalyzed by NADPH-P450 reductase and results in the formation of a free radical. In the presence of oxygen these radicals are rapidly reoxidized to the parent aromatic nitro or azo compound, concomitant with the generation of the superoxide anion radical. This futile cycling explains the toxicity of compounds, such as paraquat (Fig. B-17) or nitrofurantoin, which generate toxic superoxide under conditions in which little or no metabolism of the compound is detected. NADPH-P450 reductase is widely distributed in mammals and, consequently, these potentially toxic reactions occur in different tissues and subcellular organelles. Easily reduced compounds are readily reduced by NADPH-P450 reductase. Compounds that are more difficult to reduce, such as carbon tetrachloride, require the intact P450 monooxygenase system as a source of electrons for reduction.

Carbonyl Reductases

As mentioned previously, both alcohol and aldehyde dehydrogenases can function as reductases in the presence of NAD^+. In addition, there are a number of other carbonyl reductases that are $NADP^+$-dependent. Aldehyde reductases and carbonyl reductases are localized in the cytosolic fraction (cytosol) of broken cells, have a broad substrate specificity, have low molecular weights, and are widely distributed in extrahepatic tissues. In general, aldehyde reductases reduce only aldehydes, whereas carbonyl reductases reduce both aldehydes and ketones. Reduction of ketones can be an important metabolic pathway *in vivo*.

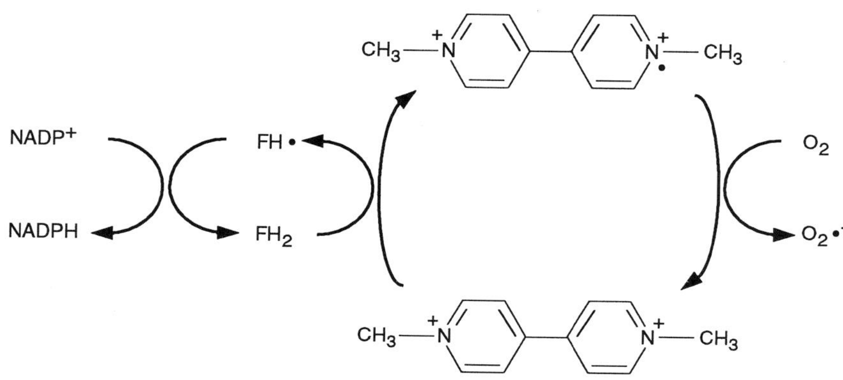

FIGURE B-17. *Futile cycle due to reaction of paraquat cation radical with molecular oxygen to generate superoxide radicals, with subsequent regeneration of paraquat. Cycle will operate as long as NADPH, required as a cofactor for P450 reductase, is present.*

Hydrolysis Reactions

When certain xenobiotics, including esters and amides, are administered to animals they are hydrolyzed. Hydrolysis reactions are important for the sequential metabolism of chemicals converted to epoxides by the P450 system. These reactions are classified as phase I because they free up functional groups (e.g., $-COOH$, $-NH_2$, $-OH$, $-SH$, and $-SO_3H$) that are important sites for conjugation (phase II) reactions.

Epoxide Hydrolase

Epoxide hydrolases catalyze the hydration of epoxides to *trans*-dihydrodiols and are very important enzymes in toxication–detoxication processes. Unsaturated aliphatic and aromatic hydrocarbons are converted to epoxides (alkene and arene oxides, respectively) by P450 monooxygenase activity. Some of these electrophilic epoxides react covalently with macromolecules, such as proteins, RNA, and DNA, resulting ultimately in acute or chronic toxicity, including necrosis, mutagenesis, carcinogenesis, and teratogenesis. In most cases, the diols produced by epoxide hydrolase are much less toxic than the epoxide substrate. With some polycyclic aromatic hydrocarbons, however, the diols are precursors for potent carcinogenic and mutagenic products. For example, benzo(a)pyrene 7,8-dihydrodiol, formed enzymatically from benzo(a)pyrene 7,8-oxide (Fig. B-18), is converted to the highly toxic benzo(a)pyrene 7,8-dihydrodio-9,10-oxide by the P450 system or by cooxidation by prostaglandin H synthase.

There are two distinct types of epoxide hydrolases, both widely distributed in mammalian tissues. One type is localized primarily in the endoplasmic reticulum, the second in the cytosol. The microsomal and cytosolic enzymes have different properties, including substrate selectivities. Several inducers of xenobiotic metabolizing enzymes, including phenobarbital, planar PCB congeners, and *trans*-stilbene oxide, selectively increase (induce) microsomal, but not cytosolic, epoxide hydrolase activity.

Carboxylesterases/Amidases

Many xenobiotic esters and amides, including a plethora of drugs, are hydrolyzed *in vivo*. These reactions are discussed together because highly purified carboxylesterases can catalyze both reactions; they cleave carboxylesters, carboxylamides, and carboxylthioesters, producing a carboxylic acid and an alcohol or phenol (Fig. B-19), amine, or mercaptan, respectively. The term carboxylesterase refers to a wide variety of enzymes with both esterase and amidase activity. There are many different esterases, some of which are important for the hydrolysis and detoxication of toxic organophosphate esters. In general, esterases are present in almost all mammalian tissues, occur as multiple isozymes, and are concentrated in the liver. The esterase activity present in plasma is normally due to release of these enzymes from liver.

Ester or amide cleavage can result in detoxication or metabolic activation, depending on the biological and chemical properties of the acids, alcohols, or amines released during hydrolysis. For example, hydroxamic acid hydrolysis has been implicated in the formation of proximate mutagens. The functional groups that become available for reaction during hydrolysis normally undergo phase II metabolism, as discussed below.

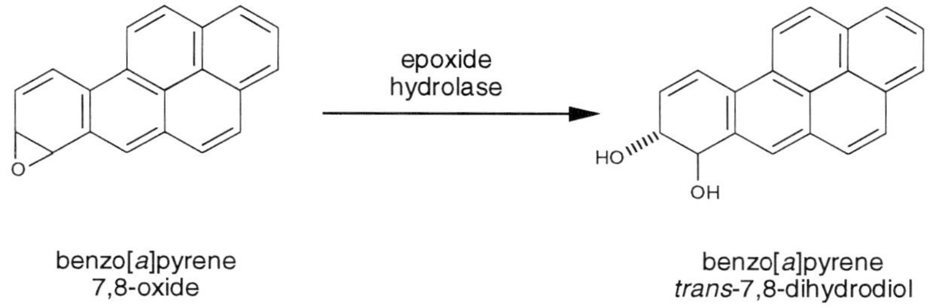

FIGURE B-18. *Conversion of benzo (a) pyrene 7,8-oxide to benzo (a) pyrene trans-7,8-dihydrodiol by microsomal epoxide hydrolase.*

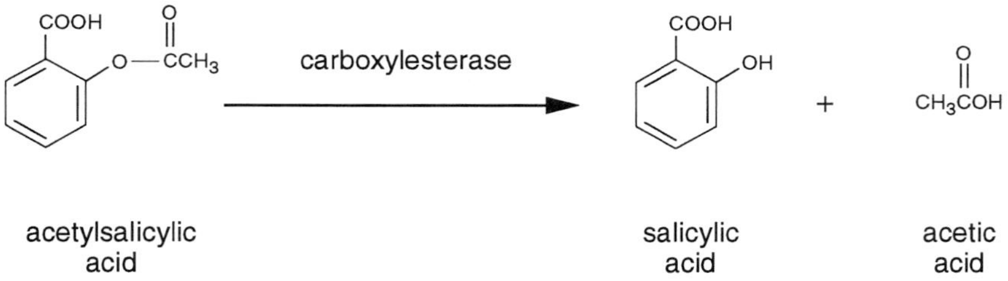

FIGURE B-19.   *Hydrolysis of acetylsalicylic acid (aspirin) to acetic acid and salicylic acid (a phenolic acid) by carboxylesterase activity.*

Conjugation Reactions

Most phase II reactions markedly increase the water solubility of xenobiotics and facilitate excretion of the chemical. Exceptions are acetylation and methylation reactions.

UDP-Glucuronosyltransferases

The most common phase II reaction is the synthesis of glucuronic acid derivatives (β-D-glucuronides) of lipophilic xenobiotics and endobiotics. Alcohols, phenols, carboxylic acids, mercaptans, primary and secondary aliphatic amines, and carbamates are converted to their β-glucuronide derivatives by UDP-glucuronosyltransferases (UDP-GT). Using molecular biology techniques, more than 30 different isozymes of UDP-GT have been identified to date. In common with the P450 monooxygenase system, UDP-GT is a microsomal enzyme, is present at highest concentrations in liver, is expressed in many extrahepatic tissues, and is induced by exposure to different classes of compounds known to modulate P450, including phenobarbital, polycyclic aromatic hydrocarbons, planar PCB congeners, and dioxins.

UDP-GT catalyzes the translocation of glucuronic acid to a substrate from the cosubstrate UDP-α-D-glucuronic acid (UDPGA) as shown in Fig. B-20.

Glucuronide conjugates excreted in the bile can be hydrolyzed to their aglycone by β-glucuronidase of the intestinal microflora. The released chemical (i.e., the aglycone) can be reabsorbed and the cycle repeated. This process is called enterohepatic circulation and accounts for the prolonged excretion of some xenobiotics that are readily glucuronidated.

Certain β-glucuronides are electrophilic in nature and may also function in toxication processes. Covalent binding of the aglycone portions of several carboxylic acid (ester) glucuronides is known to occur to nucleophilic sites on serum albumin via transacylation reactions, for example.

Sulfotransferases

Another very common phase II reaction for phenols is conjugation with sulfate to form sulfate esters (Fig. B-21). Other substrates for this pathway include alcohols, primary and secondary amines, hydroxylamines, and sulfhydryl compounds such as thiophenols. These reactions are catalyzed by a family of cytosolic enzymes, the

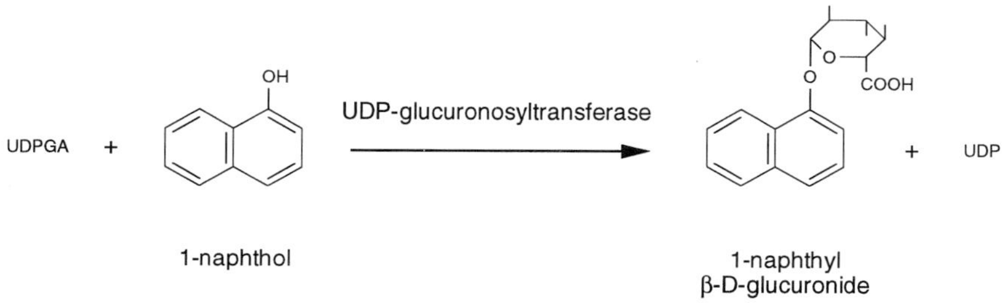

FIGURE B-20.   *Conversion of 1-naphthol to its corresponding β-D-glucuronide.*

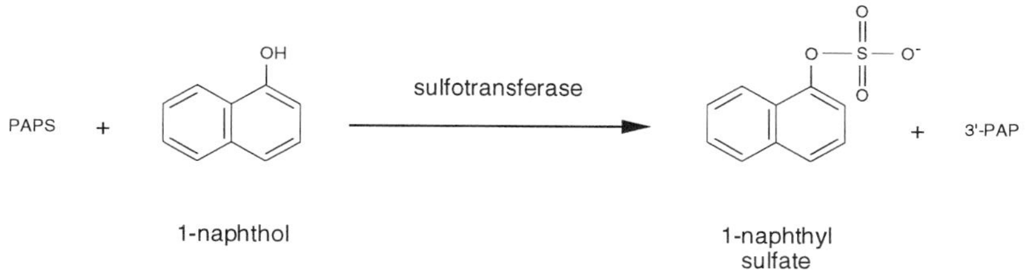

FIGURE B-21. *Conversion of 1-naphthol to 1-naphthyl sulfate by sulfotransferases.*

sulfotransferases, which require 3'-phosphoadenosine 5'-phosphosulfate (PAPS) as the cofactor.

The sulfotransferases have been divided into several groups as a result of substrate specificity determinations with purified enzymes and molecular biology studies; aryl sulfotransferases are active toward phenols, hydroxylamines, tyrosine esters, and catecholamines; alcohol sulfotransferases are active toward primary and secondary steroid alcohols; and amine sulfotransferases are active toward arylamines.

A few sulfate esters are chemically reactive and alkylate nucleophilic sites on macromolecules. This electrophilic characteristic implicates these conjugates as ultimate chemical toxicants.

Phenols, quantitatively important P450-derived metabolites of aromatic hydrocarbons, are substrates for both UDP-GT and sulfotransferases. Generally, glucuronide metabolites predominate after administration of a phenol or phenol precursor to mammals because sulfate formation is a high-affinity, low-capacity (due to sulfate depletion) system, whereas glucuronidation is a lower affinity, high-capacity system.

Glutathione *S*-Transferases

The glutathione (L-γ-glutamyl-L-cysteinylglycine; GSH) *S*-transferases are a multigene family of dimeric proteins found at relatively high concentrations in the cytosolic fraction of mammalian liver, as well as in a wide variety of extrahepatic tissues. Some GSH *S*-transferase isozymes are also localized in microsomes and within the mitochondrial matrix of the liver, at much lower concentrations than the cytosolic enzymes. A wide variety of potentially toxic, electrophilic compounds (Fig. B-22) are converted to *S*-substituted GSH adducts by this family of enzymes. These include aromatic compounds containing good leaving groups

(halogen, sulfate, sulfonate, phosphate, and nitro). Halogens are readily displaced from aromatic compounds as long as they are activated by the presence of electron-withdrawing groups (e.g., nitro). Strained three-membered rings, such as alkene and arene oxides, and four-membered lactones are readily cleaved by GSH *S*-transferases. The major factor in the transferase-catalyzed reaction of these substrates with GSH is the electrophilicity of the carbon atom where the thiol attacks. Since electrophilic chemicals are frequently very toxic the importance of the GSH *S*-transferases in detoxication cannot be overstated.

GSH *S*-transferases also catalyze a number of reactions in which an *S*-substituted GSH adduct is not formed or in which this adduct is oxidized glutathione. Examples of these reactions include the release of nitrate from nitrate esters and the release of cyanide from thiocyanates. Some GSH *S*-transferases also have glutathione peroxidase activity.

Although catalysis by GSH *S*-transferases is almost always associated with detoxication, a few substrates (e.g., the ethylene dihalides) are bioactivated to more toxic products by this pathway. Recent studies have also shown that glutathione conjugates are selectively accumulated in epithelial cells of the kidney where they are hydrolyzed. Those releasing metabolites that can undergo oxidation–reduction cycling result in cell-specific renal toxicity.

Mercapturic Acid Biosynthesis

A large variety of compounds, mostly xenobiotics, are excreted in urine as *S*-substituted *N*-acetylcysteines, also called mercapturic acids (Fig. B-23). The initial enzymatic reaction in their formation is catalyzed by the GSH *S*-transferases, as described previously. Subsequently, the glutamic acid residue is removed by γ-

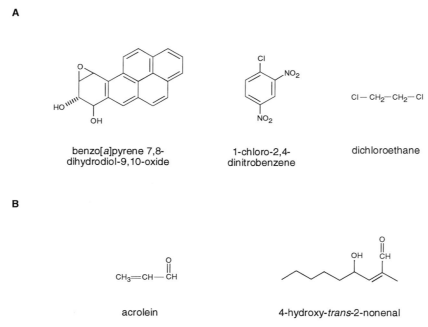

FIGURE B-22. *Structures of some common substrates of the glutathione S-transferases.*

glutamyltranspeptidase, an enzyme with very high activity in the kidney. Next, the glycine moiety is removed by dipeptidases, which have cysteinylglycinase activity. The resulting *S*-substituted cysteine is converted to the corresponding mercapturic acid by *N*-acetyltransferase activity (see below).

Although mercapturic acids are normally the major thioether products of lipophilic xenobiotics found in urine of mammals, small amounts of the corresponding *S*-cysteine conjugates are also frequently excreted. All four thioether products formed during mercapturic acid biosynthesis are routinely excreted in bile.

Cysteine Conjugate β-Lyase/Thiomethylation

In addition to being acetylated to mercapturic acids, some *S*-substituted cysteine conjugates are also hydrolyzed. The key enzyme in this reaction sequence is cysteine conjugate β-lyase, which cleaves the cysteine adduct to a free thiol, ammonia, and pyruvate (Fig. B-24).

This enzyme is present in the cytosolic fraction of rat liver and kidney and also in the microflora of the gut. Because thiols may be toxic and are more lipophilic than their cysteine conjugate precursors, β-lyase is generally a toxication pathway.

Thiols formed by mammalian or bacterial β-lyase *in vivo* are substrates for *S*-methyltransferase (Fig. B-25), an enzyme widely distributed in mammalian tissues.

This pathway accounts for the thiomethyl metabolites formed from several classes of xenobiotics. Thiomethyl metabolites can be further oxidized by the microsomal flavin-containing monooxygenases to their corresponding sulfoxide and sulfone derivatives.

Acyl-CoA:-Amino Acid *N*-Acyltransferases

Several types of xenobiotic carboxylic acids (aromatic, heteroaromatic, arylacetic, and aryloxyacetic) are conjugated with a variety of endogenous amino acids, including glycine, glutamine, or taurine, prior to excretion in mammals. An amide (peptide) bond is formed between the carboxylic acid group and the α-amino group of the amino acid during conjugation. The reactions involved in the conversion of a carboxylic acid (e.g., benzoic acid) to its glycine derivative (hippuric acid) are illustrated in Fig. B-26.

Conversion of the carboxylic acid to its CoA ester derivative is the rate-limiting step. The enzyme that catalyzes the final reaction, acyl-CoA:-amino acid *N*-

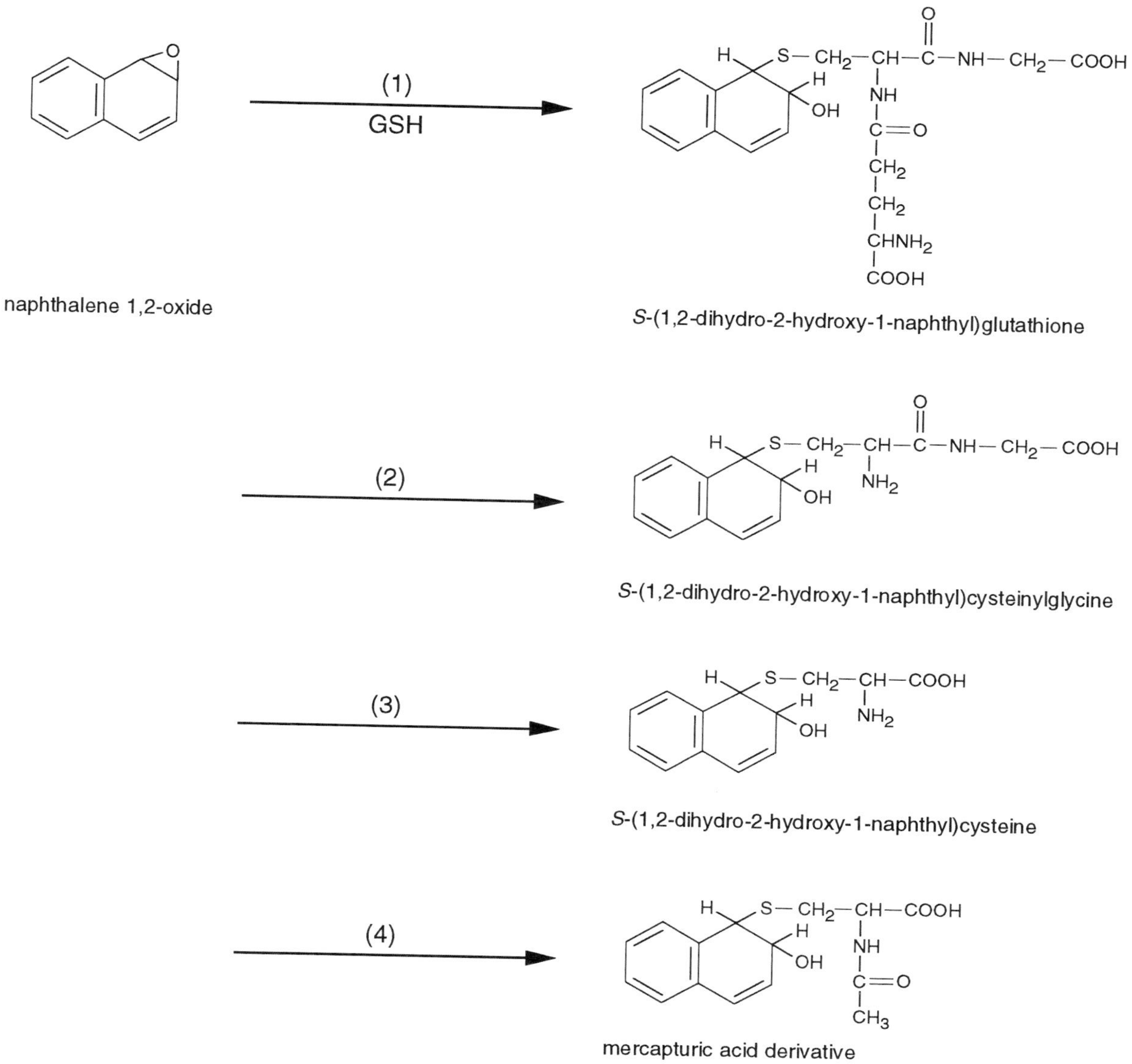

naphthalene 1,2-oxide

(1) GSH

S-(1,2-dihydro-2-hydroxy-1-naphthyl)glutathione

(2)

S-(1,2-dihydro-2-hydroxy-1-naphthyl)cysteinylglycine

(3)

S-(1,2-dihydro-2-hydroxy-1-naphthyl)cysteine

(4)

mercapturic acid derivative

FIGURE B-23. Mercapturic acid biosynthesis from naphthalene 1,2-oxide. Only one of the isomers resulting from reaction of GSH with the arene oxide is shown. (1) glutathione S-transferase, (2) γ-glutamyltranspeptidase, (3) cysteinylglycinase activity (dipeptidases), and (4) N-acetyltransferase.

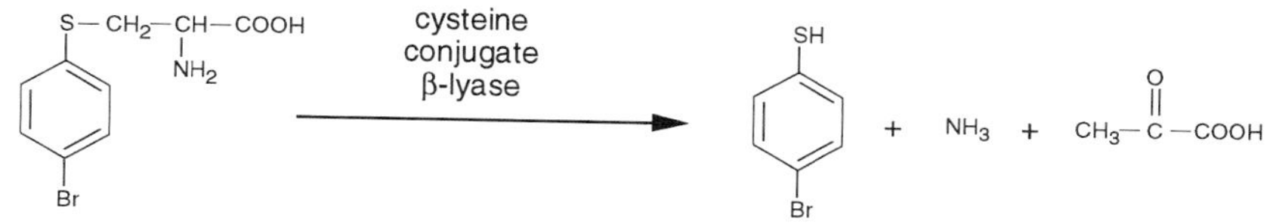

cysteine conjugate β-lyase

FIGURE B-24. Hydrolysis of S-4-bromophenyl-L-cysteine by cysteine conjugate β-lyase.

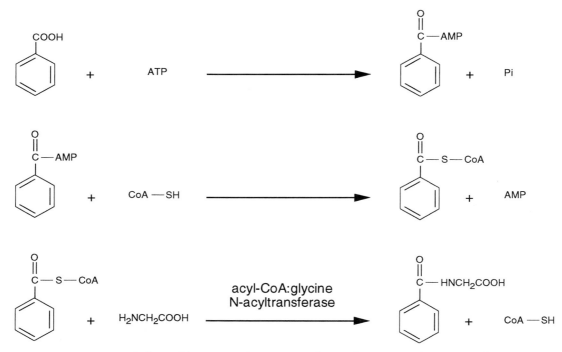

FIGURE B-25.   *S-Methylation of 4-bromothiophenol by S-methyltransferase.*

acyltransferase, is localized in the mitochondria of kidney and liver. The amino acid substrate selectivity, which varies from species to species, resides in the specific N-acyltransferase that catalyzes this reaction. In most mammalian species conjugation with glycine predominates.

N-Acetyltransferases

Acetylation of xenobiotic primary amine groups is a common metabolic pathway, whereas acetylation of xenobiotic hydroxyl and sulfhydryl groups is not. Primary aliphatic and aromatic amines, sulfonamides, hydrazines, and hydrazides are readily N-acetylated *in vivo*, and the reaction is catalyzed by various acetyl CoA:-N-acetyltransferases, commonly called N-acetyltransferases, as shown in Fig. B-27.

This family of enzymes is cytosolic in nature and is widely distributed. There are also enzymes that hydrolyze N-substituted acetamides (i.e., amidases, as described previously) and the extent to which free versus acetylated amines are present *in vivo* depends on the relative rates of the acetylation and deacetylation reactions, on the physical and chemical properties of the two products, and whether or not the amine is metabo-

FIGURE B-26.   *Metabolism of benzoic acid via its acetyl CoA derivative to hippuric acid (benzoylglycine).*

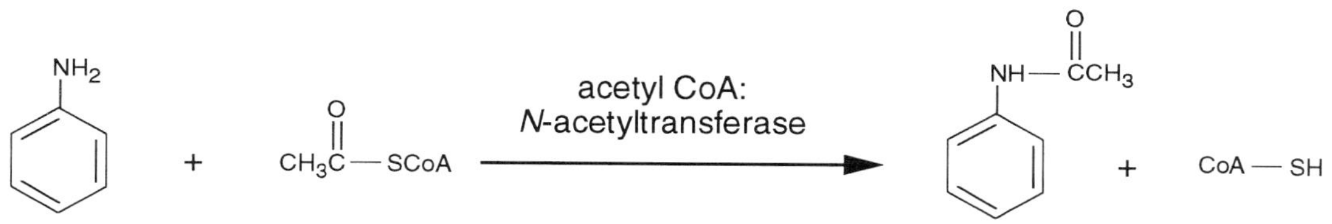

FIGURE B-27. *Acetylation of aniline by acetyl CoA:N-acetyltransferase activity.*

lized by competing pathways. Some acetylated hydroxamic acids are chemically reactive and appear to be ultimate carcinogens.

N- and O-Methyltransferases

S-adenosyl-L-methionine (SAM)-dependent methylation was briefly discussed under Thiomethylation (see Fig. B-25). Other functional groups that are methylated by this mechanism include aliphatic and aromatic amines, *N*-heterocyclics, monophenols, and polyphenols. The most important enzymes involved in these methylation reactions with xenobiotics are catechol *O*-methyltransferase, histamine *N*-methyltransferase, and indolethylamine *N*-methyltransferase—each catalyzes the transfer of a methyl group from SAM to phenolic or amine substrates (*O*- and *N*-methyltransferases, respectively). Methylation is not a quantitatively important metabolic pathway for xenobiotics, but it is an important pathway in the intermediary metabolism of both *N*-and *O*-containing catechol and amine endobiotics.

Summary

A number of enzyme systems have evolved in animals and plants which effectively convert lipophilic xenobiotics to more polar compounds that are efficiently excreted. Phase I enzymes, responsible for oxidation, reduction, and/or hydrolysis, are integrated with phase II or conjugation enzymes for reactions of both types and are normally required for the formation of products polar enough to be readily excreted. During metabolism toxic products are frequently formed, primarily but not exclusively during oxidation, and when the concentrations of these reactive metabolites exceed the capacity of detoxication systems acute (necrosis) or chronic (mutagenesis, carcinogenesis, and teratogenesis) toxicity occurs. Anything that results in the conversion of a higher proportion of the absorbed dose of a chemical to a reactive metabolite(s) increases the probability that a toxic response will occur.

Further Reading

Armstrong, R. N. (1994). Glutathione S-transferases: structure and mechanism of an archetypical detoxication enzyme. *Adv. Enzymol. Relat. Areas Mol. Biol.* **69**, 1–44. [See also other drug metabolism-related articles in this review series]

Bend, J. R., Serabjit-Singh, C. J., and Philpot, R. M. (1985). The pulmonary uptake, accumulation and metabolism of xenobiotics. *Annu. Rev. Pharmacol. Toxicol.* **25**, 97–125. [See also other drug metabolism-related articles in this review series]

Benedetti, M. S., and Dostert, P. (1994). Contribution of amine oxidases to the metabolism of xenobiotics. *Drug Metab. Rev.* **26**, 507–535. [See also other articles in this review series]

Guengerich, F. P. (Ed.) (1987). *Mammalian Cytochromes P-450*, Vols. I and II. CRC Press, Boca Raton, FL.

Kato, R., Estabrook, R. W., and Cayen, M. N. (1989). *Xenobiotic Metabolism and Disposition.* Taylor & Francis, London.

Mackenzie, P. I. (1994). The UDP glucuronosyltransferase multigene family. *Rev. Biochem. Toxicol.* **11**, 29–72. [See also other articles in this review series]

Meerman, J. H. N., Runge-Morris, M. A., Coughtrie, M. W. H. (Eds.) (1994). Sulfation of xenobiotics and endogenous compounds. *Chem.-Biol. Interact.* 92 1–476. [See also other drug metabolism-related review articles in this journal]

Sipes, I. G., and Gandolfi, A. J. (1991). Biotransformation of toxicants. In *Casarett & Doull's Toxicology: The Basic Science of Poisons* (A. M. Amdur, J. Doull, and C. D. Klaassen, Eds.), pp. 88–126. Pergamon, New York.

—John R. Bend and Christopher J. Sinal

Related Topics

Carboxylesterases
Glutathione
Kidney
Liver
Pharmacokinetics/Toxicokinetics

Bismuth (Bi)

♦ CAS: 7440-69-9

♦ SELECTED COMPOUNDS: Bismuth oxide, BiO (CAS: 1304-76-3); bismuth nitrate, $Bi(NO_3)_3$ (CAS: 10361-44-1)

♦ CHEMICAL CLASS: Metals

Uses

Several bismuth compounds have been used medicinally. Some are used for gastrointestinal distress, others are used as salves and, in rare cases, for treatment of parasites. In the past, bismuth has also been used to treat syphilis and malaria.

Commercially, bismuth is also used in the manufacturing of permanent magnets, semiconductors, and thermoelectric materials; as a catalyst in making acrylonitrile; and as an additive to improve the machinability of steels and other metals.

Exposure Pathways

The primary exposure pathway for bismuth is from medicinal preparations that are administered orally or intramuscularly. Some drinking water contains bismuth.

Toxicokinetics

As a rule, bismuth compounds are not absorbed readily (approximately 7% is absorbed) when administered orally or applied topically. Citrate enhances intestinal absorption.

Bismuth binds to one of the plasma proteins and concentrates in the kidneys, the liver (to a lesser extent),

and the skin. Very little information is available regarding the effect of bismuth on enzymes. Bismuth can displace bound lead, thus increasing the concentration of lead in the circulatory system. The urine is the major route of excretion. For some bismuth compounds, elimination may be equal between urine and feces.

Mechanism of Toxicity

The mechanism by which bismuth produces toxicity has not been identified.

Human Toxicity

Adverse reactions to bismuth include acute renal failure following ingestion of excessive concentrations. Symptoms of bismuth poisoning include fever, weakness, pain similar to rheumatism, and diarrhea. Certain people display a rash. Chronic exposure to bismuth may result in gingivitis and black spots on the gums. The bone and brain may also be targets for toxicity.

Clinical Management

There does not appear to be an antidote of choice for bismuth toxicity in humans. The newer chelating agents, *meso*-2,3-dimercaptosuccinic acid and DL-2,3-dimercapto-propane-1-sufonic acid, are being investigated experimentally as antidotes for bismuth toxicity. In mice, D-penicillamine has proven effective.

Animal Toxicity

In animals, bismuth interferes with the metabolism of copper and zinc, and it induces a metallothionein. Bismuth has not been found to be carcinogenic in animal models.

—Arthur Furst and Shirley B. Radding

Related Topics

Metallothionein
Metals

Bleach

♦ CAS: 7681-52-9

♦ PREFERRED NAME: Sodium hypochlorite

♦ SYNONYMS: Household laundry bleach (Purex, Clorox, and Dazzle); commercial laundry bleach (caustic soda bleach); Dakins solution; modified Dakins solution; sodium hypochlorite pentahydrate; surgical chlorinated soda solution

♦ CHEMICAL CLASS: Hypochlorites and related agents

Uses

Sodium hypochlorite is used in household laundry bleach, deodorizers, and antiseptics. Household laundry bleaches are approximately 5% solutions with pH adjusted to a range of 10.5–12.5. Commercial laundry bleaches contain 15% sodium hypochlorite at a pH slightly over 11.

Exposure Pathways

Ingestion is the most common route of exposure to sodium hypochlorite. Other modes of exposure are inhalation and dermal and ocular contact.

Mechanism of Toxicity

The toxicity of hypochlorite arises from its corrosive activity on skin and mucous membranes. Corrosive burns may occur immediately upon exposure to concentrated bleach products. Most of this corrosiveness stems from the oxidizing potency of the hypochlorite itself, a capacity that is measured in terms of "available chlorine." The alkalinity of some preparations may contribute substantially to the tissue injury and mucosal erosion. Sodium hypochlorite when combined with an acid or ammonia may produce chlorine or chloramine gas, respectively. An inhalation exposure to these gases may result in irritation to mucous membranes and the respiratory tract which may manifest itself as a chemically induced pneumonitis.

Human Toxicity

The resulting symptoms from an exposure to sodium hypochlorite and related compounds may range from mildly irritating to corrosive depending on the physical form and duration of exposure. It is unlikely for any symptoms to occur after ingestion of, or dermal or ocular exposure to, liquid household bleach containing sodium hypochlorite. Ocular exposure may result in irritation, burning, lacrimation, and eyelid edema. Inhalation exposures to chlorine or chloramine gas can result in dyspnea, cough, airway edema, stridor, bron-chospasm, pneumonitis, acute respiratory distress syndrome, and hypoxia.

Clinical Management

Basic and advanced life-support measures should be utilized as necessary. Treatment is generally symptomatic and supportive. Gastrointestinal evacuation procedures are generally contraindicated and unnecessary. Dilution with 4–8 oz of milk or water (not to exceed 15 ml/kg in a child) is preferable. Esophagoscopy is rarely needed following ingestion of small amounts of common household liquid bleach. Ingestion of commercial liquid bleach or bleach granules may necessitate esophagoscopy to look for burns. Neutralization with acids or basic substances is contraindicated.

For inhalation exposures, the patient should be moved to fresh air immediately. If cough or difficulty in breathing develops and is not relieved by the fresh air, the patient should be evaluated for respiratory tract irritation, bronchitis, or pneumonitis in a health care facility.

For ocular exposures to sodium hypochlorite and related agents, the eye(s) should be immediately irrigated with tap water. If an eye irrigation is delayed, the potential for injury is greater and the patient may have to be evaluated in a health care facility.

For dermal exposures, washing the affected area thoroughly with soap and water may be all that is required.

Animal Toxicity

Emesis is likely to be spontaneous. Clinical signs may include salivation, emesis, abdominal pain and tenderness, hematemesis, and bleached hair. Rats given 5–15 ml/kg of an alkaline (pH 12.0) solution containing 4.5% sodium hypochlorite died within 1–3 hr from severe local damage to the esophagus and stomach.

—*Gaylord P. Lopez*

Related Topics

Corrosives
Detergents
Surfactants, Anionic and Nonionic

Blood

Introduction

The formed elements of the blood—red blood cells (RBCs or erythrocytes), white blood cells (myeloids), immunocytes (Tand B cells), platelets (thrombocytes), and their diseases—have traditionally been the main focus of hematology. However, with the advent of increasingly sophisticated molecular investigatory tools, such as recombinant DNA technology, the appreciation of the scope and inherent complexity of the blood-forming organ has dramatically increased. The bone marrow and its formed elements can be considered as a complex organ with a total mass that is over twice as large as the liver. The cells produced by this organ provide several critical functions such as the transport of oxygen (red blood cells), hemostasis (platelets), and host resistance (immunocytes and white blood cells). Generally, each step in the intricate sequence required to produce the formed elements is vulnerable to adverse effects from a wide variety of chemicals and drugs. This entry will present a basic overview of normal bone marrow function, followed by a discussion of some of the abnormal physiologic effects that can be produced by exposure to various common drugs and chemicals.

Bone Marrow Structure and Function

In the normal adult, the marrow is found in the central hollow segment of bones. Hematopoiesis, or the production of the formed blood elements, occurs in the bone marrow. However, in the adult, it is largely restricted to scattered clusters of hemopoietic cells in the proximal epiphyses of the long bones, skull, vertebrae, pelvis, ribs, and sternum. The hematopoietic picture in adults is quite different from that seen in either prenatal or childhood time periods. Within the first 1–5 prenatal months, the liver and spleen act as the hematopoietic organs. By the fifth prenatal month, the marrow has achieved sufficient maturity to assume the dominant role in hematopoiesis. During childhood, there are high demands on the bone marrow system to produce large

quantities of the formed elements; however, with increasing chronological maturity, there is less demand on the bone marrow system and the total output of the bone marrow significantly declines.

In addition to hematopoietic cells, there are other separate and distinct cells that support and augment marrow activities. Among these cells are fibroblasts, fat cells, and reticuloendothelial and endosteal cells. In aggregate, these cells are known as the bone marrow stroma. Occasionally, the term hematopoietic microenvironment is also employed to differentiate these cells and supporting structure from the stem and progenitor cells. These cells are the focus of the next section, which presents an overview of the basic physiology of the blood-forming elements.

Hematopoiesis

In the average adult, between 200 and 400 billion blood cells are destroyed and replaced each day. This enormous turnover implies that new cells are constantly formed rather than simply released from a central storage area that contained all the cells necessary for an individual's lifetime. Hemopoiesis is the key concept that has been used to explain how the body can provide a lifetime worth of formed blood elements. Hemopoiesis is a process of cell amplification and differentiation in which a few stem cells give rise to increasingly more developed or differentiated progenitor cells, which in turn give rise to the formed blood elements. The earliest cell is known as the pluripotent stem cell or PSC. PSCs are uniquely responsible for the production of the formed elements throughout the lifetime of a human. Relatively few PSCs are required since, as these cells undergo mitosis or cell division, one replacement stem cell and one committed or daughter cell are produced. This daughter cell subsequently develops and proliferates into the various formed elements. Hence, the PSCs are considered to be self-renewing because of their ability to reproduce themselves. Figure B-28 presents the overall organization and development of the bone marrow cells. This structure is quite hierarchical and resembles a company organization chart with a single chief executive officer presiding over separate divisions, which in turn develop other specialized departments or functions. Not surprisingly, each step in the organization requires both a series of growth factors and interactions with the hematopoietic microenvironment

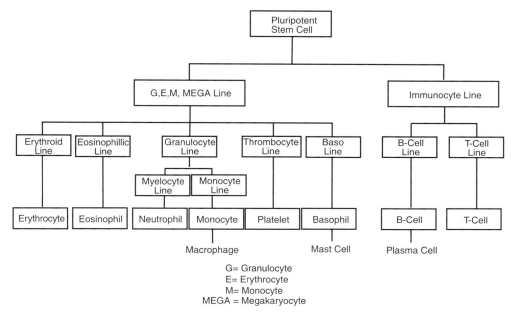

FIGURE B-28. *Bone marrow cell organization.*

to promote and control the development of each cell type. The stimulatory or growth factors are known as poietins or colony stimulating factors (CSFs). CSFs can either be lineage-specific, i.e., they act on specific cell lines, or direct acting on multipotential progenitors and stem cells. Examples of lineage-specific CSFs include (1) erythropoietin, which stimulates production of erythrocytes or red blood cells; and (2) interleukin-7, which induces the growth of B and T lymphocytes progenitors. Direct-acting CSFs include interleukins 2–6, which act on a variety of cell lines.

Formed Elements—Erythrocytes, Myeloids, and Thrombocytes

Erythrocytes

The RBC is a biconcave disk with a diameter of approximately 8 μm and a life span in the circulation of approximately 120 days. Due to its unique shape, the RBC is twice as thick at the edges (2.4 μm) as at its center. The explanation for this specialized geometry is not fully known; however, this shape tends to minimize intracellular diffusional distance and allows for easier passage through small blood vessels. The critical function of the RBC is transportation and delivery of oxygen to peripheral tissues. Approximately 30% of the wet weight of the RBC is composed of hemoglobin, the essential protein which is integral to the oxygen/carbon dioxide transport and delivery system. Hemoglobin is also capable of transporting nitric oxide (NO). NO is a unique gas that can affect the ability of blood vessels to expand or contract in addition to having a role in learning and memory.

The mature RBC is formed through a series of cell divisions that progressively increase the amount of hemoglobin in the cytoplasm. Following the last division, a special cell known as a reticulocyte is formed. The reticulocyte stays in the bone marrow for 2 or 3 days before being released into the general circulation, where over a period of 24 hr it undergoes a series of transformations that results in the appearance of a mature RBC. The reticulocyte is easy to identify in laboratory tests and the reticulocyte count or index is an important parameter that can provide information about marrow function. The reticulocyte index is equal to the reticulocyte percentage multiplied by the ratio of the patient's hematocrit (packed cell volume) to a normal hematocrit.

Hemoglobin

Hemoglobin, in the normal adult, is a protein whose main function is to transport oxygen from the lungs to tissues and to transport carbon dioxide from tissues to the lung. The hemoglobin molecule contains four separate folded peptide chains which form a hydrophobic or water "repelling" pocket around a heme group. The heme group is composed of a central iron atom complexed to four nitrogen atoms. Oxygen is capable of reversibly binding to the heme unit in a process known as oxygenation. The interactions among the subunits in a hemoglobin molecule are known as cooperativity. There are well-described regulators of the affinity of hemoglobin for oxygen that provide a control mechanism. The S-shaped graph of this oxyhemoglobin relationship is known as the oxyhemoglobin dissociation curve and represents the relationship between the partial pressure of oxygen (PO_2) in mm of mercury (Hg) and the oxygen content per 100 milliliters of blood (Fig. B-29).

The shape of this relationship is very important since it can be moved to the right, i.e., decreased affinity of hemoglobin for oxygen producing oxygen unloading, or to the left, i.e., increased affinity. These changes are produced by a variety of intracellular cofactors:

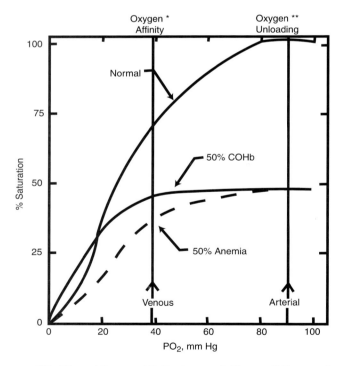

*Modifiers of Oxygen Affinity - Increase in Plasma pH, Decrease in Temperature, Decrease in 2,3-BPG.

**Modifiers of Oxygen Unloading - Decrease in Plasma pH, increase in Temperature, Increase in 2,3-BPG.

FIGURE B-29. Oxyhemoglobin dissociation curve.

hydrogen ion (pH), carbon dioxide, and the RBC enzyme 2,3 biphosphoglycerate (BPG). Molecules of 2,3-BPG bind to hemoglobin and decrease the molecules' affinity for oxygen. This causes enhanced oxygen release, or unloading, and is frequently seen in situations in which the body is responding to conditions of low oxygen supply. There are a wide variety of potential diseases and toxic exposures that can impact oxygenation and cooperativity and these will be discussed in subsequent sections.

Anemia

There are many other events that can produce a significant reduction in the RBC mass and a subsequent decrease in the oxygen-carrying capacity of the blood. Normally, the blood volume is maintained at a relatively constant level; hence, any process or event that causes a reduction in either RBCs or hemoglobin produces a condition known as an anemia. Anemias can also shift the oxyhemoglobin dissociation curve as the body attempts to compensate for reduced oxygen carrying capacity. In general, the etiology of anemias fall into three general categories: (1) acute or chronic blood loss from any source, (2) underproduction associated with a decreased reticulocyte count, and (3) hemolysis or destruction of RBCs associated with an increased reticuloctye count. There are a variety of laboratory tests that are useful for the evaluation of anemia; however, three of the most critical are the measurement of RBC size and shape (known as RBC indices), examination of the peripheral blood smear, and bone marrow examination. Each of these tests reveals information that can provide clues which lead to the etiology of the anemia. In later sections, some examples of toxins (e.g., carbon monoxide, hydrogen sulfide, and hydrogen cyanide) which cause anemias by altering the binding affinity for hemoglobin and oxygen will be presented.

In general, the amount of oxygen delivered to a given organ or tissue is directly related to three variables: (1) blood flow or cardiac output, (2) hemoglobin concentration, and (3) the difference in oxygen content (saturation) between arterial and venous blood. For example, the cardiac output can significantly increase in order to maintain adequate oxygenation of vital organs like brain and kidney at the expense of smooth muscle. Similarly, erythropoiesis can be stimulated by erythropoietin so that the overall hemoglobin levels increase. Finally, as illustrated in Fig. B-29, oxygen unloading or delivery can be augmented by a right shift in the oxygen dissociation curve facilitated by the RBC enzyme 2,3-BPG.

The converse to anemia is known as polycythemia or erythrocytosis, which is an increase above normal in the circulating quantity of RBCs. Not surprisingly, this increase in total RBCs is usually associated with a corresponding increase in hemoglobin. There are numerous causes of polycythemia, such as response to high altitude, pulmonary disease, steroids (both androgenic and glucocorticoid), stress, and smoking.

Myeloids

The myeloids or leukocytes are a highly complex and sophisticated group of cells that are primarily involved in host resistance and inflammatory response against both foreign organisms and material (e.g., chemicals and toxins). For simplicity, the leukocytes can be divided into two major groups: (1) immunocytes and (2) phagocytes. The general organizational structure and normal values are shown in Fig. B-28. These cells are thought to arise from a common PSC in the bone marrow; hence, any toxin that affects the PSC will have a potentially disastrous impact on the body's ability to respond to challenges from an external agent or foreign substance.

The immunocytes are all involved in specific types of immune response that are generally divided into two types: (1) cell-mediated, i.e., specifically sensitized T cells (derived from the thymus) which are associated with graft rejection, resistance to certain viruses, bacteria, fungi, and protozoa, and delayed-type hypersensitivity; and (2) humoral-mediated, i.e., B cells (bursa equivalent) which produce specific antibodies after the body is exposed to a specific antigen.

The phagocytes are so named because their major function is to engulf or ingest foreign organisms or material. The phagocytes include the three granulocytes known as neutrophils (54–62%), eosinophils (1–3%), and basophils (less than 1%) and the monocytes (3–7%) Monocytes circulate in the blood for several days until they migrate into the reticuloendothelial tissues (liver, spleen, and bone marrow), where they are known as macrophages. Macrophages are involved in inflammatory responses but also have a major role in the destruction and removal of old RBCs and other plasma proteins, including hemoglobin.

The phagocytes act by engulfing the foreign material/agent and produce a respiratory burst. The respiratory burst involves production of hydrogen peroxide and other highly reactive chemicals that attack the ingested material. An inflammatory response is quite commonly produced in this situation. Glucocorticoids (steroids like prednisone) tend to decrease the numbers of granulocytes that will be involved in an inflammatory reaction. This effect accounts for these drugs' beneficial impact when an antiinflammatory result is desired; however, there is also an increased susceptibility to infections that has been well documented.

There are several key terms and definitions that are given to absolute decreases or increases in the numbers of leukocytes. A fall in the total granulocyte count below 3000/mm³ is known as granulocytopenia. Granulocytopenia is commonly associated with chemically induced bone marrow damage; however, ionizing radiation and a myriad of drugs can also produce this effect. Finally, a particularly severe form of bone marrow failure is known as aplastic anemia. Aplastic anemia is diagnosed when at least two different marrow cell lines are severely depressed as demonstrated by (1) granulocytes less than 500/mm³, (2) platelets less than 20,000/mm³, (3) reticulocyte count less than 1%, or (4) a bone marrow biopsy demonstrating less than 25% cellularity.

Granulocytosis is the opposite phenomenon of decreased cellularity and refers to elevated counts over 10,000/mm³. Stress, drugs, and some bacterial toxins can produce short-term granulocytosis; however, chemical exposure is not typically associated with mild, elevated counts. Leukemias are associated with counts over 30,000/mm³ and have been associated with certain chemical exposures. This association will be presented in further detail in a subsequent section.

Thrombocytes

Platelets are produced by the fragmentation of megakaryocytes, the largest cell type in the bone marrow. Approximately one-third of the platelets are taken up by the spleen, while the other two-thirds freely circulate for 7–10 days until they are taken up by phagocytic cells. A normal platelet count is between 150,000 and 450,000/mm³. The normal platelet count is quite variable and can be affected by an individual's nutritional state or, in females, by the menstrual cycle.

Platelets are the rapid reaction troops in the situation of accidental blood loss associated with damaged blood vessels that expose collagen fibers. Normally, platelets are nonsticky; however, they rapidly and easily adhere or aggregate to exposed collagen fibers where they undergo a series of reactions that results in the formation of a thick mass known as a platelet plug. This plug acts to quickly stop bleeding; however, it must usually be reinforced by help from the clotting system so that vascular integrity is maintained. Platelet reactions are highly sensitive and vulnerable to substances that interfere with the aggregation reaction. For example, aspirin acts in a unique fashion to inhibit the aggregation reaction and has become a useful drug in the prevention of heart attacks and strokes caused by small platelet plugs.

Any disorder or agent that injures the stem cells or prevents their proliferation can drastically affect the absolute platelet count. The minimal platelet count necessary for initial hemostasis is approximately $50,000/mm^3$. If the platelet count falls below $20,000/mm^3$, a condition known as thrombocytopenia exists and the affected organism is extremely vulnerable to spontaneous bleeding episodes. Usually, thrombocytopenia due to marrow failure is also associated with reduced leukocyte and red blood cell production since chemicals or disorders that affect the megakaryocytes also impact other stem cells.

The opposite phenomenon, elevated platelet count or thrombocytosis, is diagnosed when counts are greater than $400,000/mm^3$. There are many causes of thrombocytosis, including primary [e.g., essential thrombocytosis (ET)] and secondary (e.g., response to inflammation, acute bleeding, iron deficiency, or cancers). In ET, there are colonies of megakaryocytes in the absence of any known stimulus.

Toxic Agents and Responses

Carbon Monoxide

Carbon monoxide (CO) is an odorless, tasteless, and colorless gas that is rapidly absorbed by the lungs and attaches to hemoglobin with an affinity that is 250 times greater than oxygen. Due to this extreme differential, as CO concentrations increase, the number of available sites on the hemoglobin molecule for oxygen decreases. Normally, this reaction would cause oxygen to be more freely released so that adequate tissue oxy-

genation can be maintained. This would typically produce a right shift of the oxyhemoglobin dissociation curve; however, with increasing exposure to CO and formation of carboxyhemoglobin (COHgb), there is a change in the oxyhemoglobin complex which produces a left shift in the oxygen dissociation curve (Fig. B-29). The overall effect is decreased tissue oxygenation, anaerobic metabolism, and lactic acid formation.

Exposure to CO results in a wide variety of potential adverse effects, particularly to individuals who have preexisting cardiac or lung disease. Infants, the elderly, and the developing fetus are particularly vulnerable since they have less capacity to tolerate cardiovascular compromise. An additional problem is the delayed neurological and neuropsychiatric effects that have been documented after some exposures. The incidence of delayed neurotoxicity has been reported to be between 2 and 30%.

CO poisoning is usually diagnosed by measuring the presence of (COHgb) in blood. Nonsmokers have COHgb levels of less than 1%, whereas smokers have levels of 5–10%. Unfortunately, the measured COHgb level does not always correlate with clinical findings and symptoms; therefore, the clinician should always have a high index of suspicion and aggressively evaluate and treat exposed patients. Treatment consists of removal from the source and administration of 100% oxygen and any other basic life-support measures required. In certain circumstances, i.e., COHgb levels over 25%, the use of hyperbaric oxygen is indicated (see Carbon Monoxide).

Hydrogen Cyanide and Hydrogen Sulfide

Both hydrogen cyanide (HCN) and hydrogen sulfide (H_2S) are metabolic poisons that act in relatively similar mechanistic ways. At the cellular level, the major energy source is adenosine triphosphate (ATP). ATP is primarily produced through a process known as oxidative phosphorylation, which involves the transfer of electrons to substances known as cytochromes. The cytochrome system can be viewed as a "bucket brigade" that moves critical electrons in an orderly fashion so that cellular respiration is maintained. As electrons are transferred, energy is released and used to generate ATP and water. Oxygen is the final electron acceptor in the cytochrome system and can be severely affected by metabolic toxins like HCN and H_2S. These toxins ultimately act by blocking electron transfer to molecular oxygen. This blockade produces a rise in peripheral tissue partial pressure of oxygen and a decrease in the unloading gradient for oxyhemoglobin. The net effect is the production of both high levels of oxyhemoglobin in venous return blood and significant levels of lactic acid. At high exposure concentrations, cardiopulmonary compromise is rapidly produced and death ensues.

The treatment of either HCN or H_2S toxicity is based on the use of chemicals that interrupt the binding of these materials to the cytochrome oxidase system. Sodium nitrate and amyl nitrate are both used as antidotes. These substances act by overwhelming the RBC with oxidant stress and producing a somewhat less toxic material known as methemoglobin (MetHgb). MetHgb serves as a source of circulating ferric iron (Fe^{3+}), which preferentially competes for binding by cyanide or sulfide and causes the cyanide or sulfide to dissociate from the cytochrome system and move into blood in a form complexed to methemoglobin in RBCs. This less toxic material is further detoxified by the use of another drug, sodium thiosulfate, which further enhances the conversion of cyanide to the less toxic thiocyanate. The situation with H_2S is somewhat more complex since the second step use of sodium thiosulfate is not typically recommended; however, vigorous use of 100% oxygen therapy is appropriate for treating exposure to both HCN and H_2S (see Hydrogen Sulfide).

Methemoglobin

At the molecular level, the transport of oxygen in the body is highly dependent on the maintenance of intracellular Hgb in a chemical condition known as the reduced state, or Fe^{2+}. When hemoglobin is oxidized, the Fe^{3+} state, it is known as MetHgb and is unable to bind oxygen. A small amount, less than 1%, of MetHgb is always found in normal RBCs. MetHgb can be chemically reduced by an enzyme system so that the body maintains adequate levels of Fe^{2+}. If MetHgb exceeds 10% of the total hemoglobin, then clinically observable changes such as dusky complexion can be detected in the affected individual. As MetHgb levels reach 35%, symptoms such as headache, fatigue, and shortness of breath are common. MetHgb levels over 80% are usually fatal.

There are many causes of MetHgb, including both hereditary and acquired. Drugs and toxins, such as nitrates, nitrites, nitroglycerin, aniline dyes, and sulfonamides, are associated with the production of MetHgb in certain situations. Toxic levels of MetHgb can be treated with a compound known as methylene blue, which acts to rapidly reduce the level of circulating MetHgb.

Leukemia

The leukemias are a diverse group of hematologic malignancies that arise from the malignant transformation of hematopoietic cells. These cells develop in the bone marrow and lymphoid tissue and ultimately interfere with normal cell development and immunity. Leukemias are generally divided into two groups, myeloid and lymphoid. In addition, leukemias can be further subdivided by their natural history into acute or chronic forms. The leukemias represent 3% of all malignancies and approximately 24,000 new cases a year develop in the United States.

The etiology of leukemia in most cases is unknown, although a combination of genetic and environmental factors is probably important. The most important environmental factors are drugs, radiation, and chemical exposures to a few selected substances. The most common form of leukemia associated with either chemicals or drugs is the acute nonlymphatic leukemias (ANLL), which are also referred to as acute myeloid leukemias (AML). In ANLL, large numbers of immature hematopoietic cells develop and replace the normal cells. These abnormal cells are released into the circulation and can easily be seen on peripheral blood smears. Since these cells are quite immature, the blood does not contain adequate numbers of normally functioning mature RBCs, leukocytes, and thrombocytes. AML is an aggressive and rapidly fatal disease unless appropriate therapy is begun.

The role of chemical exposure and development of ANLL has been quite controversial. This controversy is partially due to the problems associated with accurately and appropriately classifying the various leukemias. Since the mid-1980s, the nomenclature of the ANLL subtypes was established by the French–American–British Cooperative group also known as FAB. Older studies in the literature that do not use this classification scheme present a serious problem since there was a tendency to lump different categories together in order to achieve sufficient statistical power for epidemiological analysis. Nevertheless, there does appear to be sufficient evidence to link ANLL with certain exposures to benzene.

The association between benzene exposure and leukemia has been made since the late nineteenth century; however, the dose–response relationship and mechanistic explanation have been quite contentious. The most reliable evidence associating chronic benzene exposure with AML was presented in a retrospective NI-OSH study of rubber hydrochloride workers in Akron, Ohio, from 1940 to 1949. Unfortunately, the mechanism of how benzene exposure leads to the development of AML is not known. The two most frequently discussed potential mechanisms of toxicity involve either a point mutation or a chromosomal deletion. The latter is considered more likely since neither benzene nor its metabolites are mutagenic or teratogenic (see Benzene).

—*Gary R. Krieger*

Further Reading

Brooks, S. M., *et al.* (1995). *Environmental Medicine.* Mosby, St. Louis, MO.
Isselbachen, K. J., *et al.* (Eds.) (1994). *Harrison's Principles of Internal Medicine*, 13th ed. McGraw-Hill, New York.
Sullivan, J. B., Jr., and Krieger, G. R. (Eds.) (1992). *Hazardous Materials Toxicology: Clinical Principles of Environmental Health.* Williams & Wilkins, Baltimore.[2nd ed., 1998]

Related Topics

Cardiovascular System
Distribution
Immune System
Kidney
Liver

Boric Acid

♦ CAS: 10043-35-3

♦ SYNONYMS: Boracic acid; orthoboric acid; borofax; three elephant; NCI-C56417

♦ MOLECULAR FORMULA: H_3BO_3

♦ DESCRIPTION: Boric acid exists in natural deposits as a mineral, sassolite. It is also found in hot mineral water sources. The minerals are extracted with sulfuric acid and crystalline boric acid is separated.

Uses

Boric acid is used as a fireproofing agent for wood, as a preservative, and as an antiseptic. It is used in the manufacture of glass, pottery, enamels, glazes, cosmetics, cements, porcelain, leather, carpets, hats, soaps, artificial gems, and in tanning, printing, dyeing, painting, and photography. It is a constituent of nickling baths and electric condensers, and it is used for impregnating wicks and hardening steel. In laboratory procedures, boric acid is used in the preparation of buffer solutions.

Boric acid is also used as a fungicide and as an insecticide powder. Domestic use may include its application as an insecticide for crawling insects such as roaches. In medicine, it is used as a disinfectant and is a constituent of baby powders, antiseptics, diaper rash ointments, eye washes, gargles, and a variety of other consumer products for its mild antiseptic property.

Exposure Pathways

Accidental ingestion and subcutaneous routes are the primary exposure pathways. The maximum workplace concentration is 10 mg/m^3. The maximum concentration in water used in fisheries is 0.1 mg/liter.

Toxicokinetics

Water emulsifying and hydrophobic ointments containing boric acid liberate only small amounts within 24 hr compared with a near total liberation from a jelly. Boric acid is readily absorbed from the gastrointestinal tract, mucous membranes, and abraded skin. Boric acid is excreted unchanged in urine with approximately 50% excreted in the first 12 hr and the remainder excreted over a period of a few days. The half-life of boric acid given orally is estimated to be 21 hr. The fatal dose of boric acid is estimated to be about 20 g in an adult and about 5 or 6 g in an infant.

Mechanism of Toxicity

The exact mechanism of toxicity is not known. Boric acid can inhibit production of adenosine triphosphate, a cellular form of energy.

Human Toxicity

Boric acid preparations have been recognized as dangerous in the treatment of diaper rash and have been implicated in several cases of fatal poisoning. Symptoms include nausea, vomiting, bloody diarrhea, severe colic, and abdominal pain. There may be restlessness, delirium, headache, tremors, and generalized convulsions usually followed by weakness and coma. There is fever and tachypnea followed by Cheyne–Stokes-type respirations and respiratory arrest.

Changes on the skin include an erythematous skin eruption, with papules or vesicles appearing between the fingers and on the back of the hands initially and eventually becoming generalized enough to give a "boiled lobster" appearance. The skin lesions may undergo bullous formation, desquamation, excoriation, and sloughing. Hypothermia often occurs.

Renal injury, usually in the form of renal tubular necrosis, may be evidenced by oliguria, albuminuria, and eventually anuria. Signs of meningeal irritation, oliguria, and circulatory collapse may be followed by death within 5 days. Infants and young children are more susceptible to boric acid intoxication. Low levels of boric acid ingestion may lead to dry skin and mucous membranes, followed by the appearance of a red tongue, patchy alopecia, cracked lips, and conjunctivitis. Infertility among men is possible.

No major toxicological distinctions between boric acid and its salts are recognized in human beings.

Clinical Management

There is no specific antidote. Emesis should be induced or gastric lavage performed, followed by administration of activated charcoal and a laxative. Fluid and electrolyte balance, correction of acid/base disturbance, and control of seizures are essential to therapy. Dialysis, either peritoneal or extracorporeal, has been used with moderate success. Hemodialysis has been successfully used to treat acute boric acid poisoning. Sodium bicarbonate may be used for any metabolic acidosis.

Animal Toxicity

Large doses of boric acid over a period of time have been shown to cause testicular damage in animals.

—Arvind K. Agarwal

Boron (B)

♦ CAS: 7440-42-8
♦ SELECTED FORMS: Borax; kernite; tourmaline
♦ CHEMICAL CLASS: Metals

Uses

Boron is used in the form of small filaments as a reinforcing material for composites. In the nuclear industry, it is used as a neutron absorber. Boron is also used to harden metals and as an oxygen scavenger for copper and other metals. Amorphous boron is used in pyrotechnic flares to produce a green color. It is also used as a catalyst in olefin polymerization and alcohol dehydration. The principal consumption pattern in the United States for boron is for the production of glass products, soaps and detergents, and agricultural products.

Exposure Pathways

Ingestion and inhalation are the primary routes of exposure. Dermal absorption is not a factor unless the dermal barrier is compromised.

Toxicokinetics

Boron is well absorbed via the gastrointestinal tract. Systemic toxicity is more likely to result from multiple exposures rather than from single acute exposures. Boron is distributed fairly rapidly (within 30 min to 3 hr) to all tissues of the body. The apparent elimination half-life is 5–10 hr. The primary route of elimination is via the kidneys.

Mechanism of Toxicity

Boron is concentrated in the kidneys during excretion, making the kidneys a prime target organ for boron toxicity.

Human Toxicity

Single acute exposures to boron are well tolerated. Chronic exposures can lead to anorexia, weight loss, vomiting, mild diarrhea, erythematus rash, alopecia, convulsions, anemia, and kidney damage. Both vomitus and feces will be blue-green.

Clinical Management

General life support should be maintained, symptoms treated, and decontamination performed, if necessary. Emesis may be indicated in instances in which the patient has recently ingested a significant quantity of boron.

Animal Toxicity

Gastrointestinal and pulmonary disorders have been reported in lambs grazing in pastures containing a high boron content in the soil. High exposures to boron (1000–2000 ppm) for 90 days have caused oligospermia and testicular atrophy in rodents. Exposure of pregnant rats to boron has led to central nervous system abnormalities in the offspring. It is interesting to note that, compared to rodents, dogs are twice as sensitive to the toxic effects of boron.

The oral reference dose for boron is 0.09 mg/kg/day and is based on testicular atrophy and spermatogenic effect. An ambient water quality criterion of 7500 ppm boron in water has been suggested based on long-term irrigation of sensitive crops.

—*William S. Utley*

Related Topic

Metals

Botulinum Toxin

- ◆ SYNONYMS: *Clostridium botulinum;* foodborne (classic) botulism; infant botulism; wound botulism; unclassified botulism
- ◆ CHEMICAL CLASS: Foodborne toxins

Exposure Pathways

Ingestion is the primary exposure pathway for botulism. Wound botulism results from production and ab-

sorption of toxin in a contaminated wound. Foodborne botulism usually results from exposure to canned foods that are inadequately sterilized during cooking and canning. Occasional larger outbreaks occur following ingestion of contaminated food at restaurants or from commercial sources. A variety of preserved foods have been implicated, including string beans, corn, garlic, seafood, pork, and beef. Botulism has also been reported to result from ingestion of improperly prepared and stored fresh foods.

Infant botulism is the most common form of the illness. This occurs in children less than 1 year old. Most cases occur from 1 week to 11 months of age, with a peak incidence at 2–4 months of age. Breast feeding, feeding of honey or corn syrup, decreased frequency of bowel movements, and living in a rural area have been implicated as sources of *Clostridium botulinum* spores that cause infant botulism. For this reason it has been recommended to avoid feeding honey to any child 12 months of age or younger. Types A or B botulinum toxin, and rarely type F, have been responsible for all infant cases.

Mechanism of Toxicity

Despite the ubiquitous nature of botulinum spores, the incidence of disease is low. For optimal growth, *C. botulinum* requires a low acidic environment (generally pH greater than 4.5), a temperature of at least 10°C, an anaerobic environment, and a lack of competition from other bacteria. While most toxin is destroyed by boiling, only pressure cooking to 240°F will ensure destruction of spores.

The three classified types of botulism (foodborne, infant, and wound) plus one unclassified type result from infection with *C. botulinum* organisms. *Clostridium botulinum* is a strictly anaerobic, spore-forming, gram-positive rod that elaborates a potent exotoxin. Poisoning is due to this heat-labile neurotoxin produced by *C. botulinum* spores that binds irreversibly to the neuromuscular junction and impairs release of acetylcholine presynaptically. Nerve conduction is blocked, resulting in weakness.

By weight, botulinum toxin is the most potent natural poison in the world according to studies conducted in animals. While there are seven immunologically distinct toxins (A–G), the majority of poisonings in humans are caused by three toxins—A, B, and occasionally E.

Human Toxicity

Fewer than 100 cases of botulism are reported in the United States per year. Of these, one-third are food-borne botulism and two-thirds are infant botulism. The highest incidences of botulism are reported from Alaska, Washington, Oregon, and California. Symptoms usually develop 12–36 hr after toxin ingestion but have been reported as early as 3 hr and as late as 14 days. Neurologic symptoms usually appear within 3 days of generalized symptoms but may be delayed for up to 1 week. In general, the earlier the onset of symptoms, the more serious the disease and the more protracted the course. Following ingestion of food containing toxin, illness varies from a subclinical illness to a life-threatening illness. Death may ensue from respiratory failure as rapidly as 24 hr after initial exposure. Initial presenting chief complaints include ptosis, blurred vision, and a dry sore throat. Dizziness, nausea, vomiting, and abdominal cramps are not unusual; constipation is common. A descending bilateral paralysis is pathognomonic of botulism.

Infant botulism is a disease characterized by constipation, tachycardia, muscle weakness, difficulty in feeding, head lag, and diminished gag reflex and muscle tone. Infant botulism is more common during the spring and summer months and is rare during the winter months.

Clinical Management

Basic and advanced life-support measures should be utilized as necessary. Basic treatment involves supportive care and prevention of disease progression. Gastrointestinal decontamination procedures should be used as deemed appropriate to the patient's level of consciousness and the history of the ingestion. Laboratory analysis of blood, gastric contents, and food should be considered before antidotal treatment begins. An equine antitoxin is available (Trivalent ABE antitoxin) for human use. In the United States it may be obtained from the Centers for Disease Control in Atlanta, Georgia. It should be administered as soon as food or wound botulism is suspected, but it is contraindicated in infant botulism. The antitoxin exerts its effects by neutralizing blood-based toxin and prevents progression of the disease but does not reverse existing neurologic deficits. Skin testing for hypersensitivity to horse serum should precede antitoxin administration. Due to the fact that respiratory failure is the greatest threat to life, the mainstay of therapy should be early, aggressive respiratory support. Antibiotics have been recommended to eradicate the botulinum organism, but the value of this treatment is uncertain.

Recovery is often prolonged following exposure to *C. botulinum*. Mean hospitalization lasting 1–10 months and some symptoms often persisting beyond 13 months have been noted. Fatigue and dyspnea may still be problematic after 1 or 2 years. Supportive aspects of long-term intensive care are vital to a positive outcome.

Animal Toxicity

All animals are exquisitely sensitive to *C. botulinum*. Signs and symptoms in animals are similar to those observed in humans. Treatment should always be performed on the advice and with consultation of a veterinarian. Animal treatment is similar to human treatment but does vary depending on animal species.

—*Gaylord P. Lopez*

Related Topics

Aflatoxin

Mycotoxins

Neurotoxicology: Central and Peripheral

Brodifacoum

- CAS: 56073-10-0
- SYNONYMS: PP 581; WBA 8119; 3-[3-(4-bromo[1,1-biphenyl]-4-yl)-1,2,3,4-tetrahydro-1-naphthalenyl]-4-hydroxy-2H-1-benzopyran-2-one; Talon G; Ratac; Havac
- CHEMICAL CLASS: A long-acting 4-hydroxycoumarin derivative; one of the superwarfarins (also see Warfarin)

♦ CHEMICAL STRUCTURE:

Uses

Brodifacoum is used as a rodenticide (commonly 0.005% by weight).

Exposure Pathways

The most common route of exposure is oral. Transcutaneous and inhalation exposures have been implicated in workers involved the manufacture of brodifacoum.

Toxicokinetics

The metabolic fate of brodifacoum in humans is not well understood. Brodifacoum is much more lipid soluble than is warfarin, resulting in a larger volume of distribution. There is extensive hepatic sequestration and prolonged high liver concentrations in the rat. Brodifacoum may also undergo enterohepatic circulation in the rat. Based on the limited data available, the plasma half-life of brodifacoum in humans ranges from 16 to 36 days. The apparent elimination half-life in humans, using sensitive measures of enzyme inhibition, has been reported to be approximately 6 months, long after brodifacoum may be detected in the blood. This is believed to reflect hepatic accumulation of the drug. Inducers of the cytochrome P450 system have been reported to reduce the half-life of brodifacoum in animals.

Mechanism of Toxicity

Brodifacoum, like other coumarins, interferes with the production of vitamin K-dependent coagulation factors. Vitamin K_1 is a cofactor for the carboxylation of specific glutamic acid groups in coagulation factors II (prothrombin), VII, IX, and X. During this step, vitamin K_1 is oxidized to vitamin $K_{1,2,3}$ epoxide. The regeneration of vitamin K_1 by vitamin $K_{1,2,3}$ epoxide reductase is prevented by brodifacoum. As a result, dysfunctional decarboxy coagulation factors are produced and coagulation is impaired. Brodifacoum is over 100 times more potent than warfarin on a molar basis in rats.

Human Toxicity: Acute

Depletion of preformed, circulating coagulation factors must occur before any anticoagulant effects are apparent. Typically, there is a delay of 24–36 hr following ingestion before any effect is evident by measurement of the prothrombin time (PT). Significant toxicity from brodifacoum may be the result of large, one-time intentional ingestions or repeated exposures over time. Single, small accidental ingestions in children are usually benign. Bleeding may occur virtually anywhere although cutaneous, mucosal, urinary, and gastrointestinal bleeding would be expected to be most common. Fatal intracerebral hemorrhage has been reported. Poisoning due to brodifacoum has led to prolonged periods of anticoagulation, often weeks and in some cases up to 6 months or longer. The clinical effect of brodifacoum is best monitored by following the PT. Early on, the PT is more sensitive than the activated partial thromboplastin time to the effects of coumarin anticoagulants. Serum brodifacoum levels can be measured to confirm exposure, although there are no data to correlate serum levels and extent of toxicity. Factor activity can be assayed. An elevated serum ratio of vitamin K_1 epoxide to vitamin K_1 is further evidence of the presence of vitamin K_1 reductase inhibition.

Human Toxicity: Chronic

Repeated exposures over time can lead to prolonged anticoagulation.

Clinical Management

For acute, single-dose ingestions, activated charcoal should be administered. A gastric emptying procedure may be beneficial if performed soon afterward and there is no evidence of coagulopathy. Induced emesis should be avoided in the anticoagulated individual. The PT should be determined at 24–48 hr postingestion to assess the potential for toxicity. In the patient with clinical evidence of significant anticoagulation, extreme caution should be exercised with any invasive procedure. The airway should be protected if compromised by bleeding or hematoma formation. Volume resuscitation should be provided as indicated by clinical status. With active, uncontrolled, or life-threatening hemorrhage, fresh frozen plasma will provide preformed coagulation factors. Vitamin K_1 is a specific antidote for brodifacoum toxicity. Pharmacologic doses of vitamin K_1 allow the production of functional coagulation factors despite the presence of brodifacoum. The dose and route depend on the clinical setting. For rapid reversal, 5–25 mg should be administered intravenously no faster than 1 mg/min. In children, doses of 0.6 mg/kg have been recommended in warfarin poisoning, and larger doses may be necessary with brodifacoum. Clinical effects may be seen within hours. The response and duration of a single dose is variable and depends on the severity of the intoxication. Repeat doses may be necessary. In the less emergent setting, vitamin K_1 may be given subcutaneously or orally. The doses needed to maintain adequate coagulation status may be quite large; in some cases, doses of 100 mg/day or more orally have been reported, although typical doses are in the range of 10–20 mg/day. Oral vitamin K therapy may be necessary for weeks to months. Serial monitoring of the PT should be used to help guide therapy. Factor activity analysis may also be of use in assessing the adequacy of therapy. (Note: Anaphylaxis has been reported with intravenous vitamin K_1. Vitamin K_3 (menadione) is not effective therapy.)

Animal Toxicity

Mammals vary in their sensitivity to the coumarin anticoagulants. Signs of toxicity in animals include depression, anorexia, weakness, vomiting, diarrhea, bleeding, and dyspnea. Toxic effects can be monitored by following coagulation tests, especially those sensitive to prothrombin activity. Treatment in animals is as for humans. The dose of vitamin K recommended for dogs and cats is 2.5–5.0 mg/kg/day for up to 4 weeks with monitoring of coagulation parameters.

—Michael J. Hodgman

Related Topic

Coumarins

Bromine

- ◆ CAS: 7726-95-6
- ◆ SYNONYMS: Brom; brome; broom; dibromine; UN1744

Uses

Bromine was formerly used medicinally as a topical antiseptic. It is used in gold extraction, in the bleaching of fibers and silk, in shrink-proofing of wool, in photography, and in the manufacture of bromine compounds, military gas, antiknock compound (ethylene bromide), and fire-extinguishing fluid.

Exposure Pathways

Inhalation, ingestion, and eye and skin contact are the most common routes of exposure. Bromine may be absorbed through skin.

Toxicokinetics

Bromine has cumulative properties and is deposited in tissues as bromides, displacing other halogens.

Mechanism of Toxicity

Bromine causes toxicity as bromides by displacing other halogens from the body.

Human Toxicity: Acute

The respiratory system, eyes, and central nervous system are the points of attack. Bromine is extremely irritating to skin, eyes, and mucous membranes of the upper respiratory tract. Severe burns of the eye may result from liquid or concentrated vapor exposure. Liquid bromine splashed on skin may cause vesicles, blisters, and slow-healing ulcers. Inhalation of bromine is corrosive to the mucous membranes of the nasopharynx and upper respiratory tract, producing brownish discoloration of the tongue and buccal mucosa, a characteristic odor of breath, edema and spasm of the glottis, asthmatic bronchitis, and possibly pulmonary edema which may be delayed until several hours after exposure. A measle-like rash may occur. Exposure to high concentrations of bromine may lead to death due to choking caused by edema of glottis and pulmonary edema. Exposure to low concentrations results in cough, copious mucus secretion, nose bleeding, respiratory difficulty, vertigo, and headache. Usually these symptoms are followed by nausea, diarrhea, abdominal distress, hoarseness, and asthmatic-type respiratory difficulty.

Human Toxicity: Chronic

Chronic exposure may cause acne-like skin lesions and neurotoxicity.

Clinical Management

Exposure should be terminated as soon as possible by removing the victim to fresh air. The eyes and mouth should be washed with copious amounts of water. A 15- to 20-min wash may be necessary. Skin should be washed with soap. Contaminated clothing and jewelry should be removed and isolated. Contact lenses should be removed from the eyes to avoid prolonged contact of the chemical with the area. When the chemical has been swallowed, large quantities of milk should be given; if milk is not available, water should be given. Vomiting should not be given. If breathing has stopped, artificial respiration should be given. If breathing is difficult, oxygen should be given.

Animal Toxicity

Toxicity in animals is similar to that in humans.

—*Sanjay Chanda*

Bromobenzene

◆ CAS: 108-86-1

◆ SYNONYMS: Monobromobenzene; phenyl bromide

◆ CHEMICAL CLASS: Halogenated aromatic

◆ CHEMICAL STRUCTURE:

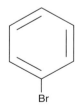

Uses

Bromobenzene is used as an industrial solvent and in organic synthesis. It is also used as an additive to motor oils.

Exposure Pathways

The primary routes of exposure are inhalation, skin contact, and accidental ingestion.

Toxicokinetics

Bromobenzene is excreted as the free and sulfate or mercapturic conjugates of the catechol derivatives. Initially bromobenzene may concentrate in the adipose tissues. Bromobenzene concentrations can be 300 times higher in the adipose tissues in first 3 hr of exposure. Bromobenzene is rapidly excreted in the urine; one report indicates that 85% of the bromobenzene may be excreted in the urine in the first 24 hr.

Mechanism of Toxicity

Bromobenzene is believed to be relatively inert, requiring metabolic activation to express toxicity to the liver and the kidney. Liver toxicity is believed to result from activation of bromobenzene to a reactive epoxide by the cytochrome P450 system. The reactive epoxides are primarily detoxified by glutathione transferase and epoxide hydratase. The involvement of glutathione transferase explains in part the decrease in glutathione levels observed following exposure to bromobenzene. It is interesting to note that in animal experiments, sulfhydryl-containing compounds, such as cysteine or methionine, partially prevented bromobenzene-induced hepatic necrosis. It has been suggested that the enterohepatobiliary cycle plays a role in the hepatic necrosis observed in bromobenzene toxicity. This is supported by the experimental findings that bromobenzene-induced hepatic necrosis can be prevented by the administration of cholestyramine.

Human Toxicity

Bromobenzene is known to be a skin irritant and is suspected of irritating the eyes and respiratory tract. The probable lethal dose is between 1 teaspoon and 1 oz for a 154-pound person. Bromobenzene is directly irritating to the skin and can act as an anesthetic when inhaled in high concentrations. It is capable of damaging the liver and the kidneys when metabolized.

Clinical Management

Treatment is generally symptomatic and supportive. General life support should be maintained, symptoms should be treated, and the patient should be decontaminated, if necessary.

Animal Toxicity

In dogs, oral exposure to high concentrations of bromobenzene leads to vomiting, diarrhea, and death. The major histopathological findings were centrilobular hepatic necrosis. The necrosis included the central hepatic veins and their respective tributaries. Bromobenzene was not found to be mutagenic in the Ames assay.

—William S. Utley

Related Topics

Kidney
Liver

Bromoform

- ◆ CAS: 75-25-2
- ◆ SYNONYM: Tribromomethane
- ◆ CHEMICAL CLASS: Halogenated hydrocarbon
- ◆ CHEMICAL STRUCTURE:

$$
\begin{array}{c}
Br \\
| \\
Br - CH \\
| \\
Br
\end{array}
$$

Uses

Bromoform is used as a chemical intermediate in the synthesis of organic chemicals and pharmaceuticals. It is used as an ingredient in fire-resistant chemicals and as an industrial solvent in liquid-solvent extractions. Bromoform is used in polymer reactions and in the vulcanization process for rubber. It is also used for medicinal purposes as a sedative, antitussive, and antiseptic.

Exposure Pathways

Ingestion is the most common form of accidental and intentional exposure. In the past, inhalation was a more common route of exposure during anesthesia. However, due to the associated toxicity, bromoform is no longer as popular as an anesthetic. Dermal absorption is possible but not likely to be a significant route of exposure in intact skin.

Toxicokinetics

Bromoform is readily absorbed from the gastrointestinal tract following ingestion and from the lungs following inhalation. Significant absorption may occur through abraded skin or open wounds. Bromoform is metabolized in the liver by the mixed-function oxidase system (cytochrome P450) to carbon monoxide and bromide. Inorganic bromide has been observed in tissues and urine following administration of bromoform. Bromoform readily distributes through the tissues of the body. It has been found in higher concentrations in the brain than in the blood following inhalation.

Mechanism of Toxicity

The wide spectrum of toxic effects elicited by bromoform suggests multiple mechanisms of actions. Bromoform is an irritant, capable of directly irritating mucosal membranes when directly exposed. Chronic exposure to bromoform may also cause fat loss in the skin leading to drying and cracking. Bromoform is capable of dissolving in phospholipid membranes, giving it the ability to produce anesthetic effects when given in high concentrations. The toxic effects on the liver and kidneys may be mediated by reactive intermediates produced by the hepatic P450 oxidative metabolism.

Human Toxicity

Bromoform's odor has been described as "chloroform-like" and the taste has been described as "sweet." Bromoform can be toxic by all routes of exposure. People appear to be able to detect bromoform at very low concentrations in liquid (0.3 ppm). Symptoms of acute exposure include severe irritation of the eyes, lacrimation, salivation, skin and respiratory tract irritation, headache, and dizziness. Excessive concentrations can lead to unconsciousness, convulsions, and possible respiratory failure. Chronic exposure may cause liver damage and memory loss. Liver damage is characterized by both fatty infiltration of the liver and necrosis. Predisposing conditions include a history of skin, liver, kidney, or chronic respiratory disease.

The OSHA PEL and ACGIH TLV is 0.5 ppm (with a notation for skin absorption). As a waste, bromoform is considered a toxic waste (No. U225).

Clinical Management

Medical surveillance may be indicated in persons with predisposing skin, liver, kidney, or respiratory conditions. Treatment is generally symptomatic and supportive. General life support should be maintained, symptoms treated, and the patient decontaminated, if necessary. The patient should be monitored for delayed liver and kidney damage. If central nervous system depression occurs, EKG and vital signs should be monitored carefully. Patients who exhibit dermal hypersensitivity may require systemic or topical antihistamines or corticosteroids.

Animal Toxicity

Bromoform was noted to be more toxic to the liver and more irritating than chloroform when given via inhalation. Bromoform was noted to produce decreased liver functions and pathologic changes to both the liver and the kidneys by both oral and inhalation routes in rodents. Single oral doses to rodents produced sedation, ataxia, piloerection, and prostration. Undiluted bromoform caused moderate irritation to the eyes of rabbits, which recovered in 1 or 2 days. Bromoform is mutagenic in the Ames assay.

—*William S. Utley*

Buckthorn

- ◆ SYNONYMS: Rhamnaceae family; *Rhamnus frangula*; alder buckthorn; frangulin; trollidora; coyotillo; wild cherry; purging buckthorn; arrow wood; berry alder; black dogwood; cascara; Hart's horn; May thorn; Persian berry; rhine berry; common buckthorn

- ◆ DESCRIPTION: Buckthorn is a shrubby tree that grows 6–12 ft tall. The leaves (up to 6 cm) are simple, ovate elliptic, with serrate margins. Some of the branches end in short thorns. The flowers have four small petals and grow solitary or in clusters from the leaf axis. They are replaced later by a berry that is green at first, turns red, and then turns black at maturity. The berry contains up to four seeds. This ornamental shrub is often found as hedges throughout the eastern United States. The shrubs are also grown along canyons and gullies in the southwestern United States and northern and central Mexico. Other species of this plant are found throughout the northern temperate zones.

Exposure Pathways

The most common route of exposure is ingestion of any parts of the plant, with the seeds being most poisonous. Anthraquinone laxatives are made from plant-derived compounds.

Toxicokinetics

Anthraquinone glycosides are poorly absorbed after ingestion. Moderate absorption occurs after hydrolysis from colonic bacteria into senna and cascara (laxative-causing agents). A bowel movement with use of senna and cascara laxatives is usually seen within 6–12 and 6–8 hr, respectively. However, this delay of symptom onset is not seen with fresh plant ingestions. After ingestion and hydrolysis, the anthraquinones are partially eliminated renally, in the feces, and in the bile. Senna, one of the anthraquinones, is not excreted into breast milk to a significant degree.

Mechanism of Toxicity

The fruit, leaves, and bark contain several anthraquinone (emodin) glycosides. After hydrolysis, a strong stimulant laxative effect is produced. Anthraquinone glycosides exhibit their effect by increasing the tone of the smooth muscle wall in the large intestine. A direct action is exhibited on the intestinal mucosa, increasing the colonic motility and colonic transit and inhibiting water and electrolyte secretion. These agents may also act on the intramural nerves and plexes of the colon. They also have a stool-softening property. The seeds/fruit of the *Karwinski humboltiana* species contains a neurotoxic c-15 polyphenol as well as anthracenones. This neurotoxin produces an ascending bilateral paralysis similar to Guillian–Barre syndrome.

Human Toxicity: Acute

Buckthorn poisonings are rare in the United States but have more often been described in Europe. Determination of a toxic dose is difficult due to various species containing different concentrations of anthraquinones. Apparently, however, the ingestion of approximately 20 berries or chewing of fresh bark is needed to induce serious symptomatology. Mild to moderate intoxication is expected with smaller, more insignificant ingestions. In these cases, the primary symptoms seen are nausea, vomiting, diarrhea, abdominal cramps, and possible palpitations. Severe poisoning has been reported from use of the bark as an abortifacient. In severe cases, kidney damage, oliguria, proteinuria, gastrointestinal hemorrhage, muscular seizures, dyspnea, and fluid depletion can result.

Ingestion of seeds or fruit of the *K. humboltiana* species can also produce neurotoxic symptoms of ascending bilateral paralysis, similar to Guillian–Barre syndrome. This may progress to respiratory paralysis. Symptom onset, including weakness, occurs after a latent period of several weeks and the paralysis may progress for a month or longer.

Human Toxicity: Chronic

Chronic ingestion of *K. humboltiana* caused progressive, symmetrical polyneuropathy that resulted in flaccid quadriplegia and respiratory insufficiency, further complicated by pneumonia. A slow but progressive improvement to an almost complete functional recovery occurred with some persistent reflex deficits.

Clinical Management

Basic and advanced life-support measures should be utilized. Emesis with syrup of ipecac is not required, but it could be administered within 30 min after ingestion if the patient is asymptomatic. Lavage may be of some use with significant ingestions. Activated charcoal without a cathartic can also be used in decontamination. Whole bowel irrigation may be indicated if large amounts are ingested. Most ingestions are self-limiting. Treatment after decontamination is symptomatic and supportive. Close monitoring of fluids and electrolytes is recommended.

A red color is seen in alkaline urine and a yellow-brown color in acid urine. This is called the Borntrager's reaction and is diagnostic for anthraquinones. The red urine must be differentiated from hematuria. No other specific laboratory tests are available to assist in diagnosis and treatment.

Animal Toxicity

Animal that have ingested anthraquinone-containing plants are likely to develop signs and symptoms similar to those of humans. These symptoms include diarrhea, vomiting, abdominal cramps, bradycardia, and fever. Death can occur with large ingestions and/or chronic exposures.

—*Lanita B. Myers*

Butadiene, 1,3-

- CAS: 106-99-0
- SYNONYMS: Biethylene; butadiene; divinyl; erythrene; pyrrolylene; vinylethylene
- MOLECULAR FORMULA: C_4H_6
- CHEMICAL CLASS: Hydrocarbon
- CHEMICAL STRUCTURE:

$$H_2C=CH-CH=CH_2$$

Uses

The primary use of butadiene is in the manufacture of synthetic elastomers (butadiene–styrene, polybutadiene, and neoprene), in food wrappings, and in the latex paint industry. About 60% of the total butadiene–styrene copolymer production is used for tires.

Exposure Pathways

Potential exposures to butadiene are likely to be limited to the industrial setting as the residual butadiene monomer content in consumer products is low and unlikely to pose a significant health threat to the general public. In the workplace, the most significant route of exposure to butadiene is inhalation during its production and use.

Toxicokinetics

Butadiene appears to be readily absorbed through the respiratory tract. Dermal absorption is anticipated to be limited due to the volatility of liquid butadiene. Butadiene is rapidly metabolized via enzyme systems located in the liver, lung, nasal mucosa, and possibly bone marrow. Initially, butadiene is metabolized (activated) to butadiene monoepoxide (BME) by cytochrome-P450 monooxygenase, an enzyme which also metabolizes BME to butadiene diepoxide (BDE). BME and BDE, which are thought to cause the DNA damage necessary for the butadiene-induced tumorigenesis, can be metabolized (inactivated) by conjugation with glutathione via glutathione *S*-transferase and by hydrolysis via epoxide hydrolase. The activation/inactivation profiles of these three enzyme systems are species specific.

Once absorbed, butadiene is rapidly distributed throughout the body. A tissue distribution study in rats indicates that the highest concentrations of butadiene are located in peripheral fat; lower concentrations are observed in the liver, brain, spleen, and kidney.

Based on a radiolabel study in mice, most of the absorbed butadiene is exhaled as the parent compound, with a lesser amount exhaled as CO_2. Smaller amounts of butadiene and/or its metabolites are detected in urine and feces, with most of the label being eliminated from the carcass within 65 hr. This is consistent with another study in rats and mice which showed that the bulk of their butadiene body burden (77–99%) is eliminated with a half-life of 2–10 hr.

Mechanism of Toxicity

The mechanism by which butadiene exerts acute toxicity is unknown. The carcinogenic potential of butadiene in rats and mice is thought to be related to the metabolism of butadiene to the DNA-reactive metabolites, BME and BDE. This hypothesis is supported by the observation that mice, which are more sensitive than rats to butadiene-induced carcinogenesis, have significantly higher concentrations of BME and BDE in their blood than do rats.

Human Toxicity

Exposures of industrial workers to butadiene concentrations of 2000–8000 ppm have been reported to cause eye, skin, and nasal irritation. High butadiene levels may cause central nervous system (CNS) depression as evidenced by blurred vision, drowsiness, fatigue, bradycardia, and hypotension. The mildly aromatic odor of butadiene, which can be detected at about 1 ppm, serves as a good warning aide. Dermal contact with liquid butadiene may produce frostbite due to cooling caused by the rapid evaporation of butadiene from the skin. It is not known if butadiene poses a carcinogenic risk in humans.

Clinical Management

The primary toxicity of butadiene is CNS depression at high concentrations. Treatment involves removal from exposure and support of respiratory function.

Animal Toxicity

Acute inhalation studies have shown that butadiene exhibits low toxicity in animals. Deep anesthesia was produced in rabbits exposed to 200,000–250,000 ppm butadiene for 8–10 min; death due to respiratory paralysis occurred within 25–35 min at 250,000 ppm. The LC_{50} in rats exposed to butadiene for 4 hr is 129,000 ppm; the LC_{50} in mice exposed to butadiene for 2 hr is 122,000 ppm. The acute oral LD_{50}s for butadiene in rats and mice are 5480 and 3210 mg/kg, respectively.

Chronic (lifetime) studies in rats and mice indicate that the inhalation of butadiene increases the incidence of tumors at various sites, with mice being significantly more susceptible to the tumorigenic effect of butadiene than rats. Fetal toxicity was observed in the pups of mated rats exposed to butadiene via inhalation, but only at concentrations that also resulted in maternal toxicity.

—*Ralph Parod*

Related Topics

Combustion Toxicology
Respiratory Tract

Butane

- ◆ CAS: 106-97-8
- ◆ SYNONYMS: *n*-Butane; diethyl; methylethylmethane; butyl hydride; butanen; butani

- CHEMICAL CLASS: Low-molecular-weight aliphatic hydrocarbon
- MOLECULAR FORMULA: C_4H_{10}

Uses

Butane is primarily available as a fuel gas. It is an abused inhalant and does not have a therapeutic use.

Exposure Pathways

Butane is inhaled by various methods. Butane lighter refills have been used for inhalant abuse by holding the nozzle between the teeth and pressing the container. Fire-breathing has also been described—this involves filling the mouth with fumes and expelling them on an open flame.

Toxicokinetics

When abused, the onset of effects is rapid, as is recovery in most cases. Signs of asphyxia are seen when oxygen is displaced to the point where the oxygen concentration ranges from 15 to 16% or less. An arterial oxygen saturation (O_2 sat) of 64–82% can impair the ability for self-rescue. Total incapacitation and unconsciousness is reached with O_2 sats of 60–70% or less.

Mechanism of Toxicity

Butane is a simple asphyxiant that causes toxicity by displacing oxygen. It is hypothesized that butane sensitizes the myocardium to endogenous catecholamines, causing ventricular fibrillation or "sudden sniffers death."

Human Toxicity

Users expect an initial elevation of mood and pleasant perceptual changes. The central nervous system (CNS) is depressed as a result of butane exposure, and users may experience disorientation, euphoria, feelings of omnipotence, excitement, illusions, hallucinations, delirium, and delusions.

Acute Toxicity

Undesirable effects can include ataxia, blurred vision, headache, tinnitus, nausea, vomiting, abdominal pain, and irritability. Other adverse effects from butane exposure include chest pains, bronchospasm, hyperemia of the rhinopharynx, tachypnea, dyspnea, and pulmonary infiltrates. Anesthetic effects, weakness, malaise,

and throat irritation have occurred. With higher inhaled doses, cardiac arrhythmias and anoxia occur. Prolonged hypoxia results in incapacitation, unconsciousness, CNS injury, metabolic acidosis, pulmonary and cerebral edema, seizures, and multiorgan damage. The practice of holding the nozzle of a butane lighter refill between the teeth and pressing the container carries an additional risk of freeze burns with resultant edema leading to severe respiratory problems. Death can result from rapid chilling of the larynx that leads to vagal inhibition. Fire-breathing has reportedly caused hemorrhagic esophagitis and gastritis.

Chronic Toxicity

There are no published reports of chronic toxicity. Tolerance develops in all cases of butane abuse. Withdrawal can result in sleep disturbances, tremors, irritability, diaphoresis, nausea, and transient illusions. Facial tics, chest pain, and abdominal cramping are also common. Withdrawal symptoms are experienced from 24 to 48 hr after discontinuation and last from 2 to 5 days.

Clinical Management

Management must focus on alleviation of anoxia and cardiac manifestations. Administration of 100% humidified oxygen with assisted ventilation may be required. The usual pharmacologic agents are employed to treat seizures, metabolic acidosis, pulmonary and cerebral edema, and cardiac arrhythmias if needed. Hyperbaric oxygen can be considered to resolve persistent neurologic deficits.

Animal Toxicity

The LC_{50} (inhalation) in mice is 680,000 mg/m^3 for 2 hr. The LC_{50} in rats is 658,000 mg/m^3 for 4 hr.

—*Janet E. Bauman*

Butter Yellow

- CAS: 60-11-7
- SYNONYMS: *p*-Dimethylaminoazobenzene (DAB); *N,N*-dimethyl-4-(phenylazo) benzenamine; methyl yellow; C.I. solvent yellow 2; C.I. 11020

- CHEMICAL CLASS: Food dye
- MOLECULAR FORMULA: $C_{14}H_{15}N_3$
- CHEMICAL STRUCTURE:

Uses
Early in this century, butter yellow was largely used as a food coloring. It is also used for determination of free HCl in gastric juice, spot test identification of peroxidized fats, as a pH indicator, and as a laboratory reagent.

Exposure Pathways
Inhalation is the most common route of exposure. When heated to decomposition it emits toxic fumes of nitrous oxides.

Toxicokinetics
Butter yellow may be rapidly absorbed by various routes including ingestion, inhalation, and dermal contact. Biotransformation involves reduction (catalyzed by at least two types of cytochrome P450) and cleavage of the azo group, demethylation, ring hydroxylation, N-hydroxylation, N-acetylation, and O-conjugation of metabolites in liver. The metabolites can bind to proteins and nucleic acids. When [^{14}C-dimethyl]-aminoazobenzene was fed to rats, most of the radioactivity was found in expired carbon dioxide. Urine of rats administered butter yellow contained 50–60% of it in the form of sulfates or glucuronides of *N*-acetylated metabolites.

Mechanism of Toxicity
It is metabolized *in vivo* to a reactive form which covalently binds to cellular macromolecules, such as proteins and DNA, to cause toxicity. Agents that prevent these bindings can decrease the toxicity.

Human Toxicity
The only occupational health observation in humans was of contact dermatitis in factory workers handling butter yellow. The target organs for toxicity are skin, liver, and bladder. Potential symptoms of overexposure are enlarged liver, hepatic and renal dysfunction, contact dermatitis, coughing, wheezing, difficulty in breathing, bloody sputum, bronchial secretions, frequent urination, hematuria, and dysuria. Butter yellow can also cause adverse reproductive effects.

In the United States, OSHA lists butter yellow as a suspected human carcinogen. Human mutation data are also reported.

Workers exposed to butter yellow should wear personal protective equipment and their work should be carried out only in restricted areas. Technical measures should prevent any contact with the skin and mucous membranes. After use, clothing and equipment should be placed in an impervious container for decontamination or disposal. Preemployment and periodic medical examination should focus on liver function.

Clinical Management
In case of contact, the eyes and skin should be flushed with water for 15–20 min. For inhalation exposure, the victim should be moved to fresh air. Oxygen and artificial respiration should be administered, if necessary. If the patient is in cardiac arrest, cardiopulmonary resuscitation should be given. Life-support measures should be continued until medical assistance has arrived. In the case of an unconscious or convulsing person, liquids should not be administered and vomiting should not be induced.

Animal Toxicity
Butter yellow is poisonous by the intravenous route. It is moderately toxic by the oral, intraperitoneal, intramuscular, and subcutaneous routes. It is an antihypertensive agent. It shows mutagenic properties after activation. It is carcinogenic by various routes in the rat and mouse (liver carcinoma). By the oral route, it causes carcinoma of the bladder and lungs. Its carcinogenic action is influenced by diet. It is shown to be teratogenic. The LD_{50} in the rat is 200 mg/kg orally and 230 mg/kg intraperitoneally. The LD_{50} in the mouse is 300 mg/kg orally and 230 mg/kg intraperitoneally.

—*Kashyap N. Thakore*

Butyl Ether

- CAS: 142-96-1
- PREFERRED NAME: Dibutyl ether

- SYNONYMS: 1-Butoxybutane; butyl oxyde; *n*-dibutyl ether
- CHEMICAL CLASS: Ether
- CHEMICAL STRUCTURE:

$$CH_3—CH_2—CH_2—CH_2—O—CH_2—CH_2—CH_2—CH_3$$

Uses

Butyl ether is used mainly as a solvent for organic materials such as resins, oils, hydrocarbons, esters, gums, and alkaloids. It is also used as an extracting agent in metal separation and as a reacting medium in organic synthesis processes. It is a common solvent found in teaching, research, and analytical laboratories.

Exposure Pathways

Exposure to butyl ether can occur through inhalation of vapor or mist, dermal contact, or oral ingestion of liquid dibutyl ether. Oral ingestion of dibutyl ether has been practiced to produce an "alcoholic" euphoria.

Toxicokinetics

Dibutyl ether is rapidly adsorbed and eliminated from the body. Dibutyl ether can cause irritation to the skin, mucus membranes, eyes, and respiratory and gastrointestinal tracts. Systematically, dibutyl ether causes central nervous system (CNS) depression and transient liver changes.

Mechanism of Toxicity

Butyl ether has the ability to dissolve lipids. As a result, they cause pain upon contact with eyes and dermal irritation and dermatitis upon contact with the skin. Damage caused by butyl ether appears to be scattered loss of epithelial cells due to solution of phospholipid cell membranes. At the CNS level, ether (like other volatile organic solvents) depresses the CNS by dissolving in the cell's lipid membrane and disrupting the lipid matrix. These effects are known as membrane fluidization. At the molecular level, membrane fluidization disrupts solute gradient homeostasis, which is essential for cell function.

Human Toxicity

Signs and symptoms of excessive exposure to dibutyl ether resemble those of ethanol intoxication except that symptoms are seen shortly after exposure and the effects are short lived. Typical symptoms include dizziness, giddiness, headache, euphoria, and CNS depression. Chronic, repeated dermal exposure may cause dermal irritation, defatting of skin, and dermatitis. Excessive consumption of dibutyl ether as an intoxicating agent has been reported to produce ether jags, respiratory depression, and death.

Clinical Management

Given the CNS and respiratory depression properties of dibutyl ether, treatment is directed at maintaining respiration and treating irritation at the site of exposure. The patient should be monitored for respiratory distress and apnea, hyperglycemia, as well as hepatic and renal dysfunction.

Animal Toxicity

Dibutyl ether is moderately toxic to animals by the oral route. The oral LD_{50} in rats has been reported to range from 3230 to 3920 mg/kg. The inhalation LC_{50} also in rats has been found to be 4000 ppm for 4 hr. The percutaneous LD_{50} in rabbits is more than 20.00 ml/kg.

—*Heriberto Robles*

Butyl Ethyl Ether

- CAS: 628-81-9
- PREFERRED NAME: Ether, butyl ethyl
- SYNONYM: Ethyl butyl ether
- CHEMICAL CLASS: Ether
- CHEMICAL STRUCTURE:

$$C_4H_9OCH_2H_5$$

Animal Toxicity

Very little is known about the toxicological properties of butyl ethyl ether. Mild irritation was produced by 500 mg of butyl ethyl ether applied to the skin of rabbits for 24 hr. It caused mild ocular effects when 500 mg was applied directly to rabbit eyes. The estimated oral LD_{50} is 1870 mg/kg in rats; the estimated LC_{50} in mice is 153 g/m³ for 15-min exposures. Animals exposed to butyl ethyl ether by the inhalation route showed central nervous system depression.

—Heriberto Robles

Butyl Methyl Ether

- ◆ CAS: 628-28-4
- ◆ PREFERRED NAME: Ether, butyl methyl
- ◆ SYNONYMS: 1-Methoxy-butane; α-methoxybutane; 1-methoxybutane
- ◆ CHEMICAL CLASS: Ether
- ◆ CHEMICAL STRUCTURE:

$$(CH_3)_3C—O—CH_3$$

Animal Toxicity

Very little is known about the toxicological properties of butyl methyl ether. In a study conducted in 1950, the inhalation LC_{50} of butyl methyl ether in mice was estimated to be 176 g/m³ for a 15-min exposure. Animals exposed to butyl methyl ether showed central nervous system depression to the point of anesthesia.

—Heriberto Robles

Butyl Nitrite

- ◆ CAS: 544-16-1
- ◆ SYNONYMS: NBN; NCI-C56553; Nitrous acid-*n*-butyl ester
- ◆ MOLECULAR FORMULA: $C_4H_9NO_2$
- ◆ CHEMICAL STRUCTURE:

$$CH_3CH_2CH_2CH_2ONO$$

Uses

Butyl nitrite is used in the manufacture of rare earth azides. It is also used as a recreational drug (for vasodilatation).

Exposure Pathways

Butyl nitrite is a poison by ingestion and intraperitoneal routes. It is mildly toxic by inhalation. When heated to decomposition it emits toxic fumes of nitrogen oxides.

Toxicokinetics

Butyl nitrite is ineffective by ingestion because it is degraded in the gastrointestinal tract. A 44% uptake of butyl nitrite was observed when rats were atmospherically exposed for 5-min periods. It is very rapidly transformed in the body. The likely products of butyl nitrite *in vivo* might be butyl alcohol, methamoglobin, nitrite ion, nitrate ion, nitrosothiols, and possibly other nitroso compounds. Butyl nitrite is also very rapidly distributed to various parts of the body such as muscles and vascular and circulating systems. The metabolites bind to hemoglobin, glutathione, and other plasma proteins. Metabolites such as nitrite ions can be eliminated in exhaled air.

Mechanism of Toxicity

Following exposure butyl nitrite causes rapid *S*-nitrosyl glutathione formation, then a concomitant decrease in protein thiols, followed by a marked adenosine triphosphate depletion. It also causes lipid peroxidation. It produces methemoglobinemia in which oxidized hemoglobin has no oxygen carrying capacity. Also in the clinical state of methemoglobinemia, the unaltered hemoglobin shows an increased affinity for oxygen resulting in symptoms of tissue hypoxia. Cyanosis occurs when methemoglobin levels are greater than 10%. Levels above 70% are potentially lethal.

Human Toxicity

Butyl nitrite is harmful if swallowed, inhaled, or absorbed through skin. It causes irritation of eyes, skin, mucous membranes, and the upper respiratory tract. Overexposure by ingestion can cause methemoglobinemia–carboxyhemoglobinemia, lowered blood pressure by vasodilatation, headache, pulse throbbing, and weakness. It can also cause behavioral changes such as altered sleep time, excitement, change in motor activity, ataxia, and rigidity. It also causes dyspnea, cyanosis, and changes in liver and kidneys. It is immunosuppressive for human lymphocytes *in vitro*.

Workers exposed to butyl nitrite should wear personal protective equipment and their work should be carried out only in restricted areas. Clothing and equipment after use should be placed in an impervious container for decontamination or disposal. Technical measures should prevent any contact with the skin and mucous membranes.

Clinical Management

In case of contact, affected eyes and skin should be flushed with water for 15–20 min. If inhaled, the affected person should be moved to fresh air. If necessary, oxygen and artificial respiration should be administered. If patient is in cardiac arrest, cardiopulmonary resuscitation should be administered. Life-support measures should be continued until medical assistance has arrived. Liquids should not be administered and vomiting should not be induced in an unconscious or convulsing person.

Animal Toxicity

The formation of butyl alcohols from butyl nitrite in experimental mice produced hepatotoxicity. It can initiate tumors via *in vivo* formation of N-nitroso compounds from butyl nitrite following exposure. The oral LD_{50} is 83 mg/kg in rats and 171 mg/kg in mice; the intraperitoneal LD_{50} is 158 mg/kg in mice. The LC_{50} is 420 ppm/4 hr in rats and 567 ppm/1 hr in mice.

—*Kashyap N. Thakore*

Related Topic

Lipid Peroxidation

Butylamines

♦ REPRESENTATIVE COMPOUNDS: *n*-Butylamine (CAS 109-73-9); *sec*-butylamine (CAS 13952-8-6); *tert*-butylamine (CAS 75-64-9)

♦ SYNONYMS:

n-Butylamine—1-butanamine, 1-aminobutane, mono-*n*-butylamine, monobutylamine

sec-Butylamine—2-butylamine, 2-butanamine, 2-aminobutane, frucote, deccotane, 1-methylpropane, 1-methylpropylamine

tert-Butylamine—isobutylamine, 2-methyl-2-propanamine, 2-aminoisobutane, 2-amino-2-methylpropane, 1,1-dimethylethylamine, 2-methyl-2-aminopropane, trimethylaminomethane.

Other butylamines, such as isobutylamine (CAS 78-81-9), diisobutylamine (CAS 110-96-3), di-*n*-butylamine (CAS 111-92-2), and tri-*n*-butylamine (CAS 102 89-9), are included in this category.

♦ CHEMICAL CLASS: Amine

♦ CHEMICAL STRUCTURE:

n-Butylamine: $CH_3CH_2CH_2CH_2NH_2$.
sec-Butylamine: $CH_3CH_2CH(NH_2)CH_3$.
tert-Butylamine: $(CH_3)_2CHCH_2NH_2$.

Uses

Butylamines are used in the manufacture of textiles, plastics, dyes, corrosion inhibitors, lubricating oil additives, antioxidants, fungicides, herbicides, rubber chemicals, and emulsifying agents. They are also used in pharmaceuticals, photographic materials, synthetic tanning agents, and miscellaneous chemicals.

Exposure Pathways

Exposure may occur through oral, dermal, or inhalation routes.

Toxicokinetics

Butylamines are well absorbed from the gut and respiratory tract. Very little is excreted in the urine. *n*-Butylamine is readily metabolized and the metabolic pathway is similar to that of other lower aliphatic amines. One of the metabolites has been reported to be acetoacetic acid. Other butylamines are also expected to be readily metabolized.

Mechanism of Toxicity

The toxic effects of butylamines are due primarily to its irritative effect on tissues. Amines also can cause a selective blockade of lysosomal degradation of protein.

Human Toxicity

Exposure of the eyes to butylamines can cause corneal burns, lacrimation, conjunctivitis, corneal edema, and loss of vision. Direct skin contact may cause severe irritation, dermatitis, and blistering. Daily exposure to 2–10 ppm has been reported to cause nose, throat, and eye irritation, headaches, and erythema, particularly about the face. Desquamation of facial skin with burning and itching may follow in several days. High vapor concentrations can cause faintness, coughing, erythema, chest pains, dizziness, depression, convulsions, narcosis, pulmonary edema, and unconsciousness. Ingestion of butylamines can result in irritation of the mouth, throat, and gastrointestinal tract with nausea and vomiting.

Clinical Management

Exposed skin and eyes should be irrigated with copious amounts of water. After inhalation exposures, the victim should be moved to fresh air and monitored for respiratory distress. Also, 100% humidified supplemental oxygen with assisted ventilation should be administered as required. If coughing or breathing difficulties are noted, the patient should be evaluated for irritation, bronchitis, or pneumonitis, including chest X-rays and determination of blood gasses. If pulmonary edema is present, positive end expiratory pressure ventilation and steroids should be considered. For ingestion exposures, copious amounts of water should be given to dilute stomach contents. Emesis or lavage should be avoided.

Animal Toxicity

Severe skin irritation with necrosis has been reported after dermal contact in the guinea pig. The LD_{50} by dermal exposure in rabbits was reported to be 850 mg/kg for *n*-butylamine and 2500 mg/kg for *sec*-butylamine. Butylamines are severely damaging to the eye when directly applied. The vapor is only mildly irritating to the eyes. At 3000–5000 ppm, *n*-butylamine produces an irritant response, labored breathing, and pulmonary edema, with death following in a matter of minutes or hours. An inhalation LC_{50} for *n*-butylamine was reported to be 800 mg/m^3 for 4 hr in mice. The oral LD_{50} in rats was reported to be 365 mg/kg for *n*-butylamine, with death due to pulmonary edema. Prior to death, the rats exhibited sedation, ataxia, nasal discharge, gasping, salivation, and convulsions. At near lethal concentrations of *n*-butylamine administered orally, rats and rabbits exhibited increased reflex excitability, increased pulse and respiration, dyspnea, convulsions, cyanosis, and coma. The LD_{50} in rats was reported to be 147 mg/kg for *sec*-butylamine and 78 mg/kg for *tert*-butylamine. A 2-year feeding study of *sec*-butylamine in rats and dogs resulted in no effects from 2500 ppm in the diet.

—*Janice M. McKee*

Butylated Hydroxyanisole

- ◆ CAS: 25013-16-5
- ◆ SYNONYMS: (1,1-Dimethylethyl)-4-methoxyphenol; 2(3)-*tert*-butyl-4-hydroxyanisole; BHA; anthracine 12
- ◆ MOLECULAR FORMULA: $C_{11}H_{16}O_2$
- ◆ CHEMICAL STRUCTURE:

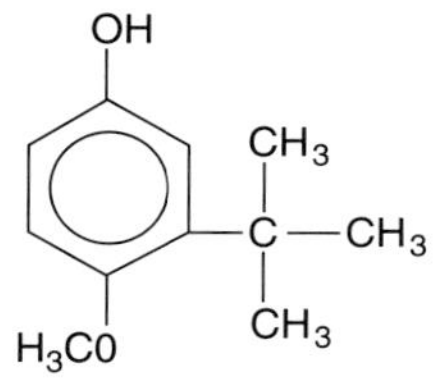

Uses

Butylated hydroxyanisole (BHA) is an antioxidant and preservative, especially in foods, cosmetics, and pharmaceuticals, and also in rubber and petroleum products.

Exposure Pathways

There is a widespread human exposure to BHA by ingestion and skin application. When heated to decomposition it emits acrid and irritating fumes and causes inhalation exposure.

Toxicokinetics

In experimental animals and in humans, BHA is absorbed rapidly after oral administration. The major metabolic pathways are conjugation (phase II) reactions, oxidative metabolism (O-demethylation) being relatively unimportant. BHA is metabolized to main metabolites, 4-O-conjugates, O-sulfates, and O-glucuronides. In dogs, oxidative metabolism is more important. It also induces both phase I and phase II drug metabolizing enzyme mRNA, protein activity, and hepatic and intestinal glutathione S-transferases.

BHA is distributed to various organs such as liver, lungs, and gastrointestinal tract. The metabolites are rapidly excreted through urine with little evidence of long-term tissue storage. When human volunteers were given a single oral dose of ^{14}C-labeled BHA (approximately 0.5 mg/kg body weight), 60–70% of the radioactivity was excreted in the urine within 2 days and 80–86.5% by Day 11. After administration of a single dose of 1000 mg BHA to New Zealand White rabbits, 46% of the dose was excreted in the urine as glucuronides, 9% as etherial sulfates, and 6% as free phenols. Excretion of glucuronides was inversely dose dependent.

Mechanism of Toxicity

The metabolites can bind to cellular macromolecules, such as proteins and DNA, to cause toxicity.

Human Toxicity

BHA is harmful if swallowed, inhaled, or absorbed through skin. It is irritating to the eyes, skin, mucous membranes, and upper respiratory tract. Prolonged or repeated exposure may cause allergic reactions in certain sensitive individuals. The target organs for toxicity are liver, lungs, and forestomach. BHA may cause cancer.

Workers exposed to BHA should wear personal protective equipment and take measures to prevent any contact with the skin and mucous membranes.

Approximately 50 countries reportedly permit the use of BHA as a food additive. BHA is classified as Generally Recognized as Safe by the U.S. FDA, when the total content of antioxidants represents not more than 0.02% w/w of the total fat or oil content of the food. It is also permitted at maximum levels of 0.001–0.02% in other specific products.

Clinical Management

In case of contact, eyes and skin should be flushed with water for 15–20 min. For inhalation exposure, the victim should be moved to fresh air. If necessary, oxygen and artificial respiration should be administered. Cardiopulmonary resuscitation should be administered if the patient is in cardiac arrest. Life-supporting measures should be continued until medical assistance has arrived. An unconscious or convulsing person should not be given liquids or induced to vomit.

Animal Toxicity

BHA induces benign and malignant tumors of the forestomach in rats and hamsters by administration through diet. It is toxic to the reproductive system and embryo in rats but not toxic to rabbits, pigs, or rhesus monkeys. It is not mutagenic to *Salmonella typhimurium*, *Drosophila melanogaster*, or Chinese hamster cells *in vitro* and did not cause chromosomal effects. In rats, the oral LD_{50} is 2000 mg/kg and the intraperitoneal LD_{50} is 881 mg/kg. The oral LD_{50} is 1100 mg/kg in mice and 2100 mg/kg in rabbits.

—*Kashyap N. Thakore*

Related Topic

Food Additives

Butylated Hydroxytoluene

◆ CAS: 128-37-0

◆ SYNONYMS: 2,6-*bis*(1,1-dimethylethyl)-4-methylphenol; 2,6-di-*tert*-butyl-*p*-cresol; BHT; Anthracine 8

◆ CHEMICAL STRUCTURE:

Uses

Butylated hydroxytoluene (BHT) is an antioxidant for food, animal feed, petroleum products, synthetic rubbers, plastics, animal and vegetable oils, and soaps. It is also used as an antiskinning agent in paints and inks.

Exposure Pathways

Ingestion is the most common route of exposure in addition to inhibition and contamination. BHT is combustible when exposed to heat or flame and can emit acrid smoke and fumes.

Toxicokinetics

In BALB/c mice, 40% of an intragastric dose of BHT was taken by the tissues within 30 min by males, whereas only 10% was absorbed in females. Oxidative metabolism (phase I reactions) mediated by the microsomal monooxygenase system is the major route for degradation; oxidation of the ring methyl group predominates in the rat, rabbit, and monkey and oxidation of the *tert*-butyl groups in man. The predominant metabolic pathway involves oxidation of the 4-methyl group. The major metabolites are 3,5-di-*tert*-butyl-4-hydroxybenzoic acid, both free and as a glucuronide, and *S*-(3,5-di-*tert*-butyl-4-hydroxybenzyl)- *N*-acetylcysteine. Moreover, BHT-quinone methide (2,6-di-*tert*-butyl-4-methylene-2,5-cyclohexadienone), a reactive metabolite, has been identified in the liver and bile of rats. Metabolites produced in mice are similar to those produced in rats, except that the major biotransformation in mice was by oxidation of *tert*-methyl groups.

Accumulation of BHT is greatest in tissues. In male and female BALB/c mice, a single intragastric dose was widely distributed to various tissues within 30 min, primarily to the small intestine, stomach, liver, kidneys, and lungs. Enterohepatic circulation of BHT has been reported in rats. BHT is also converted by cytochrome P450 monooxygenases to a chemically reactive metabolite—possibly BHT-quinone methide, which forms BHT-glutathione by nonenzymatic conjugation with glutathione.

BHT is cleared less rapidly from most species, enterohepatic circulation being partly responsible for the delay. The major metabolites of BHT in rat urine are 3,5-di-*tert*-butyl-4-hydroxybenzoic acid (BHT acid; III), both free (90% of the dose) and as a glucuronide (15%), and *S*-(3,5-di-*tert*-butyl-4-hydroxybenzyl)-*N*-acetylcysteine. The ester glucuronide and mercapturic acid were major metabolites in rat bile, while free BHT acid was the main component in the feces. In addition, 1,2-*bis*(3,5-di-*tert*-butyl-4-hydroxyphenyl)ethane has been identified in rat bile. In BALB/c mice, approximately 75% of a single oral dose was excreted in the urine during the first 24 hr; this was followed by a slower phase during which an additional 10% was excreted over the next 4 days. The total amount found in the feces was less than 1%. Female rats have greater urinary excretion of BHT than male rats, whereas male BALB/c mice excreted BHT more rapidly than females.

Mechanism of Toxicity

The metabolites can bind to cellular macromolecules, such as proteins and DNA, to cause toxicity.

Human Toxicity

BHT is harmful if swallowed, inhaled, or absorbed through skin. It causes irritation of the eyes, skin, mucous membranes, and upper respiratory tract. Prolonged or repeated contact can damage the eyes and cause nausea, dizziness, and headache. BHT is a possible carcinogen with the target organ being the lungs. It does not represent a relevant mutagenic/genotoxic risk to humans.

Approximately 40 countries reportedly permit the use of BHT as a direct or indirect food additive. BHT was approved and classified as Generally Recognized as Safe by the U.S. FDA. Regulated food products could contain a combined total of up to 0.02% BHT and butylated hydroxyanisole, based on the fat content of the food. It is also permitted at maximum levels of 0.001–0.01% in other specific products.

Workers exposed to BHT should wear personal protective equipment and take measures to prevent any contact with the skin and mucous membranes. The ACGIH recommends that occupational exposure to airborne BHT not exceed 10 mg/m^3 (TLV) as an 8-hr time-weighted average (TWA) or 20 mg/m^3.

Clinical Management

In case of contact, eyes and skin should be flushed with water for 15–20 min. For inhalation exposure, the victim should be moved to fresh air. If necessary, oxygen and artificial respiration should be administered. Cardiopulmonary resuscitation should be administered if the patient is in cardiac arrest. Life-supporting measures should be continued until medical assistance has arrived. An unconscious or convulsing person should not be given liquids or induced to vomit.

Animal Toxicity

In animals, BHT is poisonous by intraperitoneal and intravenous routes and moderately toxic by ingestion. It has produced reproductive effects in animal experiments. It is a questionable carcinogen based on experimental carcinogenic and neoplastigenic data. It induces liver tumors in long-term experiments. *In vitro* studies on bacterial, yeast and various mammalian cell lines, and primary hepatocytes demonstrate the absence of interactions with or damage to DNA. The oral LD_{50} is 890 mg/kg in rats and 1040 mg/kg in mice. In mice, the intraperitoneal LD_{50} is 138 mg/kg and the subcutaneous LD_{50} is 650 mg/kg. In the guinea pig the oral LD_{50} is 10,700 mg/kg.

—*Kashyap N. Thakore*

Related Topics

Food Additives
Respiratory Tract

Butyraldehyde, n-

- CAS: 123-72-8
- PREFERRED NAME: Butanal
- SYNONYMS: Butyraldehyde; butal;, butaldehyde; butyl aldehyde; *n*-butyl aldehyde; butyral; *n*-butyraldehyde; butyric aldehyde; butyrylaldehyde; *n*-butanal; butanaldehyde

- CHEMICAL CLASS: Aliphatic aldehyde
- CHEMICAL STRUCTURE:

$$CH_3-CH_2-CH_2-C\begin{smallmatrix}H\\\\O\end{smallmatrix}$$

Uses

Butanal is used in the manufacture of rubber accelerators, synthetic resins, solvents, and plasticizers. It has no therapeutic use at the present time.

Exposure Pathways

Butanal is a liquid at room temperature, with a relatively low vapor pressure. Limited contact could occur by exposure to butanal vapors. Butanal has appreciable solubility in water; therefore, exposure would be expected to be primarily through ingestion of, or through skin contact with, the compound or a solution of the compound.

Toxicokinetics

Butanal is readily metabolized to carbon dioxide by conversion to butyryl CoA and subsequent metabolism via the pathways of short-chain fatty acid oxidation. Detoxication by reaction with glutathione also occurs. Clearance is rapid and complete.

Mechanism of Toxicity

Butanal does not possess high acute toxicity but is a potent irritant of the skin, eyes, and upper respiratory tract. The mechanism of toxicity probably involves direct reaction between the active aldyhydic group and cellular components.

Human Toxicity

Butanal has low acute toxicity. Exposure to a large dose may have a temporary narcotic effect. Exposure to low concentrations of butanal vapors produces irritation of the eyes, nose, and throat. The compound has an unpleasant odor. Impurities (butyric acid) may be present that make the smell even more objectionable. Health effects attributed to chronic exposure to low doses of butanal vapors have not been described. Dermatitis may be expected after prolonged and repeated exposures to solutions containing butanal.

Clinical Management

Support should be given to the patient until butanal has been cleared from the body, which occurs in a relatively short time. Recovery is uneventful.

Animal Toxicity

The oral LD_{50} value for rat is 5.9 g/kg, whereas the LC_{50} is 60,000 ppm (30-min exposure). Acute exposures to butanal vapors induce inflammation of the alveolar and bronchial regions of the lung, with death due to pulmonary edema. Severe irritation of the eyes and nose are noted. Relatively high levels of butanal in the drinking water of mice for 50 days produced abnormal sperm morphology. Exposure of rodents to low concentrations of butanal allowed rapid recovery after exposure ceased.

—*Michael J. Brabec*

Butyric Acid

- ◆ CAS: 107-92-6
- ◆ SYNONYMS: Butanoic acid; ethylacetic acid; butanic acid; *n*-butyric acid; 1-propanecarboxylic acid; kyselina maselna; buttersaeure; RTECS ES5425000; UN2820
- ◆ CHEMICAL CLASS: Food additive
- ◆ CHEMICAL STRUCTURE:

$$CH_3(CH_2)_2COOH$$

Uses

Butyric acid is used in the manufacturing of esters that serve as the bases of artificial flavoring ingredients for certain liqueurs, soda-water syrups, and candies. It is used for body in butter, cheese, butterscotch, caramel, fruit, and nut flavors. It is also used for varnishes and as a decalcifier of hides.

Exposure Pathways

Skin contact or ingestion are the most common exposure pathways.

Toxicokinetics

Butyric acid is rapidly metabolized in the liver to acetic acid, acetone, acetoacetate, ketone bodies, and β-hydroxybutyrate. In humans, the elimination half-life is about 14 min.

Mechanism of Toxicity

Butyric acid causes toxicity probably by forming an acid proteinate.

Human Toxicity

Acute exposure to butyric acid may cause burns to skin and eyes. It may be harmful if inhaled. Fire may produce irritating or poisonous gases. Butyric acid is recognized as a safe food additive when used in accordance with good manufacturing practice or feeding practice.

Clinical Management

Exposure should be terminated as soon as possible by removal of the victim to fresh air. The skin, eyes, and mouth should be washed with copious amounts of water. A 15- to 20-min wash may be necessary to neutralize and remove all residual traces of the contaminant. Contaminated clothing and jewelry should be removed and isolated. Contact lenses should be removed from the eyes to avoid prolonged contact of the acid with the area. A mild soap solution may be used for washing the skin and as an aid to neutralize the acid, but should not be placed into the eyes. No cream, ointment, or dressing should be applied to the affected area. If a large quantity has been swallowed, then gastric lavage should be considered. Dilution with water may be the solution for small quantities swallowed. The victim should be kept quiet and normal body temperature should be maintained.

Animal Toxicity

The toxic effects of butyric acid in animals is similar to that in humans.

—*Sanjay Chanda*

Butyronitrile

- CAS: 109-74-0
- SYNONYMS: 1-Cyanopropane; butane nitrile; butyric acid nitrile; butanenitrile; propyl cyanide; RTECS/NIOSH ET8750000
- CHEMICAL CLASS: Nitrile
- CHEMICAL STRUCTURE:

$$CH_3CH_2CH_2CN$$

Uses

Butanenitrile is a basic intermediate in the industrial, chemical, and pharmaceutical industries. It is used as a chemical intermediate for butyric acid, butyramide, and other pharmaceuticals.

Exposure Pathways

Exposure may occur by inhalation of the vapor and by dermal contact.

Toxicokinetics

After exposure, the highest accumulations of butyronitrile and cyanides occur in the liver, stomach, intestines, kidneys, and testis. Butyronitrile was eliminated in the urine slowly and in very small quantities. Administration of nitriles in animals liberated cyanide both *in vivo* and *in vitro*. Cyanide inhibits respiration on the cellular level (see Cyanide).

Mechanism of Toxicity

The toxic mechanism in nitriles appears to be the metabolic liberation of cyanide.

Human Toxicity

Butyronitrile is considered a highly hazardous material. Inhaled butyronitrile is approximately 2.4 times as toxic as acetonitrile. Cases of severe poisoning can result in coma, seizures, palpitations, hypoventilation, hypertension, hypotension, and pulmonary edema. In addition, nausea, vomiting, and metabolic acidosis may occur. The average fatal dose of hydrogen cyanide is 50–60 mg. Exposure to 90 ppm or greater for 30 min or more may cause death. Death may result from a few minutes of exposure to 300 ppm.

Clinical Management

The victim should be treated as if exposure to cyanide has occurred. Rapid support of respiration and circulation is essential for successful treatment of cyanide intoxication. Victims should be immediately moved to fresh air and receive oxygen. Exposed eyes and skin should be rapidly and copiously flushed. Contaminated clothing should be removed. Mouth-to-mouth resuscitation should be avoided because of the danger of self-poisoning. In cases of oral exposure, gastric lavage should be performed. The Lilly Cyanide Antidote Kit should be prepared for use in symptomatic patients.

Animal Toxicity

Inhalation of butyronitrile vapor readily produced fatalities in rats with symptoms of nitrile toxicity. In acute exposure studies, rats and mice exhibited weakness, tremors, vasodialation, labored respiration, and terminal convulsions similar to other toxic nitriles.

- LD$_{50}$ (oral) rat: 50–100 mg/kg
- LD$_{50}$ (oral) rat: 0.14 g/kg
- LD$_{50}$ (oral) rat: 135 mg/kg
- LD$_{50}$ (rabbit) dermal: 400 mg/kg
- LD$_{50}$ (guinea pig) dermal: 0.1–0.5 mg/kg

—*Rhonda S. Berger and Wendy Khune*

Butyrophenones

- CAS: 495-40-9
- REPRESENTATIVE COMPOUNDS: Haloperidol (CAS: 52-86-8); droperidol (CAS: 548-73-2)

◆ SYNONYMS: Haloperidol—Haldol, 4-(4-(*p*-chlorophenyl)-4-hydroxypiperidino)-4′ fluorobutyrophenone; droperidol—dehydrobenzperidol, Inapsine, 1-(1-(3-*p*-fluorobenzoylprpyl)-1,2,3,6-tetrahydropyrid-4-yl)-2-benzimidazolinone.

◆ PHARMACEUTICAL CLASS: Neuroleptic agent, antipsychotic, major tranquilizer

◆ MOLECULAR FORMULA: $C_{10}H_{12}O$

◆ CHEMICAL STRUCTURE:

Haloperidol

Droperidol

Uses

Butyrophenones are used to treat psychosis including schizophrenia, organic psychosis, paranoid syndrome, acute idiopathic psychotic illnesses, and the manic phase of manic depressive illness. Other uses include treatment of aggressive behavior, delirium, acute anxiety, nausea and vomiting, pain, organic brain syndrome, and Tourette's syndrome.

Exposure Pathways

Haloperidol is available both in an injectable and in oral dosage form. The principal exposure pathway is intentional ingestion by adults or accidental ingestion by small children. Droperidol is available only as an injectable drug. The most common route of exposure is an accidental injection.

Toxicokinetics

Haloperidol is well absorbed orally with a bioavailability of 60% due to first-pass hepatic metabolism. Both agents are rapidly absorbed after intramuscular injection, peaking within 10 min. Butyrophenones are hepatically metabolized to inactive metabolites. Concentrations of butyrophenones are found in the liver, central nervous system, and throughout the body. Haloperidol is 92% protein bound. Haloperidol is 15% eliminated through the bile. The elimination half-life ranges from 14 to 41 hr. The half-life of droperidol is 2 hr; 10% is recovered unchanged in the urine.

Mechanism of Toxicity

Butyrophenones work primarily by blocking dopamine-mediated synaptic neurotransmission by binding to dopamine receptors. In addition to significant antidopaminergic action, butyrophenones also possess anticholinergic, α-adrenergic blockade, and quinidine-like effects.

Human Toxicity

Clinical signs of toxicity most commonly include extrapyramidal effects and sedation. Neuroleptic malignant syndrome has been reported after therapeutic use and acute intoxication. The most commonly reported dystonic reactions include akathesias, stiff neck, stiff or protruding tongue, and tremor. Children appear to be more sensitive than adults to the extrapyramidal effects of butyrophenones with facial grimacing and oculogyric crisis noted. Anticholinergic effects, including dry mouth, blurred vision, and tachycardia, may occur. Other cardiac effects include prolonged QT interval and mild hypotension. Hypokalemia has also been noted. Seizures are rarely seen. Adverse reactions following therapeutic use include sedation, dysphoria, anorexia, nausea, vomiting, constipation, diarrhea, and dyspepsia.

Clinical Management

All basic and advanced life-support measures should be implemented. Gastric decontamination should be performed. Butyrophenones are readily absorbed by activated charcoal. Aggressive supportive care should be instituted. Dystonic reactions respond well to intravenous benztropine or diphenhydramine. Oral therapy with diphenhydramine or benztropine should be continued for 2 days to prevent recurrence of the dystonic reaction. For patients suffering from neuroleptic malignant syndrome, dantrolene sodium and bromocriptine have been used in conjunction with cooling and other supportive measures. Arrhythmias should be treated with lidocaine or phenytoin. Diazepam is the drug of choice for seizures; phenytoin is used to prevent recurrence. Hemodialysis and hemoperfusion have not been shown to be effective.

Animal Toxicity

Signs of toxicity reported in animals have included sedation, dullness, photosensitivity, weakness, anorexia, fever, icterus, colic, anemia, and hemoglobinuria. Treatment consists of gastric decontamination and aggressive supportive care.

—Douglas J. Borys

Related Topics

Anxiolytics
Neurotoxicology: Central and Peripheral

Cadmium (Cd)

- CAS: 7440-43-9
- Selected Compounds: Cadmium oxide, CdO (CAS: 1306-19-0); cadmium sulfide, CdS (CAS: 1306-23-6)
- Chemical Class: Metals

Uses

Cadmium is primarily used for electroplating and galvanizing other metals because it is relatively resistant to corrosion. It is also used in electrical contacts, in soldering alloys, in nickel-cadmium storage batteries, in television phosphors, and as a stabilizer for polyvinyl chloride. Finally, given its brilliant orange color, it has been used extensively as a pigment in paints, plasters, and plastics.

Cadmium is a by-product of zinc, lead, and copper mining and smelting.

Exposure Pathways

Ingestion and inhalation are the primary routes of exposure to cadmium. Dermal contact is not a significant route of exposure. Exposure to cadmium via foodstuffs is common. Many plants and especially fish and crustaceans efficiently accumulate cadmium from soil or water. Cigarette smoke is a major source of cadmium exposure via inhalation.

Due to the wide use of cadmium-based products, cadmium is universally distributed. The cadmium content in soil (and water) has been increasing as a result of disposal of cadmium-contaminated waste and the use of cadmium-containing fertilizers (particularly on cereal crops). Commercial sludge, heavily contaminated with cadmium, has been used to fertilize agricultural fields. Cadmium concentrations in urban air are quite low, however, because of increased regulation of industrial air emissions. Lead and zinc smelters and waste incineration account for the majority of cadmium present in ambient air.

Toxicokinetics

Absorption of cadmium in the gastrointestinal tract is approximately 4–7% in adults; absorption is probably higher in children. Diets low in calcium, iron, and protein enhance cadmium absorption. Zinc is an antagonist to cadmium (decreases cadmium absorption). Cadmium absorption by the lungs is dependent on particle size and the solubility of the cadmium compound, but is generally between 15 and 30%. Dermal absorption of cadmium is insignificant.

Cadmium is a classic cumulative poison that accumulates in the kidneys over a lifetime. It is transported in the blood by erythrocytes and by albumin, and it is stored mainly in the liver and kidneys as the metallothionein (50–75% of the body burden). Cadmium binds to many proteins at the sulfate and carbonyl sites. The half-life of cadmium in these two organs may be as long as 30 years. The correlation between years of exposure and blood levels does not appear to be significant. Cadmium also accumulates in the bones and the placenta of pregnant women.

Urine is the most important excretion mechanism in humans. Urine concentration of cadmium increases with age and following kidney damage. Cadmium found on examination of hair is generally due to external contamination rather than internal absorption and distribution to the hair.

Mechanism of Toxicity

Cadmium inhibits plasma membrane calcium channels and Ca^{2+}-ATPases. It also inhibits repair of DNA damaged by various chemicals which is believed to be associated with the induction of tumors. Although cadmium forms a metallothionein, the preformed cadmium metallothionein is nephrotoxic (toxic to the kidneys); the mechanism suggested is that at some stage in the kidney the cadmium is dissociated from the metallothionein. In Itai-Itai disease (see below), patients were found to have chromosome abnormalities.

Cadmium has an affinity for sulfhydryl groups and, hence, can inhibit enzymes; however, cells treated with cadmium showed proliferation of peroxisomes which contain catalase, an enzyme.

Cadmium has no known biological function, but it is an excellent enzyme inhibitor, especially for those enzymes involved in gluconeogenesis (the generation of glycogen for energy production from noncarbohydrate precursors). It also inhibits oxidative phosphorylation (energy production) and depresses trypsin inhibitor capacity. An interesting phenomenon is that cadmium at first inhibits catalase activity and then, after a time, enhances that activity.

Human Toxicity: Acute

Acute toxicity may result from ingestion of relatively high concentrations of cadmium from contaminated food or beverages (e.g., 16 mg/liter cadmium in a beverage). Cadmium exhibits local irritant effects on the gastrointestinal tract such as nausea, vomiting, diarrhea, abdominal pain, and a choking sensation. The effects of acute toxicity are apparent immediately.

Inhalation of cadmium fumes produces local irritant effects and may result in chemical pneumonitis and pulmonary edema (possibly resulting in death).

Human Toxicity: Chronic

Chronic exposure to cadmium from any route will have adverse effects on the heart, lungs, bones, gonads, and especially the kidneys. The principal long-term effects of low-level cadmium exposure are generally chronic obstructive pulmonary disease, emphysema, and chronic renal tubular disease. Cardiovascular and skeletal effects are also possible. The initial symptoms of chronic inhalation exposure are those associated with metal fume fever (e.g., fever, headache, chest pain, sore throat, coughing, and rhinitis). Metal fume fever is most often associated with inhalation of zinc oxide but may occur following exposure to other metals such as cadmium.

Although inconclusive, there is evidence that the cadmium burden in the body can lead to hypertension, which is similar to the association of lead exposure and elevated blood pressure. Since cadmium can displace zinc, cadmium accumulation in the testes can suppress testicular function. Evidence obtained in the past several years appears to relate cadmium to prostate cancer in young men who work with cadmium. Additional investigation (such as epidemiological studies with a larger cohort) needs to be performed to verify this apparent association of cadmium with prostate cancer.

Skeletal changes due to cadmium accumulation are probably related to calcium loss, which can be influenced by diet and hormonal status. These skeletal changes include osteomalacia (softening of bone resulting from loss of minerals) and pseudofractures. In Japan, people who ate fish contaminated with cadmium experienced skeletal changes, especially in their backs. This very painful effect was called the "Itai-Itai" ("ouch-ouch") disease. Postmenopausal women with low calcium and vitamin D intake were apparently most susceptible.

Since the kidneys are the main depot for cadmium, they are of greatest concern for cadmium toxicity. Cadmium interferes with the proximal tubule's reabsorption function. This leads to abnormal actions of uric acid, calcium, and phosphorus. Amino aciduria (amino acids in the urine) and glucosuria (glucose in the urine) result and, in later stages, proteinuria (protein in the urine) results. When this happens, it is assumed that there is a marked decrease in glomeruli filtration. Long-term exposure to cadmium leads to anemia, which may result from cadmium interfering with iron absorption.

Cadmium metallothionein has also been studied extensively. This metalloprotein is high in the amino acid cysteine (approximately 30%) and is devoid of aromatic amino acids. Metallothionein itself may function to help detoxify cadmium. For some experimental tumors, cadmium appears to be anticarcinogenic (e.g., it reduces the induction of tumors).

ACGIH lists cadmium as a suspected human carcinogen. The ACGIH TLV-TWA is 0.01 mg/m^3 for elemental cadmium and inorganic compounds as total dust/particulate. The ACGIH TLV-TWA for the respirable fraction of cadmium particulate is 0.002 mg/m^3.

Clinical Management

For oral poisoning, administration of syrup of ipecac is indicated, followed by gastric lavage. The chelating agent calcium EDTA (calcium disodium salt of ethylene diamine tetraacetic acid) is indicated for acute exposure if administered shortly after cadmium exposure before new metallothionein is synthesized. BAL (British Anti-Lewisite; 2,3-dimercaptopropanol) is contraindicated as it may enhance kidney toxicity. The newer dimercapto compounds (DMSA and DMPS) are being evaluated as are derivatives of dithiocarbamates. Delayed pulmonary edema may result from inhaled cadmium dusts; therefore, supportive measures are indicated.

The apparent affinity for zinc metallothionein may someday be found to be useful as an antidote for cadmium toxicity. Antagonists to cadmium toxicity include a pretreatment with selenium and zinc. It has been postulated that this pretreatment allows cadmium to displace zinc in the zinc metallothionein. Recent evidence seems to support this hypothesis.

Animal Toxicity

Animal studies have shown cadmium to be teratogenic; the mutagenic experiments are equivocal. Cadmium produced local sarcomas in a number of rodent species when the metal, sulfide, oxide, or salts were administered subcutaneously. Intramuscular injection of cadmium powder and cadmium sulfate also produced local sarcomas. Injection of cadmium chloride into the ventral prostate resulted in a low incidence of prostatic carcinoma. Exposure via inhalation of cadmium chloride produced a dose-dependent increase in lung carcinomas in rats.

Cardiac effects (electrical and biochemical changes in the myocardium) were observed in rats exposed to cadmium in drinking water.

—*Arthur Furst, Shirley B. Radding, and Kathryn A. Wurzel*

Related Topics

Cardiovascular System
Kidney
Metallothionein
Metals
Pollution, Soil
Pollution, Water
Respiratory Tract
Sensory Organs

Caffeine

- ♦ CAS: 58-08-2
- ♦ SYNONYMS: 1,3,7-Trimethylxanthine(TMX); methyltheobromine
- ♦ PHARMACEUTICAL CLASS: A xanthine derivative that occurs naturally in coffee beans, kola nuts, and cocoa beans
- ♦ CHEMICAL STRUCTURE:

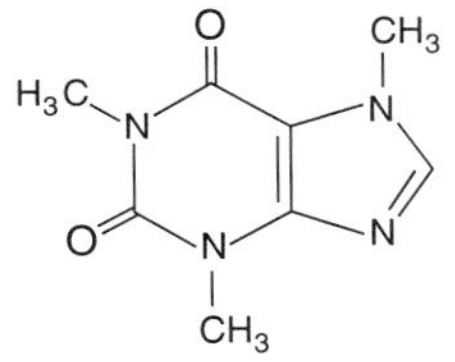

Uses

Caffeine is used as a central nervous system (CNS) stimulant, anorexiant, diuretic, and in a number of analgesic and cold medication compounds. It is also used in the treatment of spinal headaches and has been used as a respiratory stimulant in preterm infants.

Exposure Pathways

Ingestion is the most common route of exposure. Caffeine is consumed in beverages such as coffee, tea, and soda. It is available orally both singularly and as a component of a number of combination medications. It is also available for injection.

Toxicokinetics

Caffeine is rapidly absorbed after an oral dose, within 60 min, and peak levels typically occur within this same period. Food will slow absorption as may massive ingestions. The volume of distribution is 0.4–1.0 liters/kg. In adults, caffeine is extensively metabolized by the liver, primarily by N-demethylation and ring oxidation. The principal metabolites are mono- and dimethyl-urates and xanthines. Theophylline (1,3-dimethylxanthine) is a minor product of caffeine metabolism in adults. In massive overdoses, measurable serum levels of theophylline may be seen, although the level may be artifactual due to cross-reactivity with other caffeine metabolites by the assay method. The half-life of caffeine is 3–7 hr. The half-life is shorter in smokers and is prolonged by oral contraceptives, by cimetidine, and in late pregnancy. The half-life of caffeine is much longer in infants and does not approximate that seen in adults until 5 or 6 months of age. The half-life of caffeine may exceed 100 hr in preterm infants. Only 1–10% of caffeine appears unchanged in the urine. Neonates may excrete up to 85% of caffeine unchanged.

Mechanism of Toxicity

The cellular actions of caffeine include effects on translocation of intracellular calcium, inhibition of phosphodiesterase, and adenosine receptor blockade. The first two of these actions do not likely have a role at "therapeutic" caffeine doses but may play a role at toxic doses. At toxic doses caffeine causes release of catecholamines (see Catecholamines). Caffeine stimulates cardiac and skeletal muscle, gastric acid secretion, respiration, and lipolysis. It is a mild diuretic.

Human Toxicity: Acute

Acute toxicity is manifest primarily in the CNS, cardiovascular system, and gastrointestinal system. CNS signs include restlessness, tremor, nervousness, headache, insomnia, tinnitus, confusion, delirium, psychosis, and seizures. The relation of caffeine and blood pressure is complex. In caffeine-naive subjects, small to moderate doses of caffeine may cause a slight increase in blood pressure and decrease in heart rate.

In habitual users and in large doses, the effect is not predictable. Cardiac manifestations include sinus tachycardia, other dysrhythmias, asystole, and cardiovascular collapse. Other findings include tachypnea, nausea, vomiting, hematemesis, diarrhea, and fever. Case reports also include rhabdomyolysis and pulmonary edema. Laboratory findings include metabolic acidosis, respiratory alkalosis, ketosis, hypokalemia, and hyperglycemia. The estimated lethal dose in adults is 150–200 mg/kg, whereas doses of 10–15 mg/kg may produce early signs of toxicity. Serum caffeine levels correlate poorly with symptoms of caffeine toxicity, at least in part due to tolerance on the part of habitual users. Serum levels of 30 μg/ml have been associated with symptomatology and a level 80 μg/ml with death, although far higher levels have been reported in survivors.

Human Toxicity: Chronic

No definite association has been demonstrated between habitual caffeine use and hypertension, myocardial infarction, carcinogenicity, or teratogenicity. The U.S. FDA has, however, advised that pregnant women avoid or limit caffeine intake. Abrupt cessation may cause withdrawal headaches.

Clinical Management

Basic and advanced life-support measures should be instituted as indicated by clinical situation. Gastrointestinal decontamination should include activated charcoal and, if appropriate, a gastric emptying procedure. Vomiting may make activated charcoal retention difficult. Esmolol and propranolol have been used to treat tachydysrhythmias associated with caffeine toxicity. Nonselective beta-blockers may be useful for hypotension due to excessive β-2 stimulation. For seizures, standard therapy for seizures should be employed. Monitoring should be performed for fever and for fluid and electrolyte imbalances.

Animal Toxicity

Toxicity in animals is similar to that found in humans. Dehydration and hyperthermia may occur.

—*Michael J. Hodgman*

Related Topic

Developmental Toxicology

Calcium Channel Blockers

- ◆ REPRESENTATIVE COMPOUNDS: Amlodipine; bepridil; diltiazem; felodipine; isradipine; nicardipine; nifedipine; nimodipine; verapamil
- ◆ SYNONYMS: Norvasc (CAS: 88150-42-9); Vascor (CAS: 74764-40-2); cardizem (CAS: 42399-41-7); Plendil (CAS: 72509-76-3); DynaCirc (CAS: 75695-93-1); Cardene (CAS: 55985-32-5); Adalat; Procardia (CAS: 21829-25-4); Nimotop (CAS: 66085-59-4); Calan; Isoptin (CAS: 152-11-4)
- ◆ PHARMACEUTICAL CLASS: Calcium channel blockers

Uses

Calcium channel blockers are used in the management of angina pectoris, hypertension, supraventricular arrhythmias, and subarachnoid hemorrhage.

Exposure Pathways

Ingestion is the most common route for both accidental and intentional exposures. Verapamil and diltiazem are both available for parenteral administration, and toxicity can occur via a parenteral route.

Toxicokinetics

Following oral administration, absorption is rapid and almost complete (80–100%), but the ultimate bioavailability is limited and variable (15–94%) following oral administration due to significant first-pass metabolism in the liver. Protein binding is high and ranges from 70 to 99%. Volumes of distribution for some calcium channel blockers are as follows: verapamil, 5 liters/kg; diltiazem, 3.1 liters/kg; nifedipine, 0.78 liters/kg; and nicardipine, 1.1 liters/kg. Extensive hepatic metabolism occurs. Only small amounts (0–10%) are excreted unchanged in the urine. Elimination half-life ranges from 1 hr (nimodipine) to 50 hr (amlodipine).

Mechanism of Toxicity

The pharmacologic and toxicologic mechanisms of the calcium channel blockers are complex. They include interference with electrical conduction through the atrioventricular node, decreased myocardial contractility, and direct vasodilation. Calcium channel blockers also interfere with pancreatic release of insulin.

The interference with electrical conduction through the atrioventricular node is caused by interference with the influx of calcium in phase II of the action potential and manifest by bradycardia, lengthening of the PR interval, QRS widening, and QTc prolongation.

Decreased myocardial contractility is due to calcium influx into the cell, which results in increased release of calcium from the sarcoplasmic reticulum. The overall effect of this calcium influx and release is the bridging of actin and myosin and subsequent myocardial contraction. The negative inotropic effect of the calcium blockers is due to interference with this process.

Vasoconstriction occurs when calcium activates vascular myosin kinase, which in turn allows for phosphorylation of myosin and subsequent bridging with actin. Administration of calcium channel blockers will interfere with this process and produce vasodilation.

Human Toxicity

The clinical effects of the calcium channel blockers are primarily cardiovascular in nature. Due to their interference with conduction, they can cause a variety of dysrhythmias including sinus bradycardia, all degrees of atrioventricular block, junctional rhythms, pulseless electrical activity, and asystole. The negative inotropic effects of the calcium channel blockers cause significant decreases in cardiac output. Profound hypotension is observed following calcium channel blocker poisoning due to their vasodilatory properties. Renal failure secondary to decreased perfusion may be seen. The neurologic toxicities of the calcium channel blockers are most likely secondary to their cardiovascular effects. The most common neurologic effects are lethargy and coma. Neurologic deterioration can be rapid. Some patients with significant hypotension may have intact neurologic examinations initially. Seizure activity has also been observed in calcium channel blocker toxicity. The most common metabolic effects that occur in calcium channel blocker toxicity are metabolic acidosis, hyperglycemia, and hypokalemia. Hyperkalemia has also been reported.

Clinical Management

Advanced supportive care is a primary component of patient management. Emergent intubation and assisted ventilation are often necessary in these patients. Pulse oximetry should be utilized to assess respiratory status. Extensive cardiovascular monitoring is also necessary. Arterial blood gases, serum electrolytes, and glucose measurements should be obtained. Serum concentrations of specific calcium channel blockers are difficult to obtain and have limited clinical utility. Syrup of ipecac-induced emesis is contraindicated due to the rapid decreases in level of consciousness that may occur as well as emesis-induced vasovagal effects. Gastric lavage and activated charcoal can be used if warranted by the history of the ingestion and the patient's neurologic status. Whole bowel irrigation along with activated charcoal should be utilized in ingestions involving sustained-release products. Calcium salts are often administered as antidotes for calcium channel blocker toxicity although they have been used with limited success. Calcium chloride is preferred over calcium gluconate since it contains more elemental calcium on a milligram-per-milligram basis. Doses of up to 4 g of calcium have been recommended in this setting. Glucagon, which has been used in β-adrenergic blocker toxicity, has been recommended in calcium blocker toxicity. This agent has positive inotropic properties due to activation of cyclic adenosine monophosphate (AMP). It has limited beneficial effects in calcium channel blocker toxicity. Control of heart rate and rhythm present a significant challenge in this patient population. Transcutaneous pacemakers should be utilized to stabilize rate and enhance atrioventricular conduction. The vagolytic, atropine, has also been used to increase heart rate. It has limited effect since it primarily affects the sinoatrial node. The negative inotropic effects of these agents must also be treated aggressively. Positive inotropic agents, such as dopamine, dobutamine, amrinone, and isoproterenol, can be utilized to increase contractility. Isoproterenol should be used with caution due to its vasodilatory properties. Vasopressors, such as dopamine, epinephrine, and norepinephrine, may

be effective. Cardiopulmonary bypass has been used experimentally to treat patients with calcium channel blocker toxicity who do not respond to traditional therapy. Sodium bicarbonate should be administered to treat acidosis.

Seizure activity should be initially treated with benzodiazepines. If benzodiazepines are not effective, phenytoin and barbiturates can be administered. Insulin replacement may be necessary to correct hyperglycemia.

—Daniel J. Cobaugh

Related Topic

Cardiovascular System

Calomel

- ◆ CAS: 7546-30-7
- ◆ SYNONYMS: Mercurous chloride; mercury (I) chloride; mercury monochloride
- ◆ CHEMICAL CLASS: Heavy metal
- ◆ MOLECULAR FORMULA: Hg_2Cl_2

Uses

Calomel is used as a laboratory reagent, as a fungicide, and as a depolarizer in dry batteries.

Exposure Pathways

The primary routes of entry are ocular and dermal contact, inhalation, and ingestion. Calomel is found in environmental and occupational settings, such as in mercury mining operations, battery plants, paints and dyes, photography, perfumes and cosmetics, and chemical laboratories. It is poisonous by ingestion through food and intraperitoneal routes. Calomel is moderately toxic by skin contact. When heated to decomposition it emits very toxic fumes of Cl− and Hg.

Toxicokinetics

After inhalation, about 70–80% of metallic vapor is retained and absorbed. Little is taken up in the gastrointestinal tract, and less than 10% is absorbed. In the body, it is oxidized to mercuric mercury, which binds to reduced sulfhydryl groups. The kidney is the main depository following exposure to both metallic and mercuric mercury. In addition to other organs, it passes into the brain and fetus.

The metabolite is eliminated mainly in urine and feces; it is also excreted in milk. In humans, inorganic mercury compounds have two elimination half-lives: one lasts for days or weeks and the other much longer (see Mercury).

Mechanism of Toxicity

Calomel can generate reactive oxygen species and deplete glutathione levels. Both genotoxic and nongenotoxic mechanisms may contribute to renal carcinogenic effect of mercury.

Human Toxicity

Calomel is harmful and may be fatal if swallowed or inhaled. When swallowed it causes central nervous system (CNS) depression; when inhaled it causes tightness and pain in the chest, coughing, and breathing difficulties. Ocular and dermal exposure causes irritation of the eyes and skin. In cases of chronic exposure, mercury builds up in the brain, liver, and kidneys and causes headache, shakes, loose teeth, loss of appetite, skin ulceration, and impaired memory. There is inadequate evidence for carcinogenicity. Mercury concentration in urine, blood, and plasma is useful for biological monitoring.

The recommended health-based limits are 0.05 mg/m^3 for occupational exposure, 50 μg/g creatinine in urine for long-term occupational exposure to mercury vapors, and 1 μg/liter for exposure by drinking water (WHO report, 1980).

Clinical Management

In case of contact, eyes and skin should be flushed with water for 15–20 min. If inhaled, the victim should be removed to fresh air. If necessary, oxygen and artificial respiration should be administered. If the patient is in cardiac arrest, cardiopulmonary resuscitation should be provided. These life-supporting measures should be continued until medical assistance has arrived. An un-conscious or convulsing person should not be given liquids or induced to vomit.

Animal Toxicity

In animals, intense exposure causes lung damage, intestinal and renal tubular necrosis, immunosuppression, and possible cytogenetic effects. There is limited evidence for carcinogenicity. Calomel causes renal adenoma and adenocarcinoma in male mice and female rats. The oral LD$_{50}$ is 210 mg/kg in rats and 180 mg/kg in mice. The intraperitoneal LD$_{50}$ is 10 mg/kg in mice.

—*Kashyap N. Thakore*

Related Topic

Metals

Camphor

- ♦ CAS: 76-22-2

- ♦ SYNONYMS: Campho-phenique; Musterole; Ben Gay children's vaporizing rub; Vicks Vaporub; Vicks Vaposteam; Heet; Sloan's Liniment; camphorated oil; camphor spirits

- ♦ PHARMACEUTICAL CLASS: Cyclic ketone of the hydroaromatic terpene group

- ♦ CHEMICAL STRUCTURE:

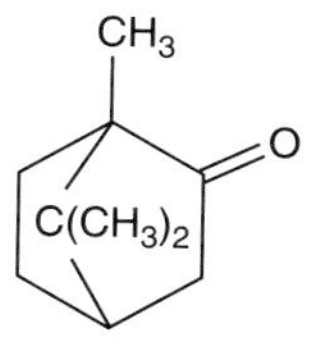

Uses

Camphor is employed externally as a rubefacient, mild analgesic, antipruritic, and counterirritant in commercially available products that contain 1.0–10% camphor. It is currently produced synthetically. It has a characteristic odor and a pungent aromatic taste.

Exposure Pathways

Ingestion is the most common route of both accidental and intentional exposure to camphor. Ocular exposures may also occur.

Toxicokinetics

Camphor in liquid form is rapidly absorbed through the skin, mucous membranes, and gastrointestinal tract. Symptoms may appear within 5–90 min following ingestion. The absorption is highly dependent on the presence of food and other chemicals which may influence the rate of camphor absorption. Camphor is metabolized to a campherol, which is conjugated with glucuronic acid in the liver. It is unclear whether camphor toxicity is attributed to the parent compound, a metabolite, or both. Camphor-related metabolites are fat soluble. Thus, significant concentrations may accumulate in fat tissue. Camphor is distributed widely in all tissues. Measurable serum levels are apparent within 15 min after ingestion of approximately 0.5–1.0 g. The volume of distribution approximates 2–4 liters/kg. The glucuronide form is excreted in the urine. The half-life of a 200-mg dose is known to be 167 min.

Mechanism of Toxicity

Camphor is both excitative and depressant to the central nervous system (CNS). Its action has been postulated to be intraneuronal on the oxidation cycle at a phase above the cytochrome b level of the cytochrome oxidase system.

Human Toxicity: Acute

Upon ingestion, an initial burning sensation may be noted in the mouth and throat. Spontaneous nausea and vomiting may occur within minutes of ingestion. Confusion, vertigo, restlessness, delirium, hallucinations, tremors, and convulsions are all directly related to the CNS involvement and may be predictors of serious toxicity. More severe intoxications may result in hepatic failure. Death may be caused by respiratory depression or may follow status epilepticus. Camphor should be considered an eye irritant.

Human Toxicity: Chronic

The chronic ingestion of camphor may produce similar toxicity but in a more insidious fashion. Liver failure is a more pronounced clinical manifestation.

Clinical Management

Basic and advanced life-support measures should be utilized as necessary. Gastrointestinal decontamination procedures should be used as deemed appropriate to the patient's level of consciousness and the history of ingestion. Activated charcoal is marginally effective in adsorbing camphor. Oils, alcohols, and other lipophilic substances enhance intestinal absorption and are contraindicated. Ocular exposures necessitate flushing with a gentle system of tepid water for a minimum of 15 min. If signs of irritation persist, an ophthalmology consult is required.

Animal Toxicity

Animal toxicity corresponds to human toxicity.

—Bonnie S. Dean

Captafol

- ♦ CAS: 2425-06-1
- ♦ SYNONYMS: 3a,4,7,7a-Tetrahydro-2-[(1,1,2,2-tetrachloroethyl)thio]-1H-isoindole-1,3(2H)-dione
- ♦ CHEMICAL CLASS: Phthalimide fungicide
- ♦ CHEMICAL STRUCTURE:

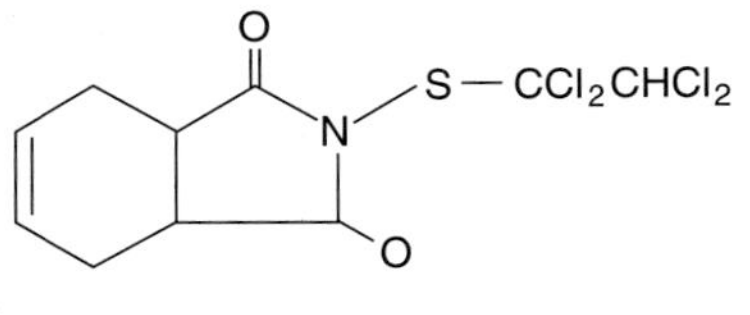

Uses

Captafol is widely used as a fungicide and seed protectant.

Exposure Pathways

Dermal and ocular exposures are the most common routes of exposure to captafol.

Toxicokinetics

Captafol is poorly absorbed from the gastrointestinal tract. The liver and the gastrointestinal tract are the primary sites of metabolism of captafol. Captafol is eliminated via urine, feces, and expired air. The major single metabolite, tetrahydrophthalimide, was detected in blood, urine, and feces, but most of the activity in the blood and urine was in the form of more water-soluble metabolites.

Mechanism of Toxicity

The primary toxicity following captafol exposure probably occurs through a hypersensitivity mechanism.

Human Toxicity

The primary symptoms of captafol exposure reported in humans include contact dermatitis and conjunctivitis. The reaction may be severe and may include stomatitis and painful bronchitis. Persons with a skin rash following exposure to captafol were found to have systemic as well as dermal disorders. Hypertension was reported in patients with marked edema. Other findings following captafol exposure include protein and urobilinogen in the urine, depression of liver function, anemia, and depression of blood cholinesterase activity. Acute oral or dermal exposure to captafol rarely results in severe toxicity. However, due to a higher level of toxicity in animal models following intraperitoneal exposure, parenteral exposure may present a greater hazard potential.

Clinical Management

Exposed eyes and skin should be flushed with copious amounts of water. In case of an inhalation exposure, the patient should be monitored for respiratory distress. Artificial ventilation may be provided and symptomatic treatment may be administered as necessary.

Animal Toxicity

The oral LD_{50} of captafol in rats varies from 2500 to 6200 mg/kg. Rats exposed to captafol at dietary levels of 1500 and 5000 ppm demonstrated enlarged livers and an increased mortality. Following exposure to 300 or 100 mg/kg of captafol, dogs suffered frequent vomiting and diarrhea during the first 4 weeks and were observed to be slightly anemic and deficient in growth during a 2-year study. Dogs at dosages of 30 mg/kg or

greater developed both absolute and relative increases in the weights of the liver and kidney. Breakdown products may contribute to the skin irritation and sensitization associated with captafol.

—Priya Raman

Related Topic

Pesticides

Captan

- ◆ CAS: 133-06-2
- ◆ SYNONYMS: *N*-(trichloromethylthio)-4-cyclohexene-1,2-dicarboximide; captane; Orthocide; Vancide 89
- ◆ CHEMICAL CLASS: Chloroalkyl thio (pthalimide) fungicide
- ◆ CHEMICAL STRUCTURE:

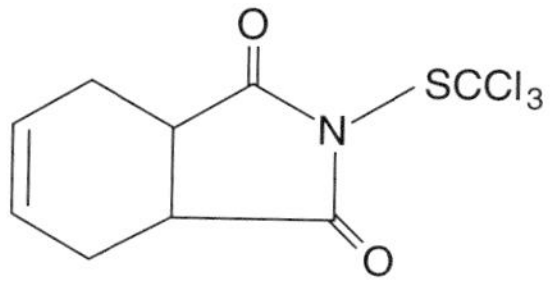

Uses

Captan is widely used as a fungicide and seed protectant. Frequently, captan is mixed with commonly used insecticides, such as diazinon, lindane, or malathion, or with other fungicides such as maneb.

Exposure Pathways

Exposure to captan may occur through oral, dermal, or inhalation exposures.

Toxicokinetics

Apparently, captan is poorly absorbed from the gastrointestinal tract and may be degraded there. Captan is also metabolized in the liver. Animal studies have indicated some elimination via urine, feces, and expired air.

Mechanism of Toxicity

The mechanism of captan toxicity has not been well defined. *In vitro*, captan has caused rapid swelling of rat liver mitochondria. The inhibition of mitochondrial function by captan is nonspecific, involving several sites. The inhibition involves an uncoupling of oxidative phosphorylation, but the mode of death of animals killed by the compound indicates that this is not the dominant mode of action.

Human Toxicity

Other than dermatitis and conjunctivitis, no signs of toxicity directly due to captan have been reported in humans. Treatment of human erythrocytes in buffered saline with $0.5 \mu M$ captan caused rapid loss of intracellular potassium. This effect was prevented by the addition of glutathione.

Clinical Management

Toxicity is unlikely following acute exposure to captan and no treatment is likely to be necessary. Any treatment is therefore symptomatic.

Animal Toxicity

Captan has a low oral toxicity to laboratory animals. In the rat, LD_{50} values of 12,600 and greater than 17,000 mg/kg have been reported. Oral LD_{50} values of 7840 and 7000 mg/kg in male and female mice, respectively, and corresponding intraperitoneal values of 518 and 462 mg/kg have been reported. Sheep and, to a lesser degree, cattle are more susceptible to captan. A dietary level of 10,000 ppm for 54 weeks produced marked growth depression in both male and female rats. Severe protein deficiency increases sensitivity to captan. If protein intake was reduced to 35% of normal, the acute toxicity of captan was barely affected, whereas 13% of normal protein intake increased toxicity 25 times while zero protein increased the toxicity of captan over 2000 times. This may be the result of decreased activity of liver microsomal enzymes or may be due to reductions in the levels of reduced sulfhydryl proteins. Due to its structural similarity to thalidomide,

the developmental toxicity of captan has been closely examined. Essentially no evidence for developmental toxicity has been found with this chemical. Captan has tested positive in a variety of *in vitro* mutagenicity assays, however.

—Todd A. Bartow

Related Topic

Pesticides

Carbamate Pesticides

Carbamate compounds are usually subdivided into at least three main groups with respect to structure and general use (see Fig. C-1): insecticides, herbicides, and thio and dithio carbamates. A variety of "R" groups may be substituted in the molecule producing, as is the case for insecticides, a variety of alkyl or aryl esters of carbamic acid. Although technically characterized as carbamate pesticides, thio and dithio carbamate compounds are not included in this brief overview because they have drastically different modes of action from the first group (see Dithiocarbamates).

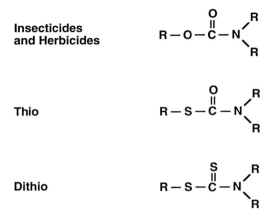

FIGURE C-1. General formulas for carbamates.

Carbamate pesticides have a colorful and interesting history of discovery and development. Oral administration of calabar bean paste, which is rich in carbamate alkaloids, was used in West Africa to reveal the guilt or innocence of people accused of witchcraft. If the alleged "witches" died after being forced to eat calabar bean paste, then they were indeed witches; if not, then they were declared innocent. Scientific investigation revealed that the active carbamate in the calabar bean was physostigmine, and the synonym for physostigmine, eserine, comes from the West Africans' word for calabar bean, esere. The first carbamate pesticides were synthesized in the mid- to late 1940s in an effort to develop new insect repellents, but the insecticidal properties of this class of compounds were quickly recognized and appreciated.

Generally, carbamate compounds, like organophosphorus pesticides, exert their primary toxic action through the inhibition of acetylcholinesterase (EC 3.1.1.7), although it is well-known that carbamates are inhibitors of many other esterases. The inhibition of acetylcholinesterase activity presumably precipitates a toxic response through the short-term increase in the concentration of acetylcholine at cholinergic junctions (e.g., central nervous system, neuromuscular junction, the parasympathetic nervous system, the sympathetic synapses, and the sympathetic innervation of the adrenal and sweat glands). Carbamates interact with acetylcholinesterase in the same manner as the natural substrate, acetylcholine, does, except that the carbamate remains in the active site for a markedly longer period of time, thereby preventing the hydrolysis of acetylcholine and resulting in a net inhibition of the enzyme's activity. The carbamylation of the active site of acetylcholinesterase is a much more labile union than is phosphorylation and does not lead to aging of the enzyme as can inhibition by organophosphate compounds. Therefore, restoration of acetylcholinesterase activity (i.e., decarbamylation or reactivation) is highly likely with the carbamate-inhibited enzyme. That is why carbamates are often labeled "reversible" inhibitors of acetylcholinesterase, i.e., because enzyme activity is restored within hours without significant *de novo* synthesis of acetylcholinesterase. Actually, in a biochemical sense, carbamates are not "reversible" inhibitors because the carbamate does not exit the active site intact; it is hydrolyzed just as acetylcholine is hydrolyzed. Because of this unstable inhibition, great care must be taken when analyzing cholinesterase

inhibition in tissues from carbamate-treated animals to prevent reactivation of the enzyme activity. Generally, carbamates do not cause peripheral neuropathy, although they may inhibit neurotoxic esterase activity. Reported effects on humans have usually been confined to the expected cholinergic overstimulation, with the exposure commonly occurring by inhalation or by the dermal route. Signs and symptoms are reported within minutes of exposure and can last for hours, but because of the "reversibility" of the inhibition, recovery is usually apparent before 24 hr. Metabolites in the urine may be used for biological monitoring. During the acute cholinergic crisis produced by these compounds, atropine (a muscarinic antagonist) may be used to counteract the effects, but oximes are contraindicated.

In addition to inhibition of acetylcholinesterase activity, carbamates have been reported to cause skin and eye irritation, hemopoietic alterations, degeneration of the liver, kidneys, and testes, as well as functional and histological changes in the nervous system after long-term, high-dose exposures. Moreover, some carbamates are known to produce reproductive and teratogenic effects. Fetuses of mothers dosed with carbamate have been reported to exhibit increased mortality, decreased weight gain, and induction of early embryonic death, so carbamates could be considered embryotoxic. Some carbamates have also been reported to be mutagenic, but with little carcinogenic potential.

As a group, carbamates have a relatively low environmental persistence, in contrast to the organochlorine pesticides. Carbamates are broken down by microorganisms, plants, animals, soil, or water, and hydrolysis is accelerated by increased light intensity, temperature, and/or alkalinity. The majority of carbamate compounds are easily absorbed through mucous membranes and respiratory and gastrointestinal tracts. Generally, metabolites are less toxic than the parent compound and the metabolites are commonly excreted in the urine. The general metabolic profile is basically the same in insects, plants, or animals. The first step in the catabolic scheme is usually hydrolysis to carbamic acid and the mechanism of hydrolysis is different for *N*-methyl and *N*-dimethyl derivatives. Carbamates, as a class, have a short half-life and are relatively rapidly excreted. Because there is very little bioaccumulation of carbamate pesticides, threat to wildlife is usually through exposure after application of the pesticides rather than through the food chain. Note that there have been reports of wildlife morbidity and mortality even if applications of certain carbamate pesticides were made at the recommended rate.

The overall toxicity profile of the carbamate pesticides covers a very wide spectrum from virtually nontoxic to some of the most highly toxic pesticides in commercial use; carbamate LD_{50} values range from 5000 to 1 mg/kg. More than 50 commercially used carbamate pesticides are in use today with the highest volume usage attributed to butylate, carbofuran, methomyl, carbaryl, and benomyl (see Carbofuran, Methomyl, Carbaryl, and Benomyl).

Further Reading

Kuhr, R. J., and Dorough, H. W. (1976). *Carbamate Insecticides: Chemistry, Biochemistry, and Toxicology.* CRC Press, Boca Raton, FL.
Smith, G. J. (1993). *Toxicology and Pesticide Use in Relation to Wildlife: Organophosphate and Carbamate Compounds.* CRC Press, Boca Raton, FL.
Vettorazzi, G, and Miles-Vettorazzi, P. (1975). *Safety Evaluation of Chemicals in Food: Toxicological Data Profiles for Pesticides. 1. Carbamate and Organophosphorus Insecticides Used in Agriculture and Public Health.* World Health Organization, Geneva.

Acknowledgments

(The research described in this article has been reviewed by the Health Effects Research Laboratory, U.S. EPA, and approved for publication. Approval does not signify that the contents necessarily reflect the views and policies of the agency nor does mention of trade names and commercial products constitute endorsement or recommendation for use.)

—*Stephanie Padilla*

Related Topics

A-esterase
Acetylcholine
Carboxylesterases
Cholinesterase Inhibition
Organochlorine Insecticides
Organophosphates
Pesticides
Pollution, Soil

Carbamazepine

- CAS: 298-46-4
- SYNONYMS: CBZ; 5H-Dibenz(b,f)-azepine-5-carboxamide; Tegretol
- PHARMACEUTICAL CLASS: Synthetic iminostilbene derivative structurally similar to imipramine, a tricyclic antidepressant. Carbamazepine is an anticonvulsant unrelated to the other anticonvulsants.
- CHEMICAL STRUCTURE:

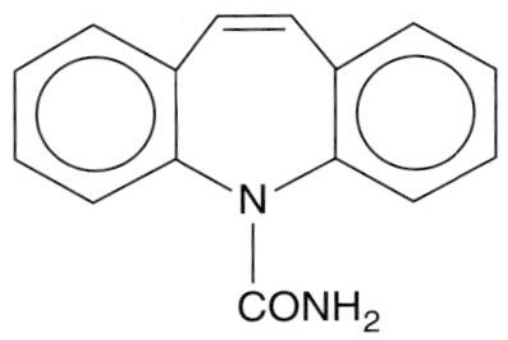

Uses

Carbamazepine is used in the treatment of epilepsy and trigeminal neuralgia. Unlabeled uses include treatment of postherpetic pain syndrome, neurogenic diabetes insipidus, bipolar disorder, alcohol withdrawal, and cocaine dependence.

Exposure Pathways

The exposure pathway for carbamazepine is exclusively oral (ingestion of tablets or suspension).

Toxicokinetics

Carbamazepine is slowly and incompletely absorbed during therapeutic use. With large ingestions, absorption may be delayed and unpredictable, producing peak levels from 4 to 72 hr after the overdose. The absorption phase in an overdose is highly variable because of carbamazepine's ability to significantly decrease gut motility and to form pharmacobezors. The primary metabolite of carbamazepine is carbamazepine-10,11-epoxide (CBZE), which also has anticonvulsant activity. A minor pathway results in iminostilbene formation. Further hydrolysis and conjugation produce six other known metabolites including 10,11-

dihydroxycarbamazepine. Protein binding is 75% for carbamazepine and 50% for CBZE. The volume of distribution is 0.8–1.9 liters/kg. The hydrolyzed and conjugated metabolites are eliminated through the kidneys, with only 1.2% free carbamazepine being found in the urine. Twenty-eight percent is eliminated unchanged in the feces. Carbamazepine induces drug metabolizing enzymes so that drug half-life is reduced in chronic use. The half-life in healthy adults ranges from 18–65 hr in a single dose to 8–17 hr in chronic administration. In newborns and children, the half-life is approximately 9 hr.

Mechanism of Toxicity

The mechanism of action is unknown. It is believed carbamazepine may decrease the conduction of sodium ions across nerve cell membranes producing a decrease in cyclic adenosine monophosphate. Additional theories include decreased turnover of γ-aminobutyric acid, a partial agonist effect at adenosine receptors, or increased firing of adrenergic neurons of the locus coreleus. It may be a combination of these.

Human Toxicity

The primary and common toxic event involves the central nervous system. Cardiac conduction delays and ventricular arrhythmias can be seen but are infrequent. In the few deaths directly attributable to carbamazepine toxicity, ventricular dysrhythmias have been the terminal event. Coma, seizures, and respiratory depression are commonly seen in adults at levels greater than 40 μg/ml (170 μmol/liter). The incidence of serious toxicity is similar in adults and children. However, serum levels are less predictive in children. Therefore, coma, seizures, and apnea may be seen at lower serum levels than in adults. Other manifestations of neurologic toxicity are nystagmus, ataxia, choreoathetoid movements, encephalopathy, absent corneal reflexes, decreased deep tendon reflexes, urinary retention, and dystonias. A cyclic clinical course can be seen, with a waxing and waning of symptoms. This may be due to the presence of a pharmacobezor in the gut or more commonly due to a decrease in gastrointestinal motility produced by the prominent anticholinergic effects of carbamazepine. Idiopathic hepatotoxicity has been reported as a rare manifestation that is not dose related.

Clinical Management

Basic and advanced life-support measures should be utilized as necessary. Gastrointestinal decontamination procedures should be used as deemed appropriate to the patient's level of consciousness and history of the ingestion. Activated charcoal effectively binds carbamazepine. Multiple dose activated charcoal (0.5 g/kg every 4 hr) has been shown to decrease the half-life of carbamazepine. Generally, supportive measures are all that is required in carbamazepine overdose. Seizures should be managed with diazepam. Ventricular arrhythmias should be managed with lidocaine. The presence of persistently high serum levels or fluctuating serum levels may suggest the presence of a pharmacobezor in the gut. Removal should be attempted, in the presence of an active bowel, with whole bowel irrigation using a polyethylene glycol-electrolyte solution.

—*Henry A. Spiller*

Carbaryl

- ◆ CAS: 63-25-2

- ◆ SYNONYMS: 1-Naphthyl *N*-methylcarbamate; Atoxan; Caprolin; Carbacide; Carbamine; Carpolin; Cekubaryl; Denapon; Denopton; Devicarb; Dicarbam; Gamonil; Hexavin; Karbaspray; Karbatox; Karbosep; Mervin; NAC; Panam; Rayvon; Septene; Sevidol; Sevin; Sevinox; Tercyl; Tricarnam; ENT 23969; UC 7744; OMS 29

- ◆ CHEMICAL CLASS: Carbamate pesticide

- ◆ CHEMICAL STRUCTURE:

Uses

Carbaryl is a broad-spectrum contact insecticide. It is also an effective systemic insecticide. Uses include control of a wide variety of pests on field crops, fruits, vegetables, nuts, ornamentals, lawns, and domestic animals. Carbaryl has also been used therapeutically in the control of human lice.

Exposure Pathways

Human exposure has occurred through ingestion, inhalation, and dermal contact. Carbaryl is available in a wide variety of formulations including dusts, wettable powders, granules, oil, and aqueous suspensions.

Toxicokinetics

The rate of dermal absorption of carbaryl in animal studies is dependent on the solvent used. Carbaryl can be hydrolyzed to 1-napthol or hydroxylated to a naphthylmethylcarbamate, either of which may be conjugated with glucuronic acid or sulfate. Nonhydrolytic pathways also play a minor role in the biotransformation of carbaryl. The major metabolite is 1-naphthol. In rats treated with carbaryl through oral gavage, the highest tissue levels of the pesticide were found in the liver, kidneys, and adipose tissue. Most mammals eliminate at least 75% of the original dose within 24–48 hr. The route of elimination is generally urinary with small amounts of certain metabolites undergoing fecal elimination.

Mechanism of Toxicity

Like aldicarb, carbaryl basically acts as a poor substrate for acetylcholinesterase, the enzyme responsible for degrading the neurotransmitter acetylcholine. Carbaryl reversibly binds to and inhibits the enzyme, resulting in accumulation of acetylcholine in the synapse, hyperstimulation of cholinergic receptors, and "cholinergic crisis." The enzyme is fairly rapidly reactivated through spontaneous decarbamoylation or via hydrolysis.

Human Toxicity: Acute

The acute effects of exposure may include the typical symptoms of cholinergic overstimulation such as the SLUDGE syndrome (salivation, lacrimation, urination, diarrhea, gastrointestinal cramping, and emesis), respiratory depression, bronchospasms, increased bronchial secretions, pulmonary edema, blurred vision, miosis, headache, tremors, muscle fasciculations, convulsions, mental confusion, and coma. Respiratory failure is the cause of death.

Human Toxicity: Chronic

An isolated human case study linked an extensive dermal exposure to a chronic polyneuropathy. Studies of oral exposure in hogs also indicate possible neuropathological effects of carbaryl.

Clinical Management

Activated charcoal may be used to reduce absorption from the gastrointestinal tract. Syrup of ipecac-induced emesis may be used if the patient is conscious and alert. Emesis is contraindicated if the patient is, or may become, convulsive. For inhalation exposure, the victim should be removed from the exposure area and observed for signs of breathing difficulty. In cases of dermal exposure, contaminated clothing should be removed and disposed of. Any exposed areas of skin should be repeatedly washed with soap and water. For eye contact, the eyes should be flushed with generous amounts of lukewarm water for a minimum of 15 min.

Clinical management is basically symptomatic and supportive. Artificial ventilation with 100% humidified oxygen is necessary in cases of respiratory paralysis. An endotracheal tube may be required to maintain the airway. Erythrocytic acetylcholinesterase (AChE) levels, arterial blood gases, and cardiac function should be monitored. Because of the rapid reactivation of AChE, significant enzyme inhibition may not be seen unless blood samples are drawn and assayed within approximately 2 hr of the exposure. Atropine should be given to relieve the muscarinic (i.e., SLUDGE) symptoms. Convulsions may be treated with intravenous diazepam. Phenytoin may be used if the convulsions are recurrent. Pralidoxime is indicated in cases of mixed exposure to both carbamates and organophosphorus compounds but is contraindicated in cases of carbamate-only exposure.

Animal Toxicity

Pigs receiving carbaryl in their diets developed incoordination, muscle contractions, and tremors cumulating in paraplegia. It is unclear if carbaryl or a metabolite, possibly unique to pigs, is responsible.

—*Paul R. Harp*

Related Topics

A-esterase
Carbamate Pesticides
Cholinesterase Inhibition
Neurotoxicology: Central and Peripheral
Pesticides

Carbofuran

- CAS: 1563-66-2
- SYNONYMS: 2,3-Dihydro-2,2-dimethyl-7-benzofuranyl methylcarbamate; Furadan; FMC 10242; Bay 70143; Chinufur; Niagra NIA-10242; OMS 864; Yaltox; NIOSH/RTECS FB 9450000; NA 2757; STCC 4921525
- CHEMICAL CLASS: *N*-methyl carbamate insecticide, acaricide, and nematocide
- CHEMICAL STRUCTURE:

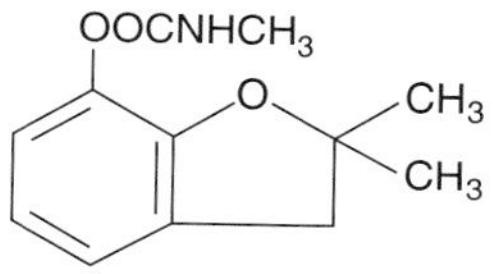

Uses

Carbofuran is used as an agricultural insecticide on tobacco, corn, alfalfa, and other field crops.

Exposure Pathways

Exposure may occur via the oral, inhalation, and dermal routes.

Toxicokinetics

Carbofuran is absorbed by the oral, inhalation, and dermal routes. It is poorly absorbed through intact skin. Approximately 75% of absorbed carbofuran is protein bound. Carbofuran is metabolized (oxidized) to 3-hydroxycarbofuran in the liver. This metabolite undergoes substantial metabolic conjugation to form a water-soluble glucaronide, which is excreted in the urine. The half-life in the rat is 20 min for the parent compound and 64 min for the 3-hydroxycarbofuran metabolite.

Mechanism of Toxicity

Carbofuran is a reversible inhibitor of acetylcholinesterase. Inhibition of acetylcholinesterase activity leads to an increase in acetylcholine at the nerve synapse resulting in excessive cholinergic stimulation. Following intravenous injection of 50 μg/kg in rats, blood acetylcholinesterase activity was depressed by 83% in 2 min. With oral exposure, acetylcholinesterase activity was depressed by 37% within 15 min of ingestion. Recovery of acetylcholinesterase activity parallels carbofuran elimination.

Human Toxicity

Exposure to carbofuran may lead to cholinergic crisis with signs and symptoms including increased salivation, lacrimation, urinary incontinence, diarrhea, gastrointestinal cramping, and emesis (SLUD syndrome). The syndrome may be indistinguishable from that seen after organophosphate poisoning. Seizures, coma, diaphoresis, muscle weakness and fasciculation, bradycardia, and tachycardia may occur. Death may be due to severe bronchoconstriction or respiratory paralysis.

Clinical Management

Rescuers and medical personnel must take precautions to avoid becoming contaminated themselves during rescue and emergency treatment. Victims should be removed from the toxic environment and 100% humidified supplemental oxygen should be administered with assisted ventilation as required. Patients with significant bronchorrhagia, pulmonary edema, convulsions, or coma may require endotracheal intubation and airway suctioning. Exposed skin and eyes should be flushed with copious amounts of water. Measures to decrease absorption may be beneficial soon after ingestion, but induced emesis should be avoided because of the potential for early development of coma or seizures. Atropine is antidotal for muscarinic symptoms and should be given in an initial dose of 2 mg and repeated every 15–30 min as required. The endpoint for atropinization is normalization of vital signs and drying of pulmonary secretions, not pupillary dilatation.

Administration of 2-PAM chloride (Protopam and pralidoxime) is generally not recommended in carbamate poisoning since it has been shown to interfere with the efficacy of atropine. Seizure control with diazepam, phenobarbital, or phenytoin may be required. Cardiovascular support and intensive supportive care may be required in serious cases.

Animal Toxicity

Dermal exposure to 1000 mg/kg caused only minimal symptoms and no deaths in rats. Inhalation exposure to 0.86 mg/m^3 caused a slight depression of the cholinesterase levels in monkeys. The no-effect levels in chronic feeding studies are 20 ppm in dogs and 25 ppm in rats. At 50 ppm chronically in the diet, significant decreases in cholinesterase activity were seen in the dog and rat. Chronic administration of 10 or 25 ppm of carbofuran in the diet for 180 days had no cumulative effect in rats.

The LD$_{50}$s in rats are 5 mg/kg (oral) and 120 mg/kg (skin). The LC$_{50}$ in rats is 85 mg/m^3 (inhalation). In dogs, the LD$_{50}$ is 19 mg/kg (oral) and the LC$_{50}$ (inhalation) is 52 mg/m^3. In rabbits the LD$_{50}$ is 885 mg/kg (skin). In wild bird species, the LD$_{50}$ is 420 μg/kg (oral) and 100 mg/kg (skin).

—Todd A. Bartow and Paul W. Ferguson

Related Topics

A-esterase
Carbamate Pesticides
Cholinesterase Inhibition
Neurotoxicology: Central and Peripheral
Pesticides

Carbon Dioxide

- CAS: 124-38-9
- SYNONYMS: Carbon ice; dry ice
- DESCRIPTION: Carbon dioxide is a colorless and odorless gas. It has a molecular weight of 44.01 and specific gravity of 1.101 at $-37°C$. It is incompatible with metals (e.g., aluminum peroxide, sodium peroxide, lithium peroxide, sodium, sodium carbide, titanium, and sodium–potassium alloy).
- MOLECULAR FORMULA: CO_2

Uses

Carbon dioxide is used in the synthesis of urea, for organic synthesis, and in the manufacture of dry ice, soft drinks, and fire extinguishers.

Mechanism of Toxicity

Carbon dioxide is a simple asphyxiant; that is, it causes toxicity by displacing oxygen from the breathing atmosphere primarily in enclosed spaces and results in hypoxia. It has been postulated that the cause of death in breathing high concentration of carbon dioxide is due to carbon dioxide poisoning and not hypoxia based on a study performed in dogs.

Human Toxicity

Carbon dioxide is a simple asphyxiant that displaces oxygen from the breathing atmosphere resulting in hypoxia. Four stages have been described (depending on the arterial oxygen saturation): (1) indifferent stage, 90% oxygen saturation; (2) compensatory stage, 82–90% oxygen saturation; (3) disturbance stage, 64–82% oxygen saturation; and (4) critical stage, 60–70% oxygen saturation or less.

Acute

Following exposure to asphyxiants, cardiovascular effects like tachycardia, arrhythmias, and ischemia are noted. Carbon dioxide exerts a direct toxic effect to the heart, resulting in diminished contractile force. It is also a vasodilator and the most potent cerebrovascular dilator known. Respiratory effects, like hyperventilation, cyanosis, and pulmonary edema, are also noted. Various neurologic effects, like dizziness, headaches, sleepiness, and mental confusion, can occur. Prolonged hypoxia may result in unconsciousness; seizures may be seen during serious cases of asphyxia. Gastrointestinal effects, like nausea and vomiting, may occur, but usually resolve within 24–48 hr following termination of exposure. Decreased vision and increased intraocular pressure may be seen with inhalation of 10% carbon dioxide. Combined respiratory and metabolic acidosis was seen in a serious exposure to dry ice. The Lake Nyos disaster in August 1986 has been postulated to have resulted from the release of carbon dioxide from rising cold deep water producing a deadly cloud of gas. Cough, headache, fever, malaise, limb swelling, and unconsciousness were noted in the victims. Inhalation of carbon dioxide is teratogenic and has caused both male and female adverse reproductive effects in rodents. Increased fetal movements have been noted in humans following inhalation with 5% carbon dioxide in air.

The lowest lethal concentration (inhalation) for humans is 100,000 ppm for 1 min. Carbon dioxide concentrations of 20–30% can cause convulsions and coma within 1 min. Unconsciousness may occur when inhaling a concentration of 12% for 8–23 min. Inhalation of 6–10% causes dyspnea, headache, dizziness, sweating, and restlessness.

Chronic
Carbon dioxide is an important component of the body and would not be expected to have a chronic toxicity. However, long-term exposures to levels as low as 0.5–1%, while being generally well tolerated, can alter the acid base and calcium–phosphorus balance resulting in metabolic acidosis and increased calcium deposits in soft tissues. Long-term exposures in the range of 1–2% can stress the adrenal cortex because of constant respiratory stimuli and this level of exposure is considered dangerous after several hours. Exposure to 2% for several hours produces headache, breathing difficulty upon exertion, and deepened respiration. Fatalities have occurred with prolonged exposure to 15–30%.

Workplace Standards
The ACGIH TLV-TWA is 5000 ppm and the ACGIH STEL is 30,000 ppm; the OSHA PEL-TWA is 500 ppm (transitional limit) and 10,000 ppm (final rule limit), and the OSHA PEL-STEL is 30,000 ppm; the NIOSH recommended exposure limit is 10,000 ppm (TWA).

Clinical Management
Victims should be moved immediately from the toxic atmosphere and receive 100% humidified supplemental oxygen with assisted ventilation as required. Patients with severe or prolonged exposure should be carefully evaluated for neurologic sequelae and provided with supportive treatment. Seizures may be controlled by administration of diazepam. If seizures cannot be controlled with diazepam or recur, phenytoin or phenobar-

bital should be administered. Rewarming has been indicated for frostbite. On ocular exposure, the eyes should be rinsed for at least 15 min.

—Swarupa G. Kulkarni and Harihara M. Mehendale

Related Topic
Combustion Toxicology

Carbon Disulfide

- ◆ CAS: 75-15-0
- ◆ SYNONYMS: Carbon bisulfide; carbon disulphide; carbon sulfide; sulphocarbonic anhydride
- ◆ CHEMICAL CLASS: Solvent
- ◆ MOLECULAR FORMULA: CS_2

Uses
The primary use of carbon disulfide is as a solvent in the manufacture of viscose rayon. It is also used as a solvent in the manufacture of resins and rubbers. Other uses include use as a pesticide, grain fumigant, and in the production of carbon tetrachloride and semiconductors.

Exposure Pathways
Inhalation and dermal contact are the most significant routes of exposure to carbon disulfide. Oral exposure may occur through accidental intake.

Toxicokinetics
Carbon disulfide is readily absorbed by oral, dermal, and inhalation routes and rapidly distributed throughout the body. Equilibrium between ambient air concentrations and exhaled air is reached within 60 min. In the brain, carbon disulfide is converted to thiocarbamate. Carbon disulfide is also metabolized by cytochrome P450 enzymes in the liver to dithiocarbamates. These

metabolites account for the neurotoxic effects of the compound. Carbon disulfide metabolites are conjugated by glutathione to form 2-thiothiazolidine-4-carboxylic acid and 2-oxythiazolidine-4-carboxylic acid and excreted by the kidneys. Unmetabolized carbon disulfide is excreted in the breath.

Mechanism of Toxicity

Carbon disulfide encephalophathy is characterized by neuronal degeneration of the central nervous system. Postmortem examination of victims revealed pallor, vacuolization, and cell loss distributed throughout many regions of the brain. Peripheral neuropathy has also been reported; however, the mechanism of toxicity is unknown. One postulated mechanism of toxicity is the ability of carbon disulfide metabolites to chelate metals necessary for neuronal health making them unavailable for normal metabolic functions.

Human Toxicity

Occupational exposure to carbon disulfide at high levels or over prolonged periods of time may result in organic brain damage, peripheral nervous system damage, neurobehavioral dysfunction, and ocular and auditory effects. Carbon disulfide exposure may also contribute to cardiovascular disease. A similar pattern of effects is observed for dermal contact and may also include skin blistering and nerve damage near the site of contact. Carbon disulfide is a severe irritant to the eyes both as a liquid splashed into the eyes and when exposure to vapors occurs. Structural and functional eye disorders have been reported in some occupationally exposed individuals. Reproductive disorders have also been reported for worker populations including decreased sperm production, menstrual irregularities, and miscarriage. The available data are inadequate to assess the carcinogenic potential of carbon disulfide.

Clinical Management

Eye and skin contact with both liquid carbon disulfide and vapors should be avoided. Contaminated areas should be immediately flushed with lukewarm water. If exposure occurs by inhalation, the source of contamination should be removed or the victim should be moved from the area to fresh air. Artificial respiration or cardiopulmonary resuscitation should be administered, if necessary. If oral exposure occurs, vomiting should not be induced. Activated charcoal and gastric lavage have been suggested to reduce absorption. Removal to a fresh air supply may increase excretion and increase the level of urinary output. Improvement in neurological functioning may gradually occur if exposure ceases. Therefore, workers who are chronically exposed should be closely monitored for onset of symptoms and moved to lower levels of exposure as needed. The effects of carbon disulfide may be related to nutritional status and a diet high in protein and minerals, such as copper and zinc, is recommended for exposed workers.

Animal Toxicity

Studies of the effects of carbon disulfide in animal models are somewhat limited due to the availability of a human data from occupational exposures. Animal studies have been used to provide histopathological and neurochemical support for the effects seen in human populations. In addition, reproductive effects have been observed in animal studies following inhalation exposure to carbon disulfide. Sperm counts and testosterone levels were reduced in rats following inhalation exposure to carbon disulfide at levels greater than 350 ppm. Pregestational exposure to levels greater than 40 ppm resulted in developmental effects in offspring of exposed rats.

The NOAEL (inhalation) for rabbits is 20 ppm (teratogenic effects). The NOAEL (oral) for rabbits is 11 mg/kg/day (converted from inhalation exposure level). The LC_{50} (inhalation) in mice is 220 ppm.

—*Linda Larsen*

Related Topic

Pollution, Air

Carbon Monoxide

- ♦ CAS: 630-08-0
- ♦ SYNONYM: Carbonic oxide

◆ CHEMICAL CLASS: Inorganic compound of carbon and oxygen

◆ MOLECULAR FORMULA: CO

◆ CHEMICAL STRUCTURE:

$$C^+ \equiv O^-$$

Uses

Carbon monoxide is used industrially as a feedstock for the production of methanol, acrylates, phosgene, and ethylene. It is also used in metallurgy applications and in industrial fuels. A major source of carbon monoxide is the incomplete combustion of carbon-containing materials.

Exposure Pathways

Exposure to this colorless, odorless gas is via inhalation. Most exposures result from incomplete combustion, especially the emissions created by internal combustion engines.

Toxicokinetics

Absorption of carbon monoxide occurs in the gas exchange region of the respiratory tract following inhalation. Most carbon monoxide binds reversibly to hemoglobin (Hb) in red blood cells; smaller amounts remain in solution or bind to cellular cytochromes. The absorption of the carbon monoxide molecule by Hb is a function of the alveolar partial pressures of carbon monoxide and oxygen, the concentrations of carbon monoxide and oxygen in blood, and a constant which describes the relative affinity of Hb for carbon monoxide compared to oxygen. This relationship is described by Haldane's first law. Experimentally, the relative affinity has been shown to be on the order of 250 times, resulting in the rapid formation of carboxyhemoglobin. Other factors that impact absorption in living organisms include ventilation rates, altitude (alveolar oxygen pressure), total blood volume, and endogenous carbon monoxide production.

Carboxyhemoglobin is completely dissociable, and carbon monoxide is liberated and eliminated through the lungs after exposure to carbon monoxide ceases. Small amounts are oxidized to carbon dioxide.

After binding to Hb to displace oxygen and form carboxyhemoglobin, carbon monoxide is transferred rapidly throughout the body, where it produces asphyxia. The majority of the body burden exists as carboxyhemoglobin, bound to hemoglobin of red blood cells, while about 10% is present in extravascular space.

Carbon monoxide is eliminated via the lungs. Carboxyhemoglobin dissociates in accordance with Haldane's law; dissociation and excretion of carbon monoxide occur rapidly after cessation of exposure but slow as carboxyhemoglobin levels decrease.

Mechanism of Toxicity

As a result of hemoglobin's high relative affinity for carbon monoxide compared to oxygen and the resulting production of carboxyhemoglobin, decreased delivery of oxygen to tissues occurs, resulting in anemic hypoxia and metabolic and functional impairment. Carbon monoxide may also exert a toxic effect by binding to cellular cytochromes. Displacement of oxygen in the tissues ultimately results in anaerobic metabolism with subsequent buildup of metabolic acids.

Human Toxicity

The effects of acute and chronic overexposures to carbon monoxide have been well documented. The effects generally result from the hypoxic action exerted on the tissues. Among the earliest and most prominent effects are central nervous system disorders, such as headaches and lightheadedness. At blood COHb levels approaching 30–40%, dizziness, incoordination, nausea, vomiting, and collapse result. At still higher levels (>40% blood saturation), syncope, convulsions, coma, and death may occur. Some studies have indicated that relatively small increments in carboxyhemoglobin levels may produce adverse cardiovascular effects. In addition, reproductive effects following exposures to carbon monoxide during pregnancy have been reported.

Clinical Management

Eye and skin contact and oral exposure are not relevant routes of exposure. If carbon monoxide is inhaled, the victim must be removed from exposure and assisted in breathing as necessary. Administration of oxygen decreases recovery time significantly: The half-life of blood carboxyhemoglobin decreases from about 6 hr in adults breathing air to less than 100 min when oxygen is administered. Hyperbaric oxygen accelerates the process of carboxyhemoglobin dissociation further and in cases of severe poisoning is preferred. Administration of 2.5–3.0 atm of oxygen for 1 or 2 hr has been recommended. Follow-up treatment should be symptomatic.

Animal Toxicity

In addition to the acute effects associated with hypoxia, carbon monoxide has been reported to cause chronic toxicity including cardiovascular, nervous system, reproductive, and developmental effects in animals. Carboxyhemoglobin levels of about 60% are lethal, with most animals dying at levels between 65 and 75% carboxyhemoglobin.

—*Daniel Steinmetz*

Related Topics

Blood
Combustion Toxicology
Indoor Air Pollution
Neurotoxicology: Central and Peripheral
Pollution, Air
Respiratory Tract
Sensory Organs

Carbon Tetrabromide

- ◆ CAS: 558-13-4
- ◆ SYNONYMS: Carbon bromide; tetrabromide methane; tetrabromo methane
- ◆ CHEMICAL CLASS: Halomethane, oxidizing agent, solvent
- ◆ MOLECULAR FORMULA: CBr_4

Uses

Carbon tetrabromide is used as an industrial solvent.

Exposure Pathways

The primary routes of entry are eye contact, skin contact, inhalation, and ingestion. When heated to decomposition, carbon tetrabromide emits toxic fumes of Br-.

Toxicokinetics

Carbon tetrabromide may be absorbed by dermal, inhalation, or oral routes. It is oxidatively metabolized by rat liver microsomes to electrophilic and potentially toxic metabolites. It is metabolized in the liver but causes primary effects on the kidneys. The electrophilic bromine derivatives formed can be excreted as such.

Mechanism of Toxicity

Carbon tetrabromide inhibits protein synthesis and causes lipid peroxidation, both of which may be involved in cell injury or death mediated by free radicals.

Human Toxicity

Carbon tetrabromide is harmful by inhalation, ingestion, or skin absorption. It causes irritation to eyes, skin, mucous membranes, and the upper respiratory tract.

In occupational settings, technical measures should prevent any contact with the skin and mucous membranes. Workers exposed to carbon tetrabromide should wear personal protective equipment and their work should be carried out only in restricted areas. After use, clothing and equipment should be placed in an impervious container for decontamination or disposal. The ACGIH TLV for carbon tetrabromide is 0.1 ppm.

Clinical Management

In case of contact, eyes and skin should be flushed with water for 15–20 min. If inhaled, the victim should be removed to fresh air. If necessary, oxygen and artificial respiration should be administered. If the patient is in cardiac arrest cardiopulmonary resuscitation should be performed. These life-supporting measures should be continued until medical assistance has arrived. An unconscious or convulsing person should not be given liquids or induced to vomit.

Animal Toxicity

Carbon tetrabromide is poisonous by subcutaneous and intravenous routes and moderately toxic by ingestion. It causes kidney toxicity and is narcotic at high concentrations. LD_{50}s in mice are 298 mg/kg (subcutaneous) and 56 mg/kg (intravenous).

—*Kashyap N. Thakore*

Carbon Tetrachloride

◆ CAS: 56-23-5
◆ SYNONYMS: Methane tetrachloride; carbon tet; carbon chloride; tetrachloromethane; perchloromethane; tetrachlorocarbon; carbona; freon 10; halon 104
◆ CHEMICAL CLASS: Halogenated hydrocarbon
◆ CHEMICAL STRUCTURE:

$$C—Cl_4$$

Uses

The carcinogenic properties of carbon tetrachloride have caused a decline in the industrial use of this chemical. It was formerly used as a solvent for fats, oils, waxes, varnishes, lacquers, and resins; as a fire extinguisher, grain fumigant, and dry cleaning agent; and in veterinary medicine as an anthelmintic. Current use of carbon tetrachloride is limited to a chemical intermediate in the industrial production of a few chlorinated, organic chemicals.

Exposure Pathways

Exposure can occur via inhalation, ingestion, and dermal contact.

Toxicokinetics

Absorbed carbon tetrachloride tends to concentrate in body fat, liver, and bone marrow. Laboratory experiments indicate that half the absorbed dose is exhaled unchanged from the lungs. The remainder of the absorbed dose is metabolized primarily in the liver and eliminated in exhaled air and in urine and feces. Carbon tetrachloride is metabolized in the liver to a trichloromethyl radical. This radical can then be subject to dimerization to hexachloroethane, reduced to chloroform, or bind to cellular macromolecules in the liver. An alternative metabolism pathway can transform carbon tetrachloride, via phosgene formation, to carbon monoxide and carbon dioxide.

Mechanism of Toxicity

Carbon tetrachloride is metabolized by cytochrome P450 to the reactive metabolites trichloromethyl free radical and trichloromethylperoxy free radical. These free radical metabolites are believed to react with cell lipids and membranes causing peroxidation of polyenoic lipids in endoplasmic reticulum. This reaction can generate secondary free radicals that are available to react with other lipids or cellular macromolecules. Lipid peroxidation and binding reactions lead to destruction of cell membranes with concomitant loss of function.

Human Toxicity

Like most organic solvents, carbon tetrachloride can cause dermatitis through its defatting action on the epidermis. Eye contact with vapor or liquid causes irritation, pain, and lacrimation. Acute ingestion or inhalation of carbon tetrachloride produces central nervous system depression. This effect is manifested by dizziness, vertigo, headache, mental confusion, and loss of consciousness. These symptoms may be accompanied by nausea, vomiting, abdominal pain, diarrhea, and hypotension. In severe cases, death may result from respiratory arrest, circulatory collapse, or ventricular fibrillation. Systemic toxicity is manifested by damage to the liver (jaundice, hepatomegaly, and ascites), kidney (acute nephrosis), and heart (ventricular arrhythmias). Liver and kidney damage can be life threatening. Carbon tetrachloride toxicity is potentiated by the consumption of alcohol.

Clinical Management

The victim should be removed from the contaminated environment and provided with supportive treatment. Contaminated clothing should be removed and the affected area should be washed with water and soap. Care should be taken to maintain respiration by giving humidified oxygen through assisted ventilation. Liver and kidney damage may be minimized by giving free radical scavengers, pyridoxine or *N*-acetylcysteine. If liver and kidney damage is apparent, supportive therapy should be provided. Renal damage may be manifested by the appearance of polyuria, which might progress to oliguria and anuria. Hematuria and proteinuria may also be seen.

Animal Toxicity

The toxicological properties of carbon tetrachloride have been extensively studied in laboratory animals. The measured median lethal doses in most experiments have ranged between 0.5 and 5.0 g/kg. Based on these results, it can be said that carbon tetrachloride is moderately toxic to laboratory animals. The lowest dose known to have caused adverse effects in rats through the oral route was 2 g/kg. The oral LD_{50} in the rat was determined to be 2800 mg/kg; the inhalation LD_{50} measured in rats was 8000 ppm for 4 hr; and the skin LD_{50} was measured to be 5070 mg/kg. Carbon tetrachloride caused mild irritation to the rabbit eye when 2.2 mg was placed in the eye for 30 sec.

—*Heriberto Robles*

Related Topics

Alkyl Halides
Mechanisms of Toxicity
Pollution, Soil
Pollution, Water

Carbonyl Sulfide

- CAS: 463-58-1
- SYNONYMS: Carbon monoxide monosulfide; carbon oxide sulfide; carbon oxysulfide; RTECS/NIOSH FG6400000
- CHEMICAL CLASS: Sulfide
- CHEMICAL STRUCTURE:

C—O—S

Uses

Carbonyl sulfide is used in the synthesis of thio organic compounds. It is a chemical intermediate for alkyl carbonates and organic sulfur compounds.

Exposure Pathways

Exposure occurs predominantly by the inhalation route.

Toxicokinetics

Carbonyl sulfide is absorbed by the inhalation route. It is metabolized to hydrogen sulfide by carbonic anhydrase. It is the hydrogen sulfide produced that is responsible for carbonyl sulfide toxicity. Hydrogen sulfide inhibits respiration on the cellular level (see Hydrogen Sulfide).

Mechanism of Toxicity

Carbonyl sulfide is an intermediate in the formation of carbon dioxide. The first step is microsomal cytochrome P450-mediated NADPH-dependent. Atomic sulfur, a common by-product of carbonyl sulfide, appears to inhibit P450. In addition, carbonyl sulfide is metabolized to hydrogen sulfide by carbonic anhydrase.

Human Toxicity

At low to moderately high vapor concentrations carbonyl sulfide can cause painful conjunctivitis, photophobia, and corneal opacity. Direct skin contact may produce erythema and pain. Gastrointestinal effects include profuse salivation, nausea, vomiting, and diarrhea. Central nervous system effects include giddiness, headache, vertigo, amnesia, confusion, and unconsciousness. Exposure to extremely high vapor concentrations will produce sudden collapse and unconsciousness and/or death from prompt respiratory paralysis.

Clinical Management

The victim should be moved to fresh air immediately. If the victim is not breathing, artificial respiration should be given. If breathing is difficult, the victim should be given oxygen. In case of ocular or dermal contact, the skin or eyes should be flushed with running water immediately. Further treatment is symptomatic. Rescuers must prevent exposure by wearing a self-contained breathing apparatus to rescue the victim.

Animal Toxicity

Exposure to carbonyl sulfide in animals produces serious nervous system effects with narcotic effects in high concentrations. Concentrations of 0.1% by volume and higher produce death within 2 hr or less. Exposure to high concentrations produces acute respiratory failure. Continuous exposure of rabbits to 50 ppm carbonyl sulfide for 1–7 weeks had no significant effect on myocardial ultrastructure, resulted in a slightly elevated mean serum cholesterol level, and did not show histopathological changes in lungs or coronary arteries.

—Rhonda S. Berger and Wendy Khune

Carboxylesterases

Carboxylesterases (CarbEs, EC 3.1.1.1; also known as aliesterases or tributyrinases) are a heterogeneous group of enzymes as they differ in substrate specificity. Despite the wide distribution of CarbEs in mammalian systems, most of their known substrates are foreign compounds that are not normally involved in intermediary metabolism. CarbEs hydrolyze xenobiotics containing an ester, thioester, or amide group, and thus play an important role in drug metabolism, carcinogenesis, and detoxification of many noxious chemicals present in our environment. The physiological function of CarbEs still remains obscure.

Physical, Chemical, and Biochemical Properties of CarbEs

The mammalian hepatic, renal, and intestinal CarbEs consist of units with a molecular weight of about 60,000. Each unit bears one active site. The amino acid sequence around the active site of several CarbEs is Gly-Glu-Ser$^+$-Ala-Gly. The pI of hepatic CarbEs is usually in the range of pH 4.7–6.5, with the pH optimum in the range of 6–10. The behavior of hepatic microsomal and cytosolic CarbEs in the *in vitro* and *in vivo* studies indicates that these enzymes are different. CarbE in hepatic microsomes has been shown to consist of three isoenzymes (RH1, molecular weight 174,000, trimer, pI 6.0; RL1, molecular weight 61,000, monomer, pI 6.5; and RL2, molecular weight 61,000, monomer, pI 5.5), which differ considerably in terms of inducibility, substrate specificity, and immunological properties.

CarbEs can catalyze hydrolytic reactions of the following types:

1. Carboxylester hydrolysis

$$R-\overset{\overset{O}{\|}}{C}-O-R' + H_2O \rightarrow R-\overset{\overset{O}{\|}}{C}-OH + HOR'$$

2. Carboxylamide hydrolysis

$$R-\overset{\overset{O}{\|}}{C}-\underset{\underset{R''}{|}}{N}-R' + H_2O \rightarrow R-\overset{\overset{O}{\|}}{C}-OH + HNR'R''$$

3. Carboxythioester hydrolysis

$$R-\overset{\overset{O}{\|}}{C}-S-R' + H_2O \rightarrow R-\overset{\overset{O}{\|}}{C}-OH + HSR'$$

The first two of these reactions are equally relevant for the biotransformation process. Amides are often more stable to enzymatic hydrolysis than the corresponding esters with similar structure. For example, phenylacetate is hydrolyzed much faster than acetanilid. In addition, CarbE can cause hydrolysis of therapeutically useful drug esters, such as chloramphenicol succinate, prednisolone succinate, procaine, and methylparaben.

Tissue Distribution of CarbEs

The activities of CarbEs have been localized and determined in almost all tissues, with the highest activity in liver. A substantial amount of the enzyme activity is present in heart, kidney, lungs, brain, skeletal muscles, testes, small intestine, pancreas, nasal mucosa, adipose tissue, and plasma. Normal values of CarbEs for some of the tissues, using tributyrin as the substrate, are given in Table C-1. No significant variability in the values of CarbEs has been found among discrete brain regions and among different fiber-dependent skeletal muscles. Studies show that CarbEs' activities of the liver, kidneys, brain, and intestinal mucosa are predominantly present in the microsomal fraction. The liver cytosolic CarbE activity is 1/20 of that present in the microsomes. The lowest CarbE activity is determined in plasma (1/70 of that in hepatic microsomes). It is suggested that at least some of the serum CarbE isoenzymes originate from the liver.

Significance of CarbE Induction or Inhibition in Metabolism and Toxicity

As mentioned before, CarbEs have a limited physiological role per se, but their induction or inhibition by some

TABLE C-1
Normal Values of CarbE in Different Tissues
of Male Sprague–Dawley Rats

Tissue	CarbE activity (μmol tributyrin/g/hr)
Brain regions	
Cortex	68 ± 2.6
Stem	68 ± 2.3
Striatum	71 ± 1.5
Hippocampus	72 ± 3.1
Muscles	
Diaphragms	67 ± 2.0
Heart	92 ± 2.9
Liver	3014 ± 6.0
Serum/plasma	30 ± 2.0

Note. Values are means $\pm$ SEM.

drugs or xenobiotics can modify the metabolism and toxicity of their own or others to a great extent.

Induction of CarbEs

Oral or parenteral administration of phenobarbital can increase the cytosolic CarbE activity more than the microsomal activity, and the activity can remain elevated for 7 days after the last phenobarbital treatment. Phenobarbital treatment has no effect on the extrahepatic CarbE activity.

Parenteral administration of *p,p'*-DDT can augment hepatic CarbE for 14 days, probably due to the sequestration of the compound in adipose tissue. The degree of enzyme induction with *p,p'*-DDT is less than that by phenobarbital. The hepatic CarbE activity can be increased up to threefold by phenobarbital and *p,p'*-DDT. Plasma CarbE activity is not altered by *p,p'*-DDT.

Hepatic and extrahepatic CarbEs activities have been studied after the exposure of rats to polycyclic aromatic hydrocarbons. In dose- and time-dependent studies, benz(*a*)anthracene, benzo(*a*)pyrene, and 3-methylcholanthrene moderately induce the hepatic cytosolic and kidney microsomal CarbEs activities, while anthracene, phenanthrene, and chrysene have no effects on these enzymes. The hepatic microsomal and kidney cytosolic enzyme activities are not altered by the polycyclic aromatic hydrocarbons. The inducibility of hepatic cytosolic CarbE by the polycyclic aromatic hydrocarbons appears to indicate that these compounds could be divided into two groups. Benz(*a*)anthracene, benzo(*a*)pyrene, and 3-methylcholanthrene are moder-

ate inducers, and anthracene, phenanthrene, and chrysene are noninducers. The commercial arochlors (i.e., polychlorinated biphenyls) have been shown to increase hepatic CarbE by twofold.

From a toxicological point of view, the induction of CarbEs may be expected to protect against toxic effects of an ester if the compound itself is directly responsible for the toxic action (e.g., the reduced toxicity of malathion or malaoxon following hexachlorobenzene exposure in rats). Conversely, CarbEs' induction may potentiate the toxic effects produced by the hydrolytic products of the compound (e.g., in the metabolism of allyl alcohol).

Inhibition of CarbEs

The most important inhibitors of CarbEs are organophosphate insecticides (parathion, paraoxon, methyl parathion, EPN, and others) and nerve agents (DFP, soman, sarin, tabun, and VX) and carbamate insecticides (carbofuran, carbaryl, aldicarb, propoxur, oxamyl, methomyl, and others). Organophosphates inhibit CarbEs irreversibly by phosphorylation and carbamates inhibit CarbEs reversibly by carbamylation; similar to the basic mechanism (i.e., acylation of the active site):

$$EH + AB \underset{K_{-1}}{\overset{K_{+1}}{\rightleftharpoons}} EHAB \xrightarrow{K_{+2}} \underset{H_2O}{EA + BH} \xrightarrow{K_{+3}} \underset{(+BH)}{EH + AOH}$$

where EH is the enzyme, AB is the inhibitor, EHAB is the enzyme–inhibitor complex, and EA is the acyl enzyme. In other words, organophosphates and carbamates inactivate CarbEs by rapid esterification of a serine residue in the active site. It is often followed by a slow hydrolysis of the new ester bond. Therefore, these compounds are not only inhibitors of CarbEs but also poor substrates. Other inhibitors of CarbEs include disulfiram (tetraethylthiuram disulfide), glucocorticoids, and dexamethasone phosphate.

Role of CarbEs in Organophosphate and Carbamate Poisoning

Depending on the involvement of one or more anticholinesterase agents, and their single or repeated exposure, CarbEs can have multiple roles, such as (1) protective role by detoxifying organophosphates or carbamates, (2) preinhibition of CarbEs as a major factor in potentiation of toxicity, and (3) role in tolerance development following repeated exposure to organophosphates.

Role of CarbEs in Detoxification of Organophosphates and Carbamates

The acute toxicity of organophosphates and carbamates is attributed to their effectiveness as inhibitors of AChE. During both acute and prolonged exposure to organophosphates and carbamates, the activities of other serine-containing esterases, such as CarbEs, are inhibited in both neuronal and nonneuronal tissues. Inhibition of CarbEs generally serves as a detoxifying mechanism by reducing the free concentration of AChE inhibitors. It has been consistently observed that low-level exposure to organophosphate or carbamate causes marked inhibition of CarbEs without inhibiting AChE, suggesting greater affinity of CarbEs than AChE to these inhibitors. CarbEs act like false targets or scavengers which bind and thereby inactivate significant amounts of these inhibitors. In recent studies, one of the oximes, HI-6, has been shown to reactivate CarbEs activity, thereby providing additional protection against soman or possibly other organophosphates poisoning by acting as endogenous scavengers.

CarbEs, in addition to serving as nonspecific binding sites, can hydrolyze the carboxylester bond (esterolytic detoxification) in malathion-type organophosphates, carbamates, pyrethroids, and benzilate insecticides, and thereby reduce the free concentration of these insecticides. It is evident, therefore, that CarbEs can detoxify organophosphates and carbamates by multiple mechanisms.

It is important to note that the diversity between toxic effects of organophosphates, carbamates, or pyrethroids could partly be due to the existing variability in the levels of CarbEs activity, which is related to species, strain, and gender differences. For example, rabbit liver CarbE is more sensitive to inhibition by malathion and isomalathion than pig liver CarbE; some strains of rats and mice have higher CarbE activity than others; and female rat plasma has higher CarbE activity than male rat plasma.

Role of CarbEs in Potentiation of Toxicity of Organophosphates or Carbamates

The toxicity of organophosphates or carbamates can be potentiated several-fold if the activity of CarbE is inhibited by pretreatment. Potentiation of malathion toxicity by EPN was an incident reported about four decades ago. The mechanism responsible for potentiation by EPN of the malathion toxicity has been explained on the basis of inhibition of the enzymatic hydrolysis of the carboxylester linkages of malathion. An impurity compound, O,O-S-trimethyl phosphorothioate, present in commercial formulations of malathion and phenthoate, has also been shown to potentiate the acute toxicity of malathion and phenthoate by inhibiting tissue CarbEs.

Toxicity of organophosphates can be potentiated 15- to 20-fold in rats and mice by pretreatment with a metabolite of tri-O-cresylphosphate, CBDP (2-O-cresyl-4H-1,3,2-benzodioxa-phosphorin-2-oxide), which is an irreversible inhibitor of CarbEs. In similar studies, tetraisopropylpyrophosphoramide (iso-OMPA), or mipafox, an organophosphate-irreversible inhibitor of CarbEs, potentiates 3- to 5-fold the toxicity of several OPs (soman, DFP, and methylparathion) and carbamates (carbofuran, aldicarb, propoxur, and carbaryl). Inhibition of CarbEs by CBDP, iso-OMPA, or mipafox pretreatment, particularly in plasma, liver, heart, brain, and skeletal muscles, is a major contributory factor in potentiation of toxicity of organophosphates and carbamates. Thus, the toxicity of any drug, pesticide, or other type of agent that is normally detoxified by CarbEs could be potentiated by an organophosphate preexposure.

Role of CarbEs in Organophosphate Tolerance Development

It has been clearly demonstrated that daily exposure of rats, mice, or guinea pigs to certain organophosphates with sublethal doses can develop severe toxicity during the first few days, but further exposure for 7–14 days can lead to development of tolerance. For example, daily dosing of DFP (0.5 mg/kg, sc) produces severe anticholinesterase signs on Day 5 (toxicity phase), but further administration results in tolerance development because on Day 14 rats are free of signs (tolerance phase). During tolerance, CarbE activity can recover up to 40% or more compared with the initial inhibition (Day 5), suggesting renewed availability of nonspecific binding sites (CarbEs). The recovery of CarbE is probably due to *de novo* synthesis since treatment with an inhibitor of protein synthesis, cycloheximide, abolishes tolerance development.

It can be concluded that the protection can be attenuated, toxicity can be potentiated, and tolerance can be abolished by preinhibition of CarbEs with iso-OMPA, mipafox, or any other CarbE inhibitor against organophosphates. In contrast to organophosphates, rats that were administered carbamates such as carbofuran or aldicarb daily for 3 or 4 weeks showed no development of tolerance to the toxicity, probably due to lack of CarbEs' recovery.

Further Reading

Boskovic, B. (1979). The influence of 2-(O-Cresyl)-4H-1 : 3 : 2-benzodioxaphosphorin-2-oxide (CBDP) on organophosphate poisoning and its therapy. *Arch. Toxicol.* 42, 207–216.

Chambers, J. E., and Chambers, H. W. (1990). Time course of inhibition of acetylcholinesterase and aliesterases following parathion and paraoxon exposures in rats. *Toxicol. Appl. Pharmacol.* 103, 420–429.

Clement, J. G. (1989). Role of aliesterases in organophosphate poisoning. *Fundam. Appl. Toxicol.* 4, S94–S105.

Cohen, S. D. (1984). Mechanisms of toxicological interactions involving organophosphate insecticides. *Fundam. Appl. Toxicol.* 4, 315–324.

Dettbarn, W. D., and Gupta, R. C. (1989). Role of esterases as false targets in organophosphorus compounds toxicity. In *Enzymes Hydrolyzing Organophosphorus Compounds* (E. Reiner, W. N. Aldridge, and F. C. G. Hoskin, Eds.), pp. 165–179. Horwood, Chichester, UK.

Gupta, R. C., and Dettbarn, W. D. (1987). Iso-OMPA-induced potentiation of soman toxicity in rat. *Arch. Toxicol.* 61, 58–62.

Gupta, R. C., and Dettbarn, W. D. (1993). Role of carboxylesterases in the prevention and potentiation of N-methylcarbamate toxicity. *Chem.-Biol. Interact.* 87, 295–303.

Gupta, R. C., and Kadel, W. L. (1989). Concerted role of carboxylesterases in the potentiation of carbofuran toxicity by iso-OMPA pretreatment. *J. Toxicol. Environ. Health* 26, 447–457.

Gupta, R. C., and Kadel, W. L. (1990). Toxic interaction of tetraisopropylpyrophosphoramide and propoxur: Some insights into the mechanisms. *Arch. Environ. Contam. Toxicol.* 19, 917–920.

Gupta, R. C., Patterson, G. T., and Dettbarn, W. D. (1985). Mechanisms involved in the development of tolerance to DFP toxicity. *Fundam. Appl. Toxicol.* 5, S17–S28.

Maxwell, D. M., and Koplovitz, I. (1990). Effect of endogenous carboxylesterase on HI-6 protection against soman toxicity. *J. Pharmacol. Exp. Ther.* 254, 440–444.

Nousiainen, U. (1984). Inducibility of carboxylesterases by xenobiotics in the rat, pp. 1–55. Dissertation submitted to the University of Kuopio.

—*Ramesh C. Gupta*

Related Topics

Biotransformation
Carbamates
Liver
Nerve Gases
Organophosphates
Pesticides

Carboxylic Acids

- ◆ SYNONYM: Organic acids
- ◆ CHEMICAL CLASS: Carboxylic acids have at least one -COOH group attached to an alkane, alkene, or aromatic hydrocarbon moiety. Examples are acetic acid, butyric acid, citric acid, and tartaric acid. These acids may contain simple aliphatic chains or branched chains, and they contain substituents such as amines or other groups (see Acetic Acid and Butyric Acid).

—*Sanjay Chanda*

Related Topic

Acids

Carcinogen Classification Schemes

Introduction

Several classification schemes have been developed for ranking the relative hazards to humans associated with chemicals that, by one or more criteria, may be considered to be potential carcinogens. The classifications represent scientific judgments that typically take into account all the data available from *in vivo* animal bioassays, *in vitro* tests for genetic toxicity, human epidemiology, and structural relationships with other known carcinogens. Classification of a chemical as a carcinogen involves the consideration of many different factors. Classification schemes provide guidance on evaluating and weighting the available evidence into defined categories that can be used to communicate the implications for risk. Factors usually taken into consideration in interpreting the results of an animal bioassay include:

- Adequacy of experimental design and conduct
- Statistical significance of any increase in tumor incidence
- The presence or absence of a dose–response relationship and correct dose selection
- Nature of tumors (benign or malignant) and relevance of tumor type to humans
- Historical control data (incidence and variability) for tumor type
- Common (spontaneous) vs uncommon tumors
- Number of organs/tissues with tumors
- Mechanistic information

Two commonly used classification schemes are one developed by U.S. EPA and one developed by IARC. The U.S. EPA classification scheme is used as a tool for the regulation of chemicals under those laws it administers (e.g., FIFRA and TSCA) as well as by many state regulatory agencies. The IARC classification scheme is commonly used in the European Community and is considered in certain U.S. regulations and laws (e.g., OSHA Hazard Communication Standard). Other respected carcinogenic classification schemes include those developed by NIOSH, OSHA, NTP, and ACGIH. Each of these schemes is described in the following sections. It should be emphasized that these classification schemes are constantly evolving and that changes may occur over time.

U.S. EPA Carcinogen Classifications

The U.S. EPA carcinogen classifications are developed using the approach detailed in the *Guidelines for Carcinogens Risk Assessment* [51 FR 33992]. The U.S. EPA

"total-weight-of-evidence" scheme classifies chemical potential carcinogens into five groups, A–E, that indicate the likelihood they are human carcinogens. These groups are described below.

Group A: Human carcinogen—
This is reserved for chemicals where there exists clear epidemiological evidence indicating an association between exposure and cancer.

Group B: Probable human carcinogen—
This group is divided into two subgroups, B1 and B2. Group B1 indicates that there is "sufficient" evidence to indicate that the material is an animal carcinogen and that there is "limited" evidence of effects in humans. Group B2 indicates that although there is sufficient evidence, the total weight of evidence for effects in humans is weaker or "inadequate."

Group C: Possible human carcinogen—
This group is used for many different chemicals where there may be limited, often marginal evidence of carcinogenicity in animals and no evidence of any effects in humans.

Group D: Not classifiable as to human carcinogenicity—
This group is used for chemicals for which no data are available.

Group E: Evidence of noncarcinogenicity for humans—
This group is used for chemicals that show no evidence of any carcinogenicity in at least two adequately conducted animal tests with different species.

Proposed U.S. EPA Classification Scheme

The U.S. EPA classification scheme was published in the 1986 cancer guidelines [51 FR 33992]. In April 1996, the U.S. EPA proposed new cancer guidelines which differ substantially from the previous guidelines. The new guidelines recommend a narrative with descriptors that replace the previous letter designations. The narrative explains the kinds of evidence available and how they fit together in drawing conclusions, along with highlighting the significant issues and strengths and limitations of the data and conclusions. The descriptors have standardized definitions. The descriptors are not meant to replace an explanation of the nuances of the biological evidence but rather to summarize it. The use of descriptors within a narrative is intended to preserve the complexity (including the gray areas) that is an essential part of the hazard classification. Risk managers are instructed to consider the entire range of information included in the narrative rather than focusing simply on the descriptor.

Each category spans a wide variety of potential data sets and weights of evidence. The three proposed categories of descriptors for human carcinogenic potential are "known/likely," "cannot be determined," and "not likely."

Known/Likely

This category of descriptors is used when the available tumor effects and other key data are adequate to convincingly demonstrate carcinogenic potential for humans. It includes cases in which agents are known human carcinogens based on either epidemiologic evidence or a combination of epidemiologic and experimental evidence, demonstrating causality between human exposure and cancer; agents that should be treated as if they were known human carcinogens, based on a combination of epidemiologic data showing a plausible causal association (not demonstrating it definitively) and strong experimental evidence; and agents that are likely to produce cancer in humans due to the production or anticipated production of tumors by modes of action that are relevant or assumed to be relevant to human carcinogenicity.

Cannot Be Determined

This category of descriptors is used when available tumor effects or other key data are suggestive or conflicting or limited in quantity and, thus, are not adequate to convincingly demonstrate carcinogenic potential for humans. It includes cases in which agents' carcinogenic potential cannot be determined but there is suggestive evidence that raises concern for carcinogenic effects; agents whose carcinogenic potential cannot be determined because the existing evidence is composed of conflicting data (e.g., some evidence is suggestive of carcinogenic effects, but other equally pertinent evidence does not confirm any concern), and agents whose carcinogenic potential cannot be deter-

mined because there are inadequate or no data to perform an assessment.

Not Likely

This category of descriptors is used when, in the absence of human data suggesting a potential for cancer effects, the experimental evidence is satisfactory for deciding that there is no basis for human hazard concern. It includes cases in which agents are not likely to be carcinogenic to humans because they have been evaluated in at least two well-conducted studies in two appropriate animal species without demonstrating carcinogenic effects; agents not likely to be carcinogenic to humans because they have been appropriately evaluated in animals and show only carcinogenic effects that have been shown not to be relevant to humans; agents not likely to be carcinogenic to humans when carcinogenicity is dose or route dependent (e.g., not likely below a certain dose range or not likely by a certain route of exposure); and agents not likely to be carcinogenic to humans based on extensive human experience that demonstrates lack of effect (e.g., phenobarbital).

IARC Carcinogen Classifications

IARC is a department of the World Health Organization (WHO). The overall classification scheme developed by IARC is similar to that used by the U.S. EPA (the U.S. EPA scheme was initially developed based on an IARC scheme). Chemicals are classified into four groups with respect to potential to cause cancer in humans. The classification reflects the strength of the evidence available from animal studies, epidemiology, and other relevant data. The IARC groups are outlined below.

Group 1: The agent is carcinogenic to humans— This group is reserved for those chemicals or agents where there is "sufficient evidence" of carcinogenicity in humans.

Group 2: The agent is probably carcinogenic to humans— This group, like the U.S. EPA group B, is divided into two subgroups, groups 2A and 2B, depending on the strength of the evidence available. Groups 2A and 2B indicate that the agent is "probably" or "possibly" carcinogenic to humans, respectively.

Group 3: The agent is not classifiable as to its carcinogenicity to humans—
This group is used for chemicals that do not fall into any of the other groups.

Group 4: The agent is probably not carcinogenic to humans—
This group is used for compounds where there exists evidence suggesting an absence of carcinogenic potential in humans.

NTP Carcinogen Classifications

NTP is responsible for preparing *Reports on Carcinogens*. The *Reports on Carcinogens* are mandated by Public Law 95-662 and are for informational purposes only. The listing of a substance in the annual report does not by itself establish that such a substance presents a risk to persons in their daily lives. Clause (I) in subparagraph (4)(A) of Section 301 (b) of the Public Health Service Act requires that a report be published which contains a list of all substances (1) "which are either known to be carcinogens or may reasonably be anticipated to be carcinogens," and (2) to which a significant number of persons residing in the United States are exposed. The conclusions regarding carcinogenicity in humans or experimental animals are based on scientific judgment, with consideration given to all relevant information. As of the September 30, 1996 update, for the purpose of *Biennial Report on Carcinogens*, the classification scheme is outlined below.

Group 1: Known to be human carcinogens—
This group is reserved for those chemicals where there is sufficient evidence of carcinogenicity from studies in humans which indicates a causal relationship between exposure to the agent, substance, or mixture and human cancer.

Group 2: Reasonably anticipated to be human carcinogens—
This group, like the U.S. EPA group B, is divided into two subgroups, groups 2A and 2B, depending on the strength of the evidence available.

Group 2A: There is limited evidence of carcinogenicity from studies in humans which indicates that causal interpretation is credible, but that alternative explanations, such as chance, bias, or confounding, could not adequately be excluded.

Group 2B: There is sufficient evidence of carcinogenicity from studies in experimental animals which indicates that there is an increased incidence of malignant tumors and/or combined benign and malignant tumors (a) in multiple species or at multiple tissue sites, (b) by multiple routes of exposure, or (c) to an unusual degree with regard to incidence, site or type of tumor, or age at onset; or there is less than sufficient evidence of carcinogenicity in humans or laboratory animals. However, the agent, substance, or mixture belongs to a well-defined, structurally related class of substances whose members are listed on a previous *Annual* or *Biennial Report on Carcinogens* as either a known to be human carcinogen or reasonably anticipated to be human carcinogen, or there is convincing relevant information that the agent acts through mechanisms indicating it would likely cause cancer in humans.

OSHA Carcinogen Classifications

The Occupational Safety and Health Act of 1970 provides for the establishment of workplace standards for toxic materials or harmful physical agents

which most adequately assures, to the extent feasible, on the basis of the best available evidence, that no employee will suffer material impairment of health or functional capacity even if such employee has regular exposure to the hazard dealt with by such standard for the period of his or her working life.

Potential occupational carcinogens regulated under OSHA are classified into two main categories based on the nature and extent of the available scientific evidence: Category I potential carcinogens and Category II potential carcinogens [29 CFR 1990].

Category I Potential Carcinogens

A substance shall be identified, classified, and regulated as a category I potential carcinogen if, upon scientific evaluation, the secretary determines that the substance meets the definition of a potential occupational carcinogen in (1) humans, or (2) a single mammalian species in a long-term bioassay in which the results are in concordance with some other scientifically evaluated evidence of a potential carcinogenic hazard, or (3) a single mammalian species in an adequately conducted long-term bioassay, in appropriate circumstances in which the Secretary of the United States Department of Labor determines the requirement for concordance is not necessary. Evidence of concordance is any of the following: positive results from independent testing in the same or other species, positive results in short-term tests, or induction of tumors at injection or implantation.

Category II Potential Carcinogens

A substance shall be identified, classified, and regulated as a category II potential carcinogen if, upon scientific evaluation, the Secretary determines that (1) the substance meets the criteria for category I potential carcinogens as described above, but the evidence is found by the secretary to be only "suggestive"; or (2) the substance meets the criteria for category I potential carcinogens as described above in a single mammalian species without evidence of concordance.

NIOSH Carcinogen Classifications

Acting under the authority of the Occupational Safety and Health Act of 1970 (Public Law 91-596), the NIOSH develops and periodically revises recommended exposure limits (RELs) for hazardous substances or conditions in the workplace. These recommendations are then published and transmitted to OSHA for use in promulgating legal standards. NIOSH may identify numerous chemicals that it believes should be treated as occupational carcinogens even though OSHA has not yet identified them as such. Generally, where OSHA has adopted the NIOSH recommendations as OSHA standards, the OSHA PELs and NIOSH RELs are equal. In cases in which the NIOSH recommendations have not been formally adopted by OSHA, the NIOSH RELs may be different from the OSHA PELs. For example, the NIOSH exposure limit for trichloroethylene (25 ppm) differs from the OSHA exposure limit (50 ppm).

The NIOSH classification scheme is one of the most simple carcinogen classification schemes. The NIOSH scheme combines into one category all carcinogens. Within this single category, NIOSH narratively de-

scribes the site of the cancer and whether the effect was seen in humans or animals. In determining their carcinogenicity, NIOSH uses a classification scheme outlined in 29 CFR 1990.103, which states in part:

> Potential occupational carcinogen means any substance, or combination or mixture of substances, which causes an increased incidence of benign and/or malignant neoplasms, or a substantial decrease in the latency period between exposure and onset of neoplasms in humans or in one or more experimental mammalian species as the result of any oral, respiratory, or dermal exposure, or any other exposure which results in the induction of tumors at a site other than the site of administration. This definition also includes any substance which is metabolized into one or more potential occupational carcinogens by mammals.

The NIOSH thresholds for carcinogens were not designed to be protective of 100% of the population. NIOSH usually recommends that occupational exposures to carcinogens be limited to the lowest feasible concentration. This perhaps is the reason why the NIOSH exposure limit for vinyl chloride is the lowest reliably detectable concentration and the OSHA exposure limit is 1 ppm.

ACGIH Carcinogen Classifications

ACGIH classifies substances associated with industrial processes that are recognized to have carcinogenic or cocarcinogenic potential. In general, the stated classification is intended to provide a practical guideline for the industrial hygiene professional to assist in control of exposures in the workplace. The classification and threshold limit values (TLVs) are not mandated by federal or state regulations, although the ACGIH classifications and values may be considered when standards are adopted by the regulatory agencies. Currently, five categories of carcinogens have been designated by the TLV Committee to recognize the qualitative differences in research results or other data. These five categories are outlined below.

A1: Confirmed human carcinogen—
The agent is carcinogenic to humans based on the weight of evidence from epidemiologic studies of exposed humans, and/or convincing clinical evidence in exposed humans.

A2: Suspected human carcinogen—
The agent is carcinogenic in experimental animals

at dose levels, by route(s) of administration, at site(s), of histologic types(s), or by mechanism(s) that are considered relevant to worker exposure. Available epidemiologic studies are conflicting or insufficient to confirm an increased risk of cancer in exposed humans.

A3: Animal carcinogen—
The agent is carcinogenic in experimental animals at a relatively high dose, by route(s) of administration, at site(s), of histologic types(s), or by mechanism(s) that are not considered relevant to worker exposure. Available epidemiologic studies do not confirm an increased risk of cancer in exposed humans. Available evidence suggests that the agent is not likely to cause cancer in humans except under uncommon or unlikely routes or levels of exposure.

A4: Not classifiable as a human carcinogen—
There are inadequate data on which to classify the agent in terms of its carcinogenicity in humans and/or animals.

A5: Not suspected as a human carcinogen—
The agent is not suspected to be a human carcinogen on the basis of properly conducted epidemiologic studies in humans. These studies have sufficiently long follow-up, reliable exposure histories, sufficiently high dose, and adequate statistical power to conclude that exposure to the agent does not convey a significant risk of cancer to humans. Evidence suggesting a lack of carcinogenicity in experimental animals will be considered if it is supported by other relevant data.

Substances for which no human or experimental animal carcinogenic data have been reported are assigned no carcinogen designation by the ACGIH.

Further Reading

American Conference of Governmental Industrial Hygienists (1991). *Documentation of the Threshold Limit Values and Biological Exposure Indices*, 6th ed. ACGIH, Cincinnati, OH.
Identification, Classification, and Regulation of Potential Occupational Carcinogens (1996). Subtitle B—Regulations Relating to Labor, Chapter XVII—Occupational Safety and Health Administration, Department of Labor, Title 29 Code of Federal Regulations (CFR) 1990. Bureau of National Affairs, Inc., Washington, DC.
NIOSH Pocket Guide to Chemical Hazards (1990, June). United States Department of Health and Human Services, Public Health Service, National Institute for Occupational Safety and Health, Washington, DC.
Seventh Annual Report on Carcinogens (1994 Summary) and September 30, 1996 Update. United States Department of Health and Human Services, Public Health Service, National Toxicology Program, Research Triangle Park, NC. [Prepared for the National Institute of Environmental Health Sciences]
U.S. EPA (1996, April 23). Proposed guidelines for carcinogen risk assessment. *Fed. Reg.* 61(79).

—*A. J. Hoffman-Kiefer and C. F. Wilkinson*

Related Topics

American Conference on Governmental Industrial Hygienists
Carcinogenesis
Dose–Response Relationship
Epidemiology
Federal Insecticide, Fungicide, and Rodenticide Act
International Agency for Research on Cancer
Levels of Effect in Toxicological Assessment
National Institute for Occupational Safety and Health
National Toxicology Program
Occupational Safety and Health Act
Occupational Safety and Health Administration
Risk Assessment, Human Health
Toxicity Testing, Carcinogenesis
Toxic Substances Control Act

Carcinogen–DNA Adduct Formation and DNA Repair

Definition

Carcinogen–DNA adducts are chemical modifications to the genetic material. They are formed in cells when an activated chemical species, usually arising from the metabolism of a xenobiotic or foreign substance, reacts with the DNA. Electrophilic chemical species form covalent addition products (adducts) in DNA because the DNA carries a negative charge. Adducts represent damage to DNA that may be fixed in the genetic code as a mutation if the DNA is incorrectly repaired or if DNA synthesis proceeds in the presence of the adduct.

Role of Adducts in Multistage Carcinogenesis

Carcinogenesis involves multiple steps that can be conceptually divided into four major stages: tumor initiation, tumor promotion, malignant conversion, and tumor progression (invasion and metastasis). The distinction between initiation and promotion was formally defined in a mouse skin chemical carcinogenesis model where the animals were treated topically with a single dose of a chemical (initiator) followed by repeated topical doses of croton oil (promoter). The sequence of molecular and biological events comprising chemical carcinogenesis has been systematically dissected and the mouse skin model has become increasingly refined. It is now recognized that carcinogenesis requires malignant conversion of hyperplastic cells from a benign or preneoplastic state and that invasion and metastasis are manifestations of further genetic and epigenetic changes. The study of this process in humans is necessarily indirect, however, and measures of age-dependent cancer incidence have shown that the rate of tumor development is proportional to the sixth power of time, suggesting that four to six independent steps are necessary.

Tumor initiation occurs as the result of irreversible genetic damage. For mutations to accumulate they must arise in cells that survive or give rise to descendants that survive the lifetime of the organism. A chemical carcinogen causes a mutation by modification of the molecular structure of the DNA. This is most often the consequence of carcinogen–DNA adduct formation. In general, a positive correlation is found between the level of carcinogen–DNA adducts that can be detected in model systems and the resulting number of tumors that develop. Thus, chemically induced tumors rarely develop in tissues that do not form carcinogen–DNA adducts. Carcinogen–DNA adduct formation is central

to theories of chemical carcinogenesis and is considered to be a necessary, but not a sufficient, prerequisite for tumor formation.

The chemical etiology of occupationally induced skin cancers was recognized as long ago as the eighteenth century. However, it was not until the 1930s that the first chemically pure carcinogens were isolated from coal–tar pitch. These chemicals were identified as polycyclic aromatic hydrocarbons, which are composed of variable numbers of fused benzene rings. Polycyclic aromatic hydrocarbons are formed in the incomplete combustion of fossil fuels and vegetable matter (including cooked foods and tobacco) and are common environmental contaminants. The polycyclic aromatic hydrocarbons are chemically unreactive, and it was not until 1951 that it was shown that enzyme-catalyzed metabolism of these compounds was necessary before they could bind covalently to cellular macromolecules to form adducts.

Metabolic Activation of Procarcinogens

Polycyclic aromatic hydrocarbons are activated in a multistep process (Fig. C-2) involving initial epoxidation, hydration of the epoxide, and subsequent epoxidation across the olefinic bond to form the ultimate carcinogenic metabolite, a vicinal diol-epoxide. The first step in this process is principally driven by a cytochrome P450, and the resulting simple arene oxide is further metabolized to a dihydrodiol by epoxide hydrolase. The second oxidation step at the site of the olefinic double bond is again catalyzed by cytochrome P450.

Metabolic activation of benzo(a)pyrene (the most extensively studied prototypical polycyclic aromatic hydrocarbon) leads to the formation of the bay region diol-epoxide (Fig. C-2, product of step 3). There are four possible stereoisomers; the reactivity of each enantiomer is variable and the enantiomeric composition of the product of metabolism is disproportionate. Biological response to the different enantiomers in mammalian systems suggests that the (+) *anti* forms, e.g., *r*-7, *t*-8-dihydroxy-*t*-9,10-epoxy-7,8,9,10-tetrahydrobenzo(a)pyrene (benzo(a)pyrene-7,8-diol 9,10-oxide), are the most active mutagens and carcinogens and the (−) *syn* forms are the least active.

The arene ring of benzo(a)pyrene-7,8-diol 9,10-oxide opens spontaneously at the 10 position giving a

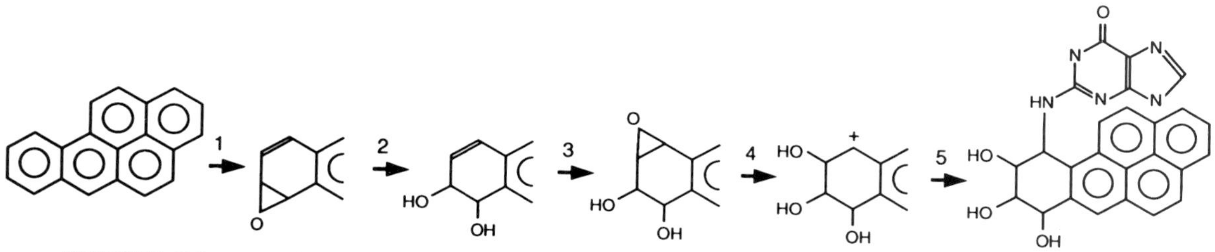

FIGURE C-2. Metabolic activation of benzo[a]pyrene (a representative polycyclic aromatic hydrocarbon). The parent hydrocarbons are chemically inert and require metabolic activation before they can exert their biological effects. Cytochrome P450 enzymes (principally CYP1A1) catalyze the formation of simple arene oxides from the parent hydrocarbons (1). The arene oxides are converted to dihydrodiols by the action of epoxide hydrolase (2). The resulting dihydrodiols are further oxidized by cytochrome P450 enzymes (principally CYP3A4) at the site of the olefinic double bond (3). Vicinal diol epoxides are highly unstable and the arene ring opening is spontaneous yielding a highly reactive carbocation (4). The eletrophilic carbocationic species can form a covalent bond with the exocyclic amino group of deoxyguanosine (5). The resulting polycyclic aromatic hydrocarbon–DNA adduct lies in the minor groove of the DNA double helix.

highly reactive carbonium ion (Fig. C-2, product of step 4) that can form a covalent addition product (adduct) with cellular macromolecules, including DNA. These adducts, for example the benzo(a)pyrene-7,8-diol 9,10-oxide binding covalently to the exocyclic amino group (N2) of guanine (Fig. C-2, product of step 5), cause the DNA to be damaged either by their persistence and consequent interference with replication or induction of mutations during DNA repair. The same basic tenet holds true for carcinogen–DNA adducts of other chemical classes that may be activated by different metabolic pathways.

Aromatic amines were first associated with increased bladder cancer among dye workers in 1895. A principal aromatic amine thought to be responsible for bladder cancer incidence among workers in the rubber industry is 4-aminobiphenyl. This compound and many related compounds are components of cigarette smoke, diesel exhaust, and the pyrolysis of certain foods. In addition, nitrated polycyclic aromatic hydrocarbons are also environmental contaminants resulting from incomplete combustion of vegetable matter and diesel fuel and are related to aromatic amines by nitroreduction.

The metabolic activation of aromatic amines is complex (Fig. C-3). They can be converted to an aromatic amide which is catalyzed by an acetyl coenzyme A-dependent N-acyltransferase (Fig. C-3, NAT). The acetylation phenotype varies among the population and persons with the rapid acetylator phenotype are at higher risk of colon cancer, whereas slow acetylators

are at risk of bladder cancer. In an alternate metabolic pathway catalyzed by cytochrome P4501A2 (Fig. C-3, CYP1A2), N-oxidation competes with the N-acyltransferase for aromatic amine metabolism, and the N-hydroxylation products when protonated form reactive electrophiles that bind covalently with DNA to produce deacetylated C8 and N2 guanosine adducts. In addition, the N-hydroxylated compounds may become glucuronidated and bind to protein. Acetylated N-hydroxy arylamines can be further activated by either an acetyl coenzyme A-dependent O-acetylase (Fig. C-3, AT) or a 3'-phosphoadenosine-5'-phosphosulfate-dependent O-sulfotransferase (Fig. C-3, ST) to N-arylhydroxamic acids that require further activation. The predominant pathway for this is through acetyltransferase-catalyzed rearrangement to a reactive N-acetoxy arylamine that produces the same deacetylated DNA adducts as previously mentioned. Sulfotransferase catalysis results in the formation of N-sulfonyloxy arylamides and ultimately results in the formation of C8 and N2 acetylated adducts. The deacetylated (or nonacetylated) adducts are by far the predominating species *in vivo*, due primarily to the absence of appropriate sulfotransferases in most target tissues.

Heterocyclic amines are formed during the preparation of cooked food, primarily from the pyrolysis (greater than 150°C) of amino acids, creatinine, and glucose. They have been recognized recently as food mutagens and have been shown to form adducts and cause liver tumors in primates. In comparison to other

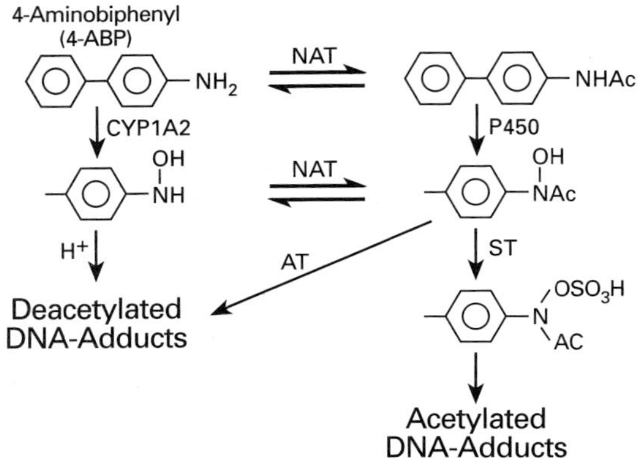

FIGURE C-3. Metabolic activation of 4-aminobiphenyl (a representative aromatic amine). The amino group of the parent aromatic amine is a substrate for competing metabolic pathways catalyzed initially by either N-acyltransferase (NAT) or cytochrome P450 (CYP1A2). The N-acyltransferase pathway results in the formation of an N-arylhydroxamic acid, whereas CYP1A2 oxidation results in the formation of an N-hydroxyaryl amine. Although covalent bonding to DNA of N-hydroxyaryl amines can be acid catalyzed (H+), they can also be further metabolized by either sulfotranserases (ST) or acyltransferases (AT). The resulting metabolites form a variety of acetylated and non- or deacetylated adducts, the most abundant being at the C8 position of deoxyguanosine. This adduct resides in the major groove of the double helix. The hydroxylamine can also form a glucuronide conjugate which may be of importance in bladder carcinogenesis; a postulated mechanism is metabolism of the aromatic amine in the liver, transport through the vascular system as the glucuronide, and hydrolysis to produce a DNA-binding species in the acidic environment of the urinary bladder.

carcinogens their metabolism is less well understood, but N-hydroxylation is considered to be a necessary step. Since they are similar in structure to the aromatic amines they are activated by cytochrome P450. The N-hydroxy metabolites of 3-amino-1-methyl-5H-pyrido[4,3-b]indole, 2-amino-6-methyldipyrido[1,2-a:3′,2′-d]imidazole, 2-amino-3-methyl-imidazo-[4,5-f]quinoline, and 2-amino-1-methyl-6-phenylimidazo-[4,5b]-pyridine, a glutamic acid pyrolysate, can react directly with DNA. However, unlike the aromatic amines this reaction is not facilitated by acid pH. Enzymic O-esterification of N-hydroxy metabolites is important in the activation of these food mutagens and the N-hydroxy metabolites are also good substrates for transacetylases. This suggests a possible etiological role

for these chemicals in colorectal cancer, in combination with the rapid acetylator phenotype.

Aflatoxins (aflatoxin B_1, B_2, G_1, and G_2) are fungal mutagens that are formed primarily as metabolites of *Aspergillus flavus*, and they contaminate cereals, grain, and nuts. A positive correlation exists between dietary aflatoxin exposure and liver cancer incidence in developing countries where grain spoilage is high. It should also be noted that a similar correlation exists between hepatitis B virus infection and incidence of hepatocellular carcinoma; chemical–viral interactive effects have been proposed. Aflatoxins are activated by cytochrome P450 enzymes. Aflatoxins B_1 and G_1 have an olefinic double bond and are more mutagenic and carcinogenic than aflatoxin B_2 and G_2,which are saturated and have an ethylenic bond at this position. This, and analysis of the DNA adducts, implies that the olefinic bond is the site of activation (Fig. C-4).

Carcinogenic *N*-nitrosamines are ubiquitous environmental contaminants and can be found in food, alcoholic beverages, cosmetics, cutting oils, hydraulic fluid, rubber, and tobacco. Tobacco-specific *N*-nitrosamines, for example, 4-(methylnitrosoamino)-1-(3-pyridyl)-1-butanone, are carcinogenic in a wide range of animal species and may account for the tumorigenic nature of snuff and chewing tobacco. Endogenous nitrosation can also occur as the result of the reaction of an amine with nitrate alone or nitrite in the presence of acid. Thus, nitrite (used in curing meats) and L-cysteine in the presence of acetaldehyde (a metabolite of alcohol) form *N*-nitrosothiazolidine-4-carboxylic acid. Although some nitrosamines can react with DNA directly, activation by cytochrome P450 enzymes also occurs.

N-nitrosodimethylamine undergoes α-hydroxylation to form an unstable α-hydroxynitrosamine. The breakdown products are formaldehyde and methyl diazohydroxide. The alkyl groups of compounds like methyl diazohydroxide are good leaving groups and so methyl diazohydroxide is a powerful methylating agent that can add a small functional group (small alkyl adduct, as opposed to the bulky aryl adducts formed by the polycyclic hydrocarbons, aromatic and heterocyclic amines, and fungal micotoxins) at more than 10 different sites in DNA. The tobacco-specific nitrosamines are not symmetrical and can also form bulky adducts. The metabolism of 4-(methylnitrosoamino)-1-(3-pyridyl)-1-butanone gives rise to either a positively charged

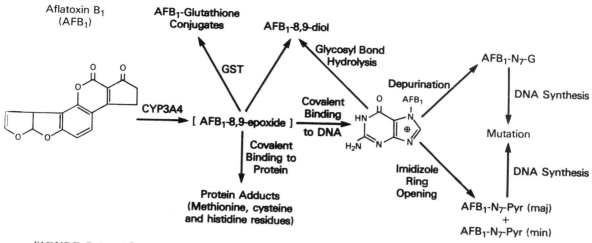

FIGURE C-4. *Aflatoxin B_1 is a fungal mycotoxin and potent human liver carcinogen. It is activated through cytochrome P450 enzyme oxidation (CYP2A3 and CYP3A4) at the olefinic bond in the 8,9 position. The resulting epoxide is highly unstable and reacts covalently at the N7 position of deoxyguanosine. Depurination of the adduct results in the formation of a promutagenic apurinic site. DNA sequencing has shown that liver tumors of individuals living in areas of Gambia and China, where fungal contamination (Aspergilus flavus) of food is extreme, have mutations that are consistent with previous genotoxic exposure (adduct formation) to aflatoxin, whereas liver tumors of individuals living in Japan, where hepatitis B virus has been identified to be the major etiological agent, generally do not.*

pyridyl-oxobutyl ion or a positively charged methyl ion, both of which are able to alkylate DNA.

Types of Carcinogen–DNA Adducts Formed in Biological Systems

There are a number of ways in which the chemical structure of DNA can be altered by a carcinogen (Fig. C-5). These include the formation of bulky aromatic adducts from (Fig. C-5a) polycyclic aromatic hydrocarbons, (Figs. C-5b and C-5c) aromatic amines, (Fig. C-5d) heterocyclic amines, (Fig. C-5e) fungal mycotoxins and other large compounds, (Figs. C-5f and C-5g) alkylation (addition of methyl and ethyl groups) by nitrosamines and other small-molecular-weight alkylating agents, and (Fig. C-5h) formation of 8-hydroxydeoxyguanosine and other adducts by oxidation. Other possibilities are dimerization and deamination (not shown in Fig. C-5). Chemical carcinogens can also cause epigenetic changes, for example, alteration in DNA methylation status. By far, the major portion of DNA damage induced by chemical carcinogens occurs on deoxyguanosine residues, and this is likely due

to the electrophilic nature of that compound. However, adducts are also formed on each of the other bases.

The presence of an adduct per se does not necessarily result in adverse biological effects. For example, mutagenesis and tumorigenesis studies have shown that specific carcinogen–DNA adducts vary in their promutagenic and tumorigenic potential. This is particularly well exemplified for alkyl adducts since the 7-alkyl-deoxyguanosine adducts are not generally as mutagenic as the 0^6-alkyl-deoxyguanosine adducts (Figs. C-5f and C-5g). In addition, the (7R)-N2-(10-{7β,8α,9α-trihydroxy-7,8,9,10-tetrahydro-benzo(a)pyrene}-yl)-deoxyguanosine in *anti* conformation (Fig. C-5a) is the adduct formed through metabolic activation of benzo(a)pyrene that is primarily associated with mutagenesis and carcinogenesis in mammalian systems. Adducts in *syn* conformation are also formed but are less mutagenic and tumorigenic.

The known range of aromatic amine adducts is more diverse than for polycyclic aromatic hydrocarbons, not only because aromatic amine DNA adducts have acetylated and nonacetylated (or deacetylated) metabolic intermediates (Figs. C-5b and C-5c) but also because they form covalent bonds at the C8, N2 and

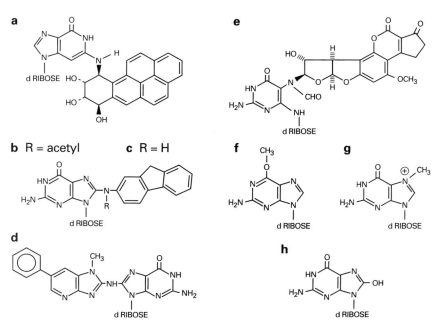

FIGURE C-5. *Molecular structures of carcinogen–adducts of deoxyguanosine. (a) (7R)-N^2-(10-[7β,8α,9α-trihydroxy-7,8,9,10-tetrahydro-benzo(a) pyrene]yl)-deoxyguanosine, formed when benzo(a)pyrene-7,8-diol 9,10-epoxide reacts with the exocyclic amino group of deoxyguanosine; (b) N-(deoxyguanosin-8-yl)-2-(acetylamino)fluorene, formed when N-hydroxylacetylaminofluorene reacts with the C8 position of the imidazole ring; (c) N-(deoxyguanosin-8-yl)-2-(amino)fluorene, formed when N-hydroxylaminofluorene reacts with the C8 position of the pyrimidine ring structure; (d) N-(deoxyguanosin-8-yl)-2-amino-1-methyl-6-phenylimidazo-[4,5b]-pyridine, formed when the N-hydroxylamine metabolite of 2-amino-1-methyl-6-phenylimidazo-[4,5b]-pyridine (PHIP), a glutamic acid pyrolysate, reacts with deoxyguanosine; (e) ring open form of N-(deoxyguanosin-7-yl)-9-hydroxyaflatoxin B$_1$, formed following the reaction of the 8,9-epoxide metabolite of aflatoxin B$_1$ at the N7 position of deoxyguanosine; (f) O^6-methyldeoxyguanosine, formed when an alkyl radical (CH$_3$$^+$), derived from an alkylating agent, reacts at the O6 position of deoxyguanosine; (g) N7-methyldeoxyguanosine, formed when an alkyl radical (CH$_3$$^+$), derived from an alkylating agent, reacts at the N7 position of deoxyguanosine; (h) 8-hydroxydeoxyguanosine, formed through exogenous or endogenous oxyradical damage (H$_2$O$_2$, ·OH, and O$_2$$^-$) at the C8 position of deoxyguanosine.*

sometimes O6 positions of deoxyguanosine as well as deoxyadenosine. However, the major adducts are C8-deoxyguanosine adducts, which reside predominantly in the major groove of DNA.

Although the evidence for activation of aflatoxins B$_1$ and G$_1$ through hydroxylation of the olefinic 8,9-position is circumstantial, the structures of the adducts are known. They are formed at the N7 position of deoxyguanosine. These adducts are relatively unstable and have a half-life of about 50 hr at neutral pH, with resulting depurination. The aflatoxin B$_1$–N7-

deoxyguanosine adduct can also undergo ring opening (Fig. C-5e) to yield two pyrimidine adducts, or alternately aflatoxin B$_1$–8,9-dihydrodiol could result. This latter possibility could restore the molecular structure of the DNA if hydrolysis of the original adduct occurs, but a potentially promutagenic lesion would result if formation of the 8,9-dihydrodiol is the result of degradation of ring-open adduct forms.

Alkylation of DNA can occur at many sites either following the metabolic activation of certain *N*-nitrosamines or directly by the action of the *N*-

alkylureas (*N*-methyl-*N*-nitrosourea) or the *N*-nitrosoguanidines. The protonated alkyl functional groups that become available to form lesions in DNA generally attack the following nucleophilic centers: adenine (N1, N3, and N7); cytosine (N3); guanine (N2, O6, and N7), and thymine (O2, N3, and O4). Some of these lesions are known to be repaired (O6-methyldeoxyguanosine), while others are not (N7-methyldeoxyguanosine). Furthermore, O6-methyldeoxyguanosine is a promutagenic lesion, whereas N7-methyldeoxyguanosine is not.

Oxyradical damage can result in the modification of DNA to form thymine glycol or 8-hydroxydeoxyguanosine adducts (Fig. C-5h). Three major pathways have been identified. Exposure to organic peroxides (catechol, hydroquinone, and 4-nitroquinoline-*N*-oxide) leads to this type of oxyradical damage. However, oxyradicals and hydrogen peroxide can be generated in the catalytic cycling of some enzymes. Cells can also be stimulated to produce peroxisomes by treatment with certain drugs and plasticizers. Exposure to tumor promoters can indirectly increase oxyradical formation; and perhaps the best known relationship is that between the phorbol esters and inflammatory cells. In this system, mediated through protein kinase C and the subsequent activation of a membrane-localized pyridine nucleotide-dependent oxidase, oxyradical formation is highly correlated with the relative potencies of the different phorbol esters. Correspondingly, promoters that do not stimulate the protein kinase C signal transduction cascade do not affect oxyradical production.

Biological Repair of Adduct Damage in DNA

DNA repair enzymes act at sites of DNA damage caused by chemical carcinogens, and five major mechanisms are known (Fig. C-6):

- Direct DNA repair
- Nucleotide excision repair
- Base excision repair
- Mismatch repair
- Post-replication repair (the recombinational mechanism is shown)

Direct DNA repair relies on the action of suicide enzymes (alkyltransferases). In the absence of DNA strand scission, these enzymes catalyze the translocation of the alkyl moiety from an alkylated base (e.g., O6-methyldeoxyguanosine) to a cysteine residue at the active site of the enzyme. Thus, one molecule of the enzyme is capable of repairing one alkyl lesion in DNA. DNA–nucleotide excision repair and base excision repair mechanisms are similar. For DNA–nucleotide excision repair, preincision recognition of the lesion is required by an endonuclease, for example, the UvrABC enzyme complex. A segment of the DNA strand, up to 100 bases, harboring an adduct is then clipped out. A patch is then synthesized in the 5′–3′ direction using the intact strand as a template and ligated to the parental DNA at its 3′ terminus. Large adducts are recognized by endonuclease complexes because of the distortions that they effect in the DNA. Base excision repair also removes a segment of DNA containing an adduct. Removal of the adducted base is brought about by a glycosylase and repair of the damaged strand is accomplished by the combined action of an exonuclease that degrades a few bases on the damaged strand and a polymerase that synthesizes a "patch" in the 5′–3′ direction using the undamaged strand as a template. The patch is then ligated as in nucleotide excision repair. Generally, small alkyl adducts, such as 3-methyladenine, are repaired in this way.

DNA mismatches occasionally occur as the result of excision repair processes. This is the result of incorporation of unmodified or conventional, but noncomplementary, Watson–Crick bases opposite each other in the DNA helix. Transition mispairs (G–T or A–C) are repaired by the mismatch repair process more efficiently than transversion mispairs (G–G, A–A, G–A, C–C, C–T, and T–T), probably due to differential recognition of the mispairings. Repair efficiency of mispairings is also dependent on their oligonucleotide environment for the same reason. Thus, mispairings in G–C-rich regions are repaired more efficiently than those in A–T-rich regions. The mechanism for correction of mispairings is essentially the same as that for excision resynthesis repair described previously, but generally involves the excision of large pieces of the DNA containing mispairings. The mismatch recognition protein simultaneously binds to unmeththylated adenine in the GATC recognition sequence and the mismatch and removes the whole intervening DNA

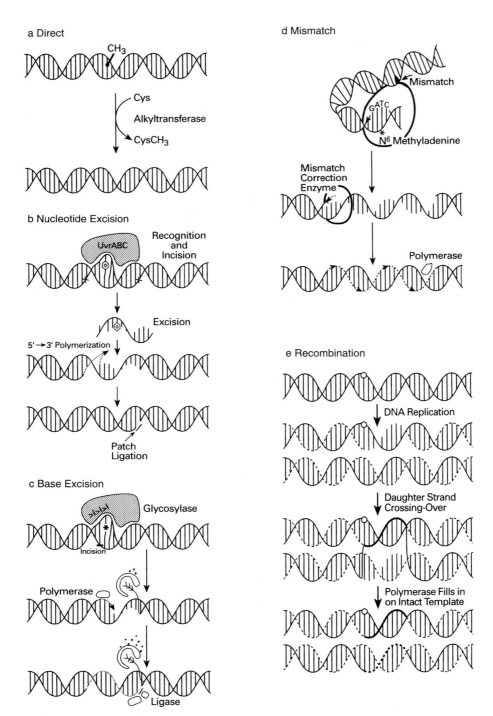

FIGURE C-6. *Mechanisms of DNA repair: (a) direct, catalyzed by an alkyltransferase, does not require strand scission; (b) nucleotide excision, catalyzed by an endonuclease complex, e.g., UvrABC; (c) base excision, catalyzed by a glycosylase and an apurinic endonuclease; (d) mismatch, catalyzed by a mismatch correction enzyme, e.g., hMLH1 and hMSH2; (e) postreplication repair through a recombination event (daughter strand gap repair).*

sequence. The parental template strand is then used by the polymerase to fill the gap.

Postreplication repair occurs by the action of a polymerase or a recombinational mechanism in response to replication of DNA on a damaged template. When the DNA polymerase reaches a replication fork in the presence of damage to the parental strand, the polymerase either stops at the lesion or proceeds past the lesion leaving a gap in the newly synthesized strand. The gap can be filled either by recombination of the daughter strand with the homologous parent strand that is mediated by the RecA protein, or, in cases in which single nucleotide gaps are left, mammalian DNA polymerases can insert an adenine residue. These mechanisms can therefore lead to recombinational events as well as base mispairing.

The rate but not the fidelity of DNA repair can be measured by adduct removal or unscheduled DNA synthesis. Substantial interindividual variations have been found in these rates. Ultraviolet light induces dimerization of thymidine residues in DNA which can be removed by excision repair, but markedly reduced rates of excision repair are found in individuals with *Xeroderma pigmentosum*. These individuals are at known risk of ultraviolet light-associated skin cancer. However, among the general population approximately 5-fold variation in rates of excision repair are found in lymphocytes treated with carcinogens *in vitro*. An association has also been found between the reduced capacity of mononuclear leukocytes *in vitro* to repair aromatic amine adducts in individuals who have first-degree relatives with cancer. Up to 40-fold variations have been reported among humans in the activity of O6-alkylguanine–DNA alkyltransferase. DNA repair rates are inhibited by aldehydes, alkylating agents, and some chemotherapeutic drugs. Decreased DNA repair capacity has also been noted in the fibroblasts of lung cancer patients compared to those of melanoma patients or noncancer controls. For benzo(*a*)pyrene-7,8-diol 9,10-epoxide–DNA adducts, a unimodal distribution of DNA repair rates has been observed in lymphocytes, but interindividual variation has been found to be substantial.

Carcinogen–DNA Adducts as Human Exposure Dosimeters

The field of human DNA adduct biomonitoring has been made possible by methods that have been devel-

oped for sensitive detection of carcinogen–DNA adducts, including immunoassays, immunohistochemistry, ^{32}P-postlabeling, fluorescence, phosphorescence spectroscopy, gas chromatography/mass spectrometry, atomic absorbance spectrometry, and electrochemical conductance. Each of these methods has its advantages and limitations. When used without preparative procedures, these techniques are generally not able to provide either quantitation of individual adducts or chemical characterization of a specific adduct. This is because humans are exposed to complex mixtures of chemical carcinogens resulting in the presence of multiple adducts in human DNA samples. An important aspect of recent approaches to human DNA adduct biomonitoring is the development of preparative strategies for sample purification that can be applied prior to the ultimate adduct quantitation to allow chemical characterization of specific DNA adducts in human tissues. Recent advances combining preparative methods (immunoaffinity chromatography, high-performance liquid chromatography, or other chromatography) with immunoassays, ^{32}P-postlabeling, synchronous fluorescence spectrometry, and gas chromatography/mass spectrometry have allowed identification and quantitation of specific DNA adducts in human tissues, potentially resulting in more precise exposure documentation in human tissues.

The determination of human DNA adduct levels, which reflect the biologically effective dose of a chemical carcinogen, integrates multiple complex genotoxic exposure variables. These include extent and frequency of chemical exposure, xenobiotic metabolism (a balance between carcinogen activation and detoxication), rates of formation of ultimate metabolites and covalent binding of ultimate metabolites to DNA, and DNA repair. Since virtually all of the enzyme systems responsible for the metabolic activation and detoxication of chemical carcinogens are polymorphic, the propensity of an individual to form carcinogen–DNA adducts in the presence of exposure is likely to vary greatly.

The majority of studies designed to monitor DNA adducts in human tissues fall into the category of exposure documentation and have concentrated on environmental and occupational exposures to agents such as polycyclic aromatic hydrocarbons, aflatoxins, 4-aminobiphenyl, styrene, nitrosamines, nitrates, and ultraviolet light. In addition, dosimetry of human DNA adduct formation has been established precisely with medicinal exposures, including cisplatin, procarbazine, dacarbazine, and 8-methoxypsoralen, as well as dietary exposure to aflatoxins. A major goal of studies in this field is the correlation of DNA adduct formation with human cancer risk assessment, and this has been accomplished in one prospective case-control study. Briefly, samples of urine and blood were obtained from more than 18,000 men in Shanghai, China, where the rate of hepatocellular carcinoma and hepatitis is very high. Within several years, 50 cancer cases were observed and these were matched to 267 controls for age, sex, and neighborhood. The blood was assayed for evidence of hepatitis virus positivity and the urine assayed for evidence of DNA adduct excretion. The final analysis showed an increased relative risk for liver cancer of 9-fold among individuals excreting DNA adducts and 60-fold among individuals positive for both biomarkers. Following the lead of this pioneering study, future human DNA adduct biomonitoring will focus on the continuing validation of DNA adduct measurements as a basis for human cancer risk assessment.

Further Reading

Beland, F. A., and Poirier, M. C. (1989). DNA adducts and carcinogenesis. In *Pathobiology of Neoplasia* (A. E. Sirica, Ed.), pp. 57–80. Plenum, New York.

Bohr, V. A., Evans, M. K., and Fornace, A. J. (1989). Biology of disease. DNA repair and its pathogenic implications. *Lab. Invest. 61*, 143–161.

Gorelick, N. J. (1993). Application of HPLC in the 32P-postlabeling assay. *Mutat. Res. 288*, 5–18.

Harris, C. C. (1991). Chemical and physical carcinogenesis. *Cancer Res. 51*, 5023s–5044s.

Poirier, M. C. (1993). Antisera specific for carcinogen–DNA adducts and carcinogen modified DNA: Application for detection of xenobiotics in biological samples. *Mutat. Res. 288*, 31–38.

Qian, G. S., Ross, R. K., Yu, M. C., Yuan, J. M., Gao, Y. T., Henderson, B. E., Wogan, G. N., and Groopman, J. D. (1994). A follow-up study of urinary markers of aflatoxin exposure and liver cancer risk in Shanghai, People's Republic of China. *Cancer Epidemiol. Biomarkers Prevention 3*, 3–10.

Singer, B., and Grunberger, D. (1983). *Molecular Biology of Mutagens and Carcinogens*. Plenum, New York.

Weston, A. (1993). Physical methods for the detection of carcinogen–DNA adducts in humans. *Mutat. Res. 288*, 19–29.

Yuspa, S. H., and Poirier, M. C. (1988). Chemical carcinogenesis: From animal models to molecular models in one decade. *Adv. Cancer Res. 50*, 25–70.

—*Ainsley Weston and Miriam C. Poirier*

Related Topics

Analytical Toxicology
Carcinogenesis
Cell Proliferation
Chromosome Aberrations
Developmental Toxicology
Dominant Lethal Tests
Host-Mediated Assay
Mechanisms of Toxicity
Molecular Toxicology
Mutagenesis
Mycotoxins
Nitrosamines
Polycyclic Aromatic Hydrocarbons
Sister Chromatid Exchange
Toxicity Testing

Carcinogenesis

Cancer, or neoplasia, which occurs in one of every four individuals and results in the death of one of every five individuals in the United States, is a complex disease with multiple causes. Many intrinsic and extrinsic factors influence the development of cancer. Intrinsic or host factors include age, sex, genetic constitution, immune system function, metabolism, hormone levels, and nutritional status. Extrinsic factors include substances eaten, drunk, or smoked; workplace and environmental (air, water, and soil) exposures; natural and medical radiation exposure; sexual behavior; and elements of lifestyle such as social and cultural environment, personal behavior, and habits. Intrinsic and extrinsic factors can interact with one another to influence the development of cancer. Because of the physical and emotional suffering associated with cancer and the immense cost to the nation in lost production and income and medical and research expenditures, considerable effort continues to be exerted to understand this complex disease so that strategies can be developed to decrease or prevent its occurrence. Current regulatory guidelines have been crafted to reduce the probability of developing cancer by lowering human exposure to agents identified as potentially capable of causing cancer.

Nomenclature of Cancer (Neoplasia)

Nomenclature associated with the study of cancer is frequently confusing because a given term often has a relatively narrow as well as a considerably broader definition based on common usage. Carcinogenesis, for example, is narrowly defined as the production of carcinoma but is more commonly used in the broadest possible sense to indicate generation of neoplasms which are new and typically abnormal growths, generally uncontrolled and becoming progressively more serious with time. Neoplasia, synonymous with carcinogenesis, refers to the process of development of neoplasms. Two important terms which relate to the clinical behavior and growth characteristics of neoplasms are (1) benign and (2) malignant, characteristic features of which are listed in Table C-2. Basically, benign neoplasms are slow-growing, localized growths frequently amenable to surgical removal with a low probability of recurrence. Malignant neoplasms have a more aggressive growth, are locally invasive, sometimes metastasize (spread to distant sites), and are difficult to remove surgically.

Two terms that have both a narrow and a broad definition are (1) tumor and (2) cancer. A tumor may be defined as and equated to a benign neoplasm but more broadly refers to any tissue enlargement or swelling. A cancer generally refers to a malignant neoplasm. Unfortunately, these two terms are frequently used interchangeably by the layperson and the professional alike. Thus, if it is said that an individual has a tumor, that individual may have a benign neoplasm (most often the case) but could have a malignant neoplasm if the term "tumor" is being used loosely. If an individual is said to have a cancer, that usually means the individual has a malignant neoplasm but, here again, loose use of the term "cancer" might include any neoplasm, including a benign one. Scientists contribute to the confusion by sometimes indicating that an agent may cause cancer, meaning either benign or malignant neoplasia. Alternatively, they may indicate that an agent is tumorigenic, which could mean that it causes tumors but frequently means that it may also cause malignant neoplasms (cancers). Common and uncritical usage of these terms is so ingrained that attempts to standardize nomenclature have been largely unsuccessful. The least ambiguous terms are "benign neoplasm" and "malignant neoplasm."

Most neoplasms are classified and named based on (1) the cell or tissue of origin and (2) benign or malignant growth characteristics. There are two basic cell types from which neoplasms may originate: mesenchymal cells and epithelial cells (Fig. C-7). Mesenchymal pertains to mesenchyma (embryonic connective tissue

TABLE C-2
Comparative Features of Benign and Malignant Neoplasms

Effect	Benign	Malignant
General effect on host	Little; not generally lethal	Will usually kill the host if not treated
Injury to host	Usually negligible but may compress or obstruct vital tissue	Can kill the host by destruction of vital tissue
Growth rate	Slow	Rapid (but slower than tissue repair); growth escapes normal control mechanisms
Extent of growth	Encapsulated; remains localized at site of origin	Infiltrates or invades and spreads to distant sites
Mode of growth	Typically grows by expansion and displaces surrounding tissues	Invades and destroys surrounding tissues
Microscopic features	Cells and structures formed by cells resemble normal tissues; may be encapsulated	Anaplastic, dysplastic, and pleomorphic; may be associated with hemorrhage, necrosis, and inflammation
Cytologic features	Mitoses rare; nucleus normal in staining and shape; nucleolus not conspicuous	Mitoses may be numerous and abnormal; nucleus often enlarged, irregular in shape, and hyperchromatic; nucleolus hyperchromatic and enlarged
Radiation sensitivity	Radiation sensitivity similar to that of normal tissues; rarely treated with radiation	Radiation sensitivity increased in approximate proportion to the degree of malignancy; frequently treated with radiation

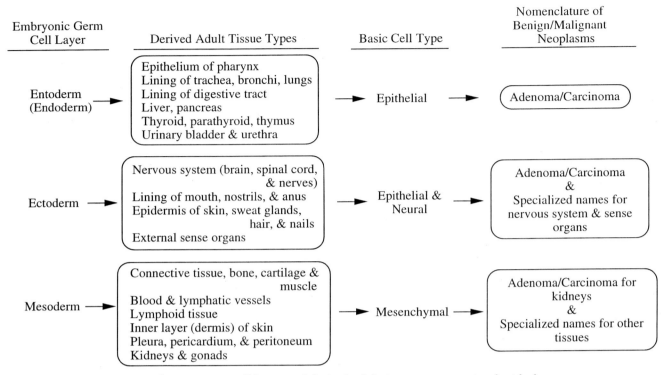

FIGURE C-7.   *Embryonic germ cell layers and derived adult tissue types associated with the two basic cell types: epithelial and mesenchymal.*

in the mesoderm) from which adult tissues such as connective tissue, blood and lymphatic vessels, and muscles and bones are formed. Epithelial cells line the internal and external surfaces of the body and form many of the major organs such as liver and lungs. Epithelial tissues are derived from the embryonic germ layers referred to as entoderm and ectoderm.

There are general guidelines used in naming neoplasms. A benign epithelial neoplasm originating within a glandular tissue is called an "adenoma," having the prefix "adeno" to designate that the origin is one of several glandular tissues and the suffix "oma" to indicate a swelling or tissue enlargement. One or more qualifiers may be added to the name to indicate the tissue of origin and various morphological features as in hepatocellular (liver cell) adenoma, thyroid follicular (forming follicles) adenoma, or renal (kidney) tubular cell adenoma. An adenoma with morphological features resembling finger-like or warty projections would be called a papillary adenoma; with cystic spaces, a cystadenoma; with both of these features, a papillary cystadenoma. Benign mesenchymal neoplasms also utilize the "oma" suffix in their name, as

in meningioma, hemangioma, and fibroma. The prefix for mesenchymal neoplasms usually identifies the specific tissue of origin such as meninges (meningioma), blood vessels (hemangioma), or fibrous connective tissue (fibroma). Nomenclature for several benign neoplasms is presented in Table C-3.

Malignant epithelial neoplasms are typically called "carcinomas" and qualified by histogenetic origin. Thus, malignant skin neoplasms are called epidermal carcinomas if they arise in the superficial layers or epidermis of the skin. If they are composed predominantly of squamous cells, they are called squamous cell carcinomas; if chiefly basal cells, basal cell carcinomas. Malignant mesenchymal neoplasms are called "sarcomas." Examples of the latter include fibrosarcoma, a malignant neoplasm of connective tissue; osteosarcoma, a malignant bone neoplasm; and leiomyosarcoma, a malignant neoplasm of smooth muscle tissue. Nomenclature for several malignant neoplasms is presented in Table C-3.

Much of the general confusion surrounding the nomenclature of neoplasms results from numerous exceptions and permutations in the general histogenetic and

TABLE C-3
Selected Nomenclature of Neoplasia

Tissue	Benign neoplasia	Malignant neoplasia
Epithelium		
Squamous	Squamous cell papilloma	Squamous cell carcinoma
Transitional	Transitional cell papilloma	Transitional cell carcinoma
Glandular		
Liver cell	Hepatocellular adenoma	Hepatocellular carcinoma
Islet cell	Islet cell adenoma	Islet cell adenocarcinoma
Connective tissue		
Adult fibrous	Fibroma	Fibrosarcoma
Embryonic fibrous	Myxoma	Myxosarcoma
Cartilage	Chondroma	Chondrosarcoma
Bone	Osteoma	Osteosarcoma
Fat	Lipoma	Liposarcoma
Muscle		
Smooth muscle	Leiomyoma	Leiomyosarcoma
Skeletal muscle	Rhabdomyoma	Rhabdomyosarcoma
Cardiac muscle	Rhabdomyoma	Rhabdomyosarcoma
Endothelium		
Lymph vessels	Lymphangioma	Lymphangiosarcoma
Blood vessels	Hemangioma	Hemangiosarcoma
Lymphoreticular		
Thymus	Not recognized	Thymoma
Lymph nodes	Not recognized	Lymphosarcoma (malignant lymphoma)
Hematopoietic		
Bone marrow	Not recognized	Leukemia
Neural tissue		
Nerve sheath	Neurilemmoma	Neurogenic sarcoma
Glioma	Glioma	Malignant glioma
Astrocytes	Astrocytoma	Malignant astrocytomas

Note. -oma, swelling; sarc-, malignant neoplasm of mesenchymal origin; carcin-, malignant neoplasm of epithelial origin.

clinical guidelines for naming neoplasms. Many of these exceptions are deeply ingrained in traditional pathology practice, and attempts at standardization have been largely unsuccessful. Examples are thymoma, lymphoma, melanoma, and neuroblastoma—neoplasms which are generally regarded as malignant despite their benign-sounding names and should more properly be called malignant thymoma or thymic sarcoma, malignant lymphoma or lymphosarcoma, malignant melanoma or melanosarcoma, and malignant neuroblastoma, respectively. Other neoplasms are named for their physical attributes such as pheochromocytoma (dark-colored neoplasms typically arising in the adrenal medulla). In addition, some neoplasms were originally named for the person first describing the lesion, and examples such as Hodgkin's disease of lymphoid tissue and Wilms' kidney tumor have persisted to this day. Neoplasms composed of mixtures of cells are named

accordingly; examples include fibroadenoma, adenosquamous carcinoma, and carcinosarcoma. To complicate matters further there are several tissue alterations that are not neoplasms but have names suggesting that they are: hamartomas (a disorganized aggregate of normal tissue components representing faulty differentiation during embryonic development) and choristomas (focal collections of normal tissue found at an abnormal site such as islands of pancreatic cells in the wall of the stomach). There are also instances in which a neoplasm is histologically malignant but clinically benign, such as in basal cell carcinoma of the skin. In addition, localized overgrowths of normal tissue components such as skin tags and vocal cord polyps are clinically recognized as tumors but are not truly neoplastic.

For brief definitions of various terms associated with carcinogenesis, refer to the glossary at the end of this entry.

Cancer (Neoplasia) as a Disease

Cancer, a multistep, progressive genetic disease involving alteration in multiple genes, is influenced by a variety of endogenous and exogenous factors. During the course of carcinogenesis, there is accumulated genetic damage ultimately resulting in the uncontrolled growth of the affected tissue. The process of carcinogenesis may be depicted schematically as in Fig. C-8 with the various steps along the pathway from normalcy to malignancy recognized by morphological and/or clinical features. It is here that the disciplines of clinical oncology and pathology are utilized to define the location of the specific neoplasm in this progressive cascade. Implicit in the process of carcinogenesis is a temporal framework that may represent months in experimental laboratory animals and years in humans. Identification of this process early in its evolution enhances the likelihood that intervention strategies such as surgical removal of a benign neoplasm may result in termination of the disease and clinical cure. By the time a neoplasm has progressed to the malignant stage and spread throughout the body, even heroic radiation and chemotherapy combined with surgery are unlikely to result in clinical cure.

Until the recent emergence of molecular biology, the multistep nature of carcinogenesis had largely been defined from structural studies of biopsy, autopsy tissue samples from humans, and experimental multistep animal models of carcinogenesis. In the past 20 years considerable attention has been focused on understanding the contributory influences of a variety of intrinsic and extrinsic factors upon carcinogenesis. It is hoped that an understanding of the role of these critical factors which impact the carcinogenic process will provide useful new intervention and prevention strategies for this important disease.

Cancer as Primarily a Multistep, Genetic Disease

During the past 40 decades of cancer research, much information has been generated indicating that cancer is a multistep, progressive disease. Support for this contention is derived from research on epidemiology and population genetics, morphological and clinical study of neoplasms, and experimental investigations in animals treated with agents that damage DNA. These studies have led to the identification of numerous endoge-

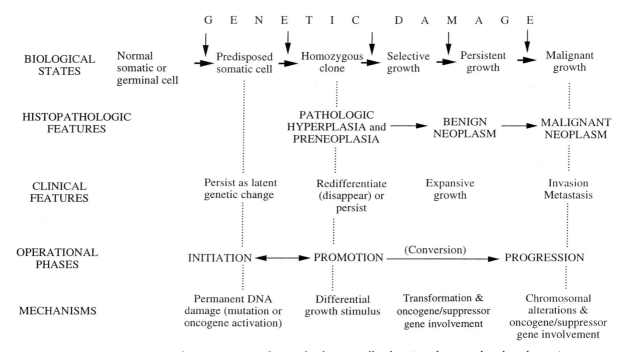

FIGURE C-8.   *Process of carcinogenesis depicted schematically showing the postulated pathway in which accumulation of genetic damage leads to malignant neoplasia.*

nous and exogenous variables that positively or negatively influence the process of carcinogenesis.

Neoplasia is frequently considered a genetic disease primarily because all known neoplasms are composed of cells with genetic alterations allowing them to escape normal homeostatic control of tissue growth. In some specific cases a critical genetic defect is inherited which predisposes individuals with such genetic traits to development of neoplasia. Examples include neurofibromatosis, retinoblastoma, and adenomatosis of the colon and rectum. In these instances, one stage in the carcinogenic process is believed to be an inherited germline mutation in the genetic material (DNA). Another inherited anomaly, an inability to repair ultraviolet light-induced DNA damage in individuals with the condition xeroderma pigmentosum, is associated with sensitivity to sunlight and a high incidence of skin neoplasia. However, the majority of genetic damage associated with carcinogenesis is acquired either *in utero* or from environmental and/or lifestyle factors to which individuals are exposed. Even for those individuals with a hereditary predisposition to neoplasia, additional DNA damage is necessary to lead ultimately to its development.

Some form of genetic alteration, typically manifested as mutations in the cellular DNA, is associated with virtually all neoplasia. Mutations in cellular DNA can arise during normal cell replication by infidelity in DNA replication (mispairing) as well as by chromosomal deletions, amplifications, or rearrangements. Just considering mispairing in nucleotide bases alone, it is estimated that spontaneous mispairing during normal cell replication can occur with a frequency of about 1.4×10^{-10} nucleotide base per cell division. Since there are approximately 10^{16} cell divisions per human life span and 2×10^9 nucleotide base pairs per genome, a total of 2.8×10^{15} mispairings could occur in a lifetime $[(1.4 \times 10^{-10}) \times (2 \times 10^9) \times 10^{16}]$. If each mispair led to a mutation that resulted in a cancer, a typical human would have billions of cancers in one average lifetime. Since such estimates of cancer frequency are clearly in excess of what is observed, it is necessary to postulate that events in addition to a single mutation are necessary for most cancers to occur. Another important consideration that explains the lower than expected cancer frequency is the existence of efficient mechanisms to repair DNA damage, thereby precluding successive accumulation of critical mutations.

Figure C-8 schematically presents as a temporal cascade a postulated pathway in which accumulation of genetic damages leads to malignant neoplasia. While it has been proposed that five or six critical events are necessary for development of many malignant neoplasms, it is important to bear in mind that there is no a priori reason to insist that a specific temporal sequence for the accumulated genetic damage is mandatory. Thus, it is entirely possible that a 1, 4, 3, 2, 5 sequence of events may be just as likely as a 1, 2, 3, 4, 5 sequence in leading to malignant neoplasia. The discussion that follows will assume the latter sequence for ease of explanation.

Some critical variables that influence the probability and rate of progressive development of neoplasia include the rate of cell proliferation, host susceptibility, the occurrence of biallelic damage, the fact that not all genetic alterations might occur within the coding region of the genome, and the amazing facility of the mammalian host to repair DNA damage prior to cell division. In recent years there has been increasing attention refocused on the role of cell proliferation in carcinogenesis. There cannot be neoplastic growth without cell proliferation, higher rates of which could contribute to a higher frequency of mutations just by chance occurrence of mistakes in DNA replication. Cell proliferation is also critical for "fixing" DNA damage since, without production of daughter cells from a damaged mother cell, there would be no inheritance of DNA damage. The cell has relatively efficient mechanisms to repair damage provided there is time prior to cell division. If a tissue is proliferating rapidly, cell division could occur before the cell has time to mend damaged DNA. While all of the above underscore the importance of cell proliferation in carcinogenesis, neoplasia does not occur exclusively or necessarily at higher frequency in tissues that have a rapid intrinsic rate of cell proliferation. Consequently, other important mechanistic factors influence the complex process of carcinogenesis.

That certain cancers occur in greater frequency within families represents primary empirical evidence for susceptibility based on some hereditary element. Some genetic predispositions exist for cancers of unknown etiology, while interactions between genetic susceptibility and environmental factors are probably responsible for a large proportion of human cancers. Hereditary predispositions include DNA repair deficiencies, inability to detoxify carcinogens, and germline loss or mutations of critical genes. Examples of genetic predispositions to cancer are listed in Table C-4. Environmental factors that would increase the risk of cancer

TABLE C-4

Examples of Genetic Predispositions to Cancer Development in Humans

Genetic predisposition	Associated cancer
Germline deletion on chromosome 13	Retinoblastoma Osteosarcoma
Germline deletion on chromosome 11	Renal nephroblastoma (Wilms' tumor) Hepatoblastoma Rhabdomyosarcoma Adrenal carcinoma
Li–Fraumeni syndrome	Soft tissue sarcomas in children Breast cancer in mothers
Von Hippel Lindau disease	Hemangiomas in the brain and retina
Von Recklinghausen's disease	Fibrosarcoma Neuroma Pheochromocytoma
Familial dysplastic nevi	Malignant melanoma
Xeroderma pigmentosa—defective ability to repair damaged DNA	Cutaneous squamous cell carcinoma
Ataxia–Telangiectasia	Leukemia Malignant lymphoma Stomach carcinoma
Familial adenomatous polyposis	Colon adenocarcinoma

development in genetically predisposed individuals include exposure to radiation and agents that stimulate cellular proliferation. Experimental systems in which to study genetic susceptibility to cancer are critically needed in order to assess the extent of gene–environmental interaction in human cancer.

For some cancers in genetically predisposed individuals data are consistent with malignant neoplasia's being associated with biallelic genetic alteration and supported by recent studies of tumor suppressor genes which prevent the development of neoplasia. Alteration or loss of a single tumor suppressor gene allele is usually insufficient to permit the development of neoplasia. In other words, the remaining functional tumor suppressor gene copy is sufficient to prevent the development of neoplasia; if it is lost or altered, however, neoplasia can develop. This situation occurs in hereditary childhood retinoblastoma, a malignant neoplasm of the retinal cells of the eye. Susceptible individuals inherit a partial loss of one copy (one allele) of chromosome 13, where the retinoblastoma tumor suppressor gene (*RB-1*) is located, and acquire an alteration or loss of the remaining *RB-1* allele during early development. The affected child subsequently develops retinoblastoma, often within the first 2 years of life.

Tissue Changes Associated with Carcinogenesis

Figure C-8 shows the interrelationship of various disciplinary aspects of carcinogenesis. Proliferative lesions which may be classified morphologically as preneoplasia, benign neoplasia, or malignant neoplasia represent a continuum of change with considerable overlap rather than discrete morphologic entities (Fig. C-9). The definitive classification of a given lesion as preneoplasia, benign neoplasia, or malignant neoplasia represents a judgment based on the experience of the diagnostic pathologist and familiarity with the species and tissue in question. These lesions are recognized by their microscopic appearance and effect on surrounding tissues and typically are a localized proliferation or hyperplasia of a specific cell type. Most neoplasms are believed to be derived from the clonal proliferation of a single initiated cell. Usually at some point early in the clonal expansion, the differentially proliferating cells become phenotypically distinguishable from the surrounding normal tissue. Although such lesions may not yet have sufficient characteristics to qualify as neoplasms, their recognition early in the process of carcinogenesis has led many to regard them as "preneoplastic."

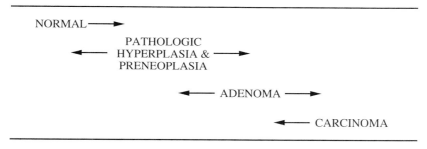

FIGURE C-9. *Morphological continuum of carcinogenesis.*

Although not all neoplasms have a preneoplastic change recognizable by the pathologist, in those instances in which presumptive alterations are observed, their occurrence documents that there is a response to tissue insult. Examples of presumptive preneoplastic lesions are presented in Table C-5. An important feature of preneoplastic lesions is their reversibility. Preneoplasia is a form of hyperplasia (an absolute increase in the number of cells in a tissue). In those experimental models of carcinogenesis in which preneoplasia is observed, it precedes the occurrence of benign neoplasia. In some instances a preneoplastic lesion represents the clonal expansion of a cell that has sustained genetic damage so that benign neoplasms arise within the preneoplastic lesion, presumably when one of the preneoplastic cells sustains additional genetic damage, giving

it a growth advantage. In other situations, the antecedent change is a localized polyclonal cellular proliferation historically associated with subsequent development of a neoplasm in the same tissue. A classical example is alcoholic cirrhosis in which chronic alcohol abuse leads to multiple, polyclonal areas of liver cell hyperplasia and an increased risk for development of hepatocellular neoplasia. In both preneoplasia and certain forms of hyperplasia, the antecedent lesions typically have a higher rate of cell proliferation than surrounding normal cells and, thus, these cells are at increased risk to sustain additional heritable genetic damage and progress to the next stage in the carcinogenic process.

There is considerable confusion regarding the significance of hyperplasia in the neoplastic process. Hyperplasia is an increase in the number of cells per unit of tissue, typically limited in amount and terminating when the stimulus that evoked it is removed. Different cell types have varying capacities to undergo hyperplasia in response to physiological or pathological stimuli. One of the most difficult judgments, even for the experienced pathologist, is whether an observed hyperplasia is part of the process of cancer development or merely an adaptive or physiologic response not likely to progress to neoplasia. This judgment is influenced by the tissue affected, whether the hyperplasia is diffuse or nodular, the age of the affected individual, the proximate cause of the hyperplastic response, and the growth pattern of the hyperplastic tissues.

A benign neoplasm is generally a localized expansive growth that compresses adjacent normal tissue but is usually not immediately life-threatening unless it physically interferes with normal function, e.g., by blocking the intestinal tract or compressing vital areas in the brain. Controversy regarding the significance of benign

TABLE C-5
Examples of Presumptive Preneoplastic Lesions

Tissue	Presumptive preneoplastic lesion
Mammary gland	Hyperplastic alveolar nodules
	Atypical epithelial hyperplasia
	Lobular hyperplasia
	Intraductal hyperplasia
	Hyperplastic terminal duct
Liver	Foci of cellular alteration
	Hepatocellular hyperplasia
	Oval cell proliferation
	Cholangiofibrosis
Kidney	Atypical tubular dilation
	Atypical tubular hyperplasia
Skin	Increase in dark basal keratinocytes
	Focal hyperplasia/hyperkeratosis
Pancreas (exocrine)	Foci of acinar cell alteration
	Hyperplastic nodules
	Atypical acinar cell nodules

neoplasia with respect to the development of malignancy is similar to that associated with preneoplastic lesions. A benign neoplasm, the clonal expansion of cells which have sustained some degree of genetic damage, is further along the spectrum of changes that precede the development of malignant neoplasia. In experimental carcinogenesis animal models, malignant neoplasia is not infrequently observed arising from or within a benign neoplasm. Features of benign neoplasms are listed in Table C-2.

Malignant neoplasms are rapidly growing, locally invasive tissue proliferations that destroy surrounding tissues and are thus life-threatening. They also have the malicious feature of spreading to distant sites in the body via the blood and lymphatic system. Although malignancy develops with greater frequency in association with (1) pathologic hyperplasia and preneoplasia, (2) qualitative alterations in cells, and (3) benign neoplasia than in association with normal tissues, these changes are not necessary precursors to malignancy. *In situ* carcinomas are malignant neoplasms that originate without evidence of antecedent benign tissue alteration. When precursor lesions are present prior to or concomitant with malignant neoplasia, it is probable that the malignancy is a consequence of the same or similar factors that produced the precursor lesions. Characteristics of malignant neoplasms are listed in Table C-2.

In addition to quantitative increases in certain cells, several qualitative cytological features help allow the morphologic classification of the spectrum of proliferative lesions that may be observed in the process of carcinogenesis. Three frequently used qualitative cytological features are metaplasia, dysplasia, and anaplasia.

Metaplasia

Metaplasia is the reversible substitution of one type of fully differentiated cell for another within a given tissue. A classic example is the replacement of the normal ciliated columnar epithelial cells in the respiratory tract airways by squamous epithelium (Fig. C-10) in situations in which there is chronic irritation from certain components of inhaled tobacco smoke. While the squamous epithelium is believed to provide functional protection against the irritant properties of the smoke, the loss of the ciliated columnar epithelium results in reduction of the functional capacity of the lungs to clear particulates from the respiratory tract. When the

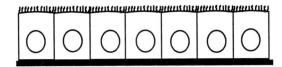

NORMAL EPITHELIUM

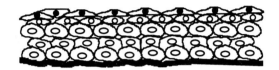

SQUAMOUS METAPLASIA

DYSPLASIA & ANAPLASIA

FIGURE C-10.   *Qualitative changes in epithelial tissues.*

irritation is removed, the squamous epithelium is replaced by normal ciliated columnar epithelium.

Dysplasia

Dysplasia is defined as abnormal formation of a tissue with respect to shape, size, and the organization of component cells. Normal cell-to-cell orientations are disorganized or disrupted, and the cells themselves vary in size and shape (Fig. C-10). When present, dysplasia may be associated with chronic irritation, occur with metaplasia, and be seen in neoplastic transformation. It is a change that is a hallmark of increased risk for development of neoplasia. Like metaplasia, dysplasia is a potentially reversible tissue alteration.

Anaplasia

Anaplasia is a qualitative alteration of cellular differentiation. Anaplastic cells are typically undifferentiated and may bear little, if any, resemblance to mature cells.

Staging and Grading of Cancers

In human oncology the experience from collective years of observation of many cancers permits the pathologist to grade and stage most cancers. The purpose of grading and staging a neoplasm is to predict its biological behavior and to help establish an appropriate therapeutic regimen. Grading is a subjective evaluation of morphological characteristics based on the extent of cellular anaplasia and the degree of proliferation evident from microscopic evaluation. Basically, neoplasms with a high degree of anaplasia, associated specific morphological patterns of growth, and evidence of numerous mitoses, some of which may be abnormal, result in a high grade of malignancy. Most grading schemes categorize neoplasms into one of three or four grades of increasing malignancy.

Staging of a cancer is independent of grading, is an index of the extent to which a cancer has spread in the body determining the patient's clinical prognosis, and influences the appropriate therapy more than grading. Criteria used for staging include the size of the primary neoplasm, the degree to which there is invasion of surrounding normal tissues, whether the cancer has spread to local lymph nodes, and the presence of spread to distant sites in the body. Thus, it is apparent that staging will have a large influence on the therapeutic approach. A small and localized breast cancer would most likely be treated by surgical excision and possibly radiation therapy, whereas a large, infiltrative breast cancer would more likely be treated by mastectomy. If the cancer has spread to lymph nodes or distant sites, more aggressive therapy is mandated.

The ultimate fate of cells or proliferative tissue masses is influenced by the amount of sustained genetic damage. Cells with minimal DNA damage may persist in a latent form, indistinguishable from surrounding normal cells. If additional damage is sustained by such a latent cell, even long after the initial insult, it may then progress further along the pathway to malignancy (see Fig. C-8). As additional genetic damage occurs, the altered cell population expands and eventually leads to irreversible uncontrolled growth that may or may not be corrected by aggressive medical intervention.

Endogenous Factors Influencing Carcinogenesis

A variety of endogenous and exogenous factors influence the development of neoplasia. Among the endogenous factors are protooncogenes, tumor suppressor genes, hormones, growth factors, age and sex, and the immune system. Exogenous factors that influence carcinogenesis include exposure to synthetic and natural chemicals, environmental exposure to ultraviolet and ionizing radiation, diet and lifestyle, and certain viruses. In some instances exogenous factors, such as diet, influence the balance of endogenous factors, such as hormones, thereby indirectly influencing carcinogenesis.

Oncogenes

Among the estimated 10,000 genes in the mammalian genome, there are approximately 50–75 genes that are classified as oncogenes because activation of these genes appears to be an essential event for the development of many, if not all, cancers. In fact, oncogenes were first discovered by studying genetic alterations in cancers. The term oncogene activation indicates a quantitative or qualitative alteration in the expression or function of the oncogene. The term oncogene is unfortunate since the unaltered (nonactivated) oncogene (usually referred to as a protooncogene) actually serves an essential function in the mammalian genome. That protooncogenes are highly conserved in evolution is evidenced by structurally and functionally similar genes in yeast, earthworms, animals, and humans. The highly conserved nature of protooncogenes is believed to be related to their essential function in normal tissue growth and differentiation. Since their normal function is to control how a tissue grows and develops, it is apparent that, if they do not function appropriately, abnormal growth and development may occur. When a primary manifestation of such abnormal growth was observed to be neoplasia, these protooncogenes were named oncogenes. This nomenclature has persisted despite the ultimate discovery that the unaltered forms of these genes are normal components of the genome.

The appearance (phenotype) and function of a tissue is a consequence of which genes are actively producing their programmed product, typically a protein, which in turn affects the structure and function of the cells comprising a given tissue. All somatic cells in the body inherit a complete complement of maternal and paternal genes. The reason that some cells form liver and produce products such as albumin while other cells form kidney tubules that function to excrete substances from the body is a consequence of which genes are expressed in those cells. In liver cells, several critical genes that are important in kidney function are not expressed and vice versa. Specific gene expression and its effect upon tissue phenotype and function are modulated by several intrinsic and extrinsic factors (Fig. C-11). Since a primary function of many oncogenes is to control cell growth, proliferation, and differentiation, inappropriate expression of these genes would be expected to influence abnormally tissue proliferation and growth. Oncogene activation is a consequence of inappropriate or excessive expression of a protooncogene.

Oncogenes can be activated by several different mechanisms (e.g., retroviral transduction, chromosomal translocation, gene amplification, point mutation, or promoter/enhancer insertion). Once activated an oncogene will either be inappropriately expressed (e.g., production of an altered message and protein) or overexpressed (e.g., production of too much of a normal message and protein). Either situation may contribute to the neoplastic process by influencing cellular proliferation and differentiation. Examples of activated or amplified oncogenes detected in human and animal neoplasms are listed in Tables C-6 and C-7. For some cancers the frequency of oncogene activation is rela-

TABLE C-6
Examples of Human Neoplasms Associated with Activated or Amplified Oncogenes

Oncogene	Type of human neoplasia
H-*ras*	Squamous cell carcinoma
	Urinary bladder carcinoma
	Lung carcinoma
	Acute myelogenous leukemia
K-*ras*	Lung adenocarcinoma
	Colon carcinoma
	Ovarian carcinoma
	Gastric carcinoma
	Renal cell carcinoma
	Acute myelogenous leukemia
	Pancreatic ductal adenocarcinoma
N-*ras*	Acute myelogenous leukemia
	Chronic myelogenous leukemia
abl	Chronic myelogenous leukemia
erb B$_2$	Breast carcinoma
	Salivary gland adenocarcinoma
myc	Small cell carcinoma of the lung
	Burkitt's lymphoma
N-*myc*	Neuroblastoma
	Carcinoma of the breast
ets-1	Acute myelomonocytic leukemia

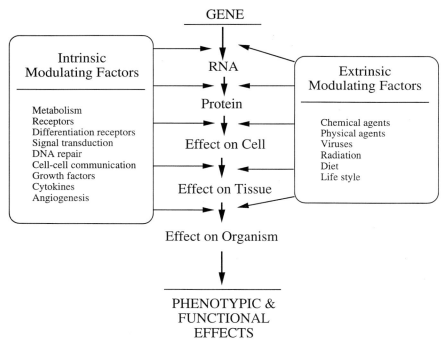

FIGURE C-11.   *Intrinsic and extrinsic factors modulating specific gene expression and its effect on tissue phenotype and function.*

TABLE C-7
Examples of Animal Neoplasms Associated with
Activated Oncogenes

Oncogene	Type of animal neoplasia
H-*ras*	Hepatocellular adenoma and carcinoma Harderian gland adenoma Mammary carcinoma Skin squamous cell carcinoma
K-*ras*	Lung adenomas and adenocarcinoma Pancreatic carcinoma Hepatocellular carcinoma
N-*ras*	Leukemia Lymphosarcoma
raf	Fibrosarcoma
neu (erb B₂)	Neuroblastoma
abl	Lymphosarcoma
c-myc	Leukemia Lymphosarcoma

tively high, while for other cancers activation of known oncogenes is uncommon. Identification of specific alterations in oncogenes in certain cancers represents a first step in determining the molecular basis of cancer and could eventually lead to development of molecular intervention and therapeutic strategies. Recent experimental evidence indicates that oncogene activation may be an early critical event in carcinogenesis, and experimental studies with known chemical carcinogens show that they produce specific alterations in certain oncogenes reflecting the manner in which the carcinogen chemically affects DNA.

Tumor Suppressor Genes

It has been shown that malignant transformation of normal cells represents a balance between genes for expression and genes for suppression of malignancy. Growth suppressor genes, sometimes called tumor suppressor genes or antioncogenes, function to suppress the development of cancers. While oncogenes must be activated to be effective, tumor suppressor genes must be inactivated or lost for cancer to develop. It has been shown that loss or mutation of both paternal and maternal copies, i.e., in both alleles, of a tumor suppressor gene must occur to ablate their effect of suppressing cancer formation. A well-known and extensively studied tumor suppressor gene is the retinoblastoma gene (*RB-1*). In hereditary retinoblastoma an affected child

is born with deletions of portions of one allele of chromosome 13 containing the *RB-1* gene. If a second event leading to a loss or alteration of the remaining *RB-1* allele occurs while retinal cells are undergoing growth during development, the ocular neoplasm, retinoblastoma, frequently present in both eyes, will occur early in life. Loss or alteration of both copies of this tumor suppressor gene is sufficient to cause retinoblastoma. Although named for the disease in which it was discovered, alterations in the *RB-1* gene have been detected in breast, lung, prostate, and bone cancers.

Growth Factors, Hormones, and Signal Transduction

While alterations in cellular DNA are critical in carcinogenesis, some cancer-causing agents, particularly those that are not genotoxic, play a major role in cancer development by indirectly influencing gene expression and growth control by altering signal transduction. While the pivotal role of hormones in the orchestration of tissue growth and development has been appreciated for decades, the recent discovery of polypeptide growth factors has added to our knowledge of the complex constellation of control mechanisms that affect normal cellular growth. Both hormones and growth factors bind to specific cellular receptors and thereby trigger a cascade of intracellular reactions that seem to be associated ultimately with cellular proliferation. This cascade of intracellular reactions is sometimes referred to as signal transduction, the process whereby a stimulus external to the cell triggers a cascade of intracellular biochemical reactions that ultimately lead to expression of specific genes. This is perhaps best exemplified by the process whereby a normal hormone stimulates a tissue to grow. An example is breast development and milk production in response to the hormone prolactin. In this example, prolactin binds to a specific prolactin receptor on the external surface of the cell, which, in turn, triggers a biochemical change inside the cell membrane via molecules that are attached to the external receptor and pass through the cell membrane. This in turn triggers a long chain of biochemical reactions ultimately resulting in a signal to specific genes in the cellular DNA so that they become active. The specific genes, in this example, initiate a program that causes breast cells to divide and secrete milk. The signal transduction pathways in mammalian cells are highly interactive with numerous positive (signal-sending) and negative (signal-blocking) feedback loops. An appropriate balance between the positive and negative feedback loops is necessary for the proper functional response to the initial stimulus.

Some forms of cancer development are believed to be facilitated by perturbations in one or more places in the signal transduction pathway. Thus, exposure to certain agents may potentially affect the balance of positive and negative feedback loops in the signal transduction pathway and make cells more susceptible to stimuli that promote growth. An example is the nongenotoxic skin tumor promoter, phorbol ester, which activates protein kinase C, a multifunctional element in the signal transduction pathway that mediates many critical cellular regulatory processes. Treatment of initiated mouse skin with phorbol ester activates protein kinase C, resulting in development of benign and malignant skin neoplasms. The complexity and pivotal importance of the signal transduction pathways help explain why multiple types of agents influence carcinogenesis, why multiple steps are involved in the carcinogenic process, and why different cancers are so heterogeneous. Signal transduction involves shifts in intracellular ion fluxes for elements such as sodium, potassium, and calcium. It also often involves activation of protein kinase C, an enzyme that phosphorylates many proteins that may be important in producing a mitogenic response. Part of the signal transduction cascade involves increased expression of cyclic adenosine monophosphate, now recognized as a mitogenic signal, and increased expression of one or more cellular protooncogenes. Current research results demonstrate that increasing numbers of protooncogenes and growth factors are integral parts of the signal transduction pathway and, when altered, influence development of cancer by subverting signal transduction. A simplified depiction of the interaction of hormones and growth factors with cellular signal transduction is presented in Fig. C-12.

The Immune System and Cancer

The proper functioning of the immune system is evidenced by recovery from common childhood diseases such as mumps and chicken pox. A properly functioning immune system recognizes the foreignness of the agents responsible for these diseases, responds to the infection, eliminates the foreign agents, and confers long-term immunity to subsequent infection by the same or similar agents. It has been proposed that cancer cells are recognized as foreign and that the immune system functions to eliminate such cells from the body before they are transformed into large, malignant neoplasms. This process involves elaboration of antibodies that bind to the cancer cells and activate a process whereby the cancer cells are killed. In addition, specific cells of the immune system, such as cytotoxic T lymphocytes, natural killer cells, and macrophages, have a mechanism for recognizing foreign cells and eliminating them from the body. The process of immune surveillance and removal of cancer cells is facilitated when the cancer cells express surface antigens that are recognized as foreign. Exposure to agents which depress the normal functioning of the immune system can lead indirectly to neoplasia by permitting early persistence and development of recently emergent cancer cells. Once a neoplasm has reached a critical size and growth rate, it may not be possible for even a properly functional immune system to eliminate effectively the neoplastic cells.

Exogenous Factors Influencing Carcinogenesis

Important exogenous factors that contribute to induction of cancer include natural and synthetic chemicals, environmental exposures to ultraviolet and medical radiation, diet and lifestyle, and viruses. Evidence for a causal association between exogenous factors and neoplasia is derived from studies of epidemiology, occupationally common cancers, and animal models.

Chemical and Physical Agents and Lifestyle Factors

Many chemicals that cause cancer interact directly with and alter DNA or are metabolized to chemical derivatives capable of doing so. Exposure to carcinogens can occur in certain occupational settings. Associations of human hepatic angiosarcomas with workplace exposure to vinyl chloride, pulmonary mesotheliomas with exposure to asbestos fibers, and leukemia with benzene are well-known examples. Exposure to other carcinogenic agents may occur in the diet or as a consequence of certain lifestyle practices, such as cigarette smoking associated with pulmonary cancer and high animal fat diets linked to breast and colon cancer. Strong associations have been made between exposure of light-skinned individuals to ultraviolet radiation and skin

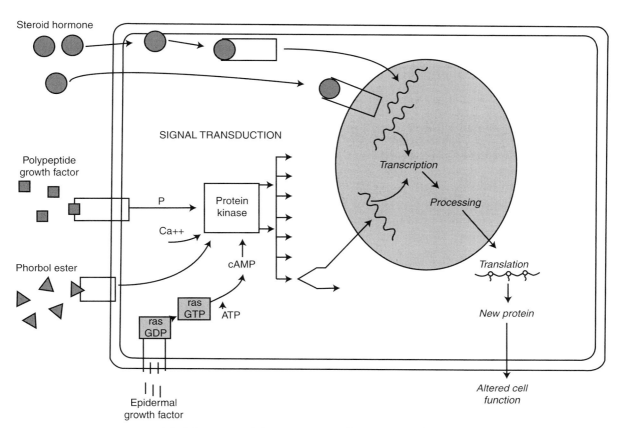

FIGURE C-12. *Simplified depiction of the interaction of hormones and growth factors with cellular signal transduction.*

cancer. Exposure to occupational ionizing radiation, X-rays, and medical use of radioisotopes have also been associated with human neoplasia. Examples include leukemias in radiologists and atom bomb victims, lung cancer in uranium mine workers, and thyroid and breast cancer following diagnostic or therapeutic use of radiation.

Viruses and Cancer

Both DNA and ribonucleic acid (RNA) viruses are strongly associated with cancer. Although viruses have been shown to cause a variety of animal neoplasms, no virus has yet been shown unequivocally to cause human neoplasia. When cells are infected by DNA viruses, the viral DNA inserts itself wholly or partially into the genome of the infected cell. It appears that such integration of viral DNA into the mammalian genome is sometimes sufficient to cause neoplastic transformation of the infected cell which is accompanied by the production of new proteins essential for the neoplastic process. RNA viruses associated with neoplasia are chiefly represented by the retroviruses. RNA viruses possess an enzyme called reverse transcriptase, which is capable of forming a DNA copy of the viral RNA when a host cell is infected by the virus. This DNA ultimately inserts itself into the host genome in much the same way as DNA viruses, possibly resulting in development of neoplasia.

The strongest link between virus infection and human neoplasia is seen for the retrovirus HTLV I associated with T cell lymphoma, human papillomavirus associated with squamous cell carcinoma of the cervix, and hepatitis B virus associated with hepatocellular carcinoma. Examples of virus-associated neoplasia in animals and humans are presented in Table C-8.

Operational Phases and Theoretical Aspects of Carcinogenesis

Carcinogenesis is a form of toxicity in which cells achieve a different steady state from normal. The al-

TABLE C-8
Viruses Causally Related to or Strongly Associated with Animal and
Human Neoplasia

Virus	Type of neoplasm	Species
DNA viruses		
Myxoma	Myxoma	Rabbit
	Myxomatosis	Rabbit
Herpes	Lymphosarcoma	Chicken
		Monkey
		Rabbit
Herpes simplex 2	Cervical carcinoma	Human
Papillomaviruses	Papillomas	Cow
		Rabbit
		Horse
		Dog
Human papillomavirus	Warts	Human
	Epidermoid carcinoma	
	Cervical carcinoma	
Woodchuck hepatitis virus	Hepatocellular carcinoma	Woodchuck
Hepatitis B virus	Hepatocellular carcinoma	Human
RNA retroviruses		
Human T cell leukemia virus (HTLV-I and -II)	T cell lymphoma	Human
Avian erythroblastosis virus	Leukemia	Chicken
	Sarcoma	
Abelson leukemia virus	Leukemia	Mouse
Harvey sarcoma virus	Sarcoma	Rat
	Leukemia	
Feline sarcoma virus	Sarcoma	Cat

tered cells do not respond to intrinsic mechanisms that normally control growth and development. Normal control mechanisms are finely tuned to permit repair of tissue alterations such as occurs when there is healing of superficial wounds or broken bones. The repair is achieved by accelerated cell division but, because of the intrinsic growth control mechanisms, once the lesion has been repaired the accelerated cell division ceases. The same homeostatic mechanisms are operational during normal growth, development, and maturation. Once adulthood has been reached, growth stops. A cancer, on the other hand, somehow escapes the influence of these homeostatic controls and is characterized by autonomous and continued growth.

In addition to being complex, the process of carcinogenesis is typically prolonged, requiring a significant portion of the life span to become clinically apparent. While perturbations in cellular DNA are essential to carcinogenesis, they alone are not sufficient to cause cancer in all cases. Thus, in some experimental situations, a few minutes of exposure to a carcinogen is sufficient to result ultimately in cancer, whereas in other situations, exposure to the same carcinogen will not result in cancer unless there is additional experimental manipulation. Smokers present a paradigm of this principle since many, but not all, ultimately develop lung cancer. In other experimental studies, simultaneous administration of a carcinogen and a second agent may enhance, reduce, or block the carcinogenic process depending on the agent employed. Many phenomena have been documented during the process of carcinogenesis in humans and experimental animal systems. These studies have elucidated some of the mechanisms and factors that influence carcinogenesis, delimited some of the specific stages in the multistep process, and continually reminded us of the complexity of this disease process.

Multistep experimental models of carcinogenesis have proven useful in defining events in the neoplastic process; have formed the foundations for current operational descriptions and hypotheses of the biological mechanisms of carcinogenesis (see Fig. C-8); exist for many organ systems including the skin, liver, urinary bladder, lung, intestine, mammary gland, and pancreas; and typically involve administration of chemical agents to laboratory animals. Operational phases of carcinogenesis include initiation, promotion, and progression.

Initiation

During the initiation phase of chemical carcinogenesis, a chemical agent or carcinogen interacts with a cell to produce an irreversible change that may ultimately be manifested by a capacity for autonomous growth. The initiated cell appears normal, and the capacity for autonomous growth may remain latent for weeks, months, or years. Initiation implies alteration in the affected cell's DNA at one or more sites, a mutational event which is by definition hereditary. Direct-acting carcinogens interact directly with cellular DNA to produce the damage while indirect-acting carcinogens must be metabolized by the cell to produce a chemical species that interacts with cellular DNA. The majority of damaged cells have the ability to repair the damaged DNA over a period of days or weeks; however, if a cell undergoes cell division with its attendant DNA replication prior to repair of the DNA damage, the DNA alteration becomes "fixed," is no longer reparable, and is inherited by all subsequent daughter cells. The operational phase of initiation is relatively short and may occur within hours or days. In contrast, the progression of an initiated cell to a fully malignant neoplasm is a prolonged process requiring months in animals and years in humans. Based on a large body of evidence that most initiators are mutagenic or genotoxic, a battery of short-term mutagenicity tests in bacteria and cell culture systems has evolved to identify chemicals with genotoxic properties. Once identified, such chemicals should be rigorously regulated to prevent human exposure. This approach is considered prudent because of the irreversible and hereditary nature of the changes that occur during initiation. Indeed, it is generally believed that even a single molecule of a mutagenic substance is potentially sufficient to damage DNA irreversibly. Thus, for practical purposes there is no threshold

or safe level of exposure to a mutagenic agent. Salient features of initiation are listed in Table C-9.

Initiators interact with host cellular macromolecules and nucleic acids in specific patterns. The majority of known initiators have both initiating and promoting (see below) activity and can thus induce neoplasms rapidly and in high yield when there is repeated or high-level exposure. When given at sufficiently low single doses, an initiated cell requires subsequent promotion for the development of any neoplasia. Thus, the dose of an initiator is a critical determinant of its carcinogenic potential.

Promotion

Promotion is classically considered that portion of the multistep carcinogenic process in which specific agents, known as promoters, enhance the development of neoplasms by providing initiated cells with a selective growth advantage over the surrounding normal cells. Characteristic features of promotion are listed in Table

TABLE C-9
Salient Features of Initiation and Promotion of Neoplasia

Initiators/initiation

- Effect is irreversible
- Only one exposure may suffice
- Multiple exposures may be additive
- Cannot identify initiated cells
- Agents are considered carcinogens
- Agents are usually mutagenic
- No measureable threshold dose
- Must be administered before the promoter
- Does not result in neoplasia unless promoter is subsequently applied
- Number of initiated cells dependent on dose of initiator

Promoters/promotion

- Nonadditive
- Agents not capable of initiation
- Modulated by diet, hormones, environment, and other factors
- Measureable threshold dose
- Measureable maximal response
- Agents not considered carcinogens but may be cocarcinogens
- Must be administered after the initiator
- Agents are usually not mutagenic
- Prolonged exposure is usually required

C-9. The promoting agents themselves are classified operationally in that a promoter is typically given at some time after chemically induced or fortuitous initiation and the usual experimental doses of promoting agent are insufficient to produce cancer without prior initiation. When classical promoters are administered at sufficiently high doses and for prolonged intervals, neoplasia can occur without evidence of prior initiation. Under these conditions, a promoting agent must be considered a complete carcinogen unless fortuitous initiation from background radiation, dietary contaminants, environmental toxins, etc. is believed to have occurred. However, under typical experimental conditions commonly employed in short- and medium-term initiation–promotion experiments, neoplasia does not typically occur in animals that are not previously initiated.

The temporal sequence of promoter administration is critical to the operational definition of promotion. The agent must be administered after initiation and cause enhancement of the neoplastic process to be considered a promoter. If an agent is given simultaneously with an initiator and results in enhancement of development of neoplasms, it is regarded as a cocarcinogen rather than a promoter. While some promoters are cocarcinogenic (e.g., phorbol esters), not all promoters (e.g., phenobarbital and phenol) possess cocarcinogenicity and, conversely, not all cocarcinogens are promoters. Under these same conditions of simultaneous administration, a diminution in the neoplasm response is considered evidence of anticarcinogenic activity. Several rodent liver tumor promoters which are active when administered after a variety of initiators prevent or delay the development of liver neoplasms when added to diets along with an active carcinogen. Finally, reversing the order of administration by giving a known promoter prior to an initiator may prevent the expression of carcinogenic activity on the part of the initiator.

While upper and lower thresholds have been demonstrated experimentally for promoters, some consider that, in an absolute sense, it is statistically impossible to prove or disprove the existence of thresholds for promoters for much the same reasons that this cannot be done for initiators. One can never be certain that an apparent no-effect-level would, indeed, be without effect if a sufficiently large enough number of animals were used. Promoters include agents such as drugs, plant products, and hormones that do not directly interact with host cellular DNA (are not genotoxic) but

somehow influence the expression of genetic information encoded in the cellular DNA. Experimental evidence suggests that regulation of gene expression is unique to the nature of the promoting agent administered. Some promoters are believed to produce their effect by interaction with receptors in the cell membrane, cytoplasm, or nucleus (e.g., hormones, dioxin, phorbol ester, and polychlorinated biphenyls). Alternatively, promoting agents may exert their effect through their molecular orientation at cellular interfaces. Other promoters may selectively stimulate DNA synthesis and enhance cell proliferation in initiated cells, thereby giving them a selective growth advantage over surrounding normal cells.

Promoters appear to have a relatively high tissue specificity. Thus, phenobarbital functions as a promoter for rodent liver neoplasia but not urinary bladder neoplasia. Saccharin, on the other hand, promotes urinary bladder neoplasia but not liver neoplasia in the rat. Similarly, 12-0-tetradecanoylphorbol-13-acetate (phorbol ester) is a potent skin and forestomach neoplasm promoter in the laboratory rodent but has no appreciable activity in the liver. Other agents, such as the antioxidants 3-*tert*-butyl-4-methoxyphenol and 2,6-di-*tert*-butyl-4-methoxyphenol, may act as promoters in one organ and antipromoters in another and have no effect in a third organ. Thus, the practical definition of a promoter must include the designation of the susceptible tissue.

Tumor promotion may be modulated by several factors such as age, sex, diet, and hormone balance. The correlation of increased rates of breast cancer in women following a "Western" lifestyle has implicated meat and fat consumption as playing an important role in breast cancer development. Experimental demonstration of the role of a high-fat diet in the promotion of mammary cancer in rats exposed to the mammary carcinogen dimethylbenzanthracene has been documented. Similarly, bile acids, as modulated by fat consumption, are known promoters of rat liver carcinogenesis and human colorectal cancer. Age- and sex-associated modulations in hormonal levels of estrogens, progesterone, and androgens have been implicated as potential promoters of breast cancer on the basis of epidemiological studies in humans. Experimental studies have repeatedly shown that these hormones, in addition to pituitary prolactin, serve to promote mammary cancer in rats initiated with mammary carcinogens.

Progression

Progression is that part of the multistep neoplastic process associated with the development of an initiated cell into a biologically malignant cell population. In common usage progression is frequently used to signify the stages whereby a benign proliferation becomes malignant or, alternatively, where a neoplasm develops from a low grade to a high grade of malignancy. During progression neoplasms show increased invasiveness; develop the ability to metastasize; and have alterations in biochemical, metabolic, and morphologic characteristics.

Expression of tumor cell heterogeneity, an important characteristic of tumor progression, includes antigenic and protein product variants, ability to elaborate angiogenesis factors, emergence of chromosomal variants, development of metastatic capability, altered metabolism, and decreased sensitivity to radiation. The development of intraneoplastic diversity may result from increasing genetic damage. Alternatively, the heterogeneity observed in tumor progression may be generated by epigenetic, regulatory mechanisms operative as a continuation of the process of promotion. More than likely, genetic and nongenetic events subsequent to initiation operate in a nonmutually exclusive manner during progression, possibly in an ordered cascade of latter events superimposed upon earlier events.

The most plausible mechanism of progression invokes the notion that, during the process of tumor growth, there is a selection that favors enhanced growth of a subpopulation of the neoplastic cells. In support of this mechanism is increased phenotypic heterogeneity observed in malignant but not benign neoplastic proliferations. Presumably, a variety of subpopulations arises, and it is only a matter of time before the emergence of a subpopulation with more malignant biological characteristics or at least an accelerated growth advantage. This can be observed occasionally during experimental hepatocarcinogenesis when a phenotypically distinguishable carcinoma can be observed arising within an existing adenoma.

Distinction between tumor promotion and tumor progression is not readily discernible in the routine histopathologic evaluation of neoplasms and may be somewhat academic because promotion may be considered part of the process of progression. In both situations the critical event is accentuated growth. What is believed to distinguish progression from promotion is

the presence of structural genomic alterations in the former and their absence in the latter. Both structural genomic changes and biochemical changes associated with tumor progression cannot be defined by conventional histopathology. Emerging technologies centered around histochemistry, immunocytochemistry, *in situ* hybridization, identification of activated oncogenes, and loss of tumor suppressor genes offer promise to distinguish various stages of progression in the evolution from benign to malignant neoplasms. Application of such technologies to human cancer has allowed development of paradigms such as that proposed by Dr. Bert Vogelstein and co-workers for human colorectal cancers (Fig. C-13). The Vogelstein model includes a mutation in one of two genes on chromosome 5. This early event in the process of colorectal carcinogenesis may occur sporadically or as an inherited mutation in individuals with a familial predisposition. A later and intermediate event in the progression of approximately half of the observed colorectal cancers involves mutations in the oncogene, K-*ras*. Later occurring events include loss and/or mutation of tumor suppressor genes (*DCC* and *p53*). Defining such critical steps in the progression of common cancers such as colorectal carcinoma provides important insights into the causes, prognosis, and potential therapies for specific cancers.

Identification of Carcinogenic Agents

There are two methods utilized to identify potential human carcinogens, the most direct of which is based on retrospective epidemiological studies in human populations using existing historical records associated with known cases of neoplasia. These records include death certificates where cause of death is indicated; hospital records; responses to questionnaires that document environmental or work-associated exposure to potential carcinogenic agents; and studies of neoplasia in culturally, ethnically, or religiously distinctive human populations. Association of cigarette smoking with lung cancer and exposure to asbestos with mesotheliomas was the result of such retrospective epidemiological work. Prospective epidemiological studies identify a given population of individuals who agree to be monitored for several years to permit identification of potential carcinogenic factors associated with neoplasms which may occur.

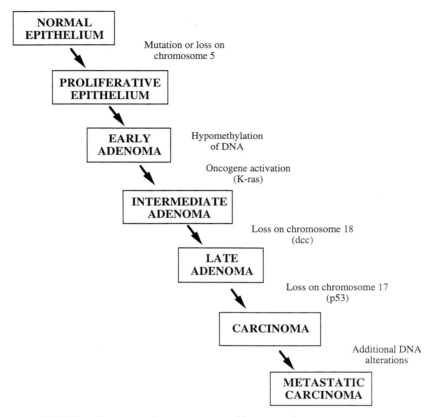

FIGURE C-13.   Multistep aspects of human colon carcinogenesis.

The second method used to identify potential human carcinogens involves testing known chemicals and agents in experimental animals. Such tests have been referred to as animal bioassays and are typically conducted using rats and mice exposed to high doses of the suspect agent for a large portion (typically 2 years) of their life span. If such agents are observed to produce neoplasia in the experimental animals, the agent is regarded as a potential human carcinogen. In countries throughout the world, legal requirements mandate that all new chemical agents and drugs be tested in animal bioassays to determine if they cause cancer in the test animals. Additionally, since the mid-1960s in the United States, the National Cancer Institute and currently the National Toxicology Program have collectively conducted animal bioassays on more than 450 chemical agents to assess their potential to cause cancer.

Interpretation of results from human epidemiological studies and animal bioassays to identify carcinogenic agents has proven difficult and controversial. Humans are rarely exposed to only one potential cancer-causing agent in their lifetime, and amount and duration of that exposure may be difficult or impossible to quantify rigorously. Many years may intervene between exposure to a potential carcinogen and ultimate development of neoplasia, making accurate assessment of cause and effect almost impossible. Despite such limitations, epidemiological studies that clearly show an association between a given chemical exposure or lifestyle habit with an enhanced rate of a specific cancer are regarded as the most relevant method for identification of human carcinogens. While animal bioassays have proven useful for identification of agents that can cause cancer in the laboratory rodent, they only identify an agent as potentially hazardous to human health. Additional facts and factors must be considered in classifying such an agent as a likely human carcinogen. The current approach for assessing the scientific relevance of either epidemiological or animal bioassay results to human health risk involves a "weight-of-evidence" procedure in which national and international panels of expert scientists from several disciplines examine all available information on the suspect agent and reach a consensus opinion. Included in this analysis are the strength of the epidemiological evidence, the dose–response curve of the animal response, comparative species metabolism and ability to extrapolate between species, likely mechanism of cancer induction for the agent in question, the genotoxicity of the agent, the amount of the agent in the environment, and the number of people potentially exposed to the agent. Based on this type of analysis, over 38 agents have been classified as known human carcinogens by the International Agency for Research on Cancer (Table C-10) and 37 more agents have been designated as likely human carcinogens. The 7th U.S. Health and Human Services *Annual Report on Carcinogens* lists 26 known human carcinogens and 154 substances

TABLE C-10

Agents or Mixtures for which There Is Sufficient Evidence of Carcinogenicity in Humans

Organic compounds	
2-Napthylamine	4-Aminobiphenyl
Aflatoxin B_1	Analgesics containing phenacetin
Azathioprine	Benzene
Benzidine	Betel quid with tobacco
bis(chloromethyl)ether	Chlorambucil
Chlornaphazine	Chloromethyl methyl ether
Cyclophosphamide	Diethylstibesterol
Melphalan	Methyl-CCNU
MOPP (and other combined therapies)	
Mustard gas	Myleran
Thiotepa	Tobacco products and tobacco smoke
Treosulfan	Vinyl chloride

Soots, tars, and oils
Coal tar pitches
Coal tars
Mineral oils, untreated and mildly treated
Shale oils
Soots

Hormones	Metals
Diethylstibesterol	Arsenic compounds
Estrogens	Chromium compounds
Oral contraceptives	Nickel and nickel compounds

Fibers	Other
Asbestos	8-Methoxypsoralen + UV radiation
Erionite	
Talc-containing asbestos fibers	

that are reasonably anticipated to be human carcinogens.

Summary and Conclusions

All of life is a balancing act of good versus evil and production versus destruction. Similar balancing factors are evident in carcinogenesis where regulatory mechanisms for tissue proliferation are balanced against those for cellular differentiation. It is well established that carcinogenesis requires the accumulation of multiple heritable alterations in the genome of the affected (cancer) cells. At the genetic level, two opposing classes of genes, oncogenes and tumor suppressor genes, have been implicated in the carcinogenic process. In addition, development of cancer is influenced by host factors such as age, sex, diet, nutrition, general health status, and inherited predispositions for cancer and by complex positive and negative intracellular signaling mechanisms. Treatment of cancer is based on our understanding of the mechanistic underpinnings of the carcinogenic process and attempts to shift the balance of critical factors in favor of patient survival. The probability of developing cancer is directly proportional to intensity, route, and duration of exposure to cancer-causing factors. Public health strategies are based on the premise that reduction or prevention of exposure to cancer-causing factors will decrease the incidence of cancer.

Glossary

Adenomatosis: a condition in which numerous adenomatous growths develop in a tissue.

Allele: one of the two gene pairs situated at the same location on a chromosome; one allele is inherited from the mother and the other from the father; characteristics such as being short or tall or having blue eyes versus brown eyes are determined by the expression of inherited alleles.

Anaplasia: lack of normal organizational or structural differentiation of a tissue; anaplastic cells are typically poorly differentiated.

Angiogenesis: the development of blood vessels.

Benign: a classification of anticipated biological behavior of neoplasms in which the prognosis for survival is good; benign neoplasms grow slowly, remain localized, and usually cause little harm to the host.

Biallelic damage: damage to both maternal and paternal copies of a gene.

Cancer: generally refers to a malignant neoplasm.

Carcinogenesis: the process of development of cancer or neoplasms.

Choristoma: a mass or collection of well-differentiated cells from one organ found within another organ, e.g., adrenal tissue present in the lung.

Clonal: pertaining to a clone; a line of cells decendent from a single cell.

Cocarcinogen: an agent that has no inherent carcinogenic activity by itself but is capable of augmenting neoplasm formation when given simultaneously with a genotoxic carcinogen.

DNA: abbreviation for deoxyribonucleic acid; the basic building block of genetic material in all organisms except RNA viruses.

Dysplasia: disordered tissue formation characterized by changes in size, shape, and orientational relationships of adult types of cells; primarily seen in epithelial cells.

Epithelial cell: cells which line internal and external surfaces of the body and form the bulk of many of the major organs of the body; they are formed from the embryonic germ layers known as entoderm and ectoderm.

Gene: the basic biological unit of heredity which is located on a chromosome.

Genome: the total complement of genes present in the set of chromosomes characteristic of a given organism.

Genotoxic: toxic to DNA; an agent or process that interacts with cellular DNA either directly or after metabolic transformation; mutagens are genotoxic agents.

Grade (grading): a subjective evaluation of the morphologic characteristics of a neoplasm based on the degree of anaplasia and proliferation evident from microscopic examination; a measure of the degree of malignancy.

Growth factors: agents that contribute to and stimulate tissue growth.

Hamartoma: a localized overgrowth of differentiated cells that have an altered growth pattern in relation to the tissue in which they are found, e.g., a nodule of disorganized striated muscle fibers within a normal skeletal muscle.

Hepatocarcinogenesis: the development of liver cancer.

Heterozygous: having different alleles at a specific position on a chromosome.

Histogenetic: pertaining to the origin, formation, or development of tissues from undifferentiated embryonic germ cell layers.

Homeostatic: pertaining to the natural state of equilibrium of the normal internal environment of the body; maintained by complex positive and negative feedback control mechanisms.

Homozygous: having identical alleles at a specific position on a chromosome.

Hyperplasia: a numerical increase in the number of normal-appearing cells within a tissue or organ.

Immune system: a primary defense system in the body capable of attacking and potentially destroying cancer cells; consists of lymphoid and related tissues from which cells are recruited to produce antibodies or to directly attack cancer cells (see Immune System).

Initiation: the first operational phase of the process of carcinogenesis during which a cell sustains a heritable alteration in DNA.

Malignant: a classification of anticipated biological behavior of neoplasms in which the prognosis for survival is poor; malignant neoplasms grow rapidly, invade, destroy tissue, and are usually fatal.

Mesenchymal cell: cells derived from embryonic mesoderm which constitute the supporting structure of tissue such as connective tissue, blood vessels, muscles, and bones.

Metaplasia: the substitution of one type of fully differentiated cells for the fully differentiated cell type normally present in a given tissue.

Metastasize: the spreading of neoplastic cells from a primary site of origin to a distant, noncontiguous site where their growth occurs.

Mitogenic: stimulating cell proliferation or division; causes mitosis.

Mutation: a structural alteration in DNA that is hereditary and may give rise to an altered phenotype (see Mutagenesis).

Neoplasia: the process of the development of neoplasms; essentially synonymous with carcinogenesis.

Neoplasm: new and typically abnormal growth which is generally uncontrolled and becomes progressively more serious with time.

Neurofibromatosis: a hereditary condition of the nervous system and other tissues of the body characterized by development of numerous neoplasms (neurofibromas) distributed over the entire body.

Nucleotide: a biochemical component of DNA that consists of a purine or pyrimidine base, a ribose or deoxyribose sugar, and a phosphate group; a basic building block of DNA.

Oncogene: a so-called cancer gene because alterations in its structure or expression are typically associated with neoplasms; an activated form of a protooncogene.

Oncogene activation: the process whereby a protooncogene is altered such that it stimulates enhanced cellular growth; several different mechanisms can lead to such activation.

Oncology: the study of neoplasia or carcinogenesis.

Phenotype: the physical appearance, biochemical makeup, and physiological behavior of an individual.

Preneoplasia: refers to the recognizable structural changes in a tissue that are sometimes antecedent to the development of neoplasia; the presence of preneoplasia indicates an increased probability for the development of neoplasia.

Progression: an operational phase of carcinogenesis associated with the development of an initiated cell into a fully malignant neoplasm; sometimes used in a more limited sense to refer to the change from a benign neoplasm to a fully malignant neoplasm.

Promotion: an operational phase of carcinogenesis in which there is enhancement of neoplasm formation when an agent (the promoter) is administered after exposure to a genotoxic carcinogen.

Protooncogene: a normal cellular structural gene that, when activated by mutations, amplifications, rearrangements, or viral transduction, functions as an oncogene and is associated with neoplasia; regulates normal processes related to cell growth and differentiation.

Retinoblastoma: an ocular neoplasm arising from germ cells of the retina.

Retroviruses: a large group of RNA viruses.

Stage (staging): a subjective assessment of the extent to which a neoplasm has spread in the body and, thus, an indication of the patient's clinical prognosis.

Threshold: the level of an agent below which no physiological, biochemical, or pathological effect can be measured.

Tumor: any tissue enlargement or swelling; frequently used as equivalent to a benign neoplasm.

Tumor suppressor gene: a gene that normally functions to suppress uncontrolled tissue growth by inhibiting the activity of oncogenes; sometimes called an antioncogene.

Weight of evidence: an approach for assessing the potential carcinogenic risk of an agent by considering all available information relative to the biological action of the agent.

Further Reading

Barrett, J. C. (Ed.) (1987). *Mechanisms of Environmental Carcinogenesis*, Vols. I and II. CRC Press, Boca Raton, FL.

Klein-Szanto, J. P., Anderson, M. W., Barrett, J. C., and Slaga, T. J. (Eds.) (1992). *Comparative Molecular Carcinogenesis*. Wiley-Liss, New York.

Maronpot, R. R. (1991). Chemical carcinogenesis. In *Handbook of Toxicologic Pathology* (W. Haschek and C. G. Rosseaux, Eds.), pp. 99–129. Academic Press, New York.

Pitot, H. C. (1986). *Fundamentals of Oncology*, 3d ed. Dekker, New York.

Tannock, I. F., and Hill, R. P. (Eds.) (1987). *The Basic Science of Oncology*. Pergamon, New York.

—*R. R. Maronpot*

Related Topics

Carcinogen Classification Schemes
Carcinogen–DNA Adduct Formation and DNA-Repair
Cell Proliferation
Chromosome Aberrations
Epidemiology
Immune System
International Agency for Research on Cancer
Mechanisms of Toxicity
Molecular Toxicology
Mouse Lymphoma Assay
Mutagenesis
Radiation Toxicology
Skin Toxicity
Toxicity Testing, Carcinogenesis

Cardiovascular System

Introduction

The scope of this entry includes the toxic effects of drugs, pesticides, other organic chemicals, metals, inorganic chemicals, and complex mixtures, such as cigarette smoke, on the human cardiovascular system (CVS). The CVS consists of the heart and the vasculature. In the latter category are the arteries, arterioles, veins, venules, and the capillaries. The focus will be on the heart and the arteries because they are where the bulk of the toxic effects occur. The first part of this entry describes the normal anatomy and the physiological processes which underlie the proper functioning of the heart. This is followed by a brief description of specific effects of selected organic and inorganic chemicals on heart function. The second part of the entry describes the normal anatomy and the physiological processes which underlie the proper functioning of the vascular system (principally, the arteries) and the responses of the vascular system to a variety of toxic agents. An extensive, but by no means complete, list of agents that are toxic to the heart and a proposed mechanism of action for each is found in Table C-11. A comparable list of agents that are toxic to the vasculature and their proposed mechanism(s) of action is found in Table C-12. (Although the point is made elsewhere in this volume, it is worth recalling here a central tenet of toxicology, as stated by the sixteenth century physician Paracelsus: "All substances are poisons. There is none that is not a poison. The right dose differentiates a poison and a remedy." Many of the substances listed in Tables C-11 and C-12 have normal and necessary, even critical, roles in the normal functioning of the body, when they are present at the proper dose. When

TABLE C-11
Cardiotoxic Agents[a]

Agent (chemical class or use category)	Proposed mechanism(s)	Associated organ/system effect(s)
Substituted aliphatic hydrocarbons		
Haloalkanes	Negative chronotropic, inotropic, and dromotropic effects that depress heart rate, contractility, and conduction	Cardiotoxicity exceeds that of similar chain-length unsubstituted hydrocarbons; maximum toxicity at 4 Cl atoms
Chloroform	Arrhythmias	Sensitizes heart to endogenous catecholamines
Cyclopropane and diethylether	Arrhythmias	Sensitize heart to catecholamines
Freons (fluorocarbons)	Reduces cardiac output and coronary flow	Reflex increases in sympathetic and parasympathetic impulses to heart via respiratory tract mucosal irritation
Haloanesthetics (halothane, methoflurane, and enflurane)	Negative chronotropic, inotropic, and dromotropic effects: possible cardiac arrest	Myocardial depression
Substituted ethanes	Negative inotropism	
Alcohols and aldehydes		
Acetaldehyde	Negative inotropic effects (after moderate ethanol intake)	Release of catecholamines and resulting sympathetic effects (at higher doses); toxicity diminishes with increasing aldehyde chain length
Ethanol	Decreases cardiac contraction; causes arrhythmias and ventricular fibrillation with sudden death (after chronic exposure); cardiomegaly	Pulmonary congestion; congestive heart failure; depression of oxidative phosphorylation in heart mitochrondria; interstitial fibrosis and increased lipid in muscle cells
PEG 500	Enhancement of the pressor effects of epinephrine	
Propylene glycol	Enhancement of arrhythmogenic effects of digitalis	
Heavy metals		
Barium	Potent arrhythmogen; produces ventricular tachycardia	Greatly prolongs action potential; antagonism of Ca^{2+} ion shortens action potential
Cadmium		
Acute	Prolongation of PR interval; heart failure in diastole	
Chronic	Cardiac hypertrophy and vacuolation in the Purkinje cells	
Cobalt	Cardiac lesions; heart failure	Antagonism of endogenous Ca^{2+}; complexes of cobalt with macromolecules
Lanthanum	Effects on sarcolemmal ion channels	Blocks Ca^{2+} channels
Lead		
Prenatal	Postnatal sensitization to the arrhythmogenic effects of norepinephrine	
Adult	Negative inotropism; EKG abnormalities and rhythm changes; deformation of T wave; prolongation of PR interval	Displacement of Ca^{2+}; interference with Ca^{2+} availability; interference with energy metabolism and ATP synthesis in heart
Manganese	Effects on sarcolemmal ion channels	Blocks Ca^{2+} channels
Nickel	Effects on sarcolemmal ion channels	Blocks Ca^{2+} channels
Vanadium	Both positive and negative inotropic effects *in vitro* depending on species; decrease of left ventricular contraction	Inotropic changes related to alteration in available surface Ca^{2+}; effects upon phosporylation reactions; inhibition of NA^+, K^+ ATPase
Gases		
Carbon disulfide	Angina pectoris	Formation of thiocarbamates; inhibition of dopamine hydroxylase; disruption of lipid thyroxine metabolism, development of coronary heart disease
Carbon monoxide (acute)	Tachycardia; bradycardia; extra systoles; increased demand for oxygen by the heart; production of angina pectoris; myocardial infarction	Interference with myocardial energy metabolism

continues

Continued

Agent (chemical class or use category)	Proposed mechanism(s)	Associated organ/system effect(s)
Drugs		
Cardioactive drugs		
Antiarrhythmics	Decreased conductivity and automaticity of the myocardium	
Quinidine and procainamide	Prolongation of QRS and QT intervals; ventricular fibrillation after iv injection; extra systoles; low doses accelerate while large doses prolong AV conduction; cardiac arrest	
Lidocaine	Sinus bradycardia; depressed automaticity of Purkinje fibers and myocardial cells; depresses myocardial contractility	Shortened action potentials of Purkinje fibers and myocardial cells
Phenytoin	Suppression of automaticity; cardiac arrest	
Adrenergic agonists		
Epinephrine and isoproterenol	Positive inatropic and chronotropic effects; ST segment deviation, ectopic beats, and subendocardial necrosis	Myocardial hypoxia; cellular Ca^{2+} overload
Isoproterenol (only)	Hypercontraction of myofibrils in apical subendocardium; appearance of donut-shaped granules in mitochondria; myocytolysis	Excessive Ca^{2+} influx
Adrenergic antagonists as well as serpine and guanethidine	Decreased cardiac contractility; production of AV block; heart failure (effect of overdose); angina and possible myocardial infarction (effects of withdrawal)	Receptor of supersensitivity; excess numbers of receptors
Glycosides of digitalis, strophanthin, and oleandrin	Increase in cardiac contractility, irritability, and arrhythmias; premature ventricular contraction; prolonged PR interval	Inhibit the sarcolemmal Na^{2+} pump (Na^+, K^+ ATPase) with elevation of intracellular Ca^{2+} via Na^+, Ca^{2+} pump; ventricular fibrillation; complete heart block
Nicotine	Arrhythmias	Suppresses K^+ conductance
Vasodilators and antihypertensives (hydralazine, diazoxide, minoxidil)	Similar effects to epinephrine, via reflex tachycardia during hypotension	Suppresses K^+ conductance
Ca^{2+} antagonists		
Bepridil	Negative chronotropic and inotropic effects	Blocks slow Ca^{2+} channels; depressed Ca^{2+} release from the SR
Papaverine	Negative chronotropic and inotropic effects	Blocks slow channels; inhibits phosphodiesterase and elevates cAMP
Verapamil and nifedipine	Negative chronotropic and inotropic effects	Excitation contraction uncoupling; block both slow Ca^{2+} and Na^+ channels; depress or block Ca^{2+} influx into myocardial cells
CNS active drugs		
Amphetamine and cocaine	Increased heart rate; blood pressure increase causing great risk when there is preexisting angina, hypertension, and atherosclerosis	Increased work load on the heart
Imipramine and amitryptyline	Low doses enhance cardia contractility, whereas high doses depress it as well as coronary flow and heart rate; prolongation of the PR, QRS, and QT interval; bundle branch block; supraventricular and ventricular arrhythmias	Cardiac arrest; catecholamines reuptake inhibition; anticholinergic effects
Lithium	Ventricular arrhythmias and, in rare instances, myocardial lesions	
MAO inhibitors	Palpitations	Exaggerated sympathomimetic effects
Marijuana	Positive inotropic and chronotropic effects; premature ventricular contractions; enhanced ventricular automaticity	Facilitation of SA and AV nodal conduction; increased workload on heart

continues

Continued

Agent (chemical class or use category)	Proposed mechanism(s)	Associated organ/system effect(s)
Methyldopa	Focal or diffuse interstitial infiltration with eosinophils, lymphocytes, and plasma cells	Hypersensitivity myocarditis
Methylsergide	Endomyocardial fibrosis; valvular defects	
Neuroleptics (phenothiazines and butyrophenones)	Tachyarrhythmias; hypotension; ventricular tachycardia and fibrillation (rare); conduction defects; prolongation of QT interval; abnormalities in T wave; sinus tachycardia; widening of QRS complex	Quinidine-type toxicity; peripheral alpha-receptor blockade; central and peripheral anticholinergic actions
Barbiturates	Depression of myocardial contractility	Inserts in lipid bilayer of membrane; stabilizes membranes
Chemotherapeutic agents		
Antimicrobial antibiotics	Weak negative inotropic effects	Depressed Ca^{2+} uptake
Antineoplastic antibiotics		
Anthracyclines (doxorubicin and daunorubicin)	Arrhythmias (acute); congestive cardiomyopathy (after chronic use); cardiac dilation, atrophy, and degeneration of the myocytes; interstitial edema and fibrosis	Possibly due to histamine release; generation of reactive oxygen; peroxidation of membrane lipids and consequent changes in membrane structure
5-Fluorouracil	Myocardial ischemia; cardiac arrest	
Cyclophosphamide (large doses)	Myocardial capillary microthrombosis; pericarditis	Cardiac failure
Emetine	Sinus tachycardia; dose-related arrhythmias; myocardial necrosis; ventricular fibrillation	Conduction disturbances; effects on K^+ ion movements
Monensin and lasalocid	Positive inotropic effect; increased cardiac output; occasional increase in heart rate and automaticity; increased coronary blood flow	Increased excitation–contraction coupling; enhanced metabolism of cardiac cells; increased sarcolemmal cationic trap for Na^+ (Iasalocid: cationic trap for K^+)
Penicillin and sulfonamide	Focal or diffuse interstitial infiltration with eosinophils, lymphocytes, and plasma cells	Hypersensitivity myocarditis
Carcinogenic agents		
1,3-Butadiene and nitrosamines	Sarcoma formation within heart	Induction of chemical carcinogenesis
Agents and drugs producing cardiovascular teratogenesis		
bis(Cichloroacetyl) diamine		Ventricular septal defects; dextrocardia; ectopia; tetralogy of Fallot; pulmonic stenosis
Caffeine		Ventricular septal defects
Cortisone		
Dextroamphetamine		Ventricular and atrial septal defects
Ethanol		Ventricular septal defects
Phenobarbital		
Salicylate and indomethacin		Ventricular septal defects
Toxins		
Batrochotoxin	Ventricular arrhythmia; fibrillation; positive inotropic effects	Increase in resting Na^+ permeability; actions upon protein constituents of Na^+ channel
Cobra venom cardiotoxin	Systolic arrest; disruption of myocardial cell membranes and myofibrils	Depression of Ca^{2+} accumulation in SR; inhibition of Ca^{2+} ATPase; SR membranes become more leaky; depression of Ca^{2+} accumulation in mitochrondria; ultimate Ca^{2+} overload

continues

Continued

Agent (chemical class or use category)	Proposed mechanism(s)	Associated organ/system effect(s)
Endotoxin	Reduced coronary perfusion; depression of contractility; negative inotropic and chronotropic responses to NE and histamine	Depression of Ca^{2+} ATPase activity; depression of Ca^{2+} uptake; reduced Ca^{2+} release by action potentials
Grayanotoxins	Positive inotropic action	Increased Na^{2+} permeability; opens voltage-dependent Na^+ channels
Sea anenome toxins (ATX-11 and CTX)	Conduction defects; negative chronotropic effect; positive inotropic effects	Greatly slows inactivation of Na^+ channels
Scorpion neurotoxin	Positive chronotropic effects; induction of fibrillation	Slows closing of Na^+ channel and opening of K^+ channels; causes neurotransmitter release
Tetrodotoxin and saxitoxin	Conduction defects	Blocks fast Na^+ channels
Vertridine and acondine	Induction of cardiac arrhythmias; increase or enhancement of automaticity; positive inotropic effect (veratridine)	Increased resting Na^+ permeability; slowed inactivation of Na^+ channel with prolonged action potentials
Volvatoxin A	Cardiac arrest in systole; increase in diastole resting tension of cardiac muscle	Makes SR membrane leaky to Ca^{2+}; alteration of ultrastructure of mitochondria, thereby inhibiting ability to accumulate Ca^{2+}

[a] From G. M. Cohen (Ed.), Target Organ Toxicity, Vol. 2, pp. 32–37, CRC Press, Boca Raton, FL, 1986.

they are present at too high a dose—a normal dose of a substance at the wrong place or wrong time is the equivalent of too high a dose—toxic responses may result.)

The Heart

The Heart as a Pump

From a simple mechanical point of view, the human heart can be viewed as two side-by-side pumps within one enclosure. Each of these pumps propels a critical fluid, i.e., blood, through a different pipeline circuit. Each pump consists of two chambers—an atrium and a ventricle (Fig. C-14). The ventricles are thicker walled and are positioned beneath the atria. The right side of the heart receives blood flow from the major veins, the superior (i.e., from the head and upper body) and inferior (i.e., from the rest of the body) venae cavae, and pumps the blood through the pulmonary circuit. As the blood passes through the lungs, carbon dioxide is exchanged for oxygen so that the blood pumped through the second (systemic) circuit to the rest of the body will be oxygenated appropriately. All four chambers of the heart are muscular. The ventricles are thicker-walled than the atria because the atria function essentially as reservoirs for blood between heart contrac-

tions, whereas the ventricles are primarily responsible for actually pumping the blood through the two circuits. The left ventricle is the largest and thickest of the four chambers since its role is to pump the blood under relatively high pressure into the aorta, the largest of the body's arteries, and from there to the rest of the body, while the right ventricle pumps blood to the lungs via the pulmonary artery. The chambers within each pump are connected to each other and to the blood vessels entering and exiting them by sets of valves. The two semilunar valves prevent leakage of blood back into (a) the right ventricle after blood has been pumped to the pulmonary artery and (b) the left ventricle after blood has been pumped to the aorta. The two atrioventricular valves prevent leakage of blood back into the atria after it has flowed to the ventricles. These two sets of valves function out of sequence with each other. The semilunar valves are closed when the ventricles are in the resting phase (diastole) of the heart beat, i.e., between contractions of the heart. The atrioventricular valves are open during diastole, thereby allowing the two ventricles to fill with blood from the corresponding atria. During the contraction phase (systole) of the heartbeat the two ventricles develop pressure and eject the blood into the pulmonary artery and aorta, respectively. At this time the atrioventricular valves are closed and the semilunar valves are open. When there is no

TABLE C-12
Vasculotoxic Agents[a]

Agent (chemical class or use category)	Vascular effect and/or primary site	Associated disease state and/or mechanism
Heavy metals		
Arsenic (arsine)	Arteriosclerosis	Peripheral vascular disease
	Pulmonary vascular lesions	Noncirrhotic portal hypertension; pulmonary edema
Beryllium	Decreased hepatic flow: hemorrhage	Occlusion of hepatic venous flow; atherosclerosis; hypertension
Cadmium	Aortic damage to endothelium allowing lipid deposition; lesions in uterine endothelial cleft; renal arteriolar thickening; effect on microcirculation	
Chromium (deficiency)	Atherosclerotic aortic plaques	Atherosclerosis; elevated serum cholesterol
Copper	Acceleration of atherosclerosis	
Chronic	Hypotension	
Acute		
Copper (deficiency)	Aortic aneurysms	
Germanium	Hemorrhage; edema in lungs and GI tract	
Indium	Hemorrhage and thrombosis in the kidney and liver	
Lead	Damage to endothelial cell with changes in blood–brain barrier permeability; changes in arterial elasticity; effects on ground substance; sclerosis of vessels in the kidney	Encephalopathy; hypertension
Mercury	Preglomerular vasoconstriction: glomerular immune complex deposits; lesions of the aorta; opening of blood–brain barrier	Glomerulonephritis; inhibition of amino acid uptake
Selenium	Atherosclerotic plaques	Atherosclerosis
Thallium	Perivascular cellular infiltration in the brain (cuffing)	
Industrial and environmental agents		
Allylamine	Renal artery lesion; intimal smooth muscle proliferation in coronary arteries	Endogenous formation of acrolein with destruction of vascular protein and nucleic acid components
β-Aminopropionitrile	Aortic lesions and atheroma formation	Damage to vascular connective tissue matrix; aneurysm
Boron	Hemorrhage; edema; increase in microvascular permeability in the lung	Pulmonary edema
Carbamylhydrazine HCl	Tumors of pulmonary blood vessels	Cancer
Carbon disulfide	Microvascular effect on ocular fundus and retina; direct injury to endothelial wall; promoter of atheroma formation	Coronary vascular disease Atherosclerosis
Chlorophenoxy herbicides		Hypertension
Dimethylnitrosamine	Decreased hepatic flow; hemorrhage; necrosis; pulmonary artery lesions; coronary vessel lesion	Occlusion of veins
4-Fluoro-10-methyl-12-benzyanthracene		
Glycerol	Strong renal vasoconstriction	Acute renal failure
Hydrochloric acid (aspiration of stomach contents)	Increased microvascular permeability	Pulmonary edema
Hydrogen fluoride	Hemorrhage; edema in the lung	Pulmonary edema
Paraquat	Vascular damage in lungs and brain	Cerebral purpura
Pyrrolizidine alkaloids	Pulmonary vasculitis; damage to vascular smooth muscle cells; proliferation of endothelium and vascular connective tissue in the liver	Pulmonary hypertension; hepatic venoocclusive disease
Organophosphate pesticides		Cerebral arteriosclerosis
Vinyl chloride	Portal hypertension; tumors of hepatic blood vessels	Cancer
Gases		
Auto exhaust	Hemorrhage and infarct in cerebral hemispheres; atheroma formation in aorta	Atherosclerosis due to CO content
Carbon monoxide	Damage to intimal layer; edema; atheroma formation	Atherosclerosis

continues

Continued

Agent (chemical class or use category)	Vascular effect and/or primary site	Associated disease state and/or mechanism
Nitric oxide	Vacuolation of arteriolar endothelial cells; edema; thickening of alveolar–capillary membranes	Pulmonary edema
Oxygen	Vasoconstriction–retinal damage; increased retinal vascular permeability–edema; increased pulmonary vascular permeability–edema	Blindness in neonate; shrinking of visual field in adults; pulmonary edema
Ozone	Arterial lesions in the lung	Pulmonary edema
Drugs and related compounds		
Antibiotic–antimitotics		
Cyclophosphamide	Lesions of pulmonary endothelial cells	
5-Fluorodeoxyuridine	GI tract hemorrhage; portal vein thrombosis	
Gentamicin	Long-lasting renal vasoconstriction	Renal failure
Vasoactive agents		
Amphetamine	Cerebrovascular lesions secondary to drug abuse	Disseminated arterial lesions similar to periarteritis nodosa
Dihydroergotamine	Spasm of retinal vessels	
Ergonovine	Coronary artery spasm	Angina
Ergotamine	Vasospastic phenomena with and without thrombosis; medial atrophy	Gangrene of peripheral tissues
Epinephrine	Peripheral arterial thrombi in hyperlipemic rats	Participates in thrombogenesis
Histamine	Coronary spasm; damage to endothelial cells in hepatic portal vein	
Methysergide	Intimal proliferation; vascular occlusion of coronary arteries	Coronary artery disease
Nicotine	Alteration of cytoarchitecture of aortic endothelium; increase in microvilli	
Nitrites and nitrates	"Aging" of coronary arteries	Repeated vasodilation
Norepinephrine	Spasm of coronary artery; endothelial damage	
Metabolic affectors		
Alloxan	Microvascular retinopathy	Diabetes; blindness
Chloroquine	Retinopathy	
Fructose	Microvascular lesions in retina	Diabetes-like condition
Iodoacetates	Vascular changes in retina	
Anticoagulants		
Sodium warfarin; warfarin	Spinal hematoma; subdural hematoma; vasculitis	Uncontrolled bleeding; hemorrhage
Radiocontrast dyes		
Metrozamide; metrizoate	Coagulation; necrosis in celiac and renal vasculature	
Cyanoacrylate adhesives		
2-Cyano-acrylate-*n*-butyl	Granulation of arteries with fibrous masses	
Ethyl-2-cyanoacrylate	Degeneration of vascular wall with thrombosis	
Methyl-2-cyanoacrylate	Vascular necrosis	
Miscellaneous drugs and compounds		
Aminorex fumarate	Intimal and medial thickening of pulmonary arteries	Pulmonary arterial hypertension
Aspirin	Endothelial damage; gastric erosion obliteration of small vessels; ischemic infarcts	Changes in the basement membrane of endothelial cells
Cholesterol; oxygenated derivatives of cholesterol; noncholesterol steroids	Atheroma formation; arterial damage	Atherosclerosis
Homocysteine	Increase of vascular fragility; loss of endothelium; proliferation of smooth muscle cells; promotion of atheroma formation	Atherosclerosis; effects on protein syntheses
Oral contraceptives	Thrombosis in cerebral and peripheral vasculature	Thromboembolic disorders

continues

Continued

Agent (chemical class or use category)	Vascular effect and/or primary site	Associated disease state and/or mechanism
Penicillamine	Vascular lesion in connective tissue matrix of arterial wall; glomerular immune complex deposits	Glomerulonephritis; inhibits synthesis of vascular connective tissue
Talc and other silicates	Pulmonary arteriolar thrombosis; emboli	
Tetradecylsulfate Na	Sclerosis of veins (used as a sclerosing agent)	Cytotoxicity
Thromboxane A_2	Extreme cerebral vasoconstriction	Cerebrovascular ischemia

[a] From M. O. Amdur, J. Doull, and C. Klassen (Eds.), *Casarett and Doulls' Toxicology—The Basic Science of Poisons* Pergamon, New York, 1991.

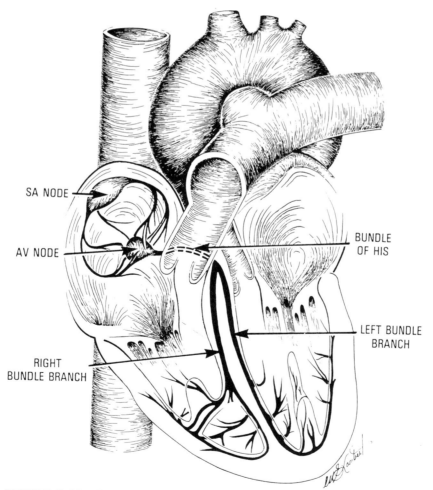

FIGURE C-14. *Diagram of the electrical pacemaking and conducting system of the heart. SA, sinoatrial node; AV, antrioventricular node (reproduced with permission from West, J. B. (Ed.), Best & Taylor's Physiological Basis of Medical Practice, 11th ed., Williams & Wilkens, Baltimore, 1985).*

injury or disease, the heart is more efficient, durable, and reliable than any man-made pump. At approximately 72 beats per minute over an 80-year life span, a heart will beat approximately 3,000,000,000 times. Two major factors underlie this organ's unique characteristics: the nature of the heart muscle and the heart's specialized electrical conduction system.

Cardiac Muscle

There are three types of muscle tissue in the body. Striated muscle tends to be attached at least at one end to the skeleton, and so is also called skeletal muscle. Since it can be contracted or relaxed voluntarily it is also called voluntary muscle. Smooth muscle is usually arranged in sheets or layers and is found in arteries and veins (see Blood Vessels), in the gastrointestinal and respiratory tracts, in the uterus, and at other sites. It is under control of the autonomic (involuntary) nervous system. The third type of muscle tissue and the only one found in the heart is cardiac muscle. Small amounts of it are also found in the superior vena cava and in the pulmonary vein. Cardiac muscle is also classified as involuntary (see Impulse Conduction). In the heart, cardiac muscle cells are joined end to end to form fibers. Gap junctions, specialized connections between these cells, are thought to facilitate the conduction of electrical impulses through the heart muscle. There are small differences between the cardiac muscle cells of the atria and the ventricles. In addition, there are other cardiac muscle cells that are specialized either to initiate or to conduct an electrical impulse for contraction through the heart.

Impulse Conduction

The continued efficiency of the heart as a pumping system is due to the synchronous contractions of the left and right sides of the heart and to all the events of the cardiac cycle following in sequence. Both of these sets of events depend on the well-controlled conduction of electrical impulses. These impulses arise in the self-firing cells of the sinoatrial (SA) node, the so-called "pacemaker." The SA node is located at the junction of the superior vena cava and the right atrium. A wave of depolarization (see below) originating at the SA node is conducted first to the cells of the right atrium, then to cells in both atria, and then converges on a second

group of specialized cardiac cells, the cells of the atrioventricular (AV) bundle. These cells conduct the impulse from the SA node to the AV node, which lies at the junction of the median wall of the right atrium and the septum which separates the two ventricles. The impulse wave next passes into the special conducting system (the bundle of His and Purkinje fibers) located in the wall between the ventricles. The depolarizing wave thus reaches the two ventricles, causing them to contract. What is the basis of the electrical excitability of these cells?

If a microelectrode is inserted into any unexcited ("resting") muscle or nerve cell, a potential difference will be recorded across the membrane of that cell. In the case of a cardiac muscle cell, this resting potential is -90 mV (inside the cell relative to the outside of the cell). In other words, the cell is polarized. In the resting cardiac muscle cell, the concentration of the positively charged potassium ion is higher inside the cell than outside, while the positively charged sodium ion is at much higher concentration outside the cell than inside. Energy is required to maintain the appropriate resting state distributions of the different ions on either side of the cell membrane. In the case of sodium and potassium ions, there is a pump, energy for which is derived from hydrolysis of the terminal phosphate from the molecule adenosine triphosphate (ATP). The enzyme responsible for this hydrolysis and for the maintenance of the proper ionic activities on either side of the membrane is called the sodium–potassium ATPase. When an electrical stimulus is received by a cardiac muscle cell, the membrane permeability to these ions changes, sodium ions move into the cell, some potassium moves out, the potential between the two sides of the cell membrane starts to disappear, and the cell becomes relatively depolarized. As depolarization progresses, the membrane potential will reach the threshold potential (approximately -70 mV for most cardiac muscle cells). Depolarization past this point results in an action potential which completely depolarizes the cell. At the peak of the action potential the inside of the cell membrane is positive relative to the outside ($+30$ mV). The cell membrane then repolarizes relatively slowly and reaches the -90 mV resting potential before it can be excited again. Not only does the wave of depolarization move very rapidly across the membrane of the cardiac muscle cell but also, because of the specialized gap junctions between the cells, the depolarization wave is propagated to adjacent cells causing their complete depolarization. The overall response is called the propagated action potential and this is what initiates the contraction of the bulk of the heart muscle cells.

The majority of cardiac muscle cells display a fast response action potential (Fig. C-15). The cells in the

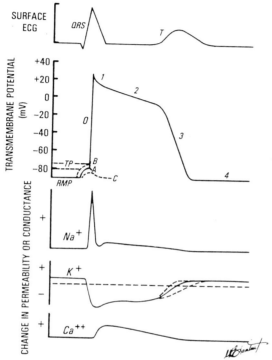

FIGURE C-15. *Relations between the surface electrocardiogram (ECG), the transmembrane potential, and diagrams of the relative changes in ion permeability (or conductance) that are responsible for the action potential. The (0–4) phases of a typical fast action potential are indicated. The resting membrane potential (RMP) is phase 4. The threshold potential (TP) is also shown. C, a subthreshold stimulus (nonpropagated); A, a threshold stimulus; B, a threshold stimulus at a lower (less negative) threshold potential (dashed line). The large rapid increase in Na^+ conductance during phase 0 is shown, together with the sustained small increase in Na^+ conductance (slow Na^+ current) during phase 2 of the action potential. The large, less rapid drop in K^+ conductance during phase 0 of the action potential is indicated and the dashed lines of altered K^+ conductance during phase 3 indicate induced changes in PK. The dashed horizontal line indicates that K^+ conductance is above zero at rest. The increased Ca^{2+} conductance during phase 2 of the action potential is shown (reproduced with permission from West, J. B. (Ed.), Best & Taylor's Physiological Basis of Medical Practice, 11th ed., Williams & Wilkens, Baltimore, 1985).*

atria and ventricles exhibit a rapid conduction velocity. The depolarization–action potential–repolarization process is characterized by five phases. In phase 0, at the threshold potential, so-called fast sodium channels in the membrane open and sodium ions rush in and initiate the action potential. These fast response cells conduct the impulses rapidly. At the end of phase 0 the cell is completely depolarized. Toward the end of phase 1 and start of phase 2 the sodium influx starts to decrease, as does the membrane potential. During the relatively long (200–300 msec) phase 2 plateau, calcium and sodium ions enter the cell through so-called slow channels. Movement of ions through these slow channels only takes place after the membrane potential has dropped to approximately −55 mV, i.e., after the fast sodium ion current has ceased. While these slow inward currents occur there is also an outward, partly balancing movement of some potassium ions which helps keep the plateau relatively steady. The calcium influx in phase 2 triggers the process known as excitation–contraction coupling, in which the myosin thick filaments slide past the thin actin filaments in the contractile unit of the muscle known as the sarcomere. This process requires energy and involves activation of a myosin ATPase which hydrolyzes ATP. The released energy is utilized to form cross-bridges between the actin and myosin molecules. Both the velocity and the force of contraction are dependent on the amount of calcium ions that reach the site of contraction. Within the resting muscle cell, calcium is sequestered in a compartment called the sarcoplasmic reticulum. During the action potential, calcium and sodium ions that enter the cell cause depolarization of the sarcoplasmic reticulum membrane resulting in release of large amounts of calcium which are needed for contraction. Between contractions, the calcium is once again sequestered in the sarcoplasmic reticulum so that the actin–myosin interaction is not favored. During the long duration of the plateau phase a new action potential cannot be initiated because the fast sodium channels are inactivated. During phase 3, membrane permeability to potassium increases and the slow calcium and sodium channels become inactivated. The membrane becomes repolarized and the resting potential returns at phase 4.

In contrast, conduction velocity is slow in muscle fibers at the SA and AV nodes. Unlike the majority of cardiac muscle cells, the pacemaker cells have an unstable resting potential (approximately −60 mV) due to a cell membrane alteration that allows sodium ions to leak into the cell while potassium ions have difficulty leaking out. This occurs through so-called slow channels. This leakage lowers the membrane potential allowing more sodium ions to move into the cell. In addition to the inward sodium movement, there is an inward movement of calcium ions, causing the pacemaker cell inner membrane to become relatively more positive. Finally, the cell becomes depolarized at approximately -40 mV. This phenomenon is called spontaneous diastolic depolarization. In pacemaker cells the phenomenon of spontaneous diastolic depolarization followed by an action potential is called automaticity. The overall effect is of pacemaker cells initiating waves of depolarization that move across the heart causing the muscle to contract. As noted previously, this phenomenon occurs approximately 72 times per minute (more during periods of excitement or physical activity and less during periods of relaxation).

Intrinsic Modulators of Cardiac Activity

The heart responds constantly to hormonal and nervous system signals. Sympathetic nervous system terminals containing norepinephrine are found in cardiac cells in the atria and ventricles. This allows for reflex regulation of heart muscle contractility. Sympathetic nerves are also present at the SA node and AV junction, where they act to increase heart rate (enhanced phase 4 depolarization) and also to improve conduction velocity by decreasing conduction time through the AV junction. Parasympathetic innervation is via fibers from the 10th cranial nerve, the vagus nerve, to the SA node, AV junction, and the atria. These fibers release the neurotransmitter acetylcholine, which slows SA node activity (decreasing the rate of phase 4 depolarization) and decreases conduction throughout the AV junction. Atrial contractility is also decreased.

Let us consider a practical example of nervous system control of activity in a healthy cardiovascular system. If blood pressure increases, pressure receptors in the carotid sinus and aortic arch will sense the stretching of the artery wall and will cause impulses to be sent to the cardiovascular regulatory centers in the medulla. Impulses from here will result in increased vagus nerve activity with resulting decreases in heart rate, peripheral vascular resistance and venous return. This will all result in a lowering of the blood pressure. On the other hand, if a sudden decrease in blood pressure is sensed,

vagal stimulation will be depressed in favor of sympathetic stimulation. This will lead to increases in heart rate, myocardial contractility, venous return, peripheral vascular resistance, and cardiac output. (The latter is the product of heart rate × stroke volume.) In addition, this overall process will be augmented by release of the naturally occurring catecholamines, epinephrine and norepinephrine, which are produced in the medulla of the adrenal gland. Three classes of molecules (adrenergic receptors) that bind these compounds are present in both the heart and the blood vessels. The response to each of these compounds is a function of the type of receptor that is involved in binding. Adrenergic receptors in the heart are of the β-1 type. Following binding to these receptors in the myocardium, catecholamines produce enhanced atrial and ventricular contraction. When they bind at the SA node they will produce an increase in heart rate, and when they bind at the AV junction, increased AV conduction occurs.

Pathologic Changes in the Heart

The major pathologic changes that occur in the heart are associated with effects on heart rate, contractility of heart muscle, or electrical conduction. Regarding heart rate changes, an arrhythmia, as the name indicates, is a loss of rhythm and here refers to an irregularity of the heart beat. Two of the more common forms are tachycardia, which is an abnormally rapid heart beat, and fibrillation, which is a rapid twitching of the muscle fibrils. Either of these can occur in the atria or the ventricles. Agents which alter ion levels and fluxes and thereby alter aspects of impulse transmission can produce arrhythmias. The most common site of arrhythmias is the SA node, which is the site of origin of the impulses transmitted through the heart. If depolarization after an action potential is accelerated or delayed an aberrant action potential can be triggered.

Another set of pathologic changes is associated with effects on the force of contraction. Heart muscle exhibits a higher rate of oxygen consumption and a greater energy requirement than any other tissue. Thus, impaired contraction can result from interference with any of the major cycles critical for proper energy metabolism or from processes that interfere with delivery or utilization of the optimum levels of oxygen. For example, if blood flow through the coronary arteries is occluded, as occurs during atherosclerosis, there will be decreased delivery of oxygen to the heart muscle.

When this occurs acutely a myocardial infarction may result. Even if death does not occur there will likely be a decrease in the force of contraction of the heart muscle. "Recreational" use of psychoactive drugs (e.g., amphetamines and cocaine) can result in profound and sudden cardiovascular responses including increases in blood pressure and heart rate. These effects can be life-threatening in individuals with underlying, and possibly previously unknown, cardiovascular problems including coronary artery disease, high blood pressure, or cerebrovascular disease.

Cells with high energy requirements, such as heart cells, have large numbers of the organelles called mitochondria which produce and supply the small molecule, ATP. When the bond connecting the terminal ATP phosphate to its neighbor is broken, a large amount of energy is released, which is now available to drive cellular processes. Enzymes are organic catalysts that interact with specific substrate molecules to help speed up chemical reactions. Enzymes that catalyze the splitting of the terminal phosphate from ATP, with its attendant release of energy, are called ATPases. The myosin ATPase involved in muscle contraction was mentioned previously and ATPases involved in the energy-driven pumping of ions, including sodium, potassium, and calcium, were mentioned previously and are noted again below. During oxidative metabolism of organic substrates, the process of electron transport to molecular oxygen in mitochondria is coupled to oxidative phosphorylation, which yields ATP. Some anesthetics and cell poisons, such as cyanide, interfere with electron transport and/or uncouple phosphorylation. This causes a direct decrease in the amount of energy available to the heart muscle and results in reduced contractility.

As noted previously, the inward calcium ion movement is vital for the contraction of cardiac muscle. This inward movement is blocked by calcium antagonists and is stimulated by the catecholamines via binding to the beta receptors, which leads to increases in the intracellular level of cyclic AMP, a compound that helps mediate numerous metabolic responses within cells. This, in turn, leads to increased availability of calcium ions for interaction with the contractile proteins. The same effect can be achieved by agents which increase the levels of free calcium ions outside the cells or by drugs that increase the levels of cyclic AMP inside the cells. Another mechanism for increasing intracellular calcium levels in cardiac cells involves the cardiac glycoside drugs, e.g., digitalis. This drug inhibits the ATPase, which pumps sodium ions out of cells. This results in elevation of sodium ion levels inside the cell, which in turn leads to increases in intracellular calcium ion levels and therefore increased contraction. There are two apparent effects of these drugs: There are increases in both the force of contraction and the rate at which the force is developed. Toxins that increase the permeability of the cardiac muscle cell membrane to sodium ion, e.g., the marine compound ciguatoxin or the poison toad active agent batrachotoxin, have a similar effect. On the other hand, agents that decrease membrane permeability to sodium ions will depress myocardial contractility. Included here are a diverse group of compounds including tetrodotoxin, from the Japanese pufferfish; the shellfish-derived poison, saxitoxin; some local anesthetics; and polyethylene glycol, the active ingredient in many antifreeze preparations. Local anesthetics, such as lidocaine and procaine, depress the fast inward sodium ion current, the slow inward current, and the potassium ion outward current. They tend to slow the heart rate and the force of contraction.

Drugs prescribed to alleviate one set of medical problems can have striking and sometimes fatal effects on the cardiac system. Antipsychotics derived from phenothiazine depress myocardial contractility and cardiac output. Chlorpromazine also can impair cardiac reflex mechanisms and cause a focal myocardial necrosis. Cyclophosphamide, an anticancer agent, also causes myocardial necrosis as well as changes in electrocardiogram patterns. Another anticancer agent, adriamycin, produces cardiomyopathies. Severe dysrhythmias and some cases of sudden death have been reported. Overdoses of the tricyclic antidepressants, e.g., amitryptaline, can result in severe cardiotoxicity, probably due to anticholinergic activity. At high doses, the antidepressant imipramine will depress contractility, lower heart rate, and depress cardiac output. Cardiac arrest may also occur. Some antibiotics, including gentamycin and neomycin, depress calcium ion uptake and therefore contractility of cardiac muscle. Although the sympathetic system transmitters, the catacholamines, are essential for maintenance of normal myocardial contractility, it has been recognized for nearly 100 years that when administered at higher than normal levels for extended periods of time, they can lead to severe myocardial necrosis.

Profound cardiotoxic responses can result from inhalation of a number of halogenated alkanes. These are low-molecular-weight hydrocarbons with some or all of the hydrogens being substituted by halogens, usually chlorine or fluorine. These agents depress heart rate, contractility of the muscle, and electrical conduction. The effects are generally more pronounced as the number of halogen atoms increases. Some of these compounds have an additional and profound effect—they sensitize the cardiac muscle cell to catecholamines, e.g., epinephrine. In people without preexisting cardiac disease, the effects of most of these compounds are reversible, although chronic exposure may cause some irreversible damage. As would be expected, the halogenated, hydrocarbon anesthetics such as halothane and enflurane have similar effects.

In contrast, low-pressure fluorocarbons, such as trichlorofluoromethane, can be particularly toxic. In most cases, the levels generally encountered in the environment are too low to have any major lasting effect and even at relatively high levels (up to 15%) fatalities are rarely, if ever, recorded. However, at levels much above this, e.g., over 20%, tragic results can ensue. Among youths who inhale these agents from closed bags to get high, fatalities can result because the levels of these agents in the bags can reach 35–40%.

There are compounds that interfere with the regular activity of calcium ions in cardiac cells, either by replacing them (as is the case with a number of heavy metals) or by altering the flux of calcium ions across the cell membrane. Among metals, lanthanum, manganese, and nickel all block calcium channels in the cell membrane. Both barium and cobalt ions antagonize endogenous calcium ion levels and tend to shorten the action potential. Lead ions have multiple effects, including displacement of calcium and interference with calcium ion availability, energy metabolism, and ATP synthesis in heart muscle cells. Among organic chemicals, the opium derivative, papaverine also blocks slow calcium ion channels. Cobra venom cardiotoxin and bacterial endotoxin both interfere with calcium ATPase activity and endotoxin also depresses calcium uptake by heart muscle cells.

Agents Causing Morphologic Changes

A number of cardiotoxic compounds have been listed to this point, including some that interfere with sodium–potassium ATPases; increase sodium or calcium influx; or depress myocardial function by replacing calcium, decreasing sodium permeability or altering contractility. These agents produce toxic responses in heart muscle often resulting in death, but do so without causing any major morphologic changes in the heart. Other cardiotoxic compounds produce characteristic morphologic lesions in heart muscle. There are three basic types of such pathological alterations. The first is toxic myocarditis. Chemicals which produce this effect cause cell damage and ultimately cell death. Whether they produce damage acutely or chronically is generally a function of the dose of toxic agent. The acute form is characterized by edema, i.e., accumulation of excess fluid, as well as by inflammatory cell responses and multiple regions of cardiac cell death. However, the inflammatory response will be attenuated or may even be absent when the toxic agent suppresses the immune system, e.g., with drugs given to prevent rejection of transplanted organs.

The second type of major morphological alteration in the heart arises from a sudden insufficiency or local arrest of the blood supply to the heart that can result in necrosis of a region of the heart. This condition is called an infarct(ion). In advanced arteriosclerosis, occlusion of the major arteries supplying the heart muscle with blood can result in myocardial infarction. Even in the absence of arteriosclerosis, infarctions can result, e.g., from amphetamine abuse, which produces severe inflammations of critical arteries (i.e., an arteritis). Intravenous drug use can cause infective endocarditis (an inflammation of the lining of the heart), which can lead to vessel occlusion with an embolus, thus resulting in an infarction. Cocaine abuse can result in ventricular tachycardia (i.e., rapid heart beat) and fibrillation, myocardial infarction, and sudden death. At higher doses, cocaine can increase the levels of catecholamines at the β-adrenergic receptors, ultimately resulting in increased calcium ion activity, accelerated heart beat, arrhythmias, etc. Chemicals that antagonize calcium ion movement through calcium-specific membrane channels prevent the ventricular arrhythmias induced by cocaine. In people free of apparent heart disease, gross myocardial infarction can result from toxic exposures to carbon monoxide, nitrates, ergot derivatives, and some potent anticancer drugs.

A third type of gross morphologic lesion in the heart muscle is hypersensitivity myocarditis. This is an inflammatory response that is the most common type of heart disease associated with drug use. There are five primary clinical criteria for diagnosis of this condition: (1) previous use of the drug(s) without deleterious incidents, (2) no apparent relationship between the size of the drug dose and the hypersensitivity response, (3) clinical symptoms consistent with responses to allergens or infectious disease agents, (4) independent confirmation of immunologic responses, and (5) persistence of the symptoms as long as drug use is continued. Histologically, there is infiltration of the heart muscle with numerous types of white blood cells. This infiltration of white cells is associated with local regions of lysis of the cardiac muscle cells. However, gross fibrosis and extensive regions of myocardial necrosis are usually absent. Among the drugs that have been reported to elicit this response are penicillin, streptomycin, ampicillin, tetracycline, sulfadiazine, and methyldopa. The specific pathological and biochemical changes that underlie these inflammatory responses have not been identified unequivocally.

The Blood Vessels

The second part of the cardiovascular system is composed of the blood vessels, which are an extensive series of tubular conduits of varying diameters. All but the narrowest of these vessels have a complex wall structure (see below). One major group of vessels, the arteries, distributes blood under various degrees of pressure to all parts of the body. A second major group of vessels, the veins, return the blood to the heart. With the exception of the pulmonary artery, which brings blood from the heart to the lungs, the arteries carry blood that is more oxygenated than the blood in their venous counterparts. The large and medium-sized arteries and veins share the same general structure, although the thicknesses of specific cell layers as well as the cell density within layers can vary considerably.

Blood Flow

Despite the system's vital function of transporting blood throughout the body, it would be incorrect to view the vascular system as merely a series of pipes of varying diameter. When the left ventricle contracts to deliver blood to the aorta, the largest artery in the body, not only is the blood pressure generated at contraction relatively high but also it is maintained at a moderately high pressure between contractions of the heart. If the

arteries were a set of rigid pipes, the pressure in the artery system would fall to zero between contractions. That this does not occur is due chiefly to the presence of numerous elastic layers (composed of the protein elastin) in the largest of the arteries. As the heart contracts, the blood pumped into these large arteries causes the elastin in the walls to stretch. Following contraction, the semilunar valves close (see description of valves) and the walls of the elastic arteries contract passively to maintain pressure within the system until the ventricles fill and contract once again. There are large, elastic arteries which function primarily to maintain the pressure within the arterial system during diastole, the resting phase of heart contraction. There are also muscular arteries, which function primarily to distribute the blood throughout the body to organs and tissues, each of which may require different amounts of blood. To help ensure that the appropriate volumes of blood are delivered on demand, the size of the lumen (the space through which the blood flows) of muscular arteries must be able to be regulated quickly and reliably. This is accomplished via innervation with sympathetic fibers of the autonomic nervous system.

Since humans stand erect, the blood must be maintained at considerable pressure in the artery system in order to overcome the force of gravity and thus provide an adequate blood supply to capillary beds, e.g., to those in the brain. However, because capillary walls are thin (to permit diffusion) the blood that is delivered to them must be delivered under reduced pressure. This is accomplished by the arterioles, which combine relatively muscular walls with a narrow lumen. The arterial blood pressure is a function of cardiac output (see Intrinsic Modulators of Cardiac Activity) and the total peripheral vascular resistance, which is primarily a function of the degree of normal tension (tonus) of the smooth muscle cells in the walls of the arterioles. If this tonus increases above the normal range for extended periods of time, hypertension (high blood pressure) will result. This tonus is under the control of the autonomic nervous system and of adrenergic hormones.

From the capillaries, blood flows first into the narrowest members of the venous system, the collecting venules, and from there into the muscular venules, whose diameter is approximately two times larger than that of the former and whose walls contain one or two layers of smooth muscle cells. Blood then flows into progressively larger veins, first to the small and then to the medium-sized veins. Veins that are located deep within tissue tend to have thinner, less muscular walls than do veins that are more exposed. [The relatively muscular medial layer (see Vessel Wall Structure) of saphenous veins from the leg made these veins a popular transplant choice for many years in coronary artery bypass surgery.] The final set of veins to receive blood before it is delivered back to the heart are the inferior and superior vena cavae. The outermost cellular layer in these veins is considerably thicker and the innermost layer is considerably thinner than those of the aorta, the first artery leaving the heart. Another difference between arteries and veins is that the latter have a more extensive vasa vasorum, an arterial blood supply to the vessel wall. Since venous blood is relatively poorly oxygenated and there may be times when cells in the outer wall of a vein may require more rapid oxygenation than could be supplied by diffusion of the low oxygen blood, this blood is supplied by the vasa vasorum. Because venous blood is under low pressure, the vasa vasorum can penetrate closer to the innermost layer of the vein, where the most oxygenated blood resides.

Pathological Changes: Veins

Approximately 90% of the pathological alterations seen in veins are associated with one of three conditions: deep-vein thrombosis, which often appears following acute myocardial infarction, thrombotic strokes, and/or major surgery; varicose veins, which usually arise secondary to sustained increases of venous pressure, usually due to venous obstruction; and superficial thrombophlebitis, which occurs in people with varicose veins as well as in some women after pregnancy. A few venotoxic responses to exogenous (i.e., from outside the body) agents are noted below. However, the great majority of vasculotoxic agents have their effects on the arteries. Therefore, only a description of artery wall structure is presented below along with listings and selected descriptions of agents that damage the arteries.

Vessel Wall Structure

There are three principal cell coats (tunics) that have been identified in the wall of large and medium-sized arteries (Fig. C-16). The outermost coat, the tunica adventitia, is composed of connective tissue cells plus extensive deposits of the proteins collagen and elastin. The adventitia in muscular arteries is approximately one-half the thickness of the middle coat, the media. In muscular arteries, the media is composed primarily of layers of smooth muscle cells. The principal extracellular protein component is elastin. In elastic arteries the tunica media also predominates, but in this case there are many layers of elastin with smooth muscle cells between the layers. The media is set off from the adventitia by a prominent elastic layer, the external elastic lamina. The adventitia of elastic arteries is thinner than that of muscular arteries. In large arteries a vasa vasorum will also be present. The innermost of the artery wall coats is the tunica intima, which is set off from the media by the internal elastic lamina. In photographs taken under a microscope, the inner elastic lamina appears fenestrated. This may serve as a relatively low resistance pathway for migration of smooth muscle cells into the intima from the media, a process thought to be involved in development of arteriosclerotic plaques. A single layer of endothelial cells borders the intima at the lumenal surface.

The intima is the most heterogeneous in composition and the most variable in size of the three major coats of the artery. The predominant cell type is the smooth muscle cell. Although some subtle differences in both appearance and behavior have been noted between smooth muscle cells in the intima versus those in the media, it is still not clear whether there is more than one type of smooth muscle cell that is present or whether the differences between smooth muscle cells in the two coats are due primarily to the differing microenvironments in these two regions of the artery wall. The intima is where many of the toxic responses of the artery wall are registered and where atherosclerosis, a major cause of death in most "civilized" societies, occurs. The characteristic lesion of this disease, the atherosclerotic plaque, is found in the intima of large and medium-sized arteries. An added problem with advanced plaques is that thrombus formation is likely to occur in regions of plaque rupture. The combination of the two events can lead to partial or even total occlusion of major arteries. If this occurs in one or more of the coronary arteries, a serious or even fatal infarction may result. A discussion of arteriosclerosis and exogenous agents that can modulate this condition is presented below.

A large variety of compounds evoke toxic responses within the arterial intima. Some will be discussed below

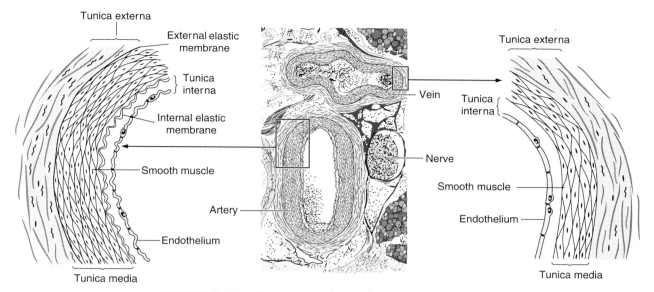

FIGURE C-16. Comparison of typical artery and typical vein.

and others are listed in Table C-12. These compounds are of interest not only because many people suffer their cardiovascular effects each year but also because an understanding of the mechanism(s) whereby some of these agents act in living organisms can provide new insights into the complex workings and controls that operate in the arterial intima. However, from the perspective of the general population, the existence of agents that can behave as vascular toxins should not be a primary public health concern, with the striking exception of cigarette smoke (discussed below). In fact, the extensive list of vasculotoxic agents listed in Table C-12 notwithstanding, it is clear that if most deaths from heart disease and stroke were due only to items on this list, then these two related conditions would quickly cease being the single greatest cause of death in the United States. As it is, there are nearly 900,000 deaths every year from these two diseases combined. The major and largely avoidable vasculotoxic agent associated with these diseases is tobacco smoke, which is discussed at the end of this entry (under Atherosclerosis).

Endothelial Damage

Maintenance of the integrity of the single layer of endothelial cells that lines all arteries is critical for normal vessel function. Until fairly recently, the endothelium was viewed largely as a mostly passive permeability barrier preventing ready access of blood-borne contaminants to the cells of the intima. It is now clear that the intact endothelium is a dynamic system which, in addition to acting as a permeability barrier, prevents adherence of white blood cells and thrombi, produces and secretes a wide range of growth regulatory molecules, and maintains vascular tone by releasing molecules that modulate dilation and constriction of blood vessels. The endothelium may even participate in its own injury on occasion. Endothelial cells are capable of oxidizing low-density lipoprotein (LDL), which is primarily responsible for transporting cholesterol through the blood to tissues. The oxidized LDL can injure the endothelium directly, attract white blood cells to the inner surface of the artery, and produce molecules that allow specific types of white blood cells to adhere to the surface. Currently, the prevailing view is that these events are critical to the early stages of arteriosclerotic plaque formation. At least one important implication should emerge from this brief summary. Since structural and metabolic integrity of endothelial cells is vital to normal arterial function, and since endothelial cell damaging agents might be present in the blood at any time, there must be efficient processes in place to repair the endothelium and maintain its integrity should it be damaged. There is growing evidence that this indeed is the case.

An observation of long standing is that blood vessels of similar anatomical structure have distinct responses to chemical stress depending on the organ system with which they are associated. This may be due to subtle differences at the cellular and subcellular levels between apparently similar cells and/or to local responses to different stimuli, e.g., due to specific hormone receptors or patterns of innervation. The blood–brain barrier, which prevents many potentially toxic blood-borne agents from reaching the brain, is a case in point. If the metabolic status of the endothelial cells in vessels at the brain level is altered, one result can be a disruption of the tight junctions between the endothelial cells, with a resulting increase in permeability. As a result, the brain, which normally is shielded from a number of toxic agents, may now be exposed to them. Lack of oxygen or markedly reduced local blood flow (ischemia) will lead to swelling of endothelial cells and a widening of the junctions. Chemicals that solubilize lipids, which are an important component of cell membranes, can also impair the barrier. These agents include alcohols and surfactants. Lead ions interact with sulfhydryl (-SH) groups that are critical to the functioning of many enzymes and structural proteins. Lead ions can produce damage to endothelial cells in blood vessels supplying the brain even before the well-known damage to nervous system cells is recognized. Chemicals that raise osmotic pressure, such as solutions of high salt or the alcohol mannitol, can cause endothelial cells to shrink, thereby causing the tight junctions between the cells to separate.

The liver is the organ largely responsible for detoxification of xenobiotic ("foreign biological") chemicals and, partly as a consequence, is also constantly at risk for damage by toxic chemicals. One such chemical, the carcinogen dimethylnitrosamine, first induces proliferation of endothelial cells, followed by increased formation of vascular connective tissue and ultimately total venous occlusion. Plant toxins of the pyrrolizidine alkaloid family, of which monocrotaline is one of the most studied compounds, can produce identical effects. Monocrotaline, which enters the body in a nontoxic form, is metabolized to its toxic form(s) by the liver. In addition to liver damage, this agent causes structural remodeling of blood vessels in the lung and a resultant increase in pulmonary arterial pressure. This effect is similar to the chronic pulmonary hypertension from which many people suffer. The monocrataline-treated rat is used as a model for this disease.

Metals

A number of metals that cause kidney damage act, at least in part, on arteries feeding this organ. Elevated levels of cadmium are associated with hypertension, at least in animal studies. Cadmium has also been implicated in thickening of the wall of arterioles and deposition of fibrotic tissue in capillaries in the testes as well as the kidneys. Agents that chelate cadmium can reverse many of these effects, as can elevation of body levels of zinc. It appears that cadmium and zinc are antagonistic and that maintenance of a cadmium/zinc ratio within fairly well-defined limits may be important in preventing cadmium-associated vessel wall changes.

Three other metal ions that have been implicated in damage to vessel walls are mercury, chromium, and arsenic. Mercury, which interferes with protein sulfhydryl groups, may cause vasoconstriction of preglomerular vessels in the kidney. There is at least one study, reported 25 years ago, which argues for a protective effect of low levels of chromium on arteries. The absence of chromium in aortas of people dying of arteriosclerotic heart disease contrasted with the presence of chromium in aortas of people dying of other causes. However, chromium deficiency does not appear in modern medical texts as a contributing factor to arteriosclerosis. Furthermore, recent studies suggest that high levels of chromium may be associated with a variety of disorders separate from heart disease.

Arsenic, though an unlikely contributor to blood vessel damage on a worldwide level, represents a striking example of how local environmental alterations can have profound effects on a large portion of a population. On the southwest coast of Taiwan, the artesian well water consumed by the local population has high levels of arsenic and approximately 1 out of every 100 people suffers from blackfoot disease, which looks and feels even worse than it sounds. In late stages of the disease, extremities can become gangrenous and spontaneous and surgical amputation of extremities often results. People suffering from this disease exhibit much higher levels not only of peripheral vascular disease but also of cardiovascular disease. The mechanism of arsenic's action on the blood vessels is still unclear.

Primary amines

Cardiotoxicity of primary amines (epinephrine, norepinephrine, and isoproterenol) was noted earlier and has been recognized for nearly 100 years. In recent years, the vascular toxicity of these and related compounds has been recognized. The effects seem to focus on medial cells of the artery wall rather than on adventitial or endothelial cells. Early changes include loss of medial cells, calcification, and loss of elastic fibers. Later, there is a compensatory proliferation of intimal cells. The vascular toxicity of two related compounds is particularly striking. One of these compounds, allylamine, will be discussed under Atherosclerosis. The second is β-aminoproprionitrile (β-APN), which is the active agent in the toxic sweet pea, *Lathyrus odoratus*. Consumption of flour derived from this plant results in a condition known as lathyrism, which is often toxic in children and young adults. Sudden death can result because of rupture of aortic aneurysms, which are ballooned and weakened segments of the aortic wall. The toxicity of β-APN has been related to its inhibition of an enzyme which normally cross-links collagen and elastin in large elastic arteries, including the aorta, thereby strengthening them.

Atherosclerosis

Arteriosclerosis (literally "artery hardening") is the general term used to describe thickening and stiffening that can occur for a variety of reasons in arteries of all sizes. From a clinical perspective, the lesion of greatest interest is the atherosclerotic plaque. This is a lipid-rich lesion associated with thickening and hardening of medium- to large-size arteries. It is the principal lesion associated with myocardial and cerebral infarction and is the primary cause of death in the United States, Canada, Europe, and Japan. Plaque development is complex, involving processes as diverse as cell proliferation, cell death (possibly of two different forms), synthesis and deposition of a variety of extracellular macromolecules (e.g., collagens, elastin, and proteoglycans), lipid accumulation, and calcification. The plaque appears in the arterial intima and a variety of cell types play major roles including: smooth muscle cells, macrophages, lymphocytes, platelets, and endothelial cells. Plaque formation has been classified sometimes as a problem of proliferation and sometimes one of degeneration, as well as an inflammatory process, a response to injury, and a process related to benign tumor formation. In truth, there is moderate to strong evidence supporting each of these views.

Although in most cases, atherosclerosis does not become manifested as a clinically serious condition until well into middle age or beyond, it is really a disease that begins early in life. Autopsy studies on U.S. soldiers killed during the Korean War revealed that many already had arterial deposits characteristic of early stages of atherosclerosis. Recent studies on children through people in the third decade of life have confirmed and expanded these findings. The good news is that while there are genetic factors which may predispose an individual to develop atherosclerosis, there is considerable evidence that individual choices and lifestyle decisions can play a large role in preventing, or at least mitigating, early onset of clinical symptoms of this disease. Furthermore, results from a limited number of animal studies suggest that it even may be possible to reverse the clinical course of the disease.

There are three areas where lifestyle modification can have profound effects on moderating development of clinically significant atherosclerosis. In addition to exercise, which is not directly relevant to this discussion, the two areas most amenable to change are diet and smoking. There is strong epidemiological evidence associating elevated levels of serum cholesterol with increasing risk of atherosclerosis and subsequent heart attacks. As noted previously, LDL is primarily responsible for solubilizing cholesterol and carrying cholesterol and its esters through the bloodstream to the tissues. Oxidized LDL can damage vessel wall cells, including endothelial cells. Oxidized LDL can act as and also generate a chemoattractant, which can attract monocytes to the endothelial surface and possibly help mediate passage of monocytes across the endothelium. Monocyte-derived macrophages act as scavengers to help remove harmful molecules such as oxidized LDL. When control mechanisms go awry, macrophages filled with oxidized LDL can become foam cells, which are considered to be critical to the formation of early stage atherosclerotic plaques. Studies on experimental animals as well as humans have shown that reduction in levels of plasma cholesterol and LDL can lead to significant widening of the arterial lumen. There is evidence that probucol, a drug touted originally for its plasma cholesterol-lowering capability, may function primarily as an antioxidant protecting the integrity of LDL.

Relaxation of blood vessels appears to be at least partially under the control of endothelial cells and their secreted products, especially endothelial cell relaxation factor (EDRF). Oxidized LDL directly inhibits the endothelial cell-associated vessel relaxation. The generation of increased reactive oxygen species in association with elevated levels of blood cholesterol has also been reported. One of these reactive oxygen species, superoxide (O_2^-), may interact with vasoactive EDRF and nitric oxide locally in the artery wall, preventing endothelial cell-dependent vasodilation. In addition, a product of the reaction of nitric oxide and superoxide, the reactive peroxynitrite, may act to stimulate lipoprotein oxidation, which, as noted previously, is regarded as an early step in atherosclerotic plaque generation. Oxidants arise from two sources. The first, which is internal, is related to various metabolic processes, including respiration and phagocytic activity to destroy bacteria- and/or virus-infected cells, and, paradoxically, attempts to detoxify foreign substances. In the process of carrying out the latter activity, toxic oxidant by-products can be produced. The second source is external. While the potential protective effects of dietary components and supplements, e.g., vitamins, are still being debated, it is reasonable to conclude that decreasing one's exposure to oxidants from external sources would be beneficial in reducing chances of not only premature atherosclerosis but also of other diseases, including cancer. By far the most common, avoidable, and dangerous source of external oxidants is cigarette smoke, which is considered a principal contributor to one-fourth of all heart disease cases, one-third of all cancers, and approximately 400,000 premature deaths in the United States every year. As economies of developing countries expand and as cigarette smoking becomes more popular throughout the world, the health problems associated with cigarette smoking will increase rapidly.

Cigarette smoke is composed of active smoke, the smoke coming from the mouth end of the cigarette and breathed in by the smoker, and passive smoke (second-hand smoke or environmental tobacco smoke), which is composed mostly of the smoke coming off the lit end of the cigarette plus a small percentage of exhaled smoke. Active and passive smoke contain many constituents in common, but often in strikingly different concentrations. Among the more than 4700 different chemicals that have been identified in cigarette smoke, prominent candidates that have been considered as vasculotoxic agents include carbon monoxide and some carcinogens. In addition to interfering with transport of well-oxygenated blood, carbon monoxide may cause endothelial cell damage directly, although if and how this happens is by no means clear. Another major class of potential vasculotoxins in cigarette smoke is the carcinogens. Most of these are found in the tar condensate fraction of cigarette smoke. Some, including benzo(*a*)pyrene, are well-known carcinogens that are found in other environmentally prominent substances including coal tar derivatives, charcoal-broiled meat, and automobile exhaust. Other smoke carcinogens include the nitrosamines, some of which are tobacco specific. Both benzo(*a*)pyrene and the parent nitrosamines require metabolic activation to become carcinogenic. The enzymes involved in these processes are members of the cytochrome P450 system. During the course of detoxifying these agents so they ultimately can be excreted readily, one or more toxic and possibly carcinogenic metabolites can be generated. Compounds such as benzo(*a*)pyrene induce the appearance of the cytochrome P450 system enzymes, and smokers are constantly exposed to the P450 inducers. Generation of endothelial cell-damaging agents during the metabolism of benzo(*a*)pyrene derived from cigarette smoke has been recently proposed, but not proven, as a mechanism to explain initiation of atherosclerotic plaques. Oxidants derived from cigarette smoke can damage lipids, an important constituent of cell membranes, as well as cellular macromolecules, including DNA. There is no direct evidence that cigarette smoke causes damage to artery wall cell DNA in either living animals or humans. However, if such damage does occur it would provide independent support for the view that DNA alterations are characteristic of atherosclerotic plaques in animal models of the disease as well as in humans. In related experimental animal studies, the chemical allylamine has been shown to cause both myocardial lesions and vascular fibrosis. The allylamine toxicity is thought to be mediated via metabolism of this compound to the reactive aldehyde, acrolein, which is also a prominent component of cigarette smoke. Studies with cultured artery wall cells indicate that the primary arterial effect of allylamine is on the smooth muscle cells. Proliferation of intimal smooth muscle cells in response to allylamine exposure has been reported to result in activation of a specific cellular DNA sequence, the H-*ras* oncogene, which has been implicated in development of certain forms of cancer. This lends further support to the contention that there may be molecular similarities between the development of the lesions of atherosclerosis and of cancer.

One of the problems researchers have faced in identifying specific health-threatening components of cigarette smoke is that while at moderate to high concentrations many of these agents can be toxic, in many cases the individual concentrations of these factors in cigarette smoke are likely too low to be able to account individually for the toxic and disease-promoting effects of cigarette smoke. The U.S. EPA side-stepped this problem in 1992 by declaring environmental tobacco smoke, with its thousands of components, to be a human class A carcinogen. The American Heart Association has classified environmental tobacco smoke as an environmental poison and as a major preventable cause of cardiovascular disease. Regarding environmental tobacco smoke, there have been estimates that as many as 60,000 excess heart disease deaths in the United States every year can be attributed directly to involuntary exposure to cigarette smoke. In support of this estimate, since 1993 at least two laboratories have reported that inhalation of sidestream cigarette smoke accelerates arteriosclerosis in two different experimental model systems of the disease. Very recent results from this author's laboratory indicate that at least one of the principal plaque-accelerating factors in environmental tobacco smoke is the volatile, unsaturated chemical 1,3-butadiene. This compound is found in the vapor phase rather than in the tar fraction, where most of the known cigarette smoke carcinogens are found. However, the concentration of butadiene in the vapor phase of environmental tobacco smoke is much higher than the concentration of most carcinogens in the tar fraction. Acrolein, referred to previously, is present in both mainstream and sidestream cigarette smoke but at concentrations that are 18 times higher in the sidestream than in mainstream smoke. Since epidemiological and autopsy evidence strongly support the view that atherosclerosis begins as early as childhood, the experimental results with environmental tobacco smoke just noted suggest that involuntary exposure of children to tobacco smoke may accelerate plaque development. In the United States, where studies show that many children are less active physically and have poorer diets than children growing up a couple of generations ago, involuntary exposure to secondhand smoke may well represent a major additional risk factor for development of atherosclerosis. Fortunately, exten-

sive epidemiological evidence from both cancer and heart disease studies indicates that as the time since cessation of smoking increases, the chances of dying prematurely from either disease decrease. Thus, the vasculotoxic effects of cigarette smoke, both active and passive, may be largely reversible.

—Arthur Penn

Further Reading

Toxic Responses of the Cardiovascular System

Baskin, S. I., and Maduh, E. U. (1993). Fundamentals of cardiotoxicology. In *General and Applied Toxicology* (B. Ballantine, T. Marrs, and P. Turner, Eds.), Vol. 1, pp. 595–618. Stockton Press, New York.

Hanig, J. P., and Herman, E. H. (1991). Toxic responses of the heart and vascular systems. In *Casarett and Doull's Toxicology, The Basic Science of Poisons* (M. O. Amdur, J. Doull, and C. Klaassen, Eds.), 4th ed., pp. 430–462. Pergamon, New York.

Van Vleet, J. F., Ferrans, V. J., and Herman, E. (1991). Cardiovascular and skeletal muscle systems. In *Handbook of Toxicologic Pathology* (W. M. Haschek and C. G. Rousseaux, Eds.), pp. 539–624. Academic Press, New York.

Atherosclerosis

Ross, R. (1993). Atherosclerosis: A defense mechanism gone awry. *Am. J. Pathol. 143,* 987–1002.

Penn, A., Chen, L. C., and Snyder, C. A. (1994). Inhalation of steady-state sidestream smoke from one cigarette promotes arteriosclerotic plaque development. *Circulation. 90,* 1361–1367.

Vascular Injury

Boor, P. J., Gotlieb, A. I., Joseph, E. C., Kerns, W. D., Roth, R. A., and Tomaszewski, K. E. (1995). Chemical-induced vasculature injury. *Toxicol. Appl Pharmacol. 132,* 177–195.

Related Topics

Blood
Metals
Tobacco Smoke

Castor Bean

- SYNONYMS: *Ricinus communis*; castor-oil plant; palma christi; koli; moy bean; mole bean; dog tick seeds
- CHEMICAL CLASS: Toxalbumins

Uses

Castor beans are used in cancer research in developing immunotoxicologic agents for chemotherapy.

Exposure Pathways

Ingestion is the most common route of exposure.

Toxicokinetics

The glycoproteins (ricin) are poorly absorbed from the gastrointestinal tract; however, once absorbed, they probably follow a distribution pattern similar to that of albumin. Many cell surfaces contain receptors specific for the ricin molecules. This molecule consists of two subunits, A and B, bound by a disulfide link. When this link is broken, the B subunit binds to galactose-containing receptors in the cell wall and is transported intracellularly. The A subunit inhibits protein synthesis. The liver, spleen, adrenal cortex, and bone marrow are the primary sites of distribution. The biotransformation and elimination of toxalbumins are poorly understood. The elimination half-life in one patient was 2 days. The reported disappearance of ricin from the plasma is according to first-order kinetics when injected intravenously into mice and human cancer patients.

Mechanism of Toxicity

The principal toxicity of the toxalbumins is protein synthesis inhibition causing hemagglutination within the first hour, adrenal insufficiency, hepatic and renal failure, endothelial damage, and, in severe cases, profound capillary hemorrhage.

Human Toxicity

The toxicity of the castor bean is variable due to erratic absorption patterns. Symptomatology can occur with the ingestion of one seed. The seeds are minimally toxic if the seed coat remains intact when ingested. Acutely, the toxalbumins cause severe gastrointestinal lesions, retinal hemorrhages, rapid and weak pulse, and possible shock due to fluid and electrolyte loss from vomiting and diarrhea. Mild to moderate central nervous system depression may be seen. Seizures can occur but are not common. Fever may be noted. Hepatic damage can occur in large overdoses with increases in alanine aminotransferase, total bilirubin, and aspartate aminotransferase. Elevated serum creatinine and hematuria is often seen. Unlike most toxalbumins, castor beans contain several allergens that can cause severe reactions in the hypersensitive individual. Late-phase complications (2–5 days postexposure) reflect the cytotoxic effects of the ricin. Patients may be asymptomatic prior to this phase, but damage to the hepatic, central nervous, renal, and adrenal systems may ensue. Laboratory radioimmunoassay is available for ricin (usually not on an emergent basis), but management must be based on symptomatology alone.

Clinical Management

No specific treatment is available for toxalbumin exposure. Aggressive gastric decontamination such as whole bowel irrigation is recommended. Supportive care primarily consists of maintaining fluid volume and electrolytes and blood products.

Animal Toxicity

The toxicity of the castor bean in animals is similar to that in humans.

—*Brenda Swanson Biearman*

Catecholamines

- DESCRIPTION: Catecholamine is the name of a group of compounds that contain a catechol nucleus (a benzene ring with two adjacent hydroxyl substituents) and an amine group. The term "catecholamine" usually implies dopamine, epinephrine, and norepinephrine. Each has its own synonyms.

◆ SYNONYMS:

Dopamine (CAS: 51-61-6)—pyrocatechol; 4-(2-aminoethyl)pyrocatechol; 3-hydroxytyramine; 3,4-dihydroxyphenethylamine; 4-(2-aminoethyl)-1,2-benzenediol; Dopastat; Intropin

Epinephrine (CAS: 51-43-4)—benzyl alcohol; adnephrine, adrenal; adrenalin; Adrine; Antiasthmatique; Asthma-Nefrin; Balmadren; Epifrin; Epirenamine; Glaucosan; Hemostatin; methylaminoethanolcatechol; Renagladin; Renalina; Renostypticin; Soladren; Simplene; Supranefran; Suprarenin; Susphrine; Sympathin; Takamine; Vasoton; Vasotonin

Norepinephrine (CAS: 51-41-2)—4-(2-amino-1-hydroxyethyl)-1,2-benzenediol; α-(aminomethyl)-3,4-dihydroxybenzyl alcohol; 2-amino-1-(3,4-dihydroxyphenyl)ethanol; 1-(3,4-dihydroxyphenyl-2-aminoethanol; Adrenor; Aktamine; Levarternol; Levonorepinephrine; Levophed; Noradrenaline; Nortrinal; Sympathin E

◆ CHEMICAL CLASS: Catecholamines are endogenous neurotransmitters or hormones. Dopamine and norepinephrine are in the monoamine class.

◆ CHEMICAL STRUCTURES:

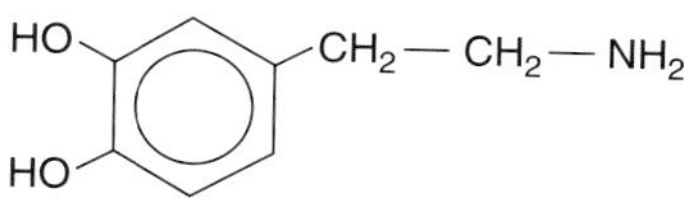

Dopamine

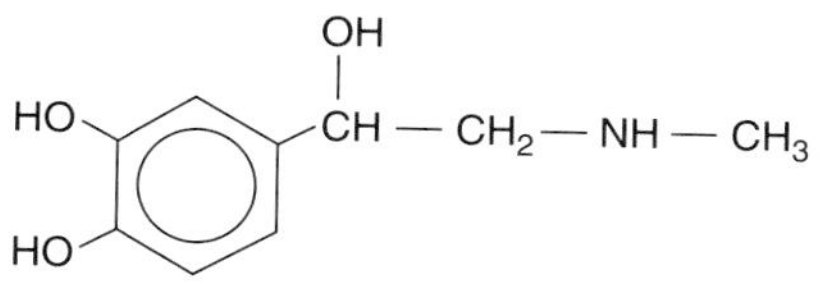

Epinephrine

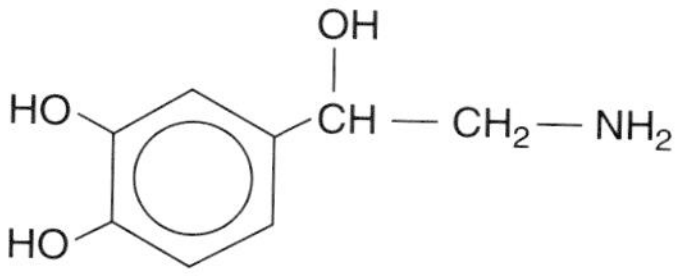

Norepinephrine

Uses

Catecholamines are sympathomimetic drugs. Dopamine and norepinephrine are used as vasopressors (antihypotensives). Epinephrine is used as a vasoconstrictor, cardiac stimulant, or bronchodilator to counter allergic reaction, anesthesia, and cardiac arrest. It is also an antiglaucoma agent.

Exposure Pathways

Catecholamines are endogenous compounds. when used therapeutically, intravenous injection or infusion is the most common route of administration. Epinephrine is available in nebulized racemic dosage form for inhalation.

Intoxication from catecholamine usually results from iatrogenic overdoses, accidental intravenous administration, and the injection of solution intended for nebulization.

Toxicokinetics

Epinephrine is well absorbed after oral administration but is rapidly inactivated in the gut mucosa. When intravenously injected or infused, the onset of drug effect is rapid (within 5 min for dopamine and 3–10 min for epinephrine) and the duration of drug effect

is short (10 min for dopamine, 1 or 2 min for norepinephrine, and 15 min to hours for epinephrine depending on route of administration). Exogenous catecholamine in the circulation is rapidly and efficiently taken up by adrenergic neurons. Catecholamine is metabolized by monoamine oxidase, which is localized largely in the outer membrane of neuronal mitochondria, and by catechol-O-methyl transferase, which is found in the cytoplasm of most animal tissues, particularly the kidneys and the liver.

The primary metabolites of dopamine are homovanillic acid and dihydroxyphenylacetic acid (75%) and norepinephrine (25%). The primary metabolites of epinephrine and norepinephrine are vanilylmandelic acid and 3-methoxy-4-hydroxy-phenethyleneglycol. Catecholamine metabolites and their conjugates are excreted in urine.

Mechanism of Toxicity

Catecholamines are sympathomimetic drugs. These drugs increase heart rate and cardiac output and may produce cardiac arrhythmias. Administration of norepinephrine also results in increased peripheral vascular resistance. Both of the previous effects may cause serious systemic hypertension, which may cause cerebral hemorrhage. Serum lactic acidosis develops because of tissue ischemia, increased glycolysis, and reduced hepatic and renal blood flow. In very high doses, a paranoid state may be induced.

Human Toxicity

At high infusion rates of dopamine, ventricular arrhythmias and hypertension may occur. Nausea, vomiting, and angina pectoris are occasionally seen. Gangrene of the extremities may occur in patients with profound shock given large doses of dopamine for long periods of time. Norepinephrine may cause dose-related hypertension (sometimes indicated by headache), reflex bradycardia, increased peripheral vascular resistance, and decreased cardiac output. High doses of norepinephrine (in excess of 8–12 mg of base/min) cause intense vasoconstriction, which results in "normal" blood pressure but decreased tissue perfusion. Local necrosis may result from perivascular infiltration and angina, mesenteric ischemia, and peripheral ischemia. Epinephrine may cause dose-related restlessness, anxiety, tremor, cardiac arrhythmias, palpitation, hypertension, weakness, dizziness, and headache. Anginal pain may occur when coronary insufficiency is present. A sharp rise in blood pressure from overdosage may cause cerebral hemorrhage and pulmonary edema.

Prolonged use and repeated injection of epinephrine may lead to tolerance and local necrosis.

Clinical Management

Basic and advanced life-support measures should be utilized as necessary. Treatment is directed at ameliorating tachycardias, shock, cardiac arrhythmias, systemic hypertension, pulmonary edema, and lactic acidosis.

—*Zhengwei Cai*

Cell Proliferation

Unicellular organisms, like yeasts, bacteria, or protozoa, have a strong selective pressure to grow and divide as rapidly as possible. The rate of cell division in these cases is limited only by the rate at which nutrients can be taken from the medium and converted into cellular materials. Multicellular organisms, on the other hand, are made up of different kinds of cells performing a variety of functions, and the survival of the organism as a whole is at stake rather than the survival or proliferation of one individual type of cell population. For the multicellular organism to survive, some cells must refrain from dividing even when nutrients are plentiful. Some cells do not to divide after certain stage of development as in the case of central nervous system. However, when need arises for new cells, as in the case of tissue injury, previously nondividing cells must be rapidly triggered to reenter the cell division cycle as part of the overall survival strategy.

Division Cycle of Cells

An adult multicellular animal must divide and supply millions of new cells just to replace the dead or dying cells. Cells go through a division cycle through a highly regulated process known as cell cycle progression (Fig. C-17). Cells divide by going through a cell cycle with the endproduct being a duplication of the contents of the mother cell in two daughter cells. In an adult animal most of the cells are in resting or in G_0 (G = gap) phase

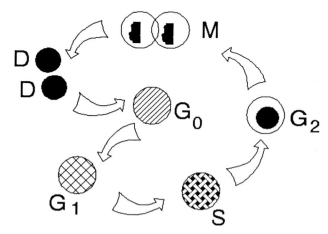

FIGURE C-17.   *In adult organisms normally cells are in resting phase (G_0) of cell division cycle. Upon appropriate stimulus the cells enter the division cycle which is characterized by G_1, S, G_2, and M Phase. After division, the daughter cells (D) may either reenter the division cycle or enter the resting phase depending on the stimulus.*

of cell cycle. When needed to divide they enter the G_1 phase of cell cycle. In most cells the DNA in the nucleus is replicated during only a limited portion of the cell cycle called the S or synthesis phase of the cell cycle. After S phase the cells go into a second interval called G_2 phase. In mitosis or M phase the contents of the nucleus condense to form visible chromosomes, which through an elaborately orchestrated series of movements are pulled apart into two equal sets and then the cell itself splits into two daughter cells. Upon loss of tissues (e.g., a population of cells and a portion of the tissue) due to injury, the division cycle of cells is stimulated in tissue- or organ-specific fashion so that the lost tissues can be replaced promptly to restore tissue function. Each of the above phases of cell division is highly regulated and orchestrated by an intricate series of signaling mechanisms. When the lost tissue is replaced the entire repertoire is brought down to the normal resting level, thereby reestablishing the cellular, organ, and tissue homeostatic mechanisms.

Genetic Control of Cell Structure and Function during and after Embryonic Development

Almost every multicellular animal is a clone of cells descended from a single original cell, the fertilized egg. Thus, the cells in the body, as a rule, are genetically alike. However, phenotypically they are different; some are specialized as muscles, others as neurons, others as hepatocytes, and so on. The different cell types are arranged in a precisely organized pattern, and the whole structure has a well-defined shape. All these features are ultimately determined by the DNA sequence of the genome, which is reproduced in every cell. Each cell must act according to the same genetic instructions, but it must interpret them with due regard to time and circumstance so as to play a proper part in multicellular organization. The development of vertebrates can be divided into three phases. In the first phase the fertilized egg cleaves to form many smaller cells and these become organized into an epithelium and perform a complex series of gastrulation and neurulation movements, whose outcome is creation of a rudimentary gut cavity and a neural tube. In the second or organogenesis phase the various organs, such as limbs, eyes, heart, and so on, are formed. In the third phase the generated struc-

tures go on to grow to their adult size. These phases are not sharply distinct but overlap considerably in time.

Terminal Differentiation and Cell Division

After the embryonic development, cells in the normal adult human body divide at very different rates. Some, such as neurons and skeletal muscle cells, do not divide at all; others, such as liver cells, normally divide once every year or two; and certain epithelial cells in the gut divide more than twice a day so as to provide constant renewal of the gut lining. Most cells in the vertebrates fall somewhere between these extremes; they can divide but normally do so infrequently. Almost all the variation lies in the time cells spend between mitosis and S phase, with slowly dividing cells remaining arrested after mitosis for weeks or even years. By contrast, the time taken for a cell to progress from the beginning of S phase through mitosis is brief (typically 12–24 hr in mammals) and remarkably constant, irrespective of the interval from one division to the next. The time cells spend in a nonproliferative, so-called G_0 state varies not only according to the cell type but also according to the circumstances. Sex hormones stimulate the cells in the wall of the human uterus to divide rapidly for a few days in each menstrual cycle to replace the tissue lost by menstruation; blood loss stimulates proliferation of blood cell precursors; and acute liver damage provokes surviving liver cells to proliferate with a cycle time of only a day or two. Similarly, epithelial cells in the neighborhood of a wound are stimulated to divide so as to repair the injured epithelium. Delicately adjusted and highly specific controls exist to govern the proliferation of each class of cells in the body according to the need.

Role of Growth Factors and Cytokines in Cell Division

When put in an artificial culture medium completely devoid of serum, vertebrate cells will not normally pass the restriction point, even though all the requisite nutrients are present in the medium, and they will halt their growth as well their progress through the chromosome cycle. Essential components of serum are some highly specific proteins (growth factors and cytokines), usually present in very small concentrations (in the order of 10^{-9}–10^{-11} M). Different cells require different sets of these proteins. Some of these proteins are involved directly in stimulating cell division and are called complete mitogens. Some can directly inhibit cell cycle progression and thus can control cell division in the body. These are called growth inhibitors. Some of the proteins can cause cell cycle progression in an indirect way and are called growth triggers. Table C-13 provides examples of some of the growth factors and cytokines with their functions.

Cell Senescence and Reluctance to Divide

Most normal cells in the body of a mammal show a striking reluctance to continue proliferating forever. Fibroblasts taken from a normal human fetus, for example, will go through about 50 population doublings when cultured in a standard growth medium; toward the end of this time, proliferation slows down and finally stops, and the cells, after spending some time in quiescent state, die. Similar cells taken from a 40-year-old stop dividing after about 40 doublings, while cells from an 80-year-old stop after about 30 doublings. Fibroblasts from animals with shorter life span stop after a smaller number of division cycles in culture. Because of the correspondence with aging of the body as a whole, this phenomenon is called cell senescence. According to one theory, cell senescence is the result

of a catastrophic accumulation of self-propagating errors in a cell's biosynthetic machinery that is unimportant under the conditions of life in the wild where most animals die from other causes long before a significant number of cells become senescent. An alternative theory is that the cell senescence is the result of a mechanism that has evolved to protect us from cancer by limiting the growth of tumors.

Cell Proliferation as a Compensatory Response to Toxic Tissue Injury

Human beings are exposed to numerous toxic insults everyday. The body has several lines of defense mechanisms to combat the toxicants. Some of the toxicants are filtered out by virtue of their particle size, even before they could enter the body. Toxicants that enter the body can be metabolized and/or conjugated to be excreted out of the body. When these first lines of defense mechanisms are overcome, then the toxic substances cause cell death in the body. The site of cell death depends on the site of action of the toxicant. At this point the tissue can respond by stimulating its healthy cells to divide and to restore tissue structure and function. In response to cell death because of toxic insult, cells in the affected tissues (with the exception of neurons) start dividing in order to replace the dead or dying cells. One surviving cell can go through several cell cycles depending on the severity of the damage. The cell division stops at the precise point when all the

TABLE C-13
Example of Growth Factors and Cytokines known to Regulate Cell Proliferation

Factor	Representative functions
Platelet-derived growth factor (PDGF)	Stimulates proliferation of connective tissue cells and neuroglial cells
Epidermal growth factor (EGF)	Stimulates proliferation of many cell types
Insulinlike growth factors I and II (IGF-I and -II)	Work with PDGF and EGF to stimulate fat cell proliferation
Fibroblast growth factor (FGF)	Stimulates proliferation of many cell types including fibroblasts, endothelial cells, and myoblasts
Interleukin-2 (IL-2)	Stimulates proliferation of T lymphocytes
Transforming growth factor β (TGF-β)	Inhibits cell cycle progression of different cell types
Interleukin-1 (IL-1)	Inhibits proliferation of hepatocytes and other cell types
Hepatocyte proliferation inhibitor	Inhibits hepatocyte proliferation
Nerve growth factor (NGF)	Promotes axon growth and survival of sympathetic and some sensory and CNS neurons
Hematopoietic cell growth factors (IL-3, GM-CSF, M-CSF, G-CSF, and erythropoietin)	Promote division of different blood cells and various other types of cells

dead cells have been replaced with new cells. At high doses of toxicants the ability of the cells to go through the cell cycle is sometimes inhibited. This leads to two consequences. First, the dead cells are not replaced and failed cell division means loss of the organ and sometimes death. Second, in the absence of compensatory cell division, injury to the tissue can progress in an unrestrained manner. The ability of the cells to go through the cell cycle as needed decreases with age. This is why an 80-year-old can be more susceptible to the same dose of a toxicant as a 40-year-old.

Tissues vary in their compensatory responses to toxic chemicals. Skin, intestine, liver, and kidney are examples of tissues that can respond to toxic injury by stimulating cells to replace the lost tissue. Epithelial cells lining the criptae of the intestines are known to renew themselves every 72 hr. Since these cells are subject to many foreign chemicals in the diet, many cells might be expected to be injured or affected in other ways. Therefore, these cells are completely renewed every 3 days. Although not as rapid, skin injury can result in replacement of injured skin rather promptly. The adult liver is normally a quiescent tissue with only an occasional cell dividing to replace dying cells to retain normal tissue homeostasis. Upon injury, however, the liver will respond promptly by stimulating its cells to divide and thereby restoring the lost tissue and function. Surgical removal of portions of the liver leads to restoration of the original liver mass through a very rapid cell proliferation and tissue repair response. Similarly, the kidney is also able to replace its cells upon toxic injury.

Stem Cells and Terminally Differentiated Cells

There are cell populations in the body that are renewed simply by duplication and then there are those that are renewed by means of stem cells. The defining properties of stem cells are (1) they themselves are not terminally differentiated cells—that is, it is not at the end of the pathway of differentiation; (2) they can divide without limit throughout the lifetime of the organism; and (3) when they divide, each daughter cell can either remain as a stem cell or it can take the path leading irreversibly to terminal differentiation.

Stem cells are required wherever there is a recurring need to replace differentiated cells that cannot them-

selves divide. There may be several reasons why a cell is terminally differentiated. The cell nucleus is digested, as in the outermost layers of the skin, or is extruded, as in the mammalian red blood cells. Alternatively, the cytoplasm may be heavily encumbered with structures, as in myofibrils of the striated muscle cells, that would hinder cell duplication. In other terminally differentiated cells the chemistry of differentiation may be incompatible with cell division. In any case, renewal must depend on stem cells.

The job of the stem cell is not to carry out the differentiated function but rather to produce the cells that will. Those stem cells that give rise to only one type of differentiated cells are called unipotent, those that give rise to a small number of cell types are called oligopotent, and those that give rise to many cell types are called pluripotent.

Cell Proliferation and Cancer

In multicellular organisms there are genes called social control genes, which are involved specifically in the social controls of cell division. A cell that undergoes a mutation or a set of mutations that disrupts the social restraint on cell division will divide without regard to the needs of the organism as a whole, and its progeny will become apparent as a tumor. Cancers, by definition, are malignant tumors; that is, the tumor cells not only divide in an ill-controlled way but also invade and colonize other tissues of the body to create widespread secondary tumors or metastases. Approximately 10^{16} cell divisions take place in a human body in the course of a lifetime. Even in an environment that is free of mutagens, mutations occur spontaneously at an estimated rate of about 10^{-6} mutations per gene per cell division—a value set by fundamental limitations on the accuracy of DNA replication and repair. Thus, in a lifetime, every single gene is likely to have undergone mutations on about 10^{10} separate occasions in any individual human being. Among the resulting mutant cells, one might expect that there would be many that have disturbances in genes involved in the regulation of cell division and consequently disobey the normal restrictions on cell proliferation. To generate cancer, a cell must undergo a number of mutations occurring together to escape the multiple controls on cell division and then accumulate further changes to become endowed with the capacity for invasion and metastasis.

From statistics it has been estimated that somewhere between three and seven independent random events, each of low probability, are typically required to turn a normal cell into a cancer cell; the smaller numbers apply to leukemia and the larger to carcinomas.

A proto-oncogene is a normal social control gene which can undergo mutation to become an oncogene. Oncogenes and protooncogenes contain DNA sequences that are closely similar but not identical. The mutations in the protooncogenes can result from spontaneous mutations or in response to chemical carcinogens or exposure to radiations. To date, more than 50 protooncogenes have been identified. It is likely, however, that many more social control genes remain to be discovered. Genes that stimulate cell division can be identified readily with current techniques, but there are several genes that have inhibitory effects on cell proliferation and recessive mutations in them are a common cause of cell transformation and cancer. The protooncogenes code for various growth-factors, growth factor receptors, and various intracellular mediators involved in signaling cells to divide.

Importance of Understanding the Mechanisms in Control of Cell Division

Understanding of the mechanisms in control of cell division offers promising opportunities for developing new avenues for therapeutic intervention, with the aim of restoring and boosting tissue repair mechanisms in cases in which tissues have been damaged or lost as seen in burn wounds, trauma cases, in cases of drug overdoses, chemical poisoning, etc. The current armamentarium of clinical treatments for patients with drug overdoses or chemical poisoning aims to prevent additional injury either by blocking further formation of toxic metabolites or by increasing clearance out of the body. While this is important in preventing further damage to the affected tissue, or damage to unaffected tissue, survival depends heavily on the remaining cells in the tissue to proliferate to replace the dead or dying cells; this, in turn, depends on how soon after the injury the patient commences active treatment. In cases in which either there is a delay in treating the patient or the initial injury itself was massive, death or loss of the organ usually occurs because the damage compromises the regenerating ability of the cells, thereby paving the way for unrestrained progression of damage. If cellular regeneration could be "actively" stimulated, even after the massive damage, by some therapeutically compatible mechanism, then it might be possible to prevent death or loss of organ. For example, animal experiments have shown that even after massive liver injury, liver failure and animal death can be obviated by stimulating tissue repair in the liver. The importance of cell division in tissue repair and recovery from injury is evident in experiments in which liver failure and animal death are observed in animals receiving an ordinarily nonlethal dose of a toxic chemical, if cell division is blocked by antimitotic agents. Perhaps carefully induced suppression of growth factors, cytokines, and protooncogenes involved in cell death or overexpression of those factors needed for cell division could stop the progression of injury and could restore the organ structure and functions. With the advent of gene therapy, specific genes could, one day, be delivered directly to the organ to induce expression/suppression of any of the factors implicated in fast recovery.

Further Reading

Alberts, B., Bray, D., Lewis, J., Raf, M., Roberts, K., and Watson, J. D. (Eds.) (1989). *Molecular Biology of the Cell*, 2d ed., pp. 728–787. Garland, New York.
Chanda, S., and Mehendale, H. M. (1996). Hepatic cell division and tissue repair: A key to survival after liver injury. *Mol. Med. Today* 2, 82–89.
Darnell, J., Lodish, H., and Baltimore, D. (Eds.) (1990). *Molecular Cell Biology*, 2nd ed., pp. 955–998. Scientific American Books, New York. [Distributed by Freeman, New York]
Mehendale, H. M. (1995). Injury and repair as opposing forces in risk assessment. *Toxicol. Lett.* 82–83, 891–899.

—*Sanjay Chanda and Harihara M. Mehendale*

Related Topics

Carcinogenesis
Carcinogen–DNA Adduct Formation and DNA-Repair
Liver
Molecular Toxicology
Mutagenesis
Tissue Repair

Centipedes

- PREFERRED NAME: Arthropoda (phylum), Chilopoda (class). There are four orders in the centipedes: Scutigeromorpha, Lithobiomorpha, Geophiulomorpha, and Scolopendromorpha.

- DESCRIPTION: Many species of centipedes of various sizes and colors are found throughout the world. Centipedes are recognized by their long, multisegmented, flattened body, with venom fangs in their first segment. The body is composed of 15–181 somites or segments, each of which has a pair of legs; a head bearing a multijointed antennae; and three pairs of mouth parts.

Exposure Pathways

A bite is the usual route of envenomation from the centipede. Accidental ingestion is another route of exposure but occurs infrequently. Occasionally, a giant centipede produces an intensely itching path of red dots over the skin where it has walked, embedding its legs partially into the skin.

Mechanism of Toxicity

Centipede venom has not been studied in detail. It contains enzymes and nonenzymatic proteins, probably small peptides. There is also a cardiotoxic protein called toxin S. Nonprotein substances in the venom include 5-hydroxytryptamine, histamine, lipids, and various polysaccharides. Anticoagulant and coagulant components have also been isolated.

Human Toxicity

About 90% of all bites are on the hands. Envenomation causes immediate local burning pain, erythema, superficial necrosis, itching, and edema. The edema may last for several hours. Lymphangitis, lymphadenopathy, anxiety, irregular pulse, headache, renal failure, rhabdomyolysis, vomiting, and dizziness are less commonly seen effects. Envenomation producing death has not been reported in the United States, but bites have caused death in other countries.

One bite may cause immediate local burning pain, erythema, bullae, superficial necrosis, rashes, and edema. There have been several cases in which bites of centipedes have produced recurring pain and swelling after a week or so and lasting for 1–3 days. It is unknown if this effect is related to the venom or to secondary phenomena such as infection.

Clinical Management

There is no first-aid measure of demonstrated value for injuries by centipedes; they are treated symptomatically. Severe pain has been treated with an injected local anesthetic. A topical corticosteroid, antihistamine, or local anesthetic combination may be of value. As infection is often a complication, the patient should be followed for possible infection and given systemic antibiotic therapy if necessary.

—*Gaylord P. Lopez*

Cephalosporins

- SYNONYMS: Cefaclor; cefadroxil; cefamandole; cefazolin; cefoperazone; cefotaxime; cefoxitin; ceftriaxone; cephalexin; cephalothin; cephradine; cephaprin; cefmetazole; cefonicid; ceforanide; cefotetan; cefprozil; loracarbef; cefperazone; cefpodoxime; cefixime; ceftazidime; ceftizoxime; moxalactam

- PHARMACEUTICAL CLASS: β-Lactam antibiotic

Uses

Cephalosporins are active *in vitro* against many gram-positive aerobic bacteria and some gram-negative aerobic bacteria. There are substantial differences among the cephalosporins in their spectra of activity as well as levels of activity against susceptible bacteria.

Exposure Pathways

The routes of exposure to cephalosporins are commonly oral, intravenous, or intramuscular. Accidental ingestion of oral dosage forms by children is the most common poisoning exposure.

Toxicokinetics

Cephalosporins are generally well absorbed following administration with bioavailability being greater than 75%. Many of these compounds are not stable in the acid environment of the stomach; therefore, only a limited number are useful for oral administration. The distribution is limited to the extracellular fluid space, with volumes of distribution for most ranging from 0.25 to 0.5 liters/kg. Protein binding is primarily to albumin. Most cephalosporins are widely distributed to tissues and fluids, including pleural fluid, synovial fluid, and bone. Some of the third-generation compounds have good distribution to the cerebrospinal fluid. The metabolites possess antibacterial activity. The cephalosporins and their metabolites are rapidly excreted by the kidneys by glomerular filtration and/or tubular secretion. Serum half-lives of these compounds range from 0.4 to 10.9 hr. Patients with immature renal systems or with renal compromise are at risk for toxicity due to decreased elimination. Cefamandole, cefmetazole, cefmenoxime, cefoperazone, and moxalactam have been associated with coagulopathies due to inhibition of platelet aggregation and prolongation of bleeding time.

Human Toxicity

Like penicillins, cephalosporins are a relatively nontoxic group of antibiotics. The primary adverse effect reported is hypersensitivity, a rare event. Cross-allergenicity with penicillins may occur. Toxicity is unlikely in children less than 6 years of age who acutely ingest less than 250 mg/kg. Nephrotoxicity is a possible, but rare, occurrence with acute ingestion. Coagulopathies have been reported following chronic intravenous use of certain cephalosporins.

Clinical Management

If a toxic or unknown amount of a cephalosporin has been ingested, gastric decontamination and the administration of activated charcoal is usually all that is needed. In the symptomatic patient, evaluation of renal function and electrolytes may be necessary. Chronic exposure usually requires discontinuation of the drug and supportive care. Anaphylaxis should be treated with epinephrine and/or diphenhydramine.

Animal Toxicity

The toxicity of cephalosporins in animals is unknown; however, it is expected to be similar to that found in humans.

—*Brenda Swanson Biearman*

Related Topic

Kidney

Charcoal

♦ CAS: 16291-96-6

♦ CHEMICAL CLASS: Carbon

Uses

Charcoal is produced by the incomplete combustion of plant or animal products. The major use of charcoal is for outdoor cooking. The second largest use of charcoal is in industrial applications in the form of activated charcoal. The activation process involves heating the charcoal, subjecting it to steam, or treating it with a chemical to both remove substances that have adhered to it and to break it down into finer particles and thus increase the surface area. Activated carbon has been used for its adsorptive properties as a "universal antidote" in cases of poisonings.

Exposure Pathways

The primary route of exposure is via inhalation of fine dust and ingestion.

Toxicokinetics

Charcoal is not absorbed through the skin or the gastrointestinal tract. When ingested it is excreted in the feces. This is readily apparent from the black color of the feces.

Mechanism of Toxicity

Charcoal is not generally considered to be toxic. It is possible to overwhelm pulmonary defense mechanisms if excessive dust is inhaled over a long period of time. The ability of charcoal to adsorb vitamins and enzymes can lead to nutritional deficiencies if ingested on a chronic basis.

Human Toxicity

There are not any reports of human toxicity.

Clinical Management

If toxicity should manifest, general life support should be maintained, symptoms treated, and decontamination performed, if necessary.

Animal Toxicity

There are not any reports of animal toxicity.

—*William S. Utley*

Chemical Industry Institute of Toxicology

The Chemical Industry Institute of Toxicology (CIIT) has a mission of developing, through research, an improved scientific basis for understanding and assessing the potential human health risks of exposure to chemicals, pharmaceuticals, and consumer products. CIIT is a not-for-profit corporation founded in the mid-1970s by 11 forward-looking chemical companies that recognized the need for a cooperative, industrywide effort to develop the scientific data needed to help ensure that chemicals could be produced and used in a safe manner. Today, CIIT is supported by dues payments from 40 major industrial firms, including most of the leaders in the chemical sector.

Early in the development of the conceptual basis for CIIT, the decision was made for the institute to conduct its own research in an in-house program as contrasted to operating by awarding external grants and contracts to other organizations. Research Triangle Park, North Carolina, was chosen as the site for the institute's laboratory because of anticipated substantial interactions with the three nearby research universities (Duke University, University of North Carolina–Chapel Hill, and North Carolina State University) and the adjacent research facilities of U.S. EPA's National Institute of Environmental Health Sciences and several corporations.

The institute began operations in February 1976 with the appointment of the late Dr. Leon Goldberg as the institute's first president. Under his direction, staff were recruited and research was initiated in temporary laboratories. In the fall of 1979, the institute moved into a newly constructed modern research facility and the pace of the research accelerated. In 1981, Dr. Goldberg retired and was succeeded by Dr. Robert A. Neal, who served as president until 1988, when he retired. Dr. Roger O. McClellan was appointed as CIIT's third president.

The research program has four basic operating tenets: (1) it addresses toxicological/risk assessment issues of concern to its member companies and society in general; (2) elucidating mechanisms of action underlying the toxicity of chemicals is essential to assessing their potential adverse effects; (3) these mechanisms of action are optimally elucidated within an exposure–dose–response paradigm through a multidisciplinary research effort; and (4) a risk assessment orientation provides an essential structure for identifying data gaps to guide the design of research to acquire critical information and for developing improved methodology for integrating and extrapolating data from molecular, cellular, animal, and human studies to likely human exposure scenarios. Chemicals are selected for study based on concerns related to the chemical and the likelihood of acquiring information that will give insight into generic issues that impact on the ability to assess human health risks of chemical exposures.

Individual research projects underway at the institute can be grouped into six research themes with resource allocations in 1996 as follows: chemical carcinogenesis, 50%; respiratory toxicology, 25%; endocrine toxicology, 15%; and reproductive/developmental toxicology, neurotoxicology, and immunotoxicology taken together, 10%. In 1996, the institute operated with a staff of 165 individuals, 35 doctoral regular staff, 30 postdoctoral fellows or visiting scientists, and 100 supporting personnel. The disciplinary background of the staff ranges from mathematics, physics, chemistry, and engineering to the biomedical sciences, such as toxicology, biochemistry, pathology, and veterinary medicine.

All of the institute's research findings are made available through regular publication in leading peer-reviewed journals, conference proceedings, monographs, and books. In addition, the institute distributes its monthly publication, *CIIT Activities*, to help keep readers informed of research underway at the institute. CIIT staff members are active in professional societies, hold adjunct appointments at nearby universities, are frequent presenters at scientific meetings, and participate extensively in advisory committees to the government and private organizations in keeping with the institute's motto of performing "science in the public interest." During the past two decades, the institute's research program has yielded numerous scientific contributions that have served to reduce the uncertainty in assessing human health risks of exposure to chemicals.

For additional information, contact the Chemical Industry Institute of Toxicology, Office of the President, Davis Drive, P.O. Box 12137, Research Triangle Park, NC 27709. Telephone: 919-558-1202; fax: 919-558-1400.

—*David M. Krentz and Harihara M. Mehendale*

(Adapted from information supplied by CIIT.)

Related Topics

    Academy of Toxicological Sciences
    International Life Sciences Institute–North
      America
    National Center for Toxicological Research
    SETAC Foundation for Environmental Education
    Society of Environmental Toxicology and
      Chemistry

Chloral Hydrate

◆ CAS: 302-17-0

◆ SYNONYMS: Noctec; "knockout drops"; "mickey finn"; chloral; hydrated chloral; chloralex;

chloralvan; novochlorhydrate; chloraldural; chloraldurat; trichloroacetaldehyde, hydrated; trichloroethylidene glycol; 2,2,2-trichlorethane-1,1-diol

◆ PHARMACEUTICAL CLASS: Chloral derivative

◆ CHEMICAL STRUCTURE:

$$CCl_3CH\,(OH)_2$$

Uses

Chloral hydrate is used as a sedative–hypnotic agent. It is also a drug of abuse.

Exposure Pathways

Ingestion of oral dosage forms is the most common route of both accidental and intentional exposures to chloral hydrate. It is available as capsules, tablets, an oral solution, and rectal suppositories.

Toxicokinetics

Chloral hydrate is rapidly absorbed from the gastrointestinal tract following oral or rectal administration. It produces its pharmacologic action within approximately 30 min. Chloral hydrate is rapidly metabolized by alcohol dehydrogenase to trichloroethanol, which is pharmacologically active. A small amount is metabolized to an inactive metabolite, trichloroacetic acid. Trichloroethanol, in turn, is either conjugated with glucuronic acid to form urochloralic acid or oxidized to trichloroacetic acid. Chloral hydrate has a half-life of elimination of only a few minutes, whereas the half-life of trichloroethanol ranges from 4 to 14 hr. The half-life of trichloroethanol may be prolonged following overdose.

Both chloral hydrate and trichloroethanol are highly lipid soluble. The apparent volume of distribution of chloral hydrate and trichloroethanol is 0.6–0.75 and 0.6–1.6 liters/kg, respectively. Trichloroethanol is approximately 40% protein bound. The active and inactive metabolites of chloral hydrate are excreted primarily in the urine. The principal metabolite excreted in the urine is trichloroacetic acid and its glucuronide conjugate.

Mechanism of Toxicity

Chloral hydrate is a central nervous system (CNS) depressant. It is probably responsible for early depressant effects, but prolonged CNS depression is largely due to trichloroethanol. The mechanism by which chloral hydrate and trichloroethanol depress the CNS is not completely known.

Human Toxicity: Acute

Chloral hydrate is an irritant to the gastrointestinal tract; as a result, ingestion may cause nausea, vomiting, and diarrhea. Gastric perforation and esophageal stricture have been reported in cases of chloral hydrate overdose. Acute ingestion of 2 g is likely to lead to toxic symptoms. The lethal dose in adults is approximately 5–10 g. Lethargy progressing to deep coma, respiratory depression, hypotension, and hypothermia are characteristic toxic manifestations of chloral hydrate overdose. Unlike most other sedative–hypnotic agents, overdose with chloral hydrate may result in serious atrial and ventricular arrhythmias. Hepatic and renal dysfunction may develop.

Human Toxicity: Chronic

Prolonged administration of chloral hydrate may lead to the development of gastritis, skin eruptions, and renal damage. Chronic use of high doses may produce psychologic and physical dependence. Abrupt discontinuation may lead to delirium and seizures.

Clinical Management

Basic and advanced life-support measures should be implemented as necessary. Gastrointestinal decontamination procedures should be used as appropriate based on the patient's level of consciousness and history of ingestion. Activated charcoal can be used to adsorb chloral hydrate. The patient's level of consciousness and vital signs should be monitored closely. Obtunded patients with reduced gag reflex should be intubated. Respiratory support including oxygen and ventilation should be provided as needed. There is no antidote for chloral hydrate. Hypotension should be treated with standard measures including intravenous fluids, Trendelenburg positioning, and dopamine hydrochloride by intravenous infusion. Cardiac arrhythmias should be managed with standard therapeutic approaches. Class IA antiarrhythmics should be avoided, however, as these may worsen cardiac conduction. Forced diuresis is of no value as a means to enhance the elimination of chloral hydrate. Hemodialysis and hemoperfusion

may be useful in severe cases in which standard supportive measures are inadequate. Withdrawal reactions should be managed with barbiturates or other sedative–hypnotic agents.

—Gregory P. Wedin

Chloramphenicol

- ◆ CAS: 56-75-7
- ◆ SYNONYMS: Chloramphen; clorolifarina; CHPC; laevomycetinum
- ◆ PHARMACEUTICAL CLASS: An antibiotic with both bacteriocidal and bacteriostatic properties
- ◆ CHEMICAL STRUCTURE:

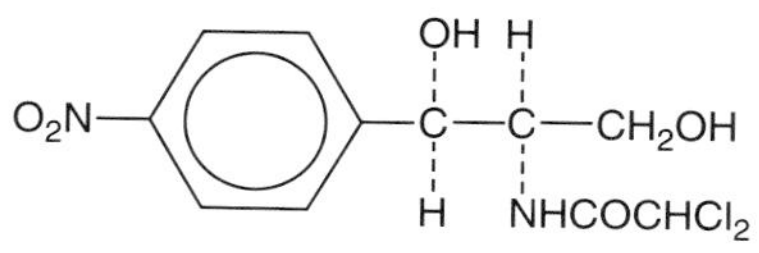

Uses

Chloramphenicol is used as an antibiotic.

Exposure Pathways

Intoxication by this agent occurs by both oral and parenteral routes.

Toxicokinetics

Chloramphenicol is well absorbed from the gastrointestinal tract; peak serum concentrations are reached 1 or 2 hr after an oral dose. Peak serum concentrations after ingestion equal those achieved after intravenous administration. Absorption after intramuscular injection is highly variable with peak concentrations achieved being 5–65% of those reached after intravenous or oral administration. The apparent volume of distribution is 0.6–1.6 liters/kg. Approximately 50%

of the drug is bound to plasma proteins (primarily albumin). Chloramphenicol diffuses into breast milk and readily crosses the placenta; fetal blood levels are 30–80% of maternal serum concentrations. Inactivation occurs primarily by hepatic glucuronidation. Metabolism is slowed in patients with hepatic insufficiency. Chloramphenicol has an elimination half-life of 1–4 hr. Urinary excretion of unchanged chloramphenicol is approximately 12% in adults and 20% in children; the remainder is eliminated as drug metabolite. Dosage modifications may be necessary in patients with renal insufficiency.

Mechanism of Toxicity

Chloramphenicol has a narrow therapeutic index with serum concentrations of 10–20 μg/ml therapeutic and >25 μg/ml toxic. Toxicity occurs through chloramphenical suppression of DNA and RNA synthesis in human as well as bacterial cells.

Human Toxicity: Acute

Acute, single overdoses of chloramphenicol produce no significant toxicity. However, in neonates, a syndrome of vomiting, irregular respirations, abdominal distension, diarrhea, cyanosis, flaccidity, hypothermia, and death may occur ("gray baby syndrome"). The gray baby syndrome typically begins 2–9 days after the initiation of chloramphenicol. It has been associated with the administration of chloramphenicol doses of >200 mg daily. Its etiology may be the combination of immature hepatic glucuronidation in conjunction with diminished urinary excretion of the parent drug, secondary to an immature renal system.

Human Toxicity: Chronic

Bone marrow suppression occurs in all individuals who take chloramphenicol regularly. This is primarily manifested by a reversible fall in reticulocyte count. Depressions in platelet count may also occur. Dose-related bone marrow suppression is associated with serum concentrations ≥25 μg/ml. Chloramphenicol can also produce severe, idiosyncratic bone marrow toxicity with the development of bone marrow aplasia. This may occur more commonly in those who undergo prolonged therapy. The reported incidence of severe bone marrow toxicity ranges from 1:30,000 to 1:100,000. Greater myelotoxicity occurs in uremic patients. Other adverse effects include hypersensitivity reactions (including rash and fever), paresthesias, and optic neuritis. Chloramphenicol may also inhibit hepatic microsomal activity, impairing the clearance of drugs including coumadin and phenytoin.

Clinical Management

Hemodialysis is ineffective. Charcoal hemoperfusion or whole blood exchange transfusion has been recommended in infants with serum concentrations of >50 μg/ml.

Animal Toxicity

Chloramphenicol is used as a veterinary antibiotic. It has a low level of animal toxicity.

—*Michael Shannon*

Related Topics

Carboxylesterases
Delayed-Type Hypersensitivity

Chlorbenzilate

- ◆ CAS: 510-15-6
- ◆ SYNONYMS: Ethyl 4,4'-dichlorobenzilate; 4,4'-chlorobenzilic acid ethyl ester; Acaraben; Benz O-chlor; Benzilan; Kop-Mite
- ◆ CHEMICAL CLASS: Organochlorine insecticide
- ◆ CHEMICAL STRUCTURE:

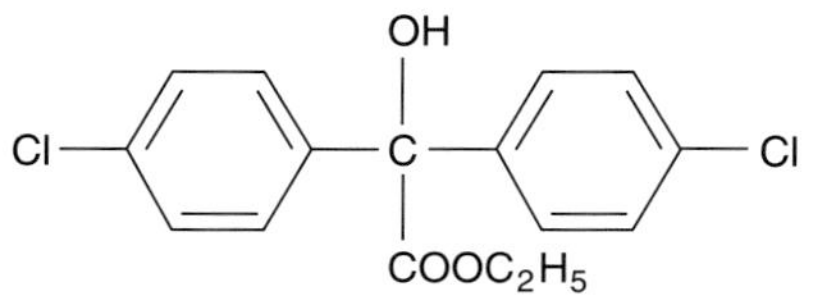

Uses

Chlorbenzilate was first introduced in the market as a technical product in 1952 and is mainly used on citrus crops as a miticide.

Exposure Pathways

The oral, dermal, and inhalation routes are the most common exposure pathways for chlorbenzilate.

Toxicokinetics

Chlorbenzilate is well absorbed from the gastrointestinal tract; the presence of absorbable lipid (animal or vegetable fat) facilitates its absorption. Organochlorines are relatively nonvolatile compounds. Absorption also occurs probably through mucocilliary trapping following inhalation exposures to sprays which may be followed by gastrointestinal absorption.

In contrast to some organochlorine insecticides, chlorbenzilate is metabolized within hours to a few days in rat liver to p,p'-dichlorobenzilic acid, p,p'-dichlorobenhydrol, p-chlorobenzoic acid, and p,p'-dichlorobenzophenone by the mixed-function oxidase system.

Pharmacokinetic studies have revealed that the excretion of chlorbenzilate by mammals does not follow first-order kinetics. The half-life for the residual compound increases with a decrease in body burden. This may be due to binding of the compound to different lipoproteins in the body to form complexes (bound forms) with different dissociation characteristics. About 40% of the total dose (12.8 mg/kg, 5 days per week for 35 weeks in dogs) was found to be excreted unchanged or as urinary metabolites. Chlorbenzilate does not tend to accumulate in fatty tissues of rats or dogs.

Mechanism of Toxicity

Chlorbenzilate primarily affects nerve cells by acting as an "axon poison" and interfering with the normal flux of sodium and potassium ions across the axonal membrane. The resulting ionic imbalance causes hyperexcitability of the nerves characterized by irritability, disturbance of the mental process, sensory aberrations, and convulsions. It can cause asphyxia by depressing the respiratory centers in the brain stem which may persist for minutes to hours depending on the degree of exposure.

Human Toxicity: Acute

Chlorbenzilate has a low acute health hazard potential. Few cases of human poisoning have been reported; however, this compound may produce tremors, ataxia, agitation, and nervousness. Symptoms exhibited by a pesticide sprayer exposed to chlorbenzilate included ataxia, delirium, fever, and muscle pains.

Clinical Management

In case of severe poisoning, life-support measures should be provided with respiratory and cardiovascular support. With inhalation exposures, the patient should be moved to fresh air immediately and may be placed on assisted ventilation along with emergency airway support and 100% humidified oxygen. In case of dermal exposure, contaminated clothes should be removed and skin, hair, and nails should be washed thoroughly with a sufficient amount of soapy water. Induction of emesis is recommended with substantial ingestion of this compound. Conventional anticonvulsant drugs may be administered to treat seizures, depending on the severity.

Animal Toxicity

The acute oral LD_{50} for mice, rats, and hamsters is approximately 700 mg/kg. Symptoms of poisoning in rodents include motor and respiratory dysfunction. Chlorbenzilate was reported to produce hepatocellular carcinoma in two strains of mice and testicular atrophy in rats. Dogs exposed repeatedly to oral doses of chlorbenzilate for a period of 35 weeks did not show any signs of overt toxicity. Rats can tolerate 500 ppm in the diet for 2 years.

—*Tamal Kumar Chakraborti*

Related Topics

Neurotoxicology: Central and Peripheral
Organochlorine Insecticides
Pesticides

Chlordane

- ◆ CAS: 57-74-9
- ◆ SYNONYM: 1,2,4,5,6,7,8,8-Octachloro-3a,4,7,7a-tetrahydro-4,7-methanoindan

- CHEMICAL CLASS: Synthetic organochlorine cyclodiene insecticide
- CHEMICAL STRUCTURE:

Uses

Chlordane is used as an insecticide.

Exposure Pathways

Exposure to chlordane may occur via ingestion, inhalation, or dermal contact. Exposure may also occur *in utero* since chlordane readily crosses the placenta.

Toxicokinetics

Chlordane is absorbed through the gastrointestinal tract, the lungs, and the skin. Absorption is potentiated by dissolution in an organic solvent. When absorbed through the respiratory tract, more than 75% of the dose was retained by rats. Absorption of oral doses is dependent on the amount of lipid-containing material in the gastrointestinal tract.

Chlordane, like other organochlorine cyclodiene insecticides, is metabolized very slowly, mainly by liver microsomal cytochrome P450. The main metabolite from a toxicity point of view is oxychlordane, which is 20- to 25-fold more toxic than the parent compound. Other metabolites, such as chlordene chlorohydrin, 1,2-dichlorochlordene, and 1-hydroxy-2-chlordene, are less toxic than the parent compound, chlordane.

Chlordane is highly lipophilic and is readily sequestered in adipose tissue and in other organs with high lipid contents. The α isomer is primarily stored in adipose tissue with the γ isomer stored to a greater extent in kidneys than in fat.

Due to the slow absorption and metabolism, approximately 80% of an oral chlordane dose is excreted in urine and feces. Chlordane may also be excreted in breast milk.

Mechanism of Toxicity

The activity of γ-aminobutyric acid, that of inducing uptake of Cl^- ions by neurons, is blocked by chlordane. This results in only a partial repolarization of activated neurons leading to an uncontrolled excited condition. Additionally, both Ca^{2+}, Mg^{2+}-adenosine triphosphatase (ATPase), and Na^+, K^+-ATPase functions are inhibited by chlordane. This leads to increased intracellular free calcium concentration in neurons and the release of neurotransmitters. These neurotransmitters potentiate depolarization of adjacent neurons in a chain-reaction fashion, propagating stimuli through the central nervous system (CNS).

Human Toxicity

Generally, nausea, vomiting, and dizziness precede CNS symptomatology. Convulsions usually follow these symptoms, although convulsions have been reported in the absence of other warning symptoms. Convulsions usually manifest themselves within 0.5–3 hr after ingestion or dermal exposure and are often accompanied by confusion, incoordination, excitability, or coma (in some cases). Respiratory arrest may result from acute exposure to high doses. Other findings include anemia, leukopenia, thrombocytopenia, and abnormal serum chemistry results. Liver toxicity is seen only in extremely high dose exposures. Recovery following convulsions has been observed in infants with dosages of 10–28 mg/kg and in adults after 32 mg/kg. Death has been observed after dermal exposure to 425 mg/kg.

Clinical Management

Treatment is symptomatic. Anticonvulsive treatment with diazepam or phenobarbital is usually effective for control of convulsions. Cholestyramine treatment has been suggested for increased elimination; this treatment has not been proven beneficial for chlordane, although contaminating heptachlor excretion was increased.

Animal Toxicity

Chlordane toxicity in animals is similar to that of other organochlorine insecticides except tremor is absent. CNS involvement produces hyperexcitability and convulsions. Immunosuppression has been reported in mice prenatally exposed to chlordane. The incidence of hepatomas is increased in mice, but not in rats, fed chlordane.

—*Benny L. Blaylock*

Related Topics

Neurotoxicology: Central and Peripheral
Organochlorine Insecticides
Pesticides
Pollution, Water

Chlordecone

- CAS: 143-50-0
- PREFERRED NAME: Kepone
- SYNONYMS: 1,3,4-Metheno-2H-cyclobuta-(cd)pentalen-2-one; 1,1α,3,3α,4,5,5,5α,5β,6-decachloroctahydro; Kepone; GC 1189; Ciba 8514; ENT 16,391; NCI-C00191; decachloroketone; decachlorotetracyclodecanone; decachlorotetrahydro-4,7-methanoindeneone
- CHEMICAL CLASS: Polycyclic chlorinated hydrocarbon
- CHEMICAL STRUCTURE:

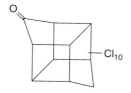

Uses

Chlordecone is used as an insecticide and fungicide.

Exposure Pathways

Ingestion is the most common route of exposure to chlordecone. Exposure to this agent may also occur via respiratory and dermal routes in industrial workers.

Toxicokinetics

Chlordecone is readily absorbed (>90%) from the gastrointestinal tract. Absorbed chlordecone rapidly (within 24–48 hr) establishes an equilibrium of distribution among most tissues. Absorption of chlordecone may also occur through skin, especially in patients with dermatitis or skin rash. Compared to other organochlorine pesticides, the ratio of chlordecone concentration in whole blood to that in fat tissue of patients exposed to chlordecone is much higher (1 : 7) and 75% chlordecone in the blood binds with albumin and high-density lipoprotein. Chlordecone is primarily stored in liver tissue followed by adipose tissue in both man and animals. In human, chlordecone is bioreduced to chlordecone alcohol in the liver followed by glucuronide conjugation of the alcohol metabolite. However, quantitatively this metabolism is very minor and not important toxicologically. Chlordecone is not subject to metabolism in animals studied so far, except Mongolian gerbils, which metabolize chlordecone similarly to humans.

Fecal excretion is the major route of elimination and only minimal amounts of chlordecone are eliminated through urine. By 84 days, 65% of the dose is excreted in the stool and only 1.6% in the urine. A substantial amount of chlordecone representing as much as 1% of the total body content enters the intestine via biliary excretion. However, the major part of biliary chlordecone (90–95%) is reabsorbed by the intestine and recirculated to the liver (enterohepatic recirculation), while the remaining 5–10% of the biliary chlordecone entering the upper intestine appears in the feces. Elimination of chlordecone from the body is slow. The half-life of chlordecone in the blood and fat tissue is 165 and 125 days, respectively. Lactating women can also excrete substantial amounts of accumulated chlordecone through breast milk.

Mechanism of Toxicity

Chlordecone inhibits brain mitochondrial and synaptisomal membrane-bound Na$^+$, K$^+$ ATPase and oligomycin-sensitive Mg^{2+} ATPase activity and thus may result in blocked cellular uptake and storage of neurotransmitters such as catecholamine and γ-aminobutyric acid, leading to neurotoxicity. Inhibition of Mg^{2+} ATPase by chlordecone and the consequently deceased hepatic mitochondrial energy production have been postulated for the mechanism of chlordecone-induced hepatic biliary dysfunction. Chlordecone is an inducer of hepatic microsomal drug metabolizing system at high doses. Dietary exposure to nontoxic levels of chlordecone (10 ppm) has been shown to cause a 70-fold increase in lethality of nonlethal dose of carbon tetrachloride in laboratory rats. The mechanism for chlordecone amplification of chloromethane hepatotoxicity and lethality is incapacitated cell division due to decreased energy owing to the disrupted intracellular calcium homeostasis. Because tissue repair cannot occur, limited liver injury progresses unabated, leading to hepatic failure and animal death. Chlordecone impairs the reproductive system by mimicking the effects of excessive estrogen.

Human Toxicity

There are no reports of death in humans exposed to chlordecone. The major target organs of chlordecone toxicity are the nervous system, the liver, and the testes. The symptoms of neuromuscular toxicity include an irregular nonpurposive waking tremor (rate, 6–8 Hz) involving the extremities, head, and trunk, and opsoclonus, an unusual oculomotor disorder consisting of chaotic eye movements causing blurred vision. Onset of tremor varies from 5 days to 8 months after initial exposure to chlordecone depending on the duration and intensity of exposure to this compound. Chlordecone poisoning causes liver and spleen enlargement, mitochondrial changes, fatty infiltration of hepatocytes, proliferation of endoplasmic reticulum, and impairment of biliary excretion of selective organic anions. Chlordecone poisoning also causes a decline in sperm number associated with abnormally low percentages of motile sperm. In most cases, these symptoms are reversible upon cessation or exposure. Removal of this chemical from the tissues is accompanied by the disappearance of clinical manifestation of toxicity.

Clinical Management

Removal of chlordecone from the body is the most important measure to take. In cases of ingestion, emesis is indicated as the treatment in other chlorinated hydrocarbon insecticide intoxications, unless the patient is comatose or has lost the gag reflex. Emesis should be followed by administration of activated charcoal and saline cathartics. Oil-based cathartics should be avoided. Administration of cholestyramine, which has been shown to bind chlordecone in the intestinal tract, is an effective way to increase fecal excretion of chlorde-

cone and to accelerate the removal of chlordecone from the blood and other tissues.

Animal Toxicity

Rats, mice, chicks, and quails are sensitive to the toxicity of chlordecone. The LD_{50} for oral administration of chlordecone in corn oil is 71, 126, 250, and 480 mg/kg for rabbits, rats, dogs, and chicks, respectively. In rabbits, the dermal LD_{50} for chlordecone is 434 mg/kg. Chlordecone tremendously potentiates lethal effects and hepatotoxicity of chloromethane in rodents. Chronic exposure causes carcinogenesis in rats and mice.

Further Reading

Mehendale, H. M. (1990). Potentiation of halomethane hepatotoxicity by chlordecone: A hypothesis for the mechanism. *Med. Hypoth.* 33, 289–299.

—*Zhengwei Cai and*
Harihara M. Mehendale

Related Topics

Liver
Neurotoxicology: Central and Peripheral
Organochlorine Insecticides
Pesticides
Psychological Indices of Toxicity

Chlordimeform

- CAS: 6164-98-3 (base); 19750-95-9 (salt)
- SYNONYMS: *N'*-(4-chloro-*o*-tolyl)-*N,N*-dimethylformamidine; Bermat; chlordimeforme; chlorodimeform; chlorophedine; chlorophenamidine; Fundal; Fundex; Galecron; Spanone. Code designations are specific for either

the base or salt form: base, ENT 27335, OMS 1209; salt, ENT 27567

♦ CHEMICAL CLASS: Organonitrogen acaricide

♦ CHEMICAL STRUCTURE:

Uses

Chlordimeform is used to control mites, ticks, and some members of the Lepidoptera order.

Exposure Pathways

The most common accidental exposure pathway is probably dermal contact. Inhalation during processing and packaging has also been reported as well as suicide attempts through ingestion of chlordimeform.

Toxicokinetics

The base formulation readily penetrates the skin but the salt form, which is much more water soluble, does not. Chlordimeform is rapidly demethylated to form demethylchlordimeform and didemethylchlordimeform, both of which are more toxic (based on acute oral LD_{50} values in mice) than the parent compound. Other active but less toxic metabolites include *N*-formyl-4-chloro-*o*-toluidine, 4-chloro-*o*-toluidine, 3-(4-chloro-*o*-tolyl)urea, 1,1-dimethyl-3-(4-chloro-*o*-tolyl)urea, and 1-methyl-3-(4-chloro-*o*-tolyl)urea. Neither chlordimeform nor any of its metabolites have been shown to accumulate in any specific tissue.

Studies using animals treated with radiolabeled chlordimeform indicated the majority of the radioactivity to be excreted in the urine within 24 hr of the last treatment. Small amounts of radioactivity were detected in the bile and feces.

Mechanism of Toxicity

Animal studies have detected a variety of pharmacological and biochemical changes in response to chlordimeform exposure. The cause of death following acute exposure appears to be cardiovascular collapse. Lethal doses of chlordimeform cause decreases in cardiac contractility and peripheral resistance resulting in severe hypotension. Respiratory arrest also occurs but is thought to be secondary to the cardiovascular effects. The effects of chlordimeform on the cardiovascular system share similarities with those seen with local anesthetics such as procaine. Chlordimeform also inhibits monoamine oxidase and acts as an uncoupler of oxidative phosphorylation (see Monoamine Oxidase Inhibitors).

Human Toxicity: Acute

In addition to the cardiovascular and respiratory effects described previously, reported signs and symptoms of chlordimeform exposure include severe hemorrhagic cystitis, gross hematuria, proteinuria, swollen liver, decreased appetite, fatigue, vertigo, and dermitis.

Human Toxicity: Chronic

A carcinogenic potential of some chlordimeform metabolites has been demonstrated.

Clinical Management

Ingestion should be treated by inducing emesis with syrup of ipecac, administering activated charcoal, or performing gastric lavage. Lavage rather than emesis is recommended in patients who are convulsing or comatose. Emesis and lavage are most effective if performed shortly after ingestion. For inhalation exposure, the victim should be removed from the exposure area and observed for signs of breathing difficulty. distress. In cases of dermal exposure, contaminated clothing should be removed and disposed of. Any exposed areas of skin should be repeatedly washed with soap and water. For eye contact, flush the eyes with generous amounts of lukewarm water for a minimum of 15 min.

Treatment is basically symptomatic and supportive; no specific antidotes are available. Artificial ventilation with 100% humidified oxygen is necessary in cases of respiratory distress. If patient is cyanotic and cyanosis does not respond to oxygen administration, methemoglobin levels should be determined. Support of cardiovascular function may also be required. Bladder damage can be determined by urinalysis. Hypotension may be treated with isotonic intravenous fluids. Dopamine or norepinephrine may be used if hypotension does not respond to infusion of fluids. Convulsions may be treated with intravenous diazepam. Phenytoin or phenobarbital may be used if the convulsions are recurrent. Because chlordimeform is a monoamine oxidase inhibitor, foods with large amounts of tryptophan or tyramine should be avoided and sympathomimetic drugs are contraindicated.

—Paul R. Harp

Related Topic

Pesticides

Chlorine

- ◆ CAS: 7782-50-5
- ◆ SYNONYMS: Bertholite; chloor; chlor; chlore; molecular chlorine; cloro; RTECS FO2100000; UN1017
- ◆ CHEMICAL CLASS: Disinfectant; bleaching agent

Uses

Chlorine is used to bleach all types of fabric, to disinfect relatively clean impervious surfaces, to purify water, and to control biofouling in cooling systems. It is used in the processing of meat, fish, vegetables, and fruits. It is also used in the manufacturing of synthetic rubber, plastics, pesticides, antifreeze, refrigerants, antiknock compounds, chlorinated hydrocarbons, polyvinyl chloride, and chlorinated lime. Chlorine is also used in detinning and dezincing iron and as an ingredient in special batteries.

Exposure Pathways

Dermal or ocular contact and inhalation are the most common exposure pathways.

Toxicokinetics

Chlorine persists as an element only at a very low pH (less than 2), and at the higher pH found in living tissue it is rapidly converted into hypochlorous acid. In this form, it apparently can penetrate the cell and form N-chloro-derivatives that can damage cellular integrity.

Mechanism of Toxicity

Chlorine reacts with body moisture to form acids. The acids form acid proteinates.

Human Toxicity: Acute

Liquid chlorine causes burns to skin and eyes and will cause frostbite. It may cause lung injury if inhaled. Chlorine causes smarting of the skin and first-degree burns on short exposure; it may cause secondary burns in long exposures. Inhalation of low concentrations causes mild mucous membrane irritation and irritation of the upper respiratory tract. Inhalation of high concentrations of the gas causes necrosis of the tracheal and bronchial epithelium as well as pulmonary edema, atelectasis, emphysema, and damage to the pulmonary blood vessels. Acute exposure may also cause anxiety and vomiting. Exposure to 500 ppm can be lethal over 30 min, while exposure to 1000 ppm can be lethal within a few minutes.

Human Toxicity: Chronic

Chronic exposure causes permanent, although moderate, reduction in pulmonary function and corrosion of teeth.

Clinical Management

Exposure should be terminated as soon as possible by removal of the patient to fresh air. The skin, eyes, and mouth should be washed with copious amounts of water. A 15- to 20-min wash may be necessary. Contaminated clothing and jewelry should be removed and isolated. Contact lenses should be removed from the eye to avoid prolonged contact of the chemical with the area. Affected areas should not be rubbed. If breathing has stopped, artificial respiration should be given. If breathing is difficult, oxygen should be given.

Animal Toxicity

Exposure of cats to a concentration of 900 mg/m^3 (300 ppm) for 1 hr may cause death after a period during which the conjunctiva is inflamed and there is coughing and dyspnea. Dogs rarely die following a 30-min exposure to 650 ppm and never die following a 30-min exposure to less than 280 ppm. The pulse rate of dogs is retarded during exposure to concentrations of 180–200 ppm or greater. In guinea pigs, the inhalation of small quantities of chlorine accelerates the course of experimental tuberculosis.

—*Sanjay Chanda*

Related Topics

Detergents
Pollution, Air
Pollution, Water
Surfactants, Anionic and Nonionic

Chlorine Dioxide

- ◆ CAS: 10049-04-4
- ◆ SYNONYMS: Chlorine-oxide; Alcide; anthium dioxcide; chlorine peroxide; Chloroperoxyl; Doxcide 50; Hallox E-100; NA 9191; Chlory Radical
- ◆ CHEMICAL CLASS: Chlorine dioxide and its by-products are collectively called oxychlorines.
- ◆ CHEMICAL STRUCTURE:

$$O=Cl=O$$

Uses

Chlorine dioxide is a strong oxidizing agent, bactericide, and antiseptic. It is used in bleaching cellulose, paper pulp, leather, flour, fats and oils, textiles, and beeswax, and in deodorizing and purifying water. It is currently considered as an alternative to chlorine as a disinfectant for public water supplies in the United States.

Exposure Pathways

Consumption of drinking water is the most probable route of exposure to chlorine dioxide and its by-products. Patients undergoing hemodialysis may be directly exposed to chlorine dioxide through a dialysis water disinfected with chlorine dioxide. Chlorine dioxide is a gas; therefore, inhalation is also an exposure pathway.

Toxicokinetics

Chlorine dioxide can be rapidly absorbed through the gastrointestinal tract. Peak blood concentration levels can be reached within 1 hr after a single dose administered orally. It can also be slowly absorbed through shaved skin with a half absorption time of 22 hr. Chlorine dioxide is metabolized to chlorite, chlorate, and mostly chloride. Most administered chlorine dioxide and its metabolites remain in plasma followed by kidneys, lungs, stomach, intestine, liver, and spleen. About 43% of orally administered chlorine dioxide is eliminated in the urine and feces within 72 hr. It is not excreted via the lungs.

Mechanism of Toxicity

The toxicity of chlorine dioxide is attributed to the oxidative stress caused by this compound and its by-products or metabolites. Animal studies and *in vitro* experiments with human red blood cells indicate that chlorine dioxide and its by-products, especially chlorite, oxidize hemoglobin to methemoglobin by inhibiting methemoglobin reductase, decreasing erythrocyte glutathione levels, stimulating erythrocyte hydrogen peroxide production, and causing hemolytic anemia.

Human Toxicity

Chlorine dioxide gas is highly irritating to the skin and mucous membranes of the respiratory tract. It may cause pulmonary edema. It is explosive in the form of concentrated vapor or solution. Daily ingestion of 1 liter of water containing 0.7 mg of chlorine dioxide has been reported to cause nausea. No other acute intoxication has been reported.

The human experience with chlorine dioxide both in controlled, prospective studies and in actual use situations in community water supplies has failed to reveal adverse health effects. However, glucose-6-phosphate dehydrogenase-deficient individuals and infants are groups thought to be at higher risk to chlorine dioxide toxicity due to their susceptibility to oxidant-induced methemoglobinemia.

The U.S. EPA has recommended standards of 0.06 mg/liter for chlorine dioxide and 0.007 mg/liter for chlorite and chlorate in drinking water.

Animal Toxicity

The deleterious effects of chlorine dioxide on red blood cells, thyroid function, and brain development have been found in animal studies with relatively high concentrations of this agent. Rats are more sensitive than mice to the developmental effects associated with chlorite-treated drinking water.

—*Zhengwei Cai*

Related Topic

Pollution, Water

Chlorobenzene

- CAS: 108-90-7
- SYNONYMS: Benzene chloride; benzene chloro-; chlorbenzene; chlorbenzol; chlorobenzol; monochlorbenzene; monochlorobenzene phenyl chloride; NCI-C54886; Caswell No. 183A; EPA Pesticide Chemical Code 056504; MCB, CP 27; I P Carrier T 40; Tetrosin SP
- MOLECULAR FORMULA: C_6H_5Cl
- CHEMICAL STRUCTURE:

C6—H5—Cl

Uses

Chlorobenzene is used as a solvent for pesticide formulations, in auto parts degreasing, and in the manufacture of adhesives, paints, polishes, waxes, pharmaceuticals, and natural rubber. It is a chemical intermediate in the production of diphenyl oxide, diisocyanates, and nitrochlorobenzene. It has also been used as a fiber swelling agent and as a dye carrier in textile processing.

Exposure Pathways

The vapor pressure of chlorobenzene is relatively high (11.8 mm Hg); therefore, inhalation is a potential route of exposure. Since chlorobenzene is soluble in water (448 ppm) and has been detected in wastewater and drinking water, there is potential for oral exposure. In addition, as a result of its solvent and degreasing properties, the potential for accidental skin contact with the material also exists.

Toxicokinetics

Data in rabbits indicate that the toxicity from a single dermal application is minimal with only slight reddening of the skin observed. Continuous skin contact with chlorobenzene for 1 week resulted in moderate erythema and slight superficial necrosis. Absorption in amounts sufficient to cause toxicity can also occur as a result of ingestion or inhalation. Because chlorobenzene is highly lipophilic and hydrophobic, it is thought to be distributed throughout the total body water, with body lipids being a major deposition site.

The kinetics of metabolism and excretion were investigated in rabbits administered a single oral dose of 0.5 mg/kg or doses of 0.5 g twice daily for 4 days. In the single-dose study, 27% of the administered dose was excreted unchanged in the expired air. The majority of the remainder was excreted in the urine as a glucuronide (25%), ethereal sulfate (27%), and mercapturic acid (20%). Similarly, rabbits administered repeated doses of chlorobenzene excreted the majority of the dose in the urine, and only small amounts were detected in the tissues and feces. It has been proposed that chlorobenzene is first oxidized to the 3,4-epoxide, which then can follow one of several pathways. One leads to the formation of the P-mercapturic acid conjugate following glutathione conjugation. A second pathway results in formation of 4-chlorocatechol, and the third pathway ends with formation of 4-chlorophenol and it conjugates. Data collected from exposed workers and volunteers indicate that for humans, the primary pathways are formation of the *p*-mercapturic acid conjugate and 4-chlorocatechol.

Mechanism of Toxicity

Similar to other volatile organic chemicals, chlorobenzene is a nervous system depressant. In addition, lesions of the liver and kidneys have also been observed following toxic doses. These sites of injury were substantiated by the observation of covalent binding to hepatic, renal, and pulmonary DNA, RNA, and other proteins in rats and mice administered a single intraperitoneal injection.

Human Toxicity

In the industrial environment, symptoms including headache, skin irritation, eye irritation, irritation of the upper respiratory tract, dizziness, somnolence, hematopoietic effects, and neuromuscular changes have been reported. Accidental ingestion of 5–10 ml of a cleaning agent containing chlorobenzene caused loss of consciousness, vascular paralysis, and heart failure in a child (approximately 2 years old).

Regulatory and recommended exposure limits for chlorobenzene include:

OSHA PEL-TWA: 75 ppm

ACGIH TLV-TWA: 10 ppm

DFG MAK: 50 ppm

Federal drinking water standard (U.S. EPA): 100 μg/liter

Minnesota drinking water standard: 100 μg/liter

Wisconsin drinking water standard: 100 μg/liter

Arizona drinking water standard: 60 μg/liter

Maine drinking water standard: 47 μg/liter

New Jersey drinking water standard: 4 μg/liter

Clinical Management

For ocular contact, the eyes should be irrigated immediately with abundant running water. If the material contacts the skin, the affected areas should be washed with soap and water promptly. If inhalation exposure occurs, the exposed person should be moved to fresh air immediately and provided with respiratory support (oxygen or artificial respiration) if necessary. If the material has been ingested, vomiting should not be in-

duced. For ingestion, gastric lavage followed by saline catharsis should be performed. Renal and hepatic function should be monitored and supported if necessary.

Animal Toxicity

The oral LD_{50} values for rats, mice, and rabbits were 2290, 2300, and 2830 mg/kg, respectively. The approximate inhalation LD_{50} (2 hr) is 4300 ppm for mice. Application of chlorobenzene to the skin of rabbits caused slight reddening; prolonged skin contact was irritating. Ocular contact in rabbits caused a transient conjunctival irritation which resolved within 48 hr.

Several repeated-exposure oral studies have been conducted in various species. Although the doses at which effects were observed are variable between species, the primary effects of chlorobenzene were observed in the liver and kidneys. Rats and mice were administered daily doses of 60–750 mg/kg, 5 days per week, for 13 weeks. Survival was lower in rats at 500 mg/kg and above and in mice at 250 mg/kg and above. Pathological changes in the liver and kidneys and changes in the hematopoietic system (spleen, bone marrow, and thymus) were observed in both species at 250 mg/kg and above. In another study, rats were administered doses ranging from 14.4 to 376 mg/kg/day, 5 days per week, over a period of 192 days. Doses of 144 mg/kg/day and above caused changes in liver and kidney weights and changes in liver morphology. Doses of 18.8 mg/kg/day and below did not cause any adverse effects. Dogs were administered oral doses ranging from 27.2 to 272.5 mg/kg/day, 5 days per week, for 93 days. There were no effects at 54.5 mg/kg/day and below. At 272.5 mg/kg/day, changes in clinical chemistry parameters were observed, four of eight dogs died, and pathological changes were observed in the liver, kidney, gastroenteric mucosa, and hematopoietic tissue. In a study to determine the carcinogenic potential of chlorobenzene, rats were administered daily doses of 0, 60, or 120 mg/kg/day, 5 days per week, for 103 weeks, and mice were similarly administered 30 or 60 mg/kg/day. No increased tumor incidences were observed in female rats or in male or female mice. Male rats administered 120 mg/kg/day had an increased incidence of hepatic neoplastic nodules (8% for untreated control, 4% for vehicle control, 8% for 60 mg/kg, and 16% for 120 mg/kg). Based on these results, the U.S.

EPA classified chlorobenzene as "D" (not classifiable as to carcinogenicity in humans).

Repeated-exposure inhalation studies have been conducted in several species. Rats, rabbits, and guinea pigs were exposed to airborne concentrations ranging from 200 to 1000 ppm for 7 hr per day, 5 days per week, for a total of 32 exposures. At 475 ppm and above, organ weight changes and histopathological changes were observed. There were no effects detected at 200 ppm. In another study, changes in hematology parameters and pathological changes in the adrenal cortex, kidney, and liver were observed in rats and rabbits exposed to airborne concentrations of 75 or 250 ppm chlorobenzene vapors for 7 hr per day, 5 days per week for 24 weeks.

Chlorobenzene was not mutagenic in several bacterial strains of *Salmonella typhimurium* or *Escherichia coli* and was negative in rat hepatic DNA repair assays; however, it was weakly positive in a mouse micronucleus assay. Exposure of rats to atmospheric concentrations up to 450 ppm did not have any adverse effects on reproductive performance or fertility of male or female rats through two consecutive generations. Chlorobenzene caused minor skeletal alterations in fetuses collected from pregnant rats exposed to atmospheric concentrations up to 590 ppm (a maternally toxic dose) for 6 hr per day during the period of organogenesis. Pregnant rabbits exposed to chlorobenzene at concentrations up to 590 ppm did not exhibit evidence of embryotoxicity or teratogenicity.

—Linda Angevine Malley and
David M. Krentz

Chloroform

♦ CAS: 67-66-3

♦ SYNONYMS: Formyl trichloride; methane trichloride; trichloromethane; NCI-C02686; R 20; RCRA waste No. U044; UN1888

♦ CHEMICAL CLASS: Chlorinated aliphatic hydrocarbon

♦ MOLECULAR FORMULA: $CHCl_3$

Uses
Chloroform was introduced in 1847 as an inhalation anesthetic. It is a central nervous system (CNS) depressant producing anesthesia, analgesia, inebriation, and narcosis. It is also used as a solvent in pesticides, as a grain fumigant, and as an intermediate in pesticides and dyes. It can be used as a rubefacient.

Exposure Pathways
Chloroform is available as emulsions, spirits, tinctures, and chloroform water. Inhalation, ingestion, and dermal contact are the most common routes of exposure.

Toxicokinetics
Chloroform is readily absorbed when inhaled. It is quickly distributed in all tissues and is highly fat soluble. It can readily cross the placental barrier.

Metabolites of chloroform include chloromethanol, hydrochloric acid, phosgene, chloride, carbon dioxide, and digluathionyl dithiocarbonate.

Chloroform is primarily excreted by the lungs as carbon dioxide. The half-life for elimination following a single oral dose of 500 mg in two subjects was 1.5 hr.

Mechanism of Toxicity
Chloroform is converted to chloromethanol by cytochrome P450s. This intermediate rapidly dechlorinates to produce hydrochloric acid and phosgene. Phosgene is a poisonous gas which can cause injury to tissues. Phosgene reacts with water to produce carbon dioxide and chloride ion. These by-products can bind to glutathione to produce diglutathione dithiocarbonate.

Human Toxicity
A dose of 10 ml of chloroform can cause CNS depression and death. The oral lethal dose is estimated to be 0.5–5 g/kg (1 oz to 1 pint) for an average 70-kg man. Chloroform can cause hepatic and renal toxicity. It is a less potent hepatotoxicant than carbon tetrachloride.

Acute

Chloroform is an irritant, CNS, and cardiovascular system depressant. Delayed kidney and liver toxicity may also occur.

Chronic

Chloroform may cause dry mouth, headache, hallucinations, dysarthria, ataxia, loss of reflexes, gastrointestinal distress, hepatotoxicity, and psychotic behavior. Chloroform is listed as a suspected carcinogen. It may also be mutagenic and can induce DNA damage, DNA repair, and sister chromatid exchanges.

Clinical Management

In cases of ingestion, ipecac-induced emesis is not recommended. Activated charcoal slurry with or without saline cathartic or sorbitol can be given in cases of oral exposures. Exposed skin should be decontaminated by repeated washing with soap. Exposed eyes should be irrigated with copious amounts of room-temperature water for at least 15 min.

Treatment is supportive and symptomatic. No specific antidote is available.

Animal Toxicity

Chloroform in animals is known to cause acute and chronic toxicity similar to that in humans. The oral LDLo in dogs is 100 mg/kg; the inhalation LCLo in cats is 35 g/m^3/4 hr; the oral LD$_{50}$ in rats is 300 mg/kg and the intravenous LDLo is 75 mg/kg.

—*Sushmita M. Chanda*

Related Topics

Neurotoxicology: Central and Peripheral
Pollution, Water

Chloromethyl Ether, bis-

- ◆ CAS: 542-88-1
- ◆ SYNONYMS: BCME; *sym*-dichloromethyl ether; dichloromethyl ether

- ◆ CHEMICAL CLASS: Alkyl organic synthetic compound
- ◆ CHEMICAL STRUCTURE:

$$ClCH_2 —O—CH_2Cl$$

Uses

bis-Chloromethyl ether is primarily used in the synthesis of plastics and ion exchange resins. It is also used as a chemical intermediate for the synthesis of other complex organic alkyl compounds. It has potential use in dental restorative materials. It is used in chloromethylating (cross-linking) reaction mixtures in anion exchange resin production.

In the textile industry, bis-chloromethyl ether is used in formaldehyde-containing reactants and resins in the finishing products of fabrics and in laminating and as adhesive in the flocking of fabrics. The nonwoven industry uses it as binders, thermosetting acrylic emulsion polymers comprising methylol acrylamide.

Exposure Pathways

The primary routes of potential human exposure to bis-chloromethyl ether are inhalation and dermal contact. The risk of potential occupational exposure to the chemical is greatest for chemical plant workers, ion exchange resin makers, laboratory workers, and polymer makers. Because bis-chloromethyl ether is highly unstable in water, it breaks down rapidly; therefore, exposure through water pollution is limited.

Toxicokinetics

bis-Chloromethyl ether is rapidly absorbed through skin and by inhalation. Since it is easily soluble in water, it interacts with cells and tissues at various levels. The absorption by the body depends on the proximity to the source of production and the industries where it is used as a starting material for certain organic substances. bis-Chloromethyl ether is metabolized mainly in the liver but also to some extent in the lung tissue. On metabolism by cytochrome P450, bis-chloromethyl ether produces epoxides, which are a highly reactive species of free radicals capable of reacting with any organic substance. They cause alkylation of DNA and bring about mutagenesis. bis-Chloromethyl ether is also converted to a potential carcinogen after metabolism. bis-Chloromethyl ether is highly soluble in water but broken down rapidly. Epoxides produced as a result of metabolism bind to cells and cell organic components easily by alkylation. Glutathione *S*-transferase, sulfotransferase, and glucuronidation help in removal of toxic metabolites.

Mechanism of Toxicity
The reactive epoxides produced as metabolites of bis-chloromethyl ether are considered to be ultimate carcinogens as well as mutagens because they are capable of bringing about DNA alkylation. The epoxide (metabolites) can also react with membrane lipid and cause autooxidation.

Human Toxicity
A study carried out by an IARC group reported that there is sufficient evidence to show the carcinogenicity of bis-chloromethyl ether in humans. Two studies of workers exposed to bis-chloromethyl ether showed an increased risk of lung cancer, mainly oval cell carcinoma. Two subsequent studies have shown a positive association between atypical cells in bronchial excretion and exposure to bis-chloromethyl ether. Several studies have demonstrated a significant incidence of lung cancer among bis-chloromethyl ether workers which was directly related to the intensity and duration of exposure. The excess respiratory cancer mortality was most marked in workers under 55 years of age.

The ACGIH TWA-TLV is 0.001 ppm (0.0047 mg/m^3), with the notation that the material is a confirmed human carcinogen.

Clinical Management
There is no antidote recommended for poisoning by bis-chloromethyl ether. Free radical scavengers should alleviate the toxicity on administration.

Animal Toxicity
When administered by subcutaneous injection, bis-chloromethyl ether induced pulmonary tumors, papillomas, and fibrosarcomas in mice of both sexes; local sarcomas in female mice; and fibromas and fibrosarcomas in female rats. The compound is also an initiator

of skin tumors in mice. bis-Chloromethyl ether induced lung tumors in mice on administration by inhalation as well as squamous cell carcinomas of the lung and esthesioneuroepitheliomas of nasal cavity in rats. When applied topically, bis-chloromethyl ether induced papillomas, most of which developed into squamous cell carcinomas in female mice.

—*Vaman C. Rao*

Related Topics
Carcinogenesis
Respiratory Tract

Chlorophenols

- REPRESENTATIVE COMPOUNDS: Pentachlorophenol (PCP); 2,4-dichlorophenol
- CHEMICAL CLASS: Aromatic alcohols
- CHEMICAL STRUCTURES:

Pentachlorophenol   2,4-dichlorophenol

Uses
Chlorophenols are used in dye synthesis and as ingredients in alcohol denaturants, fungicides, herbicides, and wood preservatives.

Exposure Pathways
Exposure to chlorophenols may occur through ingestion, inhalation, or dermal contact.

Toxicokinetics

Pentachlorophenol is rapidly absorbed following ingestion or dermal contact. Pentachlorophenol is excreted primarily unchanged. Labeled pentachlorophenol was given to rats by injection and oral routes, and 41–43% was excreted unchanged in the urine. One metabolite, tetrachlorohydroquinone (5–24%), was identified. Glucuronic acid conjugation was not seen. The primary route of elimination for pentachlorophenol is by the kidneys; 80% is thought to be excreted in the urine. The elimination half-life for pentachlorophenol may be up to 20 days in chronically exposed persons.

The volume of distribution in a surviving infant exposed to pentachlorophenol was 11.8 mg/100 ml. Also, pentachlorophenol seems to be highly protein bound in plasma.

Studies in monkeys showed the elimination half-life for pentachlorophenol to be 72–83 hr.

Mechanism of Toxicity

Chlorophenols cause an uncoupling of oxidative phosphorylation cycles in tissues, which increases basal metabolic rate and increases body temperature.

Human Toxicity

Pathologic findings in deaths by phenol or related compounds include necrosis of mucous membranes, cerebral edema, and degenerative changes in the liver and kidneys.

Acute

The most prevalent signs and symptoms after ingestion of 30–250 ml of phenol or related compounds are profuse sweating, intense thirst, nausea, vomiting, diarrhea, convulsions, pulmonary edema, cyanosis, coma, and corrosion of tissue. If death from respiratory failure is not immediate, jaundice and oliguria or anuria may occur.

Chronic (from Ingestion or Absorption from Skin)

Repeated use may cause symptoms of acute poisoning. Skin sensitivity reactions occur occasionally. Prolonged skin contact with β-naphthol may cause bladder tumors, hemolytic anemia, and lens opacities.

The OSHA PEL for pentachlorophenol is 10.5 mg/m^3. RCRA regulatory levels are 400 mg/liter for 2,4,5-trichlorophenol and 2 mg/liter for 2,4,6-trichlorophenol.

Clinical Management
Acute Poisoning

In case of ingestion, where corrosive injury is absent, the poison should be removed by ipecac emesis and use of activated charcoal. Next, milk should be given to drink. Gastric lavage and emesis are contraindicated in the presence of esophageal injury. In case of dermal exposure, the poison should be removed by washing the affected skin or mucous membrane with copious amounts of water for at least 15 min.

Chronic Poisoning

Further use of chlorophenols should be discontinued and the patient treated as for acute poisoning.

Animal Toxicity

The LD$_{50}$ for pentachlorophenol in laboratory animals ranges from 30 to 100 mg/kg.

—Stephanie E. Foster and
Paul W. Ferguson

Chlorophenoxy Herbicides

Chlorophenoxy herbicides have found widespread use in agriculture. Chemically, these herbicides exist as a phenyl moiety with one to three chlorine additions and an acidic moiety (e.g., acetic acid, propionic acid, or butyric acid). Among the most common chlorophenoxy herbicides are 2,4-dichlorophenoxyacetic acid (2,4-D), 2,4,5-trichlorophenoxyacetic acid (2,4,5-T), MCPA, and Silvex (see structures below; see 2,4-D and 2,4,5-T). These agents are formulated as metal salts, alkylamine salts, and esters.

Chlorophenoxy compounds are used as plant growth regulators by virtue of their ability to mimic the activity of naturally occurring hormones (auxins). The effects of chlorophenoxy compounds on plant growth are similar to the effects of the natural plant hormones except for exaggerated and prolonged activity. Plant death occurs because of abnormal growth and interference with nutrient transport.

2,4-D and 2,4,5-T are the most commonly used of these herbicidal agents. 2,4-D is effective against broadleaf weeds (e.g., lamb's quarter, pigweed, smartweed, and ragweed), while 2,4,5-T is more effective against woody species and is thus more useful along highways and utility rights-of-way. Common names for 2,4-D include Chloroxone, Estrone, Salvo, Weedar, and Weedone (CAS 94-75-7). Trade names for 2,4,5-T include Estron 245, Estron Brush Killer, Marks Brushwood Killer, Shellstar Brush Killer, Weedar 2,4,5-T, and Weedone (CAS: 93-76-5).

The most common exposure pathway for these agents is accidental or intentional ingestion, but dermal as well as inhalation exposures may also occur. In human studies, intravenously injected 2,4-D was completely recovered in the urine. When 2,4-D was applied to the skin, only about 6% was recovered in the urine. Following ingestion of a small dose of 2,4-D (5 mg/kg), absorption was rapid and blood levels peaked at 7–12 hr following exposure. Oral absorption of 2,4,5-T appears to follow first-order kinetics. Dermal absorption is dependent on individual characteristics of the various chlorophenoxy compounds and is significantly affected by the vehicle. 2,4-D and 2,4,5-T are highly protein bound and tend to accumulate in various tissues (liver $\geqq$ kidney $\geqq$ muscle $\geqq$ brain). 2,4-D and 2,4,5-T are primarily metabolized by acid hydrolysis in mammals and to a minor extent through conjugations. Chlorophenoxy compounds are to a great extent eliminated unchanged in the kidney through the renal organic anion secretion system and to a lesser degree by glomerular filtration.

Generally, the acute toxicity potential of chlorphenoxy herbicides is low. Myotonia is the most characteristic effect of 2,4-D overdose. 2,4-D may enhance enzyme activity of basic *p*-NPPase, which can lead to a passive increase in potassium flux and a compensatory decrease in chloride conductance. 2,4-D and 2,4,5-T both decrease cholesterol levels through the enhancement of lipid utilization in the liver. An analog, clofibrate, has been used clinically for such purposes. Chlorophenoxy compounds have no hormonal activity in mammalian species: Their mechanism of toxicity in mammalian species is not clearly understood. Acute ingestion of 2,4-D and 2,4,5-T may produce miosis, fever, hypotension, emesis, tachycardia, muscle rigidity, and coma. Complications of respiratory failure, pulmonary edema, and rhabdomyolysis may occur. Ingestion causes burning of the mouth, esophagus, and stomach. Irritation of the skin and eyes is also possible. Chloracne (most likely from contaminants) following 2,4,5-T exposure has also been reported. 2,4-D and 2,4,5-T have not proven to be teratogens, but the compounds may be linked to perinatal deaths in experimental studies. It should be noted that the defoliant "Agent Orange," a 50/50 mix of 2,4-D and 2,4,5-T, was suspected of inducing various disorders in exposed soldiers and their families. While decades of studies have failed to demonstrate conclusively long-term effects of exposure to Agent Orange, the highly toxic 2,3,7,8-tetrachlordibenzodioxin (dioxin, TCDD) contaminant in the defoliant was a possible culprit. There is "limited evidence" that chlorophenoxy herbicide exposures are associated with soft tissue sarcomas.

The acute oral LD_{50} in rodents is greater than 300 mg/kg. Dogs appear to be more sensitive to the acute toxicity of 2,4-D than other species (oral LD_{50} approximately 100 mg/kg). With lethal exposures of 2,4-D, ventricular fibrillation and death occurred shortly after exposure. At lower doses, signs of muscular dysfunction, such as myotonia, stiffness of extremities, ataxia, and paralysis, are possible. As stated before, hypothermia, reduction in metabolic rate, and coma are also possible sequelae. Long-term dietary exposure in rats (300 ppm or about 15 mg/kg/day) for over 100 days produced no consistent signs of toxicity. The toxic sequelae of 2,4,5-T are similar to those following 2,4-D exposure but spasticity may replace myotonia.

Clinical management of chlorophenoxy herbicide poisoning may include gastric emptying within 4 hr of ingestion for quantities of 2,4-D in excess of 40 mg/kg. Ipecac-induced emesis is recommended if the patient is conscious. Sodium sulfate as a cathartic is useful if no bowel movement has occurred within 4 hr after exposure. In severe cases, forced alkaline diuresis may reduce the plasma half-life of 2,4-D. For skin and eye exposure, washing of the exposed areas is necessary.

—*Thuc Pham*

Related Topic

Pesticides

Chloropicrin

- ◆ CAS: 76-06-2
- ◆ SYNONYM: Trichloronitromethane
- ◆ CHEMICAL CLASS: Fumigant insecticide
- ◆ MOLECULAR FORMULA: CCl_3NO_2

Uses

Chloropicrin is used to fumigate stored grain and to treat soil against nematodes. It is also a tear gas agent for military use.

Exposure Pathways

Dermal and eye exposures are the most common routes of chloropicrin toxicity. It may also be inhaled.

Toxicokinetics

The toxic effects of chloropicrin occur very rapidly. The liver is the primary site of metabolism of this compound.

Mechanism of Toxicity

Chloropicrin is an SN2 alkylating agent with an activated halogen group and reacts with sulfhydryl groups, "fixing" enzymes. It has the additional toxic effect of interfering with oxygen transport by its reaction with SH- groups in hemoglobin. Chloropicrin may also undergo a photochemical transformation to phosgene.

Human Toxicity

The primary signs and symptoms following inhalation exposure to chloropicrin include coughing, nasal and pharyngeal mucosal edema and erythema, lacrimation, and rhinorrhea. Fatal pulmonary edema has been reported with an onset of 3 hr postexposure. Chloropicrin is a strong eye irritant, producing ocular burning, eye pain, and lacrimation following eye exposure. These effects may last up to 30 min or longer. Redness and edema may be noted 1 or 2 days following exposure. Dermal exposure to chloropicrin produces severe skin irritation.

Clinical Management

Following an eye exposure to chloropicrin, the affected eyes should be irrigated with copious amounts of tepid water for at least 15 min. If irritation persists following decontamination, ophthalmic corticosteroids or local anesthetic ointments may be used. In case of an inhalation exposure, the patient should be monitored for respiratory distress. Emergency airway support and 100% humidified supplemental oxygen with assisted ventilation may be needed.

Following dermal exposure to chloropicrin, the exposed area must be washed thoroughly with soap and water. If dermatitis persists, topical treatment with wet dressings of Burow's solution 1 : 40, followed by corticosteroid creams or calamine lotion, may be given. Secondary infection may necessitate antibiotic therapy. Oral antihistaminics may be useful for pruritis.

Animal Toxicity

Chloropicrin is intensely irritating with an intraperitoneal LD_{50} of 25 mg/kg in mice. Mice exposed to 8 ppm chloropicrin vapor for 6 hr/day for 5 days developed moderate to severe degeneration of the respiratory and olfactory epithelium as well as fibrosing peribronchitis and peribronchiolitis of the lung. Rabbits exposed to an intravenous injection of chloropicrin at a dosage of 15 mg/kg died within 15–240 min; clinical and autopsy findings were typical of acute pulmonary edema.

—Priya Raman

Related Topic

Pesticide

Chloroquine

- ◆ CAS: 54-05-7
- ◆ SYNONYMS: SN 7,618; sanoquin; tresochin; silbesan; artichin; bipiquin; avclouor; tanakan; resochin; resoquine
- ◆ PHARMACEUTICAL CLASS: Aminoquinoline
- ◆ CHEMICAL STRUCTURE:

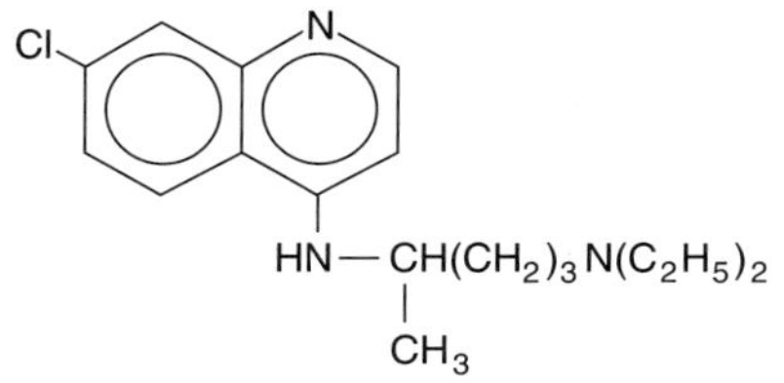

$$Cl—[quinoline ring]—N$$
$$HN—CH(CH_2)_3N(C_2H_5)_2$$
$$|$$
$$CH_3$$

Uses

Chloroquine is used as an antiinflammatory and antimalarial drug.

Exposure Pathways

Chloroquine is available in oral and intravenous forms.

Toxicokinetics

Chloroquine is absorbed rapidly and almost completely from the gut; peak serum concentrations are attained within 1 or 2 hr. Chloroquine binding to circulating plasma proteins is approximately 55%. Its volume of distribution is 116–285 liters/kg. The drug may be found in 500 times greater concentration within the liver, spleen, kidneys, lungs, and leukocytes (compared with plasma). Chloroquine appears to cross the placenta readily. A very small amount is transmitted into breast milk.

The primary route of metabolism is deethylation, producing desethylchloroquine. Elimination is significantly reduced in the presence of hepatic disease. Seventy percent of chloroquine is recovered in urine as unchanged drug. The half-life of chloroquine varies from 2 to 20 days with an average of 3 days. Urinary excretion of the drug in increased in an acid urine and decreased by urine alkalinization. Small amounts of chloroquine may be present in urine for months or even years after the drug is discontinued.

Mechanism of Toxicity

Chloroquine's mechanism of action is not completely understood but involves inhibition of DNA and RNA polymerase. The drug also possesses antiinflammatory activity. Finally, chloroquine is a direct myocardial depressant that, through its membrane-stabilizing action, impairs cardiac conduction.

Human Toxicity: Acute

Symptoms of overdosage include nausea, vomiting, visual or auditory disturbances, drowsiness, and seizures followed by severe cardiac arrhythmias, shock, or cardiorespiratory arrest. Hypotension may be severe and intractable, producing metabolic acidosis and end-organ failure. Cardiac conduction disturbances include complete atrioventricular dissociation, QRS prolongation, severe bradycardia, and ventricular fibrillation. Acute ingestion of more than 5 g of chloroquine in adults is usually fatal. Acute, nonfatal overdose of chloroquine may cause gastrointestinal upset and acute blindness or deafness.

Human Toxicity: Chronic

Chronic use of chloroquine may produce cinchonism, a syndrome characterized by headache, visual changes, and gastrointestinal disturbances. Visual disturbances are associated with retinal artery spasm. Ototoxicity may also occur. Dermatologic reactions, particularly a lichenoid skin eruption, may result from chronic chloroquine use.

Clinical Management

Immediate management is guided by the severity of toxic manifestations. In the event of depressed consciousness or seizures, airway protection should first be secured. Seizures are treated with a benzodiazepine (diazepam or lorazepam). Hypotension is treated by the establishment of vascular access followed by the administration of fluids and, if necessary, vasopressors. Cardiac conduction disturbances are treated according to advanced cardiac life-support guidelines. Gastrointestinal decontamination with consideration of gastric lavage should be considered. Activated charcoal should always be administered. Methods of extracorporeal drug removal, such as hemoperfusion and hemodialysis, are ineffective.

Animal Toxicity

Chloroquine is not used therapeutically in domestic animals. Toxic manifestations of overdose in animals are undefined.

—Michael Shannon

Chlorothalonil

- ◆ CAS: 1897-45-6
- ◆ SYNONYMS: 2,4,5,6-Tetrachlorobenzenedicarbonitrile; tetrachloroisophthalonitrile
- ◆ CHEMICAL CLASS: Industrial, agricultural, and horticultural fungicide

◆ CHEMICAL STRUCTURE:

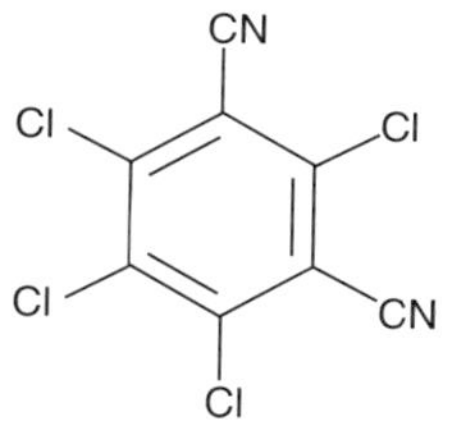

Uses

Chlorothalonil is used as a fungicide, bactericide, and nematocide. It is used as a wood preservative in some countries. Chlorothalonil is also used as a mildew-preventing agent in paints.

Exposure Pathways

Dermal and eye exposure are the most common routes of accidental exposure to chlorothalonil. It may also be ingested or inhaled.

Toxicokinetics

Chlorothalonil is rapidly absorbed orally and via inhalation. However, it is poorly absorbed by the dermal route of exposure. Hydrolysis studies indicated that the metabolism of chlorothalonil is pH dependent. Thus, 4-hydroxy-2,5,6-trichloroisophthalonitrile and 3-cyano-2,4,5,6-tetrachlorobenzamide are formed at a pH of 9 but not at a pH of 7. The metabolism of chlorothalonil was recently investigated in liver and gill cytosolic and microsomal fractions from channel catfish using high-performance liquid chromatography. The reports indicate that chlorothalonil is detoxified *in vitro* by GST-catalyzed GSH conjugation. However, no human data are currently available for the biotransformation of chlorothalonil. Chlorothalonil is primarily eliminated via the kidneys. Following administration of 1 mg/kg chlorothalonil endotracheally, orally, or dermally to rats, less than 6% was recovered in blood or urine within 48 hr.

Mechanism of Toxicity

Chlorothalonil acts as an alkylating agent and reacts with cellular sulfhydryl compounds. Alkylation of biological molecules results in effects on cellular function and viability.

Human Toxicity

Facial dermatitis has been reported in occupational exposures and can occur in the absence of direct skin contact, presumably due to the high volatility of chlorothalonil. Chlorothalonil is a strong primary skin irritant and may also cause allergic contact urticaria and anaphylaxis. Patch testing with concentrations greater than 0.01% may produce primary irritant reactions. Hypersensitivity reaction characterized by facial erythema, periorbital erythema and edema, eczema, and pruritis have been observed following chlorothalonil exposure. Photosensitivity reactions were seen in some individuals. High concentrations of chlorothalonil produce delayed irritant reactions. Delayed dermal irritant effects have also been noted 48–72 hr after cessation of exposure. Immediate respiratory reactions, such as tightness of chest and throat, may occur following inhalation exposure to chlorothalonil.

Clinical Management

One of the primary forms of treatment is to support respiratory and cardiovascular function. Dilution and dermal/eye decontamination are primary considerations. Following oral exposure, immediate dilution with 4–8 oz of milk or water is recommended. Emesis is not indicated due to the irritant properties and lack of systemic effects of chlorothalonil. In case of a dermal exposure to chlorothalonil, the exposed area should be thoroughly washed with soap and water. Allergic contact dermatitis may be treated with antihistamines, topical steroids, and/or systemic steroids. Following an eye exposure, the affected eyes should be irrigated with copious amounts of tepid water for at least 15 min.

Animal Toxicity

Undiluted chlorothalonil is a strong irritant and produces irreversible corneal, iridal, and conjunctival effects in rabbits. Weakness and sedation precedes death in animals given acute toxic doses intraperitoneally. Chronic oral administration to rats results in ataxia. Hematuria, vaginal bleeding, and epistaxis are seen in rats following chronic oral exposure. In chronic dermal exposures to chlorothalonil dissolved in acetone, the "no-effect" level for irritation is 0.001%. The 0.01% concentration is a mild irritant and 0.1% a moderate irritant. Chlorothalonil produced a dose-related increased incidence of renal tubular adenomas and adenocarcinomas in rats. The oral LD_{50} in rats is greater than 10 g/kg.

—*Priya Raman*

Chlorpheniramine

- ♦ CAS: 132-22-9
- ♦ SYNONYMS: Chlor-Trimeton; Chlo-amine; Allerchlor; Chlortab; Telachlor, Teldrin
- ♦ PHARMACEUTICAL CLASS: A propylamine-derivative (alkylamine) H-1 receptor antagonist
- ♦ CHEMICAL STRUCTURE:

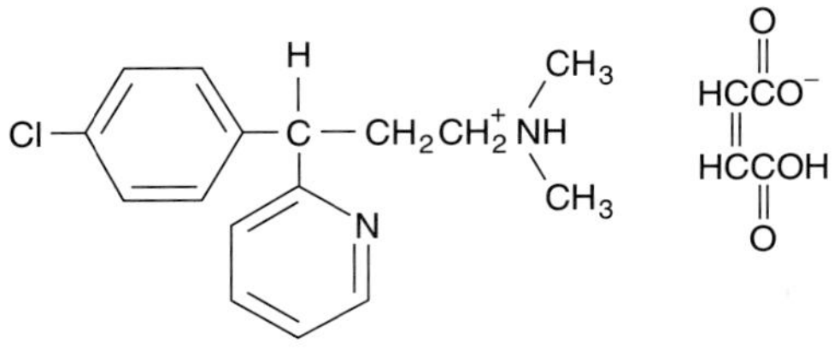

Uses

Chlorpheniramine, like other antihistamines, is most often used to provide symptomatic relief of allergic symptoms caused by histamine release.

Exposure Pathways

Ingestion and injection are the routes of both accidental and intentional exposures to chlorpheniramine.

Toxicokinetics

Chlorpheniramine is well absorbed following oral administration with peak plasma levels occurring within 2–6 hr. The drug undergoes substantial metabolism in the gastrointestinal mucosa during absorption and first pass through the liver. The volume of distribution of chlorpheniramine is 3.4–7.5 liters/kg and approximately 69–72% of the drug is bound to plasma proteins. Chlorpheniramine and its metabolites are excreted almost completely in the urine. Urinary excretion is increased with an acidic urine pH, but this is not a viable treatment option. Chlorpheniramine is eliminated more rapidly in children than adults. The drug's half-life is 9.5–13 hr for children and 14–24 hr for adults.

Mechanism of Toxicity

The toxicity of antihistamines is related to their anticholinergic (antimuscarinic) activity. The action of acetylcholine at the muscarinic receptors is blocked resulting in signs and symptoms of anticholinergic poisoning.

Human Toxicity

The alkylamine derivatives are among the most potent antihistamines producing more central nervous system (CNS) stimulation and less drowsiness than other antihistamines. In overdoses, CNS stimulation is more common in children with depression followed by excitation in adults. Anticholinergic symptoms including fixed and dilated pupils, flushed skin, dry mouth, fever, hallucinations, and seizures may also be noted. Cardiovascular effects including tachycardia, hypertension or hypotension, arrhythmias, and cardiovascular collapse may occur. Severe toxicity may result in cerebral edema, deep coma, cardiorespiratory collapse, or death.

Clinical Management

Basic and advanced life-support measures should be utilized as necessary. Appropriate gastrointestinal decontamination procedures should be administered based on the history of the ingestion and the patient's level of consciousness. Physostigmine administration may be necessary in a limited number of patients suffering from severe central and peripheral anticholinergic symptoms refractory to conventional therapy. Intravenous diazepam has been recommended for the treatment of seizures. Electrocardiogram monitoring should be considered in patients who have taken large overdoses.

Animal Toxicity

CNS changes, including sedation or hyperexcitability, salivation, and vomiting, have occurred in dogs following acute low exposures to antihistamines. Seizures and cardiac effects have occurred following acute high exposures. Symptomatic and supportive care followed by appropriate gastrointestinal decontamination procedures should be administered.

—*Carla M. Goetz*

Chlorpromazine

- CAS: 50-53-3
- SYNONYMS: 10H-Phenothiazine-10-propanamine; 2-chloro-N,N-dimethyl- (9CI CA index name); phenothiazine; 2-chloro-10-[3-(dimethylamino)propyl]-(7CI, 8CI); 2-chloro-10-[3-(dimethylamino)propyl]phenothiazine; 2-chloropromazine; 4560 R. P.; aminazin; aminazine; ampliactil; amplictil; BC 135; chlorpromanyl; chlordelazin; chlorderazin; chlorpromados; contomin; CPZ; elmarin; esmind; fenactil; fenaktyl; fraction AB; HL 5746; largactil; largactilothiazine; largactyl; megaphen; novomazina; phenactyl; proma; promactil; promazil; propaphenin; prozil; sanopron; SKF 2601-A; thorazin; thorazine; torazina; wintermin

- CHEMICAL STRUCTURE:

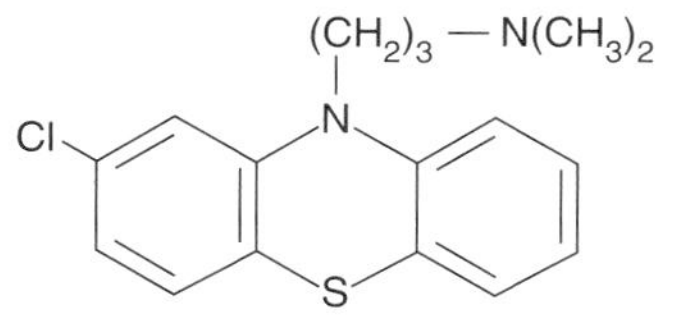

Uses

Chlorpromazine is used as a medication for both humans and animals. In humans, it is employed primarily in the treatment of psychiatric patients as an effective treatment for the management of psychotic disorders, manic depressive illness, apprehension, and anxiety, as well as for the treatment of severe behavioral problems in children. It is also used for short-term treatment of hyperactive children who exhibit excessive motor activity with accompanying conduct disorders. Chlorpromazine is also used to control nausea and vomiting, intractable hiccups, and, prior to surgery, acute intermittent porphyria. It is also used as adjunct in the treatment of tetanus.

In animals, chlorpromazine is recommended in excitable sows following farrowing, especially in those reluctant to accept their newborn; to capture African lions; as an adjunct to restraint and anesthesia; and as a neuroleptanalgesia (inducing a state of quiescence) in bears, and in reptiles prior to the administration of barbiturate anesthesia.

Exposure Pathways

Chlorpromazine is administered orally as a tablet or capsule, intravenously via injection, and via suppository. Users are exposed through oral consumption and dermal contact. Pharmacists, physicians, and nurses dispensing or administering chlorpromazine could be exposed through dermal contact. Chlorpromazine and its metabolites cross the placenta and are distributed into milk.

Toxicokinetics

Absorption of orally administered chlorpromazine is dependent on dosage form, with the elixir giving highest plasma concentration. Absorption of chlorpromazine tablets is erratic. Peak plasma levels are reached at 2–4 hr, although wide variations (at least 10-fold) in plasma concentrations occur among individuals. Chlorpromazine is 92–97% bound to plasma proteins, principally albumin, at plasma chlorpromazine concentrations of 0.01–1 μg/ml. Concentrations in the brain are four to five times higher than the plasma. Metabolites are excreted in urine for 2–6 weeks after cessation of medication.

Intramuscular administration avoids much of the first-pass metabolism in the liver (and possibly also gut) and provides measurable concentrations in plasma within 15–30 min; bioavailablity may be increased up to 10-fold.

Following a single oral chlorpromazine dose of 120 mg/sq m to four healthy men, less than 1% of the dose was excreted unchanged in the urine within 72 hr. Following continuous oral administration of chlorpromazine to a limited number of psychiatric patients in doses ranging from 0.1 to 1.4 g daily, an average of 37% of dose was excreted in urine, principally as metabolites. Although intestinal absorption is complete, oral bioavailability is 32% because of variable metabolism in intestinal wall and liver.

Chlorpromazine is strongly bound to protein, crosses blood–brain barrier, and concentrates in the brain against plasma gradient. More than 90% of the drug in plasma is bound to proteins, is metabolized in liver, and is excreted in both urine and feces. There is some evidence that chlorpromazine can cause hepatic microsomal enzyme induction, which indicates that it may accelerate its own metabolism.

Rapid placental transfer was reported in goats and mice. In goats, the fetal plasma levels approached 50% of maternal values within 10 min of the mother receiving an intravenous dose, and the fetal–maternal plasma ratio remained at 0.5 for 1 hr, whereas ratios in the liver, kidney, heart, and brain all approached one and showed a marked effect on fetal heart rate. In pregnant mice, radiolabeled chlorpromazine rapidly crossed the placenta and accumulated in eyes of both fetuses and mothers. Marked radioactivity remained in tissues of the eye for 5 months after the drug had been eliminated from other tissues.

Markedly variable half-life values reported for chlorpromazine vary from a relatively short plasma half-life of 2–6 hr to 18 hr. Disappearance of chlorpromazine from plasma includes a rapid distribution phase (half-life about 2 hr) and slower early elimination phase (half-life about 30 hr). One study reported that after 120 mg/sq m oral doses to human volunteers, chlorpromazine displayed a mean elimination half-life of approximately 18 hr (range, 6–119 hr). Half-life of elimination from human brain is unknown. The pharmacokinetics of chlorpromazine was investigated in 25 pediatric patients (aged 0.3–17 years) who received an intravenous infusion of 1 mg/kg chlorpromazine with intravenous metoclopramide administered concomitantly. Compared with previously reported values for adults, the pharmacokinetics of chlorpromazine appeared to be accelerated in children. This was especially evident for half-life and clearance values.

Chlorpromazine metabolism is exceedingly complex and varies with species; only 20 of 168 possible metabolites have been isolated. Less than 1% of the drug is excreted unchanged by the kidneys. Many of metabolites are active compounds; therefore, pharmacological half-life of these drugs is difficult to calculate.

As many as 10 or 12 chlorpromazine metabolites have been found to occur in appreciable quantities in humans. Quantitatively, the most important of these are nor2-chlorpromazine (doubly methylated), chlorphenothiazine (removal of entire side chain), methoxy and hydroxy products, and glucuronide conjugates of hydroxylated compounds. In urine, 7-hydroxylated and dealkylated (nor2) metabolites and their conjugates predominate.

In mental patients, chlorpromazine and various metabolites may be detected in urine 6–18 months after termination of treatment.

Chlorpromazine and its metabolites were found in the maternal plasma and urine, in the fetal plasma and amniotic fluid, and in neonatal urine after doses of 50–100 mg of chlorpromazine were given intramuscularly to pregnant women shortly before delivery.

Chlorpromazine undergoes considerable metabolism during absorption (in gastrointestinal mucosa) and first pass through the liver. Little or no chlorpromazine is eliminated in urine of the dog. The primary excretory product is chlorpromazine sulfoxide, but only 10–15% of the dose is eliminated as such.

Limited studies of excretion patterns have been conducted in horses following intramuscular and oral administration of chlorpromazine. After intramuscular injection, metabolites are detected in urine up to 96 hr. Following oral administration, metabolites are no longer detected after 80–96 hr. The percentage of the dose recovered in equine urine is low, with the average being 10% after intramuscular and 27% after oral administration. Unconjugated metabolites excreted in horse represented only 1–1.5% of dose after either route of administration; these were excreted entirely as sulfoxide derivatives. Glucuronide-conjugated metabolites are predominantly excreted by the horse in a ratio to unconjugated metabolites of approximately 7:1 after intramuscular injection and 18:1 after oral administration. Sulfate-conjugated metabolites make up about 5% of the total after oral administration but are detected only in trace amounts after intramuscular injection. Phenothiazine derivatives were not detected in feces with spectroscopic analytical methodology.

Mechanism of Toxicity

The action of chlorpromazine upon the brain stem reticular system is to increase reticular activity and stimulate filtering mechanisms in the reticular formation that act to reduce inflow of stimuli in a selective manner. Low doses of chlorpromazine also depress vasomotor reflexes mediated by either hypothalamus or brain stem. It inhibits release of growth hormone, perhaps by action on hypothalamus; may antagonize secretion of prolactin release-inhibiting hormone; and appears to cause reduction in secretion of corticotropin-regulatory hormone in response to certain stresses. Chlorpromazine has significant adrenergic antagonistic activity and can block the pressor effects of norepinephrine. Chlorpromazine can strongly potentiate sedatives and analgesics, alcohol, hypnotics, antihistamines, and cold remedies. It may also increase miotic and sedative effects of morphine and may increase its analgesic actions. Chlorpromazine markedly increases respiratory depression produced by meperidine. Chlorpromazine's N-demethylated metabolites may also block the antihypertensive effects of guanethidine. Chlorpromazine was also found to inhibit the reuptake of norepinephrine and 5-HT in rat cerebral cortex but does not affect reuptake of γ-aminobutyric acid. It has also been reported to be a potent competitive inhibitor of the stimulatory effects of dopamine on adenylate cyclase.

Human Toxicity

The minimum lethal or toxic dose is not well established in the literature. The acute fatal dose is thought to be in the range of 15–150 mg/kg depending on the agent. However, adults have survived ingestion of 9.7 g, and a 350-mg dose was lethal in a 4-year-old child. Four deaths were reported in children following ingestion of 20–74 mg/kg chlorpromazine.

Adverse reactions may accompany the use of products containing chlorpromazine. A representative sample is provided below; however, the *Physician's Desk Reference* should be reviewed for specific product information.

Thorazine—Some adverse effects of Thorazine may be more likely to occur, or occur in greater intensity, in patients with special medical problems, e.g., mitral insufficiency or pheochromocytoma. Adverse effects include drowsiness, jaundice, hematological disorders, agranulocytosis, and hypertension, as well as neurological disorders including dystonias, motor restlessness, pseudo-parkinsonism, and tardive dyskinesia. Other problems that may appear are endocrine disorders, autonomic disorders, changes in skin pigmentation, and ocular changes.

Acetophenazine—Major effects are peripheral anticholinergic effects. Minor effects are extrapyramidal effects and hypotension.

Chlorpromazine—Major effects are central nervous system (CNS) depression to the point of somnolence and coma; hypotension; and extrapyramidal signs (monitoring should be performed especially for neck/upper airway problems). Agitation and restlessness, seizures, pyrexia, ECG changes and arrhythmias, and autonomic signs (such as ileus) may also occur.

Ethopropazine—Primarily anticholinergic effects are expected, especially sedation; hypotension may also occur with overdose.

Mesoridazine—Major effects are altered state of consciousness, hypotension, ECG abnormalities including widened QRS complexes, ventricular tachycardia, possibly ventricular fibrillation, and seizures. Extrapyramidal signs and anticholinergic effects may also occur with mesoridazine overdose.

Regulatory levels

The U.S. FDA shall waive the requirement for the submission of evidence demonstrating the *in vivo* bioavailability of a solid oral dosage form (other than an enteric coated or controlled release dosage form) of a drug product determined to be effective for at least one indication in a Drug Efficacy Study Implementation notice or which is identical, related, or similar to such a drug product . . . if the drug product is neither the tranquilizer chlorpromazine tablets nor an identical, related, or similar drug product." [21 CFR 320.22 (4/1/90)]

In Russia, the STEL is 0.3 mg/m³.

Clinical Management

A careful patient history should be taken by physicians to determine the patient's previous and present exposure to neuroleptic agents. Emesis should not be induced. Gastric lavage should be considered; activated charcoal may be administered with or without cathartic. Positioning, fluids, and dopamine are the treatment of choice for hypotension. Treatment of seizures may be necessary in patients with CNS excitation. Ventricular tachyarrhythmias should be treated with lidocaine followed by pacing if needed. Neuroleptic malignant syndrome should be treated with dantrolene or bromocriptine along with conservative treatment. Acute renal failure may need to be addressed in patients who develop rhabdomyolysis.

Any suspected overdose patient should be transported to a health care facility as soon as possible. Any patient with clinical signs of phenothiazine overdose should be admitted.

Animal Toxicity

The oral LD$_{50}$ value for rats was 225 mg/kg. Both single and repeated oral exposures to chlorpromazine have been conducted in several species including rats, dogs, cats, and horses. The primary effects of chlorpromazine were the reduction of hematocrit, onset of cardiac arrhythmias, ocular lesions, photosensitization, stimulation of hepatic microsomal enzyme activity, delay in fetal bone ossification, and effects on the nervous system.

A single dose of 2–4 mg/kg chlorpromazine caused tachycardia, hypotension, and depression in horses and reduced the number of red blood cells in hemoglobin

for up to 2 weeks in repeated dose studies. Undesirable side effects, such as instability, lunging forward in an uncoordinated manner, stumbling, and falling, were also observed. Parenteral doses of 2.5–5 mg/kg chlorpromazine caused cardiac arrhythmias in both unanesthetized and anesthetized dogs, and dose levels of 30 mg/kg/day caused ocular lesions in two strains of dogs after 73 days. Subcutaneous doses of 1.5 mg/kg produced no signs of toxicity, but when administered intravenously caused moderate depression and ataxia which lasted between 6 and 12 hr. Marked CNS depression and ataxia were noted for 24–48 hr when the intravenous dose was increased to 3.0 mg/kg.

When a single dose of chlorpromazine hydrochloride was administered to CF rats on the 14th day of gestation, the ischium and pubis remained unossified until the 20th day of gestation; ossification of skull bones was also delayed. Ossified vertebral bodies and arches were less affected. Ossification was delayed by 1–3 days in long bones of extremities, by 1 day in scapula, and by 2 or 3 days in the ilium. The ribs were also late in maturing.

—David M. Krentz and Linda A. Malley

Related Topic

Neurotoxicology: Central and Peripheral

Chlorpyrifos

♦ CAS: 2921-88-2
♦ SYNONYMS: Dursban; Lorsban
♦ CHEMICAL CLASS: Synthetic organophosphorous insecticide in the phosphorothionate class
♦ CHEMICAL STRUCTURE:

Uses

Chlorpyrifos is a broad-spectrum insecticide useful in controlling a variety of agricultural, urban, and household insects.

Exposure Pathways

Dermal, oral, and inhalation routes are all primary exposure pathways.

Toxicokinetics

Chlorpyrifos is absorbed rapidly and effectively by the oral route but less effectively by the dermal route. Chlorpyrifos undergoes activation via the P450–mixed-function oxidase pathway to the oxygen analog, chlorpyrifos oxon. Metabolites include diethylphosphoric acid, 3,5,6-trichloro-2-pyridyl phosphate, and 3,5,6-trichloro-2-pyridinol. Differences in serum A esterase activity appear to contribute significantly to species differences in sensitivity to chlorpyrifos (see A-esterase). Major conjugates include glucuronide (80%) and glycoside (4%) metabolites. A -SCH_3 addition on the pyridinol ring has also been reported. The water-soluble glucuronide and glycoside conjugates are eliminated primarily via the urine (90%). A trace amount of the parent compound is also eliminated via the urine.

Mechanism of Toxicity

Similar to other organophosphorothionate insecticides, the toxicity of chlorpyrifos is due to inhibition of acetylcholinesterase following its oxidative metabolism to the oxon (i.e., chlorpyrifos oxon). Extensive inhibition of this enzyme results in stimulation of the central nervous system (CNS), the parasympathetic nervous system, and the somatic motor nerves (see Organophosphates and Cholinesterase Inhibition). The IC_{50} values for the insecticide and its oxon are approximately 5×10^{-3} and 5×10^{-9} M, respectively.

Human Toxicity: Acute

Ocular contact may cause pain, moderate eye irritation, and slight temporary corneal injury. Prolonged skin exposure may cause skin irritation. Toxicity is moderate for a single oral dose; swallowing larger amounts may cause serious injury, even death. If aspirated, chlorpyrifos formulations may cause airway damage or even death due to chemical pneumonia.

Human Toxicity: Chronic

Cholinesterase inhibition can sometimes persist for weeks; thus, repeated exposures to small amounts of this material may result in accumulation of acetylcholinesterase inhibition with possible sudden-onset acute toxicity.

Chlorpyrifos may be capable of causing organophosphate-induced delayed neurotoxicity in humans; a massive overdose resulted in signs characteristic of delayed neurotoxicity. A recent report suggests that a sensory neuropathy may have been caused by low-level, household chlorpyrifos exposures. Animal studies, however, indicate that doses several times higher than the LD_{50} would be required to initiate delayed neurotoxicity.

Clinical Management

For exposure to eyes, eyelids should be held open and the eyes flushed with copious amounts of water for 15 min. For exposure to skin, affected areas should be washed immediately with soap and water. The victim should receive medical attention if irritation develops and persists.

For exposure through inhalation, the victim should be removed to fresh air and, if not breathing, given artificial ventilation. The victim should receive medical attention as soon as possible.

First aid for ingestion victims would be to induce vomiting, keeping in mind the possibility of aspiration of solvents. Gastric decontamination should be performed within 30 min of ingestion to be most effective. Initial management of acute toxicity is establishment and maintenance of adequate airway and ventilation. Atropine sulfate in conjunction with pralidoxime chloride can be administered as an antidote. Atropine by intravenous injection is the primary antidote in severe cases. Test injections of atropine (1 mg in adults and 0.15 mg/kg in children) are initially administered, followed by 2–4 mg (in adults) or 0.015–0.05 mg/kg (in children) every 10–15 min until cholinergic signs (e.g., diarrhea, salivation, and bronchial secretions) decrease. High doses of atropine over several injections may be necessary for effective control of cholinergic signs. If lavage is performed, endotracheal and/or esophageal control is suggested. At first signs of pulmonary edema, the patient should be placed in an oxygen tent and treated symptomatically.

Animal Toxicity

The compound is moderately toxic in rats following oral administration (LD_{50} ~100 mg/kg) but considerably less toxic by that route in rabbits (LD_{50} ~1 g/kg).

Chlorpyrifos may inhibit preferentially plasma cholinesterase relative to erythrocyte acetylcholinesterase in laboratory rodents. At lower doses (0.08 mg/kg/day for 6 months), extensive blood cholinesterase inhibition was reported, with little effect on brain acetylcholinesterase activity.

There is considerable variation in the susceptibility of different species and the degree of toxic symptoms is also highly dependent on the route of entry. In general, birds are more sensitive to chlorpyrifos than mammals. Due to its high lipid solubility and possible sequestration into fatty tissues, under some conditions chlorpyrifos can accumulate and have persistent effects. A possible correlation between chlorpyrifos exposure and tumors of the pancreas and follicular cells of the thyroid has been noted. Chlorpyrifos has not been found to be teratogenic in experimental studies.

—Carey Pope

Related Topics

A-esterase
Cholinesterase Inhibition
Delayed Neurotoxicity
Neurotoxicology: Central and Peripheral
Organophosphates

Chlorzoxazone

- CAS: 95-25-0
- SYNONYMS: Chlorozaxone; 5-chloro-2(3*H*)-benzoxazolone; 5-chloro-2-benzoxazolol; 5-chloro-2-hydroxybenzoxazole; 2-hydroxy-5-chlorobenzoxazole; Paraflex; Biomioran; Solaxin
- CHEMICAL CLASS: Strong oxidizing agent
- MOLECULAR FORMULA: $C_7H_4ClNO_2$
- CHEMICAL STRUCTURE:

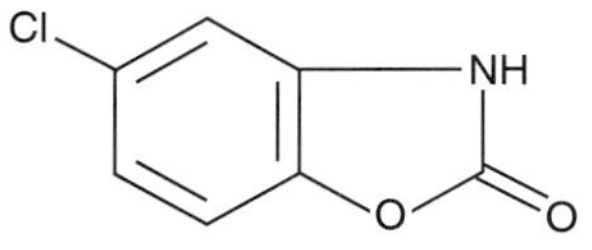

Uses

Chlorzoxazone is used as a centrally acting skeletal muscle relaxant and as an analgesic. It is also a strong oxidizing agent.

Exposure Pathways

When heated to decompose, it emits acrid smoke and irritating fumes.

Toxicokinetics

The oral suspension is rapidly absorbed and eliminated. After intravenous administration, the decay of the plasma concentration is rapid. The half-life for an oral dose is about 1 hr.

Chlorzoxazone is rapidly metabolized in the liver by carbon hydroxylation at position 6 mediated by CYP1A2 as well as by CYP2E1 and the 6-hydroxychlorzoxazone formed is conjugated with glucuronide. The concentration in fat is two times the plasma levels. Chlorzoxazone is eliminated mainly as the glucuronide conjugate by urine.

Mechanism of Toxicity

The toxic metabolite 6-hydroxychlorzoxazone is formed by action of CYP1A2 and CYP2E1.

Human Toxicity

Chlorzoxazone is harmful if swallowed, inhaled, or absorbed through skin. It causes drowsiness; central nervous system effects such as headache, dizziness, and blurred vision; nausea and vomiting; and eye, skin, and mucous membrane irritation. The target organs are the liver, nerves, and skeletal muscles. Although morbidity and mortality are low in pure compound ingestion, they may be increased in multiple ingestions.

Workers exposed to this compound should wear personal protective equipment and their work should be carried out only in restricted areas. Technical measures should prevent any contact with the skin and mucous membranes. Clothing and equipment after use should be placed in an impervious container for decontamination or disposal.

Clinical Management

In case of contact, the eyes and skin should be flushed immediately with water for at least 15 min. If the victim is not breathing, artificial respiration should be administered: if breathing with difficulty, oxygen should be given. If patient is in cardiac arrest cardiopulmonary resuscitation should be provided. If swallowed, the mouth should be washed out with water provided the person is conscious. Life-support measures should be continued until medical assistance has arrived. An unconscious or convulsing person should not be given liquids nor induced to vomit.

Animal Toxicity

In rats, the oral LD_{50} is 763 mg/kg body weight (bw) and the intraperitoneal LD_{50} is 150 mg/kg bw. In the mouse, the oral LD_{50} is 440 mg/kg bw, the intraperitoneal LD_{50} is 50 mg/kg bw, and the subcutaneous LD_{50} is 170 mg/kg bw. In the hamster, the oral LD_{50} is 662 mg/kg bw, and the intraperitoneal LD_{50} is 166 mg/kg bw.

—*Kashyap N. Thakore and Harihara M. Mehendale*

Cholesterol

- ◆ CAS: 57-88-5
- ◆ SYNONYMS: Cholest-5-en-3-ol-(3-β)- (9CI); cholest-5-en-3- β-ol; 5-cholesten-3- β-ol; cholesterin; cholestrin; dythol; provitamin D
- ◆ CHEMICAL STRUCTURE:

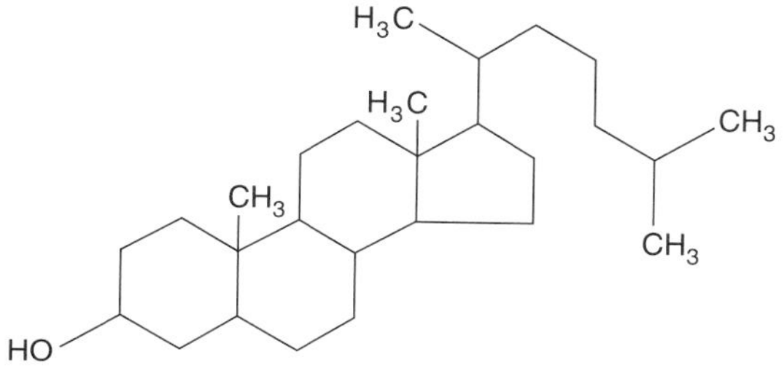

Description

A natural product, cholesterol is a fat-soluble compound that is present in our daily diet. It can form

esters with fatty acids. Most of the cholesterol found in the plasma is in the form of cholesterol esters. It can also be endogenously formed in the cells of the body. The majority of the cholesterol is produced by the liver cells; however, all other cells also form cholesterol. The membrane structure of the cells is partially made up of cholesterol. It is synthesized from multiple molecules of acetyl-CoA.

The major use of cholesterol in the body is to form cholic acid in the liver which conjugates with other substances to form bile salts. Bile salts help in digestion and absorption of fats. Cholesterol is used in small amounts to form adrenocorticoid hormones, progesterone, estrogens, and testosterone.

Adverse Effects

High blood plasma concentrations of cholesterol in the form of low-density lipoproteins is known to cause diseases such as atherosclerosis, which may lead to severe heart conditions.

—*Sushmita M. Chanda*

Related Topics

Cardiovascular System
Liver

Choline

- ◆ CAS: 62-49-7
- ◆ SYNONYMS: Bursine; ethanaminium; fagine; gossypine; luridine; sincaline; sinkalin; sinkaline; vidine
- ◆ CHEMICAL CLASS: Cholinergic agonist
- ◆ CHEMICAL STRUCTURE:

$$[HOCH_2CH_2N^+(CH_3)_3]$$

Uses

Choline is used as a direct cholinergic agonist in therapeutics and as a research tool.

Exposure Pathways

Dermal and oral contact are the most common exposure pathways.

Toxicokinetics

Choline is metabolized to trimethylamine, which is excreted in skin, lungs, and kidney.

Mechanism of Toxicity

Choline is a cholinergic agonist; therefore, it exerts toxicity by directly hyperstimulating the postganglionic cholinergic receptors. This may lead to stimulation of gastrointestinal, urinary, uterine, bronchial, cardiac, and vascular receptors.

Human Toxicity

The estimated oral lethal dose for humans is 200–400 g. Oral doses of 10 g produce no obvious effect. Vital signs may include bradycardia, hypotension, hypothermia, miosis, salivation and lacrimation, ocular pain, blurred vision, bronchospasm, muscle cramps, fasiculations, weakness, nausea, vomiting, diarrhea, and involuntary urination. No study has reported chronic toxicity of choline.

Clinical Management

Atropine sulfate is the drug of choice. Epinephrin may assist in overcoming severe cardiovascular or bronchoconstriction. Diazepam, phenytoin, and phenobarbital may be given in cases of seizures. Induction of emesis is not necessary due to spontaneous vomiting. Activated charcoal slurry with or without saline cathartic may be used. Sorbitol should not be used because it may contribute to the nausea and diarrhea. Skin decontamination should be accomplished by repeated washing with soap. Exposed eyes should be irrigated with copious amounts of room-temperature water for at least 15 min.

Animal Toxicity

Animals are known to show similar hyperstimulating effects of the cholinergic system as humans.

—*Sushmita M. Chanda*

Related Topics

Anticholinergics

Cholinesterase Inhibition
Neurotoxicology: Central and Peripheral

Cholinesterase Inhibition

Introduction and History

Cholinesterases (ChEs) are a ubiquitous groups of enzymes that hydrolyze esters of choline. A well known example is acetylcholinesterase (AChE, acetyl choline hydrolase, EC 3.1.1.7), the enzyme responsible for hydrolyzing the important neurotransmitter acetylcholine (ACh). Another ChE is butyrylcholinesterase (BuChE, acylcholine acylhydrolase, EC 3.1.1.8), also known as nonspecific cholinesterase. The preferred substrate for AChEs is ACh; BuChEs prefer to hydrolyze esters like butyrylcholine and propionylcholine. Both AChE and BuChE are inhibited by some organophosphate (OP) and carbamate (CB) esters, and also by other chemicals.

Many ChE inhibitors act at the catalytic site of the enzyme, forming enzyme-inhibitor complexes that are slow to hydrolyze. The use of ChE inhibitors as insecticides and as chemical warfare agents, their toxicity to humans, and their impact on wildlife have made them important to toxicology researchers and, public health and environmental health officials.

This entry focuses on ChE inhibitions by OPs and CBs. Other chemicals, such as tacrine, cocaine, and succinylcholine, are briefly discussed.

One of the first ChE inhibitors to be studied was a CB, physostigmine (eserine), an alkaloid from the calabar bean (*Physostigma venenosum*) used in a "trial by ordeal" in West Africa. The accused were forced to eat the poisonous beans; survivors were proclaimed innocent. The drug has been used as a treatment for glaucoma since 1877. Englehart and Loewi showed it blocked ChE activity in 1931. Shortly after, neostigmine, an analog, was shown to be effective in the symptomatic treatment of myasthenia gravis.

OPs with high toxicity were synthesized as chemical warfare agents in the late 1930s and early 1940s. During this period Schrader discovered the insecticidal properties of OPs resulting in the synthesis of tetraethyl pyrophosphate in 1941 and of parathion in 1944. Synthetic CBs developed as pesticides have been in commercial use since the 1950s. Some OPs and CBs exhibit toxicities in addition to their direct inhibitions of ChEs. These include long-term and short-term damage to nerves and muscles, mutagenicity, and effects on reproduction.

Acetylcholinesterase, Butyrylcholinesterase, and Other Esterases

AChEs and BuChEs are specialized carboxylic ester hydrolases that preferentially hydrolyze choline esters. They are classed among the B-esterases, enzymes that are inhibited by OPs. Another B-esterase is neuropathy target esterase (NTE), an enzyme implicated in organophosphate–induced delayed neuropathy (OPIDN; see Neurotoxicity, Delayed). Enzymes that actively hydrolyze OPs are known as A-esterases. They provide an important route of detoxification. Examples are paraoxonase and DFPase (Table C-14). Recently, the tertiary structure and amino acid sequences of several AChEs and BuChEs have been elucidated.

ChEs are widely distributed in the body. AChEs regulate excitation at cholinergic synapses by destroying the neurotransmitter ACh. The enzyme is one of the most active known, cycling within a few milliseconds. AChEs are found in excitable tissues at synapses, neuromuscular junctions, myotendinous junctions, central nervous system (CNS) neuron cell bodies, axons, and muscles (Table C-15). AChEs are also found in the erythrocytes [red blood cells (RBCs)] of mammals, in the serum of some birds and mammals, and in the blood platelets of rodents (rats and mice) and ruminants (sheep). (For example, the serum ChE activity of the American Kestrel, a small falcon, consists almost entirely of AChE and the serum ChE of the laboratory rat is high in both AChE and BuChE activities.) The AChE activity of human blood is localized to its RBCs. AChE activity occurs in the serum of developing mammals and birds and in precursors of formed blood elements in some species; it decreases to adult levels after birth.

BuChEs are also widely distributed. They are found at synapses, motor endplates, and muscles fibers together with AChE. BuChE activity in blood is restricted to serum.

Substrate preferences of AChE and BuChE enzymes vary with species. For example, although both mammals and bird AChEs rapidly hydrolyze ACh and its thiocholine analog acetylthiocholine (AcTh), avian AChEs also readily hydrolyze acetyl β-methylcholine and acetyl β-methylthiocholine, while mammal AChEs do not. AChEs and BuChEs respond differently to increasing substrate concentration. AChEs are inhibited

TABLE C-14
Esterase Classes

A esterases
Hydrolyze OPs to inactive products
Found in liver and HDL in plasma
High activity in mammals
Lower activity in birds
Examples: Paraoxonase and DFPase
B esterases
Widely distributed in cells and tissues
Inhibited by OPs and CBs
Slow hydrolysis of OP–enzyme complex
Relatively rapid hydrolysis of CN–enzyme complex
Examples: AChE, BuChE, CaE, and NTE

Note. Abbreviations used: OP, organophosphate ester; HDL, high-density lipoprotein; CB, carbamate; AChE, acetylcholinesterase; BuChE, butyrylcholinesterase; CaE, carboxylesterase; NTE, neuropathy target estarase.

TABLE C-15
Cholinesterase Properties

All
Hydrolyze ACh and other choline esters
AChE
Prefers ACh, is inhibited by excess substrates
Found at neural junctions and in mammal RBCs and plasma and platelets of some vertebrates
BuChE
Prefers butyrylcholine, propionylcholine
Widely distributed in vertebrate tissues and plasma

by excess substrate (often above 2 m*M*); BuChEs are less sensitive.

AChEs and BuChEs have multiple molecular forms and complicated life histories (Figs. C-18 and C-19). Some of the forms move from site to site within cells, others are secreted into body fluids. AChEs consist of asymmetric and globular forms. The asymmetric forms tend to be localized at synapses and motor end-plates; they have glycosylated heads joined together by sulfhydryl groups containing the active sites, and collagen tails that attach the enzymes to cell surfaces. The globular forms lack collagen tails; they are made up of the catalytic subunits.

AChE and BuChE subunits are synthesized within cells (e.g., nerve, muscle, liver, and some megakaryocytes), glycosylated within the Golgi apparatus, and secreted. Collagen-tailed forms become attached to the cell surface at specific binding sites. Globular forms are released into body fluids or bind to cell surfaces by ionic bonds. Antibodies have been prepared to several purified AChEs and BuChEs, and protein and nucleic acid sequences have been determined.

The three-dimensional structure of AChE from the electric organ of *Torpedo californica* has recently been established. One interesting feature is that the active site is embedded in a "gorge" of about 20 Å that reaches halfway into the protein. The postulated "anionic site," theoretically invoked to bind the quaternary ammonium ion of ACh, appears to be represented by aromatic amino acids in the gorge itself; these and charges in the active center are believed to stabilize the choline group. In addition, some inhibitions, such as that due to excess substrate, are believed to be due to a "peripheral site." Elucidation of the structure of ChE molecules opens the way to a new generation of "designer" anti-ChE agents with improved specificities of action.

Functions of Cholinesterases

$$E + AX \underset{k_{-1}}{\overset{k_{+1}}{\longleftrightarrow}} EAX \overset{k_2}{\longleftrightarrow} EA \overset{k_3}{\longleftrightarrow} E + A$$

where *E* is the enzyme, AX is the substrate (ACh) or inhibitor, EAX is the reversible enzyme complex, and *k*'s are reaction rate constants.

A hundred years of research has established that a major function of AChE is to hydrolyze the ACh released by cholinergic neurons, regulating the course of neural transmission at synapses, motor end plates, and other effector sites. The reaction is multistep: First is the formation of a reversible enzyme–substrate complex (EAX); second is the acetylation of the catalytic site of the enzyme (EA); and third is the hydrolysis of the enzyme–substrate complex yielding acetic acid, choline, and the regenerated enzyme (*E* + *A*). The generally accepted mechanism has been (a) an electrostatic attraction between the positive charge on the quaternary nitrogen atom of ACh and the negative charge on the so-called "anionic site" on the enzyme forms the enzyme substrate complex, (b) a basic imidazole moiety (histidine) and an acidic moiety (tyrosine hydroxyl) at the active site catalyze the acetylation of a serine hydroxyl, followed by (c) a rapid deacetylation restoring the enzyme and cleaving acetylcholine into acetate and choline. A similar reaction scheme is believed to apply to BuChEs. It is safe to say that the new information on the conformation of these molecules will soon result in a greater understanding of

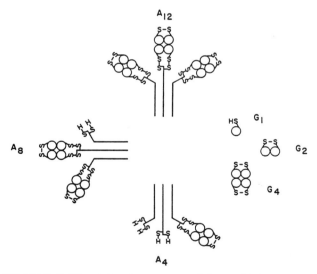

FIGURE C-18. *Subunit structure of the multiple molecular forms of ChEs. G, globular forms; A, asymmetric forms with collagen-like tails. Each circle is a catalytic subunit; disulfide bridges indicated by S–S as found in the electric organ of the electric eel (modified from Brimijoin, U.S. EPA Workshop on Cholinesterase Methodologies, 1992).*

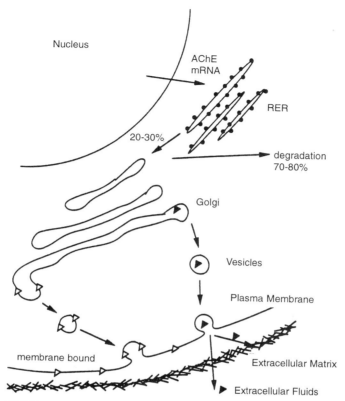

FIGURE C-19. Life cycle of ChEs. AChE is synthesized as a monomer globular form (G₁). Up to 80% is degraded by intracellular processes. Secretory forms are separated from membrane-bound forms, collagen tails are added to asymmetric forms, and the enzyme is glycosylated and becomes enzymatically active. After secretion, globular forms may escape into the body fluids, while asymmetric forms are bound to the synaptic basal lamina (modified from Brimijoin, U.S. EPA Workshop on Cholinesterase Methodologies, 1992).

the biophysical mechanisms underlying their catalytic actions.

In contrast to the functional information available for the roles of ACh and AChE, the function or functions of RBC and serum ChEs are still matters for speculation. One idea is that they protect the body from natural anti-ChE agents (e.g., phyosostigmine) encountered during the evolution of the species; another idea is that they have specific but still unknown roles in tissues. For example, there are recent reports that inhibition of BuChE activity blocks adhesion of neurites from nerve cells in culture and that AChE promotes outgrowth of neurites as if the enzymes had roles in cell adhesion and differentiation.

Toxicities

The toxicities of OPs and CBs often roughly parallel their effectiveness as inhibitors of brain AChE. For example, Fig. C-20 shows the relationship between the toxicity *in vivo* of directly acting OPs and their inhibition of AChE *in vitro*, plotting intraperitoneal LD_{50} versus the P_{50} in mice. (The LD_{50} is the dose resulting in 50% mortality; the P_{50} is the negative logarithm of the concentration of toxicant resulting in 50% inhibition of the enzyme.) Only two of the chemicals tested did not "fit" the curve.

In general, many of the physiological effects of anti-ChEs are those attributable to excess ACh at junctions

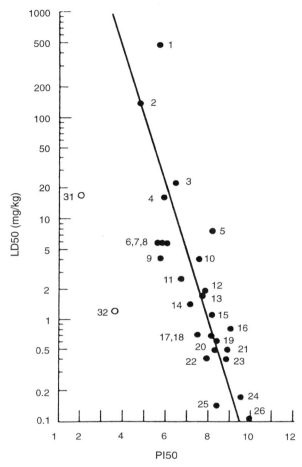

FIGURE C-20.   *Relationship of the toxicity* in vivo
(LD$_{50}$) of directly acting OPs to AChE inhibition in vivo
*(PI$_{50}$). 1, dipterex; 2, O,O-diethyl-4-chlorophenyl-
phosphate; 3, O,O-diethyl-bis-dimethylpyrophos-
phoramide(sym); 4, TIPP; 5, O,O-diethylphosphostigmine;
6, isodemeton sulfoxide; 7, isodemeton; 8, isodemeton
sulfone; 9, DFP; 10, diethylamidoethoxy-phosphoryl
cyanide; 11, O,O-dimethyl-O,O-diisopropyl pyrophosphate
(asym); 12, diethylamidomethoxyphosphoryl cyanide; 13,
tetramethyl pyrophosphate; 14, O,O-diethylo
phosphorocyanidate; 15, O,O-dimethyl-O,O-diethyl
pyrophosphate (asym); 16. soman; 17, TEPP; 18, O-
isopropyl-ethylphosphonofluoridate; 19, tabun; 20,
amiton; 21, diethylamido ispropoxyphosphoryl cyanide;
22, O,O-diethyl-S-(2-diethylaminoethyl)phosphorothioate;
23, sarin; 24, O,O-diethyl-S-(2-triethylammoniummethyl)
thiophosphate iodide; 25, echothiophate; 26,
methylfluorophosphorylcholine iodide; 27,
methylfluorophosphoryl-B-methylcholine iodide; 28, O-
ethyl-methylphosphorylthiocholine iodide; 29,
methylfluorophosphoryl-homo-choline iodide; 31,
schradan; 32, dimefox. (27–30, LD$_{50}$s 0.03–0.07) (adapted
from Gallo, L., Organophosphorus insecticides, in
Handbook of Pesticide Toxicology (W. J. Hayes, Jr., and
E. R. Laws, Jr., Eds.,. Vol. 2, p. 932, 1991).*

in the nervous system. The precise symptoms and the
time course of ChE inhibition depend on the chemicals
and the localization of the receptors affected. The properties of some cholinergic receptors are listed in Table
C-16. Cholinergic junctions are classified into several
categories based on their pharmacological sensitivities
to nicotine, muscarine, atropine, and curare. Early
symptoms of cholinergic poisoning represent stimulation of neuroeffectors of the parasympathetic system.
These effects are termed muscarinic—stimulated by
muscarine and blocked by atropine. Effects include
slowing of the heart (bradycardia), constriction of the
puplic of the eye, diarrhea, urination, lacrimation, and
salivation. Actions at skeletal neuromuscular junctions
(motor end-plates) are termed nicotinic—stimulated by
nicotine, blocked by curare, but not by atropine. Overstimulation results in muscle fasciculation (disorganized twitching), and, at higher doses, muscle paralysis.
A third site of action of anti-ChEs is the cholinergic
junctions of the sympathetic and parasympathetic autonomic ganglia. These junctions are also nicotinic—
stimulated by nicotine but not muscarine, atropine, or
curare, except at high concentrations. Their actions
affect the eye, bladder, heart, and salivary glands, with
one set often antagonizing the actions of another. Finally, there are the junctions of the CNS: Some are
stimulated by nicotine, and some are affected by atropine. They are not responsive to muscarine or curare.
CNS symptoms include hypothermia, tremors, head-

TABLE C-16
Properties of Cholinergic Receptors

Muscarinic peripheral NS
Parasympathetic nervous system
Muscarine stimulates
Atropine blocks

Nicotinic peripheral NS
Skeletal muscle motor endplates
Nicotine stimulates
Curare blocks
Atropine has no effect

Nicotinic CNS
Autonomic NS antagonist
Sympathetic and parasympathetic NS
Nicotine stimulates respiratory center

Note. Abbreviations used; NS, nervous system;
CNS, central nervous system.

ache, anxiety, convulsions, and coma. Death generally occurs when the agents extensively affect the respiratory centers in the brain. Whether or not there are consistent behavioral effects at low dose levels of OPs and CBs, such as deficits in learning and memory, is a matter of current research, especially on drugs that are under development to treat Alzheimer's disorder.

The excess ACh produced at the motor end-plate brings about a transient myopathy in experimental animals. Experiments *in vivo* and *in vitro* of Dettbarn, Wecker, Salpeter, and others using cholinergic drugs and ACh receptor blockers indicate that excess ACh leads to an influx of Ca^{++} ions and other cations into the post-synaptic cell, resulting in regions of necrosis in the muscle fiber around the motor end-plates. From 10 to 30% of the fibers may be damaged and recovery may take several weeks or more. A disorder known as Intermediate Syndrome (see Organophosphate Poisoning, Intermediate Syndrome) in humans involves prolonged muscle weakness and some muscle damage lasting several weeks or longer after exposure to high levels of some OPs, including methyl parathion, fenthion, and dimethoate.

Although most of the effects of OPs and CBs are considered to be caused by AChE inhibition, there is increasing evidence that anti-ChEs directly affect ACh receptors in the CNS and that anti-AChE pesticides depress the immune system in experimental animals.

A few OPs, such as tri-ortho cresyl phosphate (TOCP), leptophos, mipafox, methamidophos, isofenphos, and chlorpyrifos cause OPIDN, a neuropathy that results in death of motor neurons in humans and experimental animals (see Neurotoxicity, Delayed). Some, such as chlorpyrifos and isofenphos, require very high dose levels to be neuropathic—higher levels than could occur if the chemicals were used as directed. TOCP, an industrial chemical, has been responsible for the paralysis of thousands of people since the turn of the century. Inhibition of approximately 70% or more of the carboxylesterase NTE often is associated with the disorder. It is known as a "delayed" neuropathy because onset of the disorder is usually 10 days to several weeks after exposure. Discussion of this neuropathy is beyond the scope of this article, except to note that neuropathic chemicals that are the most dangerous often are those that are better NTE inhibitors than AChE inhibitors, permitting a higher dose of the chemical to be reached before cholinergic symptoms or death occurs. Agricultural chemicals are routinely screened for OPIDN using hens, since chickens are sensitive to the disorder.

The action of many toxicants, including anticholinergic compounds, often involve specific sites on molecules and cells. Such finely-tuned molecular events suggest the possibility of discovering "genocopies," genetic abnormalities that mimic chemically induced disorders. For example, patients have been reported with smaller than normal motor end-plates, defective in AChE, and suffering from muscle weakness. There are no reported AChE-less mutants; it is likely that such a genetic disaster would be lethal. There are humans with inherited differences in their serum BuChEs with decreased activity of the enzyme in their blood. Possessors of these genotypes usually are symptomless, unless they are given succinylcholine (or a similar drug) during surgery to bring about muscle relaxation. Lack of sufficient blood BuChE to speedily destroy the drug intensifies and prolongs the activity of succinylcholine, sometimes with fatal consequences. BuChEs also may play a detoxifying role in cocaine intoxication by hydrolyzing the drug. Several studies on experimental animals indicate that depressing ChEs with anti-ChEs intensifies the toxic effect of cocaine.

Organophosphorus Cholinesterase Inhibitors

OP inhibitors are substituted phosphoric acids of the form

where R_1 and R_2 are usually alkyl or aryl groups linked either directly, or via-O-or-S-groups to the P atom. According to one classification, X, termed the leaving group, may be (1) a quaternary nitrogen; (2) a fluoride; (3) a CN, OCN, SCN, or a halogen other than F; or (4) other groups. (see Fig. C-21 for representative organophosphorus cholinesterase inhibitors.)

1. OPs containing quaternary nitrogen (phosphorylcholines) are strong inhibitors of ChEs and directly acting cholinergics. One, ecothiophate iodide, is used in the treatment of glaucoma.

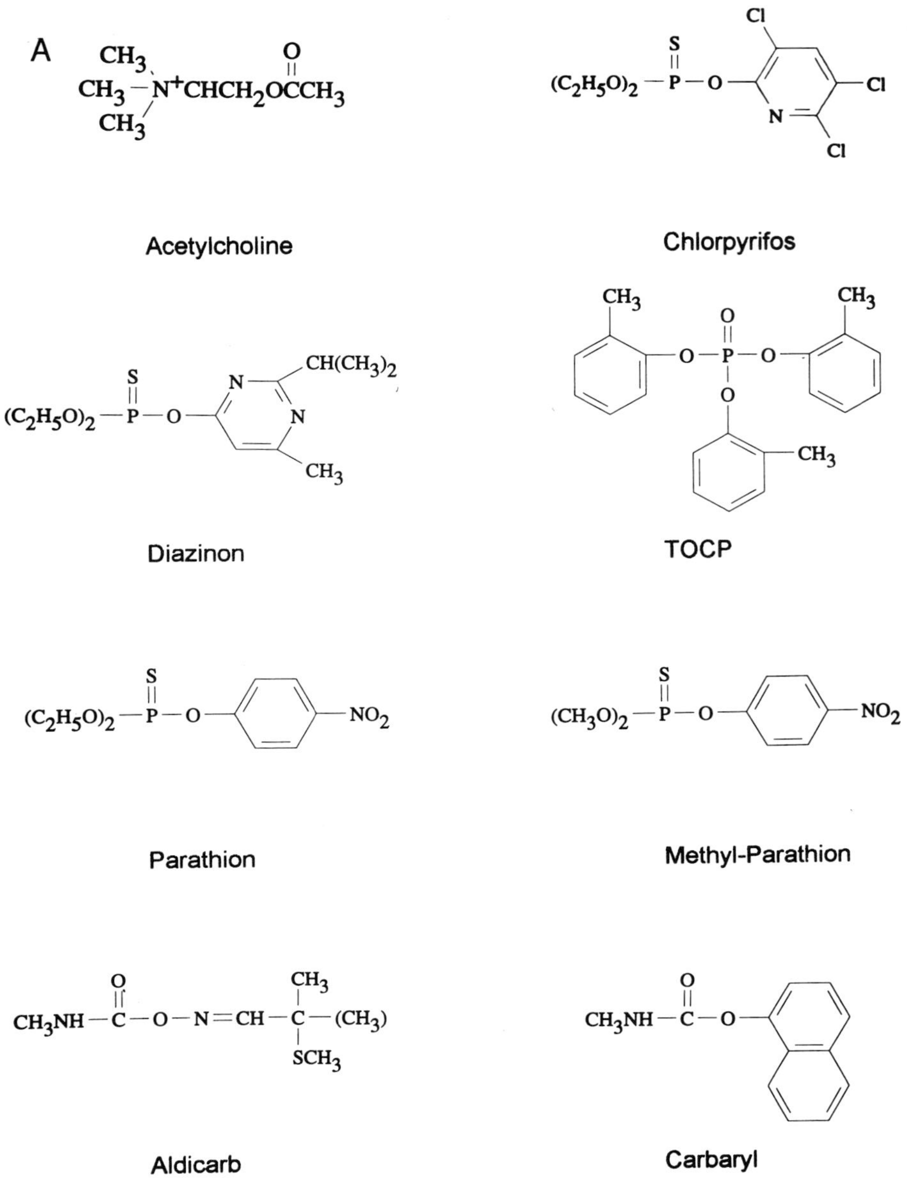

FIGURE C-21. *Representative organophosphorus and organocarbamate cholinesterase inhibitors.*

2. Fluorophosphates are also highly toxic and relatively volatile. Sarin and soman are chemical warfare agents. Diisopropyl fluorophosphate (DFP) is often used by biochemists of study serine-active enzymes. Mipafox and DFP cause OPIDN in humans and experimental animals.

3. An example of a CN containing nerve gas is Tabun.

4. Most OP pesticides are in the fourth and largest category. Many are dimethoxy or diethoxy compounds. OPs used in agriculture tend to be manufactured in the relatively stable P=S form. They are less toxic than OPs with the P=O (oxon) group. (Phosphates lack a sulfur atom, phosphorothioates have a single sulfur atom, and phosphorodithioates have two sulfur atoms.) Many pesticides, such as parathion, methyl parathion,

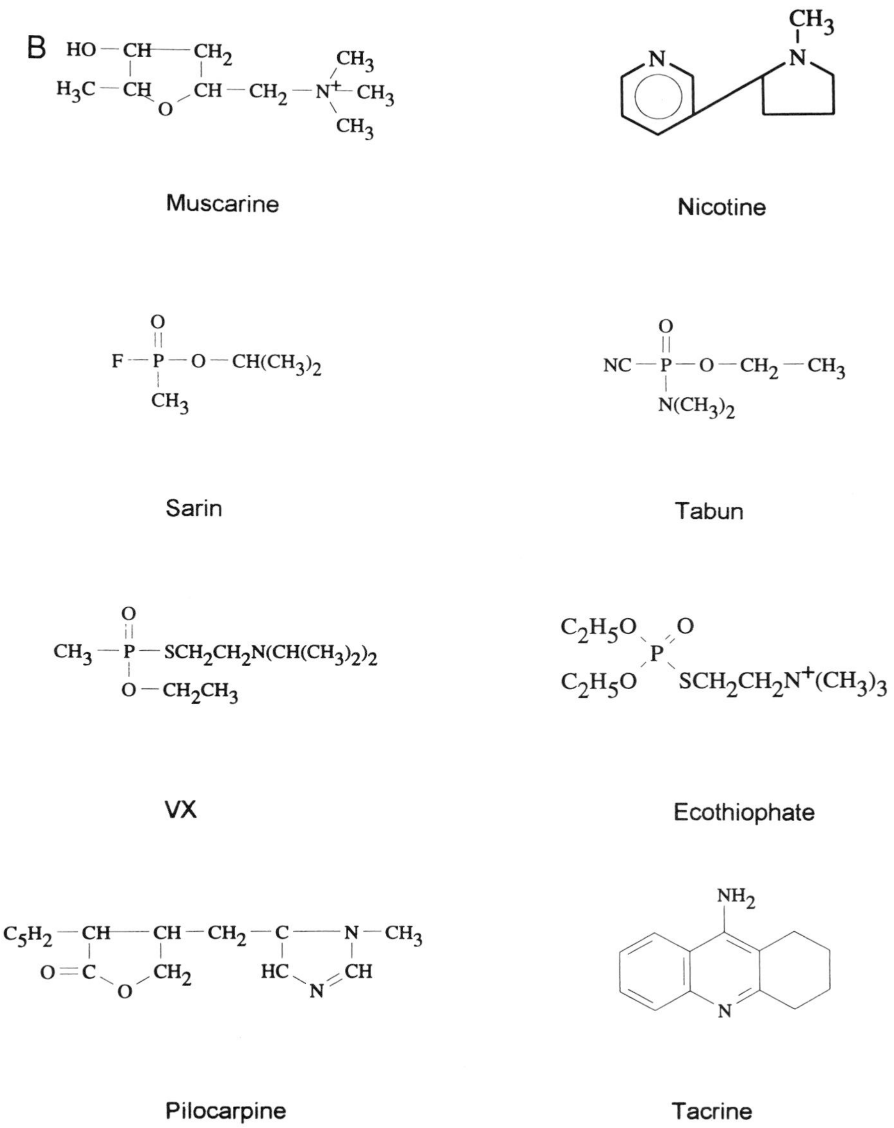

B

Muscarine

Nicotine

Sarin

Tabun

VX

Ecothiophate

Pilocarpine

Tacrine

FIGURE C-21 (continued)

diazinon, and chlorpyrifos, are phosphorothionates.

Three important chemical reactions that underlie ChE inhibitions are hydrolysis, desulfuration, and alkylation.

Hydrolysis: The rate of hydrolysis is a function of the acid and alcohol groups, pH, and temperature. It usually increases with increasing pH, temperature, and UV light.

Desulfuration: An important oxidation is the conversion of the P=S group of phosphorothionates

to P=O, the oxon form, increasing the intensity of ChE inhibition.

Alkylation: Alkyl substituents, especially methoxy groups, may act as alkylating agents. They are capable of altering nucleic acids, leading some to be concerned about OPs as mutagens.

Carbamate Cholinesterase Inhibitors

The CBs used as pesticides are N-substituted esters of carbamic acid. CBs developed in the 1950s as insect repellents were found to have insecticidal activity, leading to the development of the napthyl CBs with high anti-ChE activity and selective toxicity against insects. One example is carbaryl; it is widely used because of its low toxicity to mammals and its degradability. Aldicarb, a plant systemic, is more toxic than carbaryl. Recently, aldicarb was associated with a July 4th holiday incident when West Coast residents complained of anticholinergic symptoms after eating watermelons that may have become contaminated with aldicarb.

Most N-methyl and N,N dimethyl carbamates are better AChE inhibitors than BuChE inhibitors. However, N-carbamylated AChE spontaneously reactivates faster than N-carbamylated BuChE. AChE activity may recover as rapidly as 30 min following exposure—much faster than after exposure to OPs.

Although phosphorylation of AChE by OPs is heavily influenced by the electron withdrawing power of the leaving group, carbamylation by methyl carbamates is also greatly dependent on molecular complementarity with the conformation of the enzyme as well as reactivity of the molecule. In general, phenolic and oxime moieties are more reactive than benzyl alcohol groups.

N-methyl carbamates do not need activation to inhibit ChEs. However, at least in the case of aldicarb, inhibition increases with metabolism. Aldicarb is rapidly oxidized to the relatively stable aldicarb sulfoxide, which in turn is more slowly metabolized to aldicarb sulfone, a stronger AChE inhibitor. These products are then detoxified by conversion to oximes and nitriles, which in turn are degraded to aldehydes, acids, and alcohols. Procarbamate derivatives were developed to reduce the toxicity of CBs to mammals; the hydrogen atom on the carbamate nitrogen is replaced by a wide variety of nucleophiles—many with a sulfur atom—causing reduction in anti-ChE activity. The bond is rapidly broken in insects, restoring the activity and toxicity of the parent compound.

The rapid spontaneous reactivation of carbamates can be a problem in determining ChE activity. For example, some testing routines require that animals be put on a control diet for 24 hr before sampling. With CBs, the inhibitions may have disappeared by the time the assays are performed. In addition, the dilutions specified in some assays may reduce the inhibition and high concentrations of substrate may compete with the carbamate to further reactivate the enzyme.

Chemical Warfare Anticholinesterase Agents

Anticholinesterase chemical warfare agents have been available and stockpiled since their development immediately before and during World War II. Many countries have active research programs into their toxicity and control. P=O groups confer potent anti-ChE inhibition properties. (However, the toxicity of agents such as soman and VX may be due in part to their actions on receptor and perhaps other proteins as well as to their inhibition of AChE.) The toxicity of the nerve agents is greater than that of agricultural chemicals. For example, the dermal LD_{50} of agent VX is estimated to be 0.04–0.14 mg/kg for humans, which is at least an order of magnitude more toxic than most pesticides.

LD_{50}s for representative agricultural OPs and CBs are shown in Table C-17.

Assay Techniques

An early assay for ChE activity was a manometric method in which the change in pH due to ACh hydrolysis released CO_2 from a reaction buffer. A common technique (that of Michel) directly determines ACh hydrolysis by changes in pH. Another assay, that of Hestrin, utilizes the reaction of ACh with hydroxyl amine and ferric chloride, producing a reddish-purple complex. A test developed by Okabe and colleagues oxidizes the choline released from ACh hydrolysis and determines the H_2O_2 produced. Several assays use radioactive ACh; one method counts the acetate produced by the reaction by separating it into an organic phase, leaving the unhydrolyzed ACh behind in an aqueous phase. Another common approach utilizes thioanalogs of ACh and other esters. In the assay developed by Ellman and colleagues in 1961, hydrolysis of thiocholines such as acetylthiocholine (AcTh) is measured at 410 nm with the color reagent dithionitrobenzoate. Although assays that rely on pH or radioactivity of ACh have the advantage of using a natural substrate, assays utilizing thiocholine esters are inexpensive, readily automated, and do not require expensive disposal of radioactive wastes. Negative features are the possibility of interference of hemoglobin (Hb) in RBC samples and a nonliner reaction of the reduced glutathion in some RBCs with the color reagent. Some of

TABLE C-17
Representative Acute LD50s of Selected Organophosphates and Carbamates

Compound	LD50 (mg/kg) Oral	LD50 (mg/kg) Dermal
Organophosphates		
Dimethoxy compounds		
Azinphosmethyl (O,O-dimethyl-S-[(4-oxo-1,2,3-benzotriazin-3(4H)-yl)methyl]phosphorodithioate)	13	220
Malathion (O,O-dimethyl-S-(1,2-dicarbethoxyethyl)phosphorodithioate)	1375	>4000
Methyl parathion (O,O-dimethyl-O-(p-nitrophenyl)phosphorothioate)	14	67
Diethoxy compounds		
Parathion (O,O-diethy-O-(4-nitorphenyl)phosphorothioate)	13	21
Diazinon (O,O-diethyl-O-(2-isopropyl-6-methyl-4-pyrimidnyl)phosphorothioate)	108	200
Carbamates		
Aldicarb (2-methyl-2-(methylthio)propylideneamino-N-methylcarbamate)	0.8	3.0
Carbaryl (1-naphthyl-N-methylcarbamate)	850	>4000

Note. Adapted from Gaines, *Toxicol. Appl. Pharmacol.* **14**, 515–534, 1969.

the methods have been adapted for field use. A new generation of techniques utilize amperometric methods in which enzymes bound to electrode-like probes permit assay for the presence of anti-ChE chemicals. Whatever the assay, it is important that its conditions be validated for the species, tissues, and chemicals under study.

Biochemistry of Cholinesterase Inhibition

The inhibition of the activity of ChEs by OPs and CBs proceeds in a manner similar to the action of the enzymes on ACh. However, instead of forming a rapidly hydrolyzed acetyl–enzyme complex, the OPs and CBs respectively phosphorylate and carbamylate the catalytic sites of the enzymes. The major biochemical features of the inhibition of ChEs by OPs and CBs involve (a) activation of the inhibitors; (b) detoxification; (c) reaction of the inhibitor with the serine active site of the enzyme and loss of a "leaving group"; (d) hydrolysis of the complex and spontaneous reactivation of the enzyme; (e) loss of a second group, known as aging; and (f) recovery by synthesis of new enzyme.

One way to visualize the biochemical mechanisms underlying the toxicity of OPs and CBs is to trace the fate of an OP such as parathion from its entry into the body. Mixed function oxidases (MFO) in the liver (or in other tissues) convert parathion, a thionophosphate,

to its oxygen analog, paraoxon, increasing its anti-ChE potential by orders of magnitude. The paraoxon may exert its toxic action by inhibiting AChE, or be inactivated by conjugation with glutathione, reaction with glutathione transferases, further oxidation by MFO, or hydrolysis by A-esterases, in this case paraoxonase. Such reactions may lead to a loss in toxicity of either parathion or paraoxon. Paraoxon may also be inactivated by binding and reacting with B-esterases other than AChE, such as BuChE and carboxylesterases.

The reaction of an OP with AChE, BuChE, or other B-esterases is similar to the reaction of AChE with ACh, except that the hydrolysis step is much slower; or, in some cases, may not occur at all. Its basis is a phosphorylation of the enzyme via a nucleophilic attack. The electronegative serine hydroxyl at the catalytic site reacts with the electropositive phosphorus atom of the inhibitor to form an OP–ChE complex and loss of a side group on the phosphorus atom, known as the leaving group (X). The phosphorylated enzyme may, in time, reactivate by rehydrolysis. A similar set of reaction leads to carbamylation, except that the spontaneous reactivation tends to be more rapid than that for an OP. Spontaneous reactivation of an OP may take hours to days, whereas CBs may reactivate as soon as 30 min. In addition, OPs undergo a further reaction known as "aging" in which a second group (often an alkyl group) is lost from the phosphate, stabilizing the OP–ChE complex.

Structure/Activity

Some general rules for OPs based on their structures include the following:

1. The P=O group is more toxic than the P=S group because it is more reactive. It is more reactive due to its higher electronegativity, which causes a more electropositive P atom, facilitating its reaction with the serine hydroxyl at the active site.

2. The electron-withdrawing ability of the leaving group X is predicted by the strength of its acid. For example, fluoride is a more powerful leaving group than nitrophenol since HF is a strong acid.

3. Reactivity of the R groups is in the order methoxy $\geqq$ ethoxy $\geqq$ propoxy $\geqq$ isopropoxy $\geqq$ amino groups. The more difficult a compound is to hydrolyze, the weaker is likely to be its ChE inhibition.

4. Steric effects are also important. The longer and more branched a compound, the more reduced is its rate of inhibition, probably because of the conformation of the proteins around the catalytic site.

The terms "reversible" and "irreversible" are often misused in describing ChE inhibitions. For example, statements such as "OPs are irreversible inhibitors and CBs are reversible inhibitors" are useful insofar as they refer to the stability of the aged OP enzyme and to the more rapid hydrolysis of the CB enzyme compared to that of the unaged OP enzyme. Technically, one could argue that the term reversible should be reserved for cases in which there is an equilibrium between the substrate and the enzyme–substrate complex.

Spontaneous Reactivation of Organophosphates

Table C-18 lists the half-lives of recovery for some OP-inhibited AChEs. In general, OP–AChE complexes from dimethoxy-substituted OPs (e.g., malathion) spontaneously dephosphorylate faster than diethoxy (e.g., parathion) or diisopropoxy (e.g., DFP) complexes. Eto pointed out in 1974 that the stability of a phosphorylated AChE may be predicted from the

stability of the specific OP inhibitor itself. One possibility is that methyl groups have less steric hindrance and greater electronegativity than ethyl or isopropyl groups.

Chemical Reactivation of Organophosphates

It has been almost 40 years since I. B. Wilson and colleagues observed that nucleophiles, oximes like hydroxamic acid, reactivated OP-inhibited AChE above and beyond that occurring from spontaneous reactivation, opening the way to a treatment for OP poisoning. The oxime registered for use in the United States is 2-PAM Cl (Protopam); its methanesulfonate salt (P2S) is used in Europe. Oxime therapy should be recommended with caution for carbamate poisonings. Although it has been reported to have been beneficial in the case of aldicarb, there is evidence that 2-PAM treatment increases the toxicity of carbaryl.

The mechanism of action of oxime reactivation involves transfer of the substituted phosphate or phosphonate residue from the catalytic site of the enzyme to the oxime. In addition, 2-PAM may react directly with the free OP molecule itself. Other oximes, such as TMB-4, obidoxime, and HI-6, are reported to be superior to 2-PAM as reactivators and antidotes to chemical warfare agents. Oxime therapy should not be used in the absence of ChE inhibition since 2-PAM itself is a weak ChE inhibitor. In addition to the reactions discussed previously, direct effects of these compounds on muscle contraction and nicotinic receptors led Albuquerque and colleagues to propose that oximes also act directly on cholinergic receptors.

Aging

Research on oximes revealed an important phenomenon: The extent of reactivation of an OP–AChE complex decreased with time and depended on the OP used. This "aging" prevents both spontaneous and chemical reactivation. Evidence indicates aging is due to the loss of a second group from the phosphorus atom. Harris and colleagues in 1966 demonstrated the loss of an alkyl group from a soman–AChE complex, and showed that the percentage of enzyme losing an alkyl group was correlated to the percentage of enzyme resistant to oxime activation. In general, OP–ChE complexes that spontaneously reactivate slowly tend to age

TABLE C-18
Spontaneous Reactivation and Aging of Selected Organophosphates

Compound	Tissue	Spontaneous (hr)	Aging (hr)
Malathion	Human RBC	0.85	3.9
Methamidophos	Bovine RBC	0.13	0.54
Chlorpyrifos	Bovine RBC/mouse brain	58	36
Diazinon	Human RBC	58	41
Parathion	Rat brain/bovine RBC	103	58
Tabun	Human RBC	ND	13
Sarin	Human RBC	ND	3.0
DFP	Human RBC	ND	4.6
Soman	Human RBC	ND	0.02

Note. Adapted from Wilson *et al.,* In *Organophosphates, Chemistry, Fate, Effects* (J. E. Chambers and P. E. Levi, Eds.), Academic Press, New York, 1992.

rapidly. Exceptions are dimethoxy-phosphorylated AChEs, which both rapidly age and spontaneously reactivate. In general, agricultural chemicals (e.g., malathion, parathion, and diazinon) have half-lives of aging of hours and longer, while chemical warfare agents age rapidly (e.g., 10 min for soman).

Treatment for Anticholinesterase Poisoning

The information included here is educational; it should not be construed as specific recommendations for treatment of patients.

Inhibitions of ChEs by OPs and CBs are one of the few toxicities for which there are antidotes. The usual treatment for OP poisoning is atropine (Table C-19). The presence of atropine reduces the effectiveness of the ACh receptors, counterbalancing the excess ACh present. The recommended doses for humans are 1 g

TABLE C-19
Treatments for Anticholinesterase Poisoning

Atropine
    2 mg intravenously, at 15- to 30-min intervals as needed to
        suppress symptoms
2-Pralidoxime
    1 g either intramuscularly or intravenously two or three times
        per day or to suppress symptoms
Diazepam
    10 mg subcutaneously or intravenously, repeated as required

Note. Adapted from Environmental Health Criteria 63, 1986.

PAM Cl (intramuscular or intravenous) two or three times a day, and 2 mg atropine (intravenous) at 15- to 30-min intervals as needed. Higher doses may be used, depending on the extent of the OP intoxication. Environmental Health Criteria No. 63 describes the case of a patient who drank a large amount of dicrotophos while inebriated. Treatments were progressively increased up to 6 mg atropine intravenously every 15 min with continuous infusion of 2-PAM Cl at 0.5 g/hr. All told, 92 g of 2-PAM and 3912 mg of atropine were given to the patient, who was discharged after 33 days.

Much of the research on treatments of ChE inhibitions has concerned chemical warfare agents providing little direct information for the treatment of agricultural chemicals.

Considerable attention has been given to prophylactic treatments to protect military units and civilian populations in the event of either accidental or deliberate release of nerve gas agents. One kit contains a combination of atropine, 2-PAM, and diazepam. Another contains pyridostigmine, a carbamate with actions similar to physostigmine. Diazepam is included to lessen CNS symptoms. The use of pyridostigmine is based on the idea that if a readily rehydrolyzable carbamate will compete for AChE catalytic sites with the high-affinity binding nerve gas agents, it may reduce the percentage of AChE that becomes irreversibly inhibited. Using these agents is not without risk, since they are themselves toxic. Issuance of atropine kits to the general population of Israel during the Persian Gulf crisis led to the accidental injection of more than 200 children;

some had systemic effects but fortunately there were no fatal consequences.

The discovery of methods to isolate relatively large amounts of ChE enzymes in essentially pure form has led to a unique method of treating OP intoxication—that of adding ChEs to the blood. Several experiments indicate enough of the OPs bind to the ChEs to reduce their toxicity in experimental animals. One issue is that of possible immune responses to what might be recognized by the body as a foreign protein, but to date there is no evidence suggesting this to be a problem.

Treatments with Anticholinesterase Agents

Several anticholinesterase agents have been used to treat human disorders.

Alzheimer's Disease and Tacrine

The findings that senile dementia of the Alzheimer's type was accompanied by a loss of AChE activity (as well as other neurochemical markers for cholinergic neurons) in parts of the brain has stimulated study of cholinergic nerve activity, learning and memory, and the use of anti-ChE compounds in the treatment of Alzheimer's disease. The strategy is to increase the effective level of ACh by reducing the activity of the AChE present. One drug under test is physostigmine; others are Tacrine and its analogs. Tacrine is a weakly binding anti-ChE agent, recently approved for treatment by the U.S. FDA. The dose of Tacrine recommended (100 mg/day) was chosen on the basis of the side effects the drug has on liver function rather than on unequivocal demonstration of its effectiveness. (In some trials, up to a third of the patients were removed from the studies due to side effects of the drug.)

Glaucoma

Glaucoma is a disorder of vision accompanied by an increase in ocular pressure. Although mostly replaced by other drugs (e.g., beta-blockers, and pilocarpine), anti-ChE drugs such as ecothiopate and pyridostigmine are still used in the treatment of these common disorders.

Myasthenia Gravis

Myasthenia gravis is a progressive disorder characterized by muscle weakness; eye muscles are often the first affected. Research has shown it to be an autoimmune disease in which the victim forms antibodies to his or her nicotinic acetylcholine receptors at motor endplates. It is characterized by fatigability and weakness of skeletal muscles, especially those of the eyes. Approximately 90% of the patients have droopy eyelids and double vision. Treatments include corticosteroids and thymectomy to reduce the actions of the immune system and anti-ChE agents such as pyridostigmine to improve the effectiveness of the receptors that remain.

Wildlife and Domestic Animal Exposures

The recognition that chlorinated hydrocarbons are a persistent danger to wildlife led to a decrease in their use as agricultural chemicals and to an increase in the use of OPs and CBs. In general, OPs and CBs do not bioaccumulate as do chlorinated hydrocarbons and they are relatively biodegradable. However, they are more acutely toxic than chlorinated hydrocarbons to humans and wildlife. A thorough discussion of the comparative toxicology of OPs and CBs is outside the scope of this entry. ChE inhibitions are generally the same, regardless of the animal; differences between species are often in the overall pharmacokinetics and metabolism. For example, although birds have higher brain AChE activities than mammals, they also have less hepatic MFOs to activate OPs and less A-esterases to hydrolyze them. Much research has been done on the toxicology of OPs to wild birds from sparrows to hawks and eagles. For example, E. P. Hill and colleagues of the U.S. Fish and Wildlife Service studied the toxicity of 19 OPs and 8 CBs to 35 species of birds. In general, such studies showed that over 50% of OPs and 90% of CBs have LD_{50}s of less than 40 mg/kg for most birds.

Route of exposure may have much to do with the recovery from OPs. When pigeons were treated orally with an OP, inhibition of blood ChE was rapid, and recovery of activity occurred within a few days. However, when the treatment was conducted dermally, putting the OP on the feet, recovery of activity took several weeks, implying the presence of a depot for OPs and the possibility that birds can accumulate OPs by flying from site to site. The possibility of bioaccumulation of OPs in a food chain (usually considered to be a characteristic of chlorinated hydrocarbons) was demonstrated by the report of an eagle poisoned by an OP (Warbex) in magpies that, in turn, had obtained the OP by ingesting hair from a steer that had been treated with it for internal parasites.

Beef cattle, horses (more than sheep), goats, and swine are treated several times each year with OPs to control parasites and are fed tetrachlorvinphos to prevent fly larvae hatching in their feces. Carbaryl is commonly used for flea and tick control. Oehme states that insecticides are a common cause of poisoning of domestic animals and that "the majority of insecticide problems in domestic animals result from ignorance or mismanagement." Indeed, there is some epidemiological evidence that animal technicians in pet grooming and veterinary hospitals are exposed to the OP and CB chemicals used to control fleas and ticks while washing the animals.

Exposures in the Workplace

Worldwide, estimates of the number of humans requiring treatment due to anti-ChE chemicals run into the many thousands annually. Concern for those who manufacture and use agricultural chemicals has resulted in studies of pesticide residues, protective clothing, urinary metabolites, and blood ChE levels of farmworkers, greenhouse workers, and spray applicators. In general, the rule has been to consider decreases of blood cholinesterases of 30% or more as meaningful, signifying the worker should be removed from contact with the agent. In the United States, California requires workers to be monitored; however, even there, there is no single standard method to determine ChE activities.

Chemical Warfare and Terrorism
The use of chemical weapons, nerve gases, mustard gases, and blistering agents is banned by international treaty. Nerve agents are known to have been released from storage sites during the Persian Gulf conflict. At the time this entry was written, the role that nerve agents may have played, whether alone or in company with other chemicals, in a baffling set of symptoms

known as the Gulf War syndrome is under investigation.

Millions of pounds of chemical warfare agents are stored throughout the world. Their destruction by incineration at high temperatures, up to 2500°F (1480°C) is planned or under way in several countries. These include eight sites in the United States, such as the Tooele Army Depot in Utah and Johnston Atoll in the Pacific, which is 750 miles from Hawaii. Some of the ordinance has been stored since World Wars I and II. Complaints have been lodged by citizens groups concerned about possible risks to residents during the destruction of the chemicals.

Recently, General Schwartzkopf warned of the danger that chemical warfare weapons pose in future conflicts. Two recent episodes in which sarin was used by terrorists in Japan cast a cloud over attempts to control the use of these weapons. Sarin was released in a residential area of the city of Matsumoto on June 27, 1994, and in a crowded Tokyo subway less than a year later, on March 20, 1995. In Matsumoto, about 600 residents and rescuers were affected and 7 died. More than 5500 people were poisoned and 12 died in the Tokyo incident. Many more might have perished if it were not for the quick action and bravery of firemen, police, and others and the availability of antidotes in Japanese hospitals. (Two subway attendants died removing containers of sarin from subway cars.)

Significance of Blood Cholinesterase Levels

There has been a continuing discussion of the significance of monitoring blood ChEs of humans and wildlife. The setting of no-observable-adverse-effect levels (NOAELs) is an example. (NOAELs are the highest dose levels at which no important effect of a drug is observed). Determining NOAELS is an important step in assigning risks and safe levels for the use of a toxic chemical. Some propose that batteries of behavioral tests performed under controlled laboratory conditions provide the best data for setting safe levels of exposure. Under field conditions others propose that measurements of residues on skin and clothing, urinary metabolites of agricultural workers, and fecal metabolites of wild animals provide evidence of exposure to chemicals without invasive procedures. Proponents of the use of ChE levels point out that they represent standardized, relatively inexpensive measurements that directly demonstrate a biochemical effect of an exposure to a toxic chemical rather than merely providing evidence of the exposure itself. Recent technology permits determinations of enzyme activities on 100 μl or less of blood, obtainable by a finger prick.

Regardless, as long as millions of pounds of OPs and CBs are used annually, ChE measurements will be a useful tool in the protection of humans, domestic animals, and wildlife from overexposure to these toxic agents.

Further Reading

Ballantyne, B., and Marrs, T. C. (1992). *Clinical and Experimental Toxicology of Organophosphates and Carbamates.* Butterworth-Heinemann, Stoneham, MA.

Chambers, J. E., and Levi, P. E., (Eds.) (1992). *Organophosphates: Chemistry, Fate, Effects.* Academic Press, New York.

Hayes W. J., Jr., and Laws, E. R., Jr. (1991). *Handbook of Pesticide Toxicology.* Academic Press, New York.

Hoffman, W. E., Kramer, J., Main, A. R., and Torres, J. L. (1989). Clinical enzymology. In *The Clinical Chemistry of Laboratory Animals* (W. F. Loeb and F. W. Quimby, Eds.), pp. 237–278.

U.S. EPA (1992). Workshop on Cholinesterase Methodologies. Office of Pesticide Programs, U.S. EPA, Washington, DC.

World Health Organization (1986). *Carbamate Pesticides: A General Introduction,* Environmental Health Criteria No. 64. World Health Organization, Geneva.

—*Barry W. Wilson*

Related Topics

A-esterase
Anticholinergics
Carbamate Pesticides
Carboxylesterases
Nerve Gases
Neurotoxicity, Delayed
Neurotoxicology: Central and Peripheral
Organophosphate Poisoning, Intermediate Syndrome
Organophosphates
Pesticides
Veterinary Toxicology

Chromium (Cr)

- CAS: 7440-47-3
- SELECTED COMPOUNDS: Chromous chloride, $CrCl_2$ (CAS: 10049-05-5); chromous sulfate, $CrSO_4$ (CAS: 13825-86-0); chromic oxide, Cr_2O_3 (CAS: 1308-38-9); chromic sulfate, $Cr_2(SO_4)_3$ (CAS: 10101-53-8)
- CHEMICAL CLASS: Metals

Uses

Chromic oxide (chromium III) is an essential element. Commercially, chromium is used extensively in metal plating to obtain a shiny surface. It is a major ingredient for alloying with iron to produce special stainless steels. By treating the ore followed by electrolysis, the pure metal is obtained. Roasting the ore with lime or soda ash produces the chromate and dichromate, which have extensive commercial uses in pigments, electroplating, catalysts, corrosion inhibition, leather tanning, and wood preservation.

Exposure Pathways

Most human exposure to chromium is by ingestion, although inhalation and dermal contact are possible exposure pathways. Low levels of chromium III are found in some meats and vegetables. This biologically essential form is also found in yeasts and liver. Chromium intake in food is low, with daily human intakes less than 100 μg per day; most of this chromium intake is due to ingestion of food with minimal contribution by water supplies and ambient air.

In air and drinking water, chromium is predominately found in its hexavalent form (chromium VI), although most publications report total chromium, which includes chromium present in all valence states. The main sources of chromium VI in the ambient air are ore refining, fossil fuel combustion, cement production, and industrial operations producing fly ash.

Both chromium VI and chromium III are found in nature, but the latter predominates. Chromous (chromium II), a biologically inactive valence state, exists, but once exposed to air it rapidly oxidizes to chromium III, a biologically active form of chromium. Although the trivalent form is the most prevalent, the hexavalent form is of greater industrial importance and toxicological concern.

Toxicokinetics

Biologically, there are two forms of chromium that are of interest: chromium III and VI. Inhaled chromium can be deposited in the lungs in an insoluble form. However, some chromium is absorbed via inhalation. Ingested chromium (chromium III) is not readily absorbed from the gastrointestinal tract; approximately 1% is absorbed. Chromium VI readily crosses cell membranes and is reduced intracellularly to chromium III. The biological effects related to chromium VI may be related to the reduction of chromium III in the cell and the formation of macromolecular complexes.

Trace quantities of chromium III are essential as cofactors for insulin action, which has functions in addition to glucose metabolism. The active form of chromium III is a complex of glutamic acid and cysteine. It is called the glucose tolerance factor (GTF). The GTF is a complex of niacin and some amino acids. It is a necessary component for the functioning of insulin, which regulates sugar metabolism.

Tissue concentrations of chromium in the general population vary by geographical area. In persons without excess exposure, blood chromium concentrations are between 20 and 30 μg/ml evenly distributed between erythrocytes and plasma.

Chromium is cleared rapidly from the blood and is concentrated in the liver, spleen, and bone marrow. It is excreted mainly in the urine.

It has been suggested that the causative agent in chromium mutagenesis is chromium III (which is biologically active and binds to nucleic acids) bound to genetic material after intracellular reduction of chromium VI (which is preferentially absorbed in the cell).

Mechanism of Toxicity

Chromium is required by phosphoglucomutase for its activity. The cytochrome c reduction system also requires chromium. Chromate can inhibit a number of enzymes. Chromium toxicity mainly involves the nonessential chromium VI. This oxidation state is more easily absorbed than the essential chromium III. Chromium VI easily forms complexes with proteins. Skin dermatitis and skin sensitivity may be the result of the precipitation of proteins by chromium.

Human Toxicity

The major acute effect from excess ingestion of chromium is acute renal tubular necrosis. Ingested chromium VI affects the gastrointestinal tract—decreasing the senses of taste and smell and sometimes inducing ulcers.

Dermal contact with chromium VI can have lasting effects. For exposed workers skin ulcers result from contact with soluble chromate. In addition to contact dermatitis, a lasting sensitization may occur in workers who are not protected.

Those exposed to chromate and chromic acid mists in the local air have developed nasal membrane inflammation followed by ulceration. In extreme cases, the nasal septum perforates. Less common is the formation of polyps in the larynx and the sinuses. The lungs have been affected from chromium exposure. Pulmonary fibrosis has been reported once (without follow-up). Bronchial asthma and pulmonary congestion followed by edema are common findings. Tooth erosion is also possible due to the corrosive nature of hexavalent chromium. The trivalent form of chromium is not corrosive or irritating to tissues.

Chromium VI is considered a human carcinogen. Lung cancer has been recorded for those exposed in the chromate-producing industry and in electroplating shops with poor ventilation. Ingestion of the chromate dusts has been associated with gastric cancer as well as skin sensitization.

A number of nutritional compounds containing chromium III are available. Currently, a number of nutritional claims have been made regarding ingestion of excess chromium compounds. These claims have not been verified and are the subject of significant discussion in the nutritional and scientific communities.

According to ACGIH, chromium III compounds are not classifiable as human carcinogens and chromium VI compounds are confirmed human carcinogens. The ACGIH TLV-TWA for chromium III is 0.5 mg/m^3. The ACGIH TLV-TWA is 0.05 mg/m^3 for chromium metal and water-soluble chromium VI compounds. It is 0.01 mg/m^3 for water-insoluble chromium VI compounds.

Clinical Management

There are no specific antidotes for chromium poisoning. Since most human overexposure is by ingestion, inducing emesis followed by gastric lavage is essential. Maintaining the proper fluid balance is critical due to impact on the kidneys' ability to reabsorb fluid. It is necessary to establish that there is no impairment with breathing due to fluid accumulation in the lungs.

Animal Toxicity

Chromium VI is positive in all laboratory systems in which mutagenesis is routinely tested. It causes DNA damage to a variety of cells in both *in vitro* and *in vivo* systems. Trivalent chromium salts have little or no mutagenic activity in bacterial systems.

A large number of dichromate or chromate compounds tested positive as carcinogens in laboratory animals. Many routes of administration were used. A few researchers have shown chromium to be teratogenic; however, more research is needed.

—*Arthur Furst, Shirley B. Radding,*
and Kathryn A. Wurzel

Related Topics

Cardiovascular System
Metals
Respiratory Tract

Chromosome Aberrations

Genetic material known as deoxyribonucleic acid (DNA) is packaged with proteins into discrete units called chromosomes. The chromosome is made up of two identical arms or chromatids. There is a constriction on each chromosome, the centromere, which is involved in attachment of the spindle fiber during cell division. The centromere location is a trait of each chromosome and can be used to help identify individual chromosomes. The chromosomes of each species are characteristic in number, size, centromere location, and shape.

Any physical or chemical agent that causes chromosome aberrations, or damage, is known as a clastogen.

Chromosome aberrations can be induced by both physical agents, such as ionizing radiation, and by clastogenic chemicals. Ionizing radiation can break chemical bonds in either the DNA or protein of chromosomes; misrepair or lack of repair of these breaks results in chromosome aberrations. Chemicals can interact with the DNA to change the chemical structure, add chemical groups, or bind directly to the DNA. These chemical alterations can result in direct breakage of the chromosomes, or errors in repair of this damage may produce chromosome aberrations.

Aberrations are classified as either chromosome- or chromatid-type according to the structural changes observed. Chromosome-type aberrations occur when both chromatids are involved in formation of the aberration; in a chromatid aberration, only a single chromatid is involved in the aberration. The type of aberration produced depends on the insult and the stage of the cell cycle at the time of the damage. The cell goes through programmed stages as it divides. These stages are classified into interphase and mitosis. During interphase, the stages are further classified as G_0, G_1, S, and G_2 according to the activities of the cells. The G_0/G_1 stage represents the time in interphase between the end of mitosis and before the start of DNA synthesis (S). This stage can be very long in cells that are not dividing. The time when the cell is undergoing DNA synthesis is called S. Finally, the time after DNA synthesis and before the start of mitosis is the G_2 stage of the cell cycle.

Chromatid aberrations are produced during DNA synthesis (S) and during the time between the end of DNA synthesis and the start of mitosis (G_2). For many chemicals, the cells must go through the S phase of the cell cycle before the aberration can be expressed. Chromosome-type aberrations are produced before DNA replication, which occurs in nondividing (G_1) or resting (G_0) cells.

The frequency and type of aberrations have been studied in a number of experimental systems. Studying the presence of chromosome bridges or fragments in plant root tip cells, which were in the anaphase or telophase stages of the cell cycle, was the earliest system developed for evaluating chromosome damage. With the advent of methods to grow mammalian cells in tissue culture, to spread chromosomes in the cells with hypotonic treatment, and to block spindle fiber formation with colchicine, the use of metaphase chromosomes in mammalian cells to evaluate chromosome damage has gained wide application. Metaphase aberrations have been studied both to detect genetic changes in cancer cells and to quantitate the amount of environmentally induced genetic damage in cells from humans and experimental animal populations.

The use of cytochalasin B, which blocks cytokinesis and produces binucleated cells, makes it possible to determine which cells in a population have divided. Since only cells that have divided will contain micronuclei, the use of cytochalasin B was key to the development of the rapid micronucleus assay for detection of chromosome aberrations. Micronuclei are produced as the result of broken chromosomes, lagging chromosomes, or unbalanced exchanges.

Chromosome "banding" methods have been developed to identify individual chromosomes and to break the chromosomes into well-defined regions. The recently developed fluorescent *in situ* hybridization (FISH) technique uses DNA probes in combination with fluorescent labels to "paint" and identify specific chromosomes and aberrations associated with those chromosomes. FISH can be used to identify gene location on a chromosome, to mark individual chromosomes, and to follow specific chromosome aberrations, including exchanges.

Using a combination of these methods, it is possible to quantitate the number and type of chromosome aberrations produced in different tissues. Such quantitative evaluation allows the frequency of aberrations to be used to estimate exposure and dose from clastogenic agents. Use of aberrations as a biodosimeter has provided important exposure information in many nuclear accidents and after occupational exposures to either radiation or chemical agents. By measuring the aberration frequency under well-defined conditions, it is possible to calibrate the amount of chromosome damage against exposure or dose and to use this calibration to estimate these parameters when no physical or chemical dosimeters were present at the time of the insult.

Chromosome aberrations result in loss of and/or alterations in a cell's genetic material. This damage results in genetic imbalance, which can decrease the cell's ability to survive. In fact, most chromosomal aberrations detected by standard cytogenetic techniques are lethal to the cell. However, cells with exchange-type aberrations, which do not involve loss of genetic material, can survive cell division. Changes in gene location produced by such exchanges can result in alterations in gene regulation and expression; these changes provide a selective advantage to the cells with

the aberration. Such aberrations play an important role in both birth defects and neoplastic disease. Many chromosome aberrations are diagnostic for certain disease types. For example, the presence of a translocation between the long arms of chromosomes 9 and 22 can result in the formation of a very small chromosome (Philadelphia chromosome), which is diagnostic for chronic myelogenous leukemia. This translocation results in the fusion of two protooncogenes and causes the loss of cell cycle regulation. Protooncogenes are regulatory genes that can be activated by mutation or change in gene location to become oncogenes. Oncogenes have been shown to be involved in the formation of cancer. With the FISH technique for marking individual chromosomes, it is now possible to detect small exchange aberrations and to evaluate the role of these stable chromosome aberrations during tumor progression.

In summary, evaluation of chromosome-aberration frequency and type provides a useful method to detect and quantitate exposure to environmental insults. Chromosome aberrations can be used to diagnose some neoplastic diseases. Finally, the use of spontaneous and induced chromosome aberrations provides a useful tool to understand changes involved in abnormal development and cancer induction.

Further Reading

Bender, M. A., Awa, A. A., Brooks, A. L., Evans, H. J., Groer, P. G., Littlefield, L. G., Pereira, C., Preston, R. J., and Wachholz, B. W. (1988). Current status of cytogenetic procedures to detect and quantify previous exposures to radiation. *Mutat. Res.* 196, 103–159.

Evans, H. J. (1985). Cytogenetic and allied studies in populations exposed to radiations and chemical agents. In *Assessment of Risk from Low-Level Exposure to Radiation and Chemicals* (A. D. Woodhead, C. J. Shellabarger, V. Bond, and A. Hollaender, Eds.), pp. 429–451. Plenum, New York.

Fenech, M., and Morley, A. A. (1985). Measurement of micronuclei in lymphocytes. *Mutat. Res.* 147, 29–36.

Kurzrock, R., Gulterman, J. U., and Talpaz, M. (1988). The molecular genetics of Philadelphia chromosome-positive leukemias. *N. Engl. J. Med.* 319, 990–998.

Nowell, P. C., and Hungerford, D. A. (1960). A minute chromosome in human chronic granulocytic leukemia. *Science 132*, 1497.

Pinkel, D., Landegent, J., Collins, C., Fusco, J., Segraves, R., Lucas, J., and Gray, J. W. (1988). Fluorescent in situ hybridization with human chromosome-specific libraries: Detection of trisomy 21 and translocations of chromosome 4. *Proc. Natl. Acad. Sci. USA* **85,** 9138–9142.

Rowley, J. D. (1990). The Philadelphia chromosome translocation: A paradigm for understanding leukemia. *Cancer* **65,** 2178–2184.

—*Antone L. Brooks*

Related Topics

Ames Test
Analytical Toxicology
Carcinogenesis
Carcinogen–DNA Adduct Formation and DNA
    Repair
Developmental Toxicology
Dominant Lethal Tests
Host-Mediated Assay
Molecular Toxicology
Mutagenesis
Radiation
Sister Chromatid Exchange
Toxicity Testing, Reproductive

Chrysene

♦ CAS: 218-01-9

♦ SYNONYMS: 1,2,5,6-dibenzonaphthalene; 1,2-benzophenanthrene; 1,2-benzphenanthrene; benz(*a*)phenanthrene; benzo(*a*)phenanthrene

♦ CHEMICAL CLASS: Polynuclear aromatic hydrocarbons

♦ MOLECULAR FORMULA: $C_{18}H_{12}$

♦ CHEMICAL STRUCTURE:

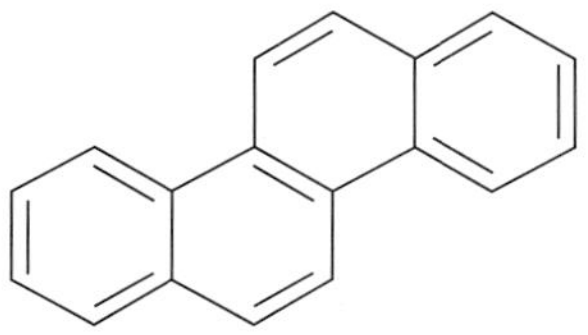

Uses

Chrysene is not produced commercially except for research purposes. However, chrysene occurs as a product of combustion of fossil fuels and has been detected in automobile exhaust. Chrysene has also been detected in air samples collected from a variety of regions nationally and internationally. The concentrations were dependent on proximity to nearby sources of pollution, such as traffic highways and industries, and was also dependent on seasons (generally higher concentrations were noted in winter months). Chrysene has also been detected in cigarette smoke and in other kinds of soot and smoke samples (carbon black soot, wood smoke, and soot from premixed acetylene oxygen flames). It has been detected as a component in petroleum products including clarified oil, solvents, waxes, tar oil, petrolatum, creosote, coal tar, cracked petroleum residue, extracts of bituminous coal, extracts from shale, petroleum asphalts, and coal tar pitch.

Exposure Pathways

Although the vapor pressure of pure chrysene is quite low (6.3×10^{7} mm Hg), it has been detected in a variety of air pollutants; therefore, exposure to chrysene by the inhalation route is possible. Chrysene has also been detected in surface water and soil samples and in a variety of cooked foods (particularly charcoal broiled/smoked); therefore, exposure to chrysene by ingestion is also possible. Dermal exposure to chrysene can also occur as a result of skin contact with soot or petroleum products.

Toxicokinetics

In general, polynuclear aromatic hydrocarbons (PAHs; the generic class name for chrysene) are highly soluble in lipids and adipose tissue and would be expected to be readily absorbed by the dermal, oral, or inhalation routes of exposure. Although specific studies investigating the absorption characteristics of chrysene have not been reported, studies in animals exposed to PAHs using the dermal, intraperitoneal, and oral routes resulted in systemic toxicity, indicating that chrysene could be absorbed by several routes.

In vitro rat liver and mouse skin preparations have been reported to metabolize chrysene to its 1,2-, 3,4-, and 5,6-dihydrodiols and to some monohydroxy derivatives including 1- and 3-phenols. In addition, the dihydrodiols have been reported to undergo further transformation to form 1,2-diol-3,4-epoxide and 3,4-diol-1,2-epoxide.

Mechanism of Toxicity

The primary toxic effect of concern for chrysene is carcinogenicity, which is most likely the result of the mutagenic activity of its metabolites, 1,2-dihydrodiol and 1,2-diol-3,4-epoxide. The 1,2-dihydrodiol and the 1,2-diol-3,4-epoxide have been shown to be mutagenic *in vitro* in bacterial and mammalian cells and have induced pulmonary adenomas when administered to newborn mice. In addition, the 1,2-dihydrodiol was active as a tumor initiating agent on mouse skin. DNA adducts in hamster cells resulting from a reaction of the DNA with 1,2-diol-3,4-expoxide have also been detected.

Human Toxicity

Chrysene is classified as A3 (animal carcinogen) by the ACGIH based on its carcinogenic effects in other species. Specific reports of toxicity or carcinogenicity in humans resulting from exposure to chrysene were not found. As a general class of compounds, PAHs have low acute toxicity. Chronic exposure to PAHs can produce a variety of effects. Exposure to the eyes can result in irritation and photosensitivity. Dermal exposure can result in erythema, burns, and "coal tar warts" (precancerous lesions enhanced by ultraviolet light exposure). Inhalation exposure may cause irritation to the respiratory tract accompanied by cough and bronchitis. Oral exposure may produce a thickening and/or whitening of the oral mucous membranes. In addition to the local effects at the site of entry, systemic toxicity may occur, which could result in hepatic or renal effects. Some PAH compounds have been noted to cause hematological effects (anemia, leukopenia, and pancytopenia) in animals and suppression of selective components of the immune system. Oncogenicity is the primary effect of concern. Increased incidences of skin, bladder, and lung tumors and tumors of the gastrointestinal tract have been reported among workers exposed to PAHs.

Clinical Management

Since PAH compounds have been noted to cause hepatic, renal, and hematopoietic abnormalities, liver function tests, renal function tests, and a complete blood count are recommended for patients with sig-nificant exposure. For an oral overdose, gastric lavage is not recommended. An oral overdose should be treated with either a single oral dose of activated charcoal slurry or a charcoal slurry mixed with a saline cathartic. If tests indicate organ function abnormalities, supportive treatment should be undertaken. Treatment for cancer should follow standard therapeutic protocols for the type and location of the cancer.

Animal Toxicity

The LD$_{50}$ of chrysene in mice, administered by intraperitoneal injection, is >320 mg/kg. Application of chrysene (0.1% in a petroleum hydrocarbon mixture known to have low embryotoxicity) to the eggshell of Mallard duck embryos resulted in embryotoxic and teratogenic effects in the ducklings. A single oral dose of chrysene administered to pregnant rats on Day 19 of gestation induced hepatic P450 enzymes in the fetal rat liver. In another study, chrysene induced benzo(*a*)-pyrene hydroxylase activity in the placenta of pregnant female rats.

The primary toxic effect elicited by chrysene is oncogenicity. Several studies have been conducted in mice in which chrysene (diluted in a variety of agents) was applied dermally either as a single dose (followed by a tumor promoting agent) or as multiple doses. Increased incidences of dermal tumors (papillomas and carcinomas) were observed in mice administered chrysene and a tumor promoting agent. Several studies were also conducted in which mice or rats received intramuscular or subcutaneous injections of chrysene in various dilutants (single or multiple doses). Most of the treatment protocols resulted in an increased incidence of tumors at the injection site. In addition, male mice which were treated on Days 1, 8, and 15 after birth with an intraperitoneal injection of chrysene (1.4 μmol diluted in dimethyl sulfoxide) had a higher incidence of pulmonary and liver tumors and lymphosarcomas compared to similarly treated controls.

The potential for mutagenic effects was determined in bacterial, fungal, and mammalian cell systems *in vitro*. Although chrysene produced negative results in *Escherichia coli* and *Saccharomyces*, it produced positive results in *S. typhimurium* in TA100. In mammalian cells, chrysene produced positive effects in Syrian hamster embryo cells *in vitro*. Administration of chrysene to Chinese hamsters by intraperitoneal injection also produced increased sister chromatid exchanges in bone marrow cells. Increased aberrations were also noted in phase II oocytes collected from NMRI mice treated orally with chrysene.

—Linda A. Malley and David M. Krentz

Related Topic

Polynuclear Aromatic Hydrocarbons

Ciguatoxin

♦ SYNONYMS: Ichthyosarcotoxism; ciguatera poisoning

♦ DESCRIPTION: Ciguatoxin is a fat-soluble oxygenated polyether compound. This toxin is sometimes contained in fish found in the Caribbean, South Pacific, and Australia and can be imported anywhere.

♦ IMPLICATED SOURCES (FISH): Amberjack; barracuda; cinnamon; coral trout; dolphin; eel; emperor; spanish mackerel; surgeon fish; grouper; kingfish; paddletail; parrot fish; red snapper; reef cord; sea bass; swordfish; yankee whiting

Exposure Pathways

Ingestion of fish containing toxin is the route of exposure. There is no reliable taste, smell, or color in fish contaminated with ciguatoxin. The toxin is not destroyed or inactivated by heating, cooking, or freezing.

Toxicokinetics

Ciguatoxin is absorbed over 30 min to 30 hr (average of 6 hr) and is distributed into the flesh, muscle, skin, mucous, and internal organs. It can also pass into the breast milk.

Mechanism of Toxicity

The primary activity of the toxin relates to its ability to act on sodium channels and cause changes in the electrical potential and permeability of cells.

Human Toxicity

Initial symptoms begin 6–24 hr after ingestion of contaminated fish and include diarrhea, vomiting, and myalgias. Since there is a delay in the onset of symptoms, patients may uncharacteristically develop signs and symptoms of ciguatera poisoning after returning from vacation. Other symptoms include blurred vision, lacrimation, transient photophobia, hypotension, and sinus bradycardia. Numbness and tingling in the extremities, paradoxical reversal of temperature perception (hot and cold reversal), vertigo, and ataxia are strongly suggestive of the diagnosis of ciguatera intoxication. These symptoms may last up to 3 weeks. Hyperkalemia is also common. Coma, respiratory depression, and death are rare and only occur in severe cases.

Clinical Management

Basic and advanced life-support measures should be utilized as necessary. Treatment is generally symptomatic and supportive. Gastrointestinal decontamination procedures should be used where appropriate. For bradycardia, atropine at doses suitable for the age and body weight of the patient would be appropriate. Use of intravenous mannitol continues to be controversial. It has been shown to be effective in the acute phase of ciguatera poisoning for some patients.

—Gaylord P. Lopez

Cimetidine

♦ CAS: 51481-01-9

♦ SYNONYM: Tagamet

♦ PHARMACEUTICAL CLASS An H-2 receptor antagonist. The drug has an imidazole ring and is structurally similar to histamine.

♦ CHEMICAL STRUCTURE:

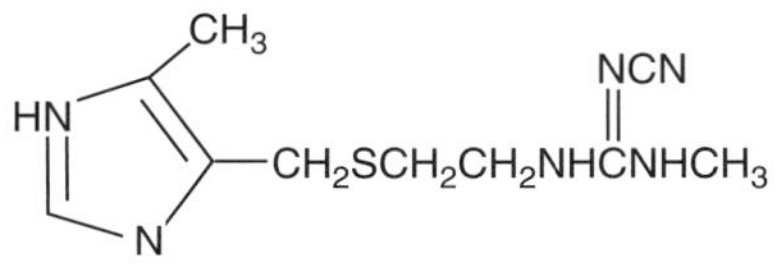

Uses

Cimetidine is indicated for the short-term and maintenance therapy of duodenal ulcers. The drug is also approved for the active treatment of benign gastric ulcers, gastroesophageal reflux disease, pathological hypersecretory conditions (i.e., Zollinger–Ellison syndrome), and prevention of upper gastroesophageal bleeding.

Exposure Pathways

Ingestion or injection are the routes of both accidental and intentional exposures to cimetidine.

Toxicokinetics

Cimetidine is rapidly and well absorbed with peak serum levels occurring 45–50 min following oral administration. Following oral administration, cimetidine is metabolized in the liver to sulfoxide and 5-hydroxymethyl derivatives. Administered intravenously, cimetidine is principally excreted unchanged in the urine. Cimetidine is distributed widely throughout the body and is 15–20% bound to plasma proteins. The principal route of excretion is the urine.

Mechanism of Toxicity

Blockade of the cardiac H-2 receptors is the postulated mechanism for the cardiovascular toxicity associated with cimetidine overdosage. Cimetidine penetrates the blood–brain barrier and has been associated with central nervous system effects.

Human Toxicity

Cimetidine overdosage is usually associated with very low morbidity and fatalities are rare. The most common symptoms reported following cimetidine overdosage are confusion, delirium with dizziness, drowsiness, slurred speech, sweating, and flushing. Cimetidine has been associated with bradycardia, hypotension, and sinus arrest. Sinus bradycardia is the most frequently reported cardiovascular effect. Other effects that have been reported following therapeutic use and overdosage of cimetidine include hematologic effects (i.e., granulocytopenia and thrombocytopenia), reversible renal failure, and endocrine dysfunction (i.e., gynecomastia). Liver enzyme elevations have been noted, probably due to an allergic reaction. Patients have survived ingestions of 12–24 g. Death has been reported in adults following acute ingestions of more than 40 g. Doses of up to 10 g have not been associated with any untoward effects, and doses up to 20 g were associated with only transient adverse effects. Severe nervous system effects have been reported following ingestion of 20–40 grams of cimetidine.

Clinical Management

Basic and advanced life-support measures should be utilized as necessary. Appropriate gastrointestinal decontamination procedures should be administered based on the history of the ingestion and the patient's level of consciousness. Electrocardiogram monitoring should be considered along with liver and renal function tests, and complete blood count.

Animal Toxicity

No data are available.

—*Carla M. Goetz*

Cisplatin

- ◆ CAS: 15663-27-1
- ◆ SYNONYMS: Biocisplatinum; CDDP; *cis*-DDP; *cis*-diaminodichloroplatinum (II); *cis*-diaminedichloroplatinum; *cis*-platin; *cis*-platinous diaminodichloride; *cis*-platinous diamine dichloride; *cis*-platinum; *cis*-platinum diaminodichloride; *cis*-platinum (II) diaminedichloride; cisplatyl; diaminedichloroplatinum; neoplatin; NSC-119875; platiblastin; platinol
- ◆ CHEMICAL STRUCTURE:

$$^-Cl-\underset{\underset{NH_3}{|}}{\overset{\overset{NH_3}{|}}{Pt}}-Cl^- \quad 2+$$

Uses

Cisplatin is primarily used for the treatment of a variety of malignancies.

Exposure Pathways

Cisplatin is only available for intravenous use. It is generally supplied in vials as a solution or as a lyophilized powder. Although instances are expected to be infrequent, the possibility exists for dermal, oral, and/ or inhalation exposure in the event of an accidental spill.

Toxicokinetics

Following an intravenous injection, several mammalian species demonstrate a similar general organ distribution. Cisplatin is rapidly distributed to all tissues, followed within the first hour by an accumulation in kidneys, liver, skin, bone, ovaries, and uterus. Approximately 60–80% is excreted in the urine within 24 hr. However, up to 4 weeks after a single dose, platinum is still detectable in kidneys, liver, skin, and lung. Following a single oral dose of cationic platinum, little absorption occurred and almost the entire dose was excreted in the feces, indicating that it would be unlikely for significant absorption to occur following oral administration of cisplatin.

The chloride atoms of the molecule may be displaced directly by reaction with neucleophiles such as thiol groups. However, hydrolysis of the chloride ion may also occur and may be responsible for formation of an active metabolite, which then reacts with nucleic acids and proteins. Investigation of metabolism of cisplatin in rat liver and kidneys following an intraperitoneal dose was conducted over a 24-hr period. Maximum platinum concentrations in kidney cortex and medulla were reached within 1 hr after dosing. In addition, the parent compound and five platinum-containing metabolites were present at 1 hr postdosing, with cisplatin being the primary species detected. Similarly, there were five platinum-containing metabolites detected in the liver at 1 hr postdosing; however, in contrast to the kidneys, cisplatin was not the primary species detected in the liver.

Mechanism of Toxicity

Cisplatin reacts with nucleosides and nucleic acids and can cross-link cellular DNA. The effects on cross-linking with DNA appear to differ among cell type; however, the effects on cross-linking are most pronounced during the S phase of the cell cycle. In addition, cisplatin inhibits a number of enzymes that contain a catalytically active sulphydryl group. Ribonucleotide reductase is extremely sensitive to the effects of cisplatin, with greater than 90% inhibition observed *in vitro* in the presence of a two-molar excess of cisplatin. The inhibition was nearly instantaneous and was irreversible.

Human Toxicity

Renal dysfunction is the major toxic effect of this drug. It reduces the single-nephron glomerular filtration rate and causes a backleak of inulin across the renal tubule. In addition, myelosuppression, peripheral neuropathy (paraesthesias), central extrapyramidal disorders, loss of deep tendon reflexes, metabolic acidosis, headaches, taste disturbance, retrobulbar neuritis, seizures, ototoxicity, nausea, vomiting, diarrhea, thirst, metallic taste, leukopenia, azotemia, hypokalemia, hypophosphatemia, hypocalcemia, and hypomagnesemia are commonly reported side affects of treatment with cisplatin. Some patients have also experienced anaphylactic reactions to treatment with platinum-containing compounds.

Clinical Management

Nephrotoxicity may be prevented or diminished by prehydration with 2 liters of normal saline administered over a 6- to 8-hr period, followed by continued hydration during and after the cisplatin infusion (total amount of saline administered should be 3–6 liters). Nausea and vomiting may be managed with antiemetics. Electrolyte concentration should be monitored and supplemented as needed. Treatment for an anaphylactic reaction would include antihistamines, administered with or without epinephrine. If accidental exposure to the eyes or skin occurs, the affected area should be washed thoroughly with soap and water, and eyes should be flushed with copious amounts of tepid water for at least 15 min. Seizures should be treated with diazepam (intravenous) or lorazepan. If seizures are uncontrollable or recur, phenobarbital and/or phenytoin should be used.

Animal Toxicity

The LD_{50} in mice following a single intraperitoneal injection was 13.0 mg/kg and following a single intravenous injection was 13.36 mg/kg for males and 12.32 mg/kg for females. The LD_{50} in male rats following a single intraperitoneal injection was 12.0 mg/kg. The toxic effects observed in rats following this treatment included leukopenia, decreased numbers of circulating platelets, lymphoid depletion, intestinal epithelial injury, bone marrow depression, and sloughing of the renal tubular epithelium. The minimum lethal dose for dogs was a single intravenous injection of 2.5 mg/kg or five daily consecutive injections of 0.75 mg/kg. Toxic symptoms in dogs included severe hemorrhagic enterocolitis, severe damage to the bone marrow and lymphoid tissue, and marked renal necrosis. Occasionally, pancreatitis was also observed in dogs. In monkeys, the minimum lethal dose was five daily intravenous injections of 2.5 mg/kg. Toxic effects observed in monkeys included nephrosis, myocarditis, and some degeneration of spermatogenic cells.

Cisplatin (13 mg/kg) administered to pregnant mice by intraperitoneal injection on Gestation Day 8 was lethal to all the fetuses. A dose of 8 mg/kg cisplatin was lethal to 98% of the fetuses, and a dose of 3 mg/kg was lethal to 31% of the fetuses. Surviving fetuses exhibited growth retardation and minor skeletal anomalies.

Female rats were administered twice weekly intraperitoneal injections of cisplatin for a cumulative dose of 15 or 34 mg/kg. Following the last dose, sensory and motor nerve conduction velocity were determined. Both doses of cisplatin significantly decreased sensory nerve conduction velocity. In addition, the level of cisplatin DNA binding in dorsal root spinal ganglion satellite cells equaled that in liver cells; however, the level of cisplatin DNA binding in spinal cord and brain was very low.

Rats were administered a single intravenous dose of 6 mg/kg cisplatin, which was either preceded (30 min prior) or was followed by (30 or 60 min) 500 mg/kg of reduced glutathione. The reduced glutathione, administered 30 min before or 30 min after the cisplatin, offered significant protection from toxicity and did not interfere with the antitumor effectiveness of cisplatin in a tumor model. However, in mice, the protective effect of reduced glutathione was only partial for some strains.

Although cisplatin is an effective chemotheraputic agent for several types of malignancies, studies in animals indicate that cisplatin can also increase the occurrence of tumors. Cisplatin was administered once weekly by intraperitoneal injection to mice for 10 weeks for a total dose of 108 (μmole/kg). Cisplatin-treated mice had a significantly higher incidence (100%) of pulmonary adenomas compared to similarly treated control mice (26%). In another study, mice received weekly intraperitoneal injections of 1.62 mg cisplatin in 5 ml/kg saline for 16 weeks, cisplatin followed by dermal application of croton oil, dermal application of croton oil alone (control), or saline alone (control). Mice treated with croton oil and cisplatin had a higher incidence of skin papillomas compared to cisplatin alone or the control groups.

Cisplatin was mutagenic in *Salmonella typhimurium* strains *his* G46/pkM101, TA100, and TA98, without metabolic activation. Cisplatin also induced increased mutations in Chinese hamster ovary cells and V79 Chinese hamster cells. Postreplication repair was induced in V79 cells and in HeLa cells; and sister chromatid exchanges were induced in V79 cells. Chromosomal damage and sister chromatid exchanges were also induced by cisplatin in human lymphocyte cultures. Similar to the *in vitro* data, intraperitoneal injection of mice with 13.85 mg/kg cisplatin induced a significant increase in sister chromatid exchanges and in chromosome aberrations.

—*Linda A. Malley and David M. Krentz*

Related Topics

Kidney
Sister Chromatid Exchanges

Clean Air Act

- ◆ TITLE: CAA
- ◆ AGENCY: U.S. EPA
- ◆ YEAR PASSED: 1970; amended 1974, 1977, 1978, 1980, 1981, 1982, 1983, 1990, and 1997
- ◆ GROUPS REGULATED: Local governments

Synopsis of Law

Section 112 of CAA provides a list of 189 hazardous air pollutants, to which U.S. EPA may add or delete pollutants. U.S. EPA must establish national emissions standards for sources that emit any listed pollutant. The original 1970 version of Section 112 required the standards of provide "an ample margin of safety to protect the public heath from such hazardous air pollutants." The implication of this language, that standards were to be set without regard to the costs of emissions control, generated interim debate from the beginning and contributed to the U.S. EPA's glacial pace of implementation.

U.S. EPA finally attempted to escape the strict interpretation of the CAA when it issued a standard for vinyl chloride in 1986. In this instance the agency claimed it could consider costs and declined to adopt a standard that would ensure safety. A court set aside this standard because U.S. EPA had improperly considered costs, but it did make clear that for determining what emissions level was safe, even for a carcinogen, U.S. EPA need not eliminate exposure. Additionally, the decision said U.S. EPA could consider costs in deciding what, if any, additional margin of protection to prescribe (*Natural Resources Defense Council, Inc. vs United States Environmental Protection Agency*, 1987).

The 1990 amendments to the CCA responded to the difficulties presented by the strict approach of Section 112. The amendments replace the health-based standard with a two-tiered system of regulation. U.S. EPA must first issue standards that are technology based, designed to require the "maximum degree of emission reduction achievable" (maximum achievable control technology; MACT) [CAA Section 112(d)(2)]. If the MACT controls are insufficient to protect human health with an "ample margin of safety," U.S. EPA must issue residual risk standards [CAA Section 112(f)]. The 1990 amendments essentially define ample margin of safety for carcinogens by requiring U.S. EPA to establish residual risk standards for any pollutant that poses a lifetime excess cancer risk of greater than 1 in 1 million.

The 1997 amendments to the Clean Air Act tightened the controls on ozone levels (to 0.08 ppm) and added strict limits on airborne particulates, particularly fin particulates. New particulate matter standards regulate particles 2.5 μm or smaller in diameter with an annual limit of 15 μg/m^3.

—*Shayne C. Gad*

Related Topics

Environmental Toxicology
Pollution, Air
Toxicity Testing, Inhalation

Clean Water Act

- ◆ TITLE: CWA
- ◆ AGENCY: U.S. EPA
- ◆ YEARS PASSED: 1972; amended 1977–1983, 1987, 1988, 1990, 1992; reauthorized in 1997; originally the Federal Water Control Act
- ◆ GROUPS REGULATED: Industry

Synopsis of Law

U.S. EPA has had responsibility for regulating toxic water pollutants since 1972. As originally enacted, Section 307 of the Federal Water Pollution Control Act required U.S. EPA to publish within 90 days and periodically add to a list of toxic pollutants for which effluent standards (discharge limits) would then be established.

Section 307(a)(4) of the Federal Water Pollution Control Act originally specified that in establishing standards for any listed pollutant U.S. EPA was to provide an "ample margin of safety," a difficult criterion to meet for most toxic pollutants and arguably impossible for any known to be carcinogenic. The law also mandated both a rapid timetable and a complex procedure for standard setting.

U.S. EPA's slow implementation of these instructions precipitated a series of lawsuits. The agency eventually reached a court-sanctioned settlement that fundamentally altered federal policy toward toxic pollutants of the nation's waterways. The settlement allowed U.S. EPA to act under other provisions of the act that allow consideration of economic costs and technological feasibility in setting limits. Congress incorporated the terms of this settlement in 1977 amendments to the statute.

In 1987, Congress again amended the Federal Water Pollution Control Act to toughen standards for toxic pollutants. Under the 1977 amendments, U.S. EPA had developed health-based "water quality criteria" for 126 compounds it had identified as toxic. These criteria described desirable maximum contamination levels which, because U.S. EPA's discharge limits were technology based, generally were substantially lower than the levels actually achieved. The 1987 amendments gave these advisory criteria real bite by requiring that states incorporate them in their own mandatory standards for water quality and impose on operations discharging into below-standard waterways additional effluent limits.

—*Shayne C. Gad*

Related Topics

Effluent Biomonitoring
Environmental Toxicology
Pollution, Water
Safe Drinking Water Act
Toxicity Testing, Aquatic

Clofibrate

- ♦ CAS: 637-07-0
- ♦ SYNONYMS: 2-(p-chlorphenoxy)-2-methylrpopionic acid ethyl ester; α-p-chlorophenoxyisobutyryl ethyl ester; Amotril; Angiokapsul; Anparton; Antilipid; Ateculon; Ateriosan; Atheropront; Atromid; Atromidin; Hyclorate; Lipavil; Liponorm; Liporil; Lipofaction; Neo-Atromid; Normet; Regelan; Serotinex
- ♦ CHEMICAL CLASS: Antihyperlipoproteinemic agent
- ♦ CHEMICAL STRUCTURE:

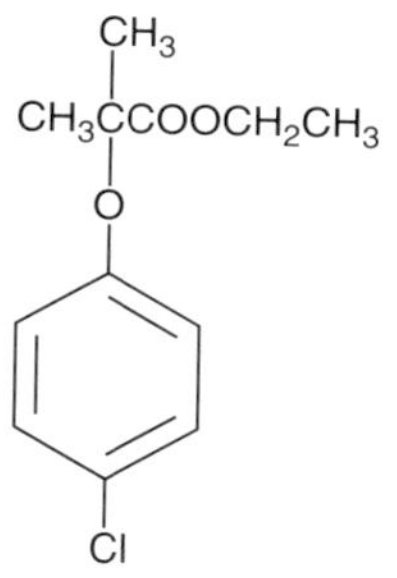

Uses

Clofibrate is a drug used to lower plasma concentration of very low density lipoprotein and also to lower plasma triglycerides concentration.

Exposure Pathways

Accidental overdose or ingestion is the most common exposure pathway.

Toxicokinetics

Clofibrate is rapidly and completely absorbed after oral administration. It is hydrolyzed to clofibric acid during absorption and in its pass through the liver. The acid binds strongly to plasma proteins. Sixty percent of it is excreted as glucuronide conjugate in the urine. Some is secreted into the bile and reabsorbed. The plasma elimination half-life is 12–25 hr. Clofibrate appears in plasma as *p*-chlorophenoxyisobutyric acid. An acyl-linked metabolite of clofibrate has been identified in human urine.

Mechanism of Toxicity

Clofibrate acyl glucuronide is an electrophilic metabolite that reacts with sulfhydryl groups and causes hepatotoxicity.

Human Toxicity: Acute

Nausea, diarrhea, skin rash, alopecia, weakness, flu-like syndrome, and severe muscle cramps are symptoms of acute toxicity.

Human Toxicity: Chronic

Chronic administration increases the incidence of cholesterolic gallstones twofold. It also causes a small increase in thromboembolic phenomenon, pulmonary embolism, intermittent claudication, and angina pectoris. The drug may increase the incidence of bowel cancer.

Clinical Management
Exposure should be terminated as soon as possible.

Animal Toxicity
Clofibrate causes hepatic tumors in rodents.

—Sanjay Chanda

Clonidine

♦ CAS: 4205-90-7

♦ SYNONYMS: Clonidine hydrochloride (CAS: 4205-90-8); 2-(2,6-dichloroanilino)-2-imidazoline hydrochloride; Catapres

♦ PHARMACEUTICAL CLASS: Imidazoline-derivative hypotensive agent; a centrally acting α2-adrenergic receptor agonist

♦ CHEMICAL STRUCTURE:

Uses
Clonidine is used in the management of hypertension or opioid withdrawal and as an aid in the diagnosis of pheochromocytoma in hypertensive patients with suggestive symptoms and borderline catecholamine values.

Exposure Pathways
Ingestion is the most common route of both accidental and intentional exposure to clonidine. Toxicity may also occur via dermal exposure. It is available in oral and transdermal dosage forms.

Toxicokinetics
Clonidine is well absorbed (75–90%) orally with peak plasma concentrations occurring in 3–5 hr. Bioavailability may decrease during chronic oral therapy. Following transdermal application, therapeutic plasma concentrations are reached within 2 or 3 days. Clonidine is metabolized by hydroxylation of the phenol ring and cleavage of the imidazole ring to four inactive metabolites. The volume of distribution is 2.1–4 liters/kg. Clonidine is distributed into the cerebral spinal fluid and into breast milk; it is not known whether the drug crosses the placenta. Approximately 65% is excreted by the kidneys and 32% as unchanged drug. The half-life is 7.5–10.8 hr (16–40 hr in patients with renal insufficiency).

Mechanism of Toxicity
Clonidine stimulates central presynaptic α2-adrenergic neuron receptors in the medulla oblongata and the locus ceruleus; this stimulation inhibits sympathetic cardiovascular discharge resulting in decreased cardiac output and heart rate and a decreased peripheral vascular resistance. At high doses, clonidine can stimulate peripheral α-adrenergic receptors resulting in vasoconstriction. Clonidine depresses the central nervous system by reducing noradrenergic activity; dopamine and dopamine metabolites increase. Clonidine can induce the release of an endogenous opioid, making it useful in treating opioid withdrawal; the hypotensive effect of this action has not been confirmed.

Human Toxicity: Acute
As little as 0.1 mg of clonidine has produced toxicity in children; adults have survived ingestions of up to 18.8 mg. Clonidine levels are not clinically useful. Features of toxicity include hypotension, bradycardia, respiratory depression, apnea, lethargy, and coma. Miosis and hypothermia and cardiac arrhythmias may be noted. Hypotension may be preceded by transient hypertension in the early stages. Symptoms generally begin within 30 min to 4 hr postingestion and resolve within 24–48 hr.

Human Toxicity: Chronic
Side effects of clonidine therapy include dry mouth, drowsiness, sedation, and constipation. Abrupt discontinuation of therapy may result in severe rebound hypertension. From 5 to 30% of patients wearing the transdermal patch will develop an allergic contact dermatitis. Since over 60% of the clonidine may be retained in the patch, even the ingestion of used patches may be associated with serious toxicity.

Clinical Management

Induction of emesis is not recommended due to the potential for rapid onset of lethargy and coma. Clonidine is adsorbed by activated charcoal. Respiratory depression, hypotension, and coma may respond to naloxone (initial dose, 0.4–2 mg intravenously repeated as necessary). Standard therapies for hypotension and cardiac arrhythmias should be utilized as clinically indicated. Paradoxical hypertension is transient and usually does not require treatment. Therapy for rebound hypertension, which follows abrupt discontinuation of clonidine therapy, includes reinstitution of clonidine or the combined administration of an α- and β-adrenergic blocking agent (e.g., phentolamine or prazosin with a beta blocker such as propranolol). Whole bowel irrigation may be useful if a transdermal patch is ingested.

Animal Toxicity

Maternal and fetal blood pressure were unaffected and the heart rate decreased in sheep injected with 300 μg of clonidine.

—*Elizabeth J. Scharman*

Related Topic

Cardiovascular System

Clostridium perfringens

- ◆ DESCRIPTION: *Clostridium perfringens* is an anaerobic, gram-positive, spore-forming bacillus. There are 12 toxins (α, β, ε, ι, δ, θ, κ, λ, μ, ν, η, and γ) produced, which are divided into five types (A–E).

Exposure Pathways

Ingestion of contaminated food is the most common route of exposure.

Mechanism of Toxicity

When there is a delay in food preparation and serving, *C. perfringens* proliferates and produces heat-stable spores. These spores release toxins in the gastrointestinal tract. The toxicity is produced by a toxin and is not an invasive infectious process. The incubation period of *C. perfringens* food poisoning is approximately 12 hr (ranging from 6 to 24 hr). The duration is approximately 24 hr (ranging from 12 to 48 hr). This organism commonly contaminates meats, meat products, gravy, and poultry. *Clostridium perfringens* type A is the most common subtype implicated in food poisoning.

Human Toxicity

Clostridium perfringens toxins produce diarrhea and abdominal cramping. The enterotoxin causes fluid, sodium, and chloride secretion into the lumen of the small intestines resulting in noninflammatory diarrheal disease. Vomiting and fever are uncommon.

Clinical Management

Clostridium perfringens-related food poisoning is a mild, self-limiting disease. Symptomatic and supportive therapy, including fluid and electrolyte replacement with careful fluid monitoring, is normally adequate intervention. The toxicity is produced by a toxin and is not an invasive infectious process; therefore, antibiotics play no role in the management of this type of food poisoning. Emetics and cathartics are not indicated since the toxin is eliminated by vomiting and diarrhea. Diagnosis is based on culture of food source.

—*Vittoria Werth*

Cobalt (Co)

- ◆ CAS: 7440-48-4
- ◆ SELECTED COMPOUNDS: Cobaltous chloride, $CoCl_2$ (CAS: 7646-79-9); cobaltous sulfide, CoS (CAS: 1317-42-6)
- ◆ CHEMICAL CLASS: Metals

Uses

Cobalt is a relatively rare metal produced primarily as a by-product of other metals, chiefly copper. It is the essential trace element found in cyanocobaltamine (vitamin B_{12}). This vitamin protects against pernicious anemia and is required in the production of red blood cells. Medicinally, cobalt salts have been used to stimulate the formation of red blood cells in individuals suffering from anemia.

Commercially, cobalt is used primarily in high-temperature alloys, in tungsten carbide tools, and (with iron and nickel) in permanent magnets. Cobalt salts are used in pigments, in paint dryers, and as catalysts in the petroleum industry.

Exposure Pathways

For the general population, ingestion is the primary exposure pathway for cobalt. For persons working in industrial settings, inhalation is a significant pathway (e.g., carbide industry emissions and airborne particulate from grinding processes) as is dermal exposure.

Toxicokinetics

Cobalt is present in small amounts in the earth's crust. Oral ingestion of cobalt salts results in ready absorption probably in the jejunum. Although cobalt is readily absorbed, increased levels do not tend to cause significant accumulation. Approximately 80% of absorbed cobalt is excreted via the urine and 15% is excreted in the feces by an enterohepatic pathway. Milk and sweat are secondary routes of excretion. The total body burden for the average person is estimated as 1.1 mg. Muscle contains the greatest mass of cobalt but the highest concentrations are found in fat. Cobalt present in the blood is associated with the red blood cells.

Mechanism of Toxicity

Cobalt most often depresses enzyme activities like catalase and amino levulinic acid synthetase. Many enzymes are depressed in the presence of cobalt (e.g., P450, enzymes involved in cellular respiration). Liver enzymes can be enhanced. Cobalt somehow interferes with and depresses iodine metabolism resulting in reduced thyroid activity. Reduced thyroid activity can lead to goiter. The Krebs citric acid cycle is blocked; thus, cellular energy production is inhibited. Cobalt can replace zinc in a number of zinc-required enzymes like alcohol dehydrogenase. Cobalt can also enhance the kinetics of some enzymes such as heme oxidase in the liver.

Human Toxicity

In industrial settings, inhalation of high concentrations of cobalt compounds has led to hard-metal pneumoconiosis, which may result in interstitial fibrosis. Workers with this condition typically develop hypersensitivity to cobalt compounds (symptoms include coughing and wheezing). A few workers have developed skin hypersensitivity after dermal contact with cobalt and its compounds.

Ingestion of cobalt may result in the production of an unusually high number of red blood cells [similar to a cancer of red blood cells (polycythemiavera)]. Finally, ingestion of cobalt salts (once added to beer as a defoaming agent) has resulted in cardiomyopathy. The signs and symptoms of cardiomyopathy due to beer consumption are similar to those of congestive heart failure. Autopsy results indicated a 10-fold increase in cobalt concentrations in heart tissue. The alcohol may have potentiated cobalt absorption or toxic effects.

The ACGIH TLV-TWA for cobalt (elemental and inorganic compounds) is 0.02 mg/m^3. ACGIH classifies cobalt as an animal carcinogen.

Clinical Management

The oil-soluble BAL (British Anti-Lewisite; 2,3-dimercaptopropanol) appears to be the antidote of choice for cobalt poisoning.

Animal Toxicity

There are significant species differences in the excretion of radiocobalt. The majority (80%) of cobalt is excreted in the feces of rats and cattle in contrast to 80% excretion via urine in humans. Some fish and mollusks bioconcentrate cobalt. When implanted intramuscularly in rats, cobalt metal produced fibrosarcomas at the site but no other routes of exposure have elicited a carcinogenic response.

—Arthur Furst, Shirley B. Radding,
and Kathryn A. Wurzel

Related Topic

Metals

Cocaine

♦ CAS: 50-36-2

♦ SYNONYMS: Ecgonine methyl ester benzoate; benzoylmethylecgonine; [1*R*-(exo,exo)]-3-(benzoyloxy)-8-methyl-8-azabicyclo[3,2,1]octane-2-carboxylic acid methyl ester; crack; rock; toot; blow; snow; dama blanca; coke; lady

♦ PHARMACEUTICAL CLASS: A naturally occurring alkaloid with local anesthetic and vasoconstrictor properties and central nervous system (CNS) stimulant and euphoric effects

♦ CHEMICAL STRUCTURE:

$$CH_3N \quad \begin{array}{c} COOCH_3 \\ OOC \end{array}$$

Uses

Cocaine is used for topical local anesthesia of mucous membranes. It is also a drug of abuse.

Exposure Pathways

Smoking the base alkaloid (as crack or freebase) and nasal insufflation of the hydrochloride salt are the most common routes of exposure in abuse. Intravenous injection and application of the hydrochloride salt to mucous membranes are also methods of abuse. In therapeutic use, a solution of hydrochloride salt is applied to mucous membranes.

Toxicokinetics

Cocaine is rapidly absorbed, in a few seconds, from the lungs as a gas, which results from volatilization of the alkaloid. Peak plasma concentrations occur within a few minutes. Absorption through mucous membranes is initially rapid, then slowed secondary to the vasoconstrictive effects of cocaine. Peak plasma concentrations occur within 1 hr after oral ingestion and nasal application. After oral administration, bioavailability is decreased secondary to presystemic hydrolysis in the gastrointestinal tract.

Cocaine is metabolized by hydrolysis to benzoylecgonine, by cholinesterase to ecgonine methyl ester, and hepatically to norcocaine and ecgonine. In the presence of ethanol, cocaine is also metabolized to cocaethylene. Cocaine metabolism is probably dose dependent at the high doses which are abused, especially with binge use. Cocaine is widely distributed in the body with an apparent volume of distribution of 1.2–1.9 liters/kg. It rapidly appears in the CNS and crosses the placenta. The elimination half-life of cocaine is approximately 1 hr at doses of less than 2 mg/kg. The measurement of low concentrations of cocaine in chronic users suggests half-lives as long as 3 days. Low doses of cocaine are excreted in the urine primarily as metabolites, with less than 10% of the dose found as unchanged cocaine.

Mechanism of Toxicity

Cocaine toxicity is primarily secondary to its ability to prevent the reuptake of neurotransmitters including serotonin, dopamine, and norepinephrine. Toxicity secondary to its local anesthetic activity is due to inhibition of sodium influx across the nerve cell membrane. The role of cocaine metabolites in producing clinical toxicity is unclear. Cocaethylene, ecgonine methyl ester, and benzoylecgonine may produce some toxicity. Additional effects of cocaine which are important in clinical toxicity include direct vasoconstriction, increased cellular oxygen consumption, increased platelet aggregation, and direct organ toxicity.

Human Toxicity: Acute

Cocaine toxicity can present after a wide range of doses, with reports of toxicity at doses of less than 1 mg/kg. Toxicity includes most organ systems, and cocaine use and culture increases the risk of trauma and infections. Acute tolerance to many CNS and cardiovascular effects of cocaine develops. Kindling, the lowering of the seizure threshold with repeated subtoxic doses, can also occur. The CNS toxicity of cocaine includes stimulation, euphoria, agitation, seizures, intracranial hemorrhage, and, with larger doses, coma. Cardiovascular toxicity can include tachycardia, hypertension, coronary artery spasm, myocardial ischemia and infarction, bradycardia, hypotension, cardiovascular collapse, dysrhythmias, and sudden death. Pulmonary toxicity after smoking the alkaloid form of cocaine includes hemorrhage, barotrauma including pneumomediastinum, pulmonary edema, and "crack lung," a hypersensitivity reaction that includes fever, productive cough, pulmonary infiltrates, and bronchospasm. Other toxicity seen with cocaine use includes hyperpyrexia, rhabdomyolysis, metabolic acidosis, and respiratory alkalosis. Cocaine use during pregnancy can result in an increased risk of abruptio placentae, spontaneous abortion, and low-birth-weight infants with congenital malformations and potentially neurobehavioral impairment. Nasal insufflation of single doses of cocaine results in plasma concentrations of 100–500 ng/ml. Blood cocaine concentrations in fatalities are described as averaging approximately 6 mg/liter, with a very wide range of reported concentrations of cocaine and metabolites.

Human Toxicitiy: Chronic

Toxicity associated with chronic use is not as well described as acute toxicity, but it appears to include cerebral atrophy, cardiomyopathy, and chronic pulmonary disease. Cocaine and its metabolites are most commonly identified in patient urine. An immunoassay directed toward identification of benzoylecgonine will frequently indicate the presence of cocaine and its metabolites for many days after use. The duration of qualitatively detected cocaine and metabolites in urine is probably dose dependent and may be up to 3 weeks in length. Chronic use of cocaine may lead to dependence.

Clinical Management

The initial management of acute cocaine intoxication should include assessment and management of the patient's airway, breathing, and circulation. Supplemental oxygen and a benzodiazepine are frequently indicated for agitation and CNS stimulation. Many findings, such as hyperpyrexia, seizures, and rhabdomyolysis, should be managed using the basic treatment approaches for the complication. Concurrent use of alcohol and other drugs is frequent and should be considered during the assessment. Treatment of cardiac toxicity should also

include supportive care, oxygen, and a benzodiazepine. Beta blockers should not be used because the unopposed alpha stimulation has been associated with hypertension and additional coronary artery constriction. Nitrates, calcium channel blockers, opiates, and thrombolytics should be used when appropriate based on the clinical and laboratory findings.

Animal Toxicity

Animal models have demonstrated acute toxicity similar to that present in humans. Dogs develop toxicity at lower doses than rats, and death appears to be associated with the development of hyperthermia. The functional status of organ enervation and the presence of anesthetics may alter cocaine toxicity in animal models.

Further Reading

Baselt, R. C., and Cravey, R. H. (1989). *Disposition of Toxic Drugs and Chemicals in Man*, 3rd ed., pp. 208–213. Yearbook Medical Publishers, Chicago.
Goldfrank, L. R., and Hoffman, R. S. (1991). The cardiovascular effects of cocaine. *Ann. Emerg. Med.* 20, 165–175.
Leikin, J. B., Krantz, A. J., Zell-Kanter, M., Barkin, R. L., and Hryhorczuk, D. O. (1989). Clinical features and management of intoxication due to hallucinogenic drugs. *Med. Toxicol. Adverse Drug Exp.* 4, 324–350.

—William A. Watson

Related Topic

Neurotoxicology: Central and Peripheral

Codeine

♦ CAS: 76-53-3

♦ PHARMACEUTICAL CLASS: Opioid agonist

♦ CHEMICAL STRUCTURE:

Uses

Codeine is used as an analgesic and an antitussive.

Exposure Pathways

Codeine is customarily ingested in the form of tablets and liquid pharmaceutical preparations.

Toxicokinetics

Codeine is well absorbed via oral and intramuscular routes of administration; it is two-thirds as effective orally as parenterally. Peak serum levels are attained in 30–60 min. Codeine is metabolized in the liver by o-demethylation and N-demethylation and partial conjugation with glucuronic acid. The volume of distribution is 3.5 liters/kg. Greater than 95% of a single dose is eliminated in 48 hr by the kidneys. The elimination half-life is 1.9–4 hr. Codeine metabolites are conjugated codeine, norcodeine, conjugated norcodeine, conjugated morphine, and hydrocodone.

Mechanism of Toxicity

Codeine acts to depress the central nervous system (CNS), thereby producing coma and cessation of respiration.

Human Toxicity: Acute

Depressant effects on the CNS are the most profound. Nausea, vomiting, and miosis may develop within 1 hr. Infants and children may demonstrate unusual sensitivity while habituated adults may have extreme tolerance to opioids. In children, greater than 1 mg/kg may produce serious symptoms and 5 mg/kg may cause a respiratory arrest. The estimated lethal adult dose of codeine is 7–14 mg/kg.

Human Toxicity: Chronic

Codeine is often subject to abuse and can cause a withdrawal syndrome after abrupt discontinuation of use.

Clinical Management

Basic and advanced life-support measures should be performed as necessary. Gastrointestinal decontamination procedures should be performed as appropriate to the patient's level of consciousness. Activated charcoal will adsorb codeine. Patients with respiratory or CNS depression can be treated with intravenous boluses of naloxone. A continuous infusion may be necessary since the toxic effects of codeine may persist longer than the duration of action of naloxone.

Animal Toxicity

Opioids have an excitatory effect on the CNS of cats and horses. Dogs will experience similar CNS depressant effects as humans. Death can occur within 12 hr. Naloxone can be used in animals as well.

—*Linda Hart*

Related Topic

Neurotoxicology: Central and Peripheral

Coke Oven Emissions

- ◆ CAS: 8007-45-2

- ◆ Synonyms: Primary synonym is coal tar pitch volatiles as benzene soluble organics. However, synonyms vary depending on the specific constituents in the emissions; RTECS No. GH0346000

- ◆ Description: Petroleum coke is a chunky powdered carbon product derived from petroleum. If petroleum coke is heated to a high temperature, it may emit volatiles such as polynuclear aromatic hydrocarbons, which could be suspect carcinogens. Such exposures can occur in coke oven workers.

The production of coke by the carbonization of bituminous coal leads to the release of chemically complex emissions from coke ovens that include both gases and particulate matter of varying chemical composition. The chemical and physical properties of coke oven emissions vary depending on the constituents. The emissions include coal tar pitch volatiles (e.g., particulate polycyclic organic matter, polycyclic aromatic hydrocarbons, and polynuclear aromatic hydrocarbons), aromatic compounds (e.g., benzene and β-naphthyl amine), trace metals (e.g., arsenic, beryllium, cadmium, chromium, lead, and nickel), and gases (e.g., nitric oxides and sulfur dioxide).

Exposure Pathways

Inhalation, dermal contact, or ocular contact are all possible routes of exposure.

Human Toxicity

Acute exposure to coke oven emissions produces irritation of the eyes, respiratory symptoms like cough, dyspnea, and wheezing.

The emissions are investigated as a carcinogen, tumorigen, and mutagen. The cancer sites include the skin, respiratory system, kidneys, and urinary bladder. Studies of coke oven workers have shown increased risk of mortality from cancer of the lung, trachea, and bronchus; cancer of the kidneys; cancer of the prostrate; and cancer at all sites combined.

Occupation-related exposure is associated with significant excess mortality from cancer of the respiratory system and of the prostate. Depending on the segment of the population considered, the respiratory cancer risk for coke oven workers was as high as 4.5 times the risk for non-oven workers. To evaluate a biologically effective exposure dose in human biomonitoring studies, DNA carcinogen adduct analysis is frequently used.

OSHA has not identified thresholds for carcinogens that will protect 100% of the population. It usually recommends that occupational exposures to carcinogens be limited to the lowest detectable concentration. To ensure maximum protection from carcinogens through the use of respiratory protection, only the most reliable and protective respirators are recommended. The OSHA PEL for benzene-soluble fraction of coke oven emissions is 0.150 mg/m^3.

Clinical Management

The exposed person should be moved to fresh air at once. If breathing has stopped, mouth-to-mouth resuscitation should be performed. The affected person should be kept warm and at rest. Exposed eyes should be washed immediately with large amounts of water; the lower and upper lids should be lifted occasionally. Medical attention should be obtained immediately.

Animal Toxicity

In animals, extracts and condensates of coke oven emissions were found to be carcinogenic in both inhalation studies and skin painting bioassays. The mutagenicity of whole extracts and condensates, as well as their individual components, provides supportive evidence for carcinogenicity. Also, several inhalation exposure studies in laboratory animals have provided evidence of the carcinogenic effect of aerosols of coal tar and its fractions. The extract was found to produce papillomas and skin carcinomas in the mice and acted as an initiating agent, although the extent to which this extract is representative of coke oven emissions is uncertain since the sample was contaminated with particulate matter from ambient air. Numerous carcinogenicity studies have shown that coal tar samples applied topically to the skin of laboratory animals produce local tumors.

—*Shashi Kumar Ramaiah
and Harihara M. Mehendale*

Related Topics

Combustion Toxicology
Pollution, Air
Polycyclic Aromatic Hydrocarbons

Colchicine

- ◆ CAS: 64-86-8
- ◆ DESCRIPTION: Colchicine is obtained from the autumn crocus, *Colchicum autumnale*, or the glory lily, *Gloriosa superba*.
- ◆ PHARMACEUTICAL CLASS: Naturally occurring alkaloid

♦ CHEMICAL STRUCTURE:

Uses

Colchicine is used in the treatment of acute gouty arthritis. Unlabeled uses include treatment of familial Mediterranean fever, neoplasms of the skin, and cirrhosis of the liver.

Exposure Pathways

Ingestion is the most common route of both accidental and intentional exposure to colchicine. It is available as an oral tablet and solution for injection.

Toxicokinetics

Colchicine is readily absorbed from the gastrointestinal tract. In therapeutic dosing, peak serum levels occur in 30–120 min. Colchicine undergoes deacetylation and hydrolysis in the liver. It has a rapid initial distribution phase, with a plasma half-life of 19 or 20 min, suggesting swift uptake by the tissues. The volume of distribution is 2.2 liters/kg. Up to 40% of colchicine is excreted in the urine, with 20–30% of this as unchanged drug. The majority of the drug undergoes enterohepatic recirculation and is excreted via bile and feces. The average elimination half-life is 20 hr.

Mechanism of Toxicity

Colchicine binds to tubulin and prevents its polymerization into microtubules, subsequently disrupting microtubule function. Consequently, it alters nuclear structure, intracellular transport, and cytoplasmic motility, ultimately causing cell death. Colchicine is a potent inhibitor of cellular mitosis.

Human Toxicity

Colchicine toxicity has been divided into three stages. The first stage, from 2 to 24 hr, is the gastrointestinal phase, notable for abdominal pain, vomiting, diarrhea, and a prominent leukocytosis. The gastrointestinal symptoms may be relieved by atropine, but this does not prevent or alter the onset of the second stage. The second stage is marked by multisystem failure. Most life-threatening symptoms occur 24–72 hr postexposure. Confusion, delirium, coma, seizures, and cerebral edema may occur. Progressive respiratory distress and pulmonary edema can occur. After an initial leukocytosis, bone marrow depression is seen with a nadir between the fourth and seventh day. Bone marrow depression, coupled with potential gastrointestinal hemorrhages and a hemolytic anemia, may produce profound anemia. Consumptive coagulopathy may also be seen. Renal function may be effected by direct organ damage as well as by decreased perfusion from profound and persistent hypotension. Cardiovascular instability along with metabolic acidosis may develop due to volume depletion, cardiac failure, and arrhythmias. Most deaths result from shock in the 24- to 72-hr period. Stage three is the recovery phase. If patients survive to this convalescent phase, the main complication is sepsis.

Clinical Management

Basic and advanced life-support measures should be utilized as necessary. Treatment of colchicine toxicity is largely supportive. Activated charcoal effectively adsorbs colchicine and should be administered. Aggressive early gastrointestinal decontamination may be life-saving. However, activated charcoal may be of use up to 24 hr postingestion. Severe anemia may require packed red blood cell replacement. Coagulopathies may respond to vitamin K and fresh frozen plasma. Hypotension may be unresponsive to fluid replacement and pressor support. Due to rapid tissue distribution and the large volume of distribution, hemoperfusion and hemodialysis are ineffective. Colchicine Fab fragments have effectively reversed hypotension and increased survival in animals and humans but are not widely available.

Animal Toxicity

Animal toxicity is primarily related to ingestion of the plant *Colchicum autumnale*. The estimated toxic dose for cows is 10 g/kg with fresh leaves or 2 or 3 g/kg with dried leaves. Symptoms may include gait disorders, hypersalivation, bloody vomitus, and diarrhea. Death within 72 hr has occurred secondary to shock.

—*Henry A. Spiller*

Combustion Toxicology

Introduction

Combustion toxicity research is the study of the adverse health effects caused by fire atmospheres. In this entry, a fire atmosphere is defined as all the effluents generated by the thermal decomposition of materials or products regardless of whether that effluent is produced under smoldering, nonflaming, or flaming conditions. The objectives of combustion toxicity research are to identify potentially harmful products from the thermal degradation of materials, to determine the best measurement methods for the identification of the toxicants as well as the degree of toxicity, to determine the effect of different fire exposures on the composition of the toxic combustion products, and to establish the physiological effects of such products on living organisms. The ultimate goals of this field of research are to reduce human fire fatalities due to smoke inhalation, to determine effective treatments for survivors, and to prevent unnecessary suffering of fire casualties caused by smoke inhalation.

Fire Death Statistics in the United States

According to the U.S. Fire Administration, fire kills more people in the United States than the combination of floods, hurricanes, tornadoes, and earthquakes. Although the reasons are not clear, the number of fire deaths per capita in the United States and Canada is about double the international average of the other countries in the industrialized world. Fire statistics collected by the National Fire Protection Association (NFPA) indicated that approximately 2 million fires occurred in the United States in 1993, the latest year for which complete statistics are available. Calculated another way, these statistics translate into a fire occurring in the United States every 16 sec, in property located outside every 35 sec, in a structure every 51 sec, in a residence every 67 sec, and in a motor vehicle every 75 sec. These fires caused approximately 4600 deaths and 30,000 reported injuries in 1993. This number for injuries is believed to be less than the actual number since many injuries are not reported. The property loss for 1993 is estimated at $8.5 billion, a value that may seem high until one realizes that the wildfires in Southern California in 1993 alone resulted in an $800,000,000 property loss.

In 1993, residential fires accounted for only 24% of the total fires but were responsible for 80% of all fire deaths and 74% of the reported injuries. Although in the years 1977–1993, the number of civilian fire fatalities in homes dropped from 6000 to 3700 per year, fires in homes still cause the greatest concern to the fire community. Statistics show that children under 5 years of age and adults over 65 years of age are the most frequent casualties of residential fires. This is attributed to their inherent difficulties in trying to escape. Statistics also show that males are more likely to die in fires than females, and Native Americans and African Americans have much higher fire death rates than the rest of the population. On the other hand, Asian Americans have one-third the average death rate. More fires and higher fire death rates occur in the South than in any of the other geographical areas of the United States; the geographical region with the next highest number of fires and fire death rate is the Northeast.

One must distinguish between the causes of fires and the causes of fire deaths. The primary causes of residential fires have been shown to be heating and cooking. Lack of central heat and the incorrect use of portable space heaters are two of the reasons given for the high fire and death rate in the South. Heating fires result in the highest property losses, primarily because cooking fires are usually noticed and extinguished before getting out of control. Fire deaths, however, usually result from fires ignited by cigarettes. The most common fire scenario leading to fire deaths is one in which a person (usually intoxicated) falls asleep in an upholstered chair while smoking. The cigarette falls into a crevice and starts the upholstered chair smoldering. The individual awakes and goes to bed. The chair can smolder for an extended period of time (in laboratory tests, 1 hr was not unusual) before bursting into flames. It is after the flaming starts that the smoke fills the room and escapes to the other rooms. It is common to find people who have died from smoke inhalation (not burns) in or near their beds indicating that the little or no effort to escape was probably due to lack of awareness of the danger. Smoke detectors in this scenario would save many lives. Statistics have shown that working smoke detectors double one's probability of escaping alive. Recent statistics have also shown that many homes have nonfunctioning smoke detectors due to being disconnected after a false alarm (usually from smoke from cooking or a wood stove) or because the beeping indicating the need for a new battery became annoying.

Since most of the deaths from fires occur in residences, the NFPA proposes the following steps to improve fire safety:

1. Increase fire safety education on fire prevention and what to do if a fire occurs.

2. Install smoke detectors in all homes and ensure that they are checked periodically for proper working order.

3. Install residential home sprinklers to prevent fires from spreading once they start.

4. Develop and use fire-safe products in the home.

5. Study the needs of the populations most at risk (the young, the elderly, and the poor) and implement preventive measures.

The U.S. Fire Administration has issued the following "Home Fire Safety Checklist":

Smoke detectors—are they (1) placed near bedrooms, (2) on every floor, (3) placed away from air vents, and (4) checked regularly for working batteries?

Electrical wiring—is it (1) replaced if frayed or cracked and (2) not placed under rugs, over nails, or in high traffic areas? Are the outlets (1) not overloaded, (2) maintained cool to the touch, not hot, and (3) not exposed (i.e., have cover plates)?

Electric space heaters—are they (1) plugged directly into wall socket with no extension cords and (2) unplugged when not in use?

Kerosene heaters—are they (1) used only where legal, (2) filled only with K-1 kerosene, never gasoline or camp stove fuel, (3) only refueled outdoors, and (4) only refueled when cool?

Woodstoves and fireplaces—are they (1) used only with seasoned wood, never green wood, artificial logs, or trash, (2) protected by screens, and (3) cleaned regularly along with the interiors, hearths, and chimneys?(Note: Never burn wood treated to resist termites or wood treated for external use such as decks because they can produce very toxic fumes.)

All alternate heaters—are they (1) used only in well-ventilated rooms, (2) stable such that they cannot be easily knocked over, (3) never used to dry clothing or other items, and (4) kept at a safe distance from curtains or furniture?

Home escape plan—is it (1) practiced every 6 months, (2) are the emergency numbers, a whistle, and a flashlight kept near the telephone, and (3) is the outside meeting place identified?

Generation of Toxic Gases in Fires: Adverse Effects of Particulates

Eighty percent of the residential fire deaths are attributed to smoke inhalation, not to burns. Smoke is defined by the American Society for Testing and Materials (ASTM) as "the airborne solid and liquid particulates and gases evolved when a material undergoes pyrolysis or combustion." The adverse effects from smoke inhalation are believed to be due mainly to the exposure to toxic gases, although the role of the particulates alone and in combination with fire gases needs further investigation. The importance, therefore, of determining the identities and concentrations of toxic gases produced from materials thermally decomposed under various fire conditions is evident. In addition, the increased variety of plastics in buildings and homes has raised the issue of whether synthetic materials may produce unusually or extremely toxic combustion products. In this entry, the phrase "extremely toxic" is a relative term indicating that the effluent from the thermal decomposition of very small quantities of a material has been noted to cause death of experimental animals (usually rats or mice) under controlled laboratory conditions. "Unusually toxic" indicates that the toxic effect cannot be totally attributable to the combustion gases (either singly or in combination) that are normally considered the main toxicants.

In 1975, the journal *Science* documented a case in which an experimental rigid polyurethane foam containing a fire retardant produced a very unusual toxic combustion product identified as 4-ethyl-1-phospha-2,6,7-trioxabicyclo[2,2,2]octane-1-oxide (commonly referred to as a bicyclic phosphate ester). Bicyclic phosphate compounds have been shown to cause seizures at very low concentrations. Because of the toxicity noted in these tests, this product never became commercially available. To a large extent, however, it was this case that generated the burgeoning interest in the field of combustion toxicology and the widespread concern about the potential formation of "supertoxicants." Although research since the 1970s has shown that this concern is largely unfounded, the bicyclophosphate ester case and at least one other product that generated extremely toxic combustion products have indicated the need to test new formulations or materials containing new combinations of compounds to ensure that extremely or unusually toxic products are not generated.

The gas composition of smoke depends on the chemical composition, the molecular structure, and the polymer formulation of the burning material, which may include a variety of additives, plasticizers, stabilizers, flame retardants, cross-linking agents, fillers, and blowing agents. In addition, the conditions of thermal degradation (e.g., temperature, oxygen availability, and ventilation) will affect the nature of the combustion atmosphere. In a series of reviews of the combustion products and toxicity of seven plastics [acrylonitrile butadiene styrenes (ABS), nylons, polyesters, polyethylenes, polystyrenes, polyvinyl chlorides (PVC), and rigid polyurethane foams] commonly found in residences, and decomposed under various thermal and atmospheric conditions, over 400 different decomposition products were noted. Many of these products were common to more than one plastic. In addition, there are probably many other combustion products that were not detected. At this time, the toxicity of most of

these individual compounds is not known and little has been done to tackle the enormous problem of determining the toxicity of combinations of these compounds. It is important to note that lack of detection of a specific combustion product from a material may only mean that the particular analytical techniques used were not suitable to detect that compound or that the investigator did not specifically analyze for that combustion product. Animal testing becomes important to ensure that an unsuspected and, therefore, undetected toxic by-product has not been formed.

Since the number of compounds one can reasonably analyze in any one test is limited, knowledge of the chemical composition, molecular structure, and formulation of the polymer can be used to provide some indication of the main gaseous products which may or may not be generated under specified experimental conditions. However, one needs to be cautious when predicting the combustion products from generic materials of unknown formulations. For example, one would expect nitrogen-containing materials (e.g., ABS, nylons, and rigid and flexible polyurethanes) to produce hydrogen cyanide (HCN) and not expect HCN from a material like PVC. However, a PVC containing zinc ferrocyanide (an additive designed to suppress smoke) as well as a vinyl chloride–vinylidene chloride copolymer were found to generate HCN. [The PVC containing zinc ferrocyanide was never made commercially available after toxicity testing indicated that its combustion products produced very rapid deaths of experimental animals (rats)]. In a similar fashion, based on the chemical composition, PVC is the only one of the seven plastics mentioned previously that would be expected to generate chlorinated combustion products. However, widespread use of halogenated fire retardants in plastic formulations makes predicting which materials will produce halogenated products extremely difficult.

Temperature also plays an important role in influencing the production of decomposition products. In general, as the temperature and thus the rate of decomposition increases, the quantity of the more complex compounds and heavier hydrocarbons decreases and the concentrations of carbon monoxide (CO), carbon dioxide (CO_2), and nitrogen dioxide (NO_2) increase. The generation of HCN has also been shown to increase as a function of temperature. Another example is hydrogen chloride (HCl), the detection of which begins when stabilized PVC is heated to approximately 200°C;

rapid dehydrochlorination then occurs at about 300°C. On the other hand, more acrolein was generated from polyethylene under lower temperature, nonflaming conditions than under higher temperature flaming conditions.

As mentioned earlier, more work is needed to examine the adverse effects of the particulate matter which is also produced when these materials are thermally decomposed. Examination of the smoke particulate and condensible matter is important for a number of reasons. First, many of the thermal degradation products may condense or be adsorbed by the soot particles and be transported along with the combustion gases into the body. Hydrogen chloride is one example of a compound that may be transported in such a fashion or can form a corrosive acid mist in moist air, such as that found in the lung. One study of the particulate matter which formed during the smoldering decomposition of rigid polyurethane foam showed that many of the compounds detected in the soot fraction were not found in the volatile fraction. Free radicals, which form in fires and are of toxicological concern due to their high reactivity, are usually considered to have very short life spans; however, if adsorbed onto soot particles, their lifetimes can be considerably longer, and if the soot particle is the correct size they can be inhaled into the individual's lungs. In addition, the particulate matter may interfere with the escape and rescue of individuals by causing the obscuration of vision, eye irritation (the eyes clamp shut and the victim is unable to see), and upper respiratory distress. An extreme case indicating the adverse effect of particulates was noted in experiments conducted at the National Institute of Standards and Technology (NIST). Rats exposed for 30 min to the smoke from polystyrene died during the exposures and the level of CO, even in combination with CO_2, was too low to account for the deaths. Pathological examination of these rats showed that their respiratory passages were completely blocked by soot and that suffocation was the likely cause of death.

Toxic Potency vs Fire Hazard vs Fire Risk

Death in a fire may be caused by

1. Carbon monoxide (CO)
2. Toxic gases in addition to CO
3. Oxygen (O_2) at levels too low to sustain life
4. Incapacitation—either physical (inability to escape) or mental (incorrect decision making)
5. Bodily burns from flame contact
6. Very high air temperatures
7. Smoke density or irritants in smoke which affect vision and interfere with ability to escape
8. Psychological effects (e.g., fear, shock, and panic)
9. Physical insults (e.g., building or ceiling collapse and lethal effects from jumping from upper floors)

Research in the field of combustion toxicology is primarily concerned with items 1–4, all of which are related to the toxic potency of the fire gas effluent. Toxic potency is defined by ASTM as "A quantitative expression relating concentration (of smoke or combustion gases) and exposure time to a particular degree of adverse physiological response, for example, death on exposure of humans or animals." This definition is followed by a discussion, which states, "The toxic potency of smoke from any material or product or assembly is related to the composition of that smoke which, in turn, is dependent upon the conditions under which the smoke is generated." It should be noted that the LC_{50} is a common endpoint used in laboratories to assess toxic potency. The LC_{50} value is the result of a statistical calculation based on multiple experiments, each with multiple animals, and indicates the concentration at which 50% of the experimental animals exposed for a specific length of time would be expected to die either during the exposure time or during the postexposure observation period. In the comparison of the toxic potencies of different compounds or materials, the lower the LC_{50} (i.e., the smaller the amount of material necessary to reach the toxic endpoint), the more toxic the material.

It is important to note that a toxicity assessment based on lethality due to toxic gases is only part of the total fire hazard that needs to be evaluated, especially when one is making choices as to the best material for a specific end use. ASTM defines "fire hazard" as the potential for harm associated with fire. The discussion that follows this definition states,

A fire may pose one or more types of hazard to people, animals or property. These hazards are associated with the environment and with a number of fire-test-response characteristics of materials, products or assemblies including but not limited to ease of ignition, flame spread, rate of heat release, smoke generation and obscuration, toxicity of combustion products and ease of extinguishment.

Other factors that need to be evaluated when considering a material for use in a given situation include the quantity of material needed, its configuration, the proximity of other combustibles, the volume of the compartments to which the combustion products may spread, the ventilation conditions, the ignition and combustion properties of the material and other materials present, the presence of ignition sources, the presence of fire protection systems, the number and type of occupants, and the time necessary to escape.

"Fire risk" is defined by ASTM as "An estimation of expected fire loss that combines the potential for harm in various fire scenarios that can occur with the probabilities of occurrence of those scenarios." The discussion following the definition of fire risk states,

> Risk may be defined as the probability of having a certain type of fire, where the type of fire may be defined in whole or in part by the degree of potential harm associated with it, or as potential for harm weighted by associated probabilities. Risk scales do not imply a single value of acceptable risk. Different individuals presented with the same risk situation may have different opinions on its acceptability.

A simple way to explain the difference between fire hazard and fire risk is to compare the fire to skydiving, a very hazardous sport; however, if one never goes skydiving, no risk is incurred.

Toxicity Assessment

Animal Exposures

In most combustion toxicology experiments, the biological endpoint has been lethality or incapacitation of experimental animals, usually rats or mice. Incapacitation in a fire can be as perilous as lethality if an individual becomes mentally incapable of correct decision making or physically unable to move. Under these circumstances, the ability to escape will be lost and death will occur unless the individual is rescued. Therefore, many fire scientists are concerned with the levels of combustion products or amounts of materials which when combusted will cause incapacitation. However, an incapacitation model for use in laboratory testing

has been especially difficult to develop. Most of the tests for incapacitation that have been designed are based on the physical motor capability of an experimental animal to perform some task (e.g., running in a motorized wheel, jumping onto a pole or lifting a paw to escape a shock, running in a maze, or pushing the correct lever to open a door to escape an irritating atmosphere). The concentration of toxic combustion products which cause the loss of these types of physical motor capabilities is usually close to the concentration which is lethal and does not usually add much additional information. Recently, however, there have been attempts at examining neurological endpoints, such as measuring the increased number of errors by humans doing mathematical problems while exposed to low levels of CO or exposing rats and pigeons to a neurobehavioral battery of 25 tests.

Whether one needs to examine incapacitation or lethality depends on the problem one is trying to solve. To determine the best material for a particular end-use application, the lethality endpoint has proven to be more definitive and will flag the materials that produce extremely toxic combustion products better than an incapacitation endpoint. There are at least two reasons for this:

1. Incapacitation is only measured during the exposure, which is usually 30 min or less, but lethality can also occur during the postexposure observation period, which can be 2 weeks or longer. A material that only causes delayed effects during the postexposure period (e.g., a material that generates HCl) can thus have an LC_{50} value that is lower than the incapacitation EC_{50} value (i.e., the amount of thermally decomposed material necessary to cause postexposure deaths is less than the amount needed to cause incapacitation during the exposure). The definition of the EC_{50} is essentially the same as that of the LC_{50}, except incapacitation rather than lethality is the endpoint and incapacitation is monitored only during the exposure.

2. In many cases in which the combustion products contain high concentrations of irritant gases, the animals would only appear to be incapacitated (i.e., they would stop responding to the test indicator due to the high-irritant quality of the

smoke), but when removed from the combustion atmosphere would immediately start responding normally.

Other delayed effects from exposures to combustion atmospheres, such as tissue or organ injury, mutagenicity, carcinogenicity, and teratogenicity, also need to be studied since they may ultimately lead to permanent disability or death. The current advances in the field of genetics provide investigators with new opportunities to examine the effects of combustion products at the molecular level. One objective could be to determine whether these toxic products cause DNA damage and/or mutations. Specific problems of interest include the following: Does the damage occur in nuclear DNA and/or mitochondrial DNA? Are certain areas of the DNA more prone to these mutations (i.e., are there hot spots)? Can we categorize the types of mutations (e.g., transitions, transversions, deletions, or insertions), How efficient are the repair mechanisms? Are these mutagens also known to be carcinogens?

Toxicity Assessment: Predictive Models

In the 1970s, there were essentially two experimental strategies to examine the issues raised by the field of combustion toxicology: the analytical chemical method and the animal exposure approach. In the analytical chemical method, investigators thermally decomposed materials under different experimental conditions and tried to determine every combustion product that was generated. This approach generated long lists of compounds. The toxicity of most of these individual compounds was unknown and the concept of examining the toxicity of all the various combinations of compounds was and still is considered an impossible task. An additional problem with the analytical method was that, as mentioned earlier, one could not be certain that every toxic product was detected and identified. This approach enabled one to identify many of the multiple products that were generated but not know the toxic potency of all the identified compounds, either singly or combined.

In the animal exposure approach, the animals (usually rats or mice) serve as indicators of the degree of toxicity of the combustion atmospheres. The materials of concern are thermally decomposed under different combustion conditions and the animals are exposed to the combined particulate and gaseous effluent. Multiple animal experiments (each with multiple animals) with different concentrations of material are conducted to determine an EC_{50} (incapacitation) or an LC_{50} (lethality) for a specific set of combustion conditions. Each material would then have a particular EC_{50} or an LC_{50} value that can be used to compare the toxicities of different materials decomposed under the same conditions. The lower the EC_{50} or LC_{50}, the more toxic are the combustion products from that material considered to be. In this approach, one knows the relative toxicity of a material compared to another material but does not know which of the toxic gases are responsible for the adverse effects.

In the 1980s, investigators began examining the possibility of combining the analytical chemical method with the animal exposure approach to develop empirical mathematical models to predict the toxicity. These predictions were based on actual experiments with animals and their response to each of the main toxic combustion gases—CO, CO_2, low O_2, HCN, NO_2, HCl, and hydrogen bromide (HBr)—and various combinations of these gases. The advantages of these predictive approaches are (1) the number of test animals is minimized by predicting the toxic potency from a limited chemical analysis of the smoke; (2) smoke may be produced under conditions that simulate any fire scenario of concern; (3) fewer tests are needed, thereby reducing the overall cost of the testing; and (4) information is obtained on both the toxic potency of the smoke (based on the mass of material burned) and the responsible gases (based on the primary toxic gases in the mixture). The prediction is checked with one or two animal tests to ensure that an unexpected gas or toxic combination has not formed. The results of using these empirical mathematical models indicated that, in most cases, one could predict the toxic potency of a combustion atmosphere with the main toxic gases and did not need to worry about the effects of minor or more obscure gases.

Primary Toxic Combustion Gases

Complete combustion of a polymer containing carbon, hydrogen, and oxygen in an atmosphere with sufficient O_2 yields CO_2 and H_2O. It is during incomplete combustion under various atmospheric conditions in either flaming or nonflaming modes that compounds of greater toxicological concern are generated. When O_2 is limited, the primary gases formed during the combustion of most materials are CO, CO_2, and H_2O. If the materials contain nitrogen, HCN and NO_2, two principal thermooxidative products of toxicological concern, are also likely to be generated. Halogenated or flame-retarded materials generally produce HCl or HBr. Other commonly found fire gases include nitrogen oxides (NO_x), ammonia (NH_3), hydrogen sulfide (H_2S), sulfur dioxide (SO_2), and fluorine compounds. One also needs to consider that in fire situations, O_2 levels drop and exposure to low O_2 atmospheres will have additional adverse physiological effects. Some of these toxic combustion gases (e.g., CO, HCN, and low O_2) produce immediate asphyxiant symptoms, while others (e.g., HCl, HBr, and NO_2) fall into an irritant category and produce symptoms following the exposures.

The N-Gas Models

The N-gas models for predicting smoke toxicity were founded on the hypothesis that a small number (N) of gases in the smoke account for a large percentage of the observed toxic potency. These predictive models were based on an extensive series of experiments conducted at NIST on the toxicological interactions of the primary gases found in fires. Both the individual gases and the complex mixtures of these gases were examined. To use these models, materials are thermally decomposed using a bench-scale method that simulates realistic fire conditions, the concentrations of the primary fire gases—CO, CO_2, low O_2, HCN, HCl, HBr, and NO_2—are measured, and the toxicity of the smoke using the appropriate N-gas model is predicted. The predicted toxic potency is checked with a small number of animal tests (using Fischer 344 male rats) to ensure that an unanticipated toxic gas was not generated or an unexpected toxicological effect (e.g., synergism or antagonism) did not occur. The results indicate whether the smoke from a material or product is extremely toxic (based on mass consumed at the predicted toxic level) or unusually toxic (the toxicity cannot be explained by the combined measured gases). These models have been shown to correctly predict the toxicity in both bench-scale laboratory tests and full-scale room burns of a variety of materials of widely differing characteristics chosen to challenge the system. The 6-gas model (without NO_2) is now included in national toxicity test method standards (ASTM E1678 and NFPA 269) approved by the American Society for Testing and Materials (ASTM) and the National Fire Protection Association, Inc. (NFPA) and an international standard (ISO 13344) that was approved by 16 member countries of the International Standards Organization (ISO). These standards were published in 1995 and 1996.

The objectives of developing the N-gas models were

- To establish the extent to which the toxicity of a material's combustion products could be explained and predicted by the interaction of the major toxic gases generated from that material in the laboratory or whether minor and more obscure combustion gases needed to be considered.

- To develop a bioanalytical screening test and a mathematical model which would predict whether a material would produce extremely toxic or unusually toxic combustion products.

- To predict the occupant response from the concentrations of primary toxic gases present in the environment and the time of exposure.

- To provide data for use in computer models designed to predict the hazard that people will experience under various fire scenarios.

The 6-Gas N-Gas Model

The 6-gas model see Eq. (1) was based on studies at NIST on the toxicological interactions of six gases—CO, CO_2, HCN, low O_2 concentrations, HCl, and HBr. First, individual gases in air were tested to determine the concentrations necessary to cause 50% of the laboratory test animals (Fischer 344 male rats) to die either during the exposure (within exposure LC_{50}) or during the exposure plus a 14-day postexposure observation period (within + postexposure LC_{50}). The studies on HCl and HBr were conducted at Southwest Research Institute (SwRI) under a grant from NIST. Similar measurements for various combinations of these gases indicated whether the toxicity of the mixtures of gases was additive, synergistic, or antagonistic.

Based on these empirical results, the following 6-gas N-gas model was developed:

$$\frac{m[CO]}{[CO_2]-b} + \frac{[HCN]}{LC_{50}HCN} + \frac{21-[O_2]}{21-LC_{50}O_2} + \frac{[HCl]}{LC_{50}HCl} + \frac{[HBr]}{LC_{50}HBr}$$
$$= N-Gas\ Value \qquad (1)$$

where the numbers in brackets indicate the time-integrated average atmospheric concentrations during a 30-min exposure period [(ppm × min)/min or, for O_2, (% × min)/min]. The other terms are defined in the following paragraphs.

Under the experimental conditions used at NIST and with Fischer 344 male rats, the 30-min LC_{50} value of CO_2 is 47% (470,000 ppm) with 95% confidence limits of 43–51%. [Caution: The values given for use in Eqs. (1) and (3) are dependent on the test protocol, on the source of test animals, and on the rat strain. It is important to verify the above values whenever different conditions prevail and, if necessary, to determine the values that would be applicable under the new conditions.] No deaths occurred in rats exposed to 26% CO_2 for 30 min. In a real fire, the highest theoretically possible concentration of CO_2 is 21%, a concentration that could only occur if all the atmospheric O_2 were converted to CO_2, which is highly improbable. Therefore, CO_2 concentrations generated in fires are not lethal. However, CO_2 is a respiratory stimulant causing an increase in both respiratory rate and tidal volume. It also increases the acidosis of the blood. When combined with any of the other tested gases, CO_2 has a synergistic toxicological effect, i.e., the toxicity of the other gases are increased in the presence of CO_2 (Table C-20). Empirically, however, it was determined that

the effect of the CO_2 can only be added into the N-gas equations once. Therefore, the CO_2 effect is included with the CO factor since there was more data on the effect of different concentrations of CO_2 on the toxicity of CO and CO is the toxicant most likely to be present in all fires. The results on the synergistic effect of CO_2 on CO indicated that as the concentration of CO_2 increases (up to 5%), the toxicity of CO increases. Above 5% CO_2, the toxicity of CO starts to revert back toward the toxicity of the CO by itself. The terms m and b in Eq. (1) define this synergistic interaction and equal −18 and 122,000, respectively, if the CO_2 concentrations are 5% or less. For studies in which the CO_2 concentrations are above 5%, m and b equal 23 and −38,600, respectively.

In rats, the 30-min LC_{50} for CO is 6600 ppm and with 5% CO_2, this value drops to 3900 ppm. Exposure to CO in air only produced deaths during the actual exposures and not in the postexposure observation period; however, exposures to CO plus CO_2 also caused deaths in the postexposure period. Carbon monoxide is a colorless, odorless, tasteless, and nonirritating poisonous gas. The toxicity of CO comes from its binding to the hemoglobin in red blood cells and the formation of carboxyhemoglobin (COHb). The presence of CO on the hemoglobin molecule prevents the binding of O_2 to hemoglobin (O_2Hb) and results in hypoxia in the exposed individual. Since the binding affinity of hemoglobin for CO is 210 times greater than its affinity for O_2, only 0.1% CO (1000 ppm) is needed to compete equally with O_2, which is normally present at 20.9% in air (20.9%/210 ≈ 0.1%). Thus, only 1000 ppm of CO in the atmosphere is enough to generate 50% COHb, a value commonly quoted (but not necessarily proven) as the concentration which is lethal to humans. The time to get to 50% COHb at 1000 ppm CO would be longer than 30 min.

The LC_{50} value of HCN is 200 ppm for 30-min exposures or 150 ppm for 30-min exposures plus the postexposure observation period. HCN caused deaths both during and following the exposures.

The 30-min LC_{50} of O2 is 5.4%, which is included in the model by subtracting the combustion atmospheric O_2 concentration from the normal concentration of O_2 in air, i.e., 21%. The LC_{50} values of HCl or HBr for 30-min exposures plus postexposures times are 3700 and 3000 ppm, respectively. HCl and HBr at levels found in fires only cause postexposure effects.

TABLE C-20
Synergistic Effects of CO_2

	LC_{50} values	
Gas[a]	Single gas	With 5% CO_2
CO_2	470,000 ppm[b]	—
CO	6,600 ppm[b]	3,900 ppm[c]
NO_2	200 ppm[d]	90 ppm[d]
O_2	5.4%[b]	6.4%[c]

[a] All gases were mixed in air; 30-min exposures of Fischer 344 rats.
[b] Deaths during the exposure.
[c] Deaths occurred during and following the exposures.
[d] Deaths during the postexposure period.

The pure and mixed gas studies showed that if the value of Eq. (1) is 1.1 ± 0.2, then some fraction of the test animals would die. Below 0.9, no deaths would be expected and above 1.3 all the animals would be expected to die. Since the concentration–response curves for animal lethalities from smoke are very steep, it is assumed that if some percentage (not 0 or 100%) of the animals die, the experimental loading is close to the predicted LC_{50} value. Results using this method show good agreement (deaths of some of the animals when the N-gas values were above 0.9) and the good predictability of this approach.

This model can be used to predict deaths that will occur only during the fire exposure or deaths during and following the fire. To predict the deaths that would occur both during and following the exposures, Eq. (1) is used as presented. To predict deaths only during the exposures, HCl and HBr, which only have postexposure effects, should not be included in Eq. (1). In small-scale laboratory tests and full-scale room burns, Eq. (1) was used successfully to predict the deaths during and following exposures to numerous materials. In the case of PVC, the model correctly predicted the results as long as the HCl was greater than 1000 ppm; therefore, it is possible that HCl concentrations under 1000 ppm may not have any observable effect on the model even in the postexposure period. More experiments are necessary to show whether a true toxic threshold for HCl does exist.

Although most of the work at NIST concentrated on deaths during or following 30-min exposures, the LC_{50}s of many of these gases both singly and mixed were determined at times ranging from 1 to 60 min, and in all the cases examined, the predictive capability of Eq. (1) holds if the LC_{50}s for the other times are substituted into the equation.

The 7-Gas Model: Addition of NO_2 to the N-Gas Model

Nitrogen dioxide is an irritant gas that will cause lachrymation, coughing, respiratory distress, increases in methemoglobin levels, and lung edema. Single brief exposures to less than lethal concentrations can cause lung damage, emphysema, or interstitial fibrosis. Low levels have been alleged to increase susceptibility to respiratory infections and aggravate reactions to allergens. Impairment of dark adaptation has also been noted. Delayed serious effects can be observed as late as 2 or 3 weeks following exposures. In the lungs, NO_2 forms both nitric (HNO_3) and nitrous (HNO_2) acids, which are probably responsible for the damage to the lung cells and connective tissue.

In fires, NO_2 may arise from atmospheric nitrogen fixation, a reaction which is material independent, or from the oxidation of nitrogen from nitrogen-containing materials. To examine the generation of NO_2 from nitrogen fixation, a small study was undertaken at NIST. In two full-scale fires of rooms in which the main source of fuel was polystyrene-covered walls, only low levels of NO_x (10 and 25 ppm) were found, indicating little nitrogen fixation under these conditions. A real example of burning nitrogen-containing materials was the 1929 Cleveland Clinic fire in which 50,000 nitrocellulose X-ray films were consumed. The deaths of 97 people in this fire were attributed mainly to NO_x. An additional 26 people died between 2 hr and 1 month following the fire, and 92 people were treated for nonfatal injuries. In laboratory tests of nitrogen-containing materials under controlled conditions, 1–1000 ppm of NO_x were measured. In military tests of armored vehicles penetrated by high-temperature ammunition, NO_2 levels above 2000 ppm were found.

Individual and Binary Mixtures

In small-scale laboratory tests of NO_2 in air, deaths of Fischer 344 male rats occur only in the postexposure period and the LC_{50} value following a 30-min exposure is 200 ppm. Carbon dioxide plus NO_2 shows synergistic toxicological effects. The LC_{50} for NO_2 following a 30-min exposure to NO_2 plus 5% CO_2 is 90 ppm (postexposure deaths) (i.e., the toxicity of NO_2 doubled).

As mentioned previously, CO produces only within-exposure deaths and its 30-min LC_{50} is 6600 ppm. In the presence of 200 ppm of NO_2, the within-exposure toxicity of CO doubled (i.e., its 30-min LC_{50} became 3300 ppm). An exposure of approximately 3400 ppm CO plus various concentrations of NO_2 showed that the presence of CO would also increase the postexposure toxicity of NO_2. The 30-min LC_{50} value of NO_2 went from 200 ppm to 150 ppm in the presence of 3400 ppm of CO. A concentration of 3400 ppm of CO was used because that concentration would not be lethal during the exposure and any postexposure effects of CO on NO_2 would become evident; the LC_{50} of CO

(6600 ppm) would have caused deaths of the animals during the 30-min exposure.

The 30-min LC_{50} of O_2 is 5.4% and the deaths occur primarily during the exposures. In the presence of 200 ppm of NO_2, the within-exposure LC_{50} of O_2 and its toxicity increased to 6.7%. In the case of O_2, increased toxicity is indicated by an increase in the value of the LC_{50} since it is more toxic to be adversely affected by a concentration of O_2 ordinarily capable of sustaining life. Exposure of the animals to 6.7% O_2 plus various concentrations of NO_2 showed that the NO_2 toxicity doubled (i.e., its LC_{50} value decreased from 200 to 90 ppm).

One of the most interesting findings was the antagonistic toxicological effect noted during the experiments on combinations of HCN and NO_2. As mentioned previously, the 30-min LC_{50} for NO_2 alone is 200 ppm (postexposure) and the 30-min within-exposure LC_{50} for HCN alone is also 200 ppm. The concentrations of either gas alone are sufficient to cause death of the animals (i.e., 200 ppm HCN or 200 ppm NO_2 would cause 50% of the animals to die either during the 30-min exposure or following the 30-min exposure, respectively). However, in the presence of 200 ppm of NO_2, the within-exposure HCN LC_{50} concentration increases to 480 ppm or, in other words, the toxicity of HCN decreases by 2.4 times.

The mechanism for this antagonistic effect is believed to be as follows: In the presence of H_2O, NO_2 forms HNO_3 and HNO_2. These two acids are the most likely suspects responsible for the lung damage leading to the massive pulmonary edema and subsequent deaths noted following exposure to high concentrations of NO_2. Nitrite ion (NO_2^-) formation occurs in the blood when the nitrous acid dissociates. The nitrite ion oxidizes the ferrous ion in oxyhemoglobin to ferric ion to produce methemoglobin (MetHb) (Eq. 2). MetHb is a well-known antidote for CN^- poisoning. MetHb binds cyanide forming cyanmethemoglobin, which keeps the cyanide in the blood and prevents it from entering the cells. In the absence of MetHb, free cyanide will enter the cells, react with cytochrome oxidase, prevent the utilization of O_2, and cause cytotoxic hypoxia. If, on the other hand, cyanide is bound to MetHb in the blood, it will not be exerting its cytotoxic effect. Therefore, the mechanism of the antagonistic effect of NO_2 on the toxicity of cyanide is believed to be due to the conversion of oxyhemoglobin [$O_2Hb(Fe^{2+})$] to methemoglobin [$MetHb(Fe^{3+})$] in the presence of nitrite (see Eq. 2).

$$2H^+ + 3NO_2^- + 2O_2Hb\,(Fe^{++}) =$$

$$2MetHb\,(Fe^{+++}) + 3NO_3^- + H_2O \qquad (2)$$

Tertiary Mixtures of NO_2, CO_2, and HCN

Earlier work indicated that the presence of 5% CO_2 with either HCN or NO_2 produced a more toxic environment than would occur with either gas alone. The antagonistic effects of NO_2 on HCN indicate that the presence of one LC_{50} concentration of NO_2 (approximately 200 ppm) will protect the animals from the toxic effects of HCN during the 30-min exposures, but not from the postexposure effects of the combined HCN and NO_2. Thus, it was of interest to examine combinations of NO_2, CO_2, and HCN. In this series of experiments, the concentrations of HCN were varied from almost 2 to 2.7 times its LC_{50} value (200 ppm). The concentrations of NO_2 were approximately equal to one LC_{50} value (200 ppm) if the animals were exposed to NO_2 alone and approximately one-half the LC_{50} (90 ppm) if the animals were exposed to NO_2 plus CO_2; the concentrations of CO_2 were maintained at approximately 5%; and the O_2 levels were kept above 18.9%. The results indicated that CO_2 does not make the situation worse, but rather provided additional protection even during the postexposure period. In each of six experiments, some or all of the animals lived through the test even though they were exposed to greater than lethal levels of HCN plus lethal levels of NO_2 and in four tests, some of the animals lived through the postexposure period even though the animals were exposed to combined levels of HCN, NO_2, and CO_2 that would be equivalent to 4.7–5.5 times the lethal concentrations of these gases. One possible reason that CO_2 seems to provide an additional degree of protection is that NO_2 in the presence of 5% CO_2 produces four times more MetHb than does NO_2 alone.

Mixtures of CO, CO_2, NO_2, O_2, and HCN

The initial design of these experiments was to look for additivity of the CO/CO_2, HCN, and NO_2 factors keeping each at about one-third of its toxic level, while keeping the O_2 concentration above 19%. When these initial experiments produced no deaths, the concentrations of CO were increased up to one-third of the LC_{50} of CO alone (6600 ppm). HCN was increased to 1.3 or 1.75 times its LC_{50} depending on whether the within-exposure LC_{50} (200 ppm) or the within- and postexposure LC_{50} (150 ppm) was being considered. NO_2 was increased up to a full LC_{50} value (200 ppm). The results indicated that just adding a NO_2 factor (e.g., $[NO_2]/LC_{50} NO_2$) to Eq. (1) would not predict the effect on the animals. A new mathematical model was developed and is shown as Eq. (3). In this model, the differences between the within-exposure predictability and the within-exposure and postexposure predictability are (1) the LC_{50} value used for HCN is 200 ppm for within-exposure or 150 ppm for within-exposure and postexposure and (2) the HCl and HBr factors are not used to predict the within-exposure lethality, only the within-exposure and postexposure lethality. According to Eq. (3), animal deaths will start to occur when the N-gas value is above 0.8 and 100% of the animals will die when the value is above 1.3. Results indicated that in those few cases in which the values were above 0.8 and no deaths occurred, the animals were severely incapacitated (close to death) as demonstrated by no righting reflex or eye reflex.

Equation (3) shows the N-gas model including NO_2:

$$N-Gas\ Value = \frac{m[CO]}{[CO_2]-b} + \frac{21-[O_2]}{21-LC_{50}(O_2)} +$$

$$\left(\frac{[HCN]}{LC_{50}(HCN)} \times \frac{0.4\,[NO_2]}{LC_{50}(NO_2)} \right) + 0.4 \left(\frac{[NO_2]}{LC_{50}(NO_2)} \right) +$$

$$\frac{[HCl]}{LC_{50}(HCl)} + \frac{[HBr]}{LC_{50}(HBr)} \tag{3}$$

For an explanation of these terms, see the paragraphs following Eq. (1). Equation (3) should be used to predict the within-exposure plus postexposure lethal toxicity of mixtures of CO, CO_2, HCN, reduced O_2, NO_2, HCl, and HBr. The LC_{50} values will be the same as those given for Eq. (1) using 150 ppm for HCN and 200 ppm for NO_2. If one wishes to predict the deaths that will occur only during the exposure, the LC_{50} value used for HCN should be 200 ppm and the HCl and

HBr factors should not be included. To predict the lethal toxicity of atmospheres that do not include NO_2, Eq. (1) is to be used.

Combustion Toxicity Test Methods

The toxicity of the combustion products from any new material formulation or product containing additives or new combinations of additives needs to be examined. Material and polymer chemists are currently trying to develop new "fire-safe" materials. The terms "fire-safe" or "fire-resistant" are not the same as noncombustible. Unless these new materials are truly noncombustible, some thermal decomposition will occur when the materials are exposed to fire conditions. Both the toxic gases and the irritants that are present in all smoke need to be considered potential dangers. The toxic products can cause both acute and delayed toxicological effects. It is the acute and extremely short-term effects that prevent escape from burning buildings by causing faulty judgment, incapacitation, and death. The irritants in the smoke can also interfere with the person's ability to escape by causing severe coughing and choking and by preventing them from keeping their eyes open long enough to find the exits. The delayed effects include tissue and organ injury, mutagenicity, carcinogenicity, and teratogenicity.

Toxicity screening tests for both the acute and delayed effects are needed to evaluate the combustion products including any irritants that may be present in newly proposed materials and products. It is imperative that the materials and products be tested under experimental conditions that simulate realistic fire scenarios of concern (e.g., flashover conditions emanating from smoldering and then flaming of upholstered furniture in homes or smoldering fires in concealed spaces in aircraft). The ideal tests should be simple, rapid, inexpensive, use the least amount of sample possible (since, in many cases, only small amounts of the new experimental material may be available), use a minimum number of test animals, and have a definitive toxicological endpoint for comparison of the multiple candidates. While faulty judgment and incapacitation are significant causes of worry since they can prevent escape and cause death, they are extremely difficult and complex endpoints to define and measure in nonhuman test subjects. Death of experimental animals (e.g., rats), on the other hand, is a more definitive and easily determined endpoint and can be used to compare the relative toxicities of alternate materials deemed suitable for the same purpose. The assumption made here is that if the combustion products of material X are significantly more lethal than those of material Y, the combustion products of X would probably cause more incapacitation and more impairment of judgment than Y. The number of experimental animals can be significantly reduced by utilizing one of the predictive mathematical models developed for combustion toxicology such as the N-gas models previously discussed.

Many test methods for the determination of the acute toxicity of combustion products from materials and products have been developed over the past two decades and continue to be developed and/or improved. In 1983, 13 of the methods published up to that time were evaluated by Arthur D. Little, Inc., to assess the feasibility of incorporating combustion toxicity requirements for building materials and finishes into the building codes of New York State. On the basis of seven different criteria, only 2 methods were found acceptable. These 2 methods were the flowthrough smoke toxicity method developed at the University of Pittsburgh and the closed-system cup furnace smoke toxicity method developed at NIST [known at that time as the National Bureau of Standards (NBS)]. Recently, standard reference materials were developed at NIST and made available to the users of these methods to provide assurance that they are performing the methods correctly. Based on the results of the Arthur D. Little report, the state of New York under Article 15, Part 1120 of the New York State Fire Prevention and Building Code decided to require that building materials and finishes be examined by the method developed at the University of Pittsburgh and that the results be filed with the state. It is important to note, however, that although the results are filed, the state of New York does not regulate any materials or products based on the results of this or any other toxicity test.

New methods that have been developed since 1983 to examine acute combustion toxicity include the University of Pittsburgh radiant furnace method II, a radiant furnace smoke toxicity protocol developed by NIST and SwRI, and the National Institute of Building Sci-ences (NIBS) toxic hazard test method. All three use radiant heat to decompose materials.

The NIST radiant test and the NIBS toxic hazard test use the same apparatus, consisting of three components: a radiant furnace, a chemical analysis system, and an animal exposure chamber. The chemical analysis system and animal exposure system are identical to that developed for the NBS cup furnace smoke toxicity method. Although the apparatuses of both methods are essentially the same, they have different toxicological endpoints. In the NIST method, an approximate LC_{50}, based on the mass of material needed to cause lethality in 50% of the test animals during a 30-min exposure and/or a 14-day postexposure period, is the determinant of toxicity. The number of animals needed to run the test is substantially reduced by first estimating the LC_{50} by the 6-gas N-gas model. This estimate is then verified with one or two animal tests to ensure that no unforeseen gas was generated. The toxicological endpoint of the NIBS toxic hazard test is the IT_{50}, the irradiation time (the time that the material is exposed to the radiant heat) that is required to kill 50% of the animals during a 30-min exposure or 14-day postexposure time. The actual results of the NIBS test with 20 materials indicated that the test animals died in very short periods of time (A. F. Grand, personal communication) and the test was unable to discriminate very well between materials. These results substantiate the thesis that mass (the smaller the mass necessary for an LC_{50}, the more toxic the material) is a better indicator of acute toxicity than time.

Both the NIST and NIBS test procedures are designed to simulate a postflashover scenario. The premise for simulating a postflashover fire is that most people that die from inhalation of toxic gases in residential fires are affected in areas away from the room of fire origin. Smoke and toxic gases are more likely to reach these distant areas following flashover. This scenario may not be relevant in certain circumstances (e.g., aircraft interior fires, where a smoldering fire in a concealed space may cause significant problems if the plane is over a large body of water and unable to land for a considerable period of time).

The NIST radiant test has recently been accepted by the ASTM and NFPA as national standards designated ASTM E1678 and NFPA 269 and titled Test Method for Measuring Smoke Toxicity for Use in Fire Hazard

Analysis and Standard Test Method for Developing Toxic Potency Data for Use in Fire Hazard Modeling, respectively. The international standard—ISO/IS 13344 titled Determination of the Lethal Toxic Potency of Fire Effluents—gives investigators the flexibility of designing or choosing a system that will simulate conditions relevant to their fire scenario rather than having to accept a designated combustion system.

Toxicant Suppressants

Fire scientists are very familiar with fire-retardant chemicals, which are defined by ASTM as "chemicals, which when added to a combustible material, delay ignition and combustion of the resulting material when exposed to fire." The discussion adds: "a fire-retardant chemical can be a part of the molecular structure, an admixture or an impregnant." The term "toxicant suppressant," however, is a new expression arising from research at NIST which demonstrated that the addition of copper compounds to flexible polyurethane foam (FPU) significantly reduced the generation of HCN as well as the toxicity of the combustion products when the foam was thermally decomposed. These experiments were designed to simulate the nonflaming and then flaming stages of a chair ignited by a cigarette (a two-phase heating system which simulates the fire scenario that results in the most fire deaths in the United States). The term toxicant suppressant may be defined as a chemical which when added to a combustible material significantly reduces or prevents one or more toxic gases from being generated when that material undergoes thermal decomposition. The resultant gas effluent should be less toxic than that from the untreated material, i.e., the toxic gas, whose concentration is being reduced, should not be converted to an equally or more toxic product.

The results of these studies at NIST indicated the following:

1. Hydrogen cyanide concentrations in the thermal decomposition products from a flexible polyurethane foam were reduced approximately 85% when the foam was treated with 0.1% or 1.0% Cu_2O and thermally decomposed via a two-phase heating system in the cup furnace smoke toxicity apparatus.

2. The copper or copper compounds could be added to the foams during or after the foams were formulated and still be operative in reducing the toxicity and HCN yield. (The BASF Corporation prepared the foams that had the Cu powder and Cu_2O added during formulation.) The addition of the copper or copper compounds during formulation did not affect the foaming process or the physical appearance on the foams except for a slight change of color.

3. Low levels of the copper compounds were effective. In particular, when cupric oxide (CuO) was used, the concentration of copper needed was only 0.08% by weight and when cuprous oxide (Cu_2O) was used, only 0.07% by weight was needed to significantly reduce the generation of HCN.

4. Full-scale room burns indicated that the presence of Cu_2O in the FPU reduced the HCN generation by approximately 50–70% when the experimental plan was designed to simulate a realistic scenario (the foams contained 1.0% Cu_2O, were covered with a cotton upholstery fabric and arranged to simulate a chair; smoldering was initiated with cigarettes, and flaming occurred spontaneously).

5. Under small-scale conditions, less than 3 ppm of NO_x was generated from the untreated foams, whereas a range of 3–33 ppm of NO_x was measured from the 0.1–1.0% Cu_2O-treated foams. About 6% of the HCN appeared to be converted to NO_x. In the full-scale room tests, approximately 23% of the HCN appeared to be converted to NO_x. Since we have shown in our laboratory that NO_2 acts as an antagonist to HCN, this amount of NO_x may also act to counteract the immediate toxic effects of any residual HCN.

6. Since the atmospheric oxygen (O_2) concentrations can reach very low levels in real fires, it was important to know if the reduction of HCN by copper would occur under low O_2 conditions. Small-scale tests with the ambient O_2 concentrations as low as 6% indicated that the HCN levels

were reduced by as much as 82% when the FPU was treated with 0.1% Cu_2O.

7. The toxicity of the gas effluent was also reduced (an indication that the HCN was not being converted into some compound that was even more toxic). Fewer animal (Fischer 344 rats) deaths occurred during the 30-min exposures to the FPU treated with the copper and copper compounds compared to the untreated FPU. Toxicity based on LC_{50} values was reduced 40–70% in the small-scale tests with 0.1% Cu_2O-treated foams. The blood cyanide levels in the animals exposed to combustion products from the CuO-treated foams for 30 min were one-half to one-fourth those measured in the animals exposed to the smoke from the same amount of untreated foam.

8. Postexposure deaths were also reduced in the animals exposed to the combustion products from the Cu and Cu_2O-treated FPU foams in the small-scale tests. These delayed postexposure deaths have not been observed in animals exposed to combustion products from flexible polyurethane foams decomposed in large-scale room fire tests. The specific cause of these postexposure deaths is not known.

9. No differences in flammability characteristics between the 0.1% Cu_2O-treated and untreated FPU foam were observed. These characteristics were examined to assure that the positive effect on toxicity was not contradicted by negative effects on the flammability properties. The flammability characteristics examined were (a) ignitability in three systems [the cup furnace smoke toxicity method, the Cone Calorimeter, and Lateral Ignition and Flame Spread Test (LIFT)], (b) heat release rates under small-scale (Cone Calorimeter) and medium-scale (furniture calorimeter) conditions, (c) heats of combustion under small-scale (Cone Calorimeter) and medium-scale (furniture calorimeter) conditions, (d) CO/CO_2 ratios under small-scale (Cone Calorimeter) and medium-scale (furniture calorimeter) conditions, (e) smoke obscuration (Cone Calorimeter), and (f) rate of flame spread (LIFT).

10. Research conducted at the BASF Corporation indicated that the physical properties of the 1.0% Cu_2O-treated FPU were not significantly different from the comparable untreated FPU. The physical properties examined were tensile strength, elongation, tear strength, resilience, indentation force deflection, support factor, compression sets, and air flow.

11. The use of melamine-treated FPU is becoming more common; it is one of two FPU foams currently allowed in Great Britain. Small-scale tests indicated that a melamine-treated FPU generated six times more HCN than an equal amount of a non-melamine-treated foam. The presence of Cu_2O reduced the HCN from the melamine foam by 90%.

Jellinek and co-workers in the late 1970s also showed that the concentrations of HCN generated from the thermal decomposition of a polyurethane at 300 and 400°C decreased when flowed through copper compounds. In their studies, the polyurethane films were usually 15 μm thick (50 mg). In some experiments, the metal powder was mixed with the polymer and, in others, copper metal films of 400–1000 Å were deposited on top of the polymer films. In most cases, the percentage of copper was 10% or greater. The lowest concentration that they tested was a 2.6% copper film which inhibited the evolution of HCN by 66%. Their experiments indicated that the copper is probably acting as an oxidative catalyst which would decompose gaseous HCN into N_2, CO_2, H_2O, and small amounts of nitrogen oxides. Further research is needed to determine if this is the actual molecular mechanism which allows copper to act as a HCN toxicant suppressant.

The research at NIST by Levin and co-workers differed from that of Jellinek in that much larger samples of FPU (including full-scale room burns of cushions and simulated chairs) were used, much smaller concentrations of copper were used, and the toxicity of the combustion products from the copper-treated FPU was also examined.

Unpublished data of Levin and Paabo also indicated that a wool fabric treated with copper would generate 50% less HCN than the untreated fabric. These results demonstrate a potentially more universal effect, namely, that treating nitrogen-containing materials with copper compounds will probably reduce the HCN generated when that material is exposed to fire conditions. Taking these results one step further, one could develop other toxicant suppressants which when added to materials and products would prevent or significantly reduce the toxic effluents that are generated when they are thermally decomposed. Since 80% of fire deaths are the result of smoke inhalation, a less toxic smoke could significantly increase the time available for escape and reduce the number of injuries and deaths from fire.

Acknowledgment

This chapter is a contribution of the National Institute of Standards and Technology (NIST) and is not subject to copyright. Certain equipment, instruments, materials, or companies are identified in this entry to specify adequately the experimental procedure. Such identification does not imply recommendation or endorsement by NIST, nor does it imply that the materials or equipment identified are the best available for the purpose.

Further Reading

Hardman, J. G., Limbird, L. E., Molinoff, P. B., Ruddon, R. W., and Gilman, A. G. (Eds.) (1996). *Goodman and Gilman's The Pharmacological Basis of Therapeutics*, 9th ed. McGraw-Hill, New York. [See also the 5th ed., 1975]

Kaplan, H. L., Grand, A. F., and Hartzell, G. E. (1983). *Combustion Toxicology, Principles and Test Methods.* Technomic, Lancaster, PA.

Karter, M. J., Jr. (1994, September/October). Fire loss in the United States in 1993. *NFPA J.*, 57–65.

Levin, B. C., Braun, E., Paabo, M., Harris, R. H., and Navarro, M. (1992). Reduction of hydrogen cyanide concentrations and acute inhalation toxicity from flexible polyurethane foam combustion products by the addition of copper compounds. In *Part IV. Effects of Combustion Conditions and Scaling on the Generation of Hydrogen Cyanide and Toxicity from Flexible Polyurethane Foam with and without Copper Compounds*, NISTIR 4989. National Institute of Standards and Technology, Gaithersburg, MD.

Nelson, G. L. (Ed.) (1995). *Fire and Polymers II; Materials and Tests for Hazard Prevention*, ACS Symposium Series 599. American Chemical Society, Washington, DC.

National Research Council, National Materials Advisory Board (1995). *Fire- and Smoke-Resistant Interior Materials for Commercial Transport Aircraft,*

Publication No. NMAB-477-1. National Academy Press, Washington, DC.

—*Barbara C. Levin*

Related Topics

Coke Oven Emissions
Indoor Air Pollution
Pollution, Air

Comprehensive Environmental Response, Compensation, and Liability Act

- ◆ TITLE: CERCLA; revised as the Superfund Amendments Reauthorization Act (SARA)
- ◆ AGENCY: U.S. EPA
- ◆ YEAR PASSED: 1981; SARA in 1986
- ◆ GROUPS REGULATED: Industry

Synopsis of Law

CERCLA, enacted by Congress in 1980, expands the authority of the federal (U.S. EPA) and state governments to investigate and respond to the release of materials into the environment. More important, the act permits the oil and chemical industries to be taxed in order to finance governmental action taken in response to the current or past release of pollutants. It also requires that all industries respond to releases immediately or reimburse the government for the costs it incurs in responding to the release of pollutants into the environment. Most commonly called superfund, SARA is best known for identifying and attempting to clean up sites identified as contaminated with toxic wastes.

—*Shayne C. Gad*

Related Topics

National Environmental Policy Act
Pollution Prevention Act
Resource Conservation and Recovery Act (RCRA)

Coniine

♦ SYNONYMS: S-2-propylpiperidine; cicutine; conicine; n-methylconine; conhydrine; psuedoconhydrine

♦ CHEMICAL CLASS: Piperidine alkaloid

♦ CHEMICAL STRUCTURE:

♦ DESCRIPTION: Poison hemlock (*Conium maculatum*) and dog parsley (*Aethusa cynapium*) are poisonous plants of the parsley family which contain coniine (see Hemlock, Poison).

Exposure Pathways

The most common route of coniine exposure is by ingestion, although there are reports of dermal and eye irritation.

Toxicokinetics

Coniine is rapidly absorbed from the gastrointestinal tract.

Mechanism of Toxicity

Coniine acts on the autonomic ganglia to produce initial stimulation of skeletal muscle followed by neuromuscular blockade. The actions of coniine are similar to those of nicotine but produce paralysis of greater numbers of central nervous system (CNS) and skeletal muscle nerve endings.

Human Toxicity

Toxic doses are difficult to determine due to differing concentrations of at least five alkaloids in each plant. A toxic dose of coniine is estimated to be 60 mg and a lethal dose is estimated to be 100–300 mg. The concentration of alkaloids varies with the age of the plant. Plants up to about 1 year old have very low alkaloid content in roots, approximately 0.15% in stems, and 0.3–0.6% in the leaves. Plants in their second year have an alkaloid content of about 1% in all parts of the plant. Geographic latitude and drying will also affect the alkaloid content.

The principal manifestations of coniine poisoning are nausea and vomiting, salivation, fever, and gradually increasing muscular weakness followed by paralysis with respiratory failure.

Clinical Management

No antidote exists for coniine. Treatment is directed at removing ingested toxin and supportive care. Gastric lavage may be used to remove the ingested plant. However, this method may not effectively remove large pieces of plant material. Administration of activated charcoal is recommended to reduce absorption into the gastrointestinal tract. Due to the rapid onset of CNS depression and seizures, emesis is generally not recommended. Respiratory failure is treated by artificial respiration with oxygen. Convulsions are controlled with diazepam.

Animal Toxicity

Certain small birds (skylarks, chaffinches, and robins) are not susceptible to coniine poisoning. Coniine toxicity has been reported in cows, goats, horses, pigs, sheep, ewes, rabbits, and chickens. The oral LD_{50} of coniine in the mouse is approximately 100 mg/kg. There are limited data in cattle, goats, and sheep suggesting developmental abnormalities to the musculoskeletal system of offspring when pregnant mothers are exposed orally to coniine (70 mg/kg for cattle and 484 mg/kg for goats and sheep).

Further Reading

Rizzi, D., Basile, C., Dimaggio, A., *et al.* (1989). Rhabdomyolysis and acute tubular necrosis in coniine (Hemlock) poisoning. *Lancet* 2, 1461–1462.

—*Todd A. Bartow and Paul W. Ferguson*

Consumer Product Safety Commission

The U.S. Consumer Product Safety Commission (CPSC) is an independent federal regulatory agency created by Congress in 1972 under the Consumer Product Safety Act. The agency's mission is to "protect the public against unreasonable risks of injuries and deaths associated with consumer products." The CPSC has jurisdiction over about 15,000 types of consumer products ranging from coffee makers to toys, lawn mowers, and fireworks.

However, some types of consumer products covered by other federal agencies are not included in CPSC's jurisdiction. For example, cars, trucks, and motorcycles are within the jurisdiction of the Department of Transportation, and food, drugs, and cosmetics are covered by the Food and Drug Administration.

Hotline Services

Consumers can contact CPSC to

Report an unsafe product

Report a product-related injury

Find out whether a product has been recalled

Learn how to return a recalled product or arrange for its repair

Get information on what to look for when buying a consumer product

Get information on how to use a consumer product safely

Receive information about ordering CPSC safety publications

The CPSC hotline number is (800) 638-2772. For the hearing- or speech-impaired, the hotline TTY number is (800)-637-8270.

The hotline staff take calls from consumers with product complaints or reports of product-related injury. The hotline is available between 8:30 AM and 5:00 PM Eastern Standard Time, Monday through Friday, except holidays.

The hotline staff speak both English and Spanish. In addition, arrangements can be made for callers to speak with someone in any of the following languages: Arabic, Burmese, Cambodian, Cantonese, Chinese, French, German, Greek, Hindi, Italian, Korean, Japanese, Punjabi, Ukrainian, Urdu, Vietnamese, and Yiddish.

How to Reach the Commission

Consumers wishing to report a dangerous product or a product-related injury and to receive information on CPSC's fax-on-demand service should call CPSC's hotline at (800) 638-2772 or CPSC's teletypewriter at (800) 638-8270. Press releases can be ordered through fax-on-demand by dialing (301) 504-0051 from the handset of a fax machine and then entering the release number. Consumers can obtain this release and recall information at CPSC's web site at http://www.cpsc.gov or via internet gopher services at cpsc.gov. Consumers can report product hazards to info@cpsc.gov.

For more information, contact: U.S. Consumer Product Safety Commission, Washington, DC 20207.

—Harihara M. Mehendale

Related Topics

Food and Drug Administration (U.S. FDA)
Food Drug and Cosmetic Act
Proposition 65
State Regulation of Consumer Products

Copper (Cu)

◆ CAS: 7440-50-8

◆ SELECTED COMPOUNDS: Cupric chloride, $CuCl_2$ (CAS: 7447-39-4); cupric sulfate, $CuSO_4$ (CAS: 7758-98-7)

◆ CHEMICAL CLASS: Metals

Uses

Copper, an essential trace element, is widely distributed in nature and widely used in industry. It is used as an electrical conductor, as a component in a variety of alloys (including gold and silver alloys), and as a constituent in paints and ceramic glazes. Because it corrodes at a very slow rate, it is used extensively for water pipes. In addition, copper sulfate mixed with lime is used as a fungicide.

Medicinally, copper sulfate is used as an emetic. It has also been used as an antihelminthic (antiparasitic agent) based on its astringent and caustic actions.

Exposure Pathways

The primary exposure pathway for copper is ingestion (e.g., food and water). Many foods contain copper, especially legumes, organ meats, and oysters. Water carried through copper pipes is also a source of this element. Inhalation is only a significant exposure pathway in industrial settings (e.g., near copper refineries).

Toxicokinetics

Approximately 50% of ingested copper is absorbed from the stomach. Although copper can be absorbed from the gastrointestinal tract, a modifying biological mechanism regulates total copper absorbed. Copper is transformed in the blood by first binding to albumin and then to a copper-specific protein (ceruloplasmin). Copper also binds to metallothionein more firmly than zinc or even cadmium. Copper is stored in the liver and bone marrow as the metallotheionein.

Copper-dependent enzymes include tyrosinase (which is involved in melanin pigment formation) and the various oxidases (i.e., cytochrome oxidase, superoxide dismutase, amine oxidase, and uricase). Copper plays a major role in the incorporation of iron into the heme of hemoglobin. Copper deficiency is characterized by hypochromic, microcytic anemia resulting from defective hemoglobin synthesis.

Copper levels in the human body vary with age. Copper levels in the brain increase with age, whereas in some tissues (e.g., liver, lungs, and spleen), copper levels are higher in newborns than in adults. Tissue levels gradually decline up to age 10 and remain relatively constant thereafter. Copper is normally excreted in bile, which plays a primary role in copper homeostasis.

Mechanism of Toxicity

Copper reduces glutathione, which is necessary for normal cell viability. The amino acid transferases are inhibited in the presence of excess copper; lipid peroxidation also occurs. Copper combines with thiol groups, which reduces the oxidation state 2 to 1 in copper and oxidizes the thiol groups to disulfides, especially in the cell membrane.

Human Toxicity

Although copper is an essential element, it is much more toxic to cells than such nonessential elements as nickel and cadmium. Acute poisoning from ingestion of excessive amounts of copper salts, most frequently copper sulfate, results in nonspecific toxic symptoms, a metallic taste, nausea, and vomiting (with vomitus possibly a blue-green color). The gastrointestinal tract can be damaged by ulceration. Severe symptoms include hypotension, coma, jaundice, and death. Liver necrosis has also been observed. In some cases, copper toxicity can result in an inability to urinate. Treatment with copper compounds can induce hemolytic anemia.

It is believed that the increased susceptibility to copper toxicity seen in infants and children is due to the normally high hepatic copper levels in early life and the fact that homeostatic mechanisms are not fully developed at birth.

Copper is associated with two genetic inborn errors of metabolism. The first, Menke's disease or Menke's kinky-hair syndrome, is a sex-linked trait characterized by peculiar hair, failure to thrive, severe neurological degradation in the brain, and death before 3 years of age. The cerebral cortex and white matter degenerates; mental retardation ensues before death. The second disease, Wilson's disease or heptolenticular degeneration, is characterized by an unusual concentration of copper in the brain, kidneys, cornea, and especially in the liver (which may become abnormally large). Mental retardation is not associated with this disease. This disease is usually treated with a chelating agent such as penicillamine or triethylene tetramine. The disease is of genetic etiology but the defect at the biochemical level is unknown.

The ACGIH TLV-TWA is 0.2 mg/m$_3$ for copper fume and 1 mg/m$_3$ for copper dusts and mists.

Clinical Management

For acute toxicity, emesis is recommended. Treatment is symptomatic. A combination of BAL (British Anti-

Lewisite; 2,3-dimercaptopropanol) and calcium–ethylene diamine tetraacetic acid has been used successfully in a poisoned infant. Penicillamine has also been used. Recently, oral administration of 2,3-dimercapto-1-propane sulfonate was found to be effective in experimental rodents. Electrolyte balance must be maintained when gastric lavage is indicated. Potassium ferrocyanide should be added to precipitate the copper.

Animal Toxicity

Copper produces lung damage by inhalation. Intratracheal administration of copper has produced lung damage in rodents; macrophages increased with degenerative membrane structure and hemoglobin values decreased. In larger animals, excess copper intake resulted in iron-deficient anemia and gastric ulcers.

No statistically significant increases in tumor formation were noted in mice fed copper for approximately 1 year. Subcutaneous and intramuscular injection of copper compounds showed a low incidence of sarcomas. The current data are adequate to assess the carcinogenicity of copper.

Mutagenesis results are dependent on the bacterial strain and copper compound evaluated. Mammalian cell tests indicate a positive mutagenic response.

—Arthur Furst, Shirley B. Radding,
and Kathryn A. Wurzel

Related Topics

Metallothionein
Metals
Pollution, Water

Corrosives

- ◆ SYNONYMS: Irritants; acids/bases
- ◆ REPRESENTATIVE COMPOUNDS: Glacial acetic acid; oxalic acid; acetic acid; nitric acid; hydrofluoric acid; bromine; potassium hydroxide; chlorine; sodium hydroxide; fluorine; sulfuric acid; hydrochloric acid (see Acetic Acid, Hydroiodic Acid, Hydrobromic Acid, Hydrochloric Acid, and Hydrofluoric Acid)

Uses

Some of the phosphoric acids are used as metal conditioners and rust removers. Acids are also used in analytical reagents, household cleaning products, fertilizers, and munitions.

Exposure Pathways

The most common route of exposure is the skin, but exposure may also occur by inhalation and ingestion.

Toxicokinetics

Acids are corrosives that penetrate the skin quickly causing a coagulation-type necrosis, with destruction of surface epithelium and submucosa. Therefore, subsequent absorption and distribution is negligible.

Mechanism of Toxicity

Corrosive acids destroy tissues by direct chemical action. The tissue protein is converted to acid proteinate, which dissolves in the concentrated acid.

Human Toxicity

After ingestion, corrosive injury to the esophagus and stomach are commonly found. With skin contact, the symptoms are severe pain and brownish or yellow stains. Burns usually penetrate the full thickness of the skin, have sharply defined edges, and heal slowly with scar formation. With eye contact, conjunctival edema and corneal destruction is prevalent. Symptoms include pain, tearing, and photophobia.

Long-term exposure to acid fumes (inhalation exposure) may cause erosion of the teeth followed by jaw necrosis. Bronchial irritation with chronic cough and frequent attacks of bronchial pneumonia are common.

The OSHA PELs for various corrosives are as follows: glacial acetic acid, 10 ppm; acetic anhydride, 5 ppm; hydrofluoric acid, 3 ppm; sulfuric acid, 1 mg/m^3; oxalic acid, 1 mg/m^3; nitric acid, 2 ppm; bromine, 0.1 ppm; chlorine, 1 ppm; fluorine, 1 ppm; hydrochloric acid, 5 ppm.

Clinical Management

In case of ingestion, neither gastric lavage nor emesis should be used. Ingested corrosives must be diluted within seconds by drinking quantities of water or milk. If vomiting is persistent, administer fluids repeatedly.

In case of eye contact, the corrosive should be diluted by irrigating the area for 30–60 min.

In case of skin contact, the corrosive should be removed by flooding the affected area with water for at least 15 min.

—Stephanie E. Foster and Paul W. Ferguson

Related Topics

Acids
Alkalies
Bases
Ocular and Dermal Studies
Sensory Organs
Skin

Corticosteroids

♦ MAJOR CLASSES: Mineralocorticosteroids (e.g., aldosterone); glucocorticoids (e.g., corticosterone)

♦ PHARMACEUTICAL CLASS: Natural or synthetic hormones

♦ CHEMICAL STRUCTURES:

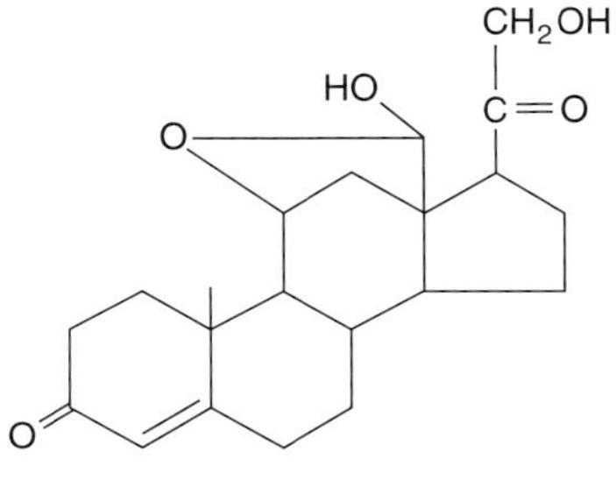

Aldosterone

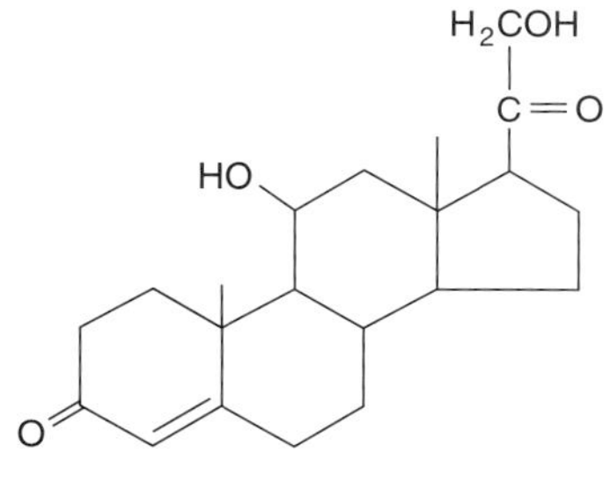

Corticosterone

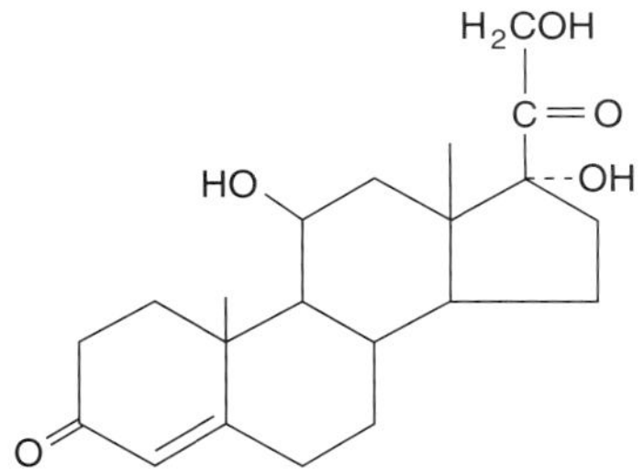

Cortisol

Uses

Corticosteroids cause a wide range of physiological effects, including impacts on protein and lipid metabolism; electrolyte and water balance; and functions of the cardiovascular system, the kidneys, skeletal muscle, the nervous system, and other organs and tissues. Corticosteroids are used to treat a wide variety of clinical conditions including adrenal insufficiency, asthma, allergic disorders, and collagen and autoimmune diseases.

Exposure Pathways

Corticosteroids are administered orally, parenterally, and topically. A certain degree of absorption into the systemic circulation occurs with all forms of topical administration. With respiratory aerosols, the total absorption is equivalent to that from parental or oral administration.

Toxicokinetics

Generally, the biological half-lives of corticosteroids can be classified as short (8–12 hr), intermediate (12–36 hr), or long (36–72 hr). Cortisone and cortisol are examples of short-lived corticosteroids. Prednisone, prednisolone, and triamcinolone are of the intermediate class. Dexamethasone and β-methasone are associated with the longer-lived class.

The adrenocortical steroids and their synthetic congeners require a double bond in the 4,5 position and a ketone group at C3 for biological activity. The reduction of the 4,5 double bond, resulting in an inactive compound, occur by both hepatic and extrahepatic metabolism. Most of the ring A-reduced metabolites can be conjugated at the 3-hydroxyl position with sulfate or glucuronic acid forming water-soluble metabolites enhancing excretion.

Mechanism of Action

The corticosteroids, like other steroid hormones, act by altering the nature of protein synthesis in target tissues. Corticosteroids interact with specific receptor proteins found in the cytoplasm of cells in many tissues to form a steroid–receptor complex. This complex then translocates into the nucleus, where it combines with DNA sequences within the regulatory region of affected genes (termed glucocorticoid response elements). Subsequently, target genes are expressed and appropriate proteins synthesized. Although there are many similarities between the mechanisms of action of the glucocorticoids and mineralocorticoids, several processes have been identified, such as tissue-restricted receptors for mineralocorticoids, to explain differences in effects of these two major corticosteroid classes.

Human Toxicity

Corticosteroids are known for two primary types of adverse effects associated with their therapeutic uses: toxicities resulting from withdrawal and those resulting from continued use of large doses (supraphysiological dose). The most significant problem of corticosteroid cessation is acute adrenal insufficiency due to long-term suppression of the hypothalamic–pituitary–adrenal axis. In the second category of adverse effects, toxicity may result from excessive overdosage of the corticosteroid. Adverse effects from long-term use of high doses of corticosteroids may include alterations in fluid and electrolyte balance causing alkalosis, edema, and hypertension; altered immune response with increased susceptibility to infection; weakness of proximal limb muscles; behavioral changes including mood swings, nervousness, and insomnia; and cataracts and osteoporosis. Growth retardation may occur from long-term administration of glucocorticoids in children.

Clinical Management

Acute overdose probably would not result in toxicity. Should oral overdosage occur, standard emergency and supportive care procedures should be followed. If anaphylaxis should occur, epinephrine may be given as 0.3–0.5 ml of a 1:1000 solution for adults (children should receive 0.01 ml/kg). Mild anaphylaxis may be treated with antihistamines alone. If chronic toxicity should occur, it is important to reduce the dosage of corticosteroid to a minimal maintenance dose at the first sign of toxicity.

—*Nancy A. Jeter, John A. McCants,
and Paul W. Ferguson*

Cosmetics

Introduction

Cosmetics are natural or synthetic products that are used to maintain hygiene and include products used to enhance appearance. This class includes dental products, bath supplies (e.g., bubble baths and bath beads), powders, lotions, lipsticks, perfumes, colognes, shampoos, and hair coloring/waving products. Most of these products contain alcohols, aromatic hydrocarbons, perborates, and anionic and nonionic surfactants. Use of cosmetics is as old as civilization itself. Women used belladonna alkaloids like atropine to dilate pupils to enhance the attractiveness of the eyes in the late nineteenth century (see Atropine, Belladonna Alkaloids, and Surfactants).

Product Formulations and Human Toxicity

Most cosmetics are nontoxic. However, caution should be exercised to prevent children from being overexposed to these products.

Hair-Coloring Products

Permanent hair colors contain an oxidizer (usually 6% hydrogen peroxide) and a dye intermediate (*p*-

phenylenediamine, resorcinol, aminophenols along with water, ammonia, glycerin, isopropanol, and propylene glycol). Semipermanent hair colors contain propylene glycol, isopropanol, fatty acids, fragrance, alkanolamines, and dyes. Some Grecian hair formulations contain lead (see Ammonia, Hydrogen Peroxide, Isopropanol, and Lead).

Large ingestions of hydrogen peroxide may produce mild gastritis due to decomposition resulting in release of oxygen.

Hair-Waving Products

Waving lotions contain thioglycolic acids and ammonia sulfides, and neutralizer solutions contain hydrogen peroxide, sodium bromate, or perborate in mildly acidic solutions. Some permanent wave fixatives contain 2–8% (weight/volume) mercuric chloride.

Sodium borate decomposes into borate and peroxide and is less toxic than potassium bromate. From 3 to 6 g and from 15 to 30 g boric acid is potentially fatal to children and adults, respectively. Cutaneous manifestations include desquamating, erythematous rash commonly over palms, soles, buttocks, and scrotum. The lesion may progress to exfoliation. Central nervous system (CNS) effects range from irritability, restlessness, and headache to coma and convulsions in severe cases. Gastrointestinal symptoms include anorexia, nausea, vomiting, and diarrhea. Acute renal tubular necrosis may lead to renal failure in moderate to severe cases (see Boric Acid).

Bromate salts are extremely toxic; they are capable of causing deafness and renal failure at doses between 240 and 500 mg/kg. Potassium bromate also used as neutralizer in cold waves is an extremely toxic compound which produces nausea, vomiting, diarrhea, deafness, acute renal failure, hypotension, CNS depression, and hemolysis. Both otic symptoms and renal impairment may be permanent. Primary tubular damage can progress to interstitial fibrosis and glomerular sclerosis.

Hair-Straightening Products

Hair straighteners contain 1–3% sodium hydroxide solution. The solution is highly caustic (see Alkalies).

Hair Sprays and Conditioners

Hair sprays contain ethanol as a solvent with resin polymers composed of vinyl acetate, acrylamide, and methyl vinyl ether (see Acrylamide). Hair conditioners contain cationic surfactants, perfumes, and alcohols. For toxicity information, see Perfumes.

Bath Preparations

Bubble baths usually contain anionic and nonionic surfactants along with alcohols and preservatives. Bath salts may contain borax, while bath oils contain vegetable and mineral oils.

Nail Polish and Removers

Nail polish contains hydrocarbon solvents (xylene, toluene, and acetone), alcohol solvents, plasticizers, and resins. Nail polish removers are solvents containing acetone or ethanol (see Acetone, Ethanol, Toluene, and Xylene).

Colognes, Perfumes, Toilet Waters

Colognes, perfumes, and toilet waters usually contain ethanol (at concentrations ranging from 50 to 95%) and volatile or essential oils (see Ethanol).

Volatile or Essential Oils

Sage, eucalyptus, turpentine, pine, pennyroyal, and cinnamon contain hydrocarbons, ethers, alcohols, esters, and ketones. These components can cause allergic contact dermatitis which begins 12 hr within sensitization and peaks at 48–72 hr. Essential oils are mucosal irritants leading to gastrointestinal distress and salivation. Concentrated formulations of essential oils can cause convulsions and CNS depression at 10-ml doses. Aspiration can cause chemical pneumonitis. Alcohol produces intoxication, which may be complicated by hypoglycemia, especially in children.

Dental Products

Toothpastes, powders, and tooth liquids contain calcium phosphates, alumina, abradants, and anionic surfactants. Mouth washes usually contain alcohol, flavoring (essential oils), and sweeteners. (For mouthwash toxicity information see Colognes, Perfumes, Toilet Waters.). Denture cleaners contain bicarbonates, borates, phosphates, and carbonates. (For toxicity information on borates, see Hair-Waving Solutions). Acrylic denture material contains methacrylate.

Deodorants

Deodorants contain aluminum and zinc (see Deodorants).

Clinical Management

Since most ingested cosmetics are nontoxic, only supportive care and dilution are required.

1. Induction of emesis depends on product toxicity, quantity, time since exposure, patient's weight, and the presence of symptoms. Cationic surfactants, perborates, and substantial ingestion of essential oils may benefit by administration of syrup of ipecac. Syrup of ipecac can be used in hydrocarbon ingestion only if the total dose of hydrocarbons exceeds 1 or 2 ml/kg.

2. Potassium borate: Lavage with 2% sodium bicarbonate solution and administer 10–50 ml of 10% sodium thiosulfate solution intravenously at the rate of 3 ml/min to reduce the bromate to the less toxic bromide ion. An alternative therapy is the administration of 100–500 ml of 1% sodium thiosulfate. Patients should be observed for development of renal toxicity and ototoxicity.

—*Prathibha S. Rao, Swarupa G. Kulkarni, and Harihara M. Mehendale*

Related Topics

Deodorants
Food, Drug and Cosmetic Act
Surfactants, Ionic and Nonionic

Cotinine

- CAS: 486-56-6
- SYNONYMS: 1-Methyl-5-(3-pyridinyl)-2-pyrrolidinone; *N*-methyl-2-(3-pyridyl)-5-pyrrolidone; CAS 486-56-6.
- CHEMICAL CLASS: Research chemical; antidepressant
- CHEMICAL STRUCTURE:

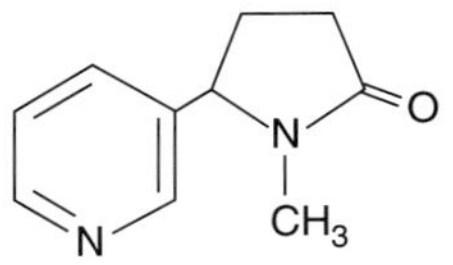

Uses
Cotinine is primarily used in research.

Exposure Pathways
Cotinine is a viscous liquid. Dermal or ocular contact are the most common exposure pathways. Because it is a predominant metabolite of nicotine, systemic exposure occurs after consumption of tobacco products.

Toxicokinetics
Cotinine is formed as a major metabolite of nicotine after tobacco smoking. The average half-life of cotinine is 19 hr (range, 10.9–37 hr). It can be detected in plasma, urine, and saliva. Cotinine is also formed in the body after intake of some vegetables (e.g., eggplant, tomato, and green pepper) primarily of the family Solanaceae. These vegetables contain nicotine as their natural defense mechanism against fungi, bacteria, insects, and animals.

It had been thought that the presence of cotinine in human urine could be used as evidence of smoking or tobacco use. However, because cotinine can also arise from consumption of some vegetables, use of cotinine as a biomarker of exposure to tobacco smoke or other forms of tobacco products may not be reliable.

Mechanism of Toxicity

Cotinine stimulates the nicotinic receptors.

Human Toxicity

Symptoms of acute toxicity include nausea, salivation, abdominal pain, vomiting, diarrhea, cold sweat, headache, dizziness, and disturbed hearing and vision.

Clinical Management

Vomiting should be induced with syrup of ipecac or gastric lavage should be performed. Respiratory assistance and treatment of shock may be necessary.

Animal Toxicity

The toxicity in animals is similar to that observed in humans. The intraperitoneal LD_{50} in mice is 930 mg/kg; the oral gavage LD_{50} in mice is 1604 mg/kg.

—*Sanjay Chanda*

Coumarins

- CAS: 91-64-5
- REPRESENTATIVE COMPOUNDS: Coumarin, warfarin, dicoumarol (see Warfarin)
- SYNONYMS: 1,2-benzopyrone; 2-*H*-benzopyran-2-one; 2,3-dihydrobenzofuran; benzo-α-pyrone; coumarinic anhydride; Cumarin; Tonka bean camphor
- CHEMICAL CLASS: Benzopyranone
- CHEMICAL STRUCTURES:

Uses

The coumarins are used as rodenticides.

Exposure Pathways

Poisonings may occur primarily through acute accidental or suicidal ingestion of these compounds. Repeated daily oral doses of these compounds have been reported to cause fatalities. Coumarin is found to be a food contaminant in vanilla from Mexico and other Caribbean countries. It was first isolated from sweet clover.

Toxicokinetics

Coumarins are rapidly absorbed from the gastrointestinal tract. The two primary metabolites of coumarin are 7-hydroxycoumarin and *O*-hydroxyphenyl acetic acid. Neither metabolite has any anticoagulant property but the latter compound was reported to be hepatotoxic. The metabolism of coumarins is species dependent. In humans, 7-OH coumarin was found to be the major metabolite (80%) and only 6% is O-OH-phenylacetic acid. 7-HCG is actively secreted through renal tubules and exclusively excreted through urine within 24 hr. The biological half-life of coumarin is 0.9 hr and that of 7-HCG is 1.3 hr.

Mechanism of Toxicity

Coumarins act as an antimetabolite of vitamin K and, therefore, inhibit the synthesis of a number of vitamin K-dependent clotting factors. Death may occur as a result of excessive hemorrhage due to depletion of prothrombin. These anticoagulants also cause capillary fragility.

Human Toxicity: Acute

The symptoms of acute human poisoning include hemoptysis, hematuria, bloody stools, widespread bruising, and bleeding into synovial spaces.

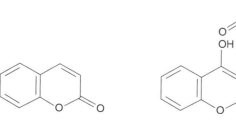

Coumarin   Warfarin

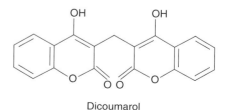

Dicoumarol

Human Toxicity: Chronic

Repeated exposure to coumarins may produce symptoms similar to those of acute poisoning. Skin necrosis is a rare complication following coumarin administration.

Clinical Management

In case of oral or parenteral exposures, gastrointestinal absorption could be retarded by inducing emesis with syrup of ipecac. Lavage is recommended to decontaminate the gut if emesis is found to be unsuccessful following two doses of ipecac. Activated charcoal in the form of an aqueous slurry can be orally administered to adsorb these compounds. The following measures should be considered if large amounts are ingested accidentally: (1) determination of prothrombin time, (2) liver function tests, and (3) central nervous system (CNS) support for depression and possible muscle paralysis.

Animal Toxicity

Considerable evidence suggests that coumarins may produce CNS depression, hepatotoxicity, muscle paralysis, and renal impairment in rats, rabbits, mice, frogs, dogs, and fish. The type of symptoms, toxic dose, and metabolite formation are species dependent. A single dose of 200–400 mg/kg can cause death in common laboratory animals. When animal data were extrapolated to humans, it was estimated that an adult human being would have to consume 1.5 lb of a warfarin concentrate or about 30 lb of a strong rat bait to cause death.

—Tamal Kumar Chakraborti

Creosote

♦ CAS: 8001-58-9

♦ SYNONYMS: Creosote; coal tar oil; brick oil; heavy oil; naphthalene oil; liquid pitch oil

♦ CHEMICAL CLASS: Petroleum distillate

♦ DESCRIPTION: Creosote is described as a brown to black oily liquid with a smoky smell and a burning caustic taste. A distillate of coal (200–250°C), creosote contains an estimated 162 different compounds. The estimated composition is as follows: aliphatic hydrocarbons (7%), polycyclic aromatic hydrocarbons (69%), and nitrogen containing polycyclic aromatic hydrocarbons (11%). Some of the polycyclic aromatic hydrocarbons identified in creosote are anthracene, benz(a)anthracene, benzo(a)pyrene, and pyrene (see separate entries for these compounds).

Uses

Creosote is primarily used as a wood preservative in the United States. It has also been used as a disinfectant, antiseptic, germicide, crop defoliant antifungal preparation, and as an animal or bird repellent.

Exposure Pathways

The primary route of exposure is dermal through handling of treated wood. Inhalation is also possible, particularly when treated wood is burned in a poorly ventilated area.

Toxicokinetics

Creosote is readily absorbed through the skin and gastrointestinal tract. As creosote is diluted, the absorption rate may actually increase. Creosote is not expected to bioconcentrate (the log of the octanol to water partition coefficient is 1). Creosote appears to be primarily excreted in the urine. Conjugation with sulfuric and hexuronic acids as well as oxidation leads to a "smoky" appearance of the urine.

Human Toxicity

Toxicity is expressed either via general depression with cardiac collapse or via the irritating/corrosive nature by irritation and burns of the skin and eyes. Large doses (7 g for adults and 1 or 2 g for children) have been associated with death 14–17 hr after ingestion. Cardiovascular collapse appears to be the primary cause of death. Nonlethal symptoms include salivation, vomiting, thready pulse, headache, and loss of pupillary reflexes.

Human Toxicity: Chronic

Chronic exposure to low levels can lead to gastrointestinal bleeding, visual disturbances, and intoxication.

Long-term self-medication has resulted in chronic symptoms of intoxication and visual disturbances. People with psoriasis who receive long-term treatment with creosote and ultraviolet radiation may be at increased risk for skin cancer.

Creosote is regulated as a combustible/flammable liquid for transport. The U.S. EPA and IARC classify creosote as a probable human carcinogen, class B1 and 2A, respectively.

Clinical Management

Acute episodes are treated similar to phenolic poisonings with initial stabilization of breathing and cardiac monitoring. Dermal decontamination is accomplished by swabbing the affected area with olive oil. For ingested material, the preferred method is administration of activated charcoal followed by a cathartic. Phenol and phenolic substances tend to exhibit an increased absorption rate at dilute concentrations and have a rapid onset of acute symptoms; therefore, there is a potential for seizures.

Animal Toxicity

Cattle have been poisoned by licking treated lumber; the estimated dose was 4–6 g/kg. A mixture of fuel oil and creosote was once widely distributed as a cure for ringworm. Excessive application of this material has caused poisoning of animals. Skin painting studies with mice have produced skin tumors. Creosote was mutagenic in the Ames test when metabolically activated with S9. The acute LD_{50} in rats is 725 mg/kg.

—*William S. Utley*

Related Topics

Petroleum Distillates
Polycyclic Aromatic Hydrocarbons

Cromolyn

- PREFERRED NAME: Cromolyn (Intal)
- SYNONYMS: Cromolyn sodium; disodium cromoglycate; disodium salt of cromolyn

- PHARMACEUTICAL CLASS: Antiinflammatory drug used in the treatment of asthma.
- CHEMICAL STRUCTURE:

Cromolyn sodium

Uses

Cromolyn is used primarily for the treatment of mild to moderate bronchial asthma and for the prevention of asthma attacks.

Exposure Pathways

For use in the treatment of asthma, cromolyn is administered by inhalation using solutions delivered by aerosol spray or nebulizer as well as a powdered drug mixed with lactose and delivered by a turbo inhaler.

Toxicokinetics

Usually, no greater than 10% of an inhaled dose of cromolyn is absorbed systemically. After complete absorption, cromolyn is excreted unchanged in urine and bile in about equal proportions. Peak plasma concentrations occur 15 min after inhalation. The distribution of cromolyn in the lung and the extent of systemic absorption is enhanced by bronchodilation during drug delivery.

The biological half-life following inhalation ranges from 45 to 100 min. By comparison, the terminal elimination half-life following intravenous administration is 20 min.

Mechanism of Action

The complete mechanism of action of cromolyn is yet to be fully defined. The major prophylactic effect of cromolyn is centered on inhibition of the degranulation of pulmonary mast cells causing a reduction in histamine release, reduced leukotriene production, and inhibition of release of inflammatory mediators from several cell types.

Human Toxicity

Because of its low toxicity, cromolyn is generally well tolerated. Adverse side effects, such as bronchospasm, cough, wheezing, laryngeal edema, joint swelling, joint pain, angioedema, headache, rash, and nausea, are rare (less than 1 in 10,000 patients). Documented instances of anaphylaxis have also been rare.

Clinical Management

Toxicity is rare, but should adverse effects occur general emergency management and supportive care procedures are indicated.

Animal Toxicity

The acute toxicity of cromolyn, measured as the LD_{50}, has been determined in the rat (6000 mg/kg, subcutaneous) and the mouse (3300 mg/kg, intravenous; 1000 mg/kg, intraperitoneal; and 4400 mg/kg, subcutaneous).

Further Reading

Kelly, H. W. (1997). Asthma pharmacotherapy: Current practices and outlook. *Pharmacotherapy* **17**(Pt. 2), 13S–21S.

—*Nancy A. Jeter and Paul W. Ferguson*

Cyanamide

- ◆ CAS: 420-04-2
- ◆ SYNONYMS: Amidocyanogen; cyanoamine; cyanogen nitride; *N*-cyanoamine; carbodiamide; carbamide
- ◆ CHEMICAL CLASS: Cyanamides
- ◆ MOLECULAR FORMULA: CH_2N_2

Uses

Cyanamide is used in the production of synthetic rubber, cyanide, fumigants, and metal cleaners. It is used as an intermediate in the chemical production industry.

Exposure Pathways

Workers handling cyanamide may be exposed through inhalation, ingestion, and ocular and dermal contact.

Toxicokinetics

The cyanide ion is not a metabolite of cyanamide biotransformation. Cyanamide is metabolized to acetylcyanamide and excreted in the urine.

Mechanism of Toxicity

Cyanamide acts by inhibiting liver aldehyde dehydrogenase.

Human Toxicity

Signs of systematic absorption and toxicity include flushing of the face and upper body, nausea, vomiting, fatigue, dyspnea, headache, and shivering. The most serious effects associated with acute, high-dose exposure are coma and cardiovascular collapse. Typical symptoms of cyanamide poisoning include vasodilation, tachycardia, bradycardia, and hypotension. These symptoms are similar to those presented following disulfiram exposure.

Cyanamide poisoning produces overactivity of the parasympathetic nervous system causing miosis, salivation, lacrimation, and twitching.

Cyanamide is irritating to living tissue. Direct contact with cyanamide is irritating to the eyes, skin, mucus membranes, and respiratory and gastrointestinal tracts. Chronic, repeated exposure may produce pneumonitis and pulmonary edema upon repeated dust inhalation; throat ulceration and esophageal irritation upon oral ingestion; dermal ulceration, dermatitis, and sensitization upon skin exposure; and keratitis, conjunctivitis, or corneal ulceration upon repeated contact with eye tissue.

Clinical Management

For irritant exposure, the treatment indicated for the specific exposure should be followed. Affected skin and eyes should be flushed with copious amounts of water. If ingested, vomiting should not be induced; instead, water or milk should be given if patient is conscious. Gastric lavage treatment may be given with caution and avoided if tracheal or esophageal ulceration is suspected. In cases of respiratory exposure, the victim should be moved to fresh air immediately and treated according to severity of irritation. The presence and severity of respiratory irritation, bronchitis, and pneumonitis should be evaluated. Assisted ventilation (100% humidified supplemental oxygen) should be provided as required, and arterial blood gases should be monitored. In any case following severe exposure, basic life support systems should be instituted.

Animal Toxicity

Cyanamide is very toxic by the oral exposure route and moderately toxic by the dermal route. The oral LD_{50} for rats is 125 mg/kg; the skin LD_{50} for rabbits is 590 mg/kg. In rats, cyanamide poisoning produces overactivity of the parasympathetic nervous system causing miosis, salivation, lacrimation, and twitching. Symptoms of severe poisoning in rats include constricted pupils at first followed by markedly dilated pupils. In addition, vessels of the iris and retina become congested. In severe cases, papilledema occurs.

—Heriberto Robles

Cyanide

- CAS: 57-12-5
- SYNONYMS: Carbon nitride ion; cyanide anion; cyanide ion; Cyanure (French); hydrocyanic acid; isocyanide; (RCRA waste No. P030)
- CHEMICAL CLASS: A cyanide is any one of a group of compounds containing the monovalent combining group CN. Inorganic cyanides are regarded as salts of hydrocyanic acid (hydrogen cyanide). Organic cyanides are usually called nitriles.
- CHEMICAL STRUCTURE:

$$-\text{C} \equiv \text{N}$$

Uses

Cyanide compounds are widely used in industry. Sodium cyanide and potassium cyanide are used extensively in the extraction of gold and silver from low-grade ores. The cyanide ion can form a wide range of complex ions with metals. These complex metal cyanide ions are extensively used in electroplating. Cyanide compounds are also used in case-hardening of iron and steel, metal polishing, photography, and the fumigation of ships and warehouses. Organic cyanide compounds are used in synthetic rubber, plastics, and synthetic fibers; they are also used in chemical synthesis. Cyanides are used in rodenticide and fertilizer production.

In addition, cyanides can be found in the seeds of the apple, peach, plum, apricot, cherry, and almond in the form of amygdalin, a cyanogenic glycoside. Amygdalin (Laetrile) has been used as an antineoplastic drug, but such beneficial effects have not been scientifically proven.

Exposure Pathways

Humans may be exposed to cyanide in a number of different forms. These include solids, liquids, and gases. Sources include industrial chemicals, natural products, medications, and combustion products. Inhalation of toxic fumes and ingestion of cyanide salts, cyanide-containing fruit seeds, and cyanide waste-contaminated drinking water are the most common exposure pathways. The respiratory route represents a potentially rapidly fatal type of exposure. Exposure to cyanides may also occur via the dermal route in industrial workers.

Toxicokinetics

Cyanide is rapidly absorbed from the skin and all mucosal surfaces; it is most dangerous when inhaled because toxic amounts are absorbed with great rapidity through the bronchial mucosa and alveoli. Once absorbed, distribution of cyanide through the body is rapid. Within a few minutes, cyanide is distributed through the body and its conversion to thiocyanate starts. Within 3 hr, 90% of the dose of cyanide is converted to thiocyanate appearing in blood. In sublethal doses, cyanide reacts with sulfane sulfur to form nontoxic thiocyanate through an enzymatic reaction involving rhodanase and mercaptopyruvate sulfur transferase. Cyanide is also trapped as cyano of vitamin B_{12}, oxidized to formate and carbon dioxide, and incorporated into cysteine. In nonfatal cases, metabolized cyanide (thiocyanate) is excreted in the urine. Although cyanide is volatile, excretion through the lungs is not a significant route of elimination of cyanide.

Mechanism of Toxicity

The cyanide ion produces cellular anoxia by reversibly inhibiting those oxidizing enzymes which contain ferric ($^{3+}$) iron, particularly cytochrome oxidase, the terminal oxidase of the mitochondrial respiratory chain. Tissue asphyxia, notably of the central nervous system (CNS), results in central respiratory arrest and death. For this reason, cyanide action has been described as "internal asphyxia." The respiratory center of the medulla ceases to function because its nerve cells can no longer obtain oxygen for their respiration. Although some cyanide combines with hemoglobin to form a stable non-oxygen-bearing compound, cyanhemoglobin, this substance is formed only slowly and in a small amount. Therefore, death is not due to cyanhemoglobin but to inhibition of tissue cell respiration.

Recent studies have shown that cyanide also inhibits the antioxidant defense enzymes (such as catalase, superoxide dismutase, and glutathione peroxidase) and stimulates neurotransmitter release. These effects of cyanide may also contribute to its acute toxicity. The prolonged energy deficit and the consequent loss of ionic homeostasis, which may result in activation of calcium signaling cascade and eventually cell injury, contribute to cyanide toxicity resulting from subacute exposure or in the postintoxication sequela.

Human Toxicity

Cyanide is a chemical asphyxiant, which renders the body incapable of utilizing an adequate supply of oxygen. Exposure to high dose of cyanide is often lethal. The lethal dose of cyanide in humans is 0.5–1.0 mg/kg. The lethal dose of hydrocyanic acid is about 50 mg for an adult; the lethal dose of the potassium or sodium salt is 200–500 mg. Inhalation of air containing a concentration of 0.2 or 0.3 mg/liter is almost immediately fatal, while 0.13 mg/liter (130 ppm) would be fatal after 1 hr.

Following the inhalation of toxic amounts of cyanide, symptoms usually appear within a few seconds, whereas it may take a few minutes for symptoms to appear following oral ingestion or skin contamination by the salts. If large amounts have been absorbed, collapse is usually instantaneous—the patient falling unconscious and dying almost immediately. With smaller doses, weakness, giddiness, headache, nausea, vomiting, and palpitation usually occur. With the rise of the blood cyanide level, ataxia develops and is followed by lactic acidosis, convulsive seizures, coma, and death. Cyanide directly stimulates the chemoreceptors of the carotid and aortic bodies to produce a brief period of hyperpnea followed by dyspnea. At higher cyanide doses, cardiac irregularities are often noted, but heart activity always outlasts the respiration.

Chronic low-level exposure to cyanide produces various signs and symptoms. The most widespread pathologic condition attributed to cyanide is tropical ataxic neuropathy associated with chronic cassava consumption. This is a diffuse degenerative neurological disease with peripheral and central signs. Cassava is the major staple food in various tropical areas; the plant has a high content of cyanogenic glycoside (linamarin). With continued ingestion over a period of time, tropical neuropathy gradually develops; the syndrome is characterized by optic atrophy, nerve deafness, and ataxia due to sensory spinal nerve involvement. Other signs include scrotal dermatitis, stomatitis, and glossitis. Chronic low-level exposure to cyanide may also lead to ultrastructural changes of heart muscle. In addition, with chronic cyanide ingestion, the thyroid may be affected due to enhanced formation of thiocyanate. Thiocyanate can block uptake of iodide by the thyroid gland, and myxedema, thyroid goiter, and cretinism may occur. This chronic effect of cyanide may pass to the fetus through maternal exposure.

Clinical Management

To be of any value, treatment of cyanide poisoning must be rapid and efficient. The rapid and early recognition of cyanide poisoning is usually difficult because most of the clinical manifestations are nonspecific. Potentially valuable cyanide blood levels are usually available for confirmation of diagnosis. Arteriolization of venous blood has been used as a significant symptom of cyanide poisoning. If cyanide was ingested, removing the unabsorbed poison by lavaging the stomach with copious amounts of water through a gastric tube is necessary. This should be continued until all odor of cyanide is gone from the lavage fluid. Artificial respiration with 100% oxygen is often used in the treatment of cyanide poisoning, although oxygen is not a specific antidote. It is theorized that oxygen therapy increases the rate of displacement of cyanide from cytochrome oxidase, and the increased intracellular oxygen tension nonenzymatically converts the reduced cytochrome to the oxidized species, enabling the electron transport system to again function. The nitrite–thiosulfate antidotal combination is still one of the most effective treatments of cyanide poisoning. Administration of sodium nitrite (300 mg for adults, intravenously) and inhalation of amyl nitrite for 15–30 sec of every minute will convert hemoglobin to methemoglobin, which has a higher affinity for cyanide than hemoglobin. A methemoglobin level of approximately 25% is desired for maintaining normal hemoglobin function and detoxification. For children weighing less than 25 kg, sodium nitrite should be dosed on the basis of their hemoglobin level and weight. Once intravenous sodium nitrite is administered, sodium thiosulfate should be immediately given to increase the rate of rhodanase-catalyzed biotransformation of cyanide to thiocyanide, which is readily eliminated in the urine. Other agents, including hydroxocobalamine or dicobalt EDTA, have been reported to be used as antidotes for acute cyanide poisoning, but they are not approved in the United States. The patient should be observed for the next 24–48 hr, and if the signs of intoxication persist or reappear, injection of nitrite thiosulfate at one-half of the recommended dose should be repeated.

Animal Toxicity

Cyanide toxicity varies with the animal species, type of cyanide compound, route of uptake, metabolic state, and other factors. The LD_{50} for cyanide has been reported in various species. Potassium cyanide, if injected, has a 24-hr LD_{50} of 6.7–7.9 mg/kg in mice. The lethal dose of potassium cyanide infused at a rate of 0.1 mg/kg/min is 2.4 mg/kg in dogs breathing room air. When hydrogen cyanide is inhaled by mice, the LD_{50} is 177 ppm with a lethal time of 29 min. The time to death is greater than 17 min for exposure to less than 266 ppm, but falls to 40 sec at 873 ppm. The LD_{50} for sodium cyanide is 4.6–15 mg/kg in rats. Male gerbils are 50-fold more sensitive to methacrylonitrile, which is metabolized to cyanide in rodents, than SD rats, and about 5-fold more sensitive than albino Swiss mice. Single and repeated low-dose cyanide intoxication can result in demyelinating lesions of the cerebral white matter in monkeys, but high doses of cyanide are required to produce similar brain lesions in rats.

—*Zhengwei Cai*

Related Topics

Blood
Neurotoxicology: Central and Peripheral
Respiratory Tract
Sensory Organs

Cyclodienes

Cyclodienes insecticides are chlorinated hydrocarbon insecticides with a polycyclic structure and, as the name implies, two unsaturated bonds. Not all of the insecticides in this class meet these criteria. Chlordane, for example, contains only one double bond in its polycyclic structure. Endrin and dieldrin are epoxides of the cyclodienes isodrin and aldrin, respectively (see Aldrin, Chlordane, Dieldrin, and Endrin).

Cyclodienes appear to act more in the central nervous system than in the peripheral nervous system. One major mode of action is the inhibition of γ-aminobutyric acid-regulated Cl^- ion flux in neurons. Cyclodienes also exert effects on membrane-bound adenosine triphosphatases (ATPases), altering Na^+, K^+, and Ca^{2+} ion transport. The result is partial depolarization of neurons rather than repolarization after activation. The accumulation of Ca^{2+} ions intracellularly in the terminal ends of neurons promotes the release of neurotransmitters from storage vesicles and the depolarization of adjacent neurons.

Symptomatology is essentially the same as that described for organochlorine insecticides. In many cases, convulsions are the first sign of toxicity without the progression of nerve hyperactivity seen in other classes of organochlorine insecticides. Clinical management is symptomatic, as described for organochlorine insecticides.

—*Benny L. Blaylock*

Related Topics

Neurotoxicology: Central and Peripheral
Organochlorine Insecticides
Pesticides

Cyclohexane

- CAS: 110-82-7
- SYNONYMS: Cicloesano; cyclohexaan; cyclohexan; exahydrobenzene; hexahydro-benzene; hexamethylene; hexanaphthene
- CHEMICAL CLASS: Saturated alicyclic hydrocarbon
- CHEMICAL STRUCTURE:

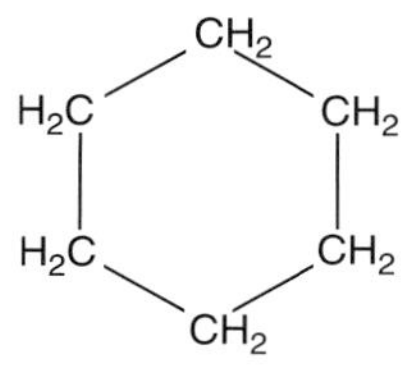

Uses

Cyclohexane is used as a solvent for fats, oils, waxes, resins, and coatings and as a paint and varnish remover. It is used in extracting essential oils and in manufacturing nylon. It is also a laboratory reagent.

Exposure Pathways

Exposure to cyclohexane may occur through dermal contact, ingestion, and inhalation routes.

Toxicokinetics

Cyclohexane is readily absorbed via inhalation and oral routes of exposure. Animal studies indicate dermal absorption to be high, probably due to the defatting action of the compound. Cyclohexane absorption into the lungs is rapid, with the concentration in the lungs reaching 42–62% of the air concentration. No information is available on the rate of absorption through the gastrointestinal tract. Cyclohexane is metabolized by cytochrome P450 enzymes in the liver and other tissues. Several metabolites have been identified including cyclohexanol and *trans*-cyclohexane-1,2-diol. These compounds have been identified in the urine of human subjects and experimental animals within 48 hr of exposure.

Mechanism of Toxicity

The precise mechanism of toxicity of cyclohexane has not been identified but is likely similar to other central nervous system (CNS) depressants and general anesthetics. These compounds are believed to exert their effects through a general interaction with the CNS and interference with neuronal membrane functions has been postulated as a mechanism of action. Disruption of membrane enzymes and the corresponding alterations in cell functions may account for the behavioral and anesthetic effects observed following exposure to various solvents.

Human Toxicity

Cyclohexane is a CNS depressant and may produce mild anesthetic effects. Inhalation exposure can cause headache, nausea, dizziness, drowsiness, and confusion. Very high concentrations may cause unconsciousness and death. Vapors may be irritating to nose and throat. Severe lung irritation, damage to lung tissues, or death may result from aspiration into the lungs. Direct dermal contact with liquid may cause mild irritation which may become more severe if exposure is prolonged. Eyes may become irritated upon exposure to vapors or liquid; however, the effect is generally mild and temporary unless exposure is prolonged. Ingestion of cyclohexane may cause sore throat, nausea, diarrhea, or vomiting. Prolonged exposure may produce liver and kidney damage.

Clinical Management

If inhalation exposure occurs, the source of contamination should be removed or the victim should be moved to fresh air. Artificial respiration should be administered or, if the heart has stopped, cardiopulmonary resuscitation provided. If dermal contact has occurred, contaminated clothing should be removed and the affected area should be washed with water and soap for at least 5 min or until the chemical is removed. Contaminated eyes should be flushed with lukewarm, gently flowing water for 5 min or until the chemical is removed. If ingestion occurs, vomiting should not be induced. Water should be given to dilute the compound. If vomiting occurs naturally, the victim should lean forward to reduce risk of aspiration. Aspiration of compound into the lungs may produce chemical pneumonitis requiring antibiotic treatment and administration of oxygen and expiratory pressure.

Animal Toxicity

The reported oral LD_{50} in rabbits is 5.5–6.0 mg/kg indicating the relatively low oral acute toxicity of cyclohexane. Lower doses (1.0–5.5 mg/kg) produced mild to extensive hepatocellular degeneration and glomerulonephristis. Vapor concentrations of 92,000 mg/m³ produced rapid narcosis and death in rabbits. In mice, concentrations of 51,000 mg/m³ caused narcosis and death occurred at 61,200–71,400 mg/m³. Microscopic changes in the liver and kidneys were observed in rabbits exposed to 2700 mg/m³ for 50 exposures. No changes were noted at 1490 mg/m³. The oral LD_{50} was reported as 12,705 mg/kg for rats and 813 mg/kg for mice.

—*Linda Larsen*

Cyclohexene

- CAS: 110-83-8
- SYNONYMS: 1,2,3,4-Tetrahydrobenzene; tetrahydrobenzene
- CHEMICAL CLASS: Cycloalkene
- MOLECULAR FORMULA: C_6H_{10}
- CHEMICAL STRUCTURE:

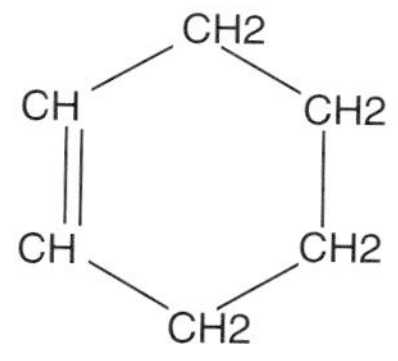

Uses

Cyclohexene is used in oil extraction and in the manufacture of adipic, maleic, and hexahydrobenzoic acids and aldehydes. It is also used as a stabilizer for high-octane gasoline and as a catalyst solvent.

Exposure Pathways

Exposure occurs most commonly through either inhalation or skin contact.

Toxicokinetics

Cyclohexene is readily hydroxylated by microsomal oxidases to the corresponding dihydroxy derivatives. These are then further conjugated and eliminated in urine.

Mechanism of Toxicity

Cyclohexene is an irritant and defats skin on direct contact. It is also an anesthetic and central nervous system (CNS) depressant on inhalation exposure. The mechanism for this toxicity is unknown.

Human Toxicity

No acute or chronic effects have been reported in humans. By analogy to effects reported with structurally similar compounds and in animals, cyclohexene is regarded as a mild respiratory irritant and CNS depressant. When ingested, it represents a low to moderate pulmonary aspiration hazard.

Clinical Management

Overexposure to vapors of cyclohexene should be treated by removing the patient to fresh air. If skin or eye contact occurs, the affected areas should be flushed with water for at least 15 min to remove residual solvent. If ingestion of cyclohexene occurs, vomiting should not be induced. This could result in aspiration of solvent into the lungs leading to chemical pneumonitis and pulmonary edema, which can be fatal.

Animal Toxicity

In general, cyclohexene appears to be only mildly toxic. Mice exposed acutely by inhalation to 8830 ppm exhibited a loss of righting reflexes; at 13,400–14,900 ppm death occurred. Dogs inhaling cyclohexene at unknown concentrations exhibited symptoms characterized by muscular quivering and incoordination. Rats, guinea pigs, and rabbits were exposed to cyclohexene vapors at 75, 150, 300, and 600 ppm for 6 hr/day, 5 days/week for 6 months. At the low doses, an increase in alkaline phosphatase was reported. At 600 ppm in rats, the same increase in alkaline phosphatase was observed along with a decrease in weight gain. Other blood and biochemical measures were within normal limits.

—*Patricia J. Beattie*

Cyclophosphamide

◆ PREFERRED NAMES: Cyclophosphamide (CAS: 50-18-0) or cyclophosphamide monohydrate (CAS: 6055-19-2)

◆ SYNONYMS:

Cyclophosphamide—2*H*-1,3,2-Oxazaphosphorin-2-amine, *N,N*-bis(2-chloroethyl) tetrahydro-, 2-oxide; 2*H*-1,3,2-oxazaphosphorine, 2-[bis(2-chloroethyl)amino] tetrahydro-, 2-oxide; Asta B 518; B 518; bis(2-chloroethyl)phosphoramide cyclic propanolamide ester; Clafen; Claphene; cyclophosphamid; cyclophosphan; cyclophosphane; cytophosphan; Cytoxan; Endoxan; Genoxal; *N,N*-bis(β-chloroethyl)-*N'*,*O*-trimethylenephosphoric acid ester diamide; *N,N*-bis(2-chloroethyl)-*N'*,*O*-propylenephosphoric acid ester diamide; NSC 26271; Procytox; Sendoxan

Cyclophosphamide monohydrate—2*H*-1,3,2-Oxazaphosphorin-2-amine, *N,N*-bis(2-chloroethyl) tetrahydro-, 2-oxide, monohydrate; 2*H*-1,3,2-oxazaphosphorine, 2-[bis(2-chloroethyl)amino] tetrahydro-, 2-oxide, monohydrate; cyclophosphamide hydrate; endoxan monohydrate

◆ CHEMICAL CLASS: Nitrogen mustard

◆ CHEMICAL STRUCTURE:

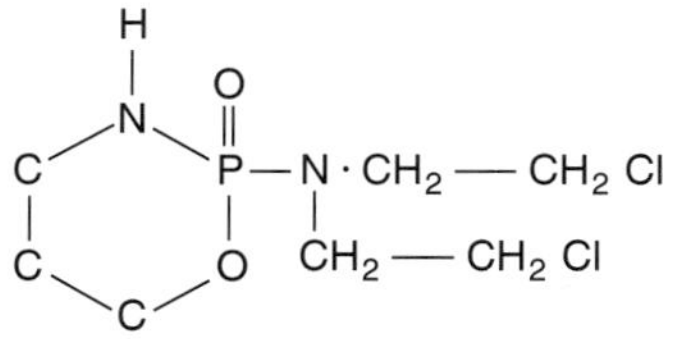

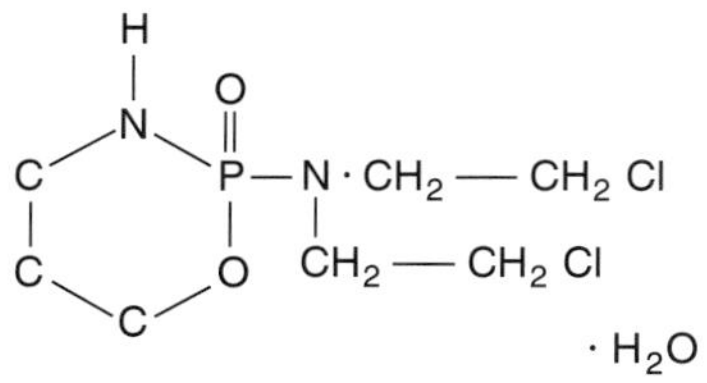

Anhydrous form

Uses

Cyclophosphamide is used in human medicine as an antineoplastic (anticancer) agent in a variety of applications. Cyclophosphamide is a potent immunosuppressive agent and is used to prevent rejection episodes following renal, hepatic, and cardiac homotransplantation; and in nonneoplastic disorders in which there is altered immune activity, such as Wegener's granulomatosis, rheumatoid arthritis, the nephrotic syndrome in children, or autoimmune ocular diseases. Cyclophosphamide has also been used in veterinary practice for defleecing sheep, and it has been tested as an insect chemosterilant.

Exposure Pathways

Exposure to this odorless, white, crystalline powder may occur during its manufacture, formulation, or distribution as an antineoplastic drug. During manufacture and experimental use, exposure may be by inhalation or skin absorption. Therapeutically, patient exposure is by the oral, intramuscular, intraperitoneal, intravenous, or intrapleural route.

Toxicokinetics

In most species, cyclophosphamide is rapidly absorbed, metabolized, and excreted. In patients, cyclophosphamide was distributed rapidly to all tissues and exhibited a half-life of 6.5–7 hr. The majority of an administered dose (50–68%) was excreted in the urine and no parent compound or metabolite was detected in expired air or feces. No radioactivity was found in expired air or feces. Carboxyphosphamide and phosphoramide mustard were detected in the urine. Cyclophosphamide is a racemer, and stereoselective metabolism by cytochrome P450 of the enantiomers has been demonstrated in mice, rats, and rabbits. The primary metabolite is the 4-hydroxy derivative, and it exists in equilibrium with aldophosphamide, its ring-opened tautomer. Either metabolite can be metabolized by mammalian enzymes to 4-ketocyclophosphamide or to the propionic derivative. Both metabolites are relatively nontoxic and represent major urinary metabolites.

Mechanism of Toxicity

An extensive study of cyclophosphamide analogs, some of which release acrolein but no phosphoramide mustard and others that cannot undergo complete metabolism, has provided strong evidence that (1) the phosphoramide mustard metabolite is responsible for the drug's antitumor activity, (2) the toxic side effects are probably due to the phosphoramide mustard and acrolein, and (3) nor-nitrogen mustard is responsible for the renal damage that occurs in some cases.

Human Toxicity

Patients treated with cyclophosphamide have been reported to exhibit various side effects such as flushing of the face, swollen lips, cardiotoxicity, pneumonitis or interstitial fibrosis, agitation, dizziness, tiredness, weakness, headache, nausea, vomiting, diarrhea, stomatitis, hemorrhagic colitis, hepatitis, hemorrhagic cystitis, fever, chills, sore throat, sweating, pancytopenia, leukopenia, alopecia, changes in the nucleoli of lymphocytes, water and sodium retention, pulmonary fibrosis, and visual blurring. Birth defects, such as limb reductions or pigmentation of the fingernails and skin, were also noted. Cystitis, hemorrhagic cystitis, and fibrosis of the bladder wall have been reported in patients treated for cystitis, rheumatoid arthritis, lupus erythematosus, and neoplasia, respectively. Fatal cardiomyopathy may result when very large doses of cyclophosphamide are given as conditioning for bone marrow transplantation. Cyclophosphamide has teratogenic and mutagenic potential and can cause sterility of either sex. It can damage germ cells in prepubertal, pubertal, and adult males and cause premature ovarian failure in females. It is most toxic to the human fetus during the first 3 months and congenital abnormalities have been detected after intravenous injections of large doses to pregnant women during this period of pregnancy. Mothers taking cyclophosphamide should avoid breast-feeding.

The U.S. FDA regulates this material as a drug.

Clinical Management

Treatment for toxic effects is supportive. Contaminated skin should be thoroughly washed with soap and water, and contaminated clothing should be discarded or left at the work site for cleaning before reuse. Hemodialysis may be beneficial in cases in which sufficient overdoses have occurred; however, the range of toxicity has not been well described and most toxicity data are from high-dose treatment regimens.

Animal Toxicity

The LD_{50} in mice ranges from 370 mg/kg (subcutaneous) to 310 mg/kg (intravenous). The LD_{50} in rats was 160 mg/kg (intravenous), 180 mg/kg (oral), and 400 mg/kg (intraperitoneal in rats bearing tumors). The intravenous LD_{50} was 400 mg/kg in guinea pigs and 40 mg/kg in dogs. In mice, rats, and dogs, the predominant hematological effect of cyclophosphamide is leucopenia; some depression of thrombocytes was also noted. Prolonged treatment of rodents with cyclophosphamide has produced pathological structural changes in a variety of organs including lung, gut, pancreas, and liver. In rats, cyclophoshamide given orally decreases mitosis in crypts, decreases the height of villi, and causes degeneration of the intestinal mucosa. A single intraperitoneal dose of cyclophoshamide caused marked necrosis of the bladder and necrosis of the renal tubular and renal pelvic epithelium in mice, rats, and dogs. Cyclophosphamide is teratogenic in the rhesus monkey when given intramuscularly for various periods between 25 and 43 days of pregnancy at doses ranging from 2.5 to 20 mg/kg body weight. Placental transfer of C^{14} has been demonstrated in mice, and a positive correlation between alkylation of embryonic DNA and the production of congenital abnormalities has been reported in mice.

—*David M. Krentz*
and Linda Angevine Malley

Related Topic

Respiratory Tract

Cypermethrin

- CAS: 52315-07-8, 69865-47-0, 86752-99-0, 86753-92-6, 88161-75-5, 97955-44-7
- SYNONYMS: (R,S)-α-cyano-3-phenoxybenzyl-2,2-dimethyl (1R,1S)-*cis,trans*-3-(2,2-dichloro vinyl)cyclopropane carboxylate; Cymbush;

Folcord; Polytrin; Ripcord; Stockade; CCN 52, PP 383, OMS 2002, NRDC 149

♦ CHEMICAL CLASS: Type II pryrethroid insecticide
♦ CHEMICAL STRUCTURE:

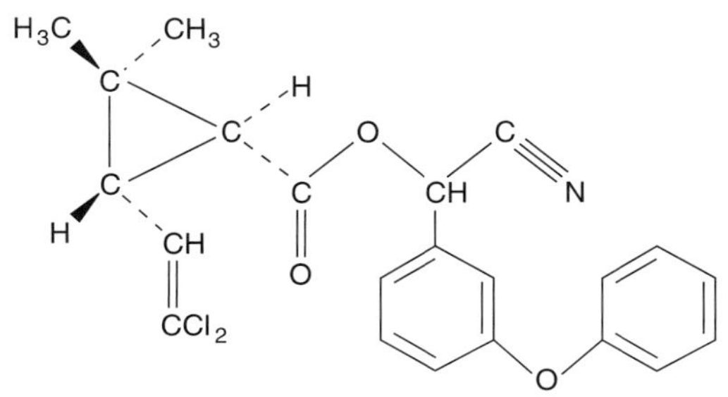

Uses

Cypermethrin is a broad-spectrum insecticide.

Exposure Pathways

Dermal contact is the most common route of exposure to cypermethrin but cases of ingestion have also been reported.

Toxicokinetics

Pyrethroids are poorly absorbed through the skin and are only moderately absorbed in the gastrointestinal tract. In one case of dermal exposure, absorption was estimated to be approximately 3%. Metabolism of cypermethrin occurs rapidly through ester cleavage and hydroxylation. Adipose tissue acts as a storage depot for cypermetherin and has varying affinities for the different isomeric forms; the elimination half-lives in laboratory rats have been reported as 3.4 and 18 days for the *trans* and *cis* isomers, respectively. Urinary excretion is the primary route of elimination.

Mechanism of Toxicity

Pyrethroids have a selective high affinity for membrane sodium channels. Closing of the channel, which ends the action potential, is slowed resulting in a prolonged "tail" current and repetitive firing of presynaptic and accompanying postsynaptic cells following a single action potential. High enough doses can cause complete depolarization and blockade of nerve conduction. Cypermethrin has also been shown to inhibit Ca^{2+}, Mg^{2+}-ATPase and calmodulin and has been suspected of interfering with the γ-aminobutyric acid-mediated chloride ionophore.

Human Toxicity: Acute

In mammals, cypermethrin produces type II motor symptoms characterized by hyperactivity, incoordination, choreoathetosis, and convulsions. In humans, extensive dermal exposure causes temporary abnormal sensations including paresthesia, numbness, and tingling. Symptoms following ingestion included nausea, vomiting, tenesmus, diarrhea, unconsciousness, and death due to respiratory failure.

Human Toxicity: Chronic

Chronic effects in humans following cypermethrin exposure have not been reported. However, studies have demonstrated possible genotoxicity in mouse spleen and bone marrow.

Clinical Management

Oral exposures may be treated with syrup of ipecac-induced emesis, activated charcoal, or gastric lavage. Emesis is contraindicated in cases of unconsciousness or convulsions. In cases of dermal exposure, contaminated clothing should be removed and disposed of. Any exposed areas of skin should be repeatedly washed with soap and water. Paresthesia may be treated with topical application of Vitamin E. For eye contact, the eyes should be flushed with generous amounts of lukewarm water for a minimum of 15 min. Seizures can be treated with intravenous diazepam; phenytoin or phenobarbital may be helpful for recurrent seizures. Atropine can be used to control excessive salivation. No specific antidotes for pyrethroid-induced neurotoxic effects have been approved for use in humans. Spontaneous recovery usually occurs with mild or moderate intoxication.

Animal Toxicity

Fish and crustaceans are extremely sensitive to pyrethroid compounds in laboratory settings. However, various factors (e.g., sediment binding) may reduce pyrethroid toxicity to these nontarget organisms in a natural environment.

—*Paul R. Harp*

Related Topics

Neurotoxicology: Central and Peripheral
Pyrethrin/Pyrethroids

Cysteine

- ◆ CAS: 616-91-1
- ◆ SYNONYMS: *N*-Acetylcysteine; L-α-acetamido-β-mercapto propionic acid; mercapturic acid; i-acetyl-3-mercaptoalanine; *N*-acetyl-L-cysteine; *N*-acetyl-*n*-cysteine; acetylcysteine; NAC
- ◆ CHEMICAL STRUCTURE:

$$^-OOC - \overset{\overset{NH_3^+}{|}}{\underset{|}{C}} - CH_2 - SH$$
$$H$$

Uses

Cysteine is a mucolytic agent for nebulization and an acetaminophen antidote.

Exposure Pathways

The most common route of exposure to cysteine is inhalation through the respiratory tract or orally. Although not approved by the U.S. FDA, it may be given intravenously in emergency situations.

Toxicokinetics

Cysteine is rapidly absorbed after oral administration, with peak plasma levels occurring in 2 or 3 hr. With intravenous administration, peak plasma levels occur immediately. Orally administered cysteine appears to distribute primarily to the kidneys, liver, and lungs. It is detectable in pulmonary secretions for at least 5 hr after the dose. Cysteine is rapidly absorbed and exists as the free species in plasma with a concomitant increase both in plasma L-cysteine levels and in protein and non-protein SH concentrations. Protein binding is more than 50%. The volume of distribution in humans is 0.337–0.47 liters/kg.

Thirty percent of intravenously administered cysteine is renally cleared. Cysteine elimination is not impaired in patients with severe liver damage. The terminal half-life of cysteine is 2–6 hr. This may be increased to 13 hr after an intravenous injection.

Mechanism of Toxicity

Fatalities from normal doses and overdoses of intravenous cysteine have not been reported. Toxicity is usually limited to anaphylactoid reactions and nausea/vomiting. The average time to onset of adverse effects following commencement of the infusion of cysteine was 30 min (range, 5–70 min). *In vivo* and *in vitro* tests indicate that cysteine is an inhibitor of allergen tachyphylaxis by inhibition of prostaglandin E synthesis. Adverse reactions are anaphylactoid in type and have been attributed to cause histamine release.

Human Toxicity

The primary toxicity of cysteine consists of nausea/vomiting, particularly after oral therapy, and an anaphylactoid reaction which may be life-threatening. Many cases of anaphylactic reactions have been reported with symptoms primarily consisting of rash, nausea, hypotension, bronchospasm, angioedema, tachycardia, and respiratory distress. Cysteine may also have some neurological toxicity which includes dizziness, intracranial hypertension, hypoactivity, ataxia, and seizures. There have been reports of mucosal damage with full strength (20%) cysteine which causes hyperemia and hemorrhages of bowel mucosa. During inhalation therapy, irritation or soreness of the mouth may occur.

Clinical Management

Basic and advanced life-support measures should be utilized as necessary. For acetaminophen overdose, a 140-mg/kg dose followed by 70 mg/kg every 4 hr for an additional 17 doses should be administered. Since cysteine has not been approved for intravenous administration, assistance is available through the Rocky Mountain Poison Center.

Animal Toxicity

Cysteine is used primarily in the treatment of acetaminophen toxicity and is also nebulized for mucolytic effects and used to treat melting corneal ulcers. Oral formulations of *n*-acetylcysteine are used intravenously in the clinical treatment of animals, although it has not been approved for this use.

—*Melissa Adams and Robert L. Judd*

2,4-D

- CAS: 94-75-7
- PREFERRED NAMES: 2, 4-Dichlorophenoxyacetic acid; dichlorophenoxyacetic acid
- CHEMICAL CLASS: Chlorinated phenoxyacetic acids. A closely related compound is 2,4,5-T (2,4,5-trichlorophenoxyacetic acid; see 2,4,5-T).
- CHEMICAL STRUCTURE:

Uses

2,4-D free acids, esters, and salts are formulated in water suspensions or solutions, or in various organic solvents, for application as systemic herbicides. Some esters are fairly volatile, whereas salts are not. The acid is corrosive. There are many commercial formulations available for weed and brush control, for certain agricultural uses, and for lawn and garden weed control.

Toxicokinetics

Rapid and complete absorption of chlorphenoxy compounds from the gastrointestinal tract has been reported. Nearly complete absorption of 2,4-D occurs within 24 hr in humans. 2,4-D is primarily metabolized by acid hydrolysis, and a minor amount is conjugated. It is highly protein bound and widely distributed. The chief organs of deposition are kidneys, liver, and the central and peripheral nervous systems.

2,4-D is primarily excreted unchanged (90%) in the kidneys via the renal organic anion secretory system. It may be conjugated to glycine or taurine. A minor fraction of 2,4-D is filtered by the glomerulus. The estimated half-life of 2,4-D is about 18 hr.

Human Toxicity

Acute ingestion can cause miosis, coma, fever, hypotension, emesis, tachycardia, muscle rigidity, possible respiratory failure, pulmonary edema, and rhabdomyolysis. Alteration in liver functions such as elevated lactate dehydrogenase and aspartate aminotransferase have also been reported. In humans the causal relationship between these effects and chlorphenoxy herbicides such as 2,4-D remains controversial and not yet proven.

Clinical Management

No specific antidote is available. The patient must be monitored for seizures; gastrointestinal irritation; possible liver, kidney, or muscle damage; arrhythmias; acidosis; dyspnea; headache; coma; hyperthermia; and hypotension. Gastric lavage and activated charcoal/cathartic are probably more useful decontamination methods.

Animal Toxicity

The oral LD_{50} in mice is 368 mg/kg. The oral LD_{50} in rats is 375 mg/kg.

—Raja S. Mangipudy and Harihara M. Mehendale

Related Topic

Pesticides

Dalapon

- ◆ CAS: 75-99-0
- ◆ PREFERRED NAME: Dalapon sodium salt
- ◆ SYNONYMS: Basfapon; Basfapon B; Basfapon/Basfapon N; BH Dalapon; Basinex; Crisapon; Ded-Weed; 2,2 dicloropropionic acid; α-dichloropropionic acid; Dowpon; Gramevin; Kenapon; Kyselina; Liropon; Proprop; Revenge; Unipon
- ◆ CHEMICAL CLASS: Pesticide (herbicide)
- ◆ CHEMICAL STRUCTURE:

$$H-\underset{\underset{H}{|}}{\overset{\overset{H}{|}}{C}}-\underset{\underset{Cl}{|}}{\overset{\overset{Cl}{|}}{C}}-\overset{\overset{O}{||}}{C}-OH$$

Uses

Dalapon is used as a herbicide.

Exposure Pathways

Dermal and inhalation exposures of liquid or vapor are the most common routes of exposure. If the chemical leaches into groundwater there is the possibility of ingestion.

Toxicokinetics

Dalapon is a polar compound that is not readily absorbed by tissues. It causes irritation to the tissue with which it comes into contact. If absorbed in the gastrointestinal tract, it is eliminated as the parent compound in the urine.

Mechanism of Toxicity

The mechanism of action of dalapon is the same as that for most acids. The tissue proteins in membranes in contact with dalapon are converted to acid proteinates, which dissolve in the concentrated acid.

Human Toxicity: Acute

Burning and irritation are the predominant acute toxicities seen with exposure to dalapon. Eye exposure may cause corneal destruction and conjunctival edema accompanied by pain and tearing. Dermal exposure causes severe burns. Ingestion causes oral, throat, and gastrointestinal irritation. Inhalation of vapors causes irritation of the eyes, nose, and throat with destruction of mucus membranes. Severe inhalation exposure may cause respiratory distress accompanied by pulmonary edema. The ACGIH TLV is 1 ppm.

Human Toxicity: Chronic

Long-term exposure may cause increased kidney and liver weights.

Clinical Management

In the event of dermal exposure to dalapon, contaminated clothing should be removed quickly and the exposed area should be flushed with copious amounts of water for 15 min. With eye exposure, the affected eye should be flushed with water for 30 min, occasionally lifting the upper and lower lids. With inhalation exposure, the victim should be moved to fresh air. If the victim is not breathing, artificial respiration should be administered. If swallowed, vomiting should not be induced. If the victim is conscious, he or she should drink plenty of water or milk.

Animal Toxicity

Dalapon toxicity in animals is similar to that observed in humans.

—Melissa Adams and Robert L. Judd

Related Topic

Pesticides

DDT

- CAS: 50-29-3

- SYNONYMS: Dichlorodiphenyl trichloroethane; *p,p'*-DDT; 1,1'-(2,2,2-trichloroethylidene)-bis-(4-chlorobenzene)

- CHEMICAL CLASS: Synthetic organochlorine insecticide

- CHEMICAL STRUCTURE:

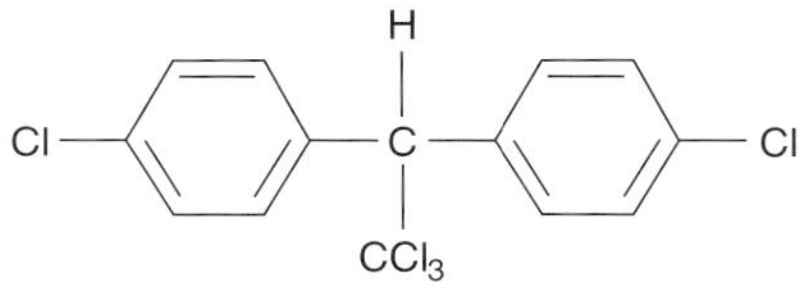

Uses

DDT is an insecticide whose use has been banned in the United States and many other parts of the world.

Exposure Pathways

Ingestion is the most common and effective route of exposure. Although absorption through the respiratory tract has been documented, the particle size of DDT is sufficiently large to be deposited in the upper airways with subsequent ingestion. Topical exposure is generally not a significant route due to poor intact epidermal absorption.

Toxicokinetics

Gastrointestinal absorption is slow with symptoms delayed several hours. Dissolved DDT (in solvents containing vegetable or animal fat) is absorbed several times faster than the undissolved compound. Due to DDT's large particle size, absorption through the respiratory tract is less important to toxicity. Skin absorption is considered almost negligible.

Metabolism of DDT proceeds at a very slow rate. Liver microsome P450 and other microsomal enzymes initially dechlorinate DDT to 1,1-dichloro-2,2-bis(*p*-chlorophenyl)ethylene (DDE) and reduction to 1,1-dichloro-2,2-bis(*p*-chlorophenyl)ethane (DDD). The conversion of DDD to bis(*p*-chlorophenyl)acetic acid (DDA) involves the formation of an acyl chloride intermediate by hydroxylation followed by hydrolysis to yield the final product.

DDT, like most other organochlorine insecticides, is highly lipophilic. It is stored in all tissues with higher levels generally found in adipose tissue. Most species store DDE more tightly than DDT.

DDA is the main form in which DDT is excreted. Most excretion takes place through bile with approximately 2% in urine and less than 1% in feces. Cows excrete approximately 10% of DDT doses in their milk; rodent females also excrete DDT in mother's milk.

Mechanism of Toxicity

The main site of toxicity for DDT is the nervous system, both central and peripheral. Neuronal membrane enzymatic and electrophysiological properties are significantly altered. In particular, sodium channels are altered such that once activated, they close slowly, prolonging the depolarization of the nerve by interfering with the active transport of Na^+ ions out of the axon. Potassium channels are also affected. DDT specifically affects Na^+,K^+-adenosine triphosphatases (ATPases) and Ca^{2+}-ATPases, which inhibit repolarization of neurons. The nerve membrane remains partially depolarized and is extremely sensitive to complete depolarization by very small stimuli. DDT also inhibits calmodulin, which is necessary for Ca^{2+} transport essential for the subsequent release of neurotransmitters.

Human Toxicity

Most human exposure cases are from ingesting very large amounts of DDT. Respiratory and/or epidermal exposure generally do not produce symptoms. Symptoms range from nausea, fatigue, and vomiting to tremor and convulsions in severe poisoning. The vomiting is not due to irritation of the gastrointestinal tract and is probably of central origin. With high oral doses of DDT, the symptoms include paresthesia of the tongue, lips, and face; apprehension and hypersensitivity to external stimuli; tremor with a gradual onset and usually mild effects; and (in extremely high doses) convulsions. Liver involvement has not been a prominent symptom in human cases of DDT poisoning but jaundice 4 or 5 days after ingestion of 5000–6000 mg has been reported. Death is unusual and is generally due to respiratory arrest.

Clinical Management

Diazepam may be beneficial to control convulsions. Sedatives, ionic calcium, and glucose are indicated based on data compiled from animal studies. Few actual cases of DDT poisoning have occurred from which to glean appropriate treatment regimens.

Animal Toxicity

The main observable toxic effect of DDT in animals is neuronal hyperactivity. Alterations in the previously mentioned ion transport channels in axon membranes produce paresthesia, hyperexcitability, irritability, fine tremors, and convulsions. In animals, hepatocyte hypertrophy and centrilobular necrosis are seen. There is also an increase in hepatic cancers.

—*Benny L. Blaylock*

Related Topics

Distribution
Environmental Hormone Disrupters
Neurotoxicity: Central and Peripheral
Organochlorine Insecticides
Pollution, Water
Psychological Indices of Toxicity
Veterinary Toxicology

Decane

- CAS: 124-18-5
- PREFERRED NAME: I-Decane; UN2247 (DOT)
- CHEMICAL CLASS: Aliphatic hydrocarbon (C10)
- MOLECULAR FORMULA: $C_{10}H_{22}$

Uses

Decane is a constituent in the paraffin fraction of petroleum and is also present in low concentrations as a component of gasoline. It is used as a solvent, in organic synthesis reactions, as a hydrocarbon standard, in the manufacture of petroleum products, in the rubber industry, in the paper processing industry, and as a constituent in polyolefin manufacturing wastes. Decane is a flammable liquid (at room temperature) that is lighter than water.

Exposure Pathways

Because decane can exist as a liquid and a vapor at normal temperature and pressure, exposure could occur by either dermal contact or inhalation; oral exposure would most likely be either incidental or accidental. Decane can be detected in urban air (up to 3 parts per billion) as a result of automobile emissions.

Mechanism of Toxicity

Decane is generally considered to be fairly nontoxic, relative to other aliphatic hydrocarbons (see Octane and Heptane). This is probably due to the fact that it is less volatile than octane or heptane and may not be as readily transferred across either the pulmonary alveoli or the blood–brain barrier. If it is aspirated into the lungs, however, decane will cause adverse effects similar to those seen with heptane or octane.

Using *in vitro* and/or microbial systems, decane has been shown to be metabolized to decanol and is thus thought to be readily biodegradable in the natural environment.

Human Toxicity

Adverse effects to humans would be expected to be similar to those seen in laboratory animals (see Animal Toxicity). There is currently no industrial air standard for occupational exposure to decane.

Clinical Management

Persons who are exposed to high concentrations of decane in air should vacate or be removed from the source and seek fresh air. Upon oral ingestion, vomiting should not be induced as pulmonary aspiration may occur, resulting in severe narcosis and/or death. In areas of expected increased concentration, extreme care must be taken to use explosion-proof apparatus and keep the areas free from ignition sources, such as sparks from static electricity.

Animal Toxicity

Decane has been shown to have narcotic effects in both mice and rats, primarily in experiments documenting acute exposure at high concentrations. One study estimated a 2-hr LC_{50} of 72,300 mg/m³ in rats. In mice, an intravenous dose of 912 mg/kg is expected to cause death in 50% of the experimental animals. Another rat study showed that a concentration of 540 ppm in air (18 hr/day, 7 days/week, 8 weeks) caused some slight adverse effects (e.g., decreased white blood cell count) but no significant toxic effects overall.

—*Stephen Clough*

DEET

- ◆ CAS: 134-62-3

- ◆ SYNONYMS: Detamide; Autan; I-Delphene; Meta-delphene; Black Flag; Tabarad; Delphene; Dieltamide; Flypel; Muskol; Naugatuck Det; Off; 612 Plus; Jungle plus; Pellit; DETA; DET; *N, N*-diethyl-3-methylbenzamide; *N, N*-diethyl-*m*-toluamide

- ◆ CHEMICAL CLASS: Methyl benzamide repellent

- ◆ CHEMICAL STRUCTURE:

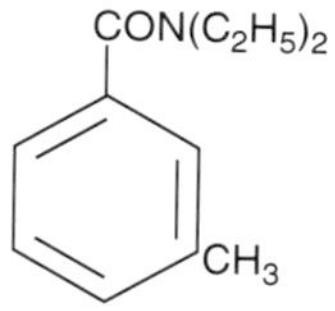

Uses

DEET is used as an insect repellent.

Exposure Pathways

DEET is available in solutions, lotions, gels, aerosol sprays, sticks, and impregnated towlettes. Dermal and ocular exposures are the most common exposure pathways.

Toxicokinetics

Approximately 50% of each topically applied dose of DEET is absorbed within 6 hr. Peak plasma levels are attained within 1 hr. Ingestion of DEET may result in symptoms within 30 min, implying very rapid absorption. DEET is metabolized by oxidative enzymes in the liver. Metabolites of DEET have not yet been characterized. Following movement through the skin, DEET is absorbed and distributed rather rapidly. Some studies indicate, however, that DEET and metabolites can remain in the skin and fatty tissues for 1 or 2 months after topical application.

Mechanism of Toxicity

DEET is primarily toxic to the central nervous system (CNS). The mechanism of toxicity is still unknown.

Human Toxicity: Acute

A toxic syndrome consisting of ataxia, hypertonicity, tremor, and clonic jerking, and progressing to coma and seizures, may occur after dermal or oral exposure. Symptoms may occur within 30 min after an acute ingestion. Dermal exposure may cause irritation, sensitization, and erythema. Toxicity is less likely following ingestion.

Human Toxicity: Chronic

Chronic application of 70% DEET solution caused paranoid psychosis and pressurized speech, flight ideas, and delusions. Repeated application causes erythema.

Clinical Management

Basic life-support measures for respiratory and cardiovascular function should be utilized. Dermal decontamination should be accomplished by repeated washing with soap. Exposed eyes should be irrigated with copious amounts of room-temperature water for at least 15 min. Ipecac-induced emesis is not recommended in cases of accidental oral exposure as coma and seizures can occur rapidly within 30 min to 1 hr of ingestion. Gastric lavage should be performed cautiously with a small-bore soft nasogastric tube with small aliquots of water or saline. Activated charcoal can also be used.

Animal Toxicity

The acute oral LD_{50} values were reported as 300 mg/kg in male rats and 2000 mg/kg in female rats. Rats given an LD_{50} dose showed signs of toxicity that included lacrimation, chromodachryorrhea, depression, prostration, tremors, asphyxial convulsions, and respiratory failure.

—Sushmita M. Chanda

DEF

- ◆ SYNONYM: Tributyl *S,S,S*-phosphorotrithioate
- ◆ CHEMICAL CLASS: Organophosphorous herbicide
- ◆ CHEMICAL STRUCTURE:

$$(C_4H_9-S)_3P=O$$

Uses

DEF is used as a cotton defoliant to facilitate mechanical harvesting.

Exposure Pathways

The major route of exposure to DEF is skin contact.

Toxicokinetics

DEF is readily absorbed through the skin. It is metabolized to *n*-butyl mercaptan in the gastrointestinal tract by hydrolysis. The metabolite is excreted in the urine.

Mechanism of Toxicity

DEF is a relatively weak inhibitor of acetylcholinesterase. The compound is hydrolyzed to a large extent in the intestine to *n*-butyl mercaptan, which is responsible for the late acute effects of DEF. The putative molecular target in neural tissue for initiation of delayed neuropathy is neurotoxic esterase or neuropathy target esterase (NTE) (see Neurotoxicity, Delayed).

Human Toxicity

Late acute poisoning from DEF is related to the release of its breakdown product, *n*-butyl mercaptan. Signs of toxicity appear within 1 hr after exposure and include general weakness, malaise, sweating, nausea, vomiting, anxiety, and drowsiness. DEF affects the lymphocyte NTE in exposed workers. Both intensity and length of exposure play important roles in determining the extent of inhibition of NTE in lymphocytes; 50% of preexposed values of NTE activity were obtained when measured 3 or 4 weeks after the beginning of DEF exposure. However, there is no direct evidence of a correlation between a high level of lymphocyte NTE inhibition and development of neuropathy in humans. Blood acetylcholinesterase and plasma butyrylcholinesterase levels remained unchanged during the study period.

Clinical Management

Supportive and symptomatic treatment should be provided to the patient following accidental or intentional exposure to DEF.

Animal Toxicity

DEF produces profound hypothermia in rats, mice, and guinea pigs by inhibition of thermogenesis. Its actions on heat conservation and motor control are minimal, however. It is effective against both shivering and nonshivering thermogenesis and completely blocks the increase in body temperature evoked by anterior hypothalamic stimulation. The toxicologic effect of DEF, the extent and permanence of injury, and the progression or improvement of clinical signs of toxicity depended on the dose, duration, and route of exposure. A subchronic administration of DEF caused three toxicologic effects in hens, depending on route of exposure: (1) an acute cholinergic effect resulting from inhibition of acetylcholinesterase, relieved by atropine, not associated with neuropathological lesions; (2) a late acute effect in chickens resulting from *n*-butyl mercaptan toxicity 4 days after oral administration of daily large doses of DEF resulting in darkening and drooping of the comb, loss of appetite and weight, weakness, emaciation, paralysis, and death, not relieved by atropine nor associated with histopathological changes in nerve tissues; and (3) delayed neurotoxicity after a delay period following topical application causing axonal and myelin degeneration resulting in ataxia, paralysis, and death.

—Priya Raman

Related Topics

Neurotoxicity, Delayed
Organophosphates
Pesticides

Deferoxamine

- CAS: 70-51-9
- PREFERRED NAME: Desferrioxamine

- SYNONYMS: *N*-benzoylferrioxamine B; deferoxaminum; deferrioxamine; deferrioxamine B; desferal; desferral; desferrin; desferrioxamine B; DF B; DFO; DFOA; DFOM; ferrioxamine B; propionohydroxamic acid; *N*-(5-(3-((5-amino-pentyl)-hydroxycarbamoyl) propionamido)pentyl)- 3-((5-(*N*-hydroxyacetamido)pentyl)-carbamoyl); 30-amino-3,14,25-trihydroxy-3,9,14,20,25-pentaazatriacotane-2,10,13,21,24-pentaone

- PHARMACEUTICAL CLASS: Deferoxamine is an iron-chelating agent. It is an iron-free derivative of ferrioxamine B, which belongs to a group of siderophores, growth factors for certain microorganisms.

- CHEMICAL STRUCTURE:

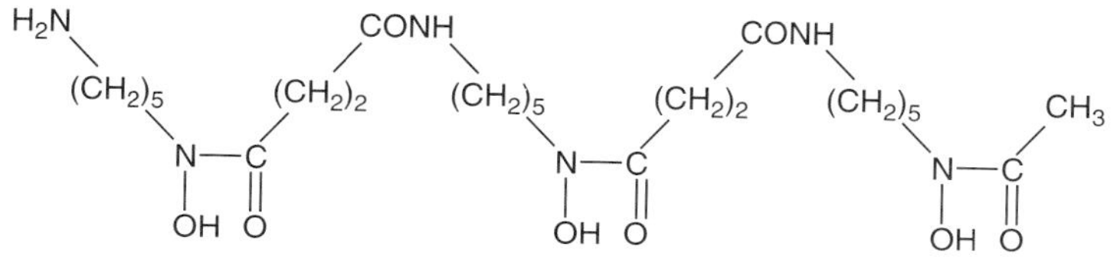

Uses

Deferoxamine is widely used for treatment of both acute iron intoxication and iron-overload anemias. Recently, it has been used in trials of malaria treatment.

Exposure Pathways

In acute cases, intravenous injection and oral ingestion are the most common routes of administration. Nightly subcutaneous infusions combined with monthly intravenous infusions are used in chronic illness.

Toxicokinetics

Deferoxamine is poorly absorbed from the gastrointestinal tract. When given orally, this drug will block absorption of small quantities of iron in both human and experimental animals. Deferoxamine is rapidly metabolized mainly in the plasma, probably by an enzyme belonging to the α_2 globulins. Of the three metabolites produced, the major one is known as metabolite C. Other organs may also have some metabolizing capacity. Deferoxamine has a very high affinity and specificity for the ferric iron and chelates it in a 1:1 molar ratio. Its elimination half-life in human plasma is between 10 and 30 min. Deferoxamine mainly distributes in blood,

but bile excretion is a possible elimination route. Renal clearance is the major elimination route of deferoxamine in humans. It accounts for about one-third of the total body clearance; 0.296 and 0.234 liters/hr/kg in healthy and hemochromatotic adults, respectively. The iron–deferoxamine complex, ferrioxamine, is also poorly absorbed from the gastrointestinal tract. Ferrioxamine does not appear to be degraded to any appreciable extent and is rapidly excreted unchanged in the urine. Its elimination half-life is 5.9 and 4.6 hr in normal and hemochromatosis adults, respectively.

Mechanism of Toxicity

Deferoxamine has some serious side effects including hypotension, renal insufficiency, neurotoxicity, growth retardation, and bacterial infections. Deferoxamine may induce venous dilation leading to poor venous return, depressed cardiac output, and eventually hypotension. An acute decrease in glomerular filtration rate and renal plasma flow is the possible mechanism underlying the nephrotoxicity induced by deferoxamine. Depletion of iron, translocation of copper, and chelation of other trace elements including zinc, due to excessive deferoxamine, may interfere with critical iron-dependent enzymes and cause oxidative damage within neural tissue. These are possible mechanisms responsible for deferoxamine-induced neurotoxicity and growth retardation.

The iron–deferoxamine complex, ferrioxamine, is a growth factor for many bacteria and by providing easily utilized iron, deferoxamine permits bacterial growth and infections. *In vitro* studies have shown that deferoxamine inhibits the synthesis of prostaglandin, hemoglobin, ferritin, collagen, and DNA.

Human Toxicity

Rapid infusion of deferoxamine over 15 min results in hypotension and tachycardia. Intravenous deferoxamine administration has been reported to cause renal insufficiency indicated by a progressive increase in serum creatinine and decrease in creatinine clearance.

Patients treated with deferoxamine chronically may develop neurotoxicity manifested as visual and hearing losses, growth retardation, and bacterial infections.

Clinical Management

When hypotension is induced by deferoxamine, the deferoxamine infusion should be discontinued and restarted at a slower rate after recovery of the blood pressure. Although the efficacy of deferoxamine in iron chelation is unquestionable, one should be cognizant of its side effects.

Animal Toxicity

The LD_{50} of deferoxamine in rats is 520 mg/kg with intravenous administration and greater than 1000 mg/kg with oral administration. Deferoxamine-induced hypotension, tachycardia, and renal insufficiency have been also reported in rats and dogs.

—*Zhengwei Cai*

DEHP

- ♦ CAS: 117-81-7
- ♦ PREFERRED NAME: Di(2-ethylhexyl) phthalate

- ♦ SYNONYMS: Dioctyl phthalate, DOP; BEHP; Compound 889; DAF 68; octyl phthalate; ethylhexyl phthalate; OCTOIL
- ♦ CHEMICAL STRUCTURE:

$$\text{COOHCH}_2\text{CH (C}_2\text{H}_5)(\text{CH}_2)_3\text{CH}_3$$
$$\text{COOHCH}_2\text{CH (C}_2\text{H}_5)(\text{CH}_2)_3\text{CH}_3$$

Uses

DEHP is a softening agent commonly used in plastics such as polyvinyl chloride. It is found in products such as telephone cords, kidney dialysis machine tubing, medical plastic bags, shower curtains, vinyl wall coverings, and children's toys.

Exposure Pathways

DEHP is readily absorbed via ingestion; it is not absorbed significantly via dermal contact. Inhalation exposure is not likely because of the compound's low vapor pressure.

Toxicokinetics

DEHP is absorbed by the oral and parenteral routes. It is not absorbed in any significant amount through intact skin. Intravenously or orally administered DEHP is rapidly metabolized to derivatives of mono-(2-ethylhexyl) phthalate (MEHP). The MEHP metabolites are mainly excreted in the urine and the bile. The estimated half-life of DEHP in humans following intravenous administration is 28 min.

Mechanism of Toxicity

DEHP is a peroxisome proliferator, so named because it causes extensive proliferation of hepatic peroxisomes in susceptible species. As peroxisome proliferators are uniformly carcinogenic in rats and mice, there is concern as to whether this effect is rodent specific or whether it can manifest in other species, including humans via drugs used in long-term therapy or through blood transfusion bags.

Human Toxicity

The effects of DEHP on humans have not been established.

Clinical Management

The potential for esophageal or gastrointestinal tract irritation following ingestion suggests that emesis should not be induced. Other measures to prevent absorption may be beneficial. Exposed skin and eyes should be copiously flushed. Liver function and blood glucose must be monitored.

Animal Toxicity

DEHP has been shown to produce liver cancer in mice and rats after lifetime exposure. It has been shown to reduce fertility and cause testicular atrophy in mice. The maternal–fetal transfer of DEHP and its major metabolite, MEHP, has been demonstrated in rats and is associated with the inhibition of brain and liver steroidogenesis in exposed offspring. In female rats, DEHP administration can suppress estradiol levels and suppress ovulation. The doses which caused cancer in animals were about 4 million times the exposure most humans are expected to receive.

*—Raja S. Mangipudy and
Harihara M. Mehendale*

Related Topics

Peroxisome Proliferators
Polymers

Delaney Clause

♦ TITLE: Delany Anti-Cancer Clause: Amendment to Food, Drug and Cosmetic Act
♦ AGENCIES: U.S. FDA and U.S EPA
♦ YEAR PASSED: 1958
♦ GROUPS REGULATED: Agricultural industry and food processors

Synopsis of Law

Substances that display carcinogenic properties in animal tests are not permitted to be intentionally added to food in any amount. The Delaney Amendment to the Food Drug and Cosmetic Act (FD&C Act), incorporated in 1958, outlines the matter as follows:

> No additive shall be deemed safe if it is found to induce cancer when ingested by man or animal or, if it is found, after tests which are appropriate for the evaluation of the safety of food additives, to induce cancer in man or animal.

Although guidance on the enforcement of this controversial amendment is quite clear with respect to substances directly added to food, there are important circumstances in which ambiguities arise, particularly in connection with indirect food additives and manufacturing by-products. Congressman James J. Delaney (Democrat, New York) chaired a select committee investigating the use of chemicals in food and cosmetics and introduced the clause into legislation.

In July 1996, Congress abruptly lifted the applicability of the Delaney clause. In passing the Quality Protection Act of 1996 (amending both the FD&C Act and the Federal Insecticide, Fungicide and Rodenticide Act), the absolute barrier of the Delaney clause was replaced with a risk assessment approach—a "health-based safety standard for pesticide residues in foods."

—Shayne C. Gad

Related Topics

Federal Insecticide, Fungicide, and Rodenticide Act
Food and Drug Administration
Food Drug and Cosmetic Act
Risk Assessment, Human Health

Deltamethrin

♦ CAS: 52918-63-5, 62229-77-0, 55700-96-4
♦ SYNONYMS: *S*-α-cyano-3-phenoxybenzyl-(1*R*)-*cis*-3-(2,2-dibromovinyl)-2,2-dimethylcyclo-propanecarboxylate; Butox; Decis; K-Othrine; NRDC 161, OMS 1998, RU 22974

♦ CHEMICAL CLASS: Type II pyrethroid insecticide

♦ CHEMICAL STRUCTURE:

Uses

Deltamethrin is a broad-spectrum insecticide.

Exposure Pathways

Exposure to deltamethrin has occurred through ingestion, inhalation, and dermal contact.

Toxicokinetics

Pyrethroids are poorly absorbed through the skin and are only moderately absorbed in the gastrointestinal tract. Metabolism of deltamethrin occurs rapidly through ester cleavage and hydroxylation. Deltamethrin is eliminated more slowly from adipose tissues than from other sites such as brain or blood. In one case of dermal exposure, absorption was estimated to be approximately 3%. Urinary excretion is the primary route of elimination.

Mechanism of Toxicity

Pyrethroids have a selective high affinity for membrane sodium channels. Closing of the channel, which ends the action potential, is slowed resulting in a prolonged "tail" current and repetitive firing of presynaptic and accompanying postsynaptic cells following a single action potential. High enough doses can cause complete depolarization and blockade of nerve conduction.

Human Toxicity: Acute

Symptoms include hyperactivity, incoordination, choreoathetosis, and convulsions. Extensive dermal exposure causes temporary abnormal sensations including paresthesia, numbness, and tingling. Symptoms following ingestion included nausea, vomiting, tenesmus, diarrhea, unconsciousness, and death due to respiratory failure.

Human Toxicity: Chronic

Chronic effects following deltamethrin exposure have not been reported.

Clinical Management

Oral exposures may be treated with syrup of ipecac-induced emesis, activated charcoal, or gastric lavage. Emesis is contraindicated in cases of unconsciousness or convulsions. In cases of dermal exposure, contaminated clothing should be removed and disposed of. Any exposed areas of skin should be repeatedly washed with soap and water. Paresthesia may be treated with topical application of vitamin E. For eye contact, flush the eyes with generous amounts of lukewarm water for a minimum of 15 min. Seizures can be treated with intravenous diazepam; phenytoin or phenobarbital may be helpful for recurrent seizures. Atropine can be used to control excessive salivation. No specific antidotes for pyrethroid-induced neurotoxic effects have been approved for use in humans. Spontaneous recovery usually occurs with mild or moderate intoxication.

Animal Toxicity

Fish and crustaceans are extremely sensitive to pyrethroid compounds in laboratory settings. However, various factors (e.g., sediment binding) may reduce pyrethroid toxicity to these nontarget organisms in a natural environment.

—Paul R. Harp

Related Topics

Neurotoxicology: Central and Peripheral
Pesticides

Deodorants

Deodorants serve either to mask human body odor or to help prevent bacterial decomposition of excreted perspiration. There are basically two types of deodorants: simple deodorants and antiperspirants. A simple deodorant consists of an antibacterial agent in a cream base; an antiperspirant contains aluminum salts, such as aluminum hydroxychloride, which reduce the flow of perspiration. These agents are used in one of four forms: cream, liquid, powder, or stick. They usually include the following ingredients: aluminum salts, titanium dioxide, oxyquinoline sulfate, zirconium salt, formaldehyde, alcohol, and antibacterial agents. Some liquid forms are propellant dispensed (aerosols). Waxes, soap, and humectants may be present in minor proportion in stick forms. Roll-on types may be added with emulsifier and thickeners. The amounts of ingredients present in these products are usually small, and unless a large quantity is ingested, no ill effect should ensue.

In case of oral ingestion, the mouth can be flushed out, and milk may be given for its soothing and diluting effect. These products are nonirritating to most people, but sensitization may occur in some individuals. For these people, the preparation should be washed off thoroughly and a substitute brand may be chosen. Discontinuing use of deodorant may be necessary. Aerosol types may cause eye irritation. The eyes should be washed carefully with lukewarm water for a few minutes. Soothing eye drops may be helpful.

—*Zhengwei Cai*

Related Topics

Detergents
Surfactants, Anionic and Nonionic

Detergent

D etergent is any of various surface-active agents (surfactants) particularly effective in dislodging foreign matter from soiled surfaces and retaining it in suspension. Soap, which is made from fats or fatty acids, is a detergent. However, in common usage the term "detergent" is not applied to soap but only to the synthetic nonsoap substances.

Dishwashing and clothes laundering are the principal applications of detergents for which the liquid bath is water. There are also nonaqueous detergents that enhance the cleaning power of organic solvents; for example, they are used in dry cleaning and in engine oil to prevent dirt and gummy decomposition products from depositing on the metal.

Because most detergents are necessary and familiar household items, they are not usually regarded as hazardous substances. However, incidental oral ingestion of detergents by children under 5 years of age is not unusual. Although severe toxic effects occur infrequently and fatalities are rare, irritation or severe injury of the oral mucosa, esophagus, and stomach may result from ingestion of these agents because of their alkaline and acidic components. Most laundry detergents are not strong enough to do significant harm, but some laundry products, automatic dishwashing detergents, wall cleaners, drain or oven cleaners, disinfectants, and ammonia can cause extensive injury. In addition, extensive eye and skin exposure to some detergents may also cause toxic effects.

Detergents vary in composition, depending on the cleaning task for which they are intended. Therefore, their potential hazardous effects also vary with composition. Detergents used for laundering usually contain several basic kinds of ingredients. There are surface-active agents, or surfactants, which are substances that greatly lower the surface tension of water. There are also builders and auxiliary components.

Surfactants have elongated molecules, one end of which is a water-insoluble nonpolar hydrocarbon, while the other end is a water-soluble polar radical, which may or may not be capable of ionizing. There are four types of surfactants: cationic, anionic, nonionic, and amphoteric. The cationic detergents or quaternary ammonium compounds are synthetic derivatives of ammonium chloride. Their toxicity has not been definitely established, but the human fatal dose by ingestion has been estimated to be from 1 to 3 g. The principal manifestations of poisoning from ingestion of these agents are vomiting, collapse, and coma due to caustic effects. Treatment consists of first giving milk, egg whites, or a mild soap solution by mouth, and then giving an emetic or performing gastric lavage with a weak soap solution. Supportive respiration should be given. If convulsions occur, short-acting barbiturates should be given parenterally.

Sodium alkylaryl sulfates, sodium alkylsulfates, and alkyl sodium isoionates are representatives of anionic detergents and generally are less toxic than cationic detergents. The LD_{50} values in animals vary from 1 to 5 g/kg. The maximum safe amount for children has been estimated at 0.1–1 g/kg. Anionic detergents may cause skin and eye irritation. Nonionic detergents, such as alkylphenyl polyethoxyethanol, polyalkalineglycol, and fatty acid alkanolamine amide, are even less toxic. Oral ingestion usually results in emesis. Amphoteric detergents contain both anionic and cationic components.

The builders of detergents are generally inorganic salts or alkalies that enhance the cleaning action of the surfactant. They are of two general types: those that hold calcium in solution as soluble chelates and those that precipitate calcium. The first category includes widely used sodium tripolyphosphate and nitrilotriacetate. The second category includes sodium carbonate, sodium bicarbonate, sodium sesquicarbonate, sodium silicate, and sodium metasilicate. Ingestion of sodium tripolyphosphate and hexametaphosphate can cause severe gastroenteritis with vomiting and diarrhea. In addition, high percentages of phosphates in detergents discharged into water systems may cause a continuous buildup of nutrients, resulting in algae and water plant growth, a complex process called eutrophication. Public concern has caused a number of states to pass legislation banning the use of phosphate detergents.

Detergents may also contain auxiliary components such as foam stabilizers, optical brighteners or whiteners, and anti-redeposition agents. Bleaching agents (chlorine-releasing agents) or bactericidal agents (mild concentrations of quaternary ammonium compounds) have been included in the formulation of a number of detergents. These additives are only moderately toxic. However, simultaneous use of acid-type toilet bowl cleaner and sodium hypochlorite bleach may produce chlorine gas, which causes respiratory irritation with coughing, labored breathing, and inflammation of the eyes and mucous membranes (see Bleach).

—*Zhengwei Cai*

Related Topics

Alkalies
Poisoning Emergencies in Humans
Sensory Organs
Skin
Surfactants, Anionic and Nonionic

Developmental Toxicology

The Problem of Birth Defects

- 1 in 33 babies is born with a serious congenital malformation.
- Birth defects are the number one cause of infant death (Fig. D-1).
- Of the 3 million births each year in the United States, 200,000 of those born live are afflicted with a structural malformation.
- Nearly 500,000 spontaneous abortions occur annually in the United States due to birth defects.
- Children born with birth defects are much more likely to die than children born without birth defects (Fig. D-2).
- Infant mortality in the United States (being nearly the highest among industrialized nations) has been increasing since the turn of the century (Fig. D-3).
- Birth defects account for more years of life lost than acquired immunodeficiency syndrome (AIDS), sudden infant death syndrome (SIDS), or prematurity (Fig. D-4).
- California leads the nation in the number of infant deaths caused by birth defects (Fig. D-5).
- Hawaii, Iowa, Kentucky, Minnesota, Montana, Nebraska, North Dakota, and New Mexico rank highest in the number of infant deaths (Fig. D-6).
- The number of white (Fig. D-7) and black (Fig. D-8) infant deaths are high in Alabama,

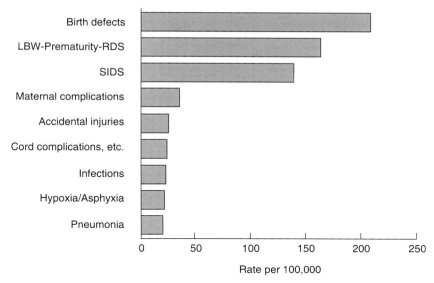

FIGURE D-1. *Leading causes of infant mortality in the United States in 1988. Rates are expressed per 100,000 births. Abbreviations: LBW, low birth weight; RDS, respiratory distress syndrome; SIDS, sudden infant death syndrome (reproduced with permission from* Birth Defects and Infant Mortality, Infant Mortality Report Series Vol. 1, No. 2, March of Dimes Birth Defects Foundation)

Louisiana, Mississippi, and Texas. Although the number of white infant deaths is low in California, the number of black infant deaths in that state is high.

• Birth defects are extraordinarily expensive. In California alone, annual public expenditure for

one malformation (spina bifida) is $18.6 million—a figure that does not include required special social or educational services, lost wages, rising cost of medical care, or payments by insurance companies to individuals. Even among infants with no obvious structural defect at birth, hospitalization costs are substantial. For babies of cocaine-positive mothers (prematurity, growth retardation, and low birth weight), labor, delivery, and postpartum care for these children averages more than 10 times ($13,222) the cost for babies born to drug-free mothers ($1,297).

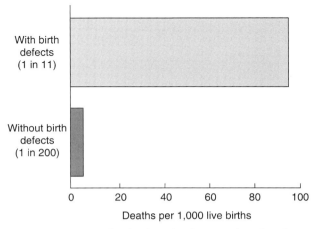

FIGURE D-2. *Risk of infant death (reproduced with permission from* Birth Defects and Infant Mortality, Infant Mortality Report Series Vol. 1, No. 2, March of Dimes Birth Defects Foundation).

The science of teratology (a word coined in 1832 by Geoffrey Saint-Hilares as literally "the study of monsters") has a history predating that of medicine as we know it today. The contemporary definition of teratology is "the science dealing with the causes, mechanisms, and manifestations of a structural or functional nature of abnormal prenatal development." Teratology can be considered a subdivision of developmental biology. Developmental toxicology encompasses embryonic and fetal death, reduced fetal growth, and other manifestations of abnormal development brought on by exposure

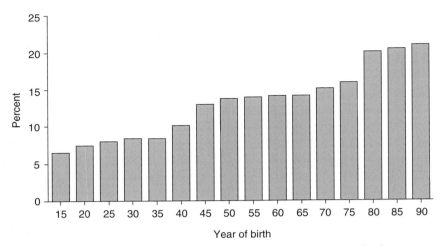

FIGURE D-3.   *National Center for Health Statistics percentage of infant mortality associated with birth defects since the early twentieth century. These data show that this percentage has been on a steady rise. This increase has occurred because infant death due to other causes, principally infectious disease, has declined at a much faster rate than has death due to birth defects (reproduced with permission from* Birth Defects and Infant Mortality, Infant Mortality Report Series Vol. 1, No. 2, March of Dimes Birth Defects Foundation).

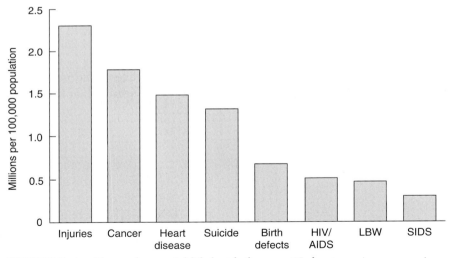

FIGURE D-4.   *Years of potential life lost before age 65 due to various causes in the United States, 1988. Abbreviations: HIV, human immunodeficiency virus; AIDS, acquired immunodeficiency syndrome; LBW, low birth weight; SIDS, sudden infant death syndrome. (reproduced with permission from* Birth Defects and Infant Mortality, Infant Mortality Report Series Vol. 1, No. 2, March of Dimes Birth Defects Foundation).

FIGURE D-5. *United States number of infant deaths caused by birth defects by state, 1988 (reproduced with permission from* Birth Defects and Infant Mortality, *Infant Mortality Report Series Vol. 1, No. 2, March of Dimes Birth Defects Foundation).*

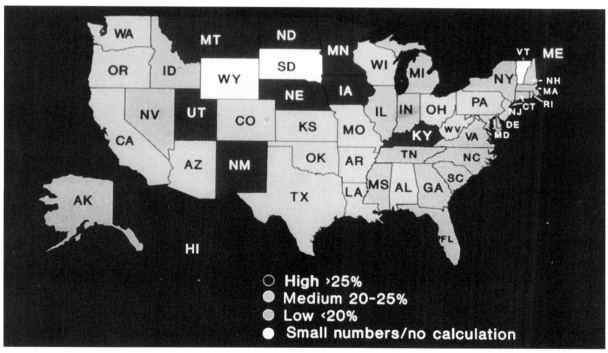

FIGURE D-6. *United States percentage of infant deaths caused by birth defects by state, 1988 (reproduced with permission from* Birth Defects and Infant Mortality, *Infant Mortality Report Series Vol. 1, No. 2, March of Dimes Birth Defects Foundation).*

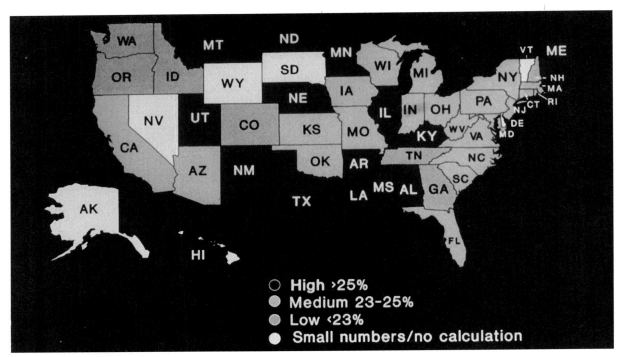

FIGURE D-7.   *United States percentage of White infant deaths caused by birth defects by state, 1988 (reproduced with permission from* Birth Defects and Infant Mortality, *Infant Mortality Report Series Vol. 1, No. 2, March of Dimes Birth Defects Foundation).*

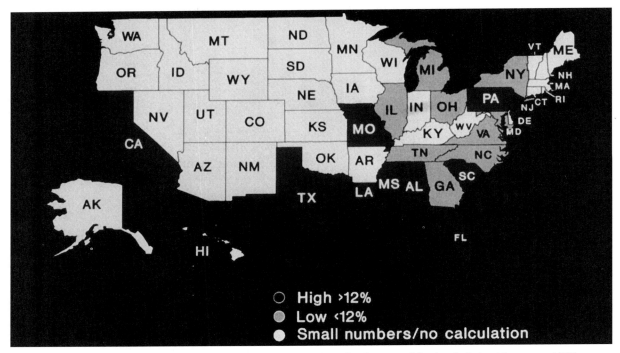

FIGURE D-8.   *United States percentage of Black infant deaths caused by birth defects by state, 1988 (reproduced with permission from* Birth Defects and Infant Mortality, *Infant Mortality Report Series Vol. 1, No. 2, March of Dimes Birth Defects Foundation).*

to xenobiotics (literally "foreign chemicals"). Virtually all chemical compounds (including common sugars like glucose or normal amino acids like phenylalanine) can induce embryotoxicity and fetotoxicity if the dose given is sufficiently large and the time and duration of exposure in pregnancy is appropriate. Maternal disease like diabetes and phenylketonuria can predispose a patient to an abnormal pregnancy outcome. Nevertheless, a compound is not usually considered a teratogen (a chemical that causes birth defects) if the dose required causes maternal signs or symptoms of poisoning in animal studies. Human teratogens are often seen where there is frank intoxication; there is no better example of this than ethanol in alcoholic mothers and their offspring who exhibit features of the fetal alcohol syndrome (FAS). Thus begins the dilemma for the clinician, the teratologist, the government regulator, and the family of an affected child—teratogens may seem to be everywhere, but on close inspection they seem to be nowhere. The birth of a malformed child has always been a matter of intense concern and sorrow; the same question always follows: "What caused it?" Ancient peoples formulated their own explanations on the cause(s) of these diseases and some remnants of their hypotheses are still with us today. As we will see in the following discussion, we have advanced from stoning or cremation of mothers of infants with birth defects. However, we have replaced the hypotheses of antiquity (termed superstition today) with more subtle (but in some instances equally preposterous) and damaging ideas in our search for responsible agents and parties.

It is often said that the cause of a particular birth defect is "multifactorial," generally taken to mean that it is the interaction of one or more environmental agents (a drug, a hazardous waste site, or a drinking water contaminant) with the genetic makeup (genotype) of the mother and her embryo. This notion of multifactorial causation stems from the writings of the French surgeon Ambroise Pare' (1510–1590) in *Chyrurgery* (1579) recounting the influence of maternal impressions (see below), demonic intervention, and environmental or mechanical factors. Actually, until the thalidomide tragedy of this century, it was generally accepted that the embryo developing in the womb was well protected and that the placenta insulated the conceptus from noxious agents. Today, the multifactorial explanation is either applied correctly, when describing the interplay between a susceptible genotype (e.g., inborn errors of metabolism) and exposures of interest or out

of frustration on the part of teratologists, obstetricians, or pediatricians who cannot otherwise account for the empirical observations at hand. This tendency taken together with the oft held view that certain birth defects are inevitable, the result of random chance, or "God's will," lead to dismay that these conditions can ever be "cured" or prevented. In contrast, the frequency of these conditions and the natural desire to ascribe one or another particular factor or agent (e.g., a drug or a workplace chemical or practice) as the cause leads today to significant legal and financial consequences that, depending on the particular circumstance, may or may not be warranted. The science of teratology has at its core the emotional distress that accompanies the birth of a malformed child. This factor is neither lost in tort adjudication nor lost in the nation's abortion ("right to life") debate. Professional scientific societies, such as the Teratology Society and the American College of Obstetricians and Gynecologists, are dedicated to the study and prevention of birth defects. Within the past several years, it has become clear that a great many of these common, costly, and deadly conditions can indeed be prevented. It is surprising that prevention can sometimes be accomplished by simple and inexpensive steps once the etiologic agent(s) has been identified.

Generally speaking, developmental toxicology focuses on abnormal morphogenesis induced by xenobiotics; however, a discussion of the extent or nature of birth defects must consider other possible causes—infectious microbes, abnormal chromosomes, radiation, hormones, and maternal chronic disease. These factors must be added to the "normal" or "background" rate of embryonic demise known from the classic studies of A. T. Hertig and J. Rock carried out during the middle of this century in couples of proven fertility under optimal conditions for pregnancy. In these studies, 15% of the oocytes failed to fertilize, 15% of the fertilized oocytes started cleavage but failed to implant, and of the 70% that implanted 58% survived. Of those surviving, 16% were abnormal.

Embryos die for any number of reasons (e.g., degeneration of the corpus luteum or a defective trophoblast) and they are aborted spontaneously with the next menstrual period—usually without producing any of the maternal signs associated with pregnancy. Thus, by the end of the first expected menstrual period more than one-half of all human eggs exposed to sperm under the best of conditions die for one reason or another.

The biological context of the word "development" covers the changes from conception through birth, neonatal life to adulthood, and to old age. The word is restricted here, however, to embryonic and fetal life ranging from subtle changes detectable only in studies of children or young laboratory animals to embryonic or fetal death. A brief discussion of functional delay or deficit (commonly referred to as behavioral teratology) is presented. It is these functional or behavioral deficits that represent insidious to overt manifestations of developmental toxicology.

To understand the causes and pathogenesis of congenital malformations, one must possess at least a working knowledge of embryology as can best be gained from completion of an undergraduate course in the subject or as is commonly taught in medical school anatomy. For purposes of the current discussion, a rudimentary understanding of biology and mammalian embryology is assumed.

Historical Lessons

While it may seem obvious, birth defects are not new. In the vast majority of cases, birth defects and their causes cannot be linked to modern consumer products, occupational exposures, therapeutic or recreational drugs, or environmental pollutants. Congenital defects are perhaps the greatest source to have influenced the myths of antiquity (second only to belief in divinity or the study of the heavens), fairy tales of Rumpelstiltskin and other dwarfs, elves and hunchbacks or otherwise twisted (arthrogryposis, torticollis, and scoliosis) trolls, or contemporary book and film scripts of the macabre.

Cyclops are first recorded as subterranean beings who serve Hephaestus (Sanskrit Yavishta and the Vedic god of fire), the Greek divine blacksmith. The sons of Uranus and Gaea (Arges, Steropes, and Brontes) are all cyclops. These cyclops forged the trident for Poseidon and the bronze helmet for Hades and were then killed in furious revenge by Apollo. The cyclops of Homer's *Odyssey* (Polyphemus, who Ulysses blinded with a sharpened, burning stake driven into his eye) inhabited the southwest coast of Sicily and lived in caves, killing and devouring any stranger who chanced upon them. According to Callimachus, the cyclops Brontes, Steropes, Acamas, and Pyracmon, who lived on Mount Etna (the active volcano near the Sicilian city of Taormina on the Ionian Sea), were

Enormous giants, big as mountains and their single eye, under a bushy eyebrow, glittered menacingly. Some made the vast bellows roar, others, raising one by one their heavy hammers struck great blows at the molten bronze and iron they drew from the furnace.

One of the colonial American explanations regarding the etiology of cyclopia was hybridization between species. The birth of a cyclopic infant or farm animal (whose mother lived near a person with features thought to resemble that of the malformed newborn) was suspect. In the *Records of the Colony and Plantation of New Haven* (1638-1648), there is recounted the story of one such unfortunate neighbor, Mr. George Spencer, who happened to live near a sow who gave birth to a cyclopic pig that had "butt one eye in the middle of the face." The jury concluded that Mr. Spencer who "had butt one eye . . . the other hath (as it is called) a pearle in it" was guilty of bestiality. The "pearle" apparently bore some superficial resemblance to the cyclopic pig's eye. The sow was "slaine in his sight, being run through with a sworde" and poor Mr. Spencer was put to death for his crime on April 8, 1642.

Cyclopia is, of course, the most severe manifestation of holoprosencephaly—a condition in which the embryonic forebrain fails to separate into right and left hemispheres. The etiologic agent(s) of human cyclopia is not known. In ruminants (sheep, cattle, and goats), however, ingestion of the plant *Veratrum californicum* on even a single day (e.g., Day 14 in ewes) reliably produces the condition. Subsequent investigations have confirmed the presence of a teratogenic alkaloid, 11-deoxojervine, in the plant. That compound is termed most appropriately cyclopamine. Cyclopamine illustrates one of the many factors that must be taken into account for interspecies extrapolation of teratology data. Cyclopamine and its congeners cannot be held accountable for human cyclopia. Cyclopamine is not teratogenic in monogastric animals (e.g., rabbits) because it is degraded by stomach acid to an inactive compound (called veratramine). Nevertheless, when cyclopamine was fed to pregnant rabbits along with sufficient alkali so as to reduce stomach acid, cyclopamine was definitely teratogenic and, in fact, induced cyclopia.

Another example of historical explanations for birth defects is the theory of maternal impression. For better or worse, such theories find their way into laws, regulations, and even (in a modified fashion) into litigation. The eighteenth-century Philadelphia surgeon, Dr. John

Morgan, described the birth of a "piebald Negro girl with splotches all over her body"—a condition attributed to her mother's habit of evening star watching. Sirenomelia and ptergium colli were attributed to the mother seeing a snake or cobra during pregnancy and anencephalus to the mother looking at monkeys. Pregnant Spartan women were legally required to concentrate on statues and pictures of beautiful gods and warriors so as to ensure strong and healthy babies. One historical account involves an Italian nobleman whose wife happened to employ a black male servant. One day during her pregnancy, she attended a cultural exhibit at the local museum. After gazing at a portrait of a Moor, she subsequently gave birth some time later to a mulatto. The nobleman protested and the legislature responded, of course, by passing a law to the effect that pregnant women were to be banned from visiting art museums.

Although it may seem odd to consider it so, even structurally normal identical (monozygotic) twins can be considered a developmental aberration. Thirty percent of the 1.08–1.36% of the normal incidence of twin births in the United States are monozygotic. The factors responsible for splitting of the single fertilized egg (zygote) into two blastocysts (each with its own inner cell mass or embryo proper) are—like the vast majority of other developmental anomalies—unknown, but maternal genotype is a predominant contributor. The stage at which splitting occurs determines whether the embryos develop each with their own placenta and amniotic cavity, whether the embryos develop having a common placenta and separate amniotic cavities, or whether the embryos share a common placenta and a common amniotic cavity. Failure of complete separation of the inner cell mass results in any of a variety of conjoined ("Siamese") twins (diploterata or "double monster").

The term "Siamese twins" was first coined by the nineteenth-century circus king, P. T. Barnum, in reference to Eng and Chang Bunker, who were joined at the sternoxiphoid. (After retiring from the circus, Eng and Chang farmed in North Carolina and married at age 44, fathered a total of 22 normal children by two sisters, and died of arteriosclerosis in 1874 at age 63—within 2 hr of one another.)

Conjoined twins can be joined at the chest (thoracopagus), lower spine (pygopagus), or skull (craniopagus), with the latter being variable in fusion at the dorsal (occipital craniopagus), parietal, or ventral (syncephalus frantalis) aspect. Symmetrical conjoined twins occur about once in every 50,000 births and craniopagus twins are born about once in 3 million births (1 in every 58 cases of conjoined twins). The recent publicized U.S. parietal craniopagus are Lori and Dori Schappell of Reading, Pennsylvania. Of interest here is that only one of these women is afflicted with spina bifida.

The most extreme occipital craniopagus (janiceps) is Janus, the Roman god of gates and doorways represented artistically with his double faces (Janus bifrons) each in opposite directions so as to observe the interior and exterior, the entrance and exit, of public buildings. Janus arose from the god Chaos when earth, air, fire, and water took form during the creation of the world. His two faces represent his confusion in his initial state; thus, not only was Janus the god of departure and return but also he was the god of daybreak and new beginnings. Janus was revered even more than Jupiter and was honored on the first day of every month. The first month of the year (Januarius) still bears his name.

Dicephalic (double-headed) monsters populate not only 1950s matinee movies but also appear in sculpture, drawings, and carvings throughout history—Catal Huyuk (6500 BC) in southern Turkey and the South Pacific (dicephalus dibrachius), clay figurines in Mexico and South America (500 BC–800 AD), and figures on Babylonian clay tablets found near the Tigris (*Cuneiform Texts from Babylonian Tablets*, British Museum, London).

Neural Tube Defects

The belief that Satan, witches, sorcerers, and other diabolic and demonic forces were responsible for congenital malformations was prevalent during the fifteenth and sixteenth centuries, and this belief found its way to the new world. Not that these concepts were new—far from it. A mummified anencephalus—a condition considered the most severe malformation compatible with intrauterine life (Fig. D-9)—was discovered in 1825 at the Egyptian catacombs of the Hermopolis sarcophagus. This individual's area cerebrovasculosa (rudimentary brain) had been ceremonially removed through the nose, as was the custom. Based on the mummy's location and condition, and the inscription on the sarcophagus, the malformed individual was considered the product of fornication between the mother and an ape. This anencephalus was brought to the Berlin

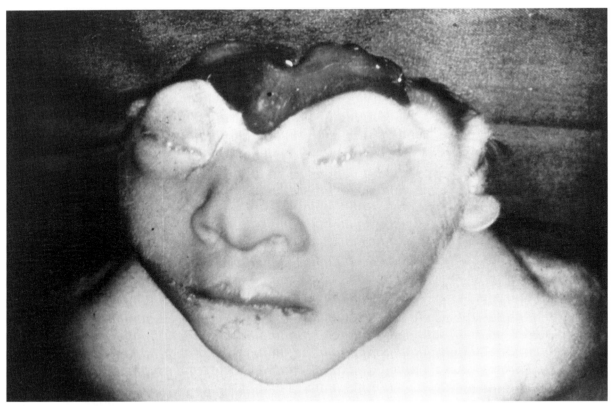

FIGURE D-9.   *Anterior view of an anencephalic human fetus. Notice the low-set ears, elevated nose and maxilla, the short neck (due to anomalies of the cervical vertebrae), and the prominent, protruding rudimentary brain.*

Museum by the King of Prussia and was unfortunately destroyed in the Second World War. By the 1600s, mothers of anencephalics were condemned in ways not dissimilar to the execution of witches in Salem, Massachusetts. One theory as to the origin of anencephalics was consanguinity with a troll—particularly dangerous being those who lived near roadways or under bridges—whose sons and daughters resembled their fathers (Fig. D-9).

Anencephalus is one of a constellation of malformations known collectively as neural tube defects (NTDs). Anencephalus is the end result of failure of neural fold elevation and fusion, a deficiency that can occur only in the most anterior region (Fig. D-9), along the entire axis (craniorachischisis totalis; Fig. D-10), or localized in areas along the spine (spina bifida; Fig. D-11). Spina bifida is a common term used to describe a range of defects of the axial skeleton, involving the vertebrae and to various degrees the cord itself. If only the verte-

brae show incomplete spinous process fusion and the subarachnoid space remains within normal limits, this is a subclinical condition known as occult spina bifida. If the vertebral arch is only rudimentary, the overlying tissues are weak, and cerebrospinal fluid pressure contributes to expansion of the subarachnoid space and the meninges herniate dorsally, this condition is diagnosed as spina bifida meningocele (cystica). If the vertebrae are so rudimentary that only the body of the bones is thickened, the spinal cord itself is displaced into the subarachnoid space (now a gross, protruding meningeal sac), the condition is classified as spina bifida with myelomeningocele (Fig. D-11). If there is complete failure of neural fold elevation in cranial, cervical, thoracic, and/or lumbar regions and the neuroectoderm is left exposed on its dorsal aspect, the spinal cord then develops with its ependymal layer in open contact with amniotic fluid and its lumen cannot be recognized. This latter condition (Fig. D-12) is termed spina bifida ap-

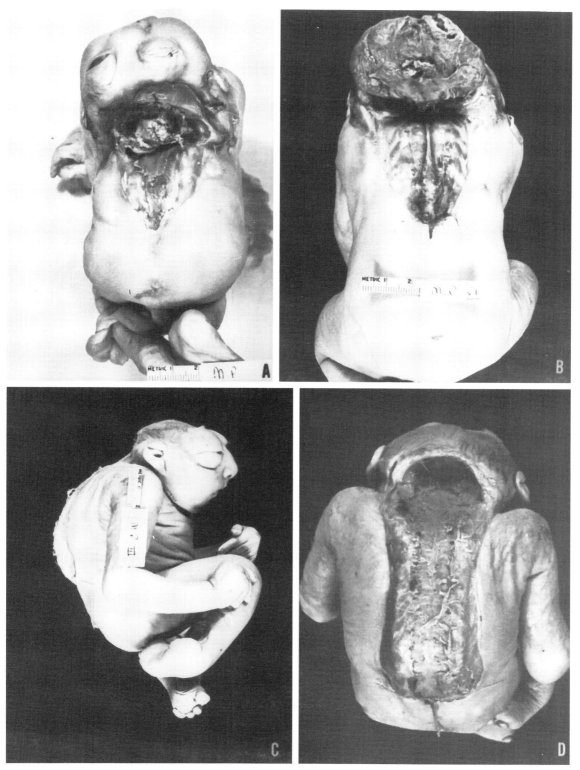

FIGURE D-10. *External appearance of various types of human craniorachischisis totalis (total myeloschisis) illustrating the severity of the dysraphic disorders. The first fetus (A) illustrates the severity of the lordosis and the shortness of the axial skeleton which can occur in these disorders. The exposed areas of the central nervous system are totally destroyed. In B, note the exencephalic brain (termed area cerebrovasculosa). C and D illustrate the lateral and posterior view. Compare C with Fig. D-9. In D, the destroyed areas of brain and spinal cord tissues have been removed to show the severity of the malformations of the vertebrae (reproduced with permission from M. Marin-Padilla, Clinical and experimental rachischisis, in* Congenital Malformations of the Spine and Spinal Cord. Volume 32. Handbook of Clinical Neurology, *North-Holland, Amsterdam, 1978).*

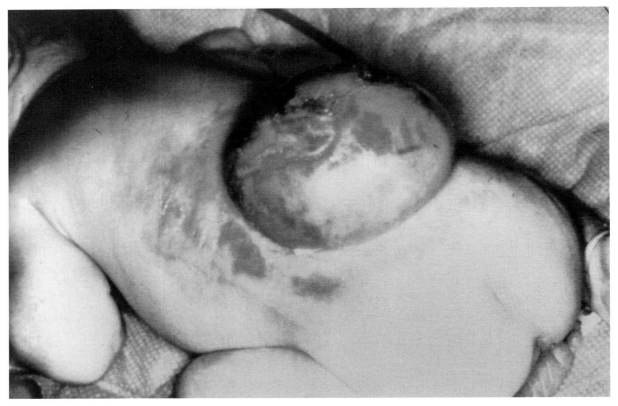

FIGURE D-11. *Newborn infant with spina bifida. Note the large meningocele on this child's back.*

erta (rachischisis or myeloschisis). Thus, spina bifida can range from a partial failure of neural tube closure, manifest as a benign subclinical condition of no practical consequence, to damage that is permanently disabling, affecting the patient's ability to walk and control normal bodily functions.

Just as partial failure of neural fold apposition and fusion can occur along the spine, it can also occur in the skull. Cephalic malformation can be complete in anencephalus (with little or no involvement of the lower spine) or incomplete as in encephalocele and Arnold–Chiari malformation (Fig. D-13). Encephalocele (Fig. D-13) is one such condition in which the brain extends through dura and membranous bone and comes into contact with the scalp.

From the early 1600s through the latter part of the nineteenth century, there was a great deal of interest on the part of physicians and surgeons in describing and classifying NTDs as well as in speculating on the cause(s) of NTDs and their pathogenesis. Dissections of fetuses with spina bifida and anencephaly were described in detail. These early studies provided a basis for understanding the morphogenesis of these malformations but gave little indication as to the actual etiologic agent(s) responsible. The 1891–1896 manuscripts by the German authors J. Arnold and H. Chiari describe hindbrain anomalies in spina bifida and provide for four different categories of herniation of the cerebellum into the foramen magnum. In that case a cervical sac or encephalocele forms; in others there is a simple cerebellar hypoplasia (termed collectively Arnold–Chiari malformations). This condition often results in hydrocephalus (accumulation of cerebrospinal fluid followed by marked expansion and thinning of the skull with subsequent compression and atrophy of the brain). Blockage of the roof of the fourth ventricle and continued production of cerebrospinal fluid by the choroid plexus leads to increased intracranial pressure. As a result, the medulla oblongata is forced into the cervical canal, the herniated cerebellum is compressed, and interrupted cerebrospinal fluid flow into the subarachnoid space produces an extensive internal hydrocephalus (Fig. D-13). In those cases of Arnold–Chiari in which the lumen of the spinal cord is open at some point

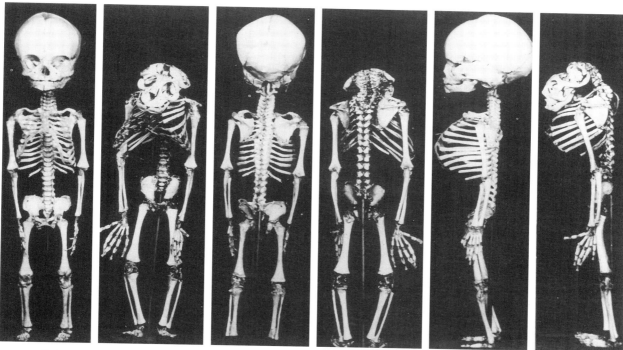

FIGURE D-12. *Posterior, anterior, and left lateral views of the entire skeletons of a 7-month premature infant and of a human anencephalic with cervical myeloschisis. Note the severity of the skeletal defects in the anencephalus, the anomalous facial bones, the short cervical column, and the absence of a proper skull (reproduced with permission from M. Marin-Padilla, Morphogenesis of anencephaly and related malformations, in* Current Topics in Pathology, *Vol. 51, pp. 145–174, Springer, New York, 1970,).*

(e.g., myeloschisis), hydrocephalus does not occur since cerebrospinal fluid either accumulates at another point (Fig. D-11) or, in cases in which the cord is exposed on the surface of the skin, cerebrospinal fluid drains to amniotic fluid and relieves increased intracranial pressure. It is by this route that α-fetoprotein (a normal serum component synthesized in the embryonic yolk sac to 12 weeks and then by fetal liver) escapes into the amniotic fluid. Radioimmunoassay of this protein is used in routine management of high-risk pregnancy where abnormally high concentrations are indicative of NTDs. Acute hydrocephalus is also a consequence of Dandy–Walker malformation, a condition characterized by defects of the ventricular system and stenosis (constriction) of the foramina of the fourth ventricle and it may present as a severe occipital encephalocele.

Anencephalus is a relatively common condition, affecting on average 1 in every 1000 births (or five or six embryos per 1000 pregnancies, given published studies of fetuses examined at 8 weeks gestation). Anencephalus occurs four times more often in males than

in females and four times more often in Caucasians than in blacks. Even lower rates occur among North American Indians (0.5 per 1000), Japanese (0.4 per 1000), and Central and South Americans (0.1–0.3 per 1000). Spontaneous early pregnancy abortion of anencephalic embryos ranges from 54% (London) to 87% (Japan).

Although the term anencephalic suggests a lack of all but the bones of the face, in fact, all of the bones of the skull are present (Fig. D-14). The anomalies of the facial bones and those of the remainder of the skull are the consequence of early disruption of the notochord, the mesoderm, and the neuroepithelium. Early deficiency in neural tube closure exposes the developing brain (neuroepithelium) to mechanical abrasion from the fourth week of gestation until birth. The hindbrain (often enclosed and therefore protected by the rudimentary neurocranium) can remain intact—containing those structures responsible for control of respiration. Thus, the newborn anencephalic can survive hours to (at most) a few days in the absence of mechanical venti-

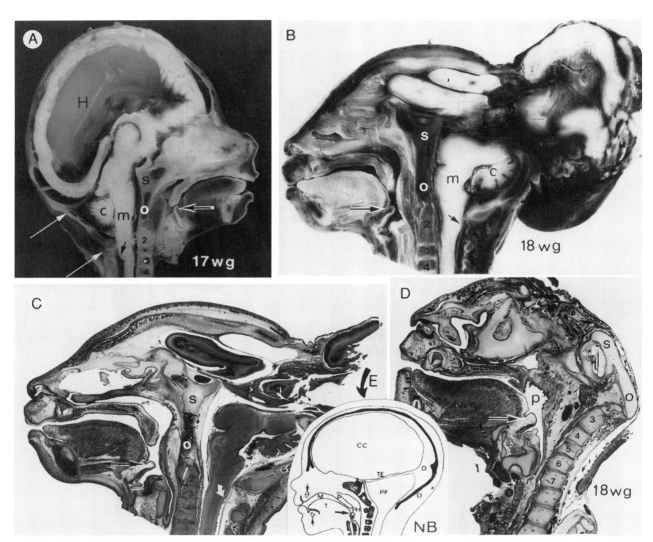

FIGURE D-13.   (A) Midsagittal section of the head of a premature infant with Arnold–Chiari malformation and secondary hydrocephalus. H, hydrocephalus; s, sphenoid; o, squama occipitalis; m, medulla; c, cerebellum; wg, weeks gestation. The cervical vertebrae are numbered. (B) Premature infant with partial failure of the anterior neural tube closure and a large occipital encephalocele. (C) Glass-slide section of the infant shown in B. (D) Premature infant with anencephalus, occipital schisis, and cervical rachischisis. The gross appearance of this infant was similar to that shown in Figs, D-9 and D-11. (E) Arrow indicates schematic ink drawing of the midsagittal section of a normal newborn's head. For comparison to A–D, locate the size and shape of the cerebral cavity (CC), the location of the tentorium (T), its angle in relation to that of the spine (D), the size and shape of the posterior fossa (PF), and the mouth (M), tongue (T), teeth (small arrows), and the nasal passage (P), pharyngeal (PH) and laryngeal (large arrow) cavities. NB, newborn. The black and white arrows in A–D point to the location of the epiglottis in relation to the base of the skull. The black arrows in A and B point to the bend in the medulla caused by the downward displacement of the subtentorial central nervous system. Note in A–C the short base of the skull, its angle to the spine, and the small posterior fossa. In D, note the angle of the spine and relatively large facial skeleton compared to that shown in E (reproduced with permission from M. Marin-Padilla, Cephalic axial skeletal-neural dysraphic disorders: Embryology and pathology, Can. J. Neurol. Sci. 18, 153–169, 1991).

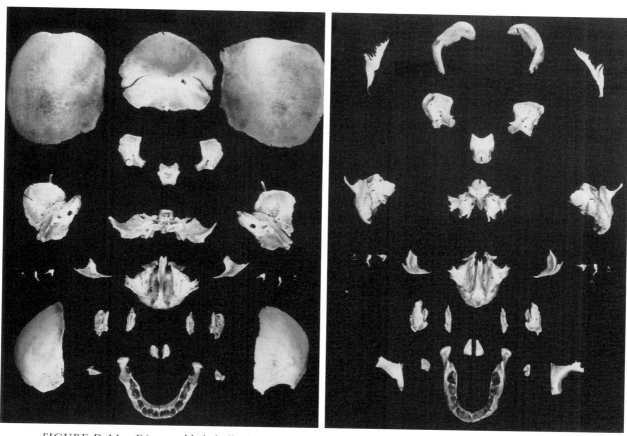

FIGURE D-14.   *Disassembled skull of a normal (left) and anencephalic (right) infant. The following bones, starting from the upper corner, are the parietals and the squama of the occipital (which in the anencephalus are represented by two small fragments); the basilar portion of the occipital with its two lateral portions; the temporals (note the rudimentary squamas in the anencephalus); the sphenoid (located in the center of each figure); the ossicles, the zygomatics, and the maxilla with the vomer and palantines; the lateral masses of the ethmoid and the turbine bones; and the frontals, the lacrimals, the mandible, and the two nasals. The rudimentary bones of the cranium and the normal (but narrowed) bones of the face are obvious (reproduced with permission from M. Marin-Padilla, Morphogenesis of anencephaly and related malformations, in* Current Topics in Pathology, *Vol. 51, pp. 145–174, Springer, New York, 1970).*

lation and aggressive life support. Anencephalic humans with larger skull bones (thus having a more representative brain) can survive for as long as a few weeks. The characteristic facial features of the anencephalus (low-set ears, protruding tongue, short maxilla, and elevated pointed nose; Fig. D-9) are direct consequences of the malformations of the bones of the base of the skull. The base of the skull (which in reality is composed of modified vertebrae) is small and, rather than forming a normal 90° angle with the spine, can be tipped to nearly 45°. The early collapse of the cephalic neural folds leads to gross malformation of the base of the skull. The normal sphenoid (with which all of these

bones articulate directly or indirectly) resembles a bird in flight (Fig. D-15A), whereas the anencephalic sphenoid resembles a bat with folded wings (Figs. D-15B–15F). Studies of disassembled normal and anencephalic skulls (Fig. D-14) show that it is the sphenoid malformations that precipitate the other gross malformations of the skull. The sphenoid is a bone that arises from mesodermal consolidation around the notochord prior to closure of the anterior neuropore. Since there is intimate communication between the embryonic brain (neuroectoderm), face (originating principally from neural crest), and the developing bones of the skull proper (neurocranium), teratogens that act on neuroec-

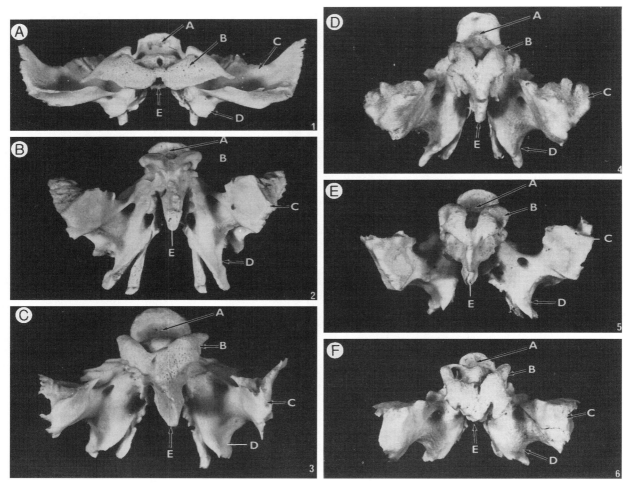

FIGURE D-15.   (A) Anterosuperior aspect of the sphenoid bone of a normal newborn infant as shown in Fig. D-14 (left) (×1.8). A, body of the sphenoid bone; B, the lesser wings; C, the greater wings; D, the pterygoid process; E, the rostrum. (B) Anterosuperior aspect of the sphenoid bone of a newborn infant with partial anterior cranioschisis (×2). (C) Anterosuperior aspect of the sphenoid bone of a newborn premature infant with complete (simple) cranioschisis (Fig. D-14, right) (×2.5). (D) Anterosuperior aspece of the sphenoid bone of a newborn premature infant with anencephalus and cervical spina bifida (×2.5). (E) Anterosuperior aspect of the sphenoid bone of a newborn premature infant with complete open spina bifida and anencephalus. The gross appearance of this infant is shown in Figs. D-10C and D-10D (×3). (F) Anterosuperior aspect of the sphenoid bone of a newborn infant with complete open spina bifida, anencephalus, a diaphragmatic hernia, and a large omphalocele (×3) (reproduced with permission from M. Marin-Padilla, Study of the sphenoid bone in human cranioschisis and craniorachischisis, Virchows Arch. Pathol. Anat. 339, 245–253, 1965).

toderm, mesoderm, or both can initiate a cascade of abnormal events which ultimately give rise to anencephalus.

The geographic distribution of NTDs is most remarkable. The incidence among people living in Ireland is four times that of people living in the United States (including those of Irish descent living in New England). Such striking differences have precipitated spec-

ulations and generated hypotheses on the cause of NTDs. One such theory, advanced in the early 1970s, was that the high incidence of anencephalus and spina bifida in Ireland was caused by ingestion of potatoes containing unidentified teratogens, potatoes infected with fungi ("blighted") after storage, or compounds produced by such potatoes in response to blight. Subsequent epidemiologic studies failed to support any of these theories; nonetheless, numerous studies have confirmed an excess number of NTD births in winter and, even more puzzling, a slow (but steady) decline in NTD births over the past five decades in the United States.

It is well known that the risk of NTDs is greatest in mothers 35 years or older (being particularly high in those who have borne several children) and that there is a marked increase in risk for mothers whose previous pregnancies ended in fetal or neonatal death (relative risk = 3.5–3.8). When one compares population prevalence of anencephalus and spina bifida (0.13 to 0.75% including New York, New England, British Columbia, Hungary, London, and South Wales), it is clear that recurrent risk is increased (from 0.1% in the general population to 1.8–7.1%) for siblings of fetuses with NTDs. In some families, as many as four anencephalics have been born to one mother. Although rare, anencephalics have been born to sisters who were daughters of women who had NTD pregnancies. Although the risk is greater among monozygotic compared to dizygotic twins, a simple genetic mechanism cannot be held responsible for NTDs. (The reader is reminded of those who advanced genetic causes for tuberculosis and Creutzfeldt–Jakob "mad cow" disease first found in the Fore cannibals of New Guinea.) More to the point, there is increased risk of NTDs in families with two or more affected pregnancies, and this increased risk for subsequent pregnancies is consistent "with a causal role of an environmental agent to which certain families are more exposed than others" (S. Yen and B. MacMahon as quoted in Elwood and Elwood, 1980). Indeed, one clue to the etiology of NTDs came from the observations of increased incidence in urban compared to rural areas—an observation consistent with a lifestyle or dietary hypothesis.

In 1992, the results of a landmark study by A. Czeizel and I. Dudas were published in the *New England Journal of Medicine*. Using a carefully controlled double-blind protocol (considered the "gold standard" by which clinical trials of new pharmaceuticals are routinely conducted), these investigators found that 0.8 mg per day of folic acid (a water-soluble vitamin) in a multivitamin preparation prevented spina bifida and anencephalus in the Hungarian women in their study. A subsequent study by M. Werler and associates of Boston University (published in 1993) confirmed that folic acid supplementation of the diet could have prevented a large proportion of spina bifida and anencephalus that occurred in the United States from 1988

through 1991. These results also confirmed those published by the United Kingdom Medical Research Council, which had conducted a randomized controlled clinical trial among women who had previously experienced a NTD pregnancy in 1991. The Hungarian trial was so successful that ethical considerations dictated its prompt discontinuation and those women who had been assigned to the placebo group were given the vitamin. These studies were extensions of previous trials on multivitamin supplements conducted in Europe and the United States in the mid-1960s and 1970s. Although it had been known since the 1950s that two cancer chemotherapeutic drugs known to induce folate deficiency (methotrexate and aminopterin) also induced terata in animals and in humans, interpretation of those data are confounded by the following facts: These folate antagonists have a number of pharmacologic actions, NTDs are not produced uniformly after exposure to either of these drugs, and NTDs are not the only malformations produced. In previous prospective studies of serum and erythrocyte folate, two important epidemiologic and clinical facts had emerged as indicators of maternal folate status: Folate deficiency can be documented in 66% of mothers of NTD pregnancies and low levels of folate mirror the woman's socioeconomic status. When matched with mothers of normal children for age, parity, time of conception, and pregnancy, 69% of mothers with NTD pregnancies were folate deficient compared with 17% of the referent controls.

In 1992 and 1993, Australian, Scottish, and Welsh departments of health and social services recommended widespread folic acid supplementation of breakfast cereals and breads. In order to avoid possible excess folate in one's diet, these groups recommended that some unfortified breads and breakfast cereals continue to be available. These groups recommended that women with spina bifida, or those who had a previous child with an NTD, consume 5 mg of folate each day if pregnancy was possible and that supplementation continue through the 12th week of gestation. It was recommended that all other women consume 0.4 mg daily, increase their consumption of folate-rich foods (fruit and vegetables), and avoid overcooking these foods. The United States Public Health Service has recommended that all women capable of becoming pregnant should consume 0.4 mg of folic acid per day for the purpose of reducing their risk of having a pregnancy affected by spina bifida or other NTDs. Nevertheless, the debate on folate supplementation of women's diets continues. Money is not an issue (400 μg of folate is currently sold at 4/100ths of a penny and food fortification for 50 servings at 200 μg per serving amounts to one U.S. cent). Rather, the benefits of folate supplementation are being weighed against the possibility of an increased risk of masking of vitamin B_{12} deficiency. This debate takes place in light of the following: Only 30% of low-income women consume the U.S. recommended daily amount of folate and fewer than 10% of women of childbearing age consume 400 μg of folate per day. Only half of the women with incomes 130% of the poverty line or less ate one serving of any vegetable over any 4 days; 18% did not eat any vegetables. With regard to fruit, fully one-third did not consume any fruit or juice and only 5% of white women 19–29 years of age and 4% of black women of that age ate two or more fruits or three or more vegetables each day. To compound the problem, cigarette smoking reduces blood levels of folate. Cigarette smoking is an increasingly popular habit among young women and a practice found more often among those in lower socioeconomic groups. Thus, numerous social considerations confront the practical implementation and the regulatory debate on folate prevention of NTDs. Of note here, it is not known how folate prevents NTDs or what role folate plays in early embryonic neural fold elevation and fusion. Whether women who have folate-preventable NTDs have defects in folate absorption, utilization, or metabolism is currently a mystery.

The Cause of Birth Defects

In English hospitals in 1914, 68% of infant deaths were due to dysentery, tuberculosis, and pneumonia. In Irish hospitals in 1915, only 2.3–3.1% of all infant deaths were due to birth defects. Since that time, infant deaths due to infectious disease have declined at rates much faster than the decline in infant deaths due to birth defects. Birth defects have assumed their increased contribution to infant death (Fig. D-3) not because of an actual increase in prevalence but because of the decline in other causes of infant death. Today, congenital malformations, prematurity, and SIDS together account for more than 50% of the total United States infant mortality (Fig. D-1). In the vast majority of cases, the cause of the birth defects is unknown (Fig. D-16).

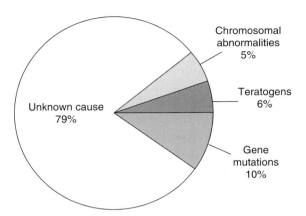

FIGURE D-16. *Pie chart of the relative percentages of the causes of human birth defects. It has been estimated by various authorities that cytogenetics contributes to no more than 5% of all malformed live births, Mendelian inheritance to no more than 15–20%, maternal infections 3 or 4%, maternal disease 3 or 4%, problems of* constraint *in utero (amniotic bands) 2%, and all drugs, chemicals, and radiation no more than 1% of the total load of structural birth defects in human beings.*

Maternal infections account for no more than 3 or 4% of the total load of congenital malformations. The most well known of these infections is rubella (German measles). Infection during various stages of gestation corresponds to the particular malformations produced. Malformation of the eye, including cataract and microphthalmia (literally small to nearly absent eye), can be induced by infection during the sixth week. Congenital deafness is induced by infection during the ninth week. Malformations of the heart (patent ductus arteriosus and ventricular septal defect), teeth, and brain (resulting in profound mental retardation) follow infection during Weeks 5–10 and Weeks 13–28, respectively. As with all known teratogens, not all children exposed to rubella *in utero* develop congenital defects, but the risk of malformation is greatest (to at least 47%) when rubella exposure occurs shortly after implantation.

Other infections clearly associated with increased risk of terata include toxoplasmosis, cytomegalovirus, and herpes simplex. Microphthalmia, blindness, hydrocephalus, and cerebral calcification can occur after infection with these organisms.

The contribution of genetic disorders accounts for no more than about 15% of the total load of congenital malformations (Fig. D-16). Inherited conditions account for (at most) 20% of the total, and abnormal cytogenetics accounts for no more than 5% of the total. An entire branch of the science of genetics is concerned with abnormally high or low numbers of chromosomes. Variable degrees of mental retardation, frank structural malformation, and sterility are common consequences of abnormal numbers of somatic chromosomes (autosomes) or of sex chromosomes. Two well-known syndromes arising from an abnormally high number of sex chromosomes are Klinefelter's (male only) and Triple X (female only), which originate as failures in normal chromosome number reduction (meiosis) in spermatogenesis and oogensis. A syndrome resulting from too few sex chromosomes is Turner's syndrome, a condition characterized phenotypically by a webbed neck and congenital absence of the ovaries.

An excess number of autosomes or the absence of one or more autosomes can either be lethal to the embryo or result in fairly well-known conditions. For example, the risk of Down's syndrome (trisomy 21 or mongolism) increases with maternal age—being 1 in 2000 for mothers aged 40 or more years. Extra chromosomes 17 and 18 result in micro- or anophthalmia (congenital absence of the eyeball), mental retardation, cleft lip, cleft palate, and deafness; this condition occurs on average once in every 5000 births.

Phenylketonuria is a genetic disease that results in abnormally high concentrations of the amino acid phenylalanine in the blood. Children of mothers with high plasma phenylalanine (>3 mg/dl) are at increased risk for microcephaly, mental retardation, heart malformations, esophageal atresia, tracheoesophageal fistula, and low birth weight. There is a clear concentration–response relationship between plasma phenylalanine and abnormal pregnancy outcome; head circumference and low birth weight are inversely (and linearly) related to maternal blood phenylalanine concentrations. Offspring of mothers with plasma phenylalanine >20 mg/dl experience a 92% incidence of congenital heart disease; those exposed *in utero* to maternal plasma phenylalanine levels higher than 3 but lower than 11 mg/dl experience a 21% incidence of congenital heart malformation. These defects can be prevented with adherence to a strict diet to control phenylalanine intake, total energy, protein, and weight gain during pregnancy; however, a normal pregnancy outcome can occur only when dietary control occurs at or before conception. Women who consume a "relaxed diet" experience a 0.6% malformation rate in their children (compared to 0.0% for those on a strict diet). Women

placed on a phenylalanine-restricted diet after conception (but before the second trimester) typically experience pregnancies with malformation rates on the order of 19–22%.

Maternal disease—not necessarily of either microbial or genetic origin–can cause or contribute to adverse pregnancy outcome. One common example is diabetes mellitus. If uncontrolled, diabetes mellitus can result in mental retardation, congenital malformation, and embryonic death. Glucose appears to be the teratogenic agent (perhaps potentiated by acetone and β-hydroxybutyrate ketone bodies) in uncontrolled diabetes. The extent of the hyperglycemia is related directly to the risk of holoprosencephaly (also usually accompanied by microcephaly, cleft lip and palate, mental retardation, and epileptiform seizures), situs inversus (complete transposition of the viscera), and ureteral duplex (complete or partial double ureter). These risks are 400, 84, and 23 times that in uncomplicated pregnancy, respectively. Sacral, vertebral, and pelvic malformation are also associated with elevated maternal glucose levels. To illustrate that stage of development determines susceptibility to teratogenic insult, women who develop diabetes relatively late in gestation (second or third trimester) do not have an increased risk of adverse pregnancy outcome. Only those women with uncontrolled diabetes prior to conception and through the first 8 weeks of pregnancy are at risk. Insulin control of maternal diabetes produces marked reduction in neonatal mortality (from 33% in the decade 1920–1930 to 6.5% from 1975 to 1979). By careful management of these mothers, the number of diabetes-induced late fetal deaths (stillbirths) and depressed or abnormally high birth weights (macrosomia) have been reduced to the point that birth defects are now the leading cause of death among offspring of diabetic mothers.

Another example of a maternal condition that contributes to birth defects is low circulating iodine. Cretinism is one of the most profound, but completely preventable, syndromes of malformation known. Characteristic consequences of prenatal iodine deficiency include pervasive mental and physical retardation, deaf-mutism (due to primary malformation of the inner ear), lack of muscle tone with a spastic or rigid walk, and failure to attain a height at maturity of less than 1 m. Today, this condition (known as endemic cretinism) is most prevalent in impoverished areas of African and East Asian countries; however, prior to implementation of a national program of iodized salt in the early part of the twentieth century, endemic cretinism was commonplace in Switzerland. After institution of iodized salt, deaf-mutism declined 50% within 8 years and no cretins have been born since 1930.

Maternal disease in addiction contributes to infant morbidity and mortality. Two of the most common agents, ethanol and tobacco smoke, are illustrative. A host of clinical and epidemiologic studies have confirmed a distinctive pattern of congenital malformations in babies born to alcoholic mothers. These malformations include microcephaly, short palpebral fissures, epicanthal folds, maxillary hypoplasia, cleft palate, micrognathia, joint disease, cardiac anomalies, capillary hemangiomata, anomalous genitalia, and retarded fetal and neonatal growth and development (a pattern referred to as FAS). FAS children have abnormal motor and psychological development and abnormal dermatoglyphic characters. Miscellaneous terata also found among these children are arthrogryposis, limb reduction defects, and gastroschisis. With alcohol, there is a definite dose–response relationship: a 9% malformation rate in light drinkers, a 14% rate for moderate drinkers, and a 32% rate for heavy drinkers. In addition, as with almost all teratogens, the severity of the malformations is greatest in those infants born to mothers consuming the highest dose. Some authors have argued that consumption of one type of alcoholic beverage (e.g., wine, beer, and schnapps) was more or less dangerous than another; however, the relationship holds true for all drinking. Pregnancy outcome is directly related to the dose of ethanol as measured by the quantity of absolute ethanol consumed per day. Ethanol doses consumed by these mothers range from "social" (1 oz of absolute ethanol/day or 350 mg/kg/day) to "heavy" (2.3–15 oz of absolute ethanol/day or 800–5600 mg/kg/day). Statistically significant reductions in birth weight (91 g for ethanol exposure before pregnancy and 160 g for exposure in late pregnancy), decreased infant length, increased stillbirth and second-trimester spontaneous abortion (0.5–1 oz of absolute ethanol/day), and increased risk of early spontaneous abortion (1 oz of absolute ethanol, twice per week) are well documented. Notwithstanding these data, results from the National Institute of Child Health and Human Development (NICHHD) study of drinking habits and pregnancy outcome in 32,870 women who had two drinks or less each day showed that those women had the same overall risk of birth defects as pregnant women who did not drink ethanol at all. NICHHD defined moderate drinking for purposes of the study on a daily basis (e.g., wine with a meal) but excluded binge drinking (foregoing drinking during the week but consuming several drinks on the weekend).

The most common addiction during pregnancy is tobacco smoking; 30.9% of all women smoke before pregnancy and 25.5% continue to smoke during pregnancy. This practice continues despite the fact that the first reports of adverse effects on the human fetus were published more than 80 years ago and despite the legally required warnings on tobacco advertisements and product packaging. At least 19 major epidemiologic studies of more than 300,000 pregnancies have been published. The results of those studies lead to the following dose–response conclusions: Smoking 10–20 cigarettes per day throughout pregnancy increases the risk of early spontaneous abortion and reduced birth weight from 92 to 316 g, and women who cease smoking during the early part of their pregnancy deliver babies with birth weights near those of babies born to mothers who have never smoked. The frequency of spontaneous abortion is directly related to the number of cigarettes smoked; the frequency for pack-a-day mothers being double that of nonsmoking mothers. The data hold true after correcting for maternal age, race, height, weight gain during pregnancy, socioeconomic status, gestational age, and parity. It appears that it is the nicotine and carboxyhemoglobin content of maternal blood that is responsible for decreased placental blood flow and anoxia leading to or contributing to the low birth weight. Perinatal mortality increases exponentially with decreasing birth weight; perinatal mortality is increased 20% in offspring of mothers who smoke less than one pack per day, and it is increased 35% in mothers who smoke more than one pack a day. Tobacco use is the single most important preventable determinant of low birth weight and its associated perinatal mortality in the United States. Of the 39,000 excess low-birth-weight babies in the United States, 5900 could be prevented each year by smoking cessation. The hospitalization and medical cost on a national basis for these babies is enormous—so much so in fact that for every $1 spent on smoking cessation during prenatal care, $6 in hospitalization and related medical care costs can be saved. This one intervention alone could double the cost savings gained by prenatal care.

Fear of Birth Defects

Two of the most questionable practices by those who are not familiar with the principles of teratology are (1) making lists of chemicals or other agents known or suspected to be teratogenic or otherwise toxic to the embryo and fetus and (2) assuming that the results of animal studies mirror what can be expected in human beings.

Perhaps the most controversial example of the first practice is California's Proposition 65, a set of laws enacted by public vote in 1986 (known as the Safe Drinking Water and Toxic Enforcement Act). Proposition 65 is now part of California's Health and Safety Code. Proposition 65's list, like other similar lists, includes chemical compounds ranging from known human teratogens to compounds having adverse effects demonstrated to occur only in laboratory animals. This type of list may reinforce the erroneous notion that compounds are either "positive" (teratogenic) or "negative" (nonteratogenic) and that providing printed or other warnings to somehow prevent exposure to those positive compounds will prevent birth defects. This is simply not the case. All compounds can cause the various manifestations of developmental toxicity provided the exposure (dose) is large enough, the route of exposure is appropriate, and the timing of exposure occurs during a susceptible period of development. One may argue that generating lists of teratogens increases public anxiety and flames the fuel of litigation.

The most obvious example of a known teratogenic compound for which warning can be effective is ethanol, but warnings alone are not the whole story. Alcohol consumption during pregnancy is a definite problem that is increasing with time (Fig. D-17). Studies of children in the general population indicate that from 0.1% to 0.2% show signs of FAS, but this figure does not represent the true measure of this problem. Among Native American tribes of the Great Plains or in northern British Columbia and the Yukon and the Southwest, studies have found 5% of the children affected. On the Pine Ridge reservation in South Dakota, more than 25% of the children show signs of FAS. Among the mothers, many are so disabled by their drinking that they are unable to care for their children. These children are difficult to place in foster homes because of their mental deficiency and inability to develop ap-

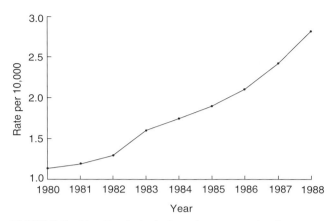

FIGURE D-17. Fetal alcohol syndrome rates in the United States, 1980–1988. These rates have shown a steady increase due to increased recognition and reporting of this condition by physicians and not necessarily due to increased alcoholism among women of childbearing age (reproduced with permission from Birth Defects and Infant Mortality, *Infant Mortality Report Series Vol. 1, No. 2, March of Dimes Birth Defects Foundation).*

propriate emotional and logical responses to everyday situations.

In the case of compounds that are merely suspected teratogens, however, there may be dark consequences. For example, in August 1973, the United States Consumer Product Safety Commission (CPSC) banned the sale of spray adhesives and published national warnings that these products caused birth defects and chromosome abnormalities. The CPSC warned all pregnant women who may have had contact with these sprays to see their physician and inquire about the chromosomes of their fetus. The minimum consequences of this regulatory action were 1273 working days logged by 130 U.S. diagnostic and genetic counseling centers on spray adhesives, at least 380 chromosome studies, 11 amniocenteses, and at least nine elective abortions out of concern for exposure to spray adhesive. Eight of these abortions were performed without first performing diagnostic amniocentesis, and one was performed in a woman who had chromosome breaks in her amniotic fluid. The genetic counselor in the latter case had informed the mother that he was unable to determine the health of her fetus with the information he had on hand; she elected abortion because of fear of possible birth defects and without telling the counselor of her decision. The aborted fetus was fixed in formalin and a detailed autopsy was performed: Not only was there no evidence for any congenital abnormality, but the chromosome change first observed was found to be due to viral contamination of her amniotic fluid sample. In those areas of the country where local newspapers had given the CPSC warnings the greatest publicity, the greatest numbers of inquiries to local genetic counseling services were made. Six months later, the CPSC withdrew the ban on these sprays because no toxicity of the substances in the spray could be demonstrated and the original observations on chromosome damage (clastogenesis) could not be confirmed.

The 1973 U.S. CPSC action and its consequence is not an isolated or rare example. Many pregnant women are very worried about birth defects. Women not exposed to any teratogenic agent appearing on any lists believe they have a one in four chance of having a child with major structural malformations—a risk equivalent to that after prenatal thalidomide exposure. Single women have a significantly higher (probability of less than 0.05) tendency to terminate their pregnancy than do married women; published studies demonstrate a greater willingness on the part of single mothers to abort their fetus when exposed to nonteratogens compared to married women in similar circumstances. The data show that the economic and social factors cited by single mothers in decisions to continue their pregnancy are compounded by a distorted perception of teratogenic risk.

That the young, the minority, and the educationally and economically disadvantaged are placed in a particularly vulnerable position with respect to this misinformation is highlighted by the following example. United States hospitalization and census data show that 61% of all pregnancies end in live birth, 26% in induced abortion, and 13% in early embryonic and late fetal death. Definite ethnic- and age-related differences underlie these overall rates. United States pregnancy rates for non-white women (80% of whom are black) average 68% higher than for white women. Although current census data show reduced numbers of births to teenage mothers (a 10% decline in the number of teenage pregnancies since the 1970s), those figures mirror the decline in the total number of teenagers (9%) over the same period. Pregnancy rates among teenagers as a group increased because of the decline in use of oral contraceptives and increased sexual activity. For U.S. teenagers 15 years of age, more than 50% of all pregnancies terminate in elective abortion; for all U.S. teenage pregnancies, 40% end in abortion and 10% in fetal loss. For women aged 15–19 years, there has been a 24% increase in elective abortion since 1976. Rates of induced abortion for U.S. non-white women are significantly greater than those for white women with the differential increasing to age 34 (after which this differential declines). These patterns underlie the fact that, in North America and Europe, there is one induced abortion for every live birth.

Among the most problematic issues raising the spectre of death and disability is radiation. After the Hiroshima and Nagasaki atomic bombs, for example, spontaneous abortion in those survivors who were pregnant increased to the point that one-third of the embryos died and of those that lived, at least 25% were afflicted with a structural malformation of one type or another (microcephaly, spina bifida, ocular defects, or oral cleft). There is no question that exposure to radiation from atomic bombs, or from X-rays or other medical procedures, has been responsible for instances of human congenital malformation. One recent report, however, illustrates an ironic and bizarre association between radiation and abortion. Following the Chernobyl meltdown and disaster in the former Soviet Union, fear and rumor were responsible for the abortion of at least 2500 otherwise wanted pregnancies in Greece. This occurred despite the fact that the radiation drifted north, to Scandinavia, and that effective exposure in Greece was 100 mrem—much less than the amount that could cause terata. In all of Western Europe, the total number of panic-induced abortions resulting from that episode has been estimated at 100,000–200,000 based on hospital records.

The second questionable practice by those unfamiliar with the principles of teratology concerns the overemphasis on animal data. Several thousand compounds have been identified as developmental toxins in animal bioassays, but only a relative handful are known human teratogens. There is a tendency among those who have not actually conducted laboratory studies, those who substitute a strength-of-evidence approach for the weight-of-evidence approach, or those who are otherwise unfamiliar with the principles of teratology to assume that the effects seen in animal (including bird) studies do or could occur in people. This is evident (1) in epidemiology studies in which investigators focus on a specific malformation in human populations after those defects have been observed in animal studies and

(2) in laboratory studies in which investigators have attempted to confirm or reproduce the human syndrome in animals. Concordance between animal and human data is the exception rather than the rule. Five teratogens are offered here to illustrate this point: acetazolamide, aspirin, caffeine, lead, and trypan blue. Finally, the retinoids (a large and diverse number of compounds of which vitamin A is a member) are presented to demonstrate that although species concordance is relatively rare, it does occur.

Acetazolamide (Diamox and Hydrazol), a prescription diuretic, is a classic example of a teratogen that produces malformations in a highly species-dependent fashion. Administration of acetazolamide to pregnant mice, hamsters, or rats causes right forelimb postaxial ectrodactyly (absent digit) and only on very rare occasions is any other malformation induced. Acetazolamide is not teratogenic in monkeys and, despite its widespread use in the 1950s in early and late human pregnancy, there has been no evidence that acetazolamide caused developmental toxicity in humans.

Acetazolamide exerts its pharmacologic action through inhibition of the enzyme carbonic anhydrase. It is believed that inhibition of this enzyme in rodent placenta results in disruption of the normal potassium ion balance, producing the malformation of the right forepaw. Replacement of the potassium lost to acetazolamide inhibition of carbonic anhydrase prevents the terata that would otherwise be induced by acetazolamide. Because the rodent fetus remains in constant orientation *in utero* with its right side against the placenta, it is thought that local disruption of potassium in that area dictates the constant malformation of only the right digit. In those rodent strains having genetic situs inversus, only the corresponding left digit is affected. In primate embryos, there is no detectable carbonic anhydrase at the stage of development where acetazolamide would be expected to induce these malformations.

Aspirin is the most widely used of any medication in the United States. Aspirin is a reliable, reproducible teratogen in rodents, cats, dogs, ferrets, and monkeys when 50–500 mg/kg oral doses are administered on a susceptible day of gestation. Malformations of all major organ systems, growth retardation, embryonic/fetal death, and behavioral deficits in the survivors are all consequences of prenatal aspirin exposure in common laboratory animals. Epidemiologic studies have failed to demonstrate any syndrome of terata; increased em-

bryonic, fetal, or neonatal mortality; or reduced birth weight that could be attributed to *in utero* aspirin exposure. Some epidemiologic data indicate that mothers of children with congenital malformations actually consumed less aspirin during the first trimester than did mothers of normal children. To be sure, there are case reports of limb reduction defects, cardiac malformations, and even cyclopia associated with prenatal aspirin exposure. When circulating aspirin concentrations in the blood of pregnant rodents or monkeys given teratogenic doses of aspirin are compared with concentrations found in human blood, the values in the animals are of the same order of magnitude (and in some instances even less than) as those found in human blood. Aspirin can be present in human umbilical cord blood and in normal newborn blood at concentrations of up to 10 times those associated with embryotoxicity in animals.

It is acknowledged that, given aspirin's widespread, unrestricted use among the 6 million pregnant women per year in the United States and given the 7.5% total malformation and minimum 13% fetal loss (spontaneous abortion and stillbirth) rates, a great many of these adverse pregnancy outcomes will have experienced aspirin exposures. It is also acknowledged that, if aspirin were a newly developed drug submitted today for regulatory evaluation, it is highly unlikely that it would be approved for marketing. The data published to date demonstrate that, at the aspirin doses usually consumed and at doses less than those causing overt intoxication (salicylism), the risk of adverse pregnancy outcome is no greater than the norm. Like acetazolamide, it appears that there is a physiologic insensitivity on the part of the human embryo compared to other species (which explains the differential teratogenic potency).

Caffeine is, without question, teratogenic in common laboratory animals. It is in the methylxanthine of chocolate and cocoa (2–20 mg/5-oz serving), coffee (74 mg/8-oz serving), soft drinks (30–58 mg/12-oz serving), prescription drugs (32–100 mg), and over-the-counter drugs (30–200 mg). In rodents and rabbits, oral doses of 80–150 mg/kg per day induce a consistent pattern of limb reduction defects and delayed development (usually measured as reduced skeletal ossification). A single, large oral bolus is more effective in producing limb malformations than is the same total dose given over several days. More than a dozen published epidemiologic and clinical studies on thousands of pregnant women and their caffeine consumption have failed to confirm any relationship between caffeine and human congenital malformation. Some studies have shown a possibility of increased risk (1.7 times normal) of early spontaneous abortion in selected subgroups consuming total doses in excess of 3 mg/kg; however, these studies failed to take into account covariates like ethanol and cigarette smoke or measured caffeine consumption before (but not during) pregnancy. A study of offspring of 1529 women who consumed three, four, or six cups of coffee per day (corresponding to 222, 296, and 444 mg caffeine per day, respectively) during early and mid-gestation at birth through age 7 and adjusted for ethanol, tobacco, aspirin, acetaminophen, dietary composition, and prescription drugs found no evidence of functional deficit. There is no question that caffeine crosses the rodent and human placenta and that caffeine has been detected in human umbilical cord and newborn blood. Here dose is the parameter that underlies the differential response in humans and laboratory animals; it is theoretically possible that daily ingestion of caffeine at concentrations equivalent to that of 75 cups of coffee, 125 cups of tea, or 200 cans of a caffeine-containing soft drink could induce terata in humans. Exposures to high doses such as these occurred in the early 1980s from ingestion of mail-order caffeine pills (each pill containing approximately 500 mg caffeine). These doses produced stroke, convulsions, tachycardia, coma, and at least 12 deaths in acute cardiopulmonary arrest. None of the individuals involved appear to have been pregnant at the time.

Lead is another example of an agent that induces terata in laboratory animals that have no direct concordance in human beings. When pregnant hamsters are given a sufficiently large dose of a water-soluble lead salt early in gestation, the otherwise normal young are born without tails. When first reported, some advocated the position that since humans do not normally grow tails, the observations in hamsters were only a laboratory curiosity. Today, however, the effects of lead on the developing human are well documented. There are at least six major prospective epidemiologic studies on prenatal lead exposure and postnatal cognitive development. Deficits in Bayley Mental Development Index scores and problems with language skills in young children are linked with maternal or umbilical cord lead concentrations in the 20–30 μg/dl range. In some children, slowed intellectual development has been noted after exposure to umbilical cord lead con-

centrations of as little as 10 μg/dl during gestation. (By comparison, normal or background blood lead in healthy adults without occupational lead exposure ranges from 2 to 6 μg/dl.) It is important to note here that these studies also show that the early postnatal delay in intelligence is not permanent and generally cannot be detected at more than 5 years of age.

These data have had a profound influence on social policy and reproductive health in the United States (Fig. D-18). For example, in February 1994, the California Department of Industrial Relations finalized a revision to the code of regulations known as Title 8, Section 1532.1—the Lead in Construction Standard. Until that revision, the General Industry Safety Orders (which promulgate maximum workplace exposure standards) had not been applied to the construction trades. The Lead in Construction Standard mandates: Workplace lead exposure assessment; designation of workplace lead containment; lead-free dressing rooms, cleaning areas, and dining areas; showers; filtered vacuum cleaning of work facilities and vehicles; prohibition of tobacco use, cosmetic application, eating, or drinking in lead-contaminated areas; protective work clothing; respirators; and periodic blood lead analyses. Promulgation of this construction standard, the institution of medical surveillance, removal protection, and assurance of protection from job discrimination are steps intended to reduce worker blood lead levels. To attain blood lead levels in the range required by these regulations, industrial hygiene and work practices (including attention to housekeeping and personal cleanliness) in traditionally filthy jobs like building demolition, abrasive blasting, welding, torch burning, cutting, and riveting must improve significantly. The quality of these practices must, in fact, approach those currently met by the food industry.

Trypan blue, a teratogenic azo dye, is yet another example of a compound which produces malformations in a highly species-specific fashion. Exposure of pregnant rodents to trypan blue during early gestation produces a uniform lumbosacral spina bifida in their offspring (Figs. D-19–D-21). These malformations are anatomically indistinguishable from those in humans with spina bifida (Fig. D-10D). After trypan blue treatment, the dye distributes to all maternal organs except the brain because it is unable to cross the blood–brain barrier. The embryo does not contain any of the dye, but the cells of the trophoblast, the parietal and the visceral yolk sac, and derivatives of the gut show visible

SUPREME COURT OF THE UNITED STATES

Syllabus

INTERNATIONAL UNION, UNITED AUTOMOBILE, AEROSPACE & AGRICULTURAL IMPLEMENT WORKERS OF AMERICA, UAW, ET AL. *v.* JOHNSON CONTROLS, INC.

CERTIORARI TO THE UNITED STATES COURT OF APPEALS FOR THE SEVENTH CIRCUIT

No. 89–1215. Argued October 10, 1990—Decided March 20, 1991

A primary ingredient in respondent's battery manufacturing process is lead, occupational exposure to which entails health risks, including the risk of harm to any fetus carried by a female employee. After eight of its employees became pregnant while maintaining blood lead levels exceeding that noted by the Occupational Safety and Health Administration (OSHA) as critical for a worker planning to have a family, respondent announced a policy barring all women, except those whose infertility was medically documented, from jobs involving actual or potential lead exposure exceeding the OSHA standard. Petitioners, a group including employees affected by respondent's fetal-protection policy, filed a class action in the District Court, claiming that the policy constituted sex discrimination violative of Title VII of the Civil Rights Act of 1964, as amended. The court granted summary judgment for respondent, and the Court of Appeals affirmed. The latter court held that the proper standard for evaluating the policy was the business necessity inquiry applied by other Circuits; that respondent was entitled to summary judgment because petitioners had failed to satisfy their burden of persuasion as to each of the elements of the business necessity defense under *Wards Cove Packing Co.* v. *Atonio*, 490 U. S. 642; and that even if the proper evaluative standard was bona fide occupational qualification (BFOQ) analysis, respondent still was entitled to summary judgment because its fetal-protection policy is reasonably necessary to further the industrial safety concern that is part of the essence of respondent's business.

FIGURE D-18. *Summary statement of the famous Johnson Controls decision by the United States Supreme Court in the matter of lead and developmental toxicity.*

accumulation of the compound. The site of trypan blue's action is the visceral yolk sac and the associated endoderm where it inhibits pinocytosis, causes increased mineral ion absorption, disrupts local osmotic balance, uncouples oxidative phosphorylation, and ultimately interferes with histiotrophic nutrition. Trypan blue does not reach or act on the embryo itself and there are no cellular or subcellular pathologies that can be attributed to the temporal presence of the dye. The interference with embryonic nutrition results in accumulation of fluid (intercellular edema) in the paraxial mesenchyme that is located above the affected endoderm. This is followed by extravasation from the blood islands of the yolk sac placenta into these expanded intercellular spaces, giving rise to failure of neural fold apposition and fusion in that area leading to myeloschisis (Fig. D-21). The cell cycle in the neuroepithelium of the affected area maintains normal generation

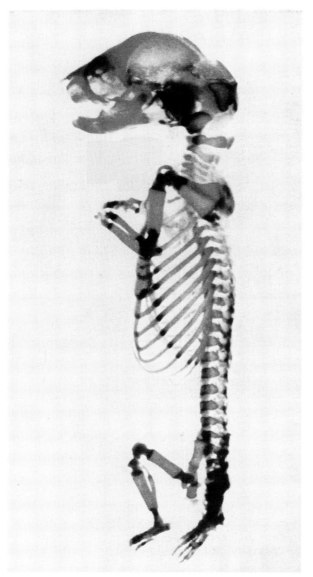

FIGURE D-19. *Skeleton of a Day 20 control rat fetus. The ossification in the skull, vertebrae, ribs, and sternebrae is nearly complete. Endochondral ossification of the pectoral and pelvic girdle and limbs is progressing (×3.2).*

times and the neuroepithelium continues to grow, only in a mechanically distorted fashion.

In rodents and rabbits, two placentas function during organogenesis: the yolk sac (choriovitelline) and the chorioallantoic placenta. In humans, the chorioallantoic placenta is the organ in which exchange between maternal and embryonic blood takes place. The choriovitelline placenta of rodents and lagomorphs is a selec-

tive adaptation in those species which have a very high reproductive potential of short gestation (hamster = 17 days) and large (6–17 littermates) litters. Trypan blue does not induce its teratogenic response in those species that do not depend on the (comparatively primitive) yolk sac placenta.

Retinoids (vitamin A and its congeners), on the other hand, are agents capable of inducing teratogenesis across species. Excess retinoic acid (RA) in human, hamster, and macaque embryos induces a nearly identical syndrome of craniofacial, cardiac, and central nervous system (CNS) terata. The craniofacial defects are due to disruption of the neural crest. The gestational stage during which embryos must be exposed to elicit craniofacial terata correspond to a time when neural crest cells still associate with the neuroepithelium. Retinoid teratogenesis displays a classic U-shaped dose–response relationship and has a definite structure–activity relationship. Experimental (proprietary) retinoids have been synthesized which are among the most powerful teratogens known, being as potent as the most active dioxin. Retinoid deficiency, however, is also teratogenic. Retinoids related to RA (including the 13-*cis* isomer and the oxidized and the glucuronide metabolites) are normal constituents of human and animal tissues; thus, there is considerable debate over what constitutes a "safe" dose or circulating concentration or what intra-embryonic retinoid concentrations are without teratogenic risk.

The effects of retinoids on embryos occur through changes in gene expression. Two families of nuclear receptors, termed retinoic acid receptor (RAR) and the related RXRs, have two endogenous retinoids, RA and its 9-*cis* isomer, as normal ligands. These receptors mediate retinoid changes in gene expression. When these receptors bind their particular ligand, they induce transcription by complex with the promoter region of the target genes (termed response element or RARE). The nucleic acid sequence GGTCA is present in at least two parts of the response element and this DNA sequence is the minimum (termed half-site) for binding of one receptor molecule. Subdivisions (RAR-α, RAR-β, and RAR-γ) of each family are recognized, each with a number of isoforms. Retinoid nuclear receptors are subsets of the steroid/thyroid hormone superfamily of nuclear receptors, all of which recognize an AGGTCA sequence, and it is the spacing of this sequence which determines response element specificity for the cognate

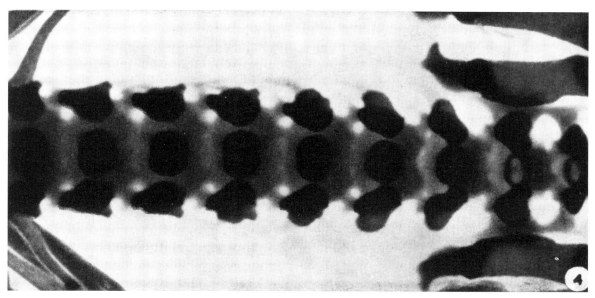

FIGURE D-20. *Higher magnification of the vertebrae, ribs, pelvic girdle, and limbs of the fetus shown in Fig. D-20. Note how the ossification centers of the bodies and arches of the vertebrae are regularly oriented (×9).*

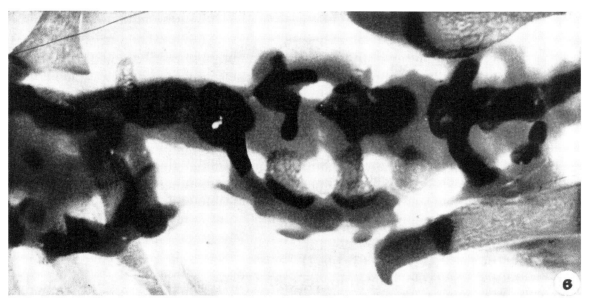

FIGURE D-21. *Magnification of the thoracic, lumber, and sacral regions of the spine of a rat fetus with trypan blue-induced spina bifida aperta on Day 21 of gestation. Note the severe malformations of the vertebral column when compared to the control (Fig. D-20). The vertebral arches and bodies are so malformed that it is difficult to identify those structures (reproduced with permission from P. W. J. Peters, A. Verhoef, A. De Liefde, and J. M. Berkvens, Development of the skeleton in normal rats and in rats with trypan blue induced spina bifida, Acta Morphol. Neerl.-Scand. 19, 21–34, 1981).*

receptor. The RXR receptor recognizes the AGGTCA half-site separated by one base pair and the RXRs form heterodimers with those nuclear receptors like RARs that bind to direct repeats of the half-site. RARs require heterodimerization with RXRs for efficient DNA binding; after RXR–RAR interaction, binding to the specific DNA sequence is stabilized. The RAR–RXR heterodimers are gene activators in the presence of ligand and are repressors in the absence of ligand.

The expression of the retinoid nuclear receptors and related proteins for cytosol retinoid transport, metabolism, and control of local concentrations of free ligand (termed the CRABPs and CRBPs) in developing embryos are carefully controlled in space and time. Embryonic CRABP is differentially expressed in those structures sensitive to retinoid teratogenesis (e.g., branchial arch mesenchyme and portions of the embryonic CNS), but this cellular binding protein is not directly involved in retinoid mechanism of action. The RAR-α serves a common "housekeeping" function, and it is present in nearly all embryonic tissues. RAR-β is abundant in the lateral nasal processes and hindbrain neuroectoderm. RAR-γ predominates in somitic mesoderm, frontonasal and precartilaginous mesenchyme, and skin. RAR-β does not co-localize in embryos with RAR-γ and neither does it occur in the limb, except in the interdigital mesenchyme (which is doomed to die in formation of normal digits). RAR-β is expressed in inner ear mesenchyme, which gives rise to structures damaged by retinoids and producing deafness in those children who survive (Fig. D-22).

It is the binding and transactivation by the retinoid nuclear receptors on target genes in embyros which is held responsible for retinoid mechanism of teratogenic action. It is the disruption of normal embryonic segmentation and alterations of normal gene expression by exogenous retinoid which leads to the defects (Fig. D-22). Homeobox (Hox) gene expression is one of the targets. The gene *Hox 2.9* (found normally in rhombomere 4 during stages sensitive to RA insult) is inappropriately expressed in rostral areas and *Krox-20* (a gene found normally in rhombomere 3) is suppressed after exposure to RA. *Hox 2.9* expression is also disrupted in neural crest and mesoderm—tissues that are known targets in retinoid teratogenesis. Normal segmented patterns of temporal gene expression are either disrupted or lost altogether in RA-treated embryos and it is these genes which have roles in normal cardiac and craniofacial and CNS morphogenesis.

The paradigm emerging is that of a cascade (a series of biochemical steps using secondary, tertiary, or more messenger molecules, each successive step amplifying by orders of magnitude those of the preceding step). After retinoid absorption from the gut, biotransformation in the liver, transport through the blood and across the placenta, and sequestration in target cells in the embryo the first biochemical actions are triggered by retinoid binding to their cognate nuclear receptor(s). The retinoid–receptor complex binds as a heterodimer to specific upstream regions of the clusters of Hox genes. Hox genes contain the information for patterning of the vertebrae, brain, and branchial arches ("Hox codes"). The proteins produced by transcription of Hox genes are themselves transcription factors (called homeoproteins). The homeoproteins have DNA-binding domains that are remarkably consistent from amphibians to birds, rodents, and primates—just as are the highly conserved regions of the retinoid nuclear receptors. (It is the highly conserved nature of the morphogenetic mechanism of action of this cascade which forms the basis for the similar teratogenic profiles of retinoids across diverse species—a reflection of the most ancient of vertebrate pattern formation maintained through prehistoric evolution.) Abnormal expression of Hox codes in target cells precipitate magnified local disruption of the ordinary careful control and coordination in vertebrate body pattern formation. Once the small window in space and time has closed and the affected progenitor cells have not arrived at their appointed station in sufficient numbers, all is lost. There is a cascade of abnormal events at the next higher level of organization—when the neural crest cells remain clustered together just under the neuroepithelium instead of migrating down to the branchial arches, insufficient numbers of these cells are available for proliferation and differentiation into bone, muscle, cartilage, thymic, and cardiac tissue and the syndrome known as retinoid embryopathy ensues.

For decades, it has been known that high levels of vitamin A during pregnancy [on the order of 40,000–250,000 international units (IU); 10,000 IU = 3.3 mg retinol] cause structural and behavioral disorders (termed functional deficits or behavioral teratology) in rodents. Prenatal hypervitaminosis A delays and disrupts motor coordination, activity, and learning-related development. In fact, vitamin A has been used for years as a reference teratogen or "positive control" in the study of potential behavioral teratology of drugs, work-

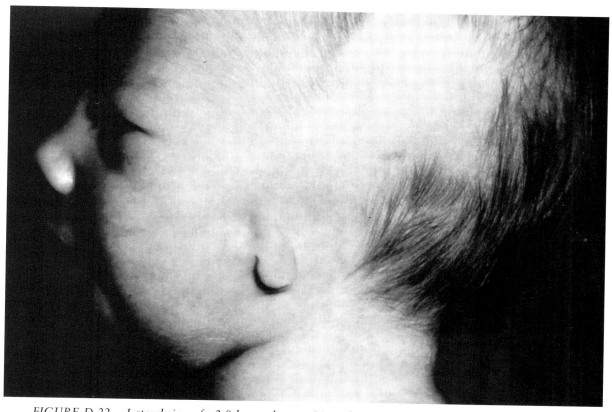

FIGURE D-22.   *Lateral view of a 2.8-kg newborn at 38 weeks gestation exposed to 80 mg/day isotretinoin (13-cis-retinoic acid) from Days 0 through 42 of pregnancy. Note the rudimentary pinna, imperforate external auditory canal, micrognathia, depressed nasal bridge, prominent occiput, and narrow sloping forehead. This child is also afflicted with paralysis of the left side of the face, abnormal visual evoked potential, hyperterlorism, latent auditory brain stem response, reduced muscle mass and tone in his legs, and eyes that cannot follow. This child also has a cleft palate, ventricular septal defect, aortic stenosis, pulmonary stenosis, dysplastic pulmonary aortic arch, and at 2 months of age he required a tracheostomy and gastrostomy. This collection of malformations is known as retinoid embryopathy. The syndrome is common in infants born to mothers ingesting from 20 to 50 mg/day isotretinoin in treatment of acne from conception (Day 0) or shortly thereafter (Days 14–16) through the first trimester (Days 35–84). Identical malformations can be induced in fetal hamsters after isotretinoin exposure during equivalent stages of embryogenesis (early primitive streak stage in hamster occurs on Day 8); in humans, retinoid embryopathy occurs after exposure during implantation (Day 7), the primitive streak stage (Day 14), closure of the anterior neuropore (at 10 somites or about Day 25), and through appearance of limb buds (Days 27 and 28). Thus, the most sensitive time for exposure occurs before the mother knows she is pregnant (i.e., from the first missed menstrual period) (reproduced with permission from C. C. Willhite, R. M. Hill, and D. W. Irving, Isotretinoin-induced craniofacial malformations in humans and hamsters, J. Craniofac. Genet. Dev. Biol. 2(Suppl.), 193–209, 1986).*

place chemicals, or other agents. After vitamin A doses that do not cause overt structural malformation, neonatal survival and weight may not be affected, but performance of the offspring on a variety of tests designed to assess mental and physical ability (e.g., rotorod balance, righting reflex and other reflex development, running wheel, water T maze, active avoidance, open field, and shock avoidance) is reduced. Thus, postnatal growth and development can be affected by exposures (dose) to retinoids less than those causing obvious structural terata. Behavioral terata induced by retinoids in rodents mirror the human experience. The risk for major craniofacial, cardiac, thymic, and CNS malformations after prenatal isotretinoin (13-*cis*-retinoic acid) exposure (Fig. D-22) is at least 25%, a value excluding those spontaneously or otherwise aborted embryos and fetuses exposed to the drug. Longitudinal follow-up of children who were exposed to isotretinoin *in utero* and survived to 5 years of age show that at least 20% are mentally retarded and that more than 50% are of substandard intelligence (Fig. D-23). The retinoids are an example of the principle that postnatal growth and development can be the most sensitive endpoint in teratogenesis. Retinoids are perhaps the best example of species concordance, dose–response in behavioral disorders at lower exposures, structural terata and growth retardation at somewhat increased exposures, and increased embryonic/fetal death at the higher exposures. The abnormal transactivation of the retinoid nuclear receptors and subsequent molecular events precipitate abnormal events at the level of the cell, leading to abnormal anatomic or mechanical disorganization in tissues (e.g., the cranial neural crest) precipitating abnormal morphogenesis of whole organs and systems relating to or surrounding those abnormal organ structures (e.g., malformation of the inner and external ear, the jaw, and CNS). As in retinoid-induced or other CNS malformation (Figs. D-12–D-15), malformation

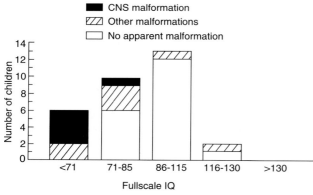

FIGURE D-23.   Relationship between congenital malformation and functional deficit in isotretinoin-exposed children at 5 years of age. All of the children classified as mentally retarded (IQ of less than 71) have major malformations. Of those with marginal intellectual ability (full-scale IQ of 71–85), 40% have major structural malformations. Among all children with significant intellectual deficits, 38% have no major malformations (reproduced with permission from J. Adams and E. J. Lammer, Relationship between dysmorphology and neuropsychological function in children exposed to isotretinoin "in utero," in Functional Neuroteratology of Short-Term Exposure to Drugs, *pp. 159–170, Teikyo Univ. Press, Tokyo, 1991).*

of the skeleton oftentimes cannot be separated in the final analysis from malformation of the brain.

Principles of Teratology

Evaluation of new pharmaceuticals, pesticides, and the like for developmental toxicity is required by law. Such testing is actually a special type of toxicity testing and the rules for these studies are based on a few generally accepted principles and a number of assumptions. The principles are listed in this section and the assumptions implicit in these studies are discussed. The overall predictive ability of the animal studies to give reliable indication of potential adverse effects in humans is then presented.

There are four general principles of developmental toxicity testing:

1. Developmental stage determines susceptibility to insult.

2. There is a dose–response continuum. The magnitude (and duration) of the exposure determines the response, which can range from subtle change to frank malformation to death.

3. Genotype influences response.

4. Paternal systems may or may not be influenced by exposure to the insult. Although the ultimate target in the conceptus, the primary site of toxic action may be elsewhere.

Developmental toxicity testing is carried out in animal bioassays. Guidelines published by domestic and international regulatory agencies specify whole animal, mammalian systems. A number of *in vitro* screening methods using invertebrate, amphibian, chick, and cultured normal or neoplastic mammalian cells have been developed, but none of these are currently accepted by regulatory agencies as adequate measures of potential developmental toxicity. These policies stem from experience; as one moves away from whole animal mammalian systems, prediction of teratogenic potential in humans becomes tenuous. *In vitro* and other systems [nematodes (*C. elegans*), fruitfly (*Drosophila*), amphibians (*Xenopus*), and Hydra] are most valuable in studies of teratogenic mechanisms of action. Nearly all the basic work in understanding homeotic gene complexes, their arrangements in tandem clusters, their segmented expression in antero-posterior domains which determine position, and the basic tenant that transcription of these genes in space and time is related to their order on the chromosome comes from *Drosophila*. Using homeobox and zinc finger probes developed in *Drosophila*, work is now under way to characterize homologous and related genes in mammals, including those homeotic gene complexes which control anteroposterior orientation and organization of the neural tube. The examples discussed in the previous section (aspirin, acetazolamide, caffeine, lead, trypan blue, and retinoids) illustrate that as knowledge of a mechanism is increased, the accuracy of interspecies extrapolation of laboratory data to human beings is increased and one can extend with confidence those conclusions to related compounds or exposure situations. Since the genes involved in embryogenesis have been conserved during evolution, alteration in homeobox sequence and expression as measured in nematode worms, sea urchins, molluscs, annelid worms, chicks, and mice can have similar consequences in human embryos.

There is no perfect animal model for accurate prediction of human teratogenic sensitivity—not all primates respond to all known or suspected human teratogens; monkeys yield the most predictive dose response in no more than half of all cases. New methods offer much promise in improving this situation. Using homologous recombination to "knock out" particular genes in embryos can yield chimeric offspring that provide excellent models of phocomelia, osteogenesis imperfecta, and congenital degeneration of Purkinje cells; inappropriate gene expression is most often lethal but, in some cases, the offspring are phenotypically normal. After insertion of bacterial genes like that for β-galactosidase (lacZ) under control of a weak constitutive promoter, those genes which fall near a strong promoter region are transcribed and the mRNA is translated into a functional enzyme detectable by histochemical methods—producing an insoluble, blue salt in those groups of cells (like neurons) expressing those DNA sequences that can be isolated. Injecting lacZ constructs into fertilized eggs and screening the transgenic embryos yields new models like "Blue mice," giving straightforward indications of disrupted genes. The tools of molecular biology and the application of rigorous interspecies dose scaling based on the principles of physiologically based pharmacokinetics (discussed in a following section) are the future of developmental toxicology and are the keys to accurate identification of those agents and levels of exposure with which the public needs to be concerned. The high degree of uncertainty currently associated with interpretation of developmental toxicity data leads to marked discord among those agencies and groups charged with protecting the public health.

Developmental toxicity testing in laboratory animals is based on the assumption that those species are, in one way or another, similar to humans in their reaction to the compound or agent of concern. The apparent concordance between human and animal data is quite poor—even when normal maternal–placenta–embryo relationships remain intact. In some cases, rodents are more sensitive to the particular chemical insult than rabbits, monkeys, or humans; in other cases, the converse is true. Drug registration, pesticide registration, and other regulatory testing protocols make use of either articulated or implicit assumptions: Metabolic fate and pharmacokinetic parameters in animals are similar to those in humans, and embryonic and fetal structural and metabolic development in animals is similar to those in humans. Since no one species is always suitable, regulatory agencies require testing in more than one species in the hope that at least one species will show some relevant adverse effect at a particular dose on which a regulatory decision can be based. In practice, this is usually the no-observed-adverse-effect level (NOAEL) or lowest-observed-adverse-effect level (LOAEL) in the most sensitive species tested.

Two practical compromises occurred at the outset of these testing protocols. First, the laboratory species used most commonly (rodents and rabbits) are relatively inexpensive, can be bred year-round, and are small, which allows for efficient husbandry, dosing, and ease of handling and permits use of the rather large numbers of individuals needed for statistical evaluation. Second, dosing is carried out for a prolonged period of gestation (usually Days 6–15 of pregnancy in rats and Days 6–18 in rabbits) to cover organogenesis and early fetal life, even though it is recognized that a single exposure during a critical period can be sufficient to induce terata. The latter practice stems from the fact that, at the outset, the investigator cannot be sure which particular gestational stage will be most sensitive; however, this approach is compromised by the possibility that some compounds (like retinoic acid and cadmium) induce or otherwise alter their own metabolism after repeated exposures so that teratogenic potency can be altered.

Methods

Of the 50 or so mixtures, compounds, or agents known to elicit human developmental toxicity under a particular regimen or exposure, the majority were first identified by astute physicians. Some human teratogens have been identified by epidemiologic studies and were only later confirmed in animals. Only a few were identified first in animal studies. Nonetheless, for every known human teratogen, there is at least one animal model. The only way the predictive ability of animal studies can be evaluated is by epidemiologic studies of sufficient statistical power. Some believe that the results of animal developmental toxicity studies have little value in predicting human response; others maintain the opposite point of view. This divergence is due to the physiologic and metabolic differences between humans and common laboratory animals and the differences in how the animal studies have been designed. For example, scientists have been unable to reproduce folic acid deficiency-induced neural tube defects in rodents, probably not because these animals are insensitive to the reproductive consequences of dietary folate deficiency but because rodents practice coprophagy (feeding on dung) from which they obtain microflora-produced nutrients.

The general application of epidemiologic methods to developmental toxicity is described below and followed by a discussion of laboratory studies in rodents and rabbits. The bulk of data available in developmental toxicology are based on these protocols. Since the difference in human and animal response appears to rest in large part on differences in behavior, physiologic parameters, and xenobiotic absorption, distribution, metabolic fate, and elimination, a brief description of transplacental pharmacokinetics is also provided.

Human Studies

The major advantage of studies of humans is that we have the data in the right species. A major disadvantage is that usually we have no reliable quantification of exposure (dose). It is, therefore, difficult if not impossible to construct a dose–response relationship.

Clinical Data

Malformations like cyclopia occur on average 60 times more often in early, aborted embryos than in infants at term. Cleft lip and palate occur 10 times more often in early abortions than at term. Study of early human embryos could be a very useful tool in identification of developmental toxins, but tabulation of anomalies in aborted embryos and fetuses is not routine. Since the total malformation rate in these spontaneous abortions ranges from 3 to 5%, it is difficult to sort out cause and effect for the comparatively small contribution of exposure to one compound or situation of relatively low teratogenic potential. It is, by comparison, relatively straightforward to identify potent, unique, or unusual teratogens in contrast to a general or common effect (e.g., ventricular septal defect). In some instances, the more rare the malformation (e.g., cyclopia), the more difficult it is to correlate etiology because of the very small number of affected infants at term; conversely, a cluster of very unique findings (e.g., thalidomide-induced phocomelia) has been the best clue historically to identify human teratogens.

Epidemiology

Birth certificates are notoriously poor sources of information because of unintentional and intentional deletion of important data. Three descriptive approaches are useful for generating hypotheses: case reports, correlation studies, and birth defect registries. Four methods are useful in testing these hypotheses: cohort studies, case control studies, cross-sectional studies, and intervention studies. The greatest limitations to all these approaches is quantifying exposure and correcting for potentially confounding factors (e.g., inherited disorders, maternal parity, disease, age, prepregnancy weight, weight gain during pregnancy, socioeconomic status, ethnicity, and diet).

Case reports can be very useful and have been the mainstay in identification of human teratogens. Usually there is no information on confounding variables, and the actual cause and effect relationship cannot be established by the case report alone. Many known human teratogens were established by a consensus of case reports, later to be confirmed by epidemiologic evaluation. Correlation studies seek relationships between geographic location, time of exposure, personal characteristics, and pregnancy outcome, but it is extremely difficult to correct for those correlations that have no actual influence on the malformation of interest. Birth defect monitoring programs (surveillance registers) can be conducted on an area-wide basis (e.g., California

Birth Defects Monitoring Program) or on an industry-wide basis (e.g., Finnish Occupational Registry); however, dose estimation and comparison to control populations are difficult These programs can provide promising leads only through laborious, expensive follow-up of the mothers with efforts to correct for recall bias. All these methods rely on identification of subgroups placed at increased risk because of their lifestyle, habits, occupations, or medical needs.

Cross-sectional studies can identify prevalence rates based on the distribution of a particular syndrome or malformation, but the study design makes it difficult to identify cause-and-effect relationships. Case-control protocols match the affected pregnancy to an unaffected pregnancy but, again, it is difficult to account for maternal recall bias and perhaps equally difficult to control for bias (even unconscious) on the part of the investigator. Cohort studies, while suffering from problems of dose determination, are usually prospective, the largest, the most expensive, the slowest, and usually the most statistically powerful to detect reliable associations. Double-blind intervention studies, like those conducted with folic acid and prevention of neural tube defects, usually yield the most conclusive data.

All these epidemiologic approaches are judged by the following criteria:

- Consistency with other epidemiologic findings
- Specificity of risk with those having highest exposure
- Strength of statistical association
- Dose–response relationship
- Biological plausibility
- Temporal relationship between exposure and outcome
- Statistical significance

Animal Bioassays

The first step in the laboratory study of developmental toxicity is determination of the substance to be tested. This might be straightforward in drug or pesticide registration studies because only a single pure compound is of concern. In the case of complex mixtures, however, selection of the test substance(s) for the particular condition(s) and route(s) of exposure can be very difficult.

For example, gasoline, diesel fuel, or aviation fuel each contain more than 250 diverse hydrocarbons, which change with source of the crude oil and are formulated differently by different refineries and change with the season of the year. While bioassays can be carried out with complex mixtures or technical-grade materials, interpretation or extension of those data to related or slightly altered materials can be most problematic.

The second step in the laboratory study of developmental toxicity is selection of the test species and strain. International (and even state) regulatory requirements differ; most look favorably on data from outbred rats and rabbits from laboratories with cumulative and comprehensive histories of malformation incidence in untreated or control animals. All regulatory agencies recommend selection of a test species that absorbs, distributes, and metabolizes the compound(s) of concern in a manner similar to that of humans; however, human pharmacokinetic data are rare and even those parameters in animals may be lacking. The animals should be housed and treated in accordance with the *Guide for Care and Use of Laboratory Animals* [U.S. National Institutes of Health (NIH) Publication No. 85-23; DHEW Publication No. 74-23]. The only difference between the treated and control groups should be exposure to the agent of concern; both groups should be provided with controlled humidity, temperature, light cycle, and the same basic, nutritionally adequate diet.

The third step in the laboratory study of developmental toxicity is the determination of dose, exposure route, and number of animals to be studied. Most regulatory agencies require at least three dose-levels in addition to a sham or vehicle control group. The highest dose is expected to produce toxicity in either the offspring or the dams; a reduction in maternal body weight gain or increased mortality (usually not more than 10% maternal death) are accepted as the highest required dose. Some agencies will allow substitution of a "limit test"; that is, if a dose of 1000 mg/kg fails to induce embryotoxicity or teratogenicity, then a full developmental toxicity bioassay may not be necessary. The lowest dose should elicit no adverse effect. Initial dose selection can be based on the anticipated therapeutic dose (or multiple thereof), on results of a range-finding toxicity study, or on an LD_{50} value in the same or a closely related species. The route of administration should be the same as the route by which humans are exposed. If the compound is to be given orally, it is preferable that intubation be used since incorporation into the feed can yield palatability problems and it provides only an approximation of dose. Regulatory agencies stipulate that 20 litters for rodents and 8–12 litters for non-rodents per dose group are necessary. Taking note that rodents can have as many as 15–18 fetuses per litter, large numbers of animals at each dose can be required for statistical evaluation. It is the litter—not the individual implantation site—that is considered the experimental unit. However, when one encounters a potent teratogen, as few as 5 litters per dose group may be sufficient to provide meaningful results.

The fourth step in the laboratory study of developmental toxicity is selection of duration of exposure. The reader should be aware that different laboratories count the days of gestation differently; some count the day after breeding as Day 0 and others count it as Day 1. Regulatory agencies stipulate that treatment begin on Day 6 (first day counted as Day 0) and continue each day through Day 15 in rodents and from Days 6 through 18 in rabbits to cover organogenesis and early fetal life. The animals should be weighed prior to dosing, regularly during treatment, and at term. Careful daily inspection for clinical signs of intoxication is needed. Just before term (e.g., Day 18 in mice and Day 15 in hamsters), the fetuses are collected by cesarean section. Conventional teratology study in rodents requires that the dams and their fetuses be killed just prior to term since malformed or otherwise defective offspring are preferentially cannibalized by the mother. Regulatory requirements for sacrifice of the dam date from 1970 U.S. Food and Drug Administration (FDA) safety evaluation guidelines. Cannibalism of malformed offspring occurs during the immediate postpartum period perhaps precipitated by the pup's inability to move about and/or nurse normally. If the dams are disturbed during this period, this will also precipitate cannibalism.

In contrast to primates, which abort dead embryos, dead rodent embryos are resorbed and the implantation site is recorded as a resorption site. The numbers of living and dead fetuses, and the number of resorption sites, are counted, and fetal weight, sex, and external malformations are recorded. Fetuses can either be inspected fresh in evaluation of internal soft tissues (known as the Staples technique) or can be placed in a fixative (usually Bouin's solution) and sectioned at a later time (Wilson technique). Other fetuses are fixed in ethanol, cleared in potassium hydroxide, and the cartilage and bone are stained with Alcian Blue and Alizarin Red, respectively. The U.S. Food and Drug Administration (FDA) recommends that one-third of the rodent fetuses be subjected to visceral examination and that two-thirds be studied for abnormalities of cartilage and bone. The U.S. Environmental Protection Agency (EPA) recommends that one-third to one-half of each litter be examined for skeletal anomalies. For rabbits, all fetuses are to be examined for both visceral and skeletal malformations.

There is no uniform and universal definition of what constitutes a malformation, a deviation, a retardation, an aberration, or a functional disorder. Only death is an unequivocal finding. All species show anatomic variations, particularly in patterns of skeletal ossification. In some cases, delays in ossification can be a reflection of retarded growth (and are often found in concert with reduced fetal body weight); however, in other cases, these changes may be a reflection of terata found after exposures to higher doses.

Just as with an LD_{50} in adult animals where, by definition, one-half of the animals die and one-half live after acute chemical insult, and the reason(s) why one particular animal (that is apparently identical to the other members of the group) lives and another dies is not known. Some fetuses in a litter can be grossly malformed and the neighboring fetus can be normal. The "litter effect" is a term applied to the finding that, at some dose, different females treated in the same way vary in the degree and even type of response.

Brief mention of *in vitro* methods is made here. Study of teratogens with chick embryos can be extremely valuable, as illustrated by the elegant mechanistic studies of retinoids in cultured wing bud; however, the chick embryo suffers from ill-defined metabolic capabilities and the numbers of false-positives due to the placement of the test substance directly on the embyro and toxic surface-actions are high. Two of the more useful *in vitro* methods are cultured whole embryos and cultured rat and mouse limb buds. For whole embryo culture, the embryos can develop normally during 48 hr of head-fold stage and precise control of exposure and experimental conditions can be achieved. The technique is inexpensive and it has been said that, with a little practice, one person can explant and culture 50 embryos in 1 day. The embryos are placed in bottles containing culture medium that are attached to a rotating drum; during incubation, oxygen and 5% CO_2 are bubbled continuously into the bottle to maintain adequate oxygen and a steady pH. The embryos are very sensitive to changes in temperature and artifacts can be induced. Although novel approaches, like including human serum from patients being treated with anticonvulsants, help in relating the rodent cultures to developmental toxicity risk assessment, the patterns of malformation induced in cultured rodent embryos can be quite different from those seen in embryos exposed to the same compound *in vivo*. A great many *in vitro* methods have been developed: cultured ova, cultured embryonic pancreas, palate, teeth, lens, kidneys, gonads, bone, thyroid, fish, sea urchin, dissociated adult invertebrate cells, micromass (dispersed embryonic limb and lung), and whole invertebrate or amphibian tail regeneration. None of these, however, is accepted by regulatory agencies in safety assessment. *In vitro* studies of developmental toxicants are generally regarded as methods for elucidation of mechanism of action and those data can be used to explain and support conclusions reached from conventional protocols.

It has become increasingly important to collect pharmacokinetic and transplacental transfer data in studies of developmental toxicants. These approaches require that quantification of parent compound and metabolites be carried out in a rigorous manner; data from total amounts of radioactivity from labeled test compounds in blood and tissues are of little utility here. The exception is use of autoradiography of early embryos, where one can obtain an approximation of compound localization in tissues of developing embryos that cannot be otherwise achieved. It may be very important to determine species-specific plasma or erythrocyte binding of the active compound since small differences here can make for large differences in the concentration of free drug and can explain what otherwise looks like a major difference in species sensitivity to the teratogen.

Classical methods for analysis of pharmacokinetic data ('box models') fit sums of exponential functions for two or three mathematical compartments, and these methods usually have no identifiable compartment other than that from which the data arise (e.g., plasma). These methods are useful first steps, yielding concentration:time curves with rates of uptake, the maximum concentration, the rates of elimination, and total dose (measured as the area under the concentration:time curve, AUC). Retinoids are examples of compounds whose teratogenic potency depends in large part on embryonic AUC values. Box models are not particularly useful in scaling dose between species as there are at least four different kinds of dose; administered dose,

absorbed dose, metabolized dose, and delivered (or target tissue) dose. Physiologically-based pharmacokinetic (PBPK) models have been developed to overcome these limitations (Fig. D-24). These methods require anatomical information (organ-specific blood flow and volume) and physiologic information (rate of metabolism), whereas traditional compartment models assume all input to a central compartment (blood) and free distribution into other compartments. The results from PBPK analyses show the influence of dose-dependent changes in drug behavior and it may be that one species appears most sensitive when compared using an administered dose basis (mg/kg-body-weight), but these species may be of equivalent sensitivity when one compares the two on a target-tissue-(delivered)-dose basis (e.g., embryonic AUC). The advantage of PBPK analyses is the inclusion of blood flow and metabolic limitations, anatomic volumes, and other characteristics (like protein binding), and these methods provide increased accuracy for interspecies extrapolation of dose.

Lessons

Three examples of mistakes known to every teratologist and the major lessons learned from each are described below. These lessons provide a warning and a caution to the reader. These lessons demonstrate how data can be misinterpreted or missed altogether and illustrate the responsibilities borne by those who make mistakes in developmental toxicology.

Diethylstilbestrol

For a great many years, the only thought given to transplacental carcinogenesis was the teratoma. Teratomas are—in a word—bizarre. Ordinarily, the primitive streak mesenchyme degenerates at its caudal terminus, but in rare cases (and mostly in female embryos and under unknown circumstances) some of these cells can persist and can become malignant either *in utero* or in infancy. In other cases, the primordial germ cells (which migrate by ameboid movements from the yolk sac along the dorsal mesentery to the gonadal ridge at Week 3 and normally differentiate into male and female germ cells) remain and give rise to teratomas in or on testes or ovaries. These uncommon tumors can arise from all three germ layers, giving rise to a disorganized mass or ball of pancreas mixed with teeth or an eyeball,

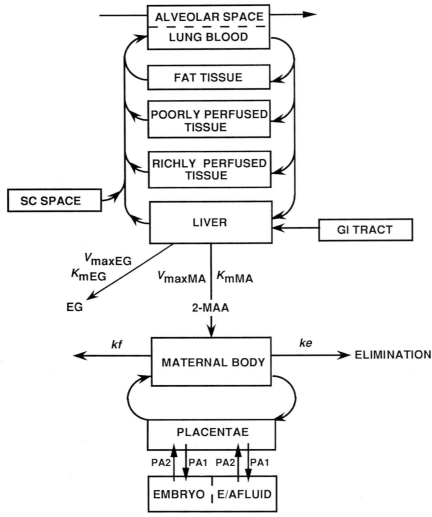

FIGURE D-24. *Schematic diagram of a physiologically based pharmacokinetic model used to describe the uptake, distribution, transplacental transfer, and elimination of a 2-methoxyethanol (2-ME) and its teratogenic metabolite, 2-methoxyacetic acid (2-MAA) in the pregnant mouse. Extraembryonic (E/A) fluid is the combined exocoelomic and amniotic fluids that surround the embryo. The results of pharmacokinetic studies show that it is the total exposure (concentration × the exposure time) of the embryo to 2-MAA rather than the peak concentration of this toxic metabolite which is most closely related to teratogenic outcome (reproduced with permission from D. O. Clarke, B. A. Elswick, F. Welsch, and R. B. Conolly, Pharmacokinetics of 2-methoxyethanol and 2-methoxyacetic acid in the pregnant mouse: A physiologically based pharmacokinetic model, Toxicol. Appl. Pharmacol. 121, 239–252, 1993).*

cartilage, and neural tube. Teratomas can also present as an independent growth, as a partial "parasitic twin" or "fetus in fetu."

Diethylstilbestrol (DES) changed forever the view of transplacental carcinogenesis. DES is a drug that was

prescribed to 1 or 2 million pregnant women at doses ranging from 1.5 to 150 mg/day from 1945 to 1970 to reduce the risk of spontaneous abortion and premature delivery. Female offspring of these women have risk for vaginal and cervical clear-cell adenocarcinoma of

1.4 per 1000. Latency is between 7 and 30 years and peak age of diagnosis is 19 years. These women may also be afflicted with structural terata of the uterus, fallopian tubes, cervix, and vagina, and may present with adenosis. Male offspring experience an increased incidence of urogenital, hypotropic, and capsular induration to testicular tissue and have associated terata of the reproductive system. The total DES dose in mothers of women with vaginal adenocarcinoma ranges from 135 to 18,200 mg.

Conventional developmental toxicity protocols can be conducted as independent studies or as part of a multigeneration or continuous breeding reproduction toxicity test. Animal bioassay for developmental toxicity has traditionally focused on gross congenital malformations, usually induced by exposure during early embryogenesis. By contrast, DES exposure late in gestation by oral dosing or subcutaneous injection in pregnant mice and rats (Days 17–21), hamsters (Days 14 or 15), and rabbits (Days 12–14) with 0.1–40 mg/kg can increase embryonic mortality, induce cryptorchidism, and cause feminization of male genitalia. Histologic examination of female and male mice or hamsters born to DES-treated mothers shows cystic metaplasia and neoplasia in segments of the reproductive tract. In animals raised to maturity, reduced fertility, ovarian tumors, endometrial hyperplasia, and uterine adenocarcinoma develop. In female rhesus monkeys born to mothers given oral DES at 1 mg/day from Day 130 to term, vaginal ridging, cervical hooding, and vaginal adenosis develop.

Three lessons are demonstrated by the DES experience:

1. Transplacental carcinogenesis can and does occur in humans and animals.

2. To focus attention on gross terata is to miss important manifestations of developmental toxicity.

3. Exposure during late gestation can have at least as great a consequence as exposure during early embryogenesis.

As a result of lessons learned from DES, an *in utero* exposure phase was added to the carcinogenicity testing requirements for food additives, drugs, and pesticides. The results of multigeneration reproduction studies must be considered together with those from carcinogenicity bioassays and conventional developmental toxicity studies in hazard evaluation.

Thalidomide

No discussion of developmental toxicity can be complete without mention of the thalidomide epidemic of the 1950s and early 1960s, which affected 10,000 children. Thalidomide ranks with methylmercury as the most infamous of all teratogens. Thalidomide was prescribed in Western Europe and Australia as a sedative anti-emetic in early pregnancy; it is still used today under careful supervision in the treatment of Hansen's disease (leprosy) and other special situations. Thalidomide phocomelia and associated limb reduction defects (oligodactyly and bone fusions) usually of the radius, humerus, and ulna occurred in at least 96% of those embryos exposed. These defects were usually bilateral and symmetrical and occurred after consumption of 0.5–1.0 mg/kg/day (corresponding to maternal circulating concentrations of 1 μg/ml) from Days 34 to 50 of pregnancy. Of the children whose mothers received thalidomide only after Day 50, 103 of 104 were normal. Of equal (but perhaps less dramatic) importance were the drug-induced malformations of the ear (anotia and microtia), the congenital deafness, epilepsy, anophthalmia, and the eye muscle and facial paralysis. Nearly one-half of these infants died in their first year, due principally to patent ductus, aortic and ventricular septal defects, and other cardiac malformations that occurred in at least 30% of all these infants.

Common laboratory rodents are refractory to thalidomide embryopathy; doses up to 4 g/kg increased the numbers of abnormal rat and mouse fetuses in only a few of the more than 60 studies. Rabbits do show limb reduction defects (classified as terminal incomplete longitudinal paraxial radial hemimelia) after thalidomide, and it is this observation that forms the basis for inclusion of rabbits in current animal testing requirements. Many species have been studied, including the armadillo, baboon (*Papio cynocephalus*), bonnet monkey (*Macaca radiata*), bushbaby, cat, cynomolgus monkey (*Macaca fascicularis*), dog, ferret, green monkey (*Cercopithecus aethiops*), guinea pig, hamster, Japanese monkey (*Macaca fuscata*), marmoset (*Callithrix jacchus*), mouse, pig, rabbit, rat, rhesus monkey (*Macaca mulatta*), and stump-tailed monkey (*Macaca arctoides*). However, only certain strains of rabbit (given 150 mg/kg/day) and non-human primates (given 4–45 mg/kg/

day) respond in a manner similar to that of humans. Not all primates respond to thalidomide (e.g., *Galago crassicaudatus*).

The mechanism of the species-specific action of thalidomide is not known. The compound crosses the placenta and has similar pharmacokinetic parameters in susceptible species (rabbit) and resistant species (rodent). The molecular structure of thalidomide and requirements for malformation are very specific. Thalidomide decomposes rapidly in water to at least 12 different products—all of which are inactive as teratogens and all of which undergo biotransformation *in vivo* to other products. These data prompted the suggestion that it is the parent thalidomide molecule that is the ultimate teratogen. Studies with supidimide analogs and other members of the phthalimide family (Fig. D-25), including side-chain or phthalimide ring-modified analogs, show precise structural requirements. Of the numerous phthalimides with their *in vivo* metabolites and related compounds examined, only thalidomide and its EM12 congener are teratogenic. Using topological parameters, geometric parameters, electronic parameters, and physiochemical parameters, several groups have attempted to derive generalized predictors for structure–activity relationships of teratogens; the structure–activity relationships with phthalimides point out the severe limitations of this approach. Regulatory agencies use structure–activity relationships to guide recommendations for collection of certain kinds of toxicity data; structure–activity relationships (even for compounds like retinoids in which these are well described) are no substitute for actual developmental toxicity data.

Four lessons are demonstrated by the thalidomide experience:

1. Terata can occur in the offspring of apparently healthy mothers.

2. Human teratogens may have little or no such activity in common laboratory animals.

3. Human terata can occur after exposures less than those used in animal studies.

4. Just because a compound has a chemical structure that appears closely related to an established teratogen does not imply that the congener is teratogenic.

Bendectin

Bendectin (Debendox) was the name given to the 1:1:1 mixture of dicyclomine hydrochloride, doxylamine succinate, and pyridoxine hydrochloride, a prescription drug first marketed in 1956 used to control nausea and vomiting in pregnancy. In 1976, the dicyclomine hydrochloride was dropped from the formulation after large clinical trials showed that it did not contribute to the efficacy of the drug. Bendectin use was very common; from 20 to 25% of all expectant mothers used the drug. Approximately 30 million pregnancies were exposed over the 27 years that the drug was available. The customary daily dose was one to four tablets/day, each containing 20 mg of active ingredient (1 or 2 mg/kg/day).

Bendectin is not teratogenic in laboratory animals (including non-human primates), but developmental toxicity (reduced fetal body weight and delayed skeletal ossification) can be induced in animals after exposures (500–800 mg/kg/day) that also cause frank maternal intoxication and/or increased maternal death. There are six published *in vitro* studies of Bendectin using cultured rodent embryos or embryonic cells; of those, only one, of mesenchymal cells, showed any indication of toxicity and that occurred after exposures to concentrations far greater than can be achieved in humans after ingestion of therapeutic doses (25 mg or more). There are at least 14 cohort and 18 case control epidemiologic investigations on Bendectin and pregnancy outcome in addition to one, by the NIH, in which the occurrence of congenital malformations was prospectively studied in 31,564 newborns. The results of the NIH study, like those of others, found that the odds ratio for any of 58 major categories of malformation and Bendectin exposure was 1.0—exactly that which is expected by chance alone. Of those categories with trends or suggestive positive associations, the magnitude of those associations was as great as that from vomiting during pregnancy without Bendectin use as with Bendectin use. There was no increase in malformation rate after exposure to Bendectin *in utero* than would otherwise be expected by chance; there are no objective data to conclude that exposure to Bendectin in animals or humans has any adverse effects on embryonic or fetal development at the doses used by these 30 million women.

As of 1987, at least 300 lawsuits had been filed alleging that Bendectin caused congenital malforma-

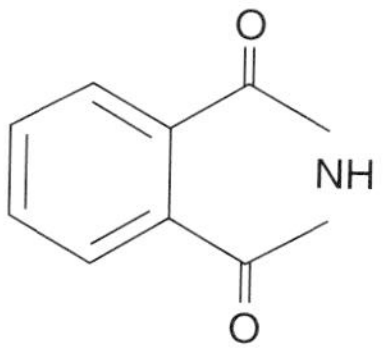

Phthalimide

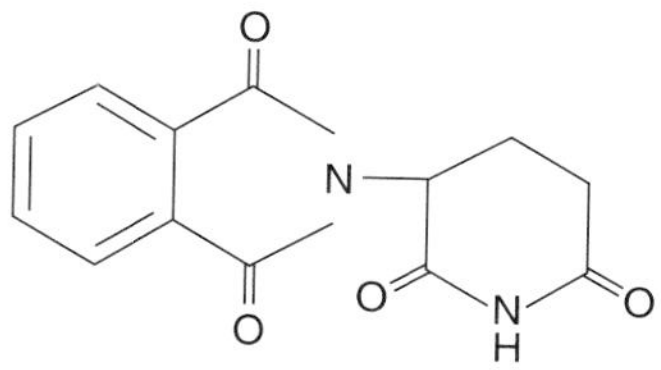

Thalidomide

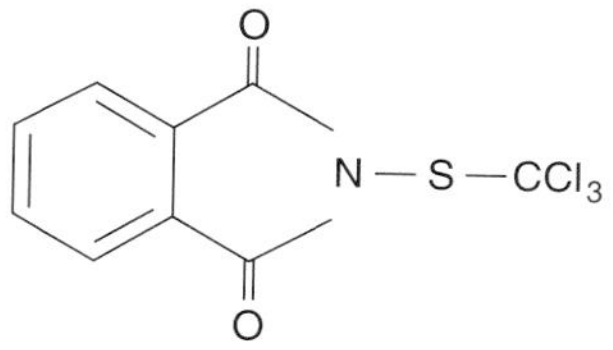

Folpet

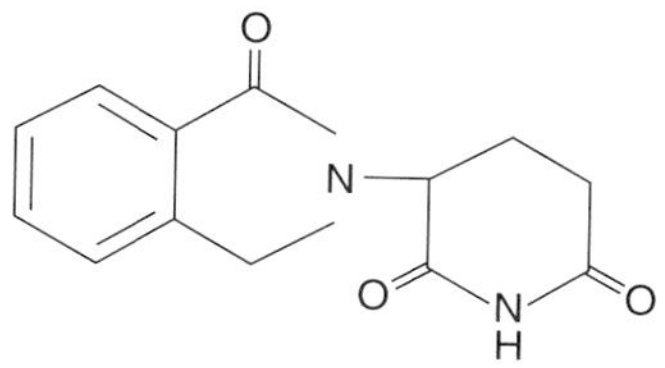

EM 12

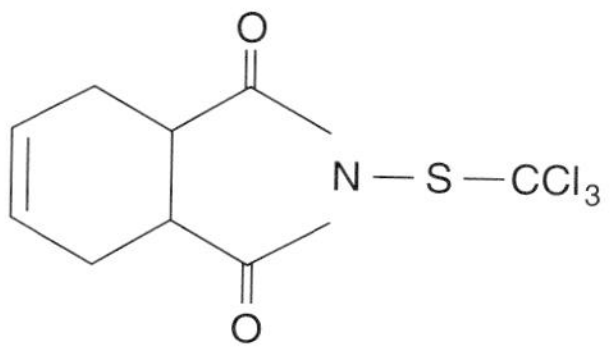

Captan

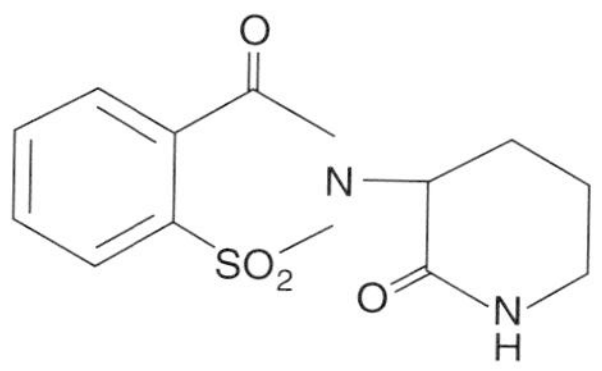

Supidimide

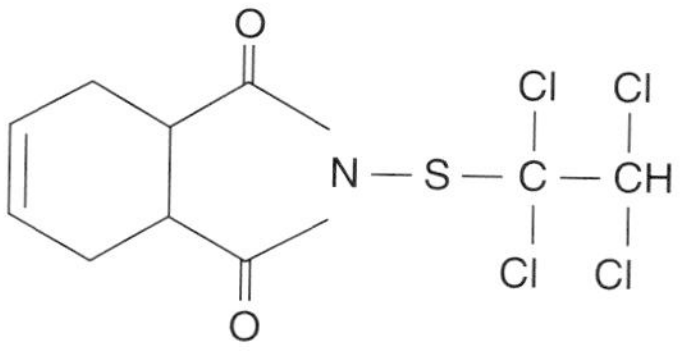

Difolatan (Captafol)

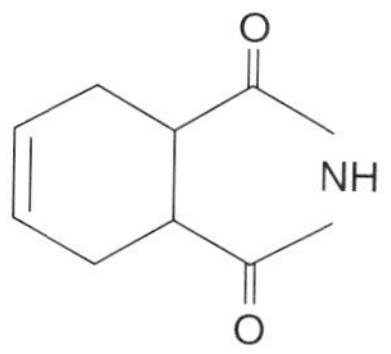

Tetrahydrophthalimide

FIGURE D-25.   The phthalimide family. In the presence of heat and ammonia, cyclic anhydrides lose a molecule of water and the two acyl groups (R–C = O) become attached in a ring to the ammonia nitrogen, forming an imide. In this way, phthalimide is produced from phthalic anhydride. Phthlalimide is acidic and when heating with potassium hydroxide in alcohol along with an alkyl halide, the corresponding N-substituted imide is produced. Neither phthalimide nor tetrahydrophthalimide are teratogenic; no derivative of tetrahydrophthalimide (captan and difolatan) is teratogenic. The N-substituted imides, folpet and supidimide, are not teratogenic. Of the phthalimides, only thalidomide and its close relative N-(2',6'-dioxopiperiden-3'-yl)-phthalimidine (EM 12) are teratogenic.

tions. The drug was dropped from commerce not because of lack of efficacy, or because it caused toxicity or because there was no market for the drug, but because of the excessive cost incurred by the manufacturer in defending the drug in litigation. The major focus here was the allegation that prenatal Bendectin exposure caused phocomelia and assorted limb reduction defects. Given the 'background' or "spontaneous" rate of major congenital malformation in the United States (3%), it would be expected that 900,000 malformed children would be born to those 30 million mothers even in the absence of any drug use. Given the U.S. background rate for limb reduction defects (1 per 3000 births), 10,000 such defects would be expected to occur even in the absence of any drug use.

How then can it be that a compound which has no detectable teratogenic activity in either animal or human studies be held responsible in the etiology of human congenital malformations? The history of Bendectin can be traced to ignorance of the principles of teratology, compounded by precedential law following the first erroneous decision and to two articles appearing in the popular press. In the September 1979 issue of the *National Enquirer,* the following was published:

> Experts Reveal . . . Common Drug Causing Deformed Babies. In a monstrous scandal that could be far larger than the thalidomide horror, untold thousands of babies are being born with hideous defects after their mothers took an anti-nausea drug (Bendectin) during early pregnancy.

At least seven women are known to have elected abortions as a consequence. In the magazine, *Mother Jones* ("The Bendectin Coverup," November, 1980), the authors counseled women to use—instead of Bendectin— "natural alternatives" including 100 mg of pyridoxine (a dose 10 times that of the same compound in the Bendectin formulation).

The Bendectin lawsuits stem from the age-old question that always follows the birth of a malformed child, "What caused it?", together with today's substantial monetary rewards that can befall successful litigation or out-of-court settlement, particularly with a large, impersonal, and wealthy corporation. Compounded by parental shock, denial, anger, and sadness, the potential astronomical medical costs and social costs for the child, the financial rewards offered by expert witness testimony, the inadequate research and reporting necessary for sensationalization in the popular press, and successful settlement or judgment in preceding cases all combine to perpetuate the myth. Juries—being people—are by nature very sympathetic to the plight of the child and his or her family who are through no fault of their own in need of financial and other assistance to address their situation. The desire to help can be overwhelming. The original Bendectin lawsuits arose from failure to understand proper interpretation of the animal data, the desire to identify a responsible agent,

and from substituting a strength-of-evidence approach for the weight-of-evidence evaluation of the data. Fundamental here was the failure to apply properly the concept of dose-response. Only after the accumulated weight of many consistent epidemiologic findings of sufficient statistical power, accounting for confounding factors and where exposures are of sufficient magnitude and during appropriate periods in gestation, can conclusions on excess risk be drawn. In the strictest sense then, the best the science of teratology can offer on the concept of safety is the statement that the exposure of concern represents no measurable excess risk.

Three lessons demonstrated by the Bendectin experience are the following:

1. Human teratogens can be confirmed only by consistent findings in epidemiologic studies, recognizing that at the 95% confidence limit, the results in 1 of 20 studies of equal statistical power can differ by chance alone.

2. Maternal intoxication in animal studies in and of itself can contribute to abnormalities in the offspring.

3. Overall weight of evidence should be used in identification of compounds, agents, and exposures of concern.

Data Interpretation and Regulatory Policy

Some have criticized those groups charged with protection of the public health for failure to designate a code or notation which can be assigned to a chemical showing it to be a reproductive or developmental toxicant (see Further Reading, U.S. EPA, 1991). Developmental toxicity is usually not the only kind of toxicity associated with exposures of concern. A 1985 U.S. EPA reevaluation of 18 pesticides (avermectin, cacodylic acid, captafol, captan, cyanazine, dinocap, EDBC, endrin, fenarimol, folpet, fusilad, nitrofen, pentachlorophenol, 2,4,5-T, silvex, TPTH, triadimefon, and warfarin) and six industrial chemicals (arsenic, two glycol ethers, lead, chloromethane, and mercury) found that in no case was embryonic or fetal toxicity the sole documented effect. Teratogenic activity almost always occurs in tandem with other adverse effects on health, including mutagenicity, male or female reproductive toxicity or carcinogenicity. Gender-specific requirements like the protective work clothing promulgated for women in the case of endrin and silvex cannot be justified, not only on the basis of U.S. Supreme Court decisions (Fig. D-18) or because gender-specific regulations concerning "women of child-bearing age" restrict the activities of those women who are not fertile but because such regulations concerning only pregnant women would most likely fail to protect the embryo since it is exposures which occur before the patient recognizes that she is pregnant which are most likely to damage her embryo. Correct determination of safe levels of exposure rests with identification of the endpoint of toxicity associated with the LOAEL regardless of whether it is developmental toxicity or another expression of toxicity.

The conventional method for interspecies extrapolation of developmental toxicity data involves empiric determination of a NOAEL in the test species followed by application of an uncertainty (or "safety") factor to calculate the safe dose for the species or individual of interest. These calculations are usually based on the administered dose and are usually expressed on a body weight basis. The default safety factor used can range from 100 to 1000 or more. Dose scaling on a body weight or surface area basis is fraught with problems. For example, scaling of a child's sulfonamide dose to the adult by body weight, surface area, age, or caloric expenditure (as a surrogate for metabolic rate) yields an adult daily and grossly excessive dose of 7 g. In the case of teratogens, the problem is even more acute: The absorbed dose can be so small (e.g., cyclopamine) or the site of drug action so restricted (e.g., trypan blue) that the quantitative differences between the species are so great that the result is judged an inherent qualitative difference. In the case of vitamin A and its metabolites, the conventional safety factor approach results in calculation of vitamin A doses so small that the risk of vitamin A deficiency-induced terata is increased.

Thirty years ago, D. Karnofsky listed the two basic tenets used today in interpretation of animal developmental toxicity data: All compounds can produce embryotoxicity if applied in sufficient dosage at an appropriate stage of development, and the purpose of evaluating chemicals for teratogenic potential is not to eliminate from use, but rather to estimate the hazard its use presents to the human embryo. Teratologists believe that developmental toxicity can be viewed most

accurately from the threshold concept; i.e., there exists an exposure below which no adverse effect will occur. This belief arises from experience with human teratogens, from pharmacokinetic considerations, and because embryos—to a point—have the remarkable ability to compensate for lost or damaged cells. For instance, early rodent and rabbit embryos can develop quite normally after surgical obliteration of at least 50% of the entire inner cell mass. Implicit here is that one must distinguish between theoretical risk and practical risk. Using statistics, arbitrary safety factors, or subjective means, theoretical risk can always be calculated and upper-bound confidence levels assigned. Multiple and consistent epidemiologic studies of high quality showing no association between developmental morbidity or mortality and the exposure of concern can never prove that no hazard exists; these data can be used only to demonstrate that the risk, if any, is so small that for all practical purposes it can be disregarded.

The weight-of-evidence evaluation takes into account the human experience and the animal data. Interpretation of human data depends on the design of the study, definition of the cohort, quantification of exposure, validity of ascertainment, control for confounding factors, size of the study population, and appropriate statistical methods. In practice, identification of human teratogens has relied on answers to two questions:

1. Is there a distinctive pattern of malformation which can be associated with the exposure of concern?

2. Have there been sufficient numbers of cases to substantiate the conclusion?

The principal disadvantage to using human data in this way is, of course, that recognition occurs after the fact. Interpretation of the animal data is carried out with the following in mind: Without exception—even with very odd, rare, or unique malformation—there is no case in which the defect has not occurred sporadically; there are no known examples of agents which cause malformation that cannot also be caused by some other agent; some species are prone to particular types of congenital malformation (e.g., cleft palate in mice); and chemicals that are not teratogenic in animals (e.g., coumarin anticoagulants) can be teratogenic in humans. Animal data can only be used to provide an approximation of risk in humans. Interpretation of the animal data rests on answers to seven questions:

1. Were the studies carried out in a species that handles the compound in a manner similar to that in humans?

2. Were the studies carried out using a route of administration applicable to anticipated human exposure?

3. Does the effect occur in more than one species?

4. Does the effect occur after doses less than those which elicit other types of toxicity?

5. Was a significant increase in the numbers of litters containing abnormal outcome observed?

6. Was a dose–response relationship identified?

7. What populations are at risk of exposure and what is the magnitude of their exposure?

Of great concern are those teratogens, like thalidomide, which are able to induce terata in the absence of any apparent disturbance in maternal well-being. A large difference in the dose required to cause embryotoxicity and a much higher dose required to cause maternal toxicity can be key in identification of those agents. Where maternal toxicity occurs after exposures equivalent to those causing increased embryonic morbidity and mortality, then estimates of safe exposure can be derived using safety factors applied to the NOAEL for adults. Even studies with low statistical power can contribute to the total weight of evidence; a strength-of-evidence (picking and choosing data to support one or another conclusion) approach cannot be justified.

For example, let us consider ribavirin, a compound found on most lists of developmental toxins. Ribavirin (1-β-D-ribofuranosyl-1,2,4-triazole-3-carboxamide; Virazole) is a nucleoside analog used in the treatment of respiratory syncytial virus where it is of clear clinical value in immunocompromised patients and against Lassa fever virus, Rift Valley fever virus, and Hantaan virus; clinical trials showing activity against influenza A and B, genital herpes, hepatitis C, and human immunodeficiency virus have been conducted in which it was given orally, intravenously, or by nasal or oral inhalation. From its approval in 1986 to 1989, approximately 50,000 U.S. infants were treated successfully. Ribavirin is a potent teratogen in several animal species

including the hamster (Fig. D-26), rat (Fig. D-27), and mouse, and it is embryolethal in rabbits at doses which do not elicit overt maternal intoxication. The hamster is more sensitive on a body weight-administered dose basis than is the rat or mouse; malformations are dose and stage dependent. Limb reduction defects range from complete amelia (Fig. D-26) to ectrodactyly (absent one or more digits). In neonatal intensive care, ribavirin is administered through ventilators; nurses and respiratory therapists are currently exposed to mean concentrations of 4.3–78 μg/m^3 with peak excursions to 95 μg/m^3. The compound was not teratogenic in the single baboon in which it was studied. In both *in vivo* and *in vitro* studies with mice, the compound inhibited embryonic DNA synthesis and induced necrosis in limb buds within 4 hr of treatment; the location of the cell death and the nature of the resulting defect depended on the concentration, stage, and duration of exposure. By 24 hr after exposure during limb development, most of the local cell debris had been cleared, but the size of the cartilage blastema was reduced. Ribavirin is metabolized *in vivo* and *in vitro* to the corresponding triphosphate, a product which inhibits the enzyme inosine monophosphate dehydrogenase which is involved in purine synthesis; it is this inhibition of RNA and DNA synthesis in target cells by which the triphosphate metabolite of ribavirin is thought to elicit the terata. Inhibition of embryonic DNA synthesis in actively dividing cells is a mechanism of action shared with other nucleoside analogs like cytosine arabinoside and 5-fluorodeoxyuridine, all of which are potent teratogens.

Although a risk–benefit analysis for ribavirin given orally or parenterally to patients with viral infections can be carried out, it is the occupational exposure to rivavirin which is controversial and problematic. Some authors have advocated the use of an arbitrary 1000-fold safety factor in evaluation of the hazard presented by aerosolized ribavirin; the compound meets all the requirements to answer the previous seven questions in the affirmative, except one. There are neither data on bioavailability after inhalation nor measures of circulating concentrations (absorbed dose) or embryonic concentrations (delivered dose) or even teratology data from inhalation exposures in animals. In the absence of accurate data on circulating and target tissue concentrations of the parent drug and its teratogenic metabolite, regulatory (NIOSH and OSHA) and advisory committees (ACGIH) are currently unable to adopt credible

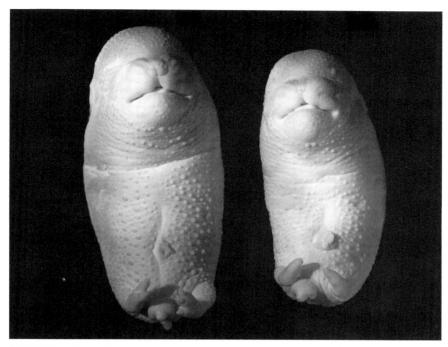

FIGURE D-26. *Hamster littermates on Day 14 of gestation from a mother given a single oral dose of 3.75 mg/kg ribavirin on Day 8 of pregnancy. Note the complete absence of the upper limbs and the reduction defects of the lower limbs. The face, skull, and brain are normal (×4.5) (reproduced with permission from V. H. Ferm, C. Willhite, and L. Kilham, Teratogenic effects of ribavirin on hamster and rat embryos, Teratology 17, 93–102, 1978).*

occupational exposure limits for this compound. Other examples of this include the teratogenic glycol ethers (Fig. D-24), in which application of default safety factors results in extremely small concentrations of the agent permitted in workplace air; in that case, however, even if workplace air is controlled to zero or near-zero values, dermal absorption of teratogenic glycol ethers through inadequate or no gloves can contribute as much if not more to dose than that which can occur at workplace air values orders of magnitude greater than those currently permitted by law.

In summarizing the results of the risk assessment, it is important that the reader look for the following three key points: a discussion of the quality of the studies supporting the concern for risk of developmental toxicity, the confidence that one can place in the NOAEL derived, and the list of uncertainties in the assesment. Animal studies (like those for aspirin and caffeine) usually employ doses much greater than those to which humans are or could be exposed. When extremely high doses are used, drug actions in the mother can elicit systemic poisoning which by itself may or may not be responsible for the effects seen. Even very reliable teratogens in animals (Figs. D-26 and D-27) may have no counterpart in humans since the dose to which people are exposed is so small.

Conclusions

What is the solution to the problem of birth defects? If not lists of developmental toxins, printed and electronic warnings, or new laws or judicial decree, then what? Although the answers to the question can be simple, true solutions to the problem are complex. Only by reductions in the numbers of infants with birth defects (Fig. D-1) can significant inroads on the problem of infant mortality (Figs. D-5–D-8) be achieved. The solutions are complex and expensive, but not as expensive as the cumulative charges for labor, postpartum care, surgery, hospitalization, prostheses, lost wages, and the extra social and educational services required by these

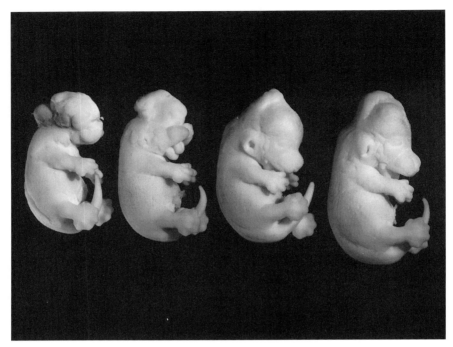

FIGURE D-27.   Rat littermates on Day 17 of gestation from a mother given a single parenteral dose of 50 mg/kg ribavirin on Day 9 of pregnancy. Note the malformations of the face, skull, and brain. The limbs are normal (×3.3) (reproduced with permission from V. H. Ferm, C. Willhite, and L. Kilham, Teratogenic effects of ribavirin on hamster and rat embryos, Teratology 17, 93–102, 1978).

patients. For every $1 spent on research into the cause and prevention of birth defects, published data show that $11 can be saved in public expenditure, insurance claims, and legal and medical payments made by individuals to address or attempt to rectify the problem.

First, only by knowing the cause of a problem can an intelligent solution be devised. Basic research into the 80% of all birth defects that are of unknown cause (Fig. D-16) sponsored by groups like the March of Dimes, the Deutsche Forschungsgemeinschaft, and the U.S. Public Health Service is key. This effort has two fronts: (1) the systematic collection and evaluation of human data and, (2) basic laboratory investigation. The first includes central birth defect registeries in Australia, California, Canada, Finland, Japan (Ishikawa, Kanagawa, Osaka, Tottori, and the National Association for Maternal Welfare), New York, Sweden, the Commission on Professional and Hospital Activities (Michigan), the U.S. Centers for Disease Control (Georgia), and the March of Dimes Nationwide Information Center (Massachusetts). Here, for example, was the first indication of valproic acid teratogenicity detected by

the Rhone-Alpes birth defect registry. The second involves understanding how embryos grow: how does the embryo know its back from front, top from bottom, and how does it control shape change from an initial ball of rapidly dividing cells to form the segmented cylinder of the torso? It is remarkable that the molecular biology of *Drosophila* has a direct bearing on understanding human embryonic development. Of equal import is understanding what the animal data are trying to tell us; as has been said by toxicologists on more than one occasion, "The animals never lie"—it is only our flimsy interpretation of the results of the animal studies upon which doubt can be cast. To make use of those data and to develop credible standards based on those data, rigorous methods for interspecies extrapolation of rodent, rabbit, and *in vitro* data are urgently needed. It makes little sense to generate animal developmental toxicity data if in the end we do not know how, if at all, those data relate to human beings. The greatest promise here rests on the development of accurate physiologically-based pharmacokinetic models (Fig. D-24) scaling from rodent to rabbit to nonhuman primate to human. Only on very rare occasion is there an opportunity to measure the compound(s) of concern in human tissue after quantifiable maternal exposure, but it has been done and every effort must be made to use the results to confirm and validate the confidence that can be placed in the PBPK results. While PBPK interspecies scaling of dose can be handled, these methods cannot scale response.

Second, once one knows the problem and has devised a solution, then the real job begins. National Center for Health Statistics data show a decline in total U.S. infant mortality from 1982 to 1992, but marked geographic and racial differences remain. The 1992 overall U.S. rate of infant death was 8.5 per 1000 live births (California, 6.9; Texas, 7.7; New York, 8.5; New Jersey, 8.5; Pennsylvania, 8.6; Ohio, 8.7; Florida, 9.1; Illinois, 10.0; Georgia, 10.4; Michigan, 10.5), a decline attributed not to reductions in the numbers of birth defects or premature births but to improved neonatal intensive care units and the introduction of synthetic pulmonary surfactants and consequent reductions in death from acute neonatal respiratory distress syndrome. Still, the years of potential life lost due to birth defects ranks fifth, just behind that of homicide and suicide (1, unintentional injury; 2, cancer; 3, cardiovascular disease); prematurity/low birth weight ranks sixth and sudden infant death syndrome seventh. Ethnic discrepancy is becoming even more pronounced; rates of white (6.3 per 1000 live births) and Latino (6.2 per 1000 live births) infant death are nearly equal, but the 1992 rate for blacks (16.7 per 1000 live births) increased from the year before.

Aggressive public health efforts to combat cigarette smoking during pregnancy can pay for themselves at least six times over. Targeting low-income and other women to increase their consumption of fruit and vegetables and improved folate status can prevent at least 50% to perhaps 80% of all cases of anencephalus and spina bifida. Alcohol has without doubt been responsible for congenital mental retardation for thousands of years; compassionate and performance-driven intervention is critical for Indian reservations where whole towns and villages are affected. Not that these measures are popular or easy; they are not attractive high-technology and they suffer from the fact that one cannot identify a villain or other responsible party upon which to place blame, extract compensation, and with whom one can take free journalistic license. That these steps are not easy can best be illustrated by iodine. Today's residents of wealthy industrialized countries give not a second thought to iodized salt; supplementary dietary iodine in prevention of goiter and cretinism is so common that these problems have disappeared. Why, then, does this completely preventable problem persist in the Third World? In impoverished countries, it is routine that a salt merchant equipped only with sack, shovel, and camel, mule, or other pack animal should go to the salt mine and return to the market to sell his wares. The consumer, whose total annual income may be only a few hundred or less U.S. dollars and who is likely illiterate, faced with the choice between the more expensive iodized variety (which looks, tastes, and for all intents and purposes works just as well as the "natural product") and the less expensive native rock salt, will select the less expensive item. The cultural resistance and practical problems faced by the World Health Organization and United Nations in widespread dissemination of iodine or vitamin A supplements are legend, but their successful introductions can make and have made tremendous improvements in the quality of life and have reduced the incidence of birth defects in those people.

Third, while it remains important to identify anthropogenic and iatrogenic contributions to the etiology of birth defects, it must be recognized that despite all the resources devoted to laboratory testings, regulation, banning, restricting or listing of drugs, pesticides, industrial chemicals, and hazardous waste, birth defects were present long before their invention analysis or laboratory testing, and despite our best efforts, only modest progress has been made in preventing birth defects (Fig. D-3). Embryogenisis is exceedingly complex—after one observes the heart form from cardiac jelly (literally by growing back on itself and tying itself into a knot), the partitioning of aortic sac and conotruncus, migration of neural crest to aorticopulmonary septum, and development of associated circulation, it is a wonder things go as well as they do as often as they do. In the final analysis, it is becoming apparent that factors like inadequate or improper nutrition, ethanol, and other lifestyle factors contribute far more to the load of human congenital malformations and other manifestations of reproductive failure than has been appreciated to date.

Further Reading

Embryology

Balinsky, B. I. (1981). *An Introduction to Embryology,* 5th ed. Saunders, Philadelphia.

Langman, J. (1975). *Medical Embryology, 3d ed.* Williams & Wilkins, Baltimore.

Moore, K. L. (1977). *The Developing Human.* Saunders, Philadelphia.

Epidemiology and Clinical Teratology

Elwood, J. M., and Elwood, J. H. (1980). *Epidemiology of Anencephalus and Spina Bifida.* Oxford Univ. Press, Oxford, UK.

Klingberg, M. A., and Weatherall, J. A. C. (Eds.) (1979). *Epidemiologic Methods for Detection of Teratogens. Contributions to Epidemiology and Biostatistics.* Karger, Basel.

Smith, D. W. (1970). *Recognizable Patterns of Human Malformation: Genetic, Embryologic and Clinical Aspects.* Saunders, Philadelphia.

Warkany, J. (1971). *Congenital Malformations. Notes and Comments.* Year Book Med. Pub., Chicago.

Laboratory Studies and Clinical Correlates

Schardein, J. (1993). *Chemically Induced Birth Defects,* 2nd ed. Dekker, New York.

Schardein, J. L., and Keller, K. A. (1989). Potential human developmental toxicants and the role of animal testing in their identification and characterization. *CRC Crit. Rev. Toxicol.* 19, 251–339.

Laboratory Protocols and Data Interpretation

U.S. Environmental Protection Agency (1991). Guidelines for developmental toxicity risk assessment; notice. *Fed. Regist.* 56(234), 63798–63826.

U.S. Food and Drug Administration (1993). *Toxicological Principles for the Safety Assessment of Direct Food Additives and Color Additives Used in Food. Redbook II.* Center for Food Safety and Applied Nutrition, Washington, DC.

World Health Organization (WHO) (1984). *Principles for Evaluating Health Risks to Progeny Associated with Exposure to Chemicals During Pregnancy,* Environmental Health Criteria No. 30. WHO, Geneva.

Pharmacokinetics/Risk Assessment

Barlow, S. M., and Sullivan, F. M. (1982). *Reproductive Hazards of Industrial Chemicals. An Evaluation of Human and Animal Data.* Academic Press, New York.

John, J. A., Wroblewski, D. J., and Schwetz, B. A. (1984). Teratogenicity of experimental and occupational exposure to industrial chemicals. In *Issues and Reviews in Teratology,* Vol. 2, pp. 267–324, Plenum, New York.

Nau, H., and Scott, W. J. (Eds.) (1987). *Pharmacokinetics in Teratogenesis,* Vols. 1 and 2. CRC Press, Boca Raton, FL.

Schardein, J. L (1983). Teratogenic risk assessment. Past, present and future. In *Issues and Reviews in Teratology,* Vol. 1, pp. 181–214. Plenum, New York.

Schardein, J. L., Schwetz, B. A., and Kenal, M. F. (1985). Species sensitivities and prediction of teratogenic potential. *Environ. Health Perspect.* 61, 55–67.

Thalidomide

Fraser, F. C., *et al.* (1988). Thalidomide retrospective: What did we learn? *Teratology* 38(3), 201–251.

Lipson, A. H., *et al.* (1992). Thalidomide retrospective: What did the clinical teratologist learn? *Teratology* 46(5), 411–418.

Retinoids

Creech-Kraft, J., and Willhite, C. C. (1996). Retinoids in abnormal and normal embryonic development. In *Environmental Toxicology and Human Development.* pp. 15–49. Taylor and Francis, Washington, D. C.

Livrea, M. A., and Packer, L. (Eds.) (1993). *Retinoids. Progress in Research and Clinical Applications.* Dekker, New York.

Morriss-Kay, G. (Ed.). (1992). *Retinoids in Normal Development and Teratogenesis.* Oxford Univ. Press, Oxford, UK.

—*Calvin C. Willhite*

Related Topics

Carcinogenesis

Chromosome Aberrations

Dominant Lethal Tests

Dose–Response Relationship

Environmental Hormone Disrupters

Epidemiology

Levels of Effect in Toxicological Assessment

Molecular Toxicology

Mutagenesis

Reproductive System, Female

Reproductive System, Male

Risk Assessment, Human Health

Toxicity Testing, Developmental

Toxicity Testing, Reproductive

Toxicology, History of

Dextromethorphan

- CAS: 125-71-3
- SYNONYMS: DM; dextromethorphan hydrobromide; demorphan; 3-methoxy-*n*-methylmorphinan; *d*-Methorphan; drug store wine; robowing
- PHARMACEUTICAL CLASS: The methyl ether of the dextrorotatory form of levorphanol, an opiate analgesic
- CHEMICAL STRUCTURE:

Uses
Dextromethorphan is used as an antitussive and cough suppressant. It is occasionally a drug of abuse.

Exposure Pathways
Dextromethorphan preparations are administered orally in tablet or liquid form. Ingestion is the most common route of accidental and intentional exposure to dextromethorphan.

Toxicokinetics
Dextromethorphan is rapidly and well absorbed from the gastrointestinal tract. Erratic and slower absorption may occur with high-dose sustained-release products. Peak plasma levels occur at 2.5 hr with conventional dosage forms and at 6 hr with sustained-release preparations. Dextromethorphan is quickly converted in the liver to an o-demethylated product, dextrorphan (3-hydroxy-N-methylmorphinan). Dextromethorphan undergoes polymorphic metabolism, depending on variations in cytochrome P450 enzyme phenotype. Five to 10% of the Caucasian population are poor metabolizers. The serum half-life of the parent drug is greatly increased in these individuals. Only minor amounts of active metabolites are formed in poor metabolizers. The volume of distribution of dextromethorphan is 5–6.4 liters/kg in dogs. Dextromethorphan and the glucuronide and sulfate ester conjugate, together with (+)-3-hydroxy-N-morphinan and traces of unmetabolized dextromethorphan, are excreted in the urine. The plasma half-life is about 2–4 hr with conventional dosage forms.

Mechanism of Toxicity
Dextromethorphan is the methylated dextro-isomer of levorphanol. Unlike the L-isomer, it has no analgesic properties. Dextromethorphan acts on the central nervous system (CNS) to elevate the cough threshold. It retains only the antitussive activity of other morphinian derivatives. Administration of dextromethorphan may be associated with histamine release. Dextromethorphan is often present in multisymptom products with a combination of ingredients. Toxic effects of concurrent agents such as antihistamines, decongestants, analgesics, and/or alcohol may be exhibited.

Human Toxicity: Acute
The potential for toxic effects, following acute overdosage of dextromethorphan, is low. Acute overdose probably will not result in severe signs and symptoms of intoxication unless massive amounts have been ingested. CNS depression may result, with amounts of 10 mg/kg or more, in children. Adults may tolerate up to 14 mg/kg with only minor effects. Ingestion of large amounts of dextromethorphan may result in lethargy, respiratory depression, nystagmus, and coma. Following a large ingestion of a long-acting preparation, symptoms may persist for 7 or 8 hr. Hemodynamic compromise and other severe symptomatology may result from concurrent ingredients in multisymptom cold products.

Human Toxicity: Chronic
Chronic abuse of dextromethorphan has resulted in psychotic behavior and hallucinations.

Clinical Management
Basic and advanced life-support measures should be instituted as necessary. Gastric decontamination may be performed as indicated by the patient's symptomatology, the specific product involved, and the history

of the ingestion. Activated charcoal may be useful to adsorb dextromethorphan and concurrent ingestants. Naloxone may be helpful in reversing the CNS and respiratory depressant effects of dextromethorphan. The level of consciousness and respiratory status should be carefully monitored. Management of the toxicological consequences of coingestants should be appropriate to the agent involved. Plasma dextromethorphan levels are not clinically beneficial to management of the overdose but may be useful in determining the metabolizer phenotype.

—*Carole Wezorek*

Diazinon

- CAS: 333-41-5
- SYNONYMS: Knox Out 2FM; Basudin; Diazitol; Dipofene; Neocidal; Nucidol; Spectracide
- CHEMICAL CLASS: Synthetic organophosphorous insecticide in the phosphorothionate class
- CHEMICAL STRUCTURE:

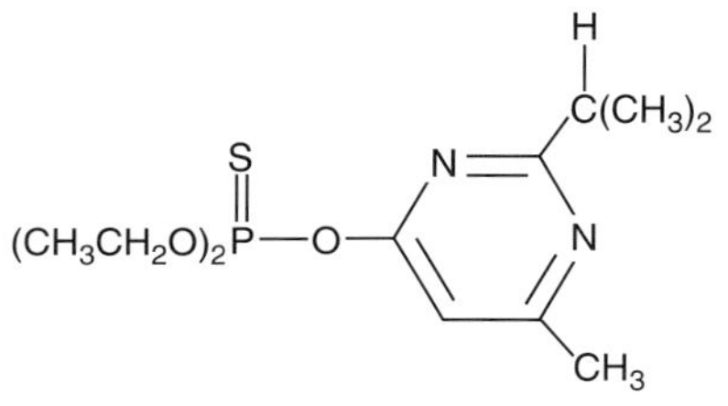

Uses

Diazinon is a broad-spectrum insecticide and nematocide for the control of a variety of household and agricultural pests.

Exposure Pathways

Dermal and inhalation routes are the primary exposure pathways. Diazinon is available as emulsifiable concentrates, wettable powders, microencapsulated formulations, and dusts.

Toxicokinetics

Diazinon is readily absorbed through the skin, lungs, and gastrointestinal tract. It is rapidly distributed to the tissues. The metabolism of diazinon by microsomal enzymes leads to both activation and detoxification (see Organophosphates). Diethyl thiophosphate and diethyl phosphate are formed in rats and cows. Some monoalkyl monoaryl derivatives may also be formed. Upon storage, diazinon can be physically converted to the oxygen analog (i.e., diazoxon) and tetraethyl monothiopyrophosphate; both are direct inhibitors of acetylcholinesterase. In contrast to most organophosphates, diazinon is relatively stable in alkaline conditions. Excretion of diazinon is rapid in laboratory rats, with approximately 50% elimination within 12 hr. About 80 and 20% is excreted in urine and feces, respectively.

Mechanism of Toxicity

As with other organophosphate insecticides, the toxicity of diazinon is due to inhibition of acetylcholinesterase (see Cholinesterase Inhibition). Typical signs of acetylcholinesterase inhibition are possible (e.g., excessive secretions, muscle fasciculations, tremors, and dyspnea).

Human Toxicity: Acute

Inhalation, dermal, or oral exposure to diazinon can produce systemic intoxication through inhibition of acetylcholinesterase. Differences in response to different formulations can be marked. Some earlier studies have reported that erythrocyte cholinesterase may be more sensitive to inhibition by diazinon than plasma cholinesterase.

Human Toxicity: Chronic

Repeated exposure may cause accumulation of acetylcholinesterase inhibition and lower the threshold for subsequent acetylcholinesterase-inhibiting exposures. Diazinon appears to be capable of inducing the intermediate syndrome (see Organophosphate Poisoning, Intermediate Syndrome).

Clinical Management

For exposure to eyes, eyelids should be held open and the eyes flushed with copious amounts of water for 15 min. For exposure to skin, affected areas should be washed immediately with soap and water. The victim should receive medical attention if irritation develops and persists.

For exposure through inhalation, the victim should be removed to fresh air and, if not breathing, given artificial ventilation. The victim should receive medical attention as soon as possible.

First aid for ingestion victims would be to induce vomiting, keeping in mind the possibility of aspiration of solvents. Gastric decontamination should be performed within 30 min of ingestion to be most effective. Initial management of acute toxicity is establishment and maintenance of adequate airway and ventilation. Atropine sulfate in conjunction with 2-PAM can be administered as an antidote. Atropine by intravenous injection is the primary antidote in severe cases. Test injections of atropine (1 mg in adults and 0.15 mg/kg in children) are initially administered, followed by 2–4 mg (in adults) or 0.015–0.05 mg/kg (in children) every 10–15 min until cholinergic signs (e.g., diarrhea, salivation, and bronchial secretions) decrease. High doses of atropine over several injections may be necessary for effective control of cholinergic signs. If lavage is performed, endotracheal and/or esophageal control is suggested. At first signs of pulmonary edema, the patient should be placed in an oxygen tent and treated symptomatically.

Animal Toxicity

The oral LD_{50} in rodents ranges from 75 to 400 mg/kg. The large differences reported in acute toxicity may be caused in part by physical changes upon storage. A large variability among species in acute response to diazinon has been reported. Some birds appear to be markedly more sensitive to diazinon than mammals. Doses of diazinon which produced maternal toxicity in rabbits were not teratogenic. Diazinon was teratogenic in chick embryos (1 mg/egg) when injected into the yolk sac on Day 4 of incubation. Following lethal doses in dogs, changes in erythrocytes (e.g., reticulocytopenia) were noted. Diazinon has little irritation or sensitization potential.

—*Thuc Pham (Clinical management section prepared by Carey Pope)*

Related Topics

Cholinesterase Inhibition
Organophosphate Poisoning, Intermediate
　　Syndrome
Organophosphates
Pesticides
Pollution, Water
Psychological Indices of Toxicity

Diazoxide

- ◆ CAS: 364-98-7
- ◆ PHARMACEUTICAL CLASS: Therapeutic thiazide
- ◆ CHEMICAL STRUCTURE:

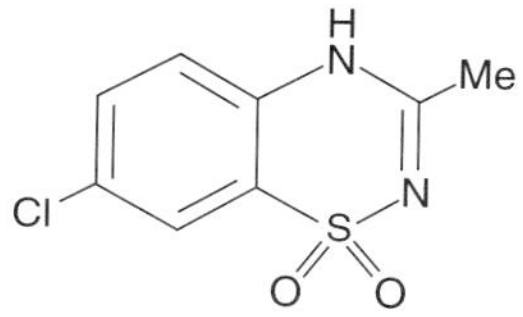

Uses

Diazoxide is administered intravenously for the acute treatment of hypertension. It is administered orally for the treatment of hypoglycemia.

Exposure Pathways

Diazoxide is therapeutically administered either intravenously or orally.

Toxicokinetics

Diazoxide is well absorbed orally. It is distributed to the plasma. Approximately 20–50% of diazoxide is eliminated unchanged by the kidneys. The remainder is metabolized by the liver to 3-carboxy and 3-hydroxymethyl derivatives.

Mechanism of Toxicity

Diazoxide acts as an antihypertensive by relaxing the arteriole smooth muscle. The cardiac output and renin secretion is increased. The result is a retention of salt and water and elevated levels of angiotensin II that eventually counteract the hypotensive action of the drug.

Human Toxicity

The effects of acute exposure include alterations of neonatal glucose levels, fetal bradycardia, and interference with labor in women treated intravenously with diazoxide. The two most frequently cited side effects are salt and water retention and hyperglycemia. Chronic exposure can lead to hypertrichosis. Diazoxide exhibits the ability to relax smooth muscle and, therefore, may be contraindicated in late pregnancy.

—William S. Utley

Dibenz[a,h]anthracene

◆ CAS: 53-70-3

◆ Synonyms: 1,2,5,6-Dibenzanthracene; DB(A,H)A; DBA

◆ Chemical Class: Polycyclic aromatic hydrocarbon

◆ Chemical Structure:

Uses

Dibenz[a,h]anthracene is a by-product of incomplete combustion and therefore is a fairly ubiquitous compound, generally strongly bound to the sediment. There are no reported industrial uses for dibenz[a,h]anthracene. It is used as a research tool.

Exposure Pathways

The primary route of exposure to dibenz[a,h]anthracene is via the skin from petroleum-based products. An additional significant route of exposure is through inhalation of cigarette smoke. While ingestion is a route of exposure, the significance of such exposure is debatable. Dibenz[a,h]anthracene is found in many food items (cereals, fruits, and vegetables) in the low parts per billion (ppb) level. It is found in cigarette smoke at 100–150 ppb and in used motor oil at approximately 14,000 ppb. It is found in petroleum products such as coal tar, mineral oil, and petroleum waxes.

Toxicokinetics

Dibenz[a,h]anthracene has an octanol to water partition coefficient (log K_{ow}) of 6.5 and will bioconcentrate in lower organism with less efficient mixed function oxidase systems. In humans, dibenz[a,h]anthracene is oxidized to the active metabolite where it is quickly eliminated or can react with target compounds such as the DNA. Dibenz[a,h]anthracene has not been isolated in the fat tissues, reflecting the effectiveness of the metabolizing enzyme system.

Mechanism of Toxicity

Dibenz[a,h]anthracene is metabolically activated by the mixed function oxidase (MFO) system of the liver (P448) to form an epoxide which subsequently covalently binds to the DNA. This interaction with the DNA is believed to result in the carcinogenicity of the material. The particular area of the compound oxidized by the MFO system will result in epoxides of varying carcinogenic potency.

Human Toxicity: Acute

As with many organic compounds, excessive acute exposure to dibenz[a,h]anthracene can lead to dizziness, nausea, and general central nervous system disturbances that resemble intoxication.

Human Toxicity: Chronic

Chronic exposure to dibenz[a,h]anthracene may produce cancer. Dibenz[a,h]anthracene is classified as a B2 probable human carcinogen based on sufficient animal data and no human data. In addition, carcinogenic polyaromatic hydrocarbons have been implicated in immunosuppresive activity. The acceptable level in drinking water is 13.3 ng/liter. Dibenz[a,h]anthracene is a RCRA hazardous waste (U063).

Predisposing Conditions

Previous exposure to polyaromatic hydrocarbons or genetic predisposition can increase the MFO system that activates dibenz[a,h]anthracene to the reactive epoxide. In addition, certain personal habits such as smoking can significantly increase a person's exposure to these compounds.

Animal Toxicity

In rats, dibenz[a,h]anthracene has been shown to cause fetal deaths when given at 5 mg/kg daily for the first day of pregnancy. Tumors in the forestomach were produced in mice given 9–19 mg over a 5- to 7-month period. Skin painting studies have produced mammary tumors. Lung tumors have been induced in rats receiving intratracheal administration of dibenz[a,h]anthracene.

—William S. Utley

Related Topics

Bioconcentration
Polycyclic Aromatic Hydrocarbons

Dibenzofuran

- ◆ CAS: 132-64-9
- ◆ SYNONYMS: 2,2′-Biphenylene oxide; 2,2′-biphenylylene oxide; dibenzo(*B*,*D*)furan; diphenylene oxide
- ◆ CHEMICAL CLASS: Strong oxidizing agent; antiestrogen
- ◆ MOLECULAR FORMULA: $C_{12}H_8O$
- ◆ CHEMICAL STRUCTURE:

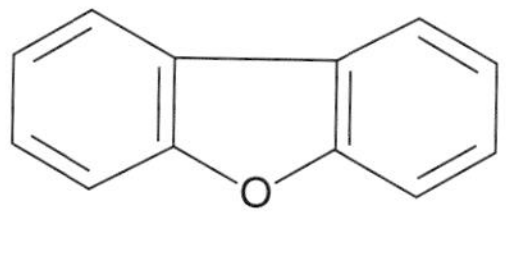

Uses

Dibenzofuran is an industrial chemical or by-product.

Exposure Pathways

Inhalation is the most common route of exposure. Dibenzofuran is present in cigarette ash and is a by-product of processes in the pharmaceutical industry. When heated to decompose, it emits acrid smoke and irritating fumes. Exposure can also occur by ingestion of contaminated food.

Toxicokinetics

Dibenzofuran may be rapidly absorbed by various routes including oral, nasal mucosal, inhalation, and dermal routes. 2,2′,3-Trihydroxy biphenyl dioxygenase is a key enzyme responsible for meta cleavage of the first aromatic ring in the degradation pathway. After intravenous or oral administration to rats, most of the compound is quickly distributed to the liver, muscle, skin, and adipose tissue and metabolized. Its metabolites may remain in the adipose tissue for a relatively long period of time. Polychlorinated dibenzofuran is highly lipophilic and is accumulated in adipose and liver tissues at a higher level and in muscle, kidneys, spleen, lungs, brain, and blood at a lower level. The metabolites are rapidly excreted mainly in bile, urine, and feces. They can also be excreted through milk.

Mechanism of Toxicity

Dibenzofuran induces hepatic, skin, and lung cytochrome P450 1A1, 1A2, and aryl hydrocarbon hydroxylase in rats. Thus, toxicity results from aryl hydrocarbon receptor signal transduction pathway. Bioactivation of many polycyclic hydrocarbon carcinogens is mediated by these enzymes.

Human Toxicity

Dibenzofuran may be harmful by inhalation, ingestion, or skin absorption and may cause irritation. It is globally distributed, is persistent in the environment, and tends to accumulate in human tissues. The major primary sources of exposure for the general population are combustion (municipal waste incineration and automobile exhaust), carbon electrode processes (smelters), chemical manufacturing wastes (chlorophenols), open-use agricultural and industrial chemicals (chlorophenols and chlorophenoxy herbicides), polychlorinted biphenyls, and aqueous chlorination (sewage sludge and kraft pulp mills). Exposure can occur through contaminated food (fish, meat, and dairy products; breast milk in the case of infants). All the chlorinated compounds have the potential to cause dermal, hepatic, and gastrointestinal toxicities. The half-life in humans is relatively long.

Clinical Management

Due to long biological half-life and lipid solubility of dibenzofurans, blood analysis may serve as an index of past cumulative occupational exposure and a means of assessing a person's exposure situation. In case of contact, the eyes and skin should be flushed immediately with water for at least 15 min. If inhaled, the victim should be removed to fresh air. If the person is not breathing, artificial respiration should be given; if breathing with difficulty, oxygen should be given. If the patient is in cardiac arrest, cardiopulmonary resuscitation should be given. In case of ingestion, the mouth should be washed out with water provided the person is conscious. These life-supporting measures should be continued until medical assistance has arrived. Liquids should not be administered to and vomiting should not be induced in an unconscious or convulsing person.

In the workplace, technical measures should prevent any contact with the skin and mucous membranes. Workers potentially exposed to this compound should wear personal protective equipment and their work should be carried out only in restricted and ventilated areas. After use, clothing and equipment should be placed in an impervious container for decontamination or disposal. Preemployment and periodic medical examination should focus on liver function.

Animal Toxicity

2,3,7,8-Tetrachlorodibenzofuran (TCDF) causes wasting syndrome, thymic atrophy and immune suppression in rodents, hair and fingernail loss in monkeys, chloracne formation in the rabbit ear, and hyperpigmentation in the rhesus monkey. It is a liver tumor promoter, teratogenic, and immunotoxic affecting natural killing cells. It is hepatotoxic and has profound effects on both steroid and growth factor receptor systems. Significant species variability is seen in the toxicity of this compound. The LD_{50} for TCDF is $5-10$ $\mu g/kg$ body weight in guinea pigs, greater than 6000 $\mu g/kg$ body weight in mice, and 1000 $\mu g/kg$ body weight in monkeys.

—*Kashyap N. Thakore and Harihara M. Mehendale*

Dibromochloropropane

- ◆ CAS: 96-2-8
- ◆ SYNONYMS: 1,2-Dibromo-3-chloropropane; DBCP; Nemafume; Nemanax; Nemaset; Nemagon; Femafume; Fumazone
- ◆ CHEMICAL CLASS: Halogenated alkane
- ◆ CHEMICAL STRUCTURE:

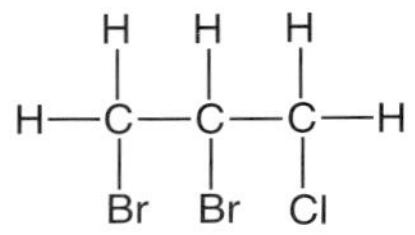

Uses

Dibromochloropropane was used as a soil fumigant and nematocide. Use of dibromochloropropane has been banned in the United States.

Exposure Pathways

Respiratory and dermal routes of exposure were most common.

Toxicokinetics

Dibromochloropropane is well absorbed by any route of exposure. Absorption is almost complete following oral exposure. Microsomal transformation leads to formation of reactive metabolites. Metabolites undergo conjugation with glutathione. Dibromochloropropane induces microsomal enzymes in the testes, liver, and kidneys. Covalently bound metabolites accumulate in the liver and kidneys. Urinary excretion is the major route of elimination.

Mechanism of Toxicity

Reactive metabolites of dibromochloropropane (e.g., epoxides) bind to cellular macromolecules.

Human Toxicity

The most marked human toxicity from dibromochloropropane exposure is infertility in males. Occupational exposure to dibromochloropropane has been associated with reduced sperm count and elevated levels of follicle-stimulating hormone and leuteinizing hormone. From a group of studies on occupational exposure to dibromochloropropane, it was estimated that approximately 15% of exposed workers were azoospermic. The effects on sperm production last years after exposure has ended, in particular with men exposed for a period of more than 4 years.

Clinical Management

Acute exposure to dibromochloropropane vapors requires removal from the source and symptomatic treatment. There is no treatment for testicular toxicity.

Animal Toxicity

The acute oral LD_{50} of dibromochloropropane in male rats and guinea pigs is about 150–300 mg/kg. The dermal LD_{50} is ~1 g/kg. Dibromochloropropane acts as a central nervous system depressant at high vapor concentrations. Early animal studies demonstrated reduced testicular weights, with testicular atrophy at higher exposure levels. Several studies have reported decreased sperm count and infertility with long-term exposure to dibromochloropropane. Dibromochloropropane was mutagenic in several bacterial assays. It is an experimental carcinogen, capable of increasing tumor incidence in a variety of tissues.

—Thuc Pham

Related Topics

Pollution, Water
Reproductive System, Male

Dibutyl phthalate

- CAS: 84-74-2
- SYNONYMS: DBP; 1,2-benzenedicarboxylic acid dibutyl ester; phthalic acid dibutyl ester; di-(*n*-butyl) phthalate; celluflex DBP; elaol; ergoplast FDB;

palatinol; polycizer DBP PX 104 RC plasticizer; genoplast B; hexaplast M/B

- CHEMICAL STRUCTURE:

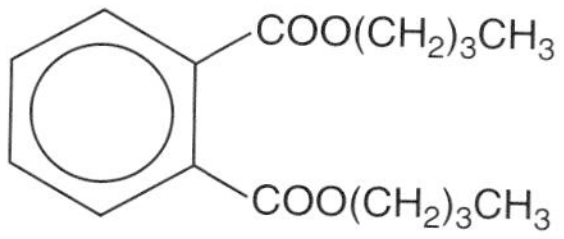

Uses

Dibutyl phthalate is used in a variety of ways, such as insect repellant, plasticizer in nitrocellulose lacquers, elastomers, explosives, nail polish, and solid rocket propellants. It can also be used as a solvent for perfume oil, perfume fixative, textile lubricating agent, safety glass, insecticides, printing inks, resin solvents, paper coatings, and adhesives.

Exposure Pathways

Exposure to dibutyl phthalate is usually by inhalation. It is known to leech out from finished plastics into blood, milk, and other food materials and, therefore, can be ingested orally. Dermal exposure is also a common route of exposure.

Toxicokinetics

Dibutyl phthalate is hydrolyzed to phthalic acid and butanol.

Mechanism of Toxicity

Dibutyl phthalate acts as an uncoupler of oxidative phosphorylation in rats.

Human Toxicity

Dibutyl phthalate has low acute toxicity based on animal studies. It can cause an immediate stinging and burning sensation upon contact by splashing. It can also cause dizziness, nausea, and contact dermatitis.

Clinical Management

Induced emesis is not recommended if the victim has any signs of esophageal or gastrointestinal tract irritation or burns, or decreased sensory response, depressed gag reflex, or impending shock. Activated charcoal slurry with or without saline cathartic or sorbitol can be given in cases of oral exposures. Skin decontamination should be done with repeated washing with soap. Exposed eyes should be irrigated with copious amounts of room-temperature water for at least 15 min. Treatment is supportive and symptomatic and no specific antidote is available.

Animal Toxicity

Exposures to dibutyl phthalate have caused photophobia, conjunctivitis, edema, and keratitis. Increase in mean liver weight and testicular atrophy have been observed in rats exposed to dibutyl phthalate. The LD_{50} in rats is 12,000 (oral) and 4000 mg/kg (intraperitoneal).

—*Sushmita M. Chanda*

Dicamba

- ◆ CAS: 1918-00-9
- ◆ SYNONYMS: Banlen; banvel 480; brush buster; compound B dicamba; velsicol compound "r"; velsicol 58-CS-11; banvel herbicide; banvel 4WS; mediben
- ◆ CHEMICAL CLASS: Benzoic acid; 3,6-dichloro-2-methoxy-
- ◆ CHEMICAL STRUCTURE:

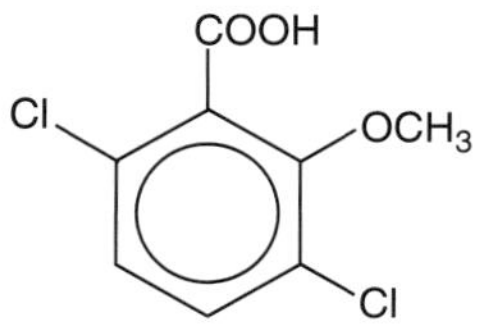

Uses

Dicamba is used as a herbicide and for pest control.

Exposure Pathways

Dicamba is available as an odorless, white or brown, crystalline solid. Ingestion and dermal contact are the most common routes of exposure.

Toxicokinetics

Dicamba is known to be well absorbed orally. Minimal absorption occurs through the skin. When given in animal feed, dicamba is readily distributed in all organs in animals. When given as a food supplement, dicamba was excreted in the rat either unchanged or as glucuronic acid conjugate. The half-life of elimination in rats was estimated to be 0.83 hr.

Mechanism of Toxicity

There is little evidence of dicamba toxicity in mammals. In plants its primary action is to act as an growth regulator.

Human Toxicity: Acute

There have been very few incidences of dicamba poisoning alone. Most exposures have occurred where it is in combination with other herbicides, especially chlorophenoxy compounds. Dicamba causes mild skin, eye, gastrointestinal, and respiratory tract irritation.

Human Toxicity: Chronic

No study has reported the chronic toxicity of dicamba.

Clinical Management

There is no specific antidote; therefore, the treatment is symptomatic and supportive. Skin decontamination should be done with repeated washing with soap. Exposed eyes should be irrigated with copious amounts of water (at room temperature) for at least 15 min. Emesis can be induced if initiated within 30 min of ingestion. Emesis is not encouraged if the patient is comatose or convulsing. In such cases, ipecac can be used to induce emesis. Activated charcoal slurry with or without saline cathartic and sorbitol may be used.

Animal Toxicity

The oral LD_{50} value is 1190 mg/kg in mice, 1039 mg/kg in rats, 2 g/kg in rabbits, and 3 g/kg in guinea pigs.

—*Sushmita M. Chanda*

Related Topics

Pesticides
Pollution, Water

Dichlone

- ◆ CAS: 117-80-6
- ◆ SYNONYMS: Dichloronaphthoquinone; 2,3-dichloro-1,4-naphthoquinone; Phygon; Algistat; Quintar; Sanquinon
- ◆ CHEMICAL CLASS: Naphthoquinone
- ◆ CHEMICAL STRUCTURE:

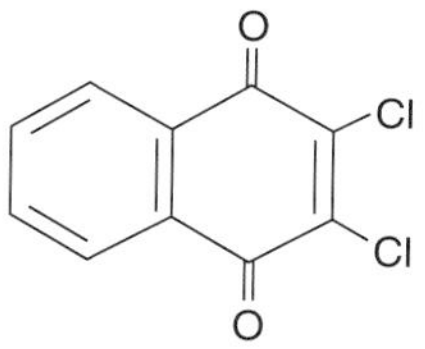

Uses

Dichlone is used as a fungicide to treat vegetable seed against seed decay. It is also used for certain apple, cherry tomato, and rose diseases.

Exposure Pathways

Dichlone exposure may occur by oral and dermal routes.

Toxicokinetics

Toxicokinetic data for dichlone are limited. It has been demonstrated that dichlone is poorly absorbed from the gastrointestinal tract.

Mechanism of Toxicity

When dichlone is heated to decomposition, it emits toxic fumes.

Human Toxicity

Dichlone has relatively low toxicity in humans. It is irritating to the skin and mucous membranes. Irritation of the cornea may occur. Ingestion of large doses usually results in prompt emesis. Large doses may cause central nervous system depression, coma, and death.

Clinical Management

Emesis may be indicated in recent substantial ingestion unless the patient is or could rapidly become obtunded, comatose, or convulsing. Emesis is most effective if initiated within 30 min. An activated charcoal cathartic may also be employed. Use of any fat should be avoided since these agents may increase the irritant effects. If clothing is contaminated, it should be removed and the skin should be washed vigorously, including hair and nails; repeated soap washings should be performed. Contaminated clothing should be discarded.

Animal Toxicity

The acute oral LD_{50} in rats is >5 g/kg. Dichlone caused swelling of rat liver mitochondria, a process that has been shown to be osmotic in nature. Consistent with these *in vitro* findings, dichlone feeding caused partial uncoupling of liver and heart mitochondria. These findings indicate that dichlone affects cellular energy metabolism by inducing oxidative stress at the organelle level and by interacting with key mitochondrial thiol groups. Dichlone produces dermal irritation in animals. Irritation is enhanced by the application of oils and other fatty substances, including perspiration. Dichlone is carcinogenic in rodents. The oral and dermal LD_{50} values for the rabbit are 1300 and 5000 mg/kg, respectively.

—*Todd A. Bartow*

Related Topic

Pesticides

Dicholorobenzene

- ◆ CAS: 106-46-7 (*p*-dichlorobenzene)
- ◆ PREFERRED NAMES: *para-* or *ortho-*dichlorobenzene
- ◆ SYNONYMS: Benzene 1,4-chloro; *p*-DCB; *o*-DCB; AI13-0050; dichloricide; evola; globol; paracide; para crystals; parazene; paradow; paramoth; paranuggets; paradi; PDC; PDCB; persia-parazol; santochlor

- CHEMICAL CLASS: Benzene, dichloro-
- CHEMICAL STRUCTURE:

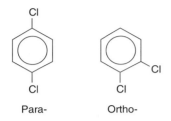

Para-   Ortho-

Uses

p-Dichlorobenzene is used as an insecticidal fumigant, disinfectant, and a chemical intermediate. It is available as mothballs, flakes, cake, crystals, and as a 100% concentrate.

Exposure Pathways

Ingestion, inhalation, and dermal or ocular contact are the most common routes of exposure.

Toxicokinetics

p-Dichlorobenzene is well absorbed orally and by inhalation. The highest concentrations are found in the adipose tissue. It is rapidly oxidized to phenolic compounds and metabolized to sulfate and glucuronic conjugates. The major metabolite is 2,5-dichlorophenol. From 91 to 97% is excreted by the kidneys within 5 days.

Mechanism of Toxicity

Unlike other halocarbons, *p*-dichlorobenzene has very low toxicity because only small amounts are converted into the epoxide intermediate.

Human Toxicity

Dichlorobenzene probably has very low toxicity. It primarily causes injury to the liver and secondarily to the kidneys. Central nervous system depression occurs at concentrations that are extremely irritating to the eyes and nose. Upon dermal contact, it may produce burning sensation.

Clinical Management

Most of the reported toxicity is due to chronic exposure. Treatment is symptomatic and supportive. Activated charcoal slurry with or without saline cathartic can be given if dichlorobenzene is ingested. Skin decon-

tamination should be done with repeated washing with soap. Exposed eyes should be irrigated with copious amounts of water (at room temperature) for at least 15 min. The victim should be moved to fresh air when exposed to dichlorobenzene through inhalation. Humidified oxygen (100%) can be supplemented with assisted ventilation.

Animal Toxicity

Dicholorobenzene is not toxic to animals. Dogs given 1.5 g/kg of *p*-dicholorobenzene did not develop toxic effects. The oral TDLo is 155 g/kg/2Y-I:CAR. The oral LD_{50} is 500 mg/kg in rats.

—*Sushmita M. Chanda*

Related Topic

Pesticides

Dichloroethanes

- SYNONYMS: Chlorinated hydrochloric ether; ethylidene chloride; ethylidine dichloride
- CHEMICAL CLASS: Chlorinated hydrocarbon
- CHEMICAL STRUCTURE:

$$Cl-\underset{\underset{Cl}{|}}{\overset{\overset{H}{|}}{C}}-\underset{\underset{H}{|}}{\overset{\overset{H}{|}}{C}}-H \quad or \quad Cl-\underset{\underset{H}{|}}{\overset{\overset{H}{|}}{C}}-\underset{\underset{H}{|}}{\overset{\overset{H}{|}}{C}}-Cl$$

Uses

1,2-Dichloroethane is utilized as a solvent, pesticide, fumigant, gasoline additive, and in the synthesis of vinyl chloride. 1,1-Dichloroethane is utilized in relatively small quantities, primarily in the chemical, agricultural, and petroleum industries. In the past, 1,1-dichlroethane was used as anesthetic.

Exposure Pathways

Inhalation through contaminated air and ingestion of contaminated water are common routes of exposure in humans.

Toxicokinetics

Dichloroethanes are readily absorbed through the lungs following inhalation exposure in both humans and experimental animals. Absorption after oral ingestion in experimental animals is rapid, complete, and essentially linear. Studies in animals have shown that dichloroethanes are well absorbed through the skin following dermal exposure. Dichloroethanes are metabolized to a variety of chlorinated metabolites, some of which (e.g. chloroacetaldehyde and acetyl chloride) are reactive species which are more toxic than the parent compounds. The relatively greater rate of metabolism of 1,1-dichloroethane relative to 1,2-dichloroethane is not consistent with the relatively higher toxicity, mutagenicity, and carcinogenicity of 1,2-dichloroethane. Therefore, it is possible that there is alternative route of metabolism. Dichloroethanes appear to be rapidly distributed in humans. In rats dichloroethanes are readily distributed throughout body tissue after inhalation or oral ingestion. The highest concentrations were found in fat. Following inhalation exposure in rats, elimination occurred primarily via the excretion of soluble metabolites and unchanged parent compound in urine and carbon dioxide in the expired breath. Urinary metabolites accounted for 84% of the absorbed dose, fecal accounted for 2%, and carbon dioxide accounted for about 7%. Following oral exposure, urinary metabolites accounted for 60%, unchanged in the breath accounted for 29%, and carbon dioxide in breath accounted for 5% of the administered 150 mg/kg dose.

Mechanisms of Toxicity

The mechanism of action of dichloroethane-induced toxicity is not fully elucidated. By most criteria 1,1-dichloroethane is less toxic than 1,2-dichloroethane. Studies of the possible mutagenicity and carcinogenicity of 1,1-dichloroethane have been negative or inconclusive. In contrast, 1,2-dichloroethane is carcinogenic in rats and mice and mutagenic in several test systems, particularly in the presence of activating enzymes, such as hepatic glutathione transferases and to a lesser extent by hepatic microsomal cytochrome P450 enzymes.

Human Toxicity

Very limited human toxicity data are available for dichloroethanes. Symptoms observed were central nervous system depression, corneal opacity, bronchitis, respiratory distress, myocardial lesions, hemorrhagic gastritis and colitis, increased blood clotting time, hepa-tocellular damage, renal necrosis, and histopathological changes in brain tissue. Death was most often attributed to cardiac arrhythmia. In the past 1,1-dichloroethane was used as an anesthetic at levels of approximately 25,000 ppm. This use was discontinued when it was discovered that cardiac arrhythmias might be induced. In persons with impaired pulmonary function, especially those with obstructive airway diseases, the breathing of dichloroethane might cause exacerbation of symptoms to its irritant properties.

The OSHA regulatory level in workplace air is 1 ppm for an 8-hr day, 40-hr week. U.S. EPA has set a limit in water of 0.005 mg/liter.

Clinical Management

Respiratory and cardiovascular function should be supported, and the victim should be moved to fresh air and given artificial respiration; if breathing is difficult, oxygen should be given. In case of contact with material, the eyes should be flushed immediately with running water for at least 15 min; the skin should be washed with soap and water. Contaminated clothing and shoes should be removed and isolated at the site. In case of oral exposure, emesis should not be induced. A charcoal slurry, aqueous or mixed with saline cathartic or sorbitol, should be administered.

Animal Toxicity

Death has also occurred in animals that ingested dichloroethanes. An acute oral LD_{50} of 680 mg/kg has been reported for rats. In mice the LD_{50} values for male (489 mg/kg) and female (413 mg/kg) are reported. No adverse clinical effects were noted in rats, rabbits, or guinea pigs exposed to 1000 ppm 1,1-dichloroethane for 13 weeks, which followed a prior 13-week exposure to 500 ppm. However, under the same conditions renal injury was apparent in cats.

—*Madhusudan G. Soni and*
Harihara M. Mehendale

Dichloroethylene, 1,1-

- ◆ CAS: 75-35-4
- ◆ SYNONYMS: 1,1-DCE; 1,1-Dichloroethene; NCI-C54262; vinylidene chloride

♦ CHEMICAL CLASS: Flammable gas, strong oxidizing agent
MOLECULAR FORMULA: $C_2H_2Cl_2$

♦ CHEMICAL STRUCTURE:

$$CH_2=CCl_2$$

Uses

1,1-Dichloroethylene is used as an intermediate in the production of "vinylidene polymer plastics" such as Saran and Velon; in interior coatings for ship tanks, railroad tank cars, and fuel storage tanks; in exterior coatings for steel piles and structures; as a copolymer latex into cement to impart strength to mortar or concrete; in barrier coatings on paper products and plastic films; and in binders for paints and nonwoven fabrics.

Exposure Pathways

Inhalation is the most common route of exposure. 1,1-Dichloroethylene has been identified in industrial and sewage treatment effluents, well water, and raw and treated drinking water. It is present in food (biscuits, potato chips, and cheese). It is a very dangerous fire hazard and explosive when exposed to heat, flame, or oxidizers, and when heated to decomposition it emits toxic fumes of Cl^-.

Toxicokinetics

In experimental rats, by inhalation or ingestion, 1,1-dichloroethylene is absorbed very rapidly and completely from the gastrointestinal tract, except in the presence of food. Peak blood levels were reached 2–8 min following administration. The biological half-life is longer for exposure via the oral route compared to intravenous route. Mice biotransform 1,1-dichloroethylene into thiodihydroxyacetic acid and *N*-acetyl-*S*-cysteinylacetyl derivatives as major metabolites with some minor metabolites which alkylate proteins, including glutathione, to a greater extent than rats. Metabolic conversion of vinylidene chloride into an epoxide which can rearrange to the corresponding acyl chloride has been proposed.

Following exposure through intraperitoneal route in mice, the compound is distributed to all tissues with peak at 6 hr and covalently binds between 6 and 12 hr with highest levels in liver, kidneys, and lungs and smaller amounts in skeletal muscle, heart, spleen, and gut. The covalent binding in the liver, kidneys, and lungs fell to 50% between 12 hr and 4 days. These three tissues showed similar spread in subcellular fractions with slight tendency to concentrate in the mitochondria. In case of rat, a single oral dose of 1,1-dichloroethylene at 25 mg/kg body weight resulted in a high concentration in the liver and kidneys after 30 min with more general distribution throughout soft tissues after 1 hr.

Elimination of 1,1-dichloroethylene is relatively rapid. Elimination is mainly pulmonary after intravenous and intraperitoneal routes but not after the intragastric route; both unchanged 1,1-dichloroethylene and its final oxidative metabolite carbon dioxide are excreted via the lungs and other metabolites are excreted via the kidneys. Part of the urinary excretion is of biliary origin. In comparison, 80% of a small intravenous dose is excreted unchanged within 1 hr of injection and more than 60% within 5 min.

Mechanism of Toxicity

1,1-Dichloroethylene undergoes biotransformation by NADPH-cytochrome P450 to several reactive species (chloroacetic acid, chloroacetyl chloride, and dichloroacetaldehyde) which conjugate with glutathione (GSH) and deplete it to cause liver, lung, and kidney toxicity. In the kidney, the presumed GSH conjugates also undergo secondary modification to reactive metabolite(s) which causes damage to proximal tubules. The reactive metabolites can activate phospholipase A_2 leading to cell death. It also covalently binds to protein and such modification may lead to tumor formation. It inhibits Ca^{2+}-ATPase in cell membrane and/or endoplasmic reticulum and causes canalicular cholestasis. However, it does not increase lipid peroxidation *in vivo* and *in vitro*.

Human Toxicity

1,1-Dichloroethylene is a central nervous system (CNS) depressant, an irritant to the skin and mucous membranes, and narcotic in high concentrations. It causes systemic effects on the liver and kidneys by inhalation. No data are available to evaluate the reproductive effects or prenatal toxicity and there is inadequate evidence of carcinogenicity.

The ACGIH TLV-TWA is 5 mg/m³. The U.S. EPA has identified the compound as a RCRA hazardous waste. In the workplace, technical measures should prevent any contact with the skin and mucous membranes.

Workers exposed to this compound should wear personal protective equipment and their work should be carried out only in restricted and ventilated areas. Clothing and equipment after use should be placed in an impervious container for decontamination or disposal.

Clinical Management

In case of contact, the eyes and skin should be flushed with water for 15–20 min. If inhaled, the exposed person should be moved to fresh air. If necessary, oxygen and artificial respiration should be administered. If patient is in cardiac arrest, cardiopulmonary resuscitation should be administered. Life-support measures should be continued until medical assistance has arrived. Liquids should not be given and vomiting should not be induced in an unconscious or convulsing person.

Animal Toxicity

At high doses, 1,1-dichloroethylene causes death from respiratory failure. Toxic effects are seen in the liver, kidneys, spleen, and lungs. 1,1-Dichloroethylene is carcinogenic in mice (kidney tumors) and rats (mammary tumors) and mutagenic in bacterial assay system. It alkylates DNA of liver and kidneys and induces DNA repair. Both toxicity and carcinogenicity are highly dose dependent. 1,1-Dichloroethylene is teratogenic and produces reproductive effects. The LC_{50} in rats is 4000 ppm/2 hr. The oral LD_{50} is 1500 mg/kg body weight in rats, 200 mg/kg body weight in mice, and 5750 mg/kg body weight in dogs. The dermal LD_{50} in rabbits is 2335 mg/kg body weight.

—Kashyap N. Thakore and
Harihara M. Mehendale

Dichloroethylene, 1,2-

◆ CAS: 540-59-0

◆ SYNONYMS:

Symmetrical—1,2-Dichloroethylene (CAS: 540-59-0)

Ethene, 1,2-dichloro-; ethylene, 1,2-dichloro-; 1,2-dichloroethene; acetylene dichloride; dioform; *sym*-dichloroethylene; DCE

cis Isomer—(*Z*)-1,2-dichloroethylene (CAS 156-59-2)

Ethylene, 1,2-dichloro-, (*Z*)-; (*Z*)-1,2-dichloroethene; 1,2-*cis*-dichloroethene; 1,2-*cis*-dichloroethylene; *cis*-1,2-dichlorethylene; *cis*-1,2-dichloroethene; *cis*-1,2-dichloroethylene; *cis*-dichloroethylene

trans Isomer—(*E*)-1,2-dichloroethylene (CAS: 156-60-5)

Ethene, 1,2-dichloro-, (*E*); ethylene, 1,2-dichloro-, (*E*)-; (*E*)-1,2-dichloroethene; 1,2-*trans*-dichloroethene; 1,2-*trans*-dichloroethylene; HCC 1130-T; *trans*-1,2-dichloroethene; *trans*-1,2-dichloroethylene

◆ CHEMICAL STRUCTURE:

Symmetrical, Cis Isomer, Trans Isomer — Cl_2 — CH=CH — Cl_2

Uses

1,2-Dichloroethylene is used as a direct solvent for perfumes, dyes, gums and waxes, oils, fats, lacquers, thermoplastics, phenols, and camphor; as a chemical intermediate for chlorinated compounds; and as an agent in retarding fermentation. 1,2-Dichloroethylene is also used as a low-temperature solvent for heat-sensitive substances in the extraction of caffeine, fats, and natural rubber, as well as in organic synthesis for polymers and telomers and as a coolant in refrigeration plants. Miscellaneous applications include use as a dry cleaning solvent, cleaning solution for printed circuit boards, and use in food packaging, adhesives, and germicidal fumigants.

Exposure Pathways

Occupational exposure to 1,2-dichloroethylene will be by inhalation or dermal and eye contact with the vapor, as well as by dermal and eye contact with the liquid during its use as a solvent. The general population is exposed to 1,2-dichloroethylene in urban air as well as from contaminated drinking water.

Toxicokinetics

1,2-Dichloroethylene is largely excreted through the lungs. *Cis* and *trans* isomers are metabolized to the same metabolites, dichloroacetic acid and dichloroethanol, in isolated perfused rat liver systems. In this system, the *cis* isomer is metabolized to a greater extent than the *trans* isomer. The metabolites are formed by an epoxide intermediate. Studies with rat liver microsomes cause a fall in microsomal cytochrome P450 content without affecting other microsomal enzymes. The decrease in cytochrome P450 only occurred in the presence of reduced nicotinamide adenine dinucleotide, suggesting that the chloroethylene must be metabolized to exert its destructive action. The loss of P450 was attributed to the destruction of heme since the fall in cytochrome P450 was always accompanied by parallel decrease in microsomal heme content.

Mechanism of Toxicity

The acute narcotic effects are due to the physical interaction of the material itself on the cells of the central nervous system (CNS). The long-term effects are most likely due to the production of an unstable reactive intermediate during biotransformation.

Human Toxicity

1,2-Dichloroethylene is toxic by inhalation, skin contact, or ingestion. OSHA considers 4000 ppm immediately dangerous to life or health. The acute toxicity of the *cis* isomer appears to be somewhat greater than that of the *trans* form. The estimated oral dose reported to cause death in humans is 3–5 ml/kg body weight. Based on a few cases of human poisoning and animal data, 1,2-dichloroethylene is considered to have toxic potential equivalent to that of trichloroethylene. Reported clinical effects from exposures to 1,2-dichloroethylene include optic neuritis, blindness, and pulmonary hemorrhage following ingestion. Respiratory depression, cyanosis, and edema can occur following inhalation. In addition, cardiac arrhythmias, hypotension, conduction defects, and myocardial injury have been reported. Nervous system effects after inhalation include CNS depression, light-headedness, dizzi-

ness, and euphoria. Mild vapor exposure may cause nausea, vomiting, weakness, tremor, and epigastric cramps. 1,2-Dichloroethylene is mildly irritating to the skin; chronic exposures may produce rash or chapped skin. 1,2-Dichloroethylene causes pain and irritation to the eye, but permanent injury is unlikely. In some cases, reversible corneal clouding will occur.

The ACGIH TLV-TWA is 200 ppm (793 mg/m^3). The excursion limit recommendation is that worker exposure levels may exceed three times the TLV-TWA for no more than 30 min during a work day, and in no circumstances should they exceed five times the TLV-TWA. Exposure limits (as of January 1993) in other countries are as follows:

OEL-Australia: TWA, 200 ppm (790 mg/m^3)

OEL-Austria: TWA, 200 ppm (790 mg/m^3)

OEL-Belgium: TWA, 200 ppm (793 mg/m^3)

OEL-Denmark: TWA, 200 ppm (790 mg/m^3)

OEL-Finland: TWA, 200 ppm (790 mg/m^3); STEL, 250 ppm (990 mg/m^3)

OEL-Germany: TWA, 200 ppm (790 mg/m^3)

OEL-Hungary: TWA, 80 mg/m^3; STEL, 160 mg/m^3

OEL-Japan: TWA, 150 ppm (590 mg/m^3)

OEL-The Netherlands: TWA, 200 ppm (790 mg/m^3)

OEL-The Philippines: TWA, 200 ppm (790 mg/m^3)

OEL-Poland: TWA, 50 mg/m^3

OEL-Russia: TWA, 150 ppm

OEL-Switzerland: TWA, 200 ppm (790 mg/m^3); STEL, 400 ppm

OEL-Turkey: TWA, 200 ppm (790 mg/m^3)

OEL-United Kingdom: TWA, 200 ppm (790 mg/m^3); STEL, 250 ppm

Check the ACGIH TLVs for OELs in Bulgaria, Colombia, Jordan, Korea, New Zealand, Singapore, and Vietnam.

Clinical Management

Persons exposed to 1,2-dichloroethylene should have their vital signs closely monitored; the heart should be monitored by EKG. Epinephrine and other catecholamines should be avoided, especially beta agonists since they may increase the risk of arrhythmias. Pulmonary edema, renal failure, and liver injury should be managed symptomatically, however; based on toxicity similarities to trichloroethylene, plasma levels following overdose/exposure are not clinically useful. Renal and liver function tests should be monitored in the presence of suspected kidney or liver injury.

Animal Toxicity

The no-adverse-effect level of 1,2-dichloroethylene in rats, rabbits, guinea pigs, and dogs is at least 1000 ppm. The most important effects of 1,2-dichloroethylene on animals appear to be CNS depression, narcosis, and irritation of the CNS. Short- and long-term chronic tests as well as multigeneration tests in mice show 1,2-dichloroethylene to have a low acute toxicity with no teratological or dominant lethal effects. However, 1,2-dichloroethylene caused apparent CNS depression and behavior changes. The liver and possibly the immune system were sensitive target organs, and male mice appeared more sensitive than females. Some dogs developed delicate corneal turbidity when repeatedly exposed to 1,2-dichloroethylene by evaporation of 10–15 ml in a chamber of 0.115 m³ volume. The turbidity was attributed to many fine gray flecks in the endothelium that cleared within 24–48 hr. *Cis* and *trans* isomers were negative in the Ames mutagenicity test; *trans*-1,2-DCE was negative in the metabolically activated test system using *Escherichia coli* strain K-12.

—*David M. Krentz and Linda Angevine Malley*

Dichloropropene, 1,3-

- ◆ CAS: 542-75-6
- ◆ SYNONYMS: 1,3-Dichloro-1-propene; 1,3-dichloropropylene; Dorlone; Nemex; Telone; Vidden D

- ◆ CHEMICAL CLASS: Halogenated hydrocarbon nematocide
- ◆ CHEMICAL STRUCTURE:

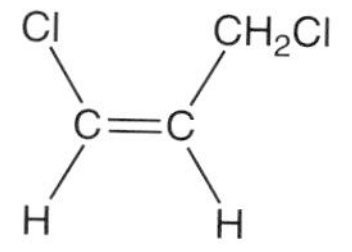

Uses

1,3-Dichloropropene is used as a soil fumigant for the control of nematodes.

Exposure Pathways

Dermal, oral, and inhalation exposures are all possible.

Toxicokinetics

1,3-Dichloropropene can be absorbed through the skin and via the respiratory and gastrointestinal tracts. It has an elimination half-life of less than 30 min from the bloodstream. Dichloropropene primarily undergoes conjugation with glutathione to form a mercapturic acid. The predominant route of excretion of 1,3-dichloropropene in rats is via the urine (about 50–80%) as mercapturic acid and sufoxide conjugates. Lesser amounts are eliminated through the feces and expired air.

Mechanism of Toxicity

1,3-Dichloropropene acts as an irritant and sensitizer. Macromolecular binding of this compound may also contribute to its toxicity.

Human Toxicity

1,3-Dichloropropene may cause headache and chest discomfort. Edema, erythema, and necrosis of the skin have also been reported.

The ACGIH TLV-TWA for 1,3-dichloropropene is 1 ppm.

Clinical Management

Treatment is symptomatic.

Animal Toxicity

The oral LD$_{50}$ of 1,3-dichloropropene in rats is about 100–700 mg/kg. The dermal LD$_{50}$ is >1200 mg/kg.

Treated animals have a "garlic" odor and exhibit eye and nasal irritation. 1,3-Dichloropropene causes both irritation and sensitization. In rabbits, marked erythema and conjunctivitis were noted following ocular exposure to 1,3-dichloropropene. Brief exposure to high concentrations (>2700 ppm) of 1,3-dichloropropene vapor may also cause lung, liver, and kidney injury. 1,3-Dichloropropene was not embryotoxic or teratogenic in inbred rats or rabbits at doses which produced maternal toxicity.

—*Jing Liu*

Related Topics

Pollution, Water
Pesticides

Dichlorvos

- ◆ CAS: 62-73-7
- ◆ SYNONYMS: DDVF; DDVP; Canogard; Dedevap; Estrosol; Herkol; Vapona
- ◆ CHEMICAL CLASS: Synthetic dimethoxy organophosphorous insecticide
- ◆ CHEMICAL STRUCTURE:

$$Cl_2C=CH-O-\overset{\displaystyle O}{\overset{\displaystyle \|}{P}}(OCH_3)_2$$

Uses

Because of its high vapor pressure, dichlorvos is useful for the control of insects in closed spaces (e.g., warehouses, greenhouses, animal shelters, homes, and restaurants). It is available in oil solutions, emulsifiable concentrates, aerosols, and baits. Therapeutically, dichlorvos is used as a broad-spectrum anthelmintic (for destroying or expelling intestinal worms). It is primarily used for insect control. Dichlorvos is also a breakdown product of the organophosphorus pesticide trichlorfon (metrifonate).

Exposure Pathways

Exposure to dichlorvos vapor can result in exposure through not only the respiratory route but also the dermal and oral routes (e.g., through contamination of feed).

Toxicokinetics

Animals exposed to dichlorvos vapor were found to absorb at least 50% of the total material by the respiratory route. Dichlorvos can also be absorbed through the oral and dermal routes. Following oral exposure, dichlorvos is rapidly detoxified in the liver. Metabolites include O,O'-dimethyl phosphate, monomethyl phosphate, O-methyl-O-2,2-dichlorovinyl phosphate (desmethyl dichlorvos), and inorganic phosphate. Detoxification processes for dichlorvos are also found in plasma.

Under most conditions, dichlorvos is not detectable in any tissues. Dichlorvos is not stored in tissues, it does not accumulate in secretions (e.g., milk), and it is below detection levels in the blood of various species at exposure levels in excess of 10 times those effective for insect control. At exceptionally high concentrations (90 mg/m^3 or about 2000 times normal exposure levels), dichlorvos was detectable in various tissues of the rat.

Dichlorvos is rapidly eliminated following any route of exposure. Desmethyldichlorvos and dimethylphosphate are rapidly excreted or further metabolized. The glucuronide conjugate of dichloroethanol is excreted in the feces. Species differences in elimination are common. For example, following oral administration, the cow eliminates approximately 50% through the feces, whereas the rat excretes only about 3% via the feces.

Mechanism of Toxicity

The toxicity of dichlorvos is due to inhibition of acetylcholinesterase and the signs of toxicity are generally similar to those caused by other organophosphorous insecticides (see Cholinesterase Inhibition). Dichlorvos is a direct inhibitor of cholinesterases; thus, toxicity rapidly follows exposure and recovery is also rapid. With inhalation exposures, airway acetylcholinesterase inhibition is possible in the absence of significant blood enzyme inhibition. The fly head acetylcholinesterase appears more sensitive to inhibition by dichlorvos relative to mammalian brain acetylcholinesterase. At high doses, dichlorvos may cause hyperglycemia and abnormal glucose tolerance.

Human Toxicity: Acute

Inhalation, dermal, or oral exposure to dichlorvos can result in systemic toxicity through inhibition of acetylcholinesterase. It has not been found to be either irritating or sensitizing. In humans, the plasma cholinesterase appears more sensitive than erythrocyte cholinesterase to inhibition by dichlorvos; thus, discrimination between these two activities may be warranted during assessment of exposures.

Human Toxicity: Chronic

Dichlorvos has little ability to produce chronic effects. Experimental studies suggest that relatively high exposures may produce delayed neurotoxicity, whereas the dibutyl analog is capable of inducing delayed neurotoxicity at relatively low doses.

Clinical Management

For exposure to eyes, eyelids should be held open and the eyes flushed with copious amounts of water for 15 min. For exposure to skin, affected areas should be washed immediately with soap and water. The victim should receive medical attention if irritation develops and persists.

For exposure through inhalation, the victim should be removed to fresh air and, if not breathing, given artificial ventilation. The victim should receive medical attention as soon as possible.

First aid for ingestion victims would be to induce vomiting, keeping in mind the possibility of aspiration of solvents. Gastric decontamination should be performed within 30 min of ingestion to be most effective. Initial management of acute toxicity is establishment and maintenance of adequate airway and ventilation. Atropine sulfate in conjunction with pralidoxime chloride can be administered as an antidote. Atropine by intravenous injection is the primary antidote in severe cases. Test injections of atropine (1 mg in adults and 0.15 mg/kg in children) are initially administered, followed by 2–4 mg (in adults) or 0.015–0.05 mg/kg (in children) every 10–15 min until cholinergic signs (e.g., diarrhea, salivation, and bronchial secretions) decrease. High doses of atropine over several injections may be necessary for effective control of cholinergic signs. If lavage is performed, endotracheal and/or esophageal control is suggested. At first signs of pulmonary edema, the patient should be placed in an oxygen tent and treated symptomatically.

Animal Toxicity

Dichlorvos is of moderate acute toxicity, with an oral LD_{50} value in rodents from 50 to 150 mg/kg. While the LC_{50} for inhibiting mammalian brain acetylcholinesterase is similar between dichlorvos and paraoxon (i.e., the active metabolite of parathion), the acute LD_{50} values for these agents are considerably different, due in part to the more rapid metabolism and elimination of dichlorvos. Dichlorvos is considered to have essentially no carcinogenic, teratogenic, or reproductive toxicity potential.

—*Carey Pope*

Related Topics

Cholinesterase Inhibition
Neurotoxicity, Delayed
Neurotoxicology: Central and Peripheral
Organophosphate Poisoning, Intermediate
    Syndrome
Organophosphates
Pesticides

Dieldrin

- CAS: 60-57-1
- SYNONYMS: 1,2,3,4,10,10-Hexachloro-6,7-epoxy-1,4,4a,5,6,7,8,8a-octahydro-*endo*-1,4-*exo*-5,8-dimethanonaphthalene; Alvit; Octalox; Panoram; Quintox
- CHEMICAL CLASS: Synthetic organochlorine insecticide
- CHEMICAL STRUCTURE:

Uses

Dieldrin is used as an insecticide.

Exposure Pathways

The most important exposure routes for dieldrin are oral and dermal.

Toxicokinetics

Dieldrin is readily absorbed through the gastrointestinal tract via the hepatic portal vein. The toxicity of dieldrin is almost as high via the dermal route indicating extensive dermal absorption. Dieldrin is metabolized by liver microsomal enzymes to less toxic metabolites, including *cis*-aldrinol and 9-hydroxy dieldrin, and photoconverted to the more toxic 2-ketodieldrin. *cis*-Aldrinol is epimerized to *trans*-aldrindiol, which is further metabolized to aldrin diacid. Like other organochlorine insecticides, adipose tissue is the major storage tissue followed by the liver, brain, and blood. The water-soluble metabolites of dieldrin detoxification are excreted to a large extent (90%) in feces and in urine. Dieldrin is also found in mothers' milk.

Mechanism of Toxicity

Dieldrin binds to the γ-aminobutyric acid receptor and inhibits chloride ion flux. The result is similar to that described for chlordane or other cyclodiene compounds.

Human Toxicity

Dieldrin toxicity is similar to that of other cyclodiene compounds. The central nervous system (CNS) is the primary target. Convulsion is the major symptom. Patients may also experience nausea, vomiting, hyperexcitability, and coma (see Chlordane).

Clinical Management

Activated charcoal has been reported to increase the speed of excretion of stored dieldrin after oral exposure has ceased. Phenobarbital or diazepam are used when anticonvulsant therapy is necessary. In severe cases of dieldrin-induced convulsions, muscle paralysis may be necessary in addition to anticonvulsant therapy.

Animal Toxicity

In addition to CNS effects, dieldrin increases hepatocarcinogenesis. Dieldrin also has been reported to be immunotoxic suppressing macrophage function and T cell-dependent humoral immune functions.

—*Benny L. Blaylock*

Related Topics

Organochlorine Insecticides
Pesticides
Pollution, Water

Diesel Exhaust

- ◆ SYNONYMS: Diesel engine emissions; diesel exhaust particulate; diesel particulate
- ◆ DESCRIPTION: Exhaust from diesel engines contains polynuclear aromatic hydrocarbons (PAHs) adsorbed onto diesel exhaust particulates as well as sulfur dioxide, nitrogen dioxide, formaldehyde, acrolein, and sulfuric acid. Diesel engines are used to power heavy machinery, locomotives, ships, buses, heavy-duty trucks, and some light-duty trucks and passenger cars.

Exposure Pathways

Inhalation is the primary route of exposure to diesel exhaust. Incidental ingestion following deposition is also a possible route of exposure.

Toxicokinetics

Clearance of diesel particles from the alveolar region of the lung (area of gas exchange) varied from approximately 2 months in rats to almost 1 year in humans. High-exposure concentrations reduce the lung clearance in animals, further increasing the lung burden. Biological fluids are relatively ineffective in extracting organics adsorbed to diesel particle surfaces. Phagocytosis by macrophages is much more effective in extracting organics. A fraction of organics was eluted in this manner within hours with the more tightly bound fraction removed with a half-life of about 1 month. Elution rates are generally more rapid than particle clearance rates so most of the organic fraction is assumed to be bioavailable even with no clearance inhibition.

Mechanism of Toxicity

Comparison of toxic responses in animals exposed to whole diesel exhaust or filtered diesel exhaust indicates that the principal etiologic agent of noncancerous health effects in animals is diesel particulate. Most of the carcinogenicity appears to be associated with the portion containing PAHs with four to seven rings. Carcinogenic effects may be a result of the formation of covalent adducts with DNA and subsequent alteration of cellular genetic information. Another proposed mechanism is based on the carcinogenic potential of the particle itself. The particle may induce increases in DNA adducts in the lungs or induce release of mediators from macrophages, many of which are considered to act via promotion.

Human Toxicity

Evidence for potential carcinogenicity of diesel particulate/exhaust is limited. Recent studies have indicated a small, but statistically significant, increase in the risk of lung cancer in occupationally exposed workers. Symptoms of acute exposure of humans to concentrations above ambient environmental concentrations are irritation of mucous membranes, eyes, and respiratory tract. Chest tightness and wheezing may occur. Neurophysiological effects include headache, nausea, heartburn, vomiting, weakness, and tingling of extremities and light-headedness. Diesel exhaust odor may cause nausea, headaches, and loss of appetite.

Animal Toxicity

Acute studies of toxic responses associated with exposure to diesel exhaust have mainly been associated with high concentrations of carbon monoxide, nitrogen dioxide, and aliphatic aldehydes. Short-term and chronic exposure studies indicate that toxic effects are related to high concentrations of particulate matter. Minimal effects on pulmonary function have been observed in short-term testing even though histological and cytological changes were noted in the lung.

Chronic studies have been performed on rats, mice, guinea pigs, hamsters, cats, and monkeys. Changes were similar to those noted in short-term studies (accumulation of particles in the lung, increase in lung weight, increase in macrophages and leukocytes, hyperplasia of alveolar epithelium, and thickening of alveolar septa). Decreased resistance to respiratory tract infections has been noted in mice exposed to diesel exhaust. Limited animal data are available indicating alteration in liver structure and function. The lowest exposure levels resulting in impaired pulmonary function varied by species.

Diesel exhaust extracts are mutagenic. Lung tumors were induced in female mice and Fischer 344 rats; however, the dose–response relationship is unclear. Dermal, skin painting, and subcutaneous injection in mice also elicited tumorigenic responses.

—*Kathryn A. Wurzel*

Related Topics

Pollution, Air
Polycyclic Aromatic Hydrocarbons
Respiratory Tract

Diesel Fuel

♦ SYNONYMS: Diesel oil (CAS: 68334-30-5); diesel fuel No. 1; diesel fuel No. 2 (CAS: 68476-34-6); diesel fuel No. 4

♦ DESCRIPTION: Diesel oil is a complex mixture produced by the distillation of crude oil. It consists of hydrocarbons having carbon numbers predominantly in the range of C9–C20 and boiling points in the range of approximately 163–357°C (325–675°F).

Diesel fuel No. 1 is a straight-run middle distillate with a boiling range consistent with that of kerosene. It contains branched-chain alkanes (paraffins), cycloalkanes (naphthenes), aromatics, and mixed aromatic cycloalkanes. The boiling point range of diesel No. 1 largely eliminates the presence of benzene and polycyclic aromatic hydrocarbons (PAHs). Kerosene contains less than 0.02% benzene and low levels of PAHs.

Diesel fuel No. 2 is a blend of straight-run and catalytically cracked streams, including straight-run kerosene, straight-run middle distillate, hydrodesulfurized middle distillate, and light catalytically and thermally cracked distillates. The boiling range is generally approximately 160–360°C (320–680°F). Diesel fuel No. 2 is similar in composition to fuel oil No. 2. Some of the PAHs contained in fuel oil No. 2, and therefore probably present in diesel fuel No. 2, include phenanthrene, fluoranthene, pyrene, benz(*a*)anthracene, chrysene, and benzo(*a*)pyrene.

Diesel fuel No. 4 is also called marine diesel fuel. It is the most viscous of the diesel fuels and contains higher levels of ash and sulfur. Diesel fuel No. 4 may contain more than 10% PAHs.

- ◆ CHEMICAL CLASS: Petroleum hydrocarbon mixture of branched-chain alkanes, cycloalkanes, aromatic compounds, and sulfurized esters.

Uses

Diesel fuel No. 1 is primarily used in city buses. Diesel fuel No. 2 has a higher specific gravity than diesel fuel No. 1, providing more energy per unit volume of fuel, and is therefore used in railcars, trucks, and boats. Diesel fuel No. 4 is used in marine vessels.

Exposure Pathways

The most common exposure pathway is dermal exposure from handling during transfer, fueling, and repair of diesel-powered vehicles. Although the constituents of diesel are not sufficiently volatile for inhalation of vapors to be an exposure route of concern, inhalation of diesel aerosols can occur. Ingestion of diesel, often associated with aspiration into the lungs, can occur as a result of accidental poisoning or suicide attempts.

Toxicokinetics

Since diesel fuel is a mixture of numerous individual substances, absorption, metabolism, and excretion are very complicated and have not been completely characterized. Systemic effects following dermal and oral exposure and inhalation of diesel aerosols have been demonstrated, indicating that absorption can occur via all routes of exposure.

The alkanes, cycloalkanes, and aromatic compounds present in diesel are lipophilic and tend to distribute to tissues with higher adipose tissue content. The reversibility and short-term nature of many effects observed during acute exposure indicate that retention of principal diesel fuel components in body tissues is limited.

The alkanes and cycloalkanes in diesel fuel are generally not readily metabolized and are mostly excreted unchanged through the lungs, with a very small fraction excreted in the urine. The aromatic constituents of diesel are subject to oxidative metabolism and are typically excreted in the urine as water-soluble metabolites.

Mechanism of Toxicity

The mechanism of action for diesel fuels is not well characterized due to the complexity of its petroleum hydrocarbon mixture. The presence of additives that improve fuel combustion or prevent microbial growth may contribute to toxicity. Based on research conducted with individual components of diesel fuels, the primary mechanism of action for central nervous system (CNS) depression from diesel fuel is the reversible, physical interaction of the aromatic and aliphatic hydrocarbons with cell membranes. Renal toxicity is possibly attributed to oxidative metabolites of some of the aromatic constituents. Eye and skin injury are attributable to direct irritant action and the high lipid solubility that may dissolve protective skin oils and allow penetration into the skin tissue. The dermal carcinogenesis observed in rodents subjected to chronic dermal exposure to diesel may be attributed to the genotoxic activity of PAHs and the promoting activity of repeated dermal injury.

Human Toxicity: Acute

Kidney toxicity has been observed in dermally exposed individuals using diesel fuel as a skin degreaser or a shampoo. In a suicide attempt, a woman ingested 1.5 liters of diesel fuel and developed a toxic lung disease and fever, which was resolved over the next 4 months.

Human Toxicity: Chronic

Little information exists on individuals chronically exposed to diesel fuel. The predominant effect reported is skin changes from chronic dermal exposure, including cutaneous hyperkeratosis. Epidemiology studies suggest a possible association with occupational diesel fuel exposure and prostate and lung cancer; however, interpretation of the studies is complicated by the populations being concurrently exposed to other petroleum mixtures and combustion products of those mixtures.

Clinical Management

Skin exposed to diesel fuel should be washed thoroughly with soap and water to minimize local irritation and prevent further absorption. Exposed eyes should be rinsed with large quantities of water for at least 15 min.

Diesel fuel may be aspirated into the lungs following vomiting, causing aspiration pneumonia. Therefore, inducing vomiting following ingestion is not indicated unless the threat of severe renal, liver, or CNS toxicity outweighs potential development of aspiration pneumonia. Inducing vomiting may be indicated if large quantities were ingested or the fuel is suspected to contain highly toxic additives. Vomiting may be induced by administering syrup of ipecac (30 ml for adults and 5 ml for children 1–12 years of age). Activated charcoal may be considered for patients who have ingested another toxic substance.

Animal Toxicity: Acute and Subacute

The principal toxicities observed in animals acutely or subacutely exposed to diesel are dermal irritation by the dermal route and renal toxicity, liver toxicity, and CNS depression from all routes of exposure.

Animal Toxicity: Oral Exposure

Diesel fuel has been demonstrated to have a low toxicity in animals following oral exposure. The LD_{50} in rats ranges from 7.5 to approximately 9 g/kg.

Animal Toxicity: Dermal Exposure

In acute irritation tests in rabbits, diesel fuel was only mildly irritating to the eyes but severely irritating to skin. Male and female B6C3F1 mice dosed dermally with 2000–40,000 mg/kg of marine diesel fuel for 14 consecutive days demonstrated skin lesions and acanthosis, parakeratosis, hyperkeratosis, and inflammatory infiltrates of the dermis. Mice receiving >20,000 mg/kg displayed 100% mortality. No treatment-related mortality was observed in mice administered 250–4000 mg/kg marine diesel fuel by dermal application 5 days per week for 13 weeks; however, the 4000 mg/kg dose group exhibited a chronic dermatitis at the site of application. Rabbits exposed to diesel fuel No. 2 for

24 hr per day, 5 days per week, for 2 weeks at doses of 4 ml/kg exhibited no mortality and 67% mortality, respectively. All animals dying exhibited signs of chronic dermal irritation, severe anorexia, and depression as the test progressed. Primary causes of death were depression and anorexia attributed to dermal irritation with infection rather than any systemic toxicity, although some liver necrosis was noted.

Rats dermally dosed with 1000 mg/kg diesel fuel per day, 5 days per week, for 2 weeks had demonstrated weight loss, reduced liver weights, serum glucose, serum protein, and serum cholesterol, as well as a reduction in hemoglobin, hematocrit, red cell count, and blood lymphocyte counts. Marine diesel fuel produced lesions in the kidneys of C3Hf/Bd mice treated dermally with 50 μl undiluted fuel three times/week for 60 weeks. Kidney lesions were not observed in a second dermal study in which B6C3F1 mice were treated with up to 500 mg/kg marine diesel fuel diluted in acetone five times per week for 103 weeks.

Animal Toxicity: Inhalation Exposure

Rats exposed to an aerosol of diesel fuel No. 2 at 100 mg/m^3 demonstrated very mild histological changes in the liver and thyroid. No other biochemical effects, hematological effects, or tissue changes were observed in exposed animals. Continuous 90-day inhalation exposure to 50 or 300 mg/m^3 of marine diesel fuels produced hyaline droplet nephropathy and reduced body weight gain in male rats. Pregnant rats exposed to 100 and 400 ppm diesel fuel No. 2 via inhalation on Days 6–15 of gestation did not produce offspring with developmental or fetotoxic effects.

Animal Toxicity: Chronic

Chronic toxicity studies in animals have been limited to dermal exposure studies. Multiple pathological disorders of the kidneys, lungs, liver, lymph nodes, and spleen were observed in greater frequencies in chronically exposed groups; however, the studies' focus has been skin changes and skin carcinogenesis.

Carcinogenicity responses are primarily dependent on the type of diesel fuel applied. Diesel fuel No. 2 did not induce a significant increase in carcinogenesis in Swiss mice when applied at 0.05 ml three times a week for 62 weeks, even in the presence of extreme skin irritation. In another study, diesel fuel No. 2 did not produce tumors by itself; however, it did promote the development of skin tumors initiated by other chemicals. In contrast, marine diesel fuel induced a significant increase in the incidence of squamous cell papillomas and carcinomas when applied to the skin of 49 or 50 male and B6C3F1 mice at doses of 250 and 500 mg/kg, 5 days per week. The 500 mg/kg group was terminated at 84 weeks due to severe skin ulcerations, and the 250 mg/kg group was carried out for 103 weeks. The chemical composition of the marine diesel fuel tested was not completely chemically characterized but consisted of a greater percentage of aromatics and a lesser percentage of alkanes compared to diesel fuel No. 2. It has been suggested that skin carcinogenesis of diesel fuels is probably promoted by chronic irritation and hyperplasia; however, the lack of carcinogenesis in the more refined diesel fuels even in the presence of marked skin irritation indicates that high concentrations of genotoxic PAHs in marine diesel fuel may be involved in the carcinogenic mechanism.

—*Shayne C. Gad*

Related Topics

Polycyclic Aromatic Hydrocarbons

Diethyl Ether

- ◆ CAS: 60-29-7
- ◆ SYNONYMS: Ether; ethyl ether; 1,1'-oxybis-ethane; diethyl oxide; ethyl oxide
- ◆ CHEMICAL CLASS: Ether
- ◆ CHEMICAL STRUCTURE:

$$CH_3CH_2OCH_2CH_3$$

Uses

Diethyl ether has a wide variety of uses in industry, medicine, and dentistry. It is used in the production of rubber, plastics, paints, and coatings. It is also utilized in the perfume, cosmetics, and toiletries industry. It can be used to denature alcohol. It can also be used as an anesthetic.

Exposure Pathways

Routes of exposure include inhalation, ingestion, and absorption through the skin.

Toxicokinetics

Diethyl ether is immediately absorbed from inhaled air into the bloodstream and passes rapidly into the brain. More than 80% will be eliminated through the lungs and another 1 or 2% excreted in the urine. The remainder may deposit in fatty tissue. Radiotracer studies in rats have shown diethyl ether can be degraded to carbon dioxide.

Mechanism of Toxicity

Ether has the ability to dissolve fats and can penetrate membranes. It causes general anesthesia.

Human Toxicity

The target organ of ether is the central nervous system (CNS). Inhalation of high concentrations may cause CNS effects including headache, dizziness, unconsciousness, and coma. It is, however, rare to find death due to an inhalation exposure. Due to defatting of the skin, dermal contact may produce signs of toxicity and a resulting dermatitis.

Clinical Management

Contact with the skin should be minimized by thoroughly washing affected areas for at least 15 min. Symptoms of dermatitis should be treated if necessary. If ingested, do not induce vomiting since ether poses an aspiration hazard and chemical pneumonitis may result. Note: CNS depression may result from ingestion. Treatment should be symptomatic. There are no known antidotes to diethyl ether.

Animal Toxicity

The reported toxic doses for mice include the following: LC_{50} (inhalation), 31,000 ppm/30 M; LD_{50} (intraperitoneal), 2.4 g/kg; LDLo (subcutaneous), 8 mg/kg; and LD_{50} (intravenous) 996 mg/kg.

—*Kathryn Kehoe*

Diethylamine

- CAS: 109-89-7
- SYNONYMS: *n*-ethylethanamine; *n,n*-diethylamine; diethamine
- CHEMICAL CLASS: Amine
- CHEMICAL STRUCTURE: $(C_2H_5)_2NH$

Uses

Diethylamine is used in the manufacture of rubber-processing chemicals, pharmaceuticals, insect repellents, resins, flotation agents, and dyes. It is used as a corrosion inhibitor in the metal industries and is used in electroplating. It is also used as a solvent for removing impurities from oils, fats, and waxes, and is used as a polymerization inhibitor.

Exposure Pathways

Exposure may occur through oral, dermal, or inhalation routes.

Toxicokinetics

Diethylamine is rapidly absorbed, with the rate of skin absorption dependent on the size of the area involved and the duration of contact. Diethylamine is primarily excreted unchanged in the urine.

Mechanism of Toxicity

The toxic effects of diethylamine are due primarily to its corrosive action on tissues.

Human Toxicity

Diethylamine is a severe skin and eye irritant. Eye exposure to diethylamine can cause edema of the corneal epithelium, generally without pain and causing colored halos around lights. This effect generally clears within 24 hr. Intense eye exposures cause blurring, photophobia, and discomfort from the roughness of the corneal epithelium. Direct contact with skin has a corrosive effect, causing erythema and blistering. Respiratory tract irritation is expected from inhalation exposures. Ingestion of diethylamine causes severe burns to the oral tissues, with emesis, abdominal pain, and diarrhea.

Clinical Management

Exposed skin and eyes should be irrigated with copious amounts of water. After inhalation exposure, the victim should be moved to fresh air and monitored for respiratory distress. Humidified supplemental oxygen (100%) should be administered with assisted ventilation as required. If coughing or breathing difficulties are noted, the patient should be evaluated for irritation, bronchitis, or pneumonitis, including chest X-rays and determination of blood gasses. If pulmonary edema is present, positive end expiratory pressure ventilation and steroids should be considered. For ingestion exposures, the use of diluents is controversial. Emesis or lavage should be avoided. A fall in blood pressure may indicate a delayed gastric or esophageal perforation.

Animal Toxicity

Diethylamine is a primary skin irritant and is an irritant to the eyes and mucous membranes. The dermal LD_{50} in rabbits was 580 mg/kg. An inhalation LC_{50} of 4000 ppm in rats was reported for a 4-hr exposure. Rabbits exposed by inhalation to 100 ppm for 6 weeks experienced irritation of the lung tissue and cornea, moderate peribronchitis, nephritis, a slight thickening of the vascular walls, and multiple punctate erosions and edema of the cornea. Changes in the liver were also noted, including parenchymatous degeneration. Parenchymatous degeneration of the heart muscle has been observed in rabbits at these concentrations but has not been confirmed in other species. The oral LD_{50} in rats has been reported to be 540 mg/kg. Intraperitoneal injection in rats resulted in a moderate inhibitory effect with respect to liver function and monoamine oxidase activity. An LC_{50} for the fathead minnow has been reported to be 855 mg/liter for 96 hr.

—Janice M. McKee

Diethylene Glycol

- CAS: 111-46-6
- SYNONYMS: DEG; ethanol, 2,2′-oxybis; bis(2-hydroxyethyl) ether; bis(β-hydroxyethyl) ether;

dicol; diglycol; dihydroxydiethyl ether; ethylene diglycol; 3-oxapentane-1,5-diol; 2,2'-oxydiethanol; 2-(2-hydroxyethyl)ethanol

♦ CHEMICAL STRUCTURE:

$$HO—CH_2—CH_2—O—CH_2—CH_2—OH$$

Uses

Diethylene glycol is used in natural gas dehydration; as a humectant for tobacco; as a lubricating and finishing agent for textiles; as a constituent of brake fluids, lubricants, mold release agents, antifreeze formulations, and inks; as a plasticizer for cork, adhesives, paper, packaging materials, and coatings; as a solvent for printing inks and textile dyes; as an intermediate in the production of the explosive diethylene glycol dinitrate; and as an intermediate in the production of some resins, morpholine, polyurethane, triethylene glycol, surfactants, and diethylene glycol esters and ethers. Diethylene glycol is a by-product of ethylene glycol production.

Exposure Pathways

Exposure to diethylene glycol by inhalation of vapor is unlikely as the vapor pressure of diethylene glycol is very low. Depending on the use, however, mists/aerosols can be generated which would be respirable and therefore a potential route of exposure. Mists/aerosols would also be deposited on the skin. Although the oral route of exposure would not be expected during proper use of the material, accidental ingestion has been reported.

Toxicokinetics

Data from rabbits indicate that the toxicity from a single dermal application is low (the LD_{50} is 13.3 g/kg); however, toxicity did occur in rats that received daily dermal doses of 2.8 g/kg/day for 2 months. Absorption in amounts sufficient to cause toxicity can also occur as a result of ingestion or inhalation of mists or aerosols.

The blood half-lives for diethylene glycol following oral doses of 6 or 12 ml/kg were 8 and 12 hr, respectively, in rats. Oxalate (the same metabolite produced for ethylene glycol) was detected in the kidneys. In addition, oxalic acid was detected in the urine of rats administered diethylene glycol in drinking water. However, in dogs, it appears that most of the material is excreted unchanged in the urine.

Mechanism of Toxicity

In large doses, diethylene glycol is a central nervous system (CNS) depressant; and lethality, which occurs within 24 hr of a large single dose, is considered to be the result of this effect. Smaller doses, which produce acute toxicity with injury or delayed lethality, primarily affect the kidneys and the liver and are associated with renal insufficiency due to swelling of the convoluted tubules and plugging of the tubules with debris.

Human Toxicity

Relatively low hazard has been associated with industrial use of diethylene glycol due to its low vapor pressure and low dermal penetration. However, the oral lethal dose in humans has been estimated as 1 ml/kg based on a number of cases of accidental ingestion or ingestion of medicinal products that contained diethylene glycol as a solvent. Effects from ingestion include CNS depression, nausea, vomiting, headache, diarrhea, abdominal pain, polyruia, oliguria, anuria, leukocytosis, ascites, hydrothorax, hydropericardium, hemorrhages, and congestion in the stomach and intestines, distention of the leptomeningeal veins, pulmonary edema, pericardial hemorrhage, liver enlargement, enlarged kidneys, cortical necrosis of the kidneys, and hyaline casts in the collecting tubules of the kidneys.

The AIHA-recommended Workplace Environmental Exposure Limit is 50 ppm (8-hr TWA) for total vapor and aerosol and 10 mg/m³ (8-hr TWA) for aerosol only.

Clinical Management

Renal and hepatic function should be monitored and supported. If acidosis occurs, treatment should begin with 1 or 2 mEq/kg (for children use 1 mEq/kg) of sodium bicarbonate intravenously, repeated every 1 or 2 hr as needed. Hemodialysis may be necessary for severe acid/base disturbances or renal failure. Treatment with ethanol (similar to that used for ethylene glycol) has been effective in animals, but efficacy data for humans are not available.

Animal Toxicity

Diethylene glycol is not a skin irritant or an eye irritant. Acute oral toxicity ranges from approximately 9 g/kg for guinea pigs to 27 g/kg for rats. Symptoms of acute toxicity were similar for rabbits, dogs, mice, and guinea pigs and consisted of thirst, diuresis, and refusal of food, followed several days later by low urine volume, proteinuria, prostration, dyspnea, bloating, coma, low body temperature, and death. In repeated dose studies, rats given a diet containing 0.25% for 30 days were not affected. Degenerative kidney lesions occurred at dietary concentrations of 1000 ppm (1%) for 30 days. Rats administered diethylene glycol via drinking water exhibited narcosis, weight depression, and mortality at concentrations of 5–20%. Diethylene glycol was administered in the diet at concentrations of 2000 ppm (2%) and 4000 ppm (4%) for approximately 2 years. Males administered the 4000-ppm diet had a higher incidence of bladder stones, and one male had a bladder tumor. These data suggest that diethylene glycol is not a primary carcinogen; however, high dietary concentrations can result in formation of bladder stones, which can lead to development of bladder tumors due to irritation. Daily oral doses of approximately 2 g/kg for 12 weeks had no effect on reproduction.

—*Linda Angevine Malley and*
David M. Krentz

Diethylstilbestrol

♦ CAS: 56-53-1

♦ SYNONYMS: DES; 4,4′-(1,2-diethyl-1,2-ethene-diyl)bisphenol; stilbestrol; stilboestrol; 3,4-bis(*p*-hydroxyphenyl)-3-hexene; Antigestil; Bio-des; Bufon; Cyren A; Distilbene; Domestrol; Estrobene; Estrosyn; Fonatol; Grafestrol; Hi-bestrol; Makarol; Micrest; Mislestrol; Neo-oestranol I; Oestrogenine; Oestromenin; Oestromensyl; Oestromon; Palestrol; Sibol

♦ PHARMACEUTICAL CLASS: A nonsteroidal, synthetic stilbene derivative with estrogenic activity

♦ CHEMICAL STRUCTURE:

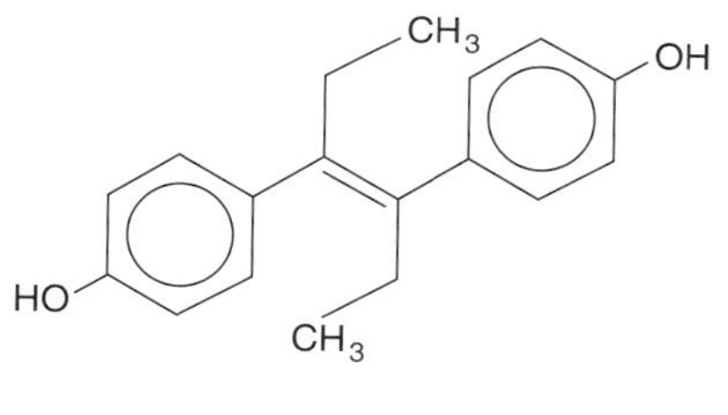

Uses

Diethylstilbestrol is an effective estrogenic agent that is used for various cancers including postmenopausal breast cancer and prostatic cancer.

Exposure Pathways

Ingestion is the most common route of both accidental and intentional exposure to diethylstilbestrol. It is available in an oral dosage form.

Toxicokinetics

Diethylstilbestrol is readily absorbed through the gastrointestinal tract. It is metabolized in the liver by oxidation and conjugation with sulfuric and glucuronic acids. A certain proportion undergoes enterohepatic circulation. The major metabolites of diethylstilbestrol are the oxides, sulfuric conjugates, and the glucuronic conjugates. Diethylstilbestrol is widely distributed throughout most body tissues with major concentrations in fat tissue. Protein binding is 50–80%. The glucuronides and sulfates of diethylstilbestrol are excreted in the urine. A portion is excreted in the bile but is mostly reabsorbed via the enterohepatic circulation.

Mechanism of Toxicity

Diethylstilbestrol stimulates estrogen receptor-containing tissue. In the 1950s and early 1960s, diethylstilbestrol was an accepted treatment for threatened miscarriages. In 1970 it was first reported that young women whose mothers had been given diethylstilbestrol during the first trimester of pregnancy had an increased incidence of vaginal dysplasia or vaginal adenocarcinoma. Approximately 25% of males exposed to diethylstilbestrol *in utero* exhibit genital lesions and low sperm counts.

Human Toxicity: Acute

Toxicity other than gastrointestinal effects is unlikely following acute ingestion.

Human Toxicity: Chronic

Purpura, edema, leg cramps, gynecomastia, porphyria cutanea tarda, and chloasma may be associated with the chronic use of diethylstilbestrol. Various cancers in premenopausal women have been shown to occur.

Clinical Management

Basic and advanced life-support measures should be utilized as necessary. Gastrointestinal decontamination procedures should be used as deemed appropriate to the patient's level of consciousness and the history of the ingestion. Activated charcoal may be used to adsorb diethylstilbestrol or concomitant ingestants.

Animal Toxicity

No known animal toxicity information is available.

—*Bonnie S. Dean*

Related Topics

Developmental Toxicology
Environmental Hormone Disrupters
Reproductive System, Female

Diflubenzuron

♦ CAS: 35367-38-5; 51026-04-1; 104790-81-0

♦ SYNONYMS: Dimilin; Difluron; 1-(4-chlorophenyl)-3-(2,6-difluorobenzoyl)urea

♦ CHEMICAL CLASS: Benzoylphenylurea

♦ CHEMICAL STRUCTURE:

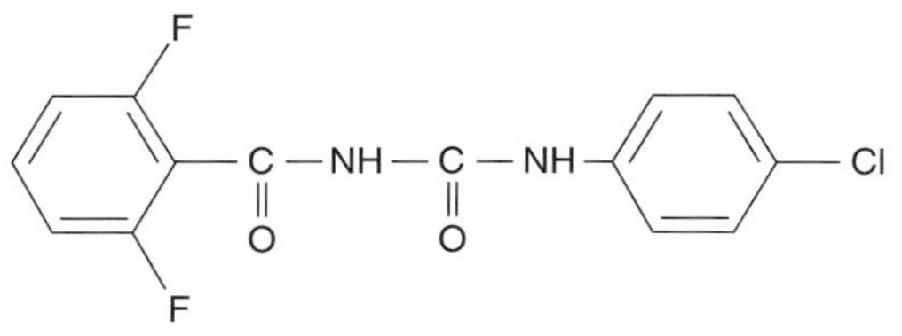

Uses

Diflubenzuron is used as an insecticide, larvicide, and insect growth regulator.

Exposure Pathways

Ingestion is route for intentional or accidental exposure. Inhalation is also a possible route of exposure.

Toxicokinetics

No literature is available on toxicokinetics in mammals.

Mechanism of Toxicity

Diflubenzuron inhibits the enzyme chitin synthase, which is required in the final step of chitin synthesis. Chitin is a polysaccharide and is a major constituent of the exoskeleton of insects. In insects, the trachea is held open by rings of chitin. The exoskeleton and the waxy covering also prevent water loss. Inhibiting chitin synthesis therefore can provide an effective means of pest control. Moreover, vertebrates and most plants do not utilize chitin thus making diflubenzuron a target-selective pesticide.

Human Toxicity

No study has reported acute or chronic toxicity of diflubenzuron in humans.

Clinical Management

Diflubenzuron has very low systemic side effects if absorbed through the skin. The exposed area should be thoroughly washed with soap and water. Eyes should be washed with copious amounts of room-temperature water for 15 min in cases of eye contamination. If small amounts are ingested, no treatment is needed. Low toxicity is seen in nontargeted species. Symptomatic treatment is recommended.

Animal Toxicity

The oral LD_{50} in rats and mice is >4 g/kg. The dermal LD_{50} in rabbits is about 2 g/kg.

—*Sushmita M. Chanda*

Related Topic

Pesticides

Difluoroethylene, 1,1-

♦ CAS: 75-38-7

♦ SYNONYMS: 1,1-Difluoroethene; NCI-C60208; vinylidene fluoride

- CHEMICAL CLASS: Flammable gas, strong oxidizing agent
- MOLECULAR FORMULA: $C_2H_2F_2$
- CHEMICAL STRUCTURE:

$$CH_2 = CF_2$$

Uses

Difluoroethylene is used in the manufacture of polyvinylidene fluoride, which is used as a thermal, chemical, and ultraviolet light-resistant agent, and as an anticorrosive agent. The monofilament form is used as filter cloth in the pulp and paper industry. It is used as an insulator due to its high melting temperature. Elastomeric copolymers are used for their heat- and moisture-resistant properties, primarily in industrial, aerospace, and automotive applications.

Exposure Pathways

The primary exposure route is inhalation.

Toxicokinetics

Absorption is very rapid after inhalation and reaches a steady state within minutes of exposure and blood levels decline rapidly at the end of exposure. Biotransformation is very slow; difluoroethylene may produce alkylating intermediate and some acetone. The tissue/air partition coefficients were determined to be 0.07, 0.18, 0.8, 1.0, and 0.29 for water, blood, liver, fat, and muscle, respectively. Difluoroethylene is eliminated as fluoride ions in urine.

Mechanism of Toxicity

Difluoroethylene may interact with the hepatic microsomal monooxygenase to form epoxide. It inhibits microsomal mixed function oxidase *in vitro*.

Human Toxicity

No data are available for occupational exposure. Acute exposure causes nausea, dizziness, and headache. Difluoroethylene is harmful if swallowed, inhaled, or absorbed through skin. It may cause heritable genetic damage and may cause tumors of sense organs, skin, and appendages. No data are available for the evaluation of carcinogenicity.

In occupational settings, technical measures should prevent any contact with the skin and mucous membranes. Workers potentially exposed to this compound should wear personal protective equipment and their work should be carried out only in restricted and ventilated areas. Clothing and equipment after use should be placed in an impervious container for decontamination or disposal. Difluoroethylene is listed as a hazardous air pollutant.

Clinical Management

In case of dermal or ocular contact, the eyes and skin should be flushed with water for 15–20 min. For inhalation exposure, the patient should be moved to fresh air. If necessary, oxygen and artificial respiration should be administered. If the patient is in cardiac arrest, cardiopulmonary resuscitation should be administered. Life-support measures should be continued until medical assistance has arrived. Do not administer liquids to or induce vomiting in an unconscious or convulsing person.

Animal Toxicity

Difluoroethylene is not acutely hepatotoxic at dose levels up to 82,000 ppm by inhalation for 3.5 hr in normal rats, whether fed or fasted. It is carcinogenic in long-term bioassays in Sprague–Dawley rats by oral administration. Evidence for human carcinogenicity is inadequate.

—*Kashyap N. Thakore and Harihara M. Mehendale*

Related Topic

Pollution, Air

Digitalis Glycosides

- REPRESENTATIVE COMPOUNDS: Digoxin; digitoxin
- SYNONYMS: Lanoxin; Lanoxicaps (CAS: 20830-75-5); Crystodigin (CAS: 71-63-6); foxglove (see Foxglove)

◆ PHARMACEUTICAL CLASS: Digoxin is a cardiac glycoside congener of digitalis with a hydroxyl group at the C12 position. Digitoxin is a cardiac glycoside congener of digitalis that does not contain a C12 hydroxyl group.

◆ CHEMICAL STRUCTURE:

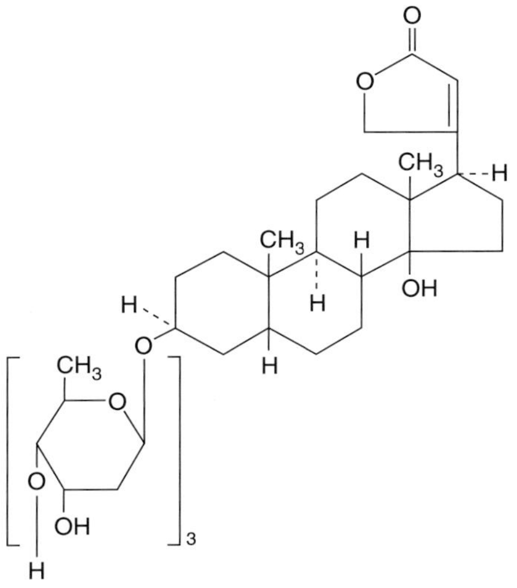

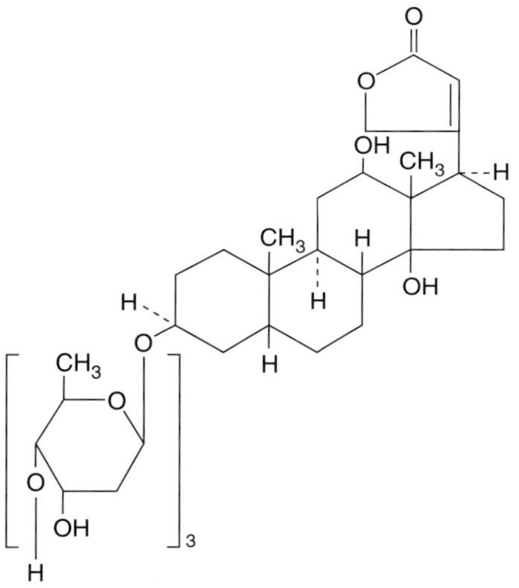

Uses

Digitalis glycosides are positive inotropic agents in congestive heart failure. They control ventricular rate in supraventricular arrhythmias including atrial fibrillation and atrial flutter.

Exposure Pathways

Ingestion is the most common exposure pathway following accidental and intentional ingestions. Digoxin is also available for parenteral administration and parenteral toxic exposures can occur.

Toxicokinetics: Digoxin

Oral administration of digoxin tablets and liquid results in 60–85% absorption from the small intestine. Liquid-filled digoxin capsules are 90–100% absorbed. The presence of food or other medications may delay oral absorption. Approximately 80% of digoxin is absorbed following intramuscular administration. Minimal metabolism occurs. Cleavage of the sugar moieties occurs in the liver and via bacteria in the large intestine. Protein binding is 20–30%. Volume of distribution approximates 4 liters/kg in adults. Digoxin is excreted in the urine primarily as unchanged drug. In healthy patients, the half-life ranges from 34 to 44 hr. The half-life can be prolonged in renal failure.

Toxicokinetics: Digitoxin

Oral absorption of digitoxin is rapid and complete. It is extensively metabolized in the liver to several active metabolites including digoxin. Protein binding is 97%. The volume of distribution approximates 0.47–0.76 liters/kg. Renal, biliary, and fecal elimination occur. The half-life can range from 4 to 14 days.

Mechanism of Toxicity

The digitalis glycosides interfere with the Na^+,K^+-ATPase pump with a resultant intracellular loss of potassium and intracellular increases in sodium and calcium. The net effects of this are increased myocardial contractility and decreased cardiac conduction.

Human Toxicity: Acute

Nausea and vomiting are frequently seen. Changes in level of consciousness may be observed. Rhythm disturbances are the most common sign of toxicity. The most common arrhythmias include bradycardia, heart block, and paroxysmal atrial tachycardia. Premature ventricular contractions and ventricular tachycardia are less common. Severe hyperkalemia can also occur following acute ingestion of a digitalis glycoside. Serum potassiums as high as 13.5 mEq/liter have been reported after digitalis ingestion. Digoxin serum concentrations can be extremely high immediately following an acute ingestion. Normal digoxin serum concentration is 0.5–2 ng/ml. Normal digitoxin serum concentrations are

18–22 ng/ml. These may decrease over 8–12 hr as distribution of the drug occurs.

Human Toxicity: Chronic

Anorexia, nausea, vomiting, and diarrhea occur after chronic exposure. Decreases in level of consciousness and delirium are observed. Visual changes including color changes and snowy vision have been described frequently. Common arrhythmias that occur during chronic toxicity include premature ventricular contractions, ventricular tachycardia, and ventricular fibrillation. Hypokalemia is often present in chronic toxicity and actually precipitates toxicity. Serum digoxin/digitoxin concentrations will be elevated but not as high as they are in acute toxicity

Clinical Management

Basic and advanced life-support measures should be utilized as necessary. A baseline 12-lead electrocardiogram should be obtained and continuous cardiac monitoring should be utilized. A digoxin/digitoxin serum concentration should be obtained as well as a serum potassium. Gastrointestinal decontamination procedures should be used as deemed necessary based on the patient's level of consciousness and history of ingestion. Syrup of ipecac-induced emesis should be avoided due to potential changes in level of consciousness and ipecac's ability to cause vagal stimulation. Multiple doses of activated charcoal can be used to enhance enterohepatic clearance of digitoxin. In chronic toxicity, hypokalemia should be treated cautiously with potassium replacement since rapid increases in serum potassium can exacerbate conduction disturbances. Ventricular arrhythmias can be treated with phenytoin or lidocaine. Overdrive pacing should also be considered. Class IA antiarrhythmics such as quinidine should be avoided since they can cause conduction disturbances. Conduction disturbances should be managed with a transvenous pacemaker. In patients with severe dysrhythmias, serum potassium concentrations >5 mEq/liter, and elevated digoxin/digitoxin serum concentrations, digoxin immune Fab (Digibind, Burroughs-Wellcome) should be considered. This sheep antibody, which binds digitalis glycosides, is effective in reversing both acute and chronic toxicities. Each 40-mg vial binds 0.6 mg of digoxin/digitoxin. Dosage should be based on serum concentration of digoxin/digitoxin or amount ingested. If these are unavailable, 10 vials can be administered.

Digibind can be administered over 30 min or it can be administered intravenous push in cardiac arrest. Since it will pull digoxin/digitoxin out of tissue sites, serum concentrations of digoxin/digitoxin will rise. These will represent bound digitalis and should not be reacted to clinically. Adverse reactions to Digibind include exacerbation of heart failure and atrial arrhythmias as well as hypokalemia. Severe allergic reactions have not been reported. The antigen/antibody complex should be eliminated within 5 days of administration. This may be delayed beyond 7 days in patients with renal failure. Hyperkalemia should not be treated with agents such as sodium polystyrene sulfonate since severe hypokalemia, due to an intracellular potassium shift, may occur following administration of Digibind. Digibind should not be used unless it is warranted since each vial costs several hundred dollars.

—*Daniel J. Cobaugh*

Related Topic

Cardiovascular System

Dimethoate

- ◆ CAS: 60-51-5
- ◆ SYNONYMS: *O,O'*-Dimethyl *S*-[2-(methylamino)-2-oxoethyl] phosphorodithioate; phosphamide; Cygon; De-fend; Rogor; Dimetate
- ◆ CHEMICAL CLASS: Synthetic organophosphorus pesticide of the phosphorothionate class
- ◆ CHEMICAL STRUCTURE:

$$(CH_3O)_2 \overset{S}{\overset{\|}{P}} - S - CH_2 - \overset{O}{\overset{\|}{C}} - \underset{\underset{H}{|}}{N} - CH_3$$

Uses

Dimethoate is a systemic and contact insecticide-acaricide used on a range of insects including mites, flies, aphids, and planthoppers. Formulations include aerosols, dusts, granules, and emulsifiable concentrates.

Exposure Pathways
Dimethoate can be absorbed by oral, dermal, or inhalation routes.

Toxicokinetics
Dimethoate is rapidly absorbed after any route of administration. It is rapidly metabolized in the liver. Like other phosphorothionate pesticides, the parent compound is activated by mixed function oxidase to the active metabolite, dimethoxon. A major route of detoxification of the parent compound is hydrolysis of the C–N bond. Pretreatment with phenobarbital increases sensitivity to dimethoate in mice. In male rats, about 60–80% of an orally administered dose of dimethoate is eliminated via the kidneys within 24 hr of exposure. Elimination was almost complete within 48 hr of exposure. Female rats appear to eliminate dimethoate at a slower rate.

Mechanism of Toxicity
Dimethoate exerts toxicity through inhibition of acetylcholinesterase (see Cholinesterase Inhibition). The oxidative metabolite (i.e., dimethoxon) is two or three orders of magnitude more potent in inhibiting acetylcholinesterase than the parent compound. The N-demethylated dimethoxon may be the most potent inhibitor of cholinesterases. The enzyme in red blood cells may be more sensitive to inhibition than plasma enzyme following dimethoate exposure.

Human Toxicity
Characteristic signs of acetylcholinesterase inhibition (e.g., diarrhea, nausea, and abdominal cramps) have been reported following dimethoate exposure (see Cholinesterase Inhibition). High exposures to dimethoate may be associated with a relapse, where the patient stabilizes and then suddenly gets much worse. Dimethoate does not cause delayed neurotoxicity but has been associated with the intermediate syndrome of organophosphate poisoning (see Organophosphate Poisoning, Intermediate Syndrome).

Clinical Management
For exposure to eyes, eyelids should be held open and the eyes flushed with copious amounts of water for 15 min. For exposure to skin, affected areas should be washed immediately with soap and water. The victim should receive medical attention if irritation develops and persists.

For exposure through inhalation, the victim should be removed to fresh air and, if not breathing, given artificial ventilation. The victim should receive medical attention as soon as possible.

First aid for ingestion victims would be to induce vomiting, keeping in mind the possibility of aspiration of solvents. Gastric decontamination should be performed within 30 min of ingestion to be most effective. Initial management of acute toxicity is establishment and maintenance of adequate airway and ventilation. Atropine sulfate in conjunction with pralidoxime chloride can be administered as an antidote. Atropine by intravenous injection is the primary antidote in severe cases. Test injections of atropine (1 mg in adults and 0.15 mg/kg in children) are initially administered, followed by 2–4 mg (in adults) or 0.015–0.05 mg/kg (in children) every 10–15 mi until cholinergic signs (e.g., diarrhea, salivation, and bronchial secretions) decrease. High doses of atropine over several injections may be necessary for effective control of cholinergic signs. If lavage is performed, endotracheal and/or esophageal control is suggested. At first signs of pulmonary edema, the patient should be placed in an oxygen tent and treated symptomatically.

After effective intervention, patients should be closely monitored for the possibility of sudden relapse.

Animal Toxicity
The acute oral LD_{50} for pure dimethoate in rodents is >500 mg/kg, but reported values using technical products range from 28 to 400 mg/kg. Early formulations contained the solvent methyl Cellosolve, which appears to have participated in chemical changes upon storage that increased mammalian toxicity. In studies comparing dermal and oral exposures, the dermal LD_{50} values were generally reported to be about twice as high. In a reproductive toxicity test, dimethoate exposure in the drinking water (about 10 mg/kg/day) was associated with >60% plasma cholinesterase inhibition in adult mice and altered pup survival and growth but was without teratogenic effects. Teratogenic effects (e.g., fused sternebrae) were reported in rats receiving 12 mg/kg/day dimethoate but were absent in animals treated with lower doses (i.e., 3 and 6 mg/kg/day).

—*Carey Pope*

Related Topics

Cholinesterase Inhibition
Neurotoxicology: Central and Peripheral
Organophosphate Poisoning, Intermediate
    Syndrome
Organophosphates

Dimethyl Sulfoxide

♦ CAS: 67-68-5

♦ SYNONYMS: Methyl sulfoxide, DMSO

♦ CHEMICAL CLASS: Sulfoxides

♦ MOLECULAR FORMULA: C_2H_6OS

♦ CHEMICAL STRUCTURE:

$$H_3C-S(=O)-CH_3$$

Uses

Dimethyl sulfoxide (DMSO) has excellent solvent properties and is, therefore, used in the topical administration of drugs, the production of synthetic fibers, the application of pesticides, and the manufacturing of industrial cleaners and paint strippers.

Exposure Pathways

Exposure via dermal absorption is common due to the use of DMSO in the drug industry. In industrial applications, exposure may occur following dermal contact, inhalation, and eye contact. Oral exposure is a less likely route.

Toxicokinetics

DMSO is readily absorbed by animals and humans by the dermal and oral routes. Higher concentrations of DMSO are more readily absorbed than more dilute solutions of DMSO in water. After dermal application, radiolabeled DMSO has been detected in blood within 5 min along with garlic halitosis resulting from the reduction metabolite, dimethyl sulfide. DMSO, due to its chemical characteristics, facilitates the absorption of many other substances through biological membranes.

The major metabolites of DMSO in humans are dimethyl sulfone and the reduction metabolite, dimethyl sulfide. Following oral administration, about two-thirds of the dose is excreted in urine as unchanged DMSO, about 20% as dimethyl sulfone, and <5% is exhaled as dimethyl sulfide. These three metabolites have also been identified in monkeys and rats. DMSO is reduced *in vitro* to dimethyl sulfide in the presence of glutathione and cysteine.

Following absorption in humans, DMSO is detected in the blood within minutes and reaches maximum levels in 4 hr. Distribution to other organs has been reported to occur within 20 min. Radiolabeled DMSO was detected in bones and teeth of animals within 1 hr. In the eyes, the highest levels appear to accumulate in the cornea; the lowest in the lens. Most absorbed DMSO is excreted in the urine unchanged or as dimethyl sulfone, the major metabolite. Studies in nonhuman primates indicate that the dimethyl sulfone metabolite appears in the blood within 2 hr and is excreted more slowly than DMSO. DMSO has a calculated half-life of 16 hr in blood; the corresponding value for dimethyl sulfone is 38 hr. About 3% of oral doses are exhaled as dimethyl sulfide in humans; the percentage appears to be slightly higher in animals.

Mechanism of Toxicity

Most physiological properties of DMSO appear to be related to its penetration properties, its potential to inhibit or stimulate enzymes *in vivo* and *in vitro*, its ability to act as a free radical scavenger, and its ability to cause histamine release from mast cells. These properties are largely based on DMSO's chemical characteristics, including its hydrogen bonding behavior, water affinity, ability to interchange with water in membranes, and ability to react with organic molecules.

Human Toxicity

DMSO is an irritant of the eyes, skin, and respiratory system. Absorption rapidly results in a garlic-like taste and odor. Overexposure may result in urticaria, headache, lethargy, nausea, and dizziness. In a few cases, eosinophilia has been reported following intravenous administration of DMSO. DMSO facilitates the penetration of other chemicals through membranes.

Clinical Management

Eye exposure should be followed by irrigation with water for at least 15 min; exposed skin should be washed thoroughly with soap and water. Resulting burns or skin irritation should be treated with standard therapy. Cases of dermal sensitization reactions may require topical antiinflammatory agents. If DMSO is swallowed, vomiting should not be induced. Charcoal in water or with a cathartic should be administered to prevent absorption. Liver and kidney function and blood parameters should be monitored.

Animal Toxicity

The acute toxicity of DMSO is generally quite low in animals. Rat oral LD_{50} values range from 14.5 to 28 g/kg and dermal LD_{50} values range up to 40 g/kg. Similar corresponding values are reported for mice. Acute toxicity in nonrodents is also low. Prolonged eye contact causes corneal injury and dermal application results in irritation and urticaria. DMSO has been reported to cause adverse reproductive effects in animals.

—*Daniel Steinmetz*

Related Topic

Sensory Organs

Dinitroanilines

♦ REPRESENTATIVE COMPOUNDS: There are four isomeric forms of dinitroaniline: 2,3-dinitroaniline (CAS No. 602-03-9); 2,4-dinitroaniline (CAS No. 97-02-9); 2,6-dinitroaniline (CAS No. 606-22-4); and 3,5-dinitroaniline (CAS No. 618-87-1).

♦ SYNONYMS:

2,3-Dinitroaniline—2,3-dinitrobenzenamine; 2,3-dinitrophenylamine

2,4-Dinitroaniline—2,4-dinitrobenzenamine; 2,4-dinitrophenylamine

2,6-Dinitroaniline—2,6-dinitrobenzenamine; 2,6-dinitrophenylamine

3,5-Dinitroaniline—3,5-dinitrobenzenamine; 3,5-dinitrophenylamine

♦ CHEMICAL CLASS: Aromatic amine

♦ CHEMICAL STRUCTURE: The structure of 2,4-dinitroaniline, the most common dinitroaniline, is shown below.

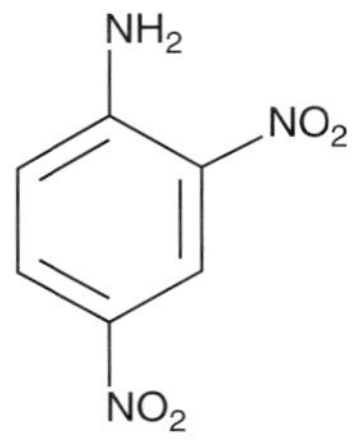

Uses

2,4-Dinitroaniline is used in the production of azo dyes.

Exposure Pathways

Exposure to dinitroanilines is most likely to occur in occupational settings. Inhalation and dermal exposure are the primary routes of exposure.

Toxicokinetics

Most of the information on the toxicokinetics of dinitroanilines pertains to 2,4-dinitroaniline. Dinitroanilines are highly toxic to humans and are well absorbed from all routes of exposure. Nine metabolites were detected in rats administered up to 90 μmol [^{14}C]2,4-dinitroaniline/kg orally or 10 μmol/kg intravenously. 2,4-Dinitrophenylhydroxylamine was the main metabolite and was excreted in the urine as the sulfate conjugate and in bile as the glucuronide. Amine hydroxylation and sulfation of 2,4-dinitroaniline are probable detoxification processes that occur rapidly and facilitate clearance.

In rats administered 0–90 μmol [^{14}C]2,4-dinitroaniline/kg orally or 10 μmol/kg intravenously, there was rapid distribution of the compound to all major tissues. Muscle, skin, and adipose tissue contained 65–70% of the ^{14}C activity in the body during the 45 min after dosing. Approximately 70–85% of the aforementioned doses were cleared from most tissues within 6 hr after administration. Three days after administration, only residual levels were detected in the major tissues. Urinary excretion of ^{14}C activity at 6 and 24 hr after dosing accounted for 30 and 63%, respectively, of the administered dose. Fecal excretion over 3 days accounted for 23% of the dose. Elimination of 2,4-dinitroaniline-derived ^{14}C activity in the bile amounted to 12.5% of the dose after 5 hr.

Mechanism of Toxicity

Much of the toxicity associated with dinitroaniline exposure is the result of methemoglobin formation in which the iron of the hemoglobin molecule is oxidized causing a deficiency in the oxygen carrying capacity of the blood. This produces the cyanosis and other signs of dinitroaniline-induced toxicity.

Human Toxicity: Acute

Little definitive information is available regarding the toxic effects of dinitroanilines. Aniline, a structurally similar compound, is a skin and eye irritant and a mild dermal sensitizer. It is rapidly absorbed by all routes of exposure and induces methemoglobinemia. Signs and symptoms of methemoglobinemia include blue skin, headache, dizziness, weakness, lethargy, loss of coordination, coma, and death. Headache and confusion occur early following poisoning, and restlessness, seizures, and coma may occur following severe poisoning. Acute exposure to 3–5 mg/kg is associated with signs and symptoms of toxicity which develop within a few hours following exposure. Liver and kidney damage may ensue within 12–72 hr postexposure and are probably secondary, hemolysis-mediated effects. As little as 1 g of aniline has caused human fatalities. The mean lethal dose for humans of the structurally related aniline has been estimated to be in the range of 15–30 g.

Human Toxicity: Chronic

Human data on the effects of chronic exposure to dinitroanilines are lacking. An increase in the severity of damage to the organs affected by acute exposure would be expected. Additionally, the adverse health effects resulting from prolonged methemoglobinemia are likely to be significant.

There are currently no regulatory or health-based guidance values for dinitroanilines.

Clinical Management

For inhalation exposures, the victim should be removed from the exposure environment and 100% humidified supplemental oxygen should be administered with assisted ventilation as required. Exposed skin and eyes should be copiously flushed with water and thoroughly decontaminated to prevent further absorption. For oral exposure, clinical management should focus on decreasing absorption. Emesis may be indicated in recent substantial ingestion unless the patient is or could rapidly become comatose or convulsive. Emesis is most effective if initiated within 30 min. Gastric lavage may be indicated if performed soon after ingestion or in patients who are comatose or at risk of convulsing. If the patient is cyanotic and symptomatic, or the methemoglobin level is greater than 30% in an asymptomatic patient, measures should be taken to correct the methemoglobinemia.

Animal Toxicity

The oral LD_{50} values for laboratory species range from 418 mg/kg (rat) to 1050 mg/kg (guinea pig). A 4-hr inhalation LClo of 17 mg/m^3 is reported for the laboratory rat. No signs of toxicity were observed in male Fischer 344 rats administered up to 90 μmol [^{14}C]2,4-dinitroaniline/kg orally or 10 μmol/kg intravenously. Animal studies have also shown varying effects on thyroid function.

Further Reading

Grayson, M., and Eckroth, D. (Eds.) (1978). *Kirk–Othmer Encyclopedia of Chemical Technology*. Wiley, New York.

Sax, N. I., and Lewis, R. J., Sr. (Eds.) (1989). *Dangerous Properties of Industrial Materials*, 7th ed. Van Nostrand Rienhold, New York.

—Robert A. Young

Dinitrophenols

♦ REPRESENTATIVE COMPOUNDS: Dinitrophenol (DNP) occurs in six different isomers—2,3-DNP,

2,4-DNP, 2,5-DNP, 2,6-DNP, 3,4-DNP, and 3,5-DNP.

◆ SYNONYMS: A number of substituted 2,4-DNPs are sold under different trade names; its analogs include DNOC, 2,4-dinitro-6-methylphenol; Binapacryl, 2-sec-butyl-4,6-dinitrophenol-3-methylcrotonate; Dinocap, 2,4-dinitro-6-(1-methy-*n*-heptyl)-phenylcrotonate, and Dinoseb, 2,4-dinitro-6-sec-butylphenol (see Dinoseb).

◆ CHEMICAL STRUCTURE:

Uses

Dinitrophenols are used as fungicides, herbicides, or insecticides. The fungicidal, herbicidal, or insecticidal properties depend on minor differences in the chemical structures of the different dinitrophenol compounds. Several dinitrophenol compounds have more than one pesticidal use.

Exposure Pathways

Dinitrophenol compounds can enter the body through inhalation, oral, or dermal routes of exposure.

Toxicokinetics

Dinitrophenols are rapidly absorbed from the gastrointestinal tract, respiratory tract, and intact skin. They can bind to plasma proteins. After absorption, they are transported through the blood to different organs and distributed in the liver, the kidneys, and the eyes.

Dinitrophenols undergo reduction in the presence of NADPH and nitroreductase and conjugation takes place at the phenolic site. Humans can slowly detoxify 2,4-DNP to 2-amino-4-nitrophenol, 2-nitro-4-aminophenol, and 2,4-diaminophenol and their glucoronic acid conjugates. The metabolism of DNP is temperature dependent (i.e., DNP metabolism is greatly diminished at low temperatures). In mice, a reduced LD_{50} and increased toxicity for dinitrophenols were observed with an increase in ambient temperature.

In humans, dogs, and rats, 2-amino-4-nitrophenol was found to be the major excretory product. Humans can slowly eliminate both the unchanged compound and the previously mentioned metabolites. Hepatic excretion is considered to be the main route of elimination of dinitrophenols. The half-life in the serum of a severely poisoned farmer was calculated to be 13.5 days. The residence half-life in humans is estimated to be 5–14 days. The elimination half-life for dinitrophenols in mice was about 6 hr.

Mechanism of Toxicity

Dinitrophenols act as uncouplers of oxidative phosphorylation. Oxygen consumption, body temperature, breathing rate, and heart rate were found to be increased following exposure to toxic levels of dinitrophenols. This is probably due to increased oxidative processes after exposure to dinitrophenols. However, the increase in oxygen consumption ability was not found after 2,5-DNP exposure. It appears that dinitrophenols interfere with the intracellular synthesis of high-energy phosphate esters. The permeability of mitochondrial membranes to hydrogen ions was found to be increased with the failure of conversion of adenosine diphosphate to adenosine triphosphate. The energy produced due to oxidation is not utilized for the synthesis of phosphates but elevates body temperature, which can lead to fatal hyperpyrexia. Inefficient circulation and respiration cannot meet the body's increased metabolic demand, resulting in anoxia and acidosis. Fat serves as an alternative fuel for metabolism. Weight loss occurs as a result of inhibition of lipogenesis from pyruvate and lactate following exposure to dinitrophenols.

Human Toxicity

Dinitrophenols are extremely toxic to humans and are well absorbed from all routes of exposure. Fatal cases of poisonings have been reported as a result of dermal exposure to dinitrophenols. Fever is a very early sign of dinitrophenol toxicity. Hepatic and renal damage were reported within 12–72 hr following exposure to dinitrophenols. Typical signs of dinitrophenol toxicity were reported to occur within a few hours following acute exposure to 3–5 mg/kg of dinitrophenol. Acute signs of toxicity include elevation of blood pressure, heart rate, and body temperature; headache; and mental confusion. Severe poisoning may cause restlessness, seizures, and coma. Cerebral edema was reported in two cases of fatal poisoning. Typical gastrointestinal symptoms may include nausea, vomiting, and abdominal cramps. Some or all of these symptoms were exhibited following repeated oral exposures to as little as 1 mg/kg/day of dinitrophenol. Repeated low-level exposures (2 mg/kg/day) can cause peripheral nerve damage. Dinitrophenols have been reported to cause cataracts (after repeated exposure).

Clinical Management

Only symptomatic treatment is available. Adequate measures should be taken to maintain fluid and electrolyte balance and keep the body temperature within tolerable limits. Measures should be taken to remove the poison from the body through gastric lavage and saline cathartic. Gastrointestinal absorption may be prevented by administering activated charcoal. Antipyretic drugs (salicylates) are not useful (they may actually potentiate the symptoms) because poisoning involves peripheral mechanisms. Therefore, control of temperature should be restricted to physical measures.

Animal Toxicity

The effects of dinitrophenol exposure in animals are very similar to those seen in humans. The oral LD_{50}s in rats, mice, guinea pigs, and dogs were reported to be 30, 20–40, 65, and 30 mg/kg, respectively.

—*Tamal Kumar Chakraborti*

Related Topic

Pesticides

Dinitrotoluene

- CAS: 25321-14-6
- REPRESENTATIVE COMPOUNDS: Dinitrotoluene (DNT) occurs in six isomeric forms: 2,3-DNT (CAS No. 602-01-7); 2,4-DNT (CAS No. 121-14-2); 2,5-DNT (CAS No. 619-15-8); 2,6-DNT (CAS No. 606-20-2); 3,4-DNT (CAS No. 610-39-9); and 3,5-DNT (CAS No. 618-85-9).

- SYNONYMS:

  [L1]2,3-DNT—1-methyl-2,3-dinitrotoluol; 2,3-dinitrotoluol

  2,4-DNT—1-methyl-2,4-dinitrotoluol; 2,4-dinitrotoluol

  2,5-DNT—1-methyl-2,5-dinitrotoluol; 2,5-dinitrotoluol

  2,6-DNT—2-methyl-1,3-dinitrotoluol; 2,6-dinitrotoluol

  3,4-DNT—1-methyl-3,4-dinitrotoluene; 3,4-dinitrotoluol

  3,5-DNT—1-methyl-3,5-dinitrotoluene; 3,5-dinitrotoluol

- CHEMICAL CLASS: Aromatic hydrocarbon
- CHEMICAL STRUCTURE: The chemical structures of the most prevalent and toxicologically important dinitrotoluenes, 2,4-DNT and 2,6-DNT, are shown below.

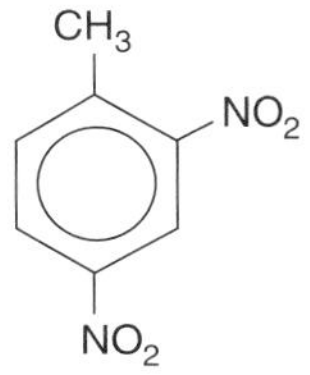

2,4-dinitrotoluene

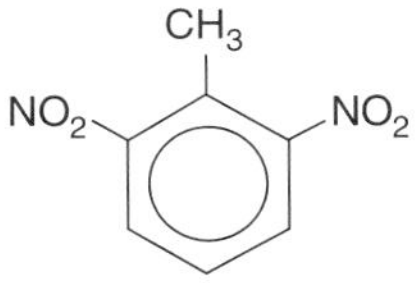

2,6-dinitrotoluene

Uses

Dinitrotoluenes are intermediates in the production of toluene diisocyanate but are also used as gelatinizing and waterproofing agents in commercial and military explosives and in the production of polyurethane foams.

Exposure Pathways

Dinitrotoluenes may occur as a contaminant of soil, surface water, and groundwater. Because DNTs are of low volatility, exposure via the air is inconsequential. The primary route of exposure to DNTs is through contaminated groundwater and, due to their mobility, in surface water as well.

Toxicokinetics

Most of the toxicokinetic data for DNTs are for the 2,4-and 2,6-isomers. Data regarding the absorption of DNT following inhalation exposure are not available, but absorption may be inferred from data on urinary metabolites in workers exposed via inhalation. Efficient absorption of various DNT isomers following oral exposure has been verified in several animal species. In animals, ingested DNT appears to be readily absorbed (55–90%) within 24 hr. Limited human data suggest that dermal exposure may result in significant absorption.

Urine from workers exposed to dinitrotoluene contained 2,4- and 2,6-DNT, 2,4- and 2,6-dinitrobenzoic acid, 2,4- and 2,6-dinitrobenzyl glucuronide, 2-amino-4-nitrobenzoic acid, and N-(acetyl)amino-4-nitrobenzoic acid. The most prevalent metabolites were 2,4-dinitrobenzoic acid and 2-amino-4-nitrobenzoic acid, collectively accounting for 74–86% of the dinitrotoluene metabolites detected. Bioactivation of dinitrotoluene in the rat is thought to occur by oxidation of the methyl group to an alcohol by a cytochrome P450-dependent pathway. The benzyl alcohol is then conjugated with glucuronic acid and excreted in the bile. Intestinal microflora hydrolyze the glucuronide and reduce one nitro group, forming an aminonitrobenzyl alcohol which can be reabsorbed from the intestine. The amino group is oxidized to an hydroxylamine by hepatic enzymes and conjugated with sulfate. Decomposition of the sulfate ester yields a highly electrophilic nitrenium (or carbonium) ion which can react with DNA and other biological nucleophiles.

Data regarding the distribution of DNT are limited to 2,4-DNT studies in animals. Following oral administration of 2,4-DNT to various laboratory species, the greatest concentrations of the chemical occurred in the liver, kidneys, and blood. Only small amounts were found in the brain, heart, and spleen. A biphasic increase in hepatic levels of 2,4-DNT in rats suggested that the chemical undergoes enterohepatic circulation.

In workers, urinary excretion of these metabolites peaked near the end of the work shift but declined to low or undetectable concentrations by the start of work the following day. The calculated elimination half-lives of total dinitrotoluene-related material detected in urine ranged from 1.0 to 2.7 hr and those of individual metabolites from 0.8 to 4.5 hr. Urinary excretion in Fischer 244 rats given 2,6-DNT accounted for half of the dose (10 mg/kg) 72 hr after administration of [^{14}C]-2,6-DNT. 2,6-Dinitrobenzoic acid, 2,6-dinitrobenzyl alcohol glucuronide, and 2-amino-6-nitrobenzoic acid accounted for 95% of the urinary ^{14}C. Fecal excretion accounted for one-fifth of the dose in 72 hr.

Mechanism of Toxicity

The most prominent toxicologic effect of DNT is the formation of methemoglobin and the subsequent effects of reduced oxygen carrying capacity of the blood which produce the cyanosis and fatigue characteristic of DNT poisoning. DNT and/or its metabolites produce this effect by oxidizing the iron in the hemoglobin molecule. This process also leads to the formation of Heinz bodies, granule-like aggregates of precipitated hemoglobin, that serve as sensitive indicators of toxic insult to the blood. Hepatotoxic effects are due, in part, to cellular damage resulting in altered hepatocytes and deficiencies in biliary excretion. DNT has also been shown to disrupt Sertoli cell function which may explain, in part, DNT's effect on the male reproductive system.

Human Toxicity

Most reports of human toxicity involve exposure to technical-grade DNT. Commercial-grade DNT is usually a combination of 2,4-DNT (approximately 76%) and 2,6-DNT (approximately 19%), with the remaining composition containing various other isomers.

Acute

The primary signs of toxicity regardless of the route of exposure are headache, fatigue, nausea, vomiting, and cyanosis resulting from methemoglobin formation. General signs and symptoms may be similar to those of alcohol intoxication. When methemoglobin levels approach 15%, cyanosis appears, and when the methemoglobin levels exceed 40% weakness and dizziness occur. Methemoglobin levels above 70% may produce muscle tremors, cardiovascular effects, and death.

Chronic

Long-term exposure to low levels of DNT will result in methemoglobinemia, the severity of which depends on the magnitude of the exposure. Although carcinogenic effects of DNT have been demonstrated in animals, there is currently no evidence of DNT carcinogenicity in humans.

The chronic oral RfDs for 2,4- and 2,6-DNT are 0.002 and 0.001 mg/kg/day, respectively. The oral slope factor for both 2,4- and 2,6-DNT is 6.8×10^{-1} (mg/kg/day)$^{-1}$. The U.S. EPA classifies both isomers as B2 carcinogens (probable human carcinogen; sufficient evidence in animals but inadequate or no evidence from epidemiologic studies). The ACGIH TLV-TWA and the NIOSH REL for DNT are 1.5 mg/m^3, with both organizations noting that the chemical is a suspected human carcinogen.

Clinical Management

For most cases of DNT poisoning, clinical management involves correction of the methemoglobinemia and associated support therapy.

Animal Toxicity

Most of the toxicity data are for the 2,4- and 2,6-DNT isomers. Oral LD$_{50}$ values for 2,4-DNT are extremely variable ranging from 177 to 609 mg/kg/day for rats and 390 to 1647 mg/kg for mice. Rat and mouse oral LD^{50} values of 216 and 607 mg/kg, respectively, have been reported for 3,5-DNT. In addition to lethality, acute oral exposures of laboratory animals to 2,4-DNT have resulted in hematologic disorders (methemoglobinemia) and toxic effects in the male reproductive system. Longer term oral exposures also induce hematologic and reproductive effects in addition to renal and neurologic disorders. For 2,6-DNT, oral LD$_{50}$ values of 665 and 714 mg/kg/day have been reported for rats and mice, respectively. The toxicologic effects of 2,6-DNT in animals are similar to those of 2,4-DNT. 2,4-DNT at an oral dose of 40 mg/kg/day for 2 years produced liver tumors in rats and at a dose of 97 mg/kg/day for 2 years produced renal tumors in male mice. 2,6-DNT at doses as low as 7 mg/kg/day produced hepatocellular carcinomas in male rats following a 1-year oral exposure. The available data also indicate strain differences in the carcinogenic response for several of the DNT isomers.

Further Reading

Agency for Toxic Substances and Disease Registry (ATSDR) (1989). *Toxicological Profile for 2,4-Dinitrotoluene and 2,6-Dinitrotoluene.* ATSDR/USPHS.

Grayson, M., and Eckroth, D. (Eds.) (1978). *Kirk–Othmer Encyclopedia of Chemical Technology.* Wiley, New York.

U.S. Air Force (1989). *The Installation Restoration Program Toxicology Guide.* Air Force Systems Command, Aerospace Medical Division.

—*Robert A. Young*

Related Topic

Pollution, Soil

Dinoseb

- ◆ CAS: 88-85-7
- ◆ SYNONYMS: Dinitrobutylphenol (DNBP); 2-sec-butyl-4,6-dinitrophenol
- ◆ CHEMICAL CLASS: Dinitrophenol (see Dinitrophenols)
- ◆ CHEMICAL STRUCTURE:

Uses

Dinoseb is used as an herbicide, insecticide, and miticide.

Exposure Pathways

Inhalation, oral, and dermal routes are the most common routes of exposure to dinoseb.

Toxicokinetics

Dinoseb is rapidly absorbed from the gastrointestinal tract, respiratory tract, and intact skin. It undergoes oxidation of either of the two methyl groups on the sec-butyl chain, conjugation of the phenolic products, and formation of many uncharacterized metabolites. Microsomal enzymes of rat liver reduce the *o*-nitro group of dinoseb. The compound is highly bound to plasma proteins. Hepatic and urinary excretion are the primary routes of elimination.

Mechanism of Toxicity

Dinoseb uncouples oxidative phosphorylation from electron transport by carrying protons across the inner mitochondrial membrane, thereby dissipating the pH gradient and membrane electrochemical potential and preventing the formation of adenosine triphosphate. Following exposure to this chemical, metabolism in all body cells is stimulated, resulting in an increase in oxygen consumption, body temperature, breathing rate, and heart rate. Dinoseb-induced weight loss may occur due to inhibition of lipogenesis from pyruvate and lactate. The body fat serves as the major fuel for the extra metabolism.

Human Toxicity

Acute exposure to dinoseb is associated with signs and symptoms of toxicity which develop rapidly within a few hours following exposure. Hyperthermia and profuse sweating are the early manifestations of toxicity. Liver and kidney damage may ensue within 12–72 hr postexposure. Early symptoms of dinoseb toxicity include headache and confusion followed by restlessness, hyperactivity, seizures, and coma following severe poisoning. The respiratory rate is usually markedly increased. Sinus tachycardia, ventricular tachycardia, and ventricular fibrillation may occur. Following ocular exposure to dinoseb, cataracts, secondary glaucoma, paresis of accommodation, and nystagmus have been reported. Other signs and symptoms following exposure to dinoseb include nausea, vomiting, abdominal pain, methemoglobinemia, and hemolytic anemia. Dinoseb has the potential to cause damage to the immune system. The U.S. EPA (1985) has set the no-observable-effect level of dinoseb for developmental toxicity at 3 mg/kg/day. Although U.S. EPA has suspended the registration of dinoseb because of its potential developmental toxicity, it is believed that there is no hazard to persons consuming food that has been treated with dinoseb. Blood glucose, liver function, and renal function tests should be monitored in symptomatic patients.

Clinical Management

Exposure to dinoseb requires symptomatic treatment. Adequate ventilation and oxygenation should be pro-vided with close monitoring of arterial blood gases. The fluid and electrolyte balances should be maintained. The body temperature should be kept within tolerable limits. Antipyretic drugs are, however, not effective because dinoseb poisoning involves peripheral metabolism, not central nervous system control of temperature. Diazepam is administered to overcome the accompanying seizure and convulsions following dinoseb exposure. In case of an oral exposure to dinoseb, gastrointestinal absorption may be prevented by gastric lavage and/or activated charcoal administration. Exposed eyes and skin should be irrigated with copious amounts of water following an ocular or dermal exposure to dinoseb.

Animal Toxicity

Neurological and skeletal malformations have been observed in laboratory animals exposed to dinoseb. Pregnant rats given 200 ppm dinoseb in their feed showed reductions in fetal survival. Surviving fetuses exhibited lower than normal birth weights. Morphologic abnormalities of the kidney have been noted in the offspring of female rats given dinoseb; however, renal function and morphology subsequently returned to normal. Dinoseb administered intraperitoneally to pregnant rats on Gestation Days 10–12 at a dose of 10.5 mg/kg/day caused a reduction in body weight in offspring. Maternal toxicity and malformations of the eye have been observed among the offspring of pregnant rats fed 200 ppm of dinoseb. Studies in laboratory animals indicate that dinoseb has the potential to cause damage to the immune system.

—Priya Raman

Related Topics

Developmental Toxicology
Pesticides

Dioctylphthalate

- ◆ CAS: 117-84-0
- ◆ PREFERRED NAME: Di-N-octylphthalate

♦ SYNONYMS: 1,2-Benzenedicarboxylic acid dioctyl ester; dioctyl *O*-benzenedicarboxylate; DNOP; *N*-octyl phthalate; *O*-benzenedicarboxylic acid dioctyl ester; octyl phthalate; phthalic acid dioctyl ester; phthalic acid dioctyl ester; benzenedicarboxylic acid di-*n*-octyl ester.

♦ CHEMICAL CLASS: Alkyl phthalate

♦ CHEMICAL STRUCTURE:

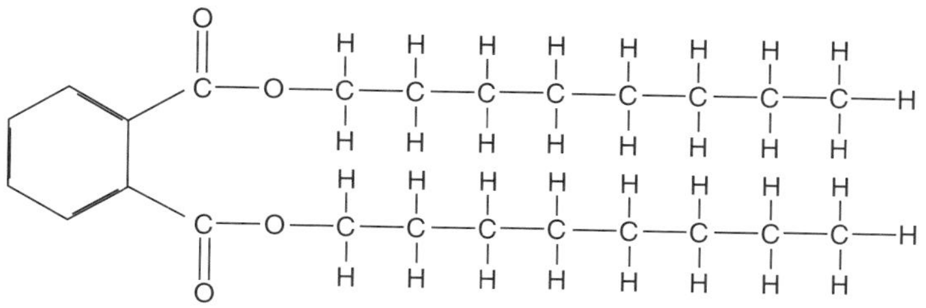

Uses

Dioctylphthalate is used as a plasticizer in cellulose ester resins and polystyrene and vinyl plastics. It is also a component of some pesticides.

Exposure Pathways

Humans may be exposed to dioctylphthalate in food (as an indirect food additive) and in drinking water. Although ingestion is the primary route of exposure for the general public, inhalation and dermal exposures may be more significant in occupational settings in which the chemical is used in industrial processes. Exposure via parenteral administration resulting from leaching of dioctylphthalate from plastic tubing and containers used in medical practice has also been documented.

Toxicokinetics

Definitive information regarding the absorption of dioctylphthalate is not available. Absorption may be inferred, however, due to systemic toxic effects following oral administration of the chemical and by analogy to absorption characteristics of similar phthalate esters. The limited information regarding the biotransformation of dioctylphthalate indicates that the chemical undergoes hydrolysis to a monoester within the intestines prior to absorption. However, it is also likely that hydrolysis may occur in the intestinal mucosal cells and in other tissues. Phthalate esters are generally widely distributed in the body. The effects observed in various organs and tissue following exposure to dioctylphthalate affirm its distribution throughout the body.

Chemical-specific elimination data for di-*N*-octylphthalate are not available. However, data from animal studies using the diisoctyl phthalate isomer have shown that it is excreted in the urine and bile as a monoester. Species-dependent quantitative and qualitative differences have been observed for excretion of this isomer as well as di-*N*-butyl phthalate and bis(2-ethylhexyl)-phthalate. Excretion half-lives of 1.2 and 5.4 hr have been reported for these compounds.

Mechanism of Toxicity

Specific data regarding the mechanism by which dioctylphthalate causes toxic responses are not available. There is some evidence that the toxic effects observed for this chemical may be due to its mono-*n*-octyl ester metabolite.

Human Toxicity: Acute

Definitive information regarding the acute toxicity of di-*N*-octylphthalate is not available. An estimated lethal oral dose in humans is between 0.5 and 15 g/kg, or between 1 oz and 1 qt in a 70-kg adult. Compounds that are structurally similar to di-*N*-octylphthalate are known to irritate mucous membranes resulting in irritation of the eyes, throat, and upper respiratory tract passages and in gastrointestinal disturbances. Generally, the acute oral toxicity of alkyl phthalates is low and the acute oral toxicity decreases as molecular weight increases.

Human Toxicity: Chronic

A case report noted the development of an asthmatic reaction in a worker continuously exposed to dioctyl phthalate during a manufacturing process. Based on the known toxic effects of di-*N*-octylphthalate in animals, chronic exposure of humans may result in liver and kidney damage.

The chronic RfD for di-*N*-octyl phthalate is 0.02 mg/kg/day. No other regulatory or health-based guideline values are currently available for di-*N*-octylphthalate. Neither the U.S. EPA nor IARC have evaluated the carcinogenicity of di-*N*-octylphthalate.

Clinical Management

The potential for esophageal or gastrointestinal tract irritation following ingestion suggests that emesis should not be induced. Other measures to prevent absorption may be beneficial. Gastric lavage may be indicated if performed soon after ingestion or if the patient is comatose or at risk of convulsing. Exposed skin and eyes should be copiously flushed with water.

Animal Toxicity

In rats fed diets containing 20,000 ppm di-*N*-octyl phthalate, an accumulation of large droplets of fat around central veins was observed that progressed to mild centrilobular necrosis and increased liver weight within 10 days. Renal toxicity has been observed in rats and mice given di-*N*-octyl phthalate in the diet (1000 ppm) for 48 weeks, and evidence of liver toxicity was noted for rats given a diet containing 3500 ppm for 7–12 months. Although many of the phthalate esters exert toxic effects on the male reproductive system, di-*N*-octyl phthalate appears to be among the least potent. Evidence for developmental toxicity are equivocal. There is currently no evidence showing that di-*N*-octylphthalate is genotoxic.

Further Reading

Grayson, M., and Eckroth, D. (Eds.) (1978). *Kirk–Othmer Encyclopedia of Chemical Technology*. Wiley, New York.

—*Robert A. Young*

Dioxane, 1,4-

- ◆ CAS: 123-91-1

- ◆ SYNONYMS: 1,4-Dioxan; diethylene oxide; *para*-dioxane; glycolethyleneether; 1,4-

diethylenedioxide; *para*-dioxan; tetrahydro-*para*-dioxin; 1,4-diethylene dioxide; dioxyethylene ether

- ◆ DESCRIPTION: Dioxane is a colorless liquid with a mild, ether-like odor. It is miscible in water and most organic solvents. It is relatively stable under normal temperature and pressure.

- ◆ CHEMICAL STRUCTURE:

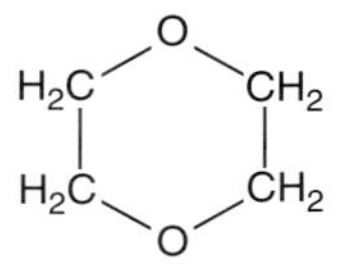

Uses

Dioxane is a solvent widely used for a wide range of organic products, including cellulose acetate, nitrocellulose, other cellulose esters or ethers, fats, oils, waxes, mineral oil, natural and synthetic resins, and polyvinyl polymers. It has been used for wetting and dispersing in textile processing, dye baths, stain, and printing compositions. It is also found in cleaning and detergent preparations, adhesives, cosmetics, deodorants, fumigants, emulsions, and polishing compositions. It has been used as an ingredient in lacquers, paints, varnishes, and paint and varnish removers. Dioxane is also used in purifying drugs and in cosmetic products such as shampoos and bath preparations. It has been used in the embedding process for the preparation of tissue sections for histology. Additionally, dioxane has been used as a stabilizer for chlorinated solvents, particularly, 1,1,1-trichloroethane, in solvent applications.

Exposure Pathways

Exposures to dioxane may occur through inhalation, ingestion, and dermal contact. Toxic gases and vapors (which may include carbon monoxide) may be released in a fire involving dioxane.

Toxicokinetics

Dioxane is readily absorbed through intact skin. In humans exposed to 50 ppm dioxane via inhalation, dioxane exhibited a half-life of 0.98 hr in plasma. Dioxane is metabolized to β-hydroxyacetic acid (HEAA) in humans and rats. Under elevated temperature, HEAA is further metabolized to 1,4-dioxane-2-one. Some believe that this 1,4-dioxane-2-one is a major metabolite and a hepatic carcinogen, however, identified by gas chromatography. One study using [^{14}C]-1,4-dioxane found that 85% of the radiolable in the urine occurred in the form of HEAA and the reminder in the form of unchanged 1,4-dioxane. The pretreatment of rats with phenobarbital or with polychlorinated biphenyls significantly increased the amount of the primary urinary metabolite excreted.

The fate of dioxane in rats is markedly dose dependent because of the limited capacity of rats to metabolize HEAA. The metabolism of dioxane is dose dependent and becomes severely nonlinear at high doses due to saturation of the enzyme systems responsible for metabolism. At saturation, maximum velocity of metabolism to HEAA was about 18 mg/kg. Plasma concentration time curves for dioxane given to rats at levels of 3–1000 mg/kg intravenously and 50 ppm vapor by inhalation for 6 hr were linear by each route with a half-life of about 1 hr. Various toxicological data indicate that dioxane toxicity occurs only after doses large enough to saturate processes for detoxification and elimination.

Mechanism of Toxicity

Pharmacokinetic and toxicological data indicate that dioxane toxicity occurs only after doses large enough to saturate processes for detoxification and elimination. Dioxane is one of many carcinogens that have not been demonstrated to react covalently with DNA. Its mode of action is not sufficiently well understood to permit assignment to a specific class of epigenetic agents.

Human Toxicity

Dioxane toxicity occurs only at high concentrations. There are five cases of fatal poisoning in men who inhaled excessive amounts of dioxane while working in a textile factory. Symptoms were irritation of the upper respiratory passage, coughing, irritation of eyes, drowsiness, vertigo, headache, anorexia, stomach pains, nausea, vomiting, uremia, coma, and death. Autopsies revealed congestion and edema of the lungs and brain and marked injury of the liver and kidneys. Blood analysis of these victims showed no abnormalities other than considerable leucocytosis. Twelve humans were exposed to 200 ppm dioxane for 15 min, considered to be the highest acceptable dose; at 300 ppm it caused irritation of the eyes, nose, and throat. Death was reported in 1 worker after 1 week on a job where the average concentration of dioxane vapor was 470 ppm. Possible skin absorption and damage to the kidneys, liver, and brain were indicated.

A mortality study was conducted on employees exposed to dioxane. Observed deaths from overall cancer were not significantly different from expected number of deaths. Dioxane can be inhaled in amounts sufficient to cause serious systemic intoxication. Injury may become apparent hours after termination of an exposure that had been erroneously considered to be negligible. Prolonged and repeated contact can cause eczema and repeated inhalation exposures to low concentrations have been fatal.

Dioxane is currently classified as B2, a probable human carcinogen, by U.S. EPA based on adequate animal studies and inadequate human studies. Three epidemiological studies on workers exposed to dioxane are available. Two of the deaths were due to cancer: one epithelial carcinoma in a 66-year-old man and one melofibrotic leukemia in a 71-year-old man. No statistically significant increase was noted based on these few cases of cancer. Among 165 production and processing workers exposed to dioxane, 12 deaths were reported. Three of these deaths were due to cancer: one stomach cancer, one alveolar carcinoma, and one mediastinal malignancy. Three deaths were not different from the expected numbers.

Clinical Management

Persons exposed to dioxane should be thoroughly medically examined before ipecac alkaloid is administered to induce emesis. If signs of oral, pharyngeal, or esophageal irritation, a depressed gag reflex, or central nervous system (CNS) excitation or depression are present, emesis should not be induced. Gastric lavage may be indicated if performed soon after ingestion or in patients who are comatose or at risk of convulsing. The airway should be protected by placing the patient in the Trendelenburg and left lateral decubitus position or by cuffed endotracheal intubation. After control of any seizures present, gastric lavage may be performed. Activated charcoal binds most toxic agents and can decrease their systemic absorption if administered soon after ingestion. In general, metals and acids are poorly bound and patients ingesting these materials will not likely benefit from activated charcoal administration. Immediate dilution with milk or water may be of benefit in caustic or irritant chemical ingestions like dioxane.

In cases of ingestion exposure, the patient should be carefully monitored for the development of any systemic signs or symptoms and symptomatic treatment should be provided as necessary. In cases of inhalation exposure, the victim should be moved to fresh air and monitored for respiratory distress. If cough or difficulty in breathing develops, the patient should be evaluated for respiratory tract irritation, bronchitis, or pneumonitis. Supplemental oxygen (100%, humidified) should be administered with assisted ventilation as required. When eyes are exposed, they should be irrigated with copious amounts of tepid water for at least 15 min. If irritation, pain, swelling, lacrimation, or photophobia persist, the patient should be seen in a health care facility. In case pulmonary edema develops due to exposure, ventilation and oxygenation should be maintained with close arterial blood gas monitoring. If PO_2 remains less than 50 mm Hg, positive end-expiratory pressure or CPAP may be necessary.

Animal Toxicity

Guinea pigs can tolerate inhalation of 2000 ppm dioxane for several hours without serious symptoms. Higher concentrations produced eye, nose, and lung irritation. Dogs given dioxane orally over a period of 9 days died after a total consumption of about 3 g/kg with severe liver and kidney damage. Single doses of 5.66, 5.17, and 3.90 g/kg to mice, rats, and guinea pigs produced symptoms progressing from weakness, depression, incoordination, and coma to death. Autopsy revealed hemorrhage areas in the pyloric region of the stomach, bladders distended with urine, enlarged kidneys, and slight proteinurea without hematuria.

Of four groups of 28–32 male Sprague–Dawley rats given 0.75, 1.0, 1.4, or 1.8% of dioxane in drinking water for 13 months, one, two, and two rats, respectively, developed nasal cavity tumors, mainly squamous cell carcinomas with adenocarcinomas in two cases. Dioxane administered at a concentration of 2 g/kg body weight orally or intraperitoneally increased liver microsomal protein content significantly in male and female mice. Male mice injected intraperitoneally showed significant increase in cytochrome b5 and P450 contents compared to female mice.

Five groups of 96 male and female Wistar rats were exposed either to air or to air containing 0.4 g/liter (111 ppm) 99.9% pure dioxane for 7 hours per day, 5 days per week, for 2 years. Fifty percent of the animals survived 20–24 months. There was no statistically significant increase in the incidence of tumors observed in 525 treated rats examined compared with 347 controls. U.S. EPA's B2 (probable human carcinogen) classification for dioxane was based on induction of nasal cavity and liver carcinoma in multiple strains of rats, liver carcinomas in mice, and gall bladder carcinomas in guinea pigs.

—*Leyna Mulholland*

Dioxins

- DESCRIPTION: There is a wide range of chlorinated dibenzo-*p*-dioxins varying in the extent of their chlorination. The nomenclature is based on the number of carbon positions that are chlorinated and include mono-, di-, tetra-, penta-, hexa-, hepta-, and octachlorinated congeners. 2,3,7,8-Tetrachlorodibenzo-*p*-dioxin (2,3,7,8-TCDD, more commonly referred to as TCDD or dioxin; CAS No. 1746-01-6) is usually of greatest concern because of high toxicity in laboratory animal models, its widespread distribution and persistence in the environment, bioaccumulation potential, and because the greatest amount of data exists for this form (see TCDD).

- CHEMICAL CLASS: Halogenated aromatic hydrocarbon

- CHEMICAL STRUCTURE: There are 74 chlorinated dibenzo-*p*-dioxin congeners. The basic structure for unsubstituted dibenzo-*p*-dioxin (showing the carbon numbering scheme that is used to name specific congeners) and the structure of 2,3,7,8-tetrachlorodibenzo-*p*-dioxin (one of 22 tetrachlorinated dibenzo-*p*-dioxins) are shown below.

Dibenzo-*p*-dioxin

2,3,7,8-tetrachlorodibenzo-*p*-dioxin

Uses

Dioxins are by-products of various chemical syntheses and are usually present as contaminants of endproducts. There is no known use.

Exposure Pathways

2,3,7,8-TCDD and other chlorinated dibenzo-*p*-dioxins are released during the combustion of many polychlorophenols and also occur as contaminants in various chemicals such as the herbicide 2,4,5-trichlorophenoxyacetic acid (see 2,4,5-T). Most significant exposure to 2,3,7,8-TCDD and other dioxins results from accidental releases or explosions in chemical plants or storage facilities for dioxin-containing chemicals. Because of the persistence of dioxin congeners in the environment and their potential for bioaccumulation, they are significant pollutants and exposure may occur via the soil, air (especially when dioxins occur as combustion products), or water. When bound to components of the soil, the health hazard from 2,3,7,8-TCDD is reduced compared to ingestion of the pure compound. However, its bioavailability varies with the specific media in which it occurs.

Toxicokinetics

Dioxins are highly lipid soluble and are efficiently absorbed by most routes of exposure although absorption will vary quantitatively depending on the route. Because dioxins are poor substrates for the enzymes typically involved with biotransformation of xenobiotics, they are very poorly metabolized. Dioxins tend to exhibit high concentrations in the liver and tend to accumulate in fatty tissue. Because of their high lipid solubility and poor metabolism, excretion of dioxins is extremely slow. The elimination half-life in humans is approximately 10 years.

Mechanism of Toxicity

Some, but not all, of the toxic effects of 2,3,7,8-TCDD are mediated by the interaction with an intracellular protein called the Ah receptor. This interaction ultimately modifies genetic components that in turn cause deleterious effects.

Human Toxicity: Acute

A TD_{Lo} of 107 μg/kg has been reported for humans although a more generally accepted minimum toxic dose for humans is 0.1 μg/kg. Nonlethal effects following short-term exposure to 2,3,7,8-TCDD include headache, fatigue, irritation of the gastrointestinal and respiratory tracts, dehydration, and skin irritation. The acneform skin irritation resulting from exposure to 2,3,7,8-TCDD or chemicals that contain TCDD is referred to as chloracne.

Human Toxicity: Chronic

The toxic effects associated with chronic exposure to 2,3,7,8-TCDD include chloracne, impaired liver function, peripheral neuropathies, and altered blood chemistry parameters. Other long-term effects may include chromosome damage, heart attacks, reproductive disorders, and cancer, although epidemiologic data regarding these effects are equivocal. Some effects of long-term low-level exposure to dioxins appear to be reversible following cessation of the exposure.

The U.S. EPA classifies 2,3,7,8-TCDD in cancer group B2 (probable human carcinogen with sufficient evidence in animals but inadequate evidence in humans), and IARC classifies the chemical as 2B (probably carcinogenic to humans; sufficient evidence in animals). The oral slope factor and unit risk for 2,3,7,8-TCDD are 1.50E + 05 (mg/kg/day)$^{-1}$ and 4.50E + 00 μg/liter, respectively. The inhalation slope factor and unit risk are 1.50E + 05 (mg/kg/day)$^{-1}$ and 3.30E-05 pg/m^3, respectively. No RfDs or RfCs have been derived for 2,3,7,8-TCDD or any other dioxin. The U.S. EPA has also issued Health Advisories (HAs) for exposure to 2,3,7,8-TCDD in drinking water. For short-term exposures the HAs are 1×10^{-3} μg/liter (1 day) and 1×10^{-4} μg/liter (10 day) for a 10-kg child, and the longer-term HAs are 1×10^{-5} and 4×10^{-5}, respectively, for a 10-kg child and 70-kg adult.

Clinical Management

There are no clinical procedures specific for dioxin intoxication, but clinical management for acute intoxication by dioxin-containing chemicals such as 2,4-D and 2,4,5-T may be applied. Basically, these procedures include decontamination of the gut and/or skin and possibly alkaline diuresis for severe overdose situations.

Animal Toxicity

Animal data have shown that 2,3,7,8-TCDD is a highly toxic chemical and capable of exerting a wide range of toxic effects. Oral LD_{50} values in animals vary considerably (e.g., 0.6, 1, 20, 114, and 1157 μg/kg for guinea pigs, dogs, rats, mice, and hamsters, respectively). Toxic effects observed in animals following acute exposure include damage of the liver, heart, thymus gland, adrenals, and immunosuppressive effects. The effects, however, vary with species. Toxic effects, including, death, have also been observed in animals following short-term dermal exposure to 2,3,7,8-TCDD. In addition to the aforementioned effects, chronic exposure in animals has also resulted in carcinogenic responses in the liver and lungs. 2,3,7,8-TCDD has been shown to be carcinogenic and teratogenic in several laboratory species.

Further Reading

Agency for Toxic Substances and Disease Registry (ATSDR) (1989). *Toxicological Profile for 2,3,7,8-Tetrachlorodibenzo-p-dioxin*, ATSDR/TP-88/23. ATSDR/USPHS.

Barnes, D. G., Kutz, F. W., and Bottimore, D. P. (1989). *Update to the Interim Procedures for Estimating Risks Associated with Exposures to Mixtures of Chlorinated Dibenzo-p-Dioxins and -Dibenzofurans (CDDs and CDFs)*, Risk Assessment Forum. U.S. EPA.

Grayson, M., and Eckroth, D. (Eds) (1978). *Kirk–Othmer Encyclopedia of Chemical Technology*. Wiley, New York.

U.S. Air Force (1989). *The Installation Restoration Program Toxicology Guide*. Air Force Systems Command, Aerospace Medical Division.

—*Robert A. Young*

Related Topics

Bioaccumulation
Distribution
Immune System
Pollution, Soil
Pollution, Water
Toxic Torts

Diphenhydramine

- CAS: 58-73-1

- SYNONYMS: Benadryl; Diphenhist; Genahist; Sominex; Nytol; Sleepinal; Caladryl; Dermarest

- PHARMACEUTICAL CLASS: An ethanolamine-derivative H-1 receptor antagonist

- CHEMICAL STRUCTURE:

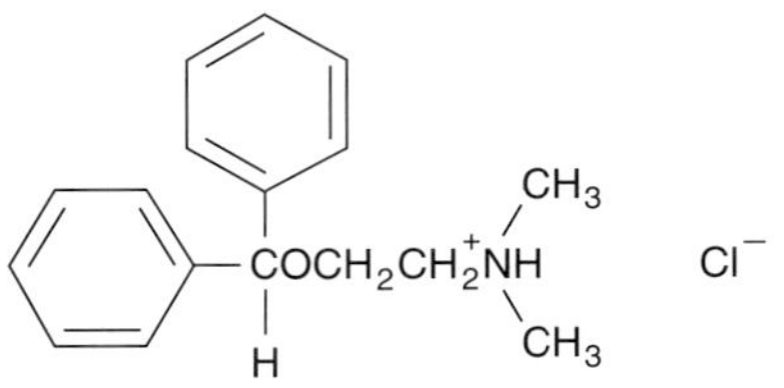

Uses

Diphenhydramine, like other antihistamines, is most often used to provide symptomatic relief of allergic symptoms caused by histamine release. The drug is also used as an antitussive, a nighttime sleep aid for the short-term treatment of insomnia, and as a preventive and treatment for motion sickness. Diphenhydramine may be useful in the treatment of parkinsonian syndrome in geriatric patients including drug-induced extrapyramidal reactions. Diphenhydramine has been used topically for the temporary relief of pruritus and pain associated with various skin conditions including minor burns, insect bites, and minor skin irritation.

Exposure Pathways

Ingestion, injection, and dermal application are the routes of both accidental and intentional exposures to diphenhydramine.

Toxicokinetics

Diphenhydramine is absorbed rapidly after an oral dose with peak plasma levels achieved within 2 hr. The drug is also absorbed through abraded skin resulting in systemic toxicity. Diphenhydramine undergoes extensive first-pass metabolism with 40–60% of an oral dose reaching systemic circulation as unchanged drug. Diphenhydramine is 98% protein bound and has an apparent volume of distribution of 3–7 liters/kg. Approximately 64% of the dose of diphenhydramine is excreted as metabolites in the urine. The serum half-life is 4–10 hr.

Mechanism of Toxicity

The toxicity of antihistamines is related to their anticholinergic (antimuscarinic) activity. The action of acetylcholine at the muscarinic receptors is blocked, resulting in signs and symptoms of anticholinergic poisoning. Diphenhydramine may produce direct toxicity unrelated to its anticholinergic properties.

Human Toxicity

Diphenhydramine overdose results in signs and symptoms of anticholinergic poisoning including dry mouth, fixed dilate pupils, flushed skin, fever, and hallucinations. Central nervous system (CNS) depression is more common in adults, whereas stimulation including tonic–clonic seizures is more common in children. Cardiovascular effects including tachycardia, hypertension or hypotension, arrhythmias, and cardiovascular collapse may occur. Symptoms of an overdosage occur within 30 min to 2 hr after an ingestion. Fatalities have occurred in children at doses under 500 mg and seizures with doses of 150 mg. A fatal adult dose is 20–40 mg/kg.

Clinical Management

Basic and advanced life-support measures should be utilized as necessary. Appropriate gastrointestinal decontamination procedures should be administered based on the history of the ingestion and the patient's level of consciousness. In a limited number of cases physostigmine administration may be necessary to treat severe central and peripheral anticholinergic symptoms refractory to conventional therapy. If physostigmine is given intravenously, the rate of administration should not exceed 2 or 3 min. Diazepam can be used to manage seizures. Continual electrocardiogram monitoring is essential.

Animal Toxicity

CNS changes including sedation or hyperexcitability, salivation, and vomiting have occurred following low exposures to antihistamines. Seizures and cardiac effects have occurred following acute high exposures. Symptomatic and supportive care followed by appropriate gastrointestinal decontamination procedures should be administered.

—*Carla M. Goetz*

Related Topics

Cholinesterase Inhibition
Delayed-type Hypersensitivity

Diphenoxylate

- CAS: 3810-80-8
- Synonym: Lomotil
- Pharmaceutical Class: Antidiarrheal
- Chemical Structure:

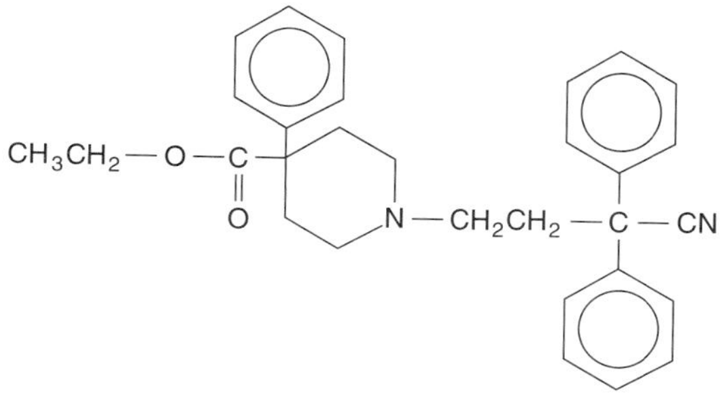

Uses

The only recognized use for diphenoxylate is in the treatment of acute and chronic diarrhea.

Exposure Pathways

Exposure is by the oral route; diphenoxylate is available in liquid and tablet form.

Toxicokinetics

Diphenoxylate is readily absorbed from the gastrointestinal tract. Peak levels occur 3 hr after a single oral dose. Diphenoxylate is metabolized rapidly in the liver to active and several inactive metabolites. Enterohepatic circulation may occur. The volume of distribution is 4.6 liters/kg. The half-life of diphenoxylate is 2.5 hr. The major metabolite difenoxin (diphenoxylic acid) is more potent and has a longer half-life (12–24 hr).

Mechanism of Toxicity

Diphenoxylate is a narcotic-like substance that slows gastrointestinal motility and depresses the central nervous system (CNS) producing coma and respiratory depression. Anticholinergic effects (secondary to the presence of atropine as an abuse deterrent) can be seen early after exposure with opioid effects occurring later. There is no correlation between the dose ingested and the severity of effects in children. Severe poisonings with coma and respiratory depression have occurred.

Human Toxicity: Acute

Single therapeutic doses produce little or no opiate-like effects in adults. Effects from larger doses (40–60 mg) are typical of opioid drugs and include miosis, ataxia, lethargy, respiratory depression, seizures, and coma. Onset of symptoms may be delayed 6–8 hr. Tachycardia, urinary retention, irritability, and cutaneous flushing may be evident before opioid symptoms. Children are more susceptible to the effects. The therapeutic index is low in children; symptoms have resulted with only one tablet.

Human Toxicity: Chronic

A morphine-like physical dependence can occur with chronic administration.

Clinical Management

Gastric lavage and activated charcoal are recommended for effective gastric decontamination. Monitoring for a minimum of 24–48 hr is recommended. Transient recovery may be observed before this time. Treat respiratory and CNS depression with intravenous administration of naloxone. Repeated boluses or continuous infusion of naloxone may be necessary.

—Linda Hart

Diphenylhydrazine

♦ PREFERRED NAMES: Diphenylhydrazine occurs as the isomers 1,1-diphenylhydrazine (CAS: 530-50-7) and 1,2-diphenylhydrazine (CAS: 122-66-7).

♦ SYNONYMS: 1,1-Diphenylhydrazine-*N,N'*-bianiline; 1,2-diphenylhydrazine hydrazobenzene; *N,N'*-diphenylhydrazine; *sym*-diphenylhydrazine

♦ CHEMICAL CLASS: Polycyclic amine

♦ CHEMICAL STRUCTURE:

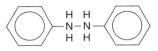

Uses

1,2-Diphenylhydrazine has been used for the production of benzidine, which, in turn, is used in the production of benzidine-based dyes. These dyes, however, are no longer produced in the United States. 1,2-Diphenylhydrazine is also used in the production of the antiinflammatory pharmaceutical agent phenylbutazone and in the production of sulfinpyrazone, a uricosuric agent.

Exposure Pathways

The primary route of exposure is likely to be via ingestion or dermal contact with dust of contaminated soil. Because of the low volatility and solubility of diphenylhydrazine, inhalation exposure or exposure via water are not likely to be significant.

Toxicokinetics

There are no human data regarding the absorption of diphenylhydrazines by any exposure route. Gastrointestinal absorption of 1,2-diphenylhydrazine in rats can be inferred by the presence of the parent compound and its metabolites in the urine and by systemic toxic effects following oral administration. No data are available regarding inhalation or dermal absorption in animals. Data are unavailable regarding the metabolism of 1,1-diphenylhydrazine. Limited data regarding the metabolism of 1,2-diphenylhydrazine by rats suggest benzidine and aniline to be major metabolites with minor metabolites including unspecified hydroxy derivatives. Conversion of 1,2-diphenylhydrazine to aniline may occur through intestinal microflora and by acid conversion in the stomach.

Data are not available regarding the distribution of either form of diphenylhydrazine. Limited data in rats indicate that urinary excretion of metabolites and unchanged parent compound occurs following oral administration of 1,2-diphenylhydrazine. No data are available regarding the excretion of 1,1-diphenylhydrazine.

Mechanism of Toxicity

The mechanism of diphenylhydrazine toxicity is not currently known. It is possible that some the toxic effects observed for diphenylhydrazine may be the result of its major metabolites, aniline and benzidine, which are known animal carcinogens.

Human Toxicity

No data are available regarding the acute or chronic toxicity of 1,1- or 1,2-diphenylhydrazine in humans.

Health-based guidance values for 1,2-diphenylhydrazine include an inhalation unit risk of 2.2×10^{-4} $(\mu g/m^3)^{-1}$ and a drinking water unit risk of 2.2×10^{-5} $(\mu g/liter)^{-1}$. The U.S. EPA classifies 1,2-diphenylhydrazine as a probable human carcinogen (B2). The cancer slope factor for 1,2-diphenylhydrazine is 8.0×10^{-1} $(mg/kg/day)^{-1}$. No regulatory values or guidance values are available for 1,1-diphenylhydrazine.

The U.S. EPA Reportable Quantity for 1,2-diphenylhydrazine is 1 lb (statutory) with a proposed Reportable Quantity of 10 lbs.

Clinical Management

Information on the clinical management of diphenylhydrazine intoxication is not available.

Animal Toxicity

Only limited data are available regarding the toxicity of diphenylhydrazine in animals. Oral LD_{50} values of 959 and 301 mg/kg have been reported for rats, and gastrointestinal hemorrhage and death have been reported for rodents following 4-week dietary exposure to a dose of 390 mg 1,2-diphenylhydrazine/kg body weight/day. Chronic exposure (78 weeks) of rats and mice to 1,2-diphenylhydrazine in the diet (equivalent to doses of 4 and 52 mg/kg/day for rats and mice, respectively) resulted in hepatocellular carcinomas. Exposure of rats to 1,2-diphenylhydrazine at doses of 5–15 mg/kg/day for 78 weeks resulted in effects ranging from death to histopathological changes in the liver and gastrointestinal tract. Other effects of chronic oral exposure of animals to 1,2-diphenylhydrazine include decreased weight gain, interstitial inflammation of the lung, and fatty degeneration and necrosis of the liver. The acute toxicity of diphenylhydrazines in animals following inhalation exposure has not been determined, and no information is available regarding the carcinogenic or noncarcinogenic effects in animals following chronic inhalation exposure to diphenylhydrazines.

Further Reading

Agency for Toxic Substances and Disease Registry (ATSDR) (1990). *Toxicological Profile for 1,2-Diphenylhydrazine*, No. TP-90-11. ATSDR/USPHS.

—*Robert A. Young*

Disc Batteries

- ◆ SYNONYM: Button batteries
- ◆ DESCRIPTION: Disc batteries are composed of two wafer-like plates that contain within their core an alkaline solution (typically up to 45% sodium or potassium hydroxide). These batteries may also contain metals including mercury, zinc, and cadmium.

Uses

Disc batteries have become ubiquitous. They are used as a power source for toys, wristwatches, cameras, and medical instruments including hearing aids. While their diameters range from 6 to 23 mm, 96% of disc batteries are 8–12 mm in diameter.

Exposure Pathways

Typical exposure to disc batteries is via ingestion. Ingestion often occurs inadvertently in those who place them in their mouths while changing the batteries of their instruments or toys. Disc battery ingestions are unique among childhood poisonings in that the mean age of victims is higher (>4 years old) than that of typical pediatric poisoning exposures. Batteries may also be placed in aural or nasal cavities.

Toxicokinetics

Following ingestion of a disc battery, the constituents of the battery do not commonly leach. However, if the battery fails to pass, metals (including mercury) can diffuse from the battery, becoming available for systemic absorption. Eighty-eight percent of ingested disc batteries are eliminated in the feces within 48 hr; 90% pass within 4 days.

Mechanism of Toxicity

Toxicity from disc battery exposure occurs through five potential mechanisms: (1) the local corrosive potential of their alkaline constituents; (2) the systemic toxicity of their individual metals; (3) potential electrochemical injury resulting from direct contact of a live battery with a mucosal surface; (4) the potential for aspiration, producing respiratory tract obstruction; and (5) potential mechanical obstruction within the gastrointestinal tract.

Human Toxicity: Acute

If a disc battery becomes lodged in the esophagus, auditory canal, or nasal cavity, its alkaline contents can leach, producing severe corrosive injury (see Alkalies). Esophageal perforation may result.

Human Toxicity: Chronic

In circumstances in which disc batteries are retained in the gastrointestinal tract for extended periods, metals may leach from the battery. Although no clinical syndromes of metal toxicity have been described after disc battery ingestion, increased serum levels of mercury have been documented (see Mercury).

Clinical Management

Because of the high rate of prompt, uncomplicated passage of disc batteries, clinical management is confined to ensuring that acute clinical toxicity does not occur. Patients with disc battery ingestion should receive a chest X-ray to ensure that the disc battery has not lodged in the esophagus. If found in the esophagus, the battery should be promptly removed. If the battery is located in the stomach and the patient does not have gastrointestinal disease associated with stenosis, stricture, or delayed gastrointestinal transit, the patient may be discharged without further intervention. If the patient has gastrointestinal disease, close outpatient monitoring should be provided.

Conservative management includes a repeat abdominal radiograph in 4 or 5 days if the battery has not been observed to pass. There is no role for gastric emptying efforts (induced emesis or gastric lavage) or use of cathartics.

Animal Toxicity

The toxicity of disc batteries in animals is present through the same mechanisms.

—*Michael Shannon*

Distribution

Distribution is the process by which absorbed chemicals are delivered to the various organs of the body in order to produce an effect, be stored, or be eliminated.

One of the major factors regulating the distribution of chemicals throughout the body is the amount of blood perfusing the various organs. Per unit of organ weight, brain and viscera, especially kidneys, are very well perfused and are therefore presented with large amounts of chemicals. Irrigation of resting skeletal muscle, skin, and bone is less important, while that of fat tissue is poor. If one also considers total tissue mass, however, the amount of blood reaching an organ can be of importance in terms of distribution. For instance, in certain individuals, fatty tissue can represent 30% of the body mass and, despite a low perfusion rate, can become an important repository for lipid-soluble chemicals.

Increasing or lowering blood perfusion will result in more or less rapid distribution of chemicals to their sites of action, storage, or removal. Muscular effort increases the amount of blood ejected by the heart per unit time. The influence of muscular activity on the overall perfusion of body organs is rather complex. For instance, the fraction of blood perfusing skeletal muscles is disproportionately increased at the expense of most of the other organs, including those responsible for elimination; skin perfusion is also markedly increased during muscular effort. On the other hand, the fraction of blood perfusing the brain remains the same. Distribution of chemicals will vary accordingly; elimination processes will be less effective allowing higher concentrations of chemicals to reach target organs, including the brain.

When the ambient temperature is elevated, the rate of skin perfusion increases; the skin then appears red and feels warm. Although skin is not an important route for transfer of chemicals, increasing the local circulation is likely to facilitate percutaneous exchange of chemicals between blood and the environment.

Another major factor determining the distribution of chemicals is their affinity for a given tissue. Affinity depends on the physicochemical properties of a chemical, the biochemical composition of the various cells in organs, and the ability of the cellular membrane acting more or less as a barrier. Chemicals may penetrate into cells by passive diffusion through the lipid-rich membrane, by special carrier-mediated transport systems, or by filtration through small water channels in the cell membrane (for more on mechanisms by which chemicals cross membranes, see Absorption). Lipid-soluble neutral molecules easily diffuse across cell membranes and tend to accumulate in lipid-rich tissues. Active transport systems (energy-requiring systems) in the liver help remove certain molecules for elimination into the bile. Distribution is therefore a dynamic process enabling chemicals to reach their sites of action, storage, or removal.

Sites of storage have considerable importance in modulating the action of a chemical or its removal. These sites are usually different from those of major action, but they could be located in organs responsible for removal (e.g., liver and kidneys). While they are stored, most chemicals are temporarily inactivated; they remain available to be released and redistributed as the concentration of free, circulating chemicals in blood decreases. Storage prolongs the residence time of chemicals in the body and helps smooth out rapid fluctuations of the concentration of circulating chemicals. On the other hand, sites of storage can represent a certain threat in a sense that deposited chemicals may be rapidly released for further redistribution to potential sites of action.

Adipose tissues (fat) can store relatively large amounts of highly lipid-soluble chemicals like chlorinated pesticides, polychlorinated biphenyls (PCBs), dioxins, furans, a number of organic solvents (benzene, trichloroethylene, and styrene), and certain drugs like anesthetics. In the past, biopsy of subcutaneous abdominal fat tissue has been used to monitor exposure to chlorinated pesticides in rural populations around the world; today, the tendency is to monitor such chemicals using blood lipids since the concentration of chemicals in the latter is in equilibrium with that in fatty tissues. Volatile chemicals, like benzene, temporarily stored in fatty tissues, are slowly released into blood and may be more easily monitored in the air exhaled by the lung. Chlorinated pesticides are also largely stored in adipose tissues of birds during summer. When birds migrate, however, adipose tissues are extensively used as a source of energy; pesticides are then mobilized and redistributed to body tissues including the central nervous system (CNS), where concentrations may reach toxic levels.

Bone is also an important site of storage for certain metals that have chemical properties similar to calcium. More than 90% of absorbed lead is incorporated into bone. Lead will stay there for years, slowly exchanging with concentrations in the blood and other tissues. When the demand for calcium in bone is high, lead may be rapidly released from its deposit and may reach toxic concentrations in target organs; this is especially true for lead workers who have accumulated large burdens of lead in bone throughout the years. Radioactive strontium is another metal with high affinity for bone. Unfortunately, bone is not only a site of deposit for strontium but also a site of action since radiations emitted by strontium induce bone cancers. Fluorides, which are also deposited in bone, may eventually cause skeletal fluorosis, a disease characterized by an increase in the density and the calcification of bone.

The affinity of liver and kidney for a number of chemicals is also considerable. A protein in the liver, ligandin, has a remarkable degree of affinity for organic acids; it plays a role in the transfer of these chemicals from blood into liver. Both liver and kidney may become storage depots for metals, like cadmium and zinc, due to the presence of small binding proteins called metallothioneins. When the binding capacity of these proteins is exceeded, local toxicity may appear, as is the case for cadmium in the kidney.

Finally, what may be considered as one of the most important storage depots in the body is plasma protein. Albumin, the most abundant protein in plasma, and other plasma proteins may bind reversibly a very large number of chemicals, many of which are therapeutic agents. The protein-bound fraction of a chemical exists in a state of equilibrium with the unbound (also called "free") fraction; only the free form of a chemical is available for biological effect and disposition. By sequestering chemicals for several hours, in certain cases, protein binding in plasma regulates the pharmacological and toxicological effects of chemicals; distribution to sites of action is delayed and access to elimination processes is slowed down. It is a fact of considerable therapeutic importance that certain drugs may be displaced from their sites of protein binding by other chemicals with considerable affinity for the same protein. Displacement of drugs that require precise dosing schedules to produce their therapeutic effect without also inducing toxicity, like anticoagulants and oral hypoglycemic agents, can lead to severe toxic manifestations.

The brain, as a site where chemicals are distributed, is a very sensitive organ. A more or less permeable membrane barrier located at the junction between the bloodstream and the brain acts as a shield to certain noxious chemicals; it is called the "blood–brain barrier."

The barrier effect is mainly due to the fact that the cells lining the walls of the capillaries present in the brain tissue are tightly joined, contrary to what prevails with capillaries in other tissues; this leaves very little space between the cells for filtration of small-size, water-soluble molecules. Moreover, the cells of brain capillaries possess very few endocytotic vesicles, which in capillaries of other tissues engulf large molecules and serve as a transfer mechanism; as a result, many neurotoxins, such as diphteria and tetanus toxins, are excluded. Furthermore, the capillaries of the brain are surrounded by prolongations of certain brain cells, thus forcing lipid-soluble chemicals to cross an additional lipid membrane. Finally, the intercellular fluid bathing the brain cells contains lower concentrations of proteins; this results in a reduction of the movement of certain water-insoluble chemicals that are more easily transported when bound to proteins.

The existence of the blood–brain barrier does not preclude the passage of chemicals into the brain. As is the case with all other cellular membranes in the body, lipid-soluble nonionized chemicals enter the brain by passive diffusion. Anesthetics, ethanol, and CNS depressants, for instance, rapidly diffuse into the brain in a matter of a few seconds or minutes. They also exit the brain rapidly when the concentration gradient between blood and brain is reversed. Elemental mercury, methylmercury, and tetraethyl lead are examples of lipid-soluble forms of metals that easily enter the brain, while the ionized, much less lipid-soluble inorganic salts of mercury and lead penetrate only poorly.

In newborn infants, the blood–brain barrier is not fully developed; certain chemicals, like lead, and some endogenous substances, like bilirubin, may therefore enter the brain more easily. Like the brain, but for different reasons, the embryo is also very sensitive to exogenous chemicals circulating in the maternal blood. The placenta is the route by which the developing embryo and fetus exchanges with maternal blood. Its main physiological function is to provide nutrients to the fetus and remove its waste products. In humans, only three layers of cells separate maternal and fetal blood and form what has been termed the placental barrier.

The placental barrier is far from being an absolute shield to the passage of foreign chemicals into the fetal circulation. Lipid-soluble, nonionized molecules cross the placenta by passive diffusion and reach equilibrium between both maternal and fetal circulations. For these chemicals, what the barrier does is simply delay the delivery of chemicals to the fetus. Large molecules and microorganisms may traverse placenta by endocytosis. Once delivered, most chemicals will diffuse back into the maternal circulation, leaving the fetus unharmed. A few chemicals, however, may have a devastating effect, killing the fertilized egg, inducing birth defects, or retarding the growth of the developing fetus.

Thus, for a number of chemicals and a few molecules of microbiological origin, the placenta does not serve as an efficient barrier. For many years, it has been known that the rubella virus (German measles) may cause human congenital anomalies. Similarly, some chemicals known as teratogens may also produce abnormalities in the developing of the human fetus. Among them are vitamins A and D taken at high dosages, certain anticancer drugs, some steroid hormones, and thalidomide, certainly the best known teratogen.

Although less spectacular than birth defects like missing limbs or cleft palate, retardation in the functional development of the fetus may be just as damaging. In this regard, the fetotoxicity of excessive alcohol consumption and tobacco smoking is well known. Governments now issue severe warnings to pregnant women concerning the danger of these actions.

Further Reading

Rozman, K., and Klaassen, C. D. (1996). Absorption, distribution, and excretion of toxicants. In *Casarett and Doull's Toxicology. The Basic Science of Poisons* (C. D. Klaassen, Ed.), 5th ed., pp. 91–112. McGraw-Hill, New York.

Roebuck, B. D. (1992). Absorption, distribution, and excretion of chemicals. In *A Primer of Environmental Toxicology* (R. P. Smith, Ed.), pp. 59–76. Lea & Febiger, Philadelphia.

—Jules Brodeur and Robert Tardif

Related Topics

Absorption
Blood
Developmental Toxicology
Excretion
Gastrointestinal System
Kidney
Liver
Metallothionein
Neurotoxicology: Central and Peripheral
Pharmacokinetics/Toxicokinetics
Skeletal System

Disulfiram

- ◆ CAS: 97-77-8
- ◆ SYNONYMS: Tetraethylthiuram disulfide; abstensil; alcophobin; alk-aubs; Antabuse; antadix
- ◆ PHARMACEUTICAL CLASS: Thiuram derivative with an unclassified therapeutic class
- ◆ CHEMICAL STRUCTURE:

$$(C_2H_5)_2NC \overset{\underset{\displaystyle S}{\|}}{{}} -S-S- \overset{\underset{\displaystyle S}{\|}}{C} N(C_2H_5)_2$$

Uses

Disulfiram is used to produce hypersensitivity to ethanol and serve as a deterrent to ethanol abuse.

Exposure Pathways

Disulfiram is only available in an oral form.

Toxicokinetics

Disulfiram is rapidly absorbed although up to 20% is excreted unchanged in the feces. However, clinical effects do not appear for 3–15 hr after ingestion. Disulfiram has a large volume of distribution secondary to its high lipid solubility. Its protein binding is approximately 50%. Disulfiram is metabolized in the liver to produce diethyldithiocarbamate, diethylamine, and carbon disulfide. Metabolites are excreted in the urine although a small amount is excreted from the lungs as carbon disulfide.

Mechanism of Toxicity

Disulfiram has multiple mechanisms of action. Its most well-defined action is inhibition of aldehyde dehydrogenase, which thereby retards the breakdown of acetaldehyde. The inhibition of aldehyde dehydrogenase after disulfuran may persist for 120 hr. Disulfuran has additional toxicologic actions including inhibition of the enzymes dopamine β-hydroxylase and those of the hepatic microsomal systems.

Human Toxicity: Acute

Acute overdose of disulfiram, in the absence of concomitant ethanol ingestion, may produce hypotension. When taken with ethanol, a constellation of severe reactions including flushing, vasodilation, pulsating headache, vomiting, and chest pain may occur. Less commonly, severe reactions including hypotension with shock, coma, seizures, and myocardial infarction may occur. An ethanol level as low as 5–10 mg/dl may produce this reaction with fully developed symptoms appearing when ethanol concentrations exceed 50 mg/dl. These toxic manifestations correlate with increased serum concentrations of acetaldehyde and may persist for 1 to 2 weeks after cessation of disulfiram use.

Human Toxicity: Chronic

In the absence of ethanol ingestion, chronic disulfiram use may produce adverse effects including fatigue, impotence, headache, dermatitis, and a metallic or garlic aftertaste. Neurologic complaints including vertigo, irritability, insomnia, slurred speech, and personality changes may occur. Less commonly, peripheral neuropathy, optic neuritis, delirium, and bizarre behavior may occur. Hematologic and gastrointestinal toxicity include blood dyscrasias and cholestatic hepatitis, respectively. Clinically important drug interactions include impaired metabolism of barbiturates, warfarin, and phenytoin; this may result in toxicity from these agents.

Clinical Management

Management of disulfiram ingestion is focused on gastrointestinal decontamination. After ingestion, gastric evacuation should be performed in patients presenting within 1 hr of ingestion. Activated charcoal should be administered. Supportive care should be provided as needed.

Animal Toxicity

Disulfiram is not used therapeutically in domestic animals. Its toxicity when ingested in overdose is undefined.

—Michael Shannon

Dithiocarbamates

- ◆ PREFERRED NAMES: Thiram (CAS: 137-26-8); ziram (CAS: 137-30-4); maneb (CAS: 12427-38-2); zineb (CAS: 12122-67-7)
- ◆ REPRESENTATIVE COMPOUNDS: Methyldithiocarbamates (metham); dimethyldithiocarbamates (DDC, ferbam, thiram, and ziram); diethyldithiocarbamates (sulfallate); ethylenebisdithiocarbamates (anobam, maneb, nabam, and zineb).
- ◆ SYNONYMS:

   Thiram—Arasan; Fernasan; Nomersan; Puralin; Tersan; Thiosan

   Ziram—Corozate; Fuclasin; Karbam White; Mathasan; Milbam; Nibam; Zimate

   Maneb—Dithane M-22; Manzate

   Zineb—Dithane Z-78; Lodacol; Parzate

- ◆ CHEMICAL STRUCTURE:

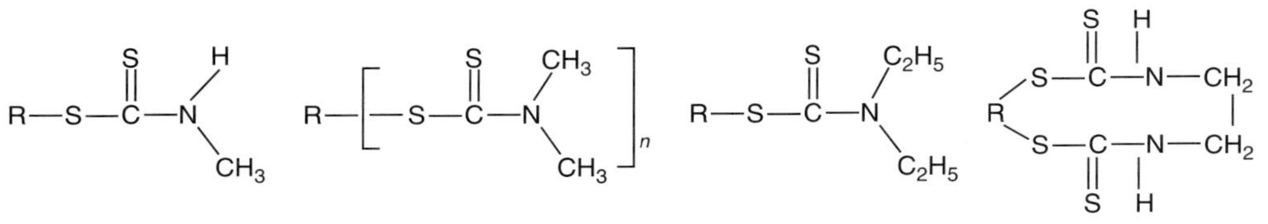

R=Na$^+$ (Metham)   R=Zn$^+$ (Ziram)   R=2—Chloroallyl (Sulfallate)   R=Mn (Maneb)   R=Zn (Zineb)

Uses

Most dithiocarbamates are used as fungicides. Dithiocarbamates are also used as herbicides and, at least one compound, Metham, is used as nematocide.

Exposure Pathways

Human poisonings have been reported following oral, inhalation, and dermal exposures to these compounds.

Toxicokinetics

Dithiocarbamates are well absorbed from the gastrointestinal tract and skin. The rate of dermal and gastroin-

testinal absorption is slower compared to that of the organochlorine and ester insecticides. Ethylenebisdithiocarbamates and ethylenethiourea, metabolic breakdown products of ethylenebisdithiocarbamates, may be absorbed through the lung, the gastrointestinal tract, and the skin. The absorption of ethylenebisdithiocarbamate from the gastrointestinal tract may be altered by the presence of cations occurring naturally in food.

Very little information is available regarding metabolism of these compounds. Indirect evidence suggests, however, that these chemicals undergo rapid metabolism and excretion in humans, usually within hours or days following absorption. The metal moiety of each ethylenebisdithiocarbamate is eliminated in the metabolic process. The initial metabolites of these ethylenebisdithiocarbamates are not identical to those of dimethyldithiocarbamates. A common metabolic product of all dithiocarbamate fungicides is carbon disulfide. Carbon disulfide may undergo further metabolism to form thiourea (an antithyroid substance), which may partially explain the tendency of different dithiocarbamates to affect thyroid function. Ethylene thiourea is reported to be metabolized *in vivo* and *in vitro* to ethylene urea with release of atomic sulfur. This sulfur atom can bind to macromolecules in the liver and may alter the activity of some enzymes in the endoplasmic reticulum. It is suggested that binding of this reactive sulfur atom in the thyroid gland may cause a decrease in iodination of tyrosine leading to thyroid dysfunction.

One study reported that the elimination half-life for ethylenethiourea was approximately 100 hr. Generally, dithiocarbamates are rapidly excreted through the kidneys. However, the effect of thiram on acetaldehyde dehydrogenase tends to persist for 10–14 days.

Mechanism of Toxicity

Thiram can precipitate an Antabuse (disulfiram) reaction in persons who have consumed a substantial amount of alcohol by inhibiting the enzyme, acetaldehyde oxidase. The ethylenebisdithiocarbamates have the same potential. Thiram and ethylenebisdithiocarbamates are metabolized partially to carbon disulfide in the body as mentioned before. This may partially explain the neurotoxicity of these compounds following high doses. The ethylenebisdithiocarbamates may be degraded in the environment and also in mammalian tissues to form ethylenethiourea, which may act as a goitrogen, mutagen, and carcinogen. The central nervous system (CNS) toxicity of maneb may be partially attributed to its manganese content.

Human Toxicity

As opposed to carbamate pesticides (see Carbamate Pesticides), exposure to dithiocarbamates does not precipitate symptoms of cholinergic crisis. Historically, systemic poisoning by dithiocarbamates has been rare. However, an Antabuse-like reaction (flushing, sweating, headache, weakness, hypotension, and tachycardia) may occur when ethanol is consumed following exposure to thiram and metallobisdithiocarbamates.

Interestingly, this is not typically observed following carbamate, monothiocarbamate, or ethylenedithiocarbamate exposure.

Ataxia, weakness, hypothermia, and ascending paralysis are possible neurologic symptoms after exposure to thiram and ethylenedithiocarbamate. Gastrointestinal signs following dithiocarbamate exposure include nausea, vomiting, and diarrhea. Renal failure has been reported following maneb exposure. Exposure to sprays, solutions, suspensions, and powders of these agents may cause skin and mucous membrane irritation. Thiuram, the ethyl analog of thiram, was reported to cause peripheral neuropathy in humans (characterized by pain, numbness, and weakness in the extremities). In one case, coma, seizures, and right hemiparesis have been reported following exposures to maneb and zineb. Occupational exposure to maneb-containing fungicides caused extrapyramidal symptoms in two agricultural workers. In addition to the previously mentioned symptoms, mental confusion, drowsiness, lethargy, and flaccid paralysis were reported to occur with thiram poisoning.

Clinical Management

Symptomatic treatment is recommended as there is no specific antidote available for poisoning by these compounds. In case of accidental oral poisoning, gastric lavage should be performed soon after ingestion. Absorption of these compounds may be prevented by administering activated charcoal slurry. Conventional anticonvulsant drugs may be used to treat seizures.

Animal Toxicity

All dithiocarbamtes have moderate to extremely low acute toxicity. The oral LD_{50} values of these agents vary from 285 to 7500 mg/kg (average; 2500 mg/kg). Large doses of thiram caused ataxia, and hyperactivity followed by clonic convulsion, loss of muscle tone, and dyspnea in rats and mice. Decreased fertility and impaired thyroid function were reported in cattle after repeated exposure to 200 mg/kg of zineb for 80 days. Dogs given ziram for 5–9 months (25 mg/kg/day) exhibited convulsions and some lethality.

—Tamal Kumar Chakraborti

Related Topics

Carbamate Pesticides
Pesticides

Diuron

- CAS: 330-54-1
- SYNONYMS: 3-(3,4-Dichlorophenyl)-1,1-dimethylurea; DCMU; DMU; Cekiuron; Crisuron; Dailon; Diater; Diurex; Duirol; Karmex
- CHEMICAL CLASS: Substituted urea herbicide
- CHEMICAL STRUCTURE:

Uses

Diuron is used for weed control and as a soil sterilant.

Exposure Pathways

Dermal and inhalation routes are the primary exposure pathways in occupational settings. Ingestion is also a possible route of accidental exposure.

Toxicokinetics

Diuron is absorbed from the gastrointestinal and respiratory tracts. It undergoes hydroxylation and dealkylation with the urea moiety generally unchanged. The metabolites are mainly excreted in the urine.

Mechanism of Toxicity

Diuron is a selective inhibitor of the Hill reaction of photosynthesis.

Human Toxicity

Diuron produces little toxicity in humans except irritation of the skin, eyes, and nose.
The ACGIH TLV-TWA is 10 mg/m^3.

Clinical Management

Treatment is symptomatic.

Animal Toxicity

Diuron exhibits low acute toxicity. The oral LD$_{50}$ in rats is >3 g/kg. Dietary protein levels may influence the toxicity of diuron. High acute dosage of diuron (about 3 g/kg) in rats may cause drowsiness and ataxia; animals that survived were irritable and hyperexcitable. Diarrhea, hypothermia, and clinical signs of renal dysfunction were reported. Subacute exposure of diuron may cause growth retardation and increase erythropoiesis.

—*Jing Liu*

Related Topic

Pesticides

Dominant Lethal Tests

Dominant lethal mutations are generally considered to be the result of mutations in germinal tissue which do not cause dysfunction of the gametes but which result in prenatal death of embryos heterozygous for the mutations. Thus, dominant lethal tests, usually conducted in rodents, assess the heritability of genomic mutations (i.e., mutations that can be passed to the next generation). Obviously, since embryos heterozygous for dominant lethal mutations do not survive, the mutations that result in dominant lethality cannot be passed to succeeding generations. However, the assumption is made that if dominant lethal mutations are present, other dominant and recessive nonlethal mutations could also be present and inherited by future generations.

Pioneering studies that provided the scientific foundation for the dominant lethal test were conducted in the 1930s, but the only known mutagen before the end of World War II was radiation. It was not until the late 1950s and the 1960s that, as a result of genetic research, scientists expressed concerns that chemicals might be hazardous to the germ-line of humans and suggested that routine toxicity testing of chemicals should include assays for mutagenicity. The dominant lethal test was one of the first genetic toxicology tests to be developed to assess the potential hazards of chemicals. In the early 1970s, the dominant lethal test (together with the host-mediated assay and the *in vivo* cytogenetic assay) was one of the original three screening tests recommended for evaluating the mutagenic effects of chemicals and, in the ensuing time, numerous approaches for examining dominant lethal mutations in rodents have been developed and evaluated. The dominant lethal test is no longer used as an initial, or first-tier, test largely because of the time and expense that are involved as well as the numbers of animals that are required to assess the results for statistical significance. However, the dominant lethal test has gained acceptance as a second-tier test for national and international regulatory submissions.

At one time, it was thought that the dominant lethal test was a relatively insensitive test in which few first-tier mutagens would be found positive, which led to recommendations that the numbers of animals utilized be increased to enhance the sensitivity of the test. Although contrary to efforts to reduce animal usage, if a test is conducted with too few animals for defensible results, it is even more wasteful of animals. However, because of the time and expense involved, if a newly developed chemical is found to be mutagenic in a first-tier test, a decision may be made to discontinue further development and plans to market the chemical rather than to initiate additional testing. Hence, dominant lethal testing for regulatory submissions is usually reserved for registration or reregistration of products with a wide or potentially wide usage for which there are no acceptable replacements.

Dominant lethal tests are usually conducted in mice or rats. The mouse dominant lethal test is more economical, but the rat dominant lethal test is more informative, as corpora lutea may be accurately enumerated in pregnant female rats, but not in mice, to assess preimplantation loss. Although dominant lethal tests may be performed with treated females, the tests are commonly performed with treated males in order to identify stages sensitive to mutation induction during germ cell development because the knowledge of stage sensitivity is important for risk evaluation. In a typical dominant lethal test, males are dosed with three levels of the test chemical with the objective of the highest dose being one that will exhibit some signs of toxicity but that will be low enough for a sufficient number of males to survive through the duration of the test. Following dosing, each treated male is mated to one or more

virgin females over each of a series of mating cycles. The females are euthanized mid-gestational term, and the uterine contents are examined to enumerate the number of live implantations (fetuses), early and late dead implantations (dominant lethals), and total implantations. If the test is conducted in rats, the corpora lutea are also enumerated to determine the number of ovulated eggs that fail to develop into fetuses (preimplantation loss). Data, including fertility indices, are then analyzed statistically based on groups of treated males, and each parameter and each mating group is evaluated independently.

Dominant lethal protocols can vary in the route and duration of exposure of the treated males, the number of males per dose level, the number of females per male per mating interval, and the number of mating intervals. However, it is necessary that dominant lethal tests be conducted with sufficient numbers of animals and mating intervals to maximize the test's sensitivity for detecting genomic mutational events and to provide information on germ cell-stage sensitivity of any mutagenic effects that may be observed. Thus, it is necessary to define protocols for each test material that will address these concerns as well as current regulatory requirements. Economies of testing may be achieved by using a laboratory's recently obtained (e.g., within approximately 12 months) historical positive control values for the same species and strain, with the same number of mating cycles, rather than using a concurrent group of positive control animals. However, even then, one dominant lethal test may involve more than 2000 animals.

Although current evidence indicates that the dominant lethal test is less sensitive than first-tier tests for assessing gene and chromosomal mutations *in vitro*, and although the correlation between positive dominant lethal results and carcinogenicity is apparently low, which would be expected, it has been found that dominant lethal results are highly predictive of the outcome of the even more expensive and time-consuming third-tier specific locus and heritable translocation tests for genetically transmissible mutations. Thus, the apparent lack of sensitivity of the dominant lethal test in comparison to *in vitro* tests and the lack of concordance of dominant lethal test results with carcinogenicity may reflect differences between *in vitro* and *in vivo* exposures of target cells, differences in the cellular and genetic mechanisms of carcinogenesis in somatic tissues and the induction of heritable mutations in germinal tissues, and/or the failure of some mutagenic chemical metabolites to reach germinal tissues because the dominant lethal test has high concordance with other mammalian germ cell mutagen tests.

—Ann D. Mitchell

Related Topics

Analytical Toxicology
Carcinogen–DNA Adduct Formation and DNA Repair
Carcinogenesis
Chromosome Aberrations
Developmental Toxicology
Host-Mediated Assay
Molecular Toxicology
Mouse Lymphoma Assay
Mutagenesis
Reproductive Toxicology
Sister Chromatid Exchange
Toxicity Testing

Dose–Response Relationship

Introduction

What toxicologists call "the dose–response relationship" is probably the single most important principle of toxicology as a scientific discipline. The concept in its simplest form states that as one modifies the amount (the "dose") of potentially toxic material to which a living organism is exposed, the occurrence and severity of the adverse effect (the "response") will also be modified. In human medicine, the dose-response relationship is clearly evident with most prescribed therapeutic agents: If the amount of drug administered is too small (below the therapeutic level), no biological effect occurs; if the dose is increased and

the amount administered lies in the therapeutic range, the beneficial actions of the medication appear; if, however, the dose is too large (above the therapeutic range), deleterious signs and symptoms (toxicity) may become evident. Toxicologists hold that the dose–response relationship applies not only to therapeutic agents but also to all chemical substances. The underlying principle is that the biological effects (beneficial or deleterious) of chemicals are due to the amount of active material at the site (or sites) of action and that the concentration or the amount of the substance at the site (the "internal" dose) is related to the amount of chemical administered (the "external" dose).

The relationship between the dose and the adverse toxic event can be described in either of two ways. In one situation, it may be viewed as a deleterious action observed in a single individual, but where the severity of the effect increases (or the number of toxic actions increases) in that individual as the dose increases; this is referred to as a "graded" relationship. In another set of circumstances, the toxic event may be described in terms of the numbers of individuals (or a percentage) of a defined population affected, where the frequency increases as the exposure (dose) increases; this is referred to as an "all-or-none" or "quantal" relationship. The demonstration of a dose–response relationship suggests causality between the degree of exposure and the adverse event.

Toxicity versus Hazard

While the dose–response relationship is of paramount importance in modern toxicology, the concept of a dose-dependent causal association in toxicology is not a modern idea. In the sixteenth century, Paracelsus, a Swiss physician, elaborated the thesis that all chemicals possess toxic properties, but that the amount of material (dose) to which the subject is exposed determines whether or not toxicity will occur. Traditionally toxicologists have made a clear distinction between the terms "toxicity" and "hazard." Toxicity is a qualitative term that denotes deleterious effects: All chemicals possess toxic properties. For some substances, the adverse property is observed when very small quantities are absorbed by the target species; for others, large amounts must be absorbed to exert an adverse effect. Hazard, on the other hand, is the likelihood that toxicity will occur given the manner in which the chemical is being used (the likelihood that a dangerous situation exists). To assess hazard, one needs to know the potential toxic properties of the chemical substance as well as the amount of substance available to the organism. The conditions of exposure will determine the amount of material absorbed by the organism. Thus, living organisms, including humans, can be exposed to potentially toxic substances without hazard (or danger), if the dose absorbed is sufficiently small and the exposure is nonrepetitive.

There is some confusion, unfortunately, in the traditional use of the terminology toxicity and hazard by toxicologists, as described previously, and the use of somewhat similar terms by other groups of professionals. In 1983, the National Research Council of the National Academy of Sciences of the United States published a report titled *Risk Assessment in the Federal Government: Managing the Process*; this work has had a marked influence on the risk assessment process used by regulatory agencies worldwide. The risk assessment process, in this report, consists of four components: hazard identification, dose–response assessment, exposure assessment, and risk characterization. "Hazard identification" in the context of the report is concerned with evaluating the potential adverse health effects of a chemical, mixture of chemicals, or process; thus, it is very similar to the traditional term toxicity used by toxicologists discussed in the previous paragraph. It is unfortunate that this confusion occurs since it detracts from the traditional use of the terminology toxicity and hazard used by toxicologists for many years before the publication of the NRC-NAS report.

Dose–Response Curves

The typical dose–response curve is usually sigmoidal in shape (Fig. D-28, top), when the response, expressed as a percentage of the frequency, is plotted against the dose, expressed as a logarithmic scale. The sigmoidal curve represents the cumulative curve of a normal (Gaussian) distribution of the response, where the response for each individual dose level, expressed as a percentage of the total minus the percentage responding at the preceding lower dose, is plotted against the log dose (Fig. D-28, middle). The resulting bell-shaped, normal distribution curve reflects the variation in susceptibility among individual subjects in a given population to the effects of the chemical. The subjects respond-

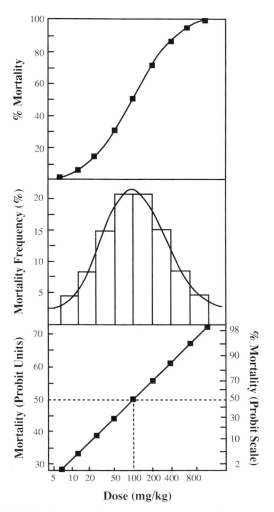

FIGURE D-28. *Graphical representation of a typical dose–response relationship for assessing mortality in laboratory animals receiving varying dosages of a toxicant. In all panels, the dosage (mg/kg) is plotted as a logarithmic scale. (Top) The response (% mortality) is plotted as a cumulative percentage of the total number of dead animals (the number of animals killed at a specific dosage level and all dosages below it are added together and the percentage of the total number is calculated). (Middle) Each bar (mortality frequency %) represents the percentage of the total number of animals that died at each dosage minus the percentage that died at the immediately lower dosage. The curve that joins the bars is the bell-shaped relationship known as the normal frequency distribution curve. (Bottom) The cumulative percentage of the total number of dead animals seen in the top panel is expressed in probit units on the left ordinate [mortality (probit units)] and as a probit scale on the right ordinate [% mortality (probate scale)] (reproduced with permission from C. D. Klaassen and D. L. Eaton, Principles of toxicology, in* Casarett and Doull's Toxicology *(M. O. Amdur, J. Doull, and C. D. Klaassen, Eds.), 4th ed., Pergamon, New York, 1991).*

ing at the lower doses (left side of curve) represent hypersusceptible subjects, whereas those responding at the higher doses (right side of curve) are considered resistant subjects; those situated at the middle doses represent the average responders.

Mathematically, if one plots the cumulative response along a probability unit ("probit") scale, the cumulative sigmoidal dose–response curve is transformed into a straight line. Such a transformation is illustrated in Fig. D-28 (bottom). Converting the data to straight lines is essential if one wishes to compare the dose–response characteristics of several chemicals because mathematical procedures exist for determining various parameters of such a linear relationship. One can assess the degree of change in response to the degree of change in dose; this is called the "slope" of the dose–response curve. With some agents, the change in response may be extremely abrupt (a relatively small change in dose results in a large change in response); this type of chemical is said to possess a "steep" dose–response curve. With other agents, the change may be quite small (a relatively large change in dose is required to elicit a small change in response); such a substance is said to exhibit a "flat" dose–response relationship. Linear curves can also be compared mathematically to determine if they are parallel to each other.

While the typical dose–response curve can usually be described by a sigmoidal or linear curve, it may be convex, concave, or even bimodal. These other configurations are usually the exception rather than the rule and depend on the mechanism of action of the material in question and even the presence of multiple toxicity sites. In some instances, the curve may be "J-shaped" or "U-shaped". This type of dose–response relationship is observed in a phenomenon known as "hormesis," a situation in which exposure to small amounts of material actually confers resistance to the agent before frank toxicity begins to appear following exposures to larger amounts.

The Threshold Concept

The dose–response relationship is crucial to another concept of major importance in modern toxicology— the existence of "thresholds." Toxicologists hold that a toxicant must be present at its cellular site of action in sufficient amounts to exert its deleterious effects. When the concentration is too small, it is said that the "threshold" has not been reached; therefore, the material does not exert an adverse action. An interesting stochastic analysis performed by Dinman in 1972 suggests that thresholds (critical concentrations) are essential for all mammalian processes; a concentration of about 10^4 molecules per cell has been estimated as the limiting concentration for biological activity. Even for the most potent toxicant known, botulinum toxin, calculations indicate that a dosage exerting no toxic effect in mice would yield about 10^7 molecules/kg body weight. It has been estimated that a 70-kg human consists of about 6×10^{13} cells; a limiting level of activity has been calculated to be about 9×10^{15} molecules/kg body weight, or about 1×10^{-8} M when expressed as a concentration. For very active substances, one has to take into consideration the fact that distribution in the body is not uniform, and that certain cells can exhibit preferentially high affinities for particular agents. Thus, the concept of thresholds exists for the most active of compounds.

There are a number of mechanisms that can lead to the demonstration of the existence of thresholds. The classical receptor agonist–antagonist relationships observed in pharmacology represent one example. Pharmacokinetic thresholds also exist, where absorption, distribution, biotransformation, and excretion can determine the effective "dose" of the potential toxicant at its biological target site. Even the tissue response itself can act as a type of threshold. For normal physiological processes, critical concentrations of vitamins or hormones are needed to avoid abnormal deficiency states.

From a regulatory standpoint, the threshold concept exerts important practical consequences. Safe exposure conditions for humans coming in contact with various chemical agents can be established. With pharmaceutical products, beneficial therapeutic interventions are possible without accompanying toxic effects. The use of ventilation systems to avoid buildup of harmful concentrations of solvents in ambient air in the workplace is another example. The establishment of permissible chemical residues in food products is commonplace. The application of this approach to drinking water is also reasonably established.

The applicability of the threshold concept to chemically induced cancers, however, has been questioned. Some scientists maintain that the nature of the carcinogenic process is such that one cannot assume the presence of a threshold. With the so-called genotoxic carcinogens, agents that act on DNA, this has been particularly the case. Most toxicologists would agree that even with such a type of carcinogen, a dose level exists below which one would not observe neoplasia; there would be, however, considerable division of opinion on its predictability. Therefore, in terms of prudence, thresholds are usually not applied in the risk assessment of this type of carcinogen by some regulatory bodies. On the other hand, with nongenotoxic carcinogens, a consensus favors the position that dosages which do not result in neoplasia can be reasonably estimated. Here, one sees more application of the threshold concept in risk assessment by regulatory bodies. As mechanisms of action become better understood, scientists are more likely to apply the concept of thresholds in the regulatory process.

Specific Dose-Dependent Values

Specific dose-dependent values on the linear dose–response curves can be estimated and statistically compared as well as the degree of variability (confidence limits) representative of the data being analyzed. Perhaps the most common specific toxicity value so determined in laboratory animals is the "median lethal dosage" or the LD_{50}. This value is the estimated dosage that would be expected to kill 50% of a given population of animals under the conditions of a particular laboratory test. With medicinal agents, another useful value related to the LD_{50} is the "therapeutic index." Here one is interested in the "median effective dosage" (ED_{50}) for a beneficial pharmacological therapeutic effect and how it compares to the toxic potency of the agent. One way of assessing this situation is to calculate an LD_{50}/ED_{50} ratio; the larger the ratio, the greater the relative safety of the chemical. Another way of assessing the relative safety of medicinal agents is to compare the ED_{99} (the dose that is effective in 99% of a given population) to the LD_1 (the dose that is lethal to 1% of the same population. The ratio LD_1/ED_{99} is called the "margin of safety"; the larger the ratio, the greater the relative safety of the medicinal agent.

Chronicity Index

The result of repetitive exposures to a given chemical may be different than that when exposure to the mate-

rial only occurs once or twice. The cumulative effects of repetitive exposures over time may render the agent more hazardous. This property can be estimated in animals by comparing the lethal potency (LD_{50}) of an agent, given only once, to its lethal potency when administered repetitively. The "chronicity index" is a term that has been applied to a ratio of the "1-dose LD_{50}" (animals receive the material only once) and the "90-dose LD_{50}" (animals receive the material repetitively each day for 90 days). If the 1-dose:/90-dose ratio is close to 1, this is an indication that repetitive administration does not result in cumulative effects or cumulative retention, whereas if the ratio increases and is larger than 1, it is quite likely that the agent exerts cumulative effects or is retained over the repetitive exposures. While this type of toxicological assessment is not performed routinely, the principle behind it is reasonably well understood. When the chronicity index is close to 1, this means that with repetitive exposures, it takes about the same amount of total material to be absorbed and retained to exert the toxic effect as it does when the total dose is given over a much smaller period of time. In contrast, when the chronicity index is larger than 1, it takes less total material to be absorbed and retained over time to exert the toxic effect than when the total dose is given over a much smaller period of time. Chemicals that are slowly degraded in the body and have high lipid solubility tend to accumulate and are retained in the body; they exhibit chronicity indices >1. Sodium chloride and caffeine are examples of two chemicals with chronicity indices of about 1 in laboratory rats, whereas the indices for mirex and dieldrin, two pesticides that are slowly degraded and very soluble in fat, are over 10.

Dose–Response Limits in Regulatory Toxicology

Although mathematical ratios are not usually derived from subchronic or chronic toxicity studies conducted with laboratory animals for regulatory purposes, the dose–response relationship is a very important part of such studies. Different dose levels are utilized in such experiments, and it is desirable to have at least one dose level where no biological effects occur following exposure. One finds various terms used to describe the severity of biological effects observed or extrapolated from such studies. The "no-observable-effect level"

(NOEL) is described as the highest dosage that exerts no significant difference between the exposed animals and the unexposed control group. The "lowest-observable-effect level" (LOEL) is the lowest dose used in a study that results in the appearance of some statistically significant mild, biological effect (beneficial or deleterious). The "no-observable-adverse-effect level (NOAEL) is the highest dosage where a mild biological effect is observed, but the effect is not considered deleterious to the well being of the animal. The "lowest-observable-adverse-effect level" (LOAEL) is the lowest dose used that results in the appearance of a mild, but deleterious effect. Finally, the "frank-effect level" (FEL) is a treatment level that results in the appearance of overt toxic effects. Figure D-29 is a graphical representation of where these various dosage levels might appear in a typical dose–response curve. Not all of them are observed in each subchronic or chronic study. Nevertheless, one usually sees an FEL, a LOAEL, and either a LOEL or a NOAEL. One can also extrapolate a NOAEL or a NOEL from the data derived in a subchronic or chronic study because these dosage levels can be important for establishing safety guidelines for humans.

The dose–response relationship is the basis by which regulatory bodies define under what limits humans can be exposed to potentially toxic chemicals and yet not suffer adverse effects. A number of different government bodies establish regulations to define safe exposure conditions. The "Acceptable Daily Intake" (ADI) is defined as the daily intake of a chemical which, during an entire lifetime, appears to be without appreciable risk; the ADI is used by the U.S. FDA for calculating permissible levels of nonfood ingredient residues in food (for instance, food additives or pesticides). The U.S. OSHA estimates "Permissible Exposure Limits" for chemical contaminants in occupational environments, to which workers are exposed for given periods of time during normal working conditions. Finally, the U.S. EPA makes use of the "Reference Dose" and "Reference Concentration" to estimate levels of noncarcinogenic environmental chemicals to which humans can be exposed during a lifetime without deleterious effects. All of these various safety indicators use NOAEL or NOEL values as part of the calculation.

Further Reading

Derelanko, M. J. (1995). Risk assessment. In *CRC Handbook of Toxicology* (M. J. Derelanko and M. A. Hollinger, Eds.), pp. 591–676. CRC Press, Boca Raton, FL.

Dinman, B. D. (1972). "Non-concept" of "no-threshold": chemicals in the environment. *Science 175*, 495–497.

Ecobichon, D. J. (1992). *The Basis of Toxicity Testing*. CRC Press, Boca Raton, FL.

Hayes, W. J., Jr. (1975). *Toxicology of Pesticides*, pp. 37–106. Williams & Wilkins, Baltimore, MD.

Klaassen, C. D., and Eaton, D. L. (1991). Principles of toxicology. In *Casarett and Doull's Toxicology* (M. O. Amdur, J. Doull, and C. D. Klaassen, Eds.), 4th ed., pp. 12–49. Pergamon, New York.

Lu, F. C. (1991). *Basic Toxicology*, 2d ed. Hemisphere, New York.

Ottoboni, M. A. (1984). *The Dose Makes the Poison*, 1st ed. Vincente Books, Berkeley, CA.

Plaa, G. L. (1995). Environmental toxicology. In *Planet Earth* (J. A. Leith, R. A. Price, and J. H. Spencer, Eds.), pp. 149–159. McGill-Queen's Univ. Press, Montreal.

—*Gabriel L. Plaa*

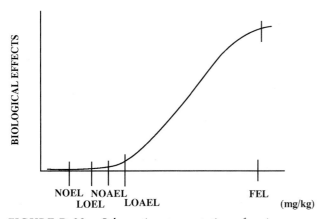

FIGURE D-29. Schematic representation of various dosage limits used in regulatory toxicology and where they usually appear in a typical dose–response curve. NOEL, no-observable-effect level; LOEL, lowest-observable-effect level; NOAEL, no-observable-adverse-effect level; LOAEL, lowest-observable-adverse-effect level; FEL, frank effect level (reproduced with permission from D. J. Ecobichon, The Basis of Toxicity Testing, CRC Press, Boca Raton, FL, 1992).

Related Topics

Exposure Assessment
Exposure Criteria
Hazard Identification
Hormesis

LD_{50}/LC_{50}
Levels of Effect in Toxicological Assessment
Maximum Allowable Concentration
Maximum Tolerated Dose
Median Lethal Dose
Permissible Exposure Level

Pharmacokinetics/Toxicokinetics
Risk Assessment, Human Health
Risk Characterization
Structure–Activity Relationships
Threshold Limit Values
Toxicity Testing

E. coli

- PREFERRED NAME: *Escherichia coli*
- DESCRIPTION: *Escherichia coli* are gram-negative, motile rod-like bacteria that vary in size.

Mechanism of Toxicity

Escherichia coli are complex, toxin-producing organisms. Most strains of *E. coli* require adhesion to the gut to produce disease, except those that are pathogenic and produce hemolytic–uremic syndrome.

Toxicokinetics

The onset of symptoms is 24–72 hr with a duration of approximately 1 week. This organism produces four distinct syndromes: enterotoxigenic *E. coli*, infantile enteropathogenic *E. coli*, enteroinvasive strains, and enterohemorrhagic *E. coli*.

Enterotoxigenic *E. coli*—this syndrome is produced by a heat-stable enterotoxin that causes travelers' diarrhea. This enterotoxin produces its effects by activating adenyl guanase and therefore increasing intracellular guanosine monophosphate.

Infantile enteropathogenic *E. coli*—this strain causes diarrhea in infants.

Enteroinvasive strains—these organisms penetrate and multiply within the epithelial cells.

Enterohemorrhagic *E. coli*—this strain of *E. coli* can produce hemorrhagic colitis and hemolytic–uremic syndrome. Damage to endothelial cells (caused by the toxin) of the glomerular capillaries and small renal arterioles may initiate the hemolytic–uremic syndrome.

Human Toxicity

Four distinct clinical syndromes are caused by pathogenic strains of *E. coli*: a dysentery-like picture (vomiting and diarrhea), a cholera-like illness, a hemolytic–uremic syndrome, and a hemorrhagic colitis.

Clinical Management

Travelers' diarrhea can be treated with antibiotic therapy. Pepto-Bismal reduces the incidence of travelers' diarrhea. *Escherichia coli* (0157 : H7) produces severe diarrhea and cramping. It resolves in about 1 week. When the type of food poisoning is identified, hospitalization is essential so that renal functions can be monitored.

—*Vittoria Werth*

Ecological Toxicology

Ecological toxicology is the study of harmful effects of substances on organisms, populations, communities, and ecosystems. Any substance, man-made or natural, is a potential subject of concern and study in the field of ecological toxicology. Substances of concern are those present in the environment at concentrations (or show signs of increasing to sufficiently high concentrations) that may elicit toxicological effects. The effects may be found in individual organisms, in populations made up of individuals, in communities made up of interacting populations, and in ecosystems. At the community and ecosystem levels of assessment, subtle impacts on individual organisms may become very difficult to discern. Impacts on individuals may be very relevant, but they may also pose insignificant impacts at the community and ecosystem levels. As the organization of the environmental components becomes more complex, ecological properties begin to emerge that were not readily discernible when only individual organisms were examined.

The development of ecological toxicology as a recognized field of specialization is a recent phenomenon which arose from its fledgling beginnings after the publication of *Silent Spring* by Rachel Carson in 1962. However, it had been understood for a long time that excessive levels of chemicals in the environment could have severe consequences. Excessive levels of salt were known to inhibit plant growth in biblical times. Agricola discussed the effects of various types of air pollution on miners in 1556. There had been numerous reports on effects on individual organisms of selected species of fish, birds, and mammals prior to 1962. These were then summarized and presented in an easy-to-read manner in *Silent Spring*.

In addition to concerns about toxic substances of human origin, the environment abounds with animal and plant toxins that point to a veritable chemical warfare among certain species. For instance, toxins have been used by snakes, spiders, and jellyfish as tools in capturing prey. Certain species of amphibians secrete potent toxins in their skin, making them virtually immune to predation. Many plant alkaloids represent effective defenses against being eaten by herbivores. Some of the naturally occurring toxins are among the most toxic substances known to humans.

The basic goal of the evaluation of effects of pollutants on the environment is usually stated in terms of an assessment of the impacts of a stressor on the structure and function of ecosystems. However, direct evaluations of the structure and function of ecosystems are extraordinarily complex and difficult to accomplish. The fundamental properties of ecosystems are often evaluated indirectly through the measurement of the abundances of individual species by assessing changes in species composition, biomass, diversity, primary productivity, colonization rates, and similar parameters.

Further Reading

Bourdeau, P., and Treshow, M. (1978). Ecosystem response to pollution. In *Principles of Ecotoxicology* (G. C. Butler, Ed.), SCOPE Report No. 12, pp. 313–330. Wiley, New York.

Carson, R. (1962). *Silent Spring*. Houghton-Mifflin, New York.

Moriarty, F. (1988). *Ecotoxicology*, 2nd. ed. Academic Press, New York.

National Research Council (1981). *Testing for Effects of Chemicals on Ecosystems. Report by the Committee to Review Methods for Ecotoxicology*, pp. xv, 103. National Academy Press, Washington, DC.

Peterle, T. J. (1991). *Wildlife Toxicology*, pp. xxi, 322. Van Nostrand Reinhold, New York.

Sheehan, P. J., Miller, D. R., Butler, G. C., and Bourdeau, P. (Eds.) (1984). *Effects of Pollutants at the Ecosystem Level*, SCOPE Report No. 22, pp. xv, 443. Wiley, New York.

—Rolf Hartung

Related Topics

Bioaccumulation
Bioconcentration
Biomagnification
Biomarkers, Environmental
Ecological Toxicology, Experimental Methods
Effluent Biomonitoring
Environmental Processes
Environmental Toxicology
Pollution, Air
Pollution, Soil
Pollution, Water
Population Density
Risk Assessment, Ecological

Ecological Toxicology, Experimental Methods

It is easiest to study effects on individuals in the laboratory. It is also comparatively easy to look for effects in individuals collected from the environment. Consequently, a large mass of information has been collected for effects observed on individual organisms. Many authors have suggested far-reaching conclusions regarding the health of the environment based on the effects that have been found in individuals. Although it is clear that effects found in individuals can represent information that may have significant consequences in an ecosystem setting, the specific effects have to be interpreted with great caution because the specific effects can range from adaptive responses that may be of little consequence to the individual to responses that signal incipient life-threatening damage to the individual. Therefore, efforts have been made to study interacting groups of individuals belonging to several species in microcosms, mesocosms, and field studies.

These indicators of ecosystem properties reflect the structure and functioning of the individual communities that are part of the ecosystem. In turn, the structure and functioning of the communities reflect the success of the populations of individual species, including their interspecific interactions in the form of competition, predator–prey relationships, etc.

The success of individual populations in an ecosystem context is ultimately determined by the aggregate success of the individuals that make up the populations, and the success of the individual is ultimately determined by its ability to function, to compete, and to reproduce within this convoluted population–community–ecosystem complex.

In this intricate organizational matrix, the individual organisms have been most susceptible to detailed study; and because the system is made up of diverse collections of individuals, adverse effects on individuals can have obvious consequences for the overall system.

However, there are some important difficulties in trying to make conclusions for the total structure and function of ecosystems based on the health status of individuals. The basic reason for this is that each level of organizational complexity introduces new interactive properties and compensatory processes that are not evident from the study of a lower level of organization by itself. For instance, while large changes in reproductive success of individuals will have a significant impact on the success of a population, slight or even moderate changes may not have any significant impacts on populations because of the multitude of other factors that operate in the environment that are also important determinants of the success of a population. At higher levels of organization, studies in microcosms that are

able to integrate many interspecific interactions are generally relatively poor predictors of ultimate effects in ecosystems, unless the impacts on the model systems are severe.

It became apparent fairly early that it was necessary to include interactions among species in any assessment of potential ecotoxicological effects. Thus, there were studies that looked at the influence of chemicals on predator–prey relationships, on competition, on parasitism, etc. (National Research Council, 1981). However, the experimental conditions to study effects on these interactions are usually so artificial that it becomes difficult to transfer the information to field conditions.

Microcosms

A considerable number of tests have been conducted using small subunits of natural systems, formerly called microecosystems. This term has been largely abandoned in favor of the term "microcosm." For aquatic systems, such microcosms may be assembled using known species at known densities (gnotobiotic microcosms), or they may be collected from the environment as they would normally exist as "transplanted microcosms." The aquatic microcosms are usually between 1 and 1000 liters in size and perform best with pelagic phytoplankton and zooplankton communities. Most of these systems cannot accommodate even small fish without losing all environmental realism. The reproducibility of these tests is strongly influenced by the size of the system and the species and their number of individuals that are initially included in the microcosm studies.

Highly synthetic micrososms have also been used to investigate the fate and transport of chemicals, especially pesticides. In this application the microcosms were able to provide qualitative and semi-quantitative information on the potential behavior of these substances in the environment.

Terrestrial microcosms are usually constructed to isolate a segment of the terrestrial environment (e.g., a plant with its intact root system and the natural growth medium). In this case the microcosm allows a more natural assessment of various factors (e.g., nutrient requirements, gas exchange, and growth). It generally does not allow for the assessment of plant–plant or plant–herbivore interactions.

Mesocosms

The major distinction between microcosms and mesocosms is one of size. This allows for a greater number of species, a greater number of organisms, a more varied set of physical and chemical characteristics, and therefore greater complexity. Mesocosms may range from 1000 liters and larger containers, usually located outdoors, to artificially constructed earthen ponds that can be 0.1 ha in size with a volume of 1 million liters. The distinction in sizes of microcosms and mesocosms overlaps, and there is no clear boundary. Likewise, the distinction between pond mesocosms and field studies lies only in the fact that the mesocosm is usually artificially constructed, and the field may not be. These distinctions become especially fuzzy when one tries to compare various types of mesocosm and field study designs.

Field Studies

Field studies fall into two major classes. The better defined studies have bounded areas which receive intentional measured treatments. There is no general agreement on the definition of field enclosures and the range of designs that are included under the definition of mesocosm studies.

Field studies also include investigations of contamination incidents where neither the treatment nor the extent of the affected area can be strictly defined. Many authors characterize these as studies at the ecosystem or landscape levels.

Bounded Field Studies or Enclosures

Enclosures of the air space above small segments of terrestrial plant communities have been used to measure the production and consumption of carbon dioxide and oxygen in relation to available light. Similar enclosures have also been used to look at impacts of air pollutants on community respiration and photosynthesis. Pesticides are commonly tested on selected species of wildlife in large pens and in field trials under the conditions of intended use.

Enclosed field studies have been used much more extensively for the aquatic environment. Artificial streams, constructed ponds, and littoral enclosures have been used extensively to assess impacts on experimentally introduced aquatic life and on other aquatic life that has been allowed to naturally colonize these water bodies.

Open Field Studies

There is some overlap between bounded and open field studies. Thus, a few studies have taken advantage of upstream–downstream comparisons in natural streams by dosing the stream with a test substance at an artificial weir so that downstream effects on biota could be assessed in comparison to unexposed upstream controls. In all open field studies, just as much attention needs to be paid to the measurement of exposure as to the measurement of effect.

Upstream–downstream comparisons of exposure and effects on biota are more commonly undertaken in the course of a point source discharge of effluents. A great deal of care must be taken to match the physical environments that are being compared in such studies because they have such a significant influence on habitat and therefore on local species distributions.

When the upstream–downstream comparisons are not possible, as is the case with terrestrial or lacustrine environments, then one needs to establish exposure gradients and examine the biota for associated gradients.

Endpoints

The ecological endpoints can be classified as structural endpoints, such as:

- Biodiversity (species richness)
- Abundance or population density
- Similarity (species and number of individuals per species present among sites)
- Biomass (total and per species)

These structural endpoints require the identification of the species that are present and their enumeration. This can become an overwhelming task when one has to deal with open boundary field studies.

Other ecological endpoints can be classified as functional:

- Primary productivity
- Respiration rates
- Mineralization rates
- Nutrient cycling
- Fluctuations in abundance
- Recovery after insult (resiliency)

The first four of these endpoints are difficult to evaluate without the use of isotopes, which limits the extent of their applicability. The last two endpoints require repeated species and population assessments through time.

Scaling

These indicators of ecosystem properties reflect the structure and functioning of the individual communities that are part of the ecosystem. In turn, the structure and functioning of the communities reflect the success of the populations of individual species, including their interspecific interactions in the form of competition, predator–prey relationships, etc.

The success of individual populations in an ecosystem context is ultimately determined by the aggregate success of the individuals that make up the populations, and the success of the individual is ultimately determined by its ability to function, to compete, and to reproduce within this convoluted population–community–ecosystem complex.

In this intricate organizational matrix, the individual organisms have been most susceptible to detailed study; and because the system is made up of diverse collections of individuals, adverse effects on individuals can have obvious consequences for the overall system.

However, there are some important difficulties in trying to make conclusions for the total structure and function of ecosystems based on the health status of individuals. The basic reasons for this are that each level of organizational complexity introduces new interactive properties and compensatory processes that are not evident from the study of a lower level of organization by itself. For instance, while large changes in reproductive success of individuals will have a significant impact on the success of a population, slight or even moderated decreases may not have any significant impacts on populations because of the multitude of other factors that operate in the environment that are also important determinants of the success of a population. At higher levels of organization, studies in microcosms that are able to integrate many interspecific interactions are generally relatively poor predictors of ultimate effects in ecosystems, unless the impacts on the model systems are severe.

Because it is easiest to study effects on individuals in the laboratory, and because it is easiest to look for effects in individuals collected from the environment, a large mass of information has been collected for such effects. Many authors have suggested far-reaching conclusions regarding the health of the environment based on the effects that have been found in individuals. Although it is clear that effects found in individuals can represent information that reflects potentially significant consequences in an ecosystem setting, the specific effects have to be interpreted with great caution because the specific effects can range from adaptive responses that may be of little consequence to the individual to responses that signal incipient life-threatening damage to the individual.

Further Reading

Cairns, J., Jr., and Niederlehner, B. R. (Eds.) (1995). *Ecological Toxicity Testing*. Lewis, Boca Raton, FL.

National Research Council (1981). *Testing for Effects of Chemicals on Ecosystems*, pp. xv, 103. National Academy Press, Washington, DC.

Rand, G. M. (Ed.) (1995). *Fundamentals of Aquatic Toxicology*, 2nd ed., pp. xxi, 1125. Taylor & Francis, Washington, DC.

Sheehan, P. J., Miller, D. R., Butler, G. C., and Bourdeau, P. (Eds.) (1984). *Effects of Pollutants at the Ecosystem Level*, SCOPE Report No. 22, pp. xv, 443. Wiley, New York.

Suter, G. W., II (1993). *Ecological Risk Assessment*, pp. xiv, 538. Lewis, Boca Raton, FL.

—Rolf Hartung

Related Topics

Bioaccumulation
Bioconcentration
Biomagnification
Biomarkers, Environmental
Ecological Toxicology
Effluent Biomonitoring
Environmental Processes

Environmental Toxicology
Pollution, Air
Pollution, Soil
Pollution, Water
Population Density
Risk Assessment, Ecological

EDTA

- CAS: 60-00-4
- PREFERRED NAME: Ethylenediamine-tetra-acetic acid
- SYNONYMS: Ethylenedinitrilo-tetra-acetic acid; Celon A; Cheelox; Edetic; Nullapon B Acid; Trilon B; Versene
- CHEMICAL CLASS: Organic chelating agent
- DESCRIPTION: EDTA is a white, odorless, crystalline (sugar or sand-like) material. It has a molecular weight of 292.28 and its melting point is at 2400°C. It is water insoluble.
- CHEMICAL STRUCTURE:

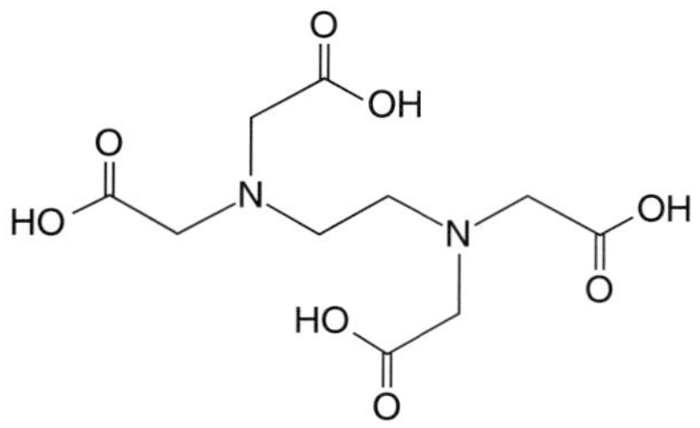

Uses

EDTA is used as a food additive, in herbicides, in pharmaceuticals, and in a variety of consumer products.

Exposure Pathways

The most probable routes of human exposure to EDTA would be ingestion and dermal contact. Workers in-volved in the manufacture or use of EDTA may be exposed by inhalation and dermal contact.

Toxicokinetics

EDTA is essentially not metabolized by the human body and it is rapidly excreted in the urine. About 50% of EDTA administered intravenously is excreted within 1 hr and 90% within 7 hr. EDTA and its metal chelates do not permeate the cellular membrane to a significant extent; thus, most of the EDTA remains in the extracel-lular fluid until excreted in the urine.

Mechanism of Toxicity

All known pharmacological effects of EDTA result from formation of chelates with divalent and trivalent metal ions in the body. Also, the effects on rat liver glucocorticoid receptor in vitro have been studied. At 4°C, 10 mmol EDTA had a stabilizing effect on un-bound hepatic glucocorticoid receptors. Apparently, endogenous metal ions are involved in the processes of glucocorticoid–receptor complex stabilization and transformation. Furthermore, EDTA increases the ab-sorption of a number of agents. This effect is nonspe-cific because EDTA increases the absorption of bases, acids, and neutral compounds. It appears that by che-lating calcium, EDTA causes a general increase in mem-brane permeability.

Human Toxicity

The following acute (short-term) health effects may occur immediately or shortly after exposure to EDTA: contact may irritate the skin causing a rash or burning feeling; contact with high concentrations may irritate the eyes; and inhalation of EDTA dust may irritate the nose and throat.

The FAO/WHO acceptable daily intake for EDTA is 0–2.5 mg/kg body weight.

Clinical Management

In case of contact with EDTA, the eyes should be flushed immediately with running water for at least 15 min. Affected skin should be washed with soap and water. Contaminated clothing and shoes should be removed and isolated at the site.

—Cephas C. Barton and Harihara M. Mehendale

Effluent Biomonitoring

Introduction

The Clean Water Act mandates the reduction of water pollution sources, largely through control of point source discharges. Under the National Pollutant Discharge Elimination System (NPDES), municipal and industrial entities that discharge wastewater (e.g., sewage, pulp, and paper) into national waterways are required to obtain a permit and meet imposed effluent limitations. These limitations, designed to be protective of water uses, human health, and aquatic life, are generally expressed in terms of a numerical limitation for a specific chemical (e.g., 10 μg/liter of copper). However, one shortcoming to chemical-specific requirements is that bioavailability of multiple chemical species in complex effluents is not directly evaluated; thereby providing little information on an organism's ability to integrate the effects of prolonged exposure to multiple contaminants.

In addition to chemical-specific requirements, toxicity requirements can also be imposed to ensure that the discharge is meeting the goal of protecting aquatic biota. NPDES permit requirements for aquatic toxicity may be numerical limitations or merely biomonitoring requirements. Whole-effluent toxicity (WET) testing monitors whether the discharge is likely to be toxic to aquatic life by exposing test organisms to various concentrations of the effluent and observing their response. The principal advantage to WET testing over chemical-specific requirements is the direct and integrated biological assessment of antagonistic, synergistic, or additive effects from multiple chemical interactions. During both acute (short-term) and chronic (long-term) WET exposures, generally invertebrate and fish species are challenged with various concentrations of effluent under controlled conditions, and lethal or sublethal endpoints are measured. In compliance with biomonitoring, if toxicity is observed above a certain magnitude, defined in terms of acute or chronic "toxic units" (TU_a and TU_c), duration (e.g., time averaged instream concentration), and frequency (e.g., not to be exceeded more than once every 3 years), additional studies may be required to evaluate, treat, and reduce effluent toxicity. [The TU_a is defined as $100/LC_{50}$, and the TU_c is defined as $100/$(no-observed-effect concentration (NOEC), or $100/$chronic value which is defined as the geometric mean of the LOEC and NOEC.]

This entry is intended to provide an introduction and practical analysis of the current standardized aspects of aquatic toxicity biomonitoring of industrial and municipal effluents. These tests and monitoring programs are needed to determine whether management requirements or regulatory criteria are being met as well as to assess the temporal and spatial trends in water quality.

Acute and Chronic Tests

A battery of aquatic biological tests have been developed to evaluate the toxic effects elicited from effluents. For all the types of toxicity tests, selection of exposure concentrations and duration, test species and strains, and monitored water quality parameters are critical. The short- and long-term methods for evaluating the potentially toxic discharges have been well characterized.

Acute Toxicity Tests

Acute bioassays are designed to assess the effects of toxic substances that occur within a relatively short period after exposure (48–96 hr). The effluent toxicity elicited by test organisms is often fatal and rarely reversible. The relevant information to be gained from this type of test is the distribution of the exposure–response relationship and the nature or potency of the toxic effects, such as immobility, percentage mortality (LC_{50}), and time interval to mortality.

Ideally, the exposure duration should be long enough to allow for steady-state conditions (body burden and elimination of toxics) to occur, although this is not always achieved.

The responses monitored are generally binary variables—meaning "all or none" (e.g., mobile or immobile)—with known sample size and unknown probability, versus more graded or continuous observations.

Although crude as an endpoint, mortality is highly visible, clearly defined, and measurable and has been utilized as a first tier in hazard prediction. The acute procedure is often applied in situations in which the effluent is diluted at least 80 times; if there is less dilution of the effluent by the receiving water, then the 7-day chronic test often is performed.

Acute tests can be further divided with respect to flow regime: typically either static or static renewal. During the static procedure, test organisms are exposed to the environmental sample in relatively nontoxic and nonreactive cups, glass beakers, or aquaria and the test solution is not changed. Advantages to this design include ease of operation, minimal space needed, and minimal waste generated. This system is relatively simple and cost-effective for evaluating large numbers of samples or where only limited volumes of effluent are available. However, the concentrations of the effluent components—dissolved oxygen, metabolic by-products, and hydrogen ions—may change throughout the test due to complex biochemical events: uptake by the test organism, adsorption on the organism or on the walls of the test container, biodegradation, vaporization, and precipitation. The potential accumulation of metabolic or other wastes may lead to undue stress on the test organism and variable results. These difficulties may be minimized by using the static-renewal test.

During the static-renewal acute test, the exposure solutions are periodically renewed (24 or 48 hr) either by carefully transferring the test organisms to a freshly prepared exposure solution or by gently decanting and refilling test containers. Abrupt renewal of test concentrations may also introduce stress to the test organisms. In contrast to the static procedure, the static-renewal system is designed to mimic more natural intermittent exposure scenarios (e.g., acid rain and agrochemical applications) and mitigate the changes induced by unstable, volatile, or high oxygen-demanding effluents.

Weaknesses in the results of acute lethality tests include:

- Lack of ecological relevance due to laboratory culture conditions (e.g., test organisms more or less sensitive, more or less crowded, than indigenous species)

- Restriction to monitoring a single life stage of uniform size under a set of controlled conditions (versus varying temperatures, seasons, food supply, predation, etc.)

- Lack of direct relevance to the susceptibility and bioavailability of other life stages.

In addition, laboratory-prepared water used for diluting the effluent is relatively free of suspended solids, humic matter, and other varying components (e.g., hardness level) that may serve as nutrients, provide sorption sites, and have direct effects on the functional expression of toxicity. Thus, toxicity as expressed by the LC_{50} should not be viewed as a biological constant but as a value that varies with age and physical, chemical, and biological factors.

Chronic Toxicity Tests

Chronic whole-effluent toxicity tests involve exposing organisms (usually emphasizing presumably sensitive early life stages, such as neonates or newly hatched larvae) to various effluent concentrations for a period of typically 7 days, during which the exposure solutions are renewed daily. The 7-day cladoceran survival and reproduction test is one of several procedures employed for estimating the chronic toxicity of effluents. These longer term tests provide information on effluent effect endpoints such as fecundity, growth, reproduction, life span, behavior, and mortality.

Specific exposure concentrations for effluent dilution are often evenly spaced in a linear or geometric series (e.g., 100, 50, 25, 12.5, and 6.25% effluent). Replicate exposures are required, with test vessels arranged in the incubator in a randomized block design. At the end of the test, statistical calculations of NOECs for survival and reproduction or growth are typically performed as a multiple comparison procedure of effluent-treated groups. The induced effects of effluent pollutants on growth or reproduction and on survival may or may not be equal, suggesting information on the relative sensitivities of the test organisms. The test endpoints and duration are assumed to demonstrate that the aggregated substances in the effluent have or have not been protective of aquatic life.

Advantages to the chronic test include its attempts to detect longer term, and perhaps more subtle, sublethal, toxic effects by studying relatively susceptible neonates or larvae and their reproduction or growth. The 7-day chronic procedure may mimic intermittent and fluctuating pollutant exposures in nature through daily renewals, thereby allowing for organism recovery, adaptation, acclimatization, or simple stressor avoidance.

In general, a greater proportion of aquatic organisms are exposed to sublethal concentrations of toxics compared to acutely lethal concentrations. There is general acceptance that chronic exposure of fish to sublethal levels of toxics makes them more prone to disease states. Although scientifically controversial, intuitively, one can grasp how individual responses such as growth, reproduction rates, and survival probability, as a function of age, could relate to population level responses.

One limitation to chronic WET tests is that they do not indicate which stressor(s) is causing the observed effects. Furthermore, they are generally subject to false positives (type I error) and false negatives (type II error) because of weak correlation between WET test results and in-stream effects. For instance, during the 7-day fathead minnow survival and growth test, growth rate endpoints (dry weights compared to controls) may not be a reliable indicator of latent toxic effects since there is potential for reproductive failure in successive generations even in the absence of statistically poor growth.

A related general weakness in WET testing schemes involves the natural "variability" of effluents, and whole-effluent tests, which may be unrelated to the actual effluent toxicity but related to short-term spatial, temporal, and seasonal variation at a site. More frequent effluent testing may identify these atypical toxicity responses.

Another major disadvantage of the chronic WET tests is referred to as "simulated toxicity," whereby the observed toxicity is attributable to adverse interactions from biological growth of freshwater pathogenic (e.g., *Aeromonas hydrophilia*) or sheathed (*Sphaerotilus natans*) bacteria on the test organisms and is not necessarily a manifestation of chemical constituents present in the effluent. The microbial growth may be a result of contaminated culture conditions (e.g., masses of bacterial growth surrounding brood pouch eggs) or microbial proliferation due to nutrients associated with the effluent.

Statistical disadvantages in chronic testing data analyses have been presented, suggesting that NOECs are misleading and should be phased out of regulatory use. This research suggests that NOEC concepts do not consider basic variability and are artifacts of the test design in terms of effluent concentrations and intervals chosen. As an alternative, regression analysis has been proposed as a more robust procedure than traditional hypothesis testing to mitigate the problems of violation of assumptions, experimental error, and variability in test protocols.

Physical, Chemical, and Biological Test Factors

Although variability is inherent in all environmental measurements, a variety of physical, chemical, and biological factors should be noted in compliance biomonitoring to enhance the consistency and defensibility of toxicity test results. These various parameters may affect the effluent toxicity to aquatic biota, and it is important that the investigator take them into account. Test conditions should mimic receiving water conditions, whenever feasible, to allow accurate assessment of the in-stream effect of an effluent. However, the use of upstream water for dilution should be avoided due to the potential variability in quality over the testing period.

Physical considerations of particular importance in aquatic toxicity testing include exposure temperature, periodicity and intensity of light, test organism loading, test duration, laboratory equipment, test methods, and data recording. For instance, control and test solution volume should be sufficient to allow organisms free mobility, to provide adequate supply of dissolved oxygen (e.g., >4 mg/liter), and to prevent buildup of metabolic waste products such as ammonia. Chemical considerations of interest that contribute to observed WET toxicity include pH, alkalinity, hardness, salinity, dissolved gases, organics, inorganics, sediments, and humic substances. For instance, the bioavailability and toxicity of metals to aquatic biota are known to differ with pH, hardness, and oxidation state. Fluctuations of pH levels of a receiving water can result in over- or underpredicted effluent toxicity. Biological factors inherent to biomonitoring include test organism age class, size, genetic state, acclimation to culture conditions, nutritional health, parental care, and food quality and quantity. For instance, dietary factors such as algal digestibility and availability affect test organism body size, sensitivity, and performance, as evidenced in *Ceriodaphnia* reproduction tests.

Laboratory Dilution Water

Water selected for diluting the effluent should be of uniform quality, free of contaminants (e.g., benzene and pesticides), and available in large enough volumes for culture maintenance, acclimation, and testing. Based on these minimal requirements, synthetic laboratory water prepared with reagent-grade chemicals and deionized water is particularly useful for most small laboratory operations and where precise water quality parameters are desired (e.g., hardness). Water quality parameters such as alkalinity, hardness, pH, and specific conductance should be measured to check for consistencies between batches, as part of good laboratory practice. However, if a variety and numerous types of toxicity tests are performed routinely, then a high-quality surface water such as a large lake may be desirable. Generally, lake water is less prone to changes in quality (suspended solids, organic matter, and runoff contaminants) that may affect results compared to river water. A municipal water supply is least desirable for laboratory diluent due to potential interferences from the constant oxidation of residuals needed for safe and potable water.

Test Organism Selection and Culture

Invertebrates and fish are typically used in toxicity testing of effluents and are the primary focus of most historical and present methods (ASTM, U.S. EPA, and OECD). For acute and chronic effluent testing, the most typical invertebrate species is the water flea (*Daphnia magna*, *D. pulex*, or *Ceriodaphnia dubia*), and the routine vertebrate test species is the fathead minnow (*Pimephales promelas*). These test species have held regulatory acceptance and meet regulatory guidelines. Another reason these test organisms have been favored is their relatively small size, lending themselves to practical considerations such as availability of the effluent and sensitivity to toxics. Aquatic criteria often require acute and chronic toxicity data on a range of organisms representing many families. Standardized culturing techniques are required to assure the body size, performance, health, and sensitivity of test organisms and are continually being refined. The quality and quantity of food for cultures–microalgae (*Selenastrum capricornutum*), cereal leaves, trout chow, yeast, rotifers, or brine shrimp–are critical for individual energy assimilation and balance and for precise and quantifiable data.

Good Laboratory Practice

Good laboratory practice regulations govern the planning, experimental design, and conducting of whole effluent toxicity studies and are described in federal publications and toxicity testing manuals (e.g., Code of Federal Regulations, U.S.). In laboratories, much cost and effort are associated with the handling of samples and information as well as the test itself. Good laboratory quality assurance begins before a sample is accepted, carries through during sample log-in and testing, and continues after completion of the analysis.

This system of quality assurance was introduced to ensure that toxicity tests are competently performed and that data are not fabricated. A sound, written quality assurance program is an essential basis for laboratory operation and is often found in the form of standard operating procedures. The quality assurance program functions not only to monitor the reliability of data recorded and reported but also to control data quality in order to meet regulatory requirements. Clearly written protocols for each procedure employed eliminate or reduce errors in laboratory operation caused by factors such as personnel (qualifications and technical competence), supplies (e.g., purity of reagents), equipment calibration and maintenance, sample storage and handling, and analytical methods. Quality control describes the set of quantifiable measures, including established protocols and standard equipment, used to define daily laboratory activity.

The good laboratory practice criteria for whole effluent toxicity tests include species acceptability, exposure system conditions, physical and chemical conditions, and statistical data analysis methods. For instance, the test acceptability criteria for the larval fathead minnow 7-day chronic tests involves having 80% or greater survival of controls and an average dry weight of surviving control fish equal to or greater than 0.25 mg.

Sample Collection and Handling

Care should be taken that the sample collected is representative and undergoes minimal changes prior to toxicity evaluation. Hence, a 24-hr composite sample obtained with a refrigerated, proportional flow sampling device is most ideal. This type of sample could then be readily shipped to the contracting laboratory in a cooler packed with ice to maintain sample integrity. A "chain-of-custody" form must accompany the shipment indicating the source and type of sample, time of collection, whether pre- or postchlorination, and the name of the individual who collected the sample.

Upon arrival, the sample should be logged in the laboratory record book indicating time of arrival, sample temperature, and pH. The sample should in all cases be analyzed for residual chlorine and, if present, oxidized with sodium thiosulfate before it is employed for toxicity evaluation. A portion of the sample should also be removed for alkalinity and hardness analyses. Often it is necessary to coarse filter the sample to remove floc or suspended debris before testing; however, this practice may reduce the sample's toxicity. The remainder of the sample should be kept at 4°C for a period not to exceed 72 hr after initial sample collection. It is desirable to employ two separate 24-hr composite samples for performing a 96-hr acute larval fathead minnow test. This would allow renewal after 48-hr exposure. In the 7-day tests with *Ceriodaphnia* and *Pimephales*, three separate 24-hr composite samples should be employed for the daily renewal of the various exposure solutions. Toxicity data summary sheets should include daily routine physical–chemical measurements and sample information. It is essential that good laboratory practices are used in all aspects of sample collection, treatment, and analysis to obtain quality and defensible results.

Data Analysis

Statistical methods in effluent toxicity evaluations enable the investigator to quantify the observed exposure–response relation, with reference to the desired endpoint. The resulting statistical confidence in the data may then be used for comparison to replicate tests and for regulatory decision making.

Acute Methods

Many models have been used to calculate LC_{50} results, such as graphical interpolation, moving average, probability unit (probit), logistic unit (logit), Litchfield–Wilcoxon, and Spearman–Karber (often trimmed) methods. Commonly used statistical software includes ToxStat, U.S. EPA Probit Analysis, and SAS Institute. The challenge then arises as to which model to choose, given the toxicity data. The most obvious choice is the model that holds the most biological support. However, the answer is that there is not much biological basis for these models, and the investigator is left to choose the most appropriate model based on the test results and standard methods.

For instance, the probit and logit models assume that the organism tolerances have a lognormal distribution, with a typical exposure–response curve given by a sigmoidal-shaped curve. Both methods utilize transformations and curve fitting of the data and require varying observations of partial mortalities. The criterion for acceptable acute tests in terms of mortality in the controls is often specified to be ≤10%.

Chronic Methods

In terms of the statistical methods of the sublife cycle whole effluent tests, survival and reproduction data from the 7-day exposure are analyzed using hypothesis testing to determine "acceptable" concentrations. In preliminary statistical analyses, the data are tested for normality and homogeneity of variance. Dunnett's test is typically used to compare the treatment mean with the control mean in order to obtain the NOEC and the lowest-observed-effect level for each endpoint. If either the test for normality or homogeneity fails, then the nonparametric Steel's Many One Rank test may be used with the cladoceran test data.

Test Failures

Whole effluent toxicity tests may fail for primarily two reasons: an aborted test due to a lack of good quality data generated (e.g., organism health or effluent sample exceeded its holding time of 72 hr), or whole effluent toxicity is demonstrated to exceed permitted levels.

In terms of the second case, whole effluent toxicity may be the direct result of unidentified and persistent contaminants discharged to a wastewater treatment facility. These influent sources of test failures include refractory substances ("pass through toxics") in effluents, such as floc and coagulating agents, or pesticides such as diazinon, which cause observed patterns of strong toxicity to waterfleas. Furthermore, ammonia may be present in concentrations which cause WET test failures even though the effluent discharge meets NPDES permit numerical limitations. Although the federal regulations do not define the "failure" of an effluent toxicity test as an automatic permit violation (subject to fine), in practice, single-species WET tests have been treated in this manner. In fact, the significance of episodic exceedance of WET limits should depend on a host of additional factors, such as the receiving water conditions, mixing zone, and duration of events.

Often, regulatory requirements are such that test failures trigger additional evaluations. As part of the toxicity-based approach to effluent permitting, toxicity identification and reduction evaluations (TI/RE) schemes were implemented to identify and reduce the toxic components of complex effluents (e.g., un-ionized ammonia, hexavalent chromium, and chlorides) and assist managers in controlling toxics. The TIE process involves the stepwise chemical manipulation (pH adjustment, filtration, aeration, C18 solid-phase extraction column, EDTA chelation, and sodium thiosulfate treatments) of the effluent to identify specific compounds causing toxicity and render them biologically unavailable.

However, before implementing potentially time-consuming and expensive toxicity reduction studies, it is important to understand their capabilities and limitations. For instance, past protocols have had the greatest success with effluent constituents that cause acute toxicity and have questionable usefulness for chronic toxicity. In addition, no method ensures success for very complex waste mixtures (e.g., polar organic compounds), which are exceedingly difficult to analyze.

Furthermore, a TRE may vary between states. The U.S. EPA's TRE is based on compliance with whole effluent toxicity, but the state of Pennsylvania's toxics reduction evaluation addresses compliance with chemical specific water quality standards and implements action-oriented programs providing technological solutions (e.g., pretreatment facilities) for pollutant removal. Thus, various toxicity reduction options may be available and should be evaluated before initiating time-consuming and expensive studies.

Alternative Test Methods

There are major efforts in support of replacing *in vivo* vertebrate tests, refining existing approaches, and reducing test organism mortality. In terms of the adequacy or inadequacy of current *in vivo* designs, questions have arisen as to their accuracy in predicting hazards to humans and aquatic life, inter- and intralaboratory variation, and ethical concerns on how animals are used. For instance, in the United Kingdom, fish are protected under the Animal Scientific Procedures Act of 1986 as soon as they are capable of independent feeding. Furthermore, there has been a general evolution in toxicology studies from a merely descriptive science to one describing actual mechanisms of action.

The bacterial luminescence toxicity assay Microtox (*Vibrio fischeri*) has achieved international recognition for its usefulness in detecting acute and chronic aqueous toxicity that correlates with traditional invertebrate and fish species. Also, microalgae (*Selenastrum*), brine shrimp (*Artemia*), plants (*Lemna*), mysid shrimp (*Mysidopsis*), and fish cells (both primary and lineages) have gained acceptance as useful alternatives for measuring aquatic toxicity. In conjunction with chemical-specific analysis, a general consensus is that a battery of biological tests should be utilized to broadly characterize toxicity because each test organism and system responds and characterizes toxicity uniquely. Overall, water quality biomonitoring must harbor goals of cost-effectiveness, sensitivity, relevance, and precise results.

Further Reading

ASTM (1993). *ASTM Standards on Aquatic Toxicology and Hazard Evaluation*. ASTM, Philadelphia.

Australian and New Zealand Environment and Conservation Council (ANZECC) (1992). *Australian Water Quality Guidelines for Fresh and Marine Waters*. ANZECC, Canberra, Australia.

OECD (1987). The use of biological tests for water pollution assessment and control. In: Proceedings of the International Seminar on the Use of Biological Tests for Water Pollution Assessment and Control. Varese, Italy: ISPRA Research Centre (OECD Environmental Monograph No. 11).

Rand, G. M. (Ed.) (1995). *Fundamentals of Aquatic Toxicology*, 2nd ed. Taylor & Francis, Washington, DC.

Society of Environmental Toxicology and Chemistry (SETAC) (1996). *Whole Effluent Toxicity Testing*, Workshop proceedings, Pellston, Michigan, September 1995. SETAC, Pensacola, FL.

U.S. EPA (1991). *Technical Support Document for Water Quality-Based Toxics Control*, EPA/505/2-90/001. U.S. EPA, Office of Water, Washington, DC.

U.S. EPA (1993). *Methods for Measuring the Acute Toxicity of Effluents and Receiving Waters to Freshwater and Marine Organisms*, EPA/600/4-90/027F. U.S. EPA, Washington, DC.

U.S. EPA (1994). *Short-Term Methods for Estimating the Chronic Toxicity of Effluents and Receiving Water to Freshwater Organisms*, EPA/600/4-91/002. U.S. EPA, Washington, DC.

—Peter G. Meier and Leonard I. Sweet

Related Topics

Bioaccumulation
Bioconcentration
Biomagnification
Biomarkers, Environmental
Clean Water Act
Ecological Toxicology
Ecological Toxicology, Experimental Methods
Environmental Processes
Environmental Toxicology
Laboratory Practices
Microtox
Pollution, Water
Population Density
Risk Assessment, Ecological
Toxicity Testing, Aquatic

Emergency Response

The tragic release of methyl isocyanate at Bhopal, India, in 1984, which killed an estimated 2000 people, raised the awareness of the world to the potential hazards of chemical accidents. Previous to this, the release of 2,3,7,8-tetrachlorodibenzodioxin at Seveso, Italy, contaminated a large area and resulted in community evacuation and 175 cases of chloracne. Although these major incidents are quite rare considering the amounts of chemicals used worldwide, adequate planning and risk management can prevent such occurrences.

Furthermore, experience has shown that a well-managed emergency response program can minimize the consequences of a chemical release. In-depth planning is essential to ensure that actions by the chemical facility and the community mitigate the impact of any release on people and the environment.

Exposure Guidelines

To evaluate the effects of possible accidental chemical exposures, toxic endpoints are needed for each chemical of interest. Historically, chemical exposure guidelines have been developed for occupational exposures and not for emergency situations. The major guidelines for occupational exposure in the United States are the ACGIH TLVs, the OSHA PELs, the AIHA Workplace Environmental Exposure Levels, and the NIOSH Recommended Exposure Limits. None of these guidelines are designed for emergency situations or even single, acute exposures in general. Furthermore, they are designed for mainly healthy workers and do not normally consider more susceptible members of the population. Temporally, they are designed for long-term exposures such as 8–10 hr per day, 5 or 6 days per week for several years up to a working lifetime. In contrast, an emergency exposure is, by definition, a rare event of short duration, and possibly at a high or unknown concentration which could involve a heterogeneous population.

Currently, there are two major reference sources available specifically for use in emergency exposures: the National Research Council (NRC) and the American Industrial Hygiene Association (AIHA). Emergency Exposure Guidance Levels (EEGLs) have been developed by the NRC for use by the military. They have been developed for a population of healthy young adults and are not applicable to the general public. They provide a single value which is the estimate for the highest exposure which will not interfere with one's ability to perform specific tasks. While Short-Term Public Emergency Guidance Levels have also been developed by the NRC to address exposure to civilians living in and near military installations, these cover only a few chemicals.

Emergency Response Planning Guidelines (ERPGs), published by the AIHA, are specifically designed for use in chemical emergencies and are applicable to the general population including susceptible (but not hypersusceptible) populations. The values are intended to provide estimates of concentration ranges above which one could reasonably anticipate observing adverse effects as described in the definitions for ERPG-1, ERPG-2, and ERPG-3 (Fig. E-1) as a consequence of exposure to the specific substance. Currently, there are ERPGs on the most commercially important highly hazardous chemicals (about 70; see Table E-1), and work is under way to cover the remaining important chemicals of concern.

In 1993, NRC recommended that Community Emergency Exposure Limits be developed to cover

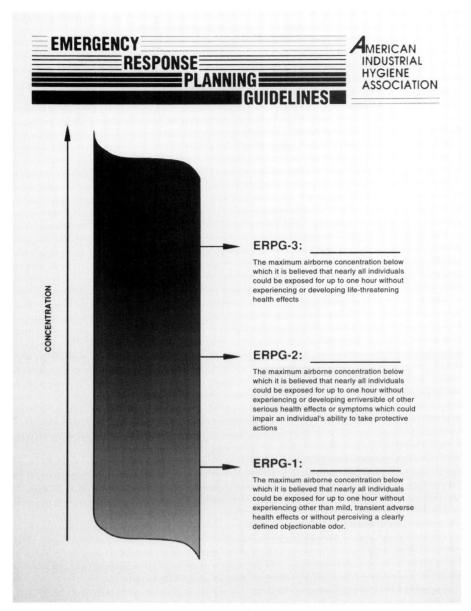

FIGURE E-1. AIHA emergency response planning guidelines (reproduced with permission from AIHA).

chemicals for which emergency planning is recommended by the U.S. EPA, OSHA, and other government groups. Since then, the EPA has organized a committee of experts to develop Acute Exposure Guideline Levels for use by government agencies. The guidelines are similar in intent to the ERPGs and would similarly consist of three guidance levels and be targeted to the general population.

Applications of ERPGs

The three-level format of the ERPGs was developed specifically to address the problems of chemical releases. Each level has a specific use in planning.

ERPG-1 is a concentration that may produce some discomfort or simply an odor in the community but essentially no deleterious health effects. Above this concentration, discomfort or objectionable odor may become increasingly common depending on the characteristics of the chemical. Actions taken at this level by the emitter or the local emergency personnel may be to simply communicate to the community that there is not a problem and that the odor or other effects will be transient. In some circumstances, this may be a level at which emergency response personnel may be alerted to prepare for further action.

ERPG-2 is a concentration below which escape is not impaired and there are no irreversible or serious effects. Above this concentration irreversible effects, escape impairment, and disability would become increasingly common. This concentration is the most important in an actual emergency because it is at this concentration where mitigating steps must be taken. In the affected geographical areas these steps could involve sheltering in place, evacuation, or donning of respirators.

ERPG-3 is a concentration below which death or life-threatening effects would be unlikely in the exposed population. Above this concentration death and life-threatening effects would become increasingly common. This level should be used only for advanced planning; emergency actions should be taken at ERPG-2, not ERPG-3. This level is used for worst-case scenarios (consequence analysis), sizing of vessels in plant design, plant siting decisions, insurance purposes, etc.

Integration of Toxic Endpoints with Other Factors

In determining the geographical zone where toxic effects might be produced during actual emergency releases, meteorological conditions and plume behavior are very important. It is common for industrial sites to have wind direction and speed indicators to aid in monitoring release plumes. The degree of air mixing or dilution of the chemical is a complex phenomenon dependent on factors such as ambient temperature, degrees of reaction of the chemical with water in the air, atmospheric stability class, and local terrain. Determining the amount of the release and the release rate from a broken or leaking pipe (the most common scenario) is not simple, and confounding factors such as whether there is an entrained aerosol can add large uncertainties.

TABLE E-1
AIHA ERPGs 1997

Chemical	ERPG-1	ERPG-2	ERPG-3
Acetaldehyde	10 ppm	200 ppm	1000 ppm
Acrolein	0.1 ppm	0.5 ppm	3 ppm
Acrylic acid	2 ppm	50 ppm	750 ppm
Acrylonitrile	10 ppm	35 ppm	75 ppm
Allyl chloride	3 ppm	40 ppm	300 ppm
Ammonia	25 ppm	200 ppm	1000 ppm
Benzene	50 ppm	150 ppm	1000 ppm
Benzyl chloride	1 ppm	10 ppm	25 ppm
Bromine	0.2 ppm	1 ppm	5 ppm
1,3-Butadiene	10 ppm	200 ppm	5000 ppm
n-Butyl acrylate	0.05 ppm	25 ppm	250 ppm
n-Butyl isocyanate	0.01 ppm	0.05 ppm	1 ppm
Carbon disulfide	1 ppm	50 ppm	500 ppm
Carbon tetrachloride	20 ppm	100 ppm	750 ppm
Chlorine	1 ppm	3 ppm	20 ppm
Chlorine trifluoride	0.1 ppm	1 ppm	10 ppm
Chloroacetyl chloride	0.1 ppm	1 ppm	10 ppm
Chloropicrin	NA[a]	0.2 ppm	3 ppm
Chlorosufonic acid	2 mg/m^3	10 mg/m^3	30 mg/m^3
Chlorotrifluoroethylene	20 ppm	100 ppm	300 ppm
Crotonaldehyde	2 ppm	10 ppm	50 ppm
Diborane	NA	1 ppm	3 ppm
Diketene	1 ppm	5 ppm	50 ppm
Dimethylamine	1 ppm	100 ppm	500 ppm
Dimethyldichlorosilane	0.8 ppm	5 ppm	25 ppm
Dimethyl disulfide	0.01 ppm	50 ppm	250 ppm
Dimethylformamide	2 ppm	100 ppm	200 ppm
Dimethyl sulfide	0.5 ppm	500 ppm	2000 ppm
Epichlorohydrin	2 ppm	20 ppm	100 ppm
Ethylene oxide	NA	50 ppm	500 ppm
Formaldehyde	1 ppm	10 ppm	25 ppm
Furfural	2 ppm	10 ppm	100 ppm
Hexachlorobutadiene	3 ppm	10 ppm	30 ppm
Hexafluoroacetone	NA	1 ppm	50 ppm
Hexafluoropropylene	10 ppm	50 ppm	500 ppm
Hydrogen chloride	3 ppm	20 ppm	100 ppm
Hydrogen cyanide	NA	10 ppm	25 ppm
Hydrogen fluoride	2 ppm	20 ppm	50 ppm
Hydrogen peroxide	10 ppm	50 ppm	100 ppm
Hydrogen sulfide	0.1 ppm	30 ppm	100 ppm
Isobutyronitrile	10 ppm	50 ppm	200 ppm
2-Isocyanatoethyl methacrylate	NA	0.1 ppm	1 ppm
Lithium hydride	25 μg/m^3	100 μg/m^3	500 μg/m^3
Methyl alcohol	200 ppm	1000 ppm	5000 ppm
Methyl bromide	NA	50 ppm	200 ppm
Methyl chloride	NA	400 ppm	1000 ppm

(continues)

TABLE E-1 (continued)

Chemical	ERPG-1	ERPG-2	ERPG-3
Methyl iodide	25 ppm	50 ppm	125 ppm
Methyl isocyanate	0.025 ppm	0.5 ppm	5 ppm
Methyl mercaptan	0.005 ppm	25 ppm	100 ppm
Methylene chloride	200 ppm	750 ppm	4000 ppm
Methyltrichlorosilane	0.5 ppm	3 ppm	15 ppm
Monomethylamine	10 ppm	100 ppm	500 ppm
Perchloroethylene	100 ppm	200 ppm	1000 ppm
Perfluoroisobutylene	NA	0.1 ppm	0.3 ppm
Phenol	10 ppm	50 ppm	200 ppm
Phosgene	NA	0.2 ppm	1 ppm
Phosphorus pentoxide	5 mg/m^3	25 mg/m^3	100 mg/m^3
Propylene oxide	50 ppm	250 ppm	750 ppm
Styrene	50 ppm	250 ppm	1000 ppm
Sulfur dioxide	0.3 ppm	3 ppm	15 ppm
Sulfuric acid (oleum, sulfur trioxide, and sulfuric acid)	2 mg/m^3	10 mg/m^3	30 mg/m^3
Tetrafluoroethylene	200 ppm	1000 ppm	10000 ppm
Titanium tetrachloride	5 mg/m^3	20 mg/m^3	100 mg/m^3
Toluene	50 ppm	300 ppm	1000 ppm
Trichloroethane	350 ppm	700 ppm	3500 ppm
Trimethoxysilane	0.5 ppm	2 ppm	5 ppm
Trimethylamine	0.1 ppm	100 ppm	500 ppm
Uranium hexafluoride	5 mg/m^3	15 mg/m^3	30 mg/m^3
Vinyl acetate	5 ppm	75 ppm	500 ppm

[a] NA, Not appropriate.

Obviously, these calculations cannot be performed at a moment's notice during an emergency unless prepared systems are in place. Devices such as provided by Safer Systems present a computerized terrain map that shows population centers and the chemical release point. Data from on-line weather stations is coupled to release rates and preentered physical/chemical parameters from the released chemical to develop a temporal forecast of the plume behavior. When this is coupled to toxicity endpoints such as ERPGs, a map overlay can be produced showing the geographical zones (isopleths) where ERPG concentrations might occur (Fig. E-2). This information may then be used to trigger a preplanned set of actions such as evacuation or sheltering in place in areas where the ERPG-2 concentration is predicted to occur.

Consequence Analysis

Using the previous systems, various scenarios can be developed to plan for theoretical emergencies. One approach is to start with a "worst-case scenario" in which, for instance, the largest (no matter how probable) chemical release would occur, coupled to the worst weather conditions (based on local historical data), in the direction of the closest population. If this first very conservative analysis shows that there is no potential problem with the chemical process then the analysis can end. If, however, there is a projected significant effect to the local population, a more in-depth analysis would be applied. The in-depth analysis would use more credible release scenarios coupled to a more sophisticated analysis of the physical/chemical parameters involved in the plume formation and might take into account mitigating processes such as scrubbing or flare towers and automatic chemical line shut-offs. Other factors such as local terrain effects and degree of population sheltering might also be considered.

The results of these analyses can be shared with local emergency planning committees so that the chemical facility and the community can work together to form contingency plans. A more formalized approach to consequence analysis is required for certain stationary emissions sites in the EPA's Risk Management Program rule, published in June 1996. This rule applies to facilities having over a certain threshold quantity of listed "highly hazardous" chemicals. By June 21, 1999, the rule requires facilities to develop and implement appropriate risk management plans to minimize the frequency and severity of chemical plant accidents and to share the plans with the EPA and the general public.

Further Reading

Alexeeff, G., Lewis, D., and Lipsett, D. (1992). Use of toxicity information in risk assessment for accidental releases of toxic gases. *J. Hazardous Mater.* 29, 387–403.

Cavender, F. L., and Gephart, L. A. (1994). Emergency response planning. *Proc. Air Waste Manage. Assos. 11,* 93-RA-111.01.

National Research Council, Committee on Toxicology (1992). *Guidelines for Developing Spacecraft Maximum Allowable Concentrations for Space Station Contaminants.* National Academy Press, Washington, DC.

National Research Council, Committee on Toxicology (1993). *Guidelines for Developing Community*

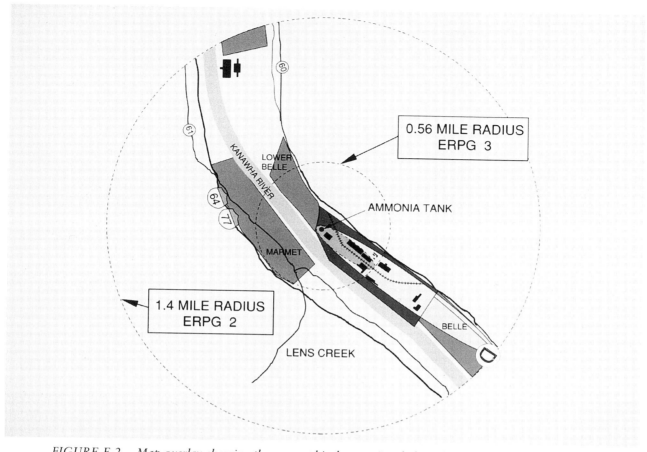

FIGURE E-2. *Map overlay showing the geographical zones (isopleths) where ERPG concentrations might occur.*

Emergency Exposure Levels for Hazardous Substances. National Academy Press, Washington, DC.

Rusch, G. M. (1993). The history and development of Emergency Response Planning Guidelines. *J. Hazardous Mater. 33,* 193–202.

Woudenberg, F., and Van Der Torn, P. (1992). Emergency exposure limits: A guide to quality assurance and safety. *Qual. Assurance Good Practice Regul. Law 1,* 249–293.

—*David P. Kelly and Finis L. Cavender*

Related Topics

Exposure Criteria
Levels of Effect in Toxicological Assessment
National Institute for Occupational Safety and Health
Occupational Safety and Health Administration

Endocrine System

Overview of Endocrine System

The life of a multicell organism requires coordination between organs and tissues. The endocrine system, composed of ductless secretory organs and structures, maintains this coordination by producing hormones in response to physiological and environmental changes. Hormones are discharged into the blood and lymphatic system and transported to other parts of the body, including other hormone-secreting organs, in order to elicit characteristic cellular responses in target organs. The major endocrine organs include the hypothalamus, pituitary, thyroid, parathyroid, pancreas, adrenal, ovary, and testis. Other endocrine organs and tissues include the placenta, liver, kidneys, and cells throughout the gastrointestinal tract.

Hormones can be glycoproteins, polypeptides, peptides, steroids, or modified amino acids. They function as messengers traveling through the bloodstream to target tissues and organs, where they bind to surface or nuclear receptors and regulate gene expression, ion channels, or enzyme activities. The major target organs and tissues include mammary glands, reproductive organs, bone, muscle, and the nervous system.

Endocrine organs function to maintain a relative balance of cellular parameters in the body through the hormones they produce. Hormones are produced in response to changes in the external environment and the internal physiological status. Examples of external signals include light cycles, temperature, nutrient availability, and toxicants. Disease, growth, and reproduction are examples of internal factors that accompany changes in hormonal balance. The intensity of the endocrine effect on the target organ depends on the amount of hormones produced and the binding property, amount, and response of receptors in the target cells.

Target cells differ from nontarget cells by the presence of hormone-specific cell receptors. A receptor, therefore, is a signal discriminator capable of binding to a specific hormone and translating its message into specific cellular responses. These responses regulate protein phosphorylation, cell growth, gene expression, enzyme activity, nutrient metabolism, mineral release/retention, and cell death. Hormones also modulate responses of the immune and central nervous systems.

The production of any hormone in the endocrine system is the result of an entire chain of events involving precisely choreographed interactions of many other endocrine organs. For example, the initiation of testosterone and estrogen production can be traced to the release of gonadotropin releasing hormone (GnRH) from the hypothalamus in the brain. GnRH stimulates the pituitary gland to produce luteinizing hormone (LH) and follicle stimulating hormone (FSH), which in turn act on the testis and ovary to stimulate testosterone and estrogen production, respectively.

When there has been a sufficient cellular response in the target cell, negative feedback, a control mechanism, relates this information back to the organ that produced the hormone, inhibiting further hormone production. This mechanism prevents overstimulation and

modulates the stability of the cellular status. For example, in the case of testicular function, sufficient amounts of testosterone in the blood will send feedback signals to the hypothalamus and the pituitary gland. This will reduce the production of GnRH, LH, and FSH, thus decreasing the signals for further testosterone production and release.

Mechanisms of Endocrine Disorders

In general, due to precise yet adaptable control mechanisms, and the intertwined nature of the hormonal balance, moderate amounts of chemical effects on hormones seldom compromise normal physiological functions. Fluctuations of hormone concentration and receptor activities, by design, absorb environmental and physiological challenges in order to maintain functional equilibrium in the body. Only when the equilibrium control mechanisms are overwhelmed do deleterious effects occur.

Another consequence of the interdependent nature of the endocrine system is that manifestation of an endocrine disorder is virtually always associated with changes in synthesis or concentration of multiple hormones. For example, in 2,3,7,8-tetrachlorodibenzo-*p*-dioxin (TCDD)-treated rats, the decrease in thyroid hormone 3,5,3′,5′-tetraiodothyroxine (T_4) is always associated with an increase in the blood concentration of thyroid stimulating hormone, which is secreted by the pituitary gland in response to the low blood T_4.

Altered hormone concentrations in response to chemical exposure could be an adaptation response. It alone, therefore, is not a sufficient indicator of toxicity. Endocrine toxicity is characterized by disease conditions in the host in addition to hormonal changes.

Dysfunction of the endocrine system could be due to either hyperfunction (excessive hormone production or responses) or hypofunction (insufficient hormone production or responses). Environmental chemicals that have the potential to perturb the endocrine system are known as endocrine or, synonymously, hormone disruptors. Endocrine disruptors are defined as exogenous agents that interfere with synthesis, secretion, transport, binding, cellular response mechanisms, metabolism, or elimination.

Excessive amounts of hormones in circulation may be due to overproduction of hormones in the endocrine organs, rapid release of hormones from storage, decreased hormone metabolism, or decreased clearance and excretion of hormones. On the other hand, cell injury in hormone-producing tissues, inhibition of synthetic enzymes, and induction of metabolic enzymes are causes of hormone deficiency.

Examples of Alterations in Endocrine System

Hormone Synthesis

Hormone synthesis can be altered by modification of the activity of hormone synthesizing enzymes, lack of precursors, or interference with enzyme cofactors such as divalent cations. For example, the fungicide fenarimol inhibits the enzyme aromatase, which converts testosterone to estrogen.

Cadmium decreases testosterone production by preventing the synthesis of cholesterol, a precursor of all steroid hormones. Other chemicals that interfere with steroid hormone synthesis include aminoglutethimide, cyanoketone, and ketoconazole. Copper chelating compounds, such as dithiocarbamates, metam sodium, and carbon disulfide, suppress the conversion of dopamine to norepinephrine and subsequently to epinephrine.

Storage and Release

In addition to synthesis, the release of hormones from the storage compartment in cells also controls the amount of hormones in circulation. Reserpine and amphetamine are examples of compounds that affect hormone storage in granular vesicles. Compounds that activate LH receptors could potentially cause hypersecretion of testosterone from the leydig cells (site of testosterone synthesis in the testes). On the other hand, direct cell injury of secretory cells may cause hyposecretion.

Carrier Proteins

Most lipid-soluble hormones in the blood are bound to specialized carrier proteins. The availability of hormones for physiological functions depends on the total concentration of the hormone as well as the amount of hormone existing in the free state; protein-bound hormones are not readily available for receptor binding. While lack of carrier proteins could impair the transport of hormones to target organs, excessive amounts may decrease the availability of free hormones.

Some estrogenic compounds (compounds with estrogen activity) are known to increase the amount of testosterone–estrogen-binding globulin (TEBG), a sex hormone carrier protein, while high doses of androgens and glucocorticoids may decrease the TEBG concentration in plasma. Salicylates and diphenylhydantoin have been shown to cause changes in thyroxine-binding globulin, a thyroid hormone carrier protein, thus modifying the amount of free circulating thyroxine.

Receptor and Ligand Interactions

Endocrine disruptors often are structural analogs of endogenous hormones (hormones produced naturally in the host). Hormone analogs may act like the endogenous hormone if the analog–receptor complex in the target cell mimics the function of the hormone–receptor complex. Hydroxy metabolites of both o,p'-DDT and methoxychlor bind to estrogen receptors and cause estrogenic effects in birds and reptiles. Alkyl phenols, the biodegradation products of alkyl phenol ethoxylates, bind to estrogen receptors in fish and human cells *in vitro* and induce estrogenic effects in rats. Some hydroxy metabolites of polychlorinated biphenyls (PCBs) are also estrogenic. The receptor-binding property of diethylstilbestrol (DES), a synthetic estrogen, is implicated in the detrimental effects on the reproductive system of men and women exposed to DES *in utero* (see Diethylstilbestrol).

Hormone analogs can also act like antagonists, if they compete with endogenous hormones at the receptor binding site, but elicit no cellular response. An example of such an estrogen antagonist is tamoxifen, which binds competitively to the estrogen receptor and alters the effectiveness of the hormone–receptor complex in regulating gene expression. Vinclozolin, a dicarboximide fungicide, is an androgen antagonist; its metabolite blocks the androgen receptor (see Androgens). Similarly, the major DDT metabolite, 1,1-dichloro-2,2-bis(p-chlorophenyl)ethylene (p,p'-DDE), acts as an antiandrogen in rats, causing abnormalities in male sexual development.

Many endocrine disruptors can bind to more than one type of receptor. For example, o,p-DDT and chlordecone bind to both estrogen and progesterone receptors, while nonylphenol and EPTE, a metabolite of methoxychlor, bind with the same affinity to estrogen, progesterone, and androgen receptors.

Endocrine disruptors may also interfere with hormone function by altering the nature of the receptors or interfering with interactions between the hormone–receptor complex and genes or other cellular components. For example, TCDD and some of the PCBs that resemble the chemical structure of TCDD act as antiestrogens by decreasing the sensitivity of estrogen receptors to estrogen. In such cases, despite adequate estrogen production, organisms respond as if in an estrogen-deficient condition. In addition, under normal conditions, the hormone–receptor complex elicits cellular responses by regulating gene expression. Another type of endocrine disruptor interferes with the interaction between this hormone–receptor complex and DNA.

Metabolism and Clearance

Chemicals that cause either induction (increased synthesis) or degradation of hormones are also potential disruptors of the endocrine system. Several liver enzymes, including cytochrome P450 enzymes, inducible by drugs and environmental pollutants, are involved in hormone clearance. For example, DDT and similar compounds are potent inducers of cytochrome P450-dependent monooxygenases, an enzyme system that degrades endogenous androgens. These compounds, therefore, potentially have antiandrogenic activity. Likewise, lindane has been reported to decrease the amount of circulating estrogen by increasing estrogen clearance.

It has also been hypothesized that long-term hormone imbalance can induce cancers in endocrine-sensitive organs, such as gonads, adrenals, thyroid, prostate, and breast. Lipid-soluble compounds are of special concern because they are retained in the body, therefore causing a long-term effect. Some of these chemicals are known cytochrome P450 enzyme inducers, such as PCBs, DDT, and butylated hydroxytoluene, and have been implicated in cancers in the adrenals, uterus, and thyroid.

UDP-glucuronosyltransferase, an enzyme that conjugates UDP-glucuronic acid with T_4 and other steroid hormones, can be induced by TCDD via an aryl hydrocarbon receptor-dependent mechanism. In rats, TCDD exposure has been shown to increase the rate of removal of T_4 from the blood.

Structurally similar compounds can also compete with endogenous hormones for the binding sites of metabolic enzymes and make the enzyme unavailable for normal hormone degradation. This would lead to a decreased clearance rate and prolong the half-life of circulating endogenous hormones.

Cell Differentiation

Other hypotheses suggest that *in utero* and neonatal exposure to estrogenic compounds may affect cell differentiation and alter cellular responses to sex hormones later in life, resulting in cancer. This mechanism is probably involved in the development of vaginal cancer in women whose mothers were given large doses of diethylstilbestrol during pregnancy. In this case, the manifestation of vaginal cancer was not evident until years after exposure.

Endocrine Organs as Targets of Radiation

Endocrine organs can also become the target of physical agents, such as radiation. This is especially true if a radioactive compound is actively taken up through normal mechanisms and concentrated in an organ. For example, the inorganic iodine in the body is largely taken up by the thyroid in connection with the synthesis of thyroid hormone. More than 20% of iodine in the body is found in the thyroid, an organ which weighs less than 0.005% of the total body. Almost 10 years after the radioactive fallout from the 1986 Chernobyl nuclear power plant explosion, a more than 10-fold increase in the incidence of childhood thyroid cancer in Belarus, the Ukraine, and the Russian Federation was observed. These countries received most of the radioactive fallout. Although many toxic and radioactive compounds were released through the explosion, the geographical distribution of the cancers most closely matches the pattern of fallout from the radioactive iodine. It is also evident through this accident that children are much more susceptible to thyroid cancer caused by radioactive iodine than adults.

Radiation can also cause male infertility, impotency, decreases in sperm fertilizing ability, and damage to the process of sperm formation. These disorders have been observed in victims of the Chernobyl accident as well as in patients treated with radiotherapy. In laboratory rats, the damage to sperm formation can be alleviated by treatment with testosterone and estrogen, indicating an etiology of sex hormone imbalance.

—Karen Chou and Barbara Salem

Further Reading

Amdur, M. O., Doull, J., and Klaassen, C. D. (1991). *Casarett and Doull's Toxicology, The Basic Science of Poisons*, 4th ed. Pergamon, New York.
Thomas, J. A., and Colby, H. D. (1996). *Endocrine Toxicology*, 2nd ed. Taylor & Francis, Washington, DC.

Related Topics

Environmental Hormone Disruptors
Radiation
Reproductive System, Female
Reproductive System, Male
Toxicity Testing, Reproductive

Endrin

- CAS: 72-20-8
- SYNONYMS: 1,2,3,4,10,10-Hexachloro-6,7-epoxy-1,4,4a,5,6,7,8,8a-octahydro-1,4-*endo,endo*-5,8-dimethanonaphthalene; Endrex; Hexadrin
- CHEMICAL CLASS: Synthetic organochlorine insecticide
- CHEMICAL STRUCTURE:

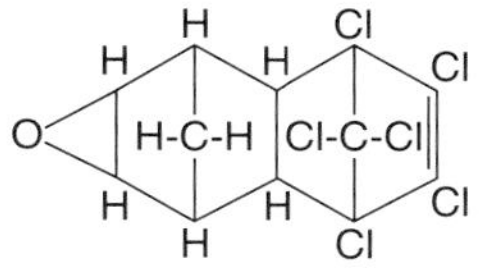

Uses

Endrin is used as an insecticide.

Exposure Pathways

The most important exposure routes for endrin are oral and dermal.

Toxicokinetics

Endrin is absorbed through the gastrointestinal tract and through intact skin.

Endrin is the result of epoxidation of isodrin. Endrin is metabolized by liver microsomal enzymes to more water-soluble metabolites including the hydroxylation products *syn*-12-hydroxyendrin and *anti*-12-hydroxyendrin, both of which are somewhat more toxic than the parent compound. The oxidative metabolite 12-ketoendrin is significantly more toxic than the parent compound and exerts its toxic effects earlier than endrin.

Because of the efficiency of its oxidative metabolism, endrin does not accumulate as effectively as other organochlorine insecticides. Higher concentrations may be found in the liver than adipose tissue.

Greater than 90% of the compound and its metabolites are excreted in feces and/or bile.

Mechanism of Toxicity

Endrin binds to the γ-aminobutyric acid receptor and inhibits chloride ion flux. The result is similar to that described for chlordane or other cyclodiene compounds.

Human Toxicity

Endrin is more highly toxic than other organochlorine insecticides. Symptomatology is exemplified by rapid onset of violent epileptiform convulsions in severe poisoning cases. The onset may be rapid (within 0.5 hr) or delayed as much as 10 hr after ingesting contaminated material. Following the convulsion, a state of semiconsciousness or coma follows. Significant hyperthermia with temperatures as high as 107°F have been reported. Death may follow as rapidly as 0.5 hr after ingestion of contaminated material.

Lower limb weakness, dizziness, and nausea are common in milder intoxications but vomiting is not as common as with other cyclodiene insecticides.

Clinical Management

Management of endrin poisoning is symptomatic. Diazepam or phenobarbital are used to control convulsions. In severe cases, mechanically assisted breathing may be necessary as well as administration of succinylcholine for muscle relaxation and control.

Animal Toxicity

Animal toxicity with endrin is similar to that of other organochlorine cyclodiene insecticides (see Dieldrin).

—*Benny L. Blaylock*

Related Topics

Organochlorine Insecticides
Pesticides
Pollution, Water

Environmental Hormone Disruptors

Alligators in Florida's Lake Apopka are failing to reproduce; many males have reduced genitalia. Female–female pairs of gulls in California are building nests, and some young males in the same population show partially feminized reproductive tracts. Male rainbow trout in Great Britain living near sewer outlets are generating substances in their tissues normally found only in females' eggs. Meanwhile, several studies suggest that over the past 40 years human males have suffered a worldwide, steep decline in the numbers and quality of their sperm cells. Incidences of abnormal or incomplete genital development, such as undescended testes and malformed penises, are reported to be increasing in certain Scandinavian countries. Also, a variety of human cancers known to be affected by sex hormones appear to be on the rise, including breast cancer, testicular cancer, and prostate cancer.

There is increasing concern among some scientists that these phenomena represent the diverse manifestations of a common, emerging environmental problem: accumulation in the environment of certain persistent chemicals that mimic the actions of the body's natural sex hormones.

Natural hormones produced within our bodies, including estrogens and androgens (such as testosterone), control the normal processes of development, growth, and control of sexual differentiation, sexual behavior, and reproductive function. Appropriately timed changes in the circulating concentrations of these hormones exert this control through binding of the hormone molecules to certain specific receptor molecules in cells of the tissues to be affected, triggering the appropriate biological effects.

It is becoming clear that a number of foreign chemicals can bind to these hormone receptors as well, although generally more weakly than do the genuine hormone molecules. At least in the laboratory, some of these chemicals in sufficient concentrations appear to be able to trigger the receptors' biological actions (i.e., they act as "agonists"). Others may act to block access of genuine hormone molecules to their receptors (i.e., they act as "antagonists"), perhaps preventing needed modulation of biological function by the body's own hormones. Still others affect the metabolism, and hence the concentrations, of the body's natural hormones. A variety of compounds, representing different compositions and chemical classes, appear to have such properties. These include certain chlorinated organic compounds (principally some pesticides such as DDT, kepone, and others, but also certain polychlorinated biphenyls), some plasticizers and breakdown products of polycarbonate plastic, and some pharmaceuticals, such as diethylstilbestrol (DES).

Such compounds are coming to be termed "xenoestrogens," referring to the fact that they are agents foreign to the body's own metabolism yet have properties that mimic those of endogenous sex hormones. (Since androgens as well as estrogens are at issue, and since blocking as well as stimulation is of concern, the broader term "environmental hormone disruptors" is also used.) The fear is that exposure to xenoestrogens, by producing hormonal stimuli that are of inappropriate timing, magnitude, or biological context, may result in unwanted biological effects such as the proliferation of mammary tissue cells (increasing cancer risk) or the diversion of the normal paths of sexual differentiation of developing embryos. The matter is complicated by the existence of environmental compounds that act as anti-estrogens—that is, compounds that tend to reduce the response to sex-hormonal stimuli. Some agents may act as antiestrogens at low doses and as estrogen agonists at high doses.

What is less clear is whether the small concentrations of xenoestrogens usually experienced by most humans are capable of having any significant biological effect. The quantities involved are small; they are typically dwarfed by our intake of naturally occurring estrogenic compounds that are found in a variety of vegetables in our diet. These plant compounds, often termed phytoestrogens, have been consumed in significant quantities for centuries by some cultures. For instance, soy products, long consumed in the Orient, are a major source of certain estrogenic flavonoids. While phytoestrogens can provide estrogenic stimuli in test tube studies, the effect of low exposures in living organisms can sometimes be to suppress responses to estrogen.

Exposure to phytoestrogens is in turn dwarfed by the doses of estrogens that many humans (at least female humans) receive in birth control pills or postmenopausal hormone replacement therapy. Modern birth control pills have not been associated with elevated cancer risks, although older formulations and postmenopausal hormone replacement therapy have shown some increase in breast cancer risk in epidemiological studies. Some scientists have questioned whether the low levels of man-made xenoestrogens in the environment can plausibly be thought to affect the risks of hormonally influenced cancers when the exposure levels are over 100-fold less than exposures to naturally occurring phytoestrogens—and millions of times smaller than exposures to estrogens in birth control pills. These critics point out that natural variation in estrogen concentrations from one individual woman to another—variation that appears to have little if any bearing on health—is much larger than the increase that could come from environmental xenoestrogens, and that hormonal control is biologically designed to produce correct response despite these differences. They state that, aside from specific instances of high local contamination (such as the spill of the pesticide dicofol thought to be responsible for the Lake Apopka alligator problem), there is little indication of any widespread health problems associated with exposure to xenoestrogenic chemicals.

The wide span of opinion on the xenoestrogen issue is typical of the early stages of an emerging scientific question, when possibilities of great concern are raised but existing information (and, perhaps more important, scientific consensus about the meaning of that information) is insufficient to resolve whether emerging fears are well founded. If the environment is indeed accumulating compounds that can have widespread and serious effects on the health, development, and fertility of wildlife and humans, then we face a great problem; such compounds will be difficult to control and even more difficult to remove from the environment. On the other hand, if low exposure levels pose little real risk, then a crash program of xenoestrogen control and cleanup will divert precious resources from other pressing environmental problems and could divert regulatory and research attention away from other true causes of breast cancer, birth defects, and wildlife toxicity.

Research efforts on xenoestrogens and related issues of environmental hormone disruption have increased markedly during the past few years in government, industry, and academic laboratories, and further increases in research efforts are likely. The National Academy of Sciences has convened a Committee on Hormone-Related Toxicants in the Environment to prepare a report assessing the known and suspected mechanisms and impacts on wildlife and humans, identifying the significant uncertainties, and recommending a scientific framework with which to approach the problem. There will also be much discussion at meetings of professional societies and specially convened colloquia.

As the field matures, research will move beyond its current foci of identifying compounds that can lead to estrogen-mimicking effects in isolated cells in the test tube and documenting apparent associations of reproductive and developmental difficulties and xenoestrogen exposure in field studies. These descriptions of phenomena of concern will increasingly be dissected by controlled analytical studies that aim to determine the specific combinations of dose level, exposure duration, and sensitive periods that can elicit toxic effects in whole, living animals. Key questions will be whether hormone disruption has an exposure threshold below which no ill effects are caused, whether simultaneous exposure to several different agents at levels that are individually without effect can cause responses by their joint action, whether the large exposures to phytoestrogens (relative to artificial compounds) play either a protective or an exacerbating role, and attempts to define the conditions under which laboratory responses can predict the existence and magnitude of true risks among wildlife species and humans exposed to low levels of potential xenoestrogens in the environment. There will also be a large emphasis on development and validation of batteries of screening tests so that potential new estrogenic compounds can be identified before they enter the environment. Finally, the conclusions about the ability of low exposures to environmental compounds must be reconciled with studies on the health effects of variation among individuals in their concentrations of genuine sex hormones, including the effects of drugs, oral contraceptives, and estrogen replacement therapy, which lead to changes in estrogenic stimulation that are much larger than seems plausible for environmental agents.

If the xenoestrogen problems are real, they have taken some time to develop and will take a good deal more time to cure. Policy-makers face a great challenge, given the complexity of hormonal action and the unanswered questions regarding exposure thresholds for biological effect, the relation of test-tube studies to effects in living organisms, the unknown net effect of low levels of estrogenic and anti-estrogenic agents experienced in combination, and the role of natural and artificial sources. It is likely to be difficult to design policies that are reasonably assured of improving rather than worsening the situation, much less ones that will solve the problem. There is a clear need for research to illuminate these questions and for vigorous scientific debate and examination of the issues to discern the appropriate interpretation of the information we now have on hand.

Acknowledgment

This entry was adapted by the author from an earlier essay appearing in *Risk in Perspective*, Vol. 4(3) (April 1996), published by the Harvard Center for Risk Analysis.

Further Reading

Colborn, T., Peterson-Myers, J., and Dumanoski, D. (1996). *Our Stolen Future: How We Are Threatening Our Fertility, Intelligence and Survival—A Scientific Detective Story*. Dutton/Signet, New York.
Safe, S. (1995). Environmental and dietary estrogens and human health: Is there a problem? *Environ. Health Perspect.* 103(4), 346–351.

—Lorenz Rhomberg

Related Topics

Developmental Toxicology
Endocrine System
Reproductive System, Female
Reproductive System, Male

Toxicity Testing, Developmental
Toxicity Testing, Reproductive

Environmental Processes

Transport Mechanisms

The predominant abiotic environmental process for inorganic and organic substances is dispersion following the concentration gradient. The major components of this are diffusion and turbulent mixing. In air and water the mixing component predominates under most conditions. In soils and groundwater, diffusion plays an important role. The transport of materials is greatly influenced by changes of state. Therefore, molecular weight, polarity, vapor pressure, and water solubility are major determinants of the ease of vaporization, water solubility, sorption to various types of particles, and lipid solubility.

Thus, the ease with which substances can be sorbed or desorbed greatly influences the fraction that can appear in free vapor form in the air, or freely dissolved in water, and this in turn affects the biological availability. If the particles on which a substance has become strongly adsorbed are large, then transport in water or in air would be retarded.

If substances have low water solubility and high lipid solubility then they tend to be able to pass through cell membranes and concentrate in lipids, the basis of much of the observed bioaccumulation. While at first this appears to be a movement of a chemical against the concentration gradient, it is actually moving along a gradient of increased entropy so that no external energy needs to be supplied for this process to take place.

The concentration that may be found locally due to distal releases depends on the release rate, air and water flows and turbulence, transfer among media (e.g., water to sediment), rate of transformations, and the proximity of other similar sources.

Environmental transport processes can be more or less successfully evaluated using a large series of computer models. Long-term and short-term atmospheric diffusion models are generally based on Gaussian plume assumptions. Surface water models for lakes need to consider thermal stratification, wind, waves, and currents. For streams, they need to consider cross section, stream velocity, bottom roughness, and sediment interactions. Groundwater modeling is particularly difficult, in part because of the difficulty of gathering appropriate data that will form a representative construct of the soils, water volumes and flows, hydraulic gradients, porosity, sorption, and ion exchange capacities.

When the transport and equilibrium tendencies of organic compounds are considered, the fugacity approach has been particularly helpful. Fugacity modeling estimates the tendency of a molecule to leave a particular phase (e.g., water) and enter a different one (e.g., air). In its simplest form such a model can estimate the fugacity equilibrium among several compartments. For more detailed analyses, such a model can be expanded to take transfer rates into account.

Fate

Inorganic Chemicals: Metals

The individual elements are clearly persistent. However, that does not mean that the biological availability of the metal is invariant. Factors that affect deep burial or sedimentation affect availability. Also, the behavior and the toxicities of the individual metals and their compounds are strongly affected by their physical and chemical forms. A particular example to illustrate this point is mercury. Its toxicity as the metal differs drastically whether it is inhaled as the vapor or whether it is ingested as the liquid metal. Some mercury salts, such as mercurous chloride, used to be given as a drug, but methyl mercury compounds are exceedingly toxic by any route. The interconversions of mercury and its compounds in the environment are extremely complex, and they have major effects on biological absorption efficiency and toxicity (see Mercury).

The physical and chemical forms of all of the metals exert strong influences on transport, biological availability, and toxicity so that in many cases it makes no sense to speak of the toxicities of chromium, copper, mercury, lead, etc. without specifying these conditions.

Organic Chemicals

The persistence of organic chemicals varies greatly, ranging from very labile free radicals and epoxides, such as ethylene oxide, to very stable compounds, such as decachlorobiphenyl. However, none can equal the perfect persistence of the elements that make up the inorganic chemicals, and the predominant processes for organic chemicals in the environment are degradative.

Oxidation is a particularly important process. It occurs in air, in surface and groundwater, and in soils, as long as adequate oxygen concentrations are present. Reaction rates with hydroxyl radicals in air are often known or can be calculated. Peroxy radicals, singlet oxygen, and hydroxyl radicals can also be present in water and participate in oxidation reactions.

Photochemical reactions can be very important, especially if the organic molecule is able to absorb light energy efficiently. Photolytic reactions may still occur by indirect pathways involving the oxidative radicals cited previously or by the actions of, for example, nitrate radicals. Photolysis reactions also occur in water, depending on light penetration and quantum yield. An example is the photolysis of the hexacyanoferrate ion to yield a cyanide ion.

Perhaps the most important degradative processes are microbial aerobic and anaerobic degradation. These biodegradation processes have been studied in special protocols, such as activated sludge systems, anaerobic digesters, and others.

Further Reading

Bodek, I., Lyman, W. J., Reehl, W. F., and Rosenblatt, D. H. (1988). *Environmental Inorganic Chemistry*. Pergamon, New York.

Conway, R. A. (Ed.) (1982). *Environmental Risk Analysis for Chemicals*, pp. xxiv, 558. Van Nostrand Reinhold, New York.

Fischer, H. B., List, E. J., Koh, R. Y., Imberger, J., and Brooks, N. H. (1979). *Mixing in Inland and Coastal Waters*, pp. xiv, 483. Academic Press, New York.

Howard, P. H. (1989). *Handbook of Environmental Fate and Exposure Data for Organic Chemicals*. Lewis, Chelsea, MI.

Lyman, W. J., Reehl, W. F., and Rosenblatt, D. H. (1990). *Handbook of Chemical Property Estimation Methods*. American Chemical Society, Washington, DC.

National Research Council (1990). *Ground Water Models*, pp. xv, 303. National Academy Press, Washington, DC.

Samiullah, Y. (1990). *Prediction of the Environmental Fate of Chemicals*, pp. xi, 285. Elsevier, London. 1990

—Rolf Hartung

Related Topics

Bioaccumulation
Bioconcentration
Biomagnification
Biomarkers, Environmental
Ecological Toxicology
Ecological Toxicology, Experimental Methods
Effluent Biomonitoring
Environmental Toxicology
Metals
Microtox
Photochemical Oxidants
Pollution, Air
Pollution, Soil
Pollution, Water
Population Density
Risk Assessment, Ecological

Environmental Protection Agency, National Health and Environmental Effects Research Laboratory

The U.S. EPA National Health and Environmental Effects Research Laboratory (NHEERL or "National Effects Lab"), with headquarters located in Research Triangle Park, North Carolina, was created because of the need to address environmental health issues from an integrated multidisciplinary perspective. The goals of NHEERL are to identify and resolve, through short-term and long-term ecological and health research, the scientific issues that are responsive and relevant to U.S. EPA's regulatory programs. To achieve these goals research is designed, within a risk assessment context, to answer basic scientific questions and reduce major uncertainties about effects produced by chemicals and human activities on human health and the environment.

The NHEERL research program is focused on the potential effects of environmental pollutants and other anthropogenic stresses on human health and the ecosystems in which we live. NHEERL is the focal point for toxicological, clinical, epidemiological, ecological, and biogeographic research within the U.S. EPA. Through high-quality peer-reviewed studies, NHEERL scientists create and apply biological assays and toxicologic assessment methods, predictive pharmacokinetic/pharmacodynamic models, ecosystem function theory, and advanced extrapolation methods to improve the scientific underpinnings of the agency's risk assessments and regulatory/policy decisions. NHEERL fosters cooperative research projects with academic and other scientific institutions to complement its intramural research.

As good scientific and corporate citizens, NHEERL scientists serve as scientific advisors/reviewers in providing technical assistance to program offices, regions, states, tribes, other U.S. EPA offices of research and development (ORD) national laboratories, senior agency managers, agency workgroups, and interagency task forces. NHEERL also provides national and international leadership in identifying studies and resolving important human and ecological issues and by influencing the research planning/priorities of other research organizations.

NHEERL achieves its mission through the integrated activity of nine research divisions, five of which focus on human health issues and four of which focus on ecological issues. Opportunities for coordination, collaboration, and cooperation among these divisions are pursued with vigor to provide a unique synergy of effort. The five health research divisions are located in Research Triangle Park, North Carolina, and include developmental toxicology, environmental carcinogenesis, experimental toxicology, human studies, and neurotoxicology.

In addition to the nine research divisions, NHEERL also directs the Environmental Monitoring and Assessment Program (EMAP). EMAP activities cut across the ecology divisions and include designing and implementing a comprehensive, long-term nationwide environmental research, monitoring, and assessment program to assess and to document periodically the condition of the nation's ecological resources; providing a scientifically valid process for combining the ecosystem-specific data into comprehensive ecological risk assessments of major environmental conditions on a regional and national basis; and ensuring extensive involvement of other EPA laboratories, other federal agencies, regional offices, states, and interested international communities to ensure a well-coordinated and maximally leveraged research program.

For additional information, contact U.S. EPA National Health and Environmental Effects Research Laboratory, Research Triangle Park, NC 27711. Telephone: 919-541-2281; fax: 919-541-4324.

—*David M. Krentz and Harihara M. Mehendale*

(Adapted from information supplied by U.S. EPA)

Related Topics

Agency for Toxic Substances and Disease Registry
Food and Drug Administration
National Center for Toxicological Research
National Institute for Occupational Safety and Health
National Institute of Environmental Health Sciences
National Institutes of Health
National Toxicology Program

man health may be central, peripheral, or even excluded. Concepts that address specific topics within this very broadly defined scope of environmental toxicology can include bioaccumulation, biodiversity, biomarkers, ecological toxicology, ecological risk assessment, effluent testing, pesticides, transport, and fate of chemicals.

—*Rolf Hartung*

Related Topics

Bioaccumulation
Bioconcentration
Biomagnification
Biomarkers, Environmental
Ecological Toxicology
Ecological Toxicology, Experimental Methods
Effluent Biomonitoring
Environmental Processes
Pollution, Air
Pollution, Soil
Pollution, Water
Population Density
Risk Assessment, Ecological

Environmental Toxicology

Environmental toxicology is the study of harmful effects of chemicals on biological organisms examined in relation to environmental influences. This definition is extremely broad and encompasses many aspects of toxicology. Individual perceptions of the concept of "environment" range from the environment external to individual cells to global considerations. Indoor environments or occupational environments may be included or excluded. Considerations for hu-

Eosinophilia– Myalgia Syndrome

Introduction

During the summer and fall of 1989, an epidemic of a seemingly new disease occurred in the United States. The illness was characterized by blood eosinophilia (raised numbers of a type of white blood cell) and myalgia (severe muscle pain) and termed the eosinophilia–myalgia syndrome (EMS). It was initially recognized in October 1989 when physicians in New Mexico identified three women with similar clinica findings; all three had consumed L-tryptophan prior to onset of illness. These findings were publicized by the local news media. Soon after, additional cases were recognized throughout the United States and in several other countries.

Epidemiological studies initiated in early November 1989 by the health departments of New Mexico and Minnesota demonstrated a strong association between antecedent tryptophan consumption and EMS. A national surveillance program was initiated by the U.S. Centers for Disease Control (CDC) to investigate the new disease. On November 11, 1989, the U.S. FDA issued a nationwide warning that advised consumers to discontinue use of tryptophan food supplements. Six days later, the agency requested a nationwide recall of all dietary supplements that would provide a daily dose of more than 100 mg of tryptophan. The recall was expanded on March 22, 1990, to include all products containing tryptophan at any dose (with the exception of protein supplements, infant formulae, and intravenous solutions that incorporated small amounts of tryptophan for nutritional requirements).

With the removal of tryptophan from the consumer markets, the number of new EMS cases diminished rapidly. Nevertheless, over 1500 persons were affected by the illness; to date, 37 deaths have been attributed to EMS. Many patients remain in a chronic phase of the disease. The toll would certainly have been higher were it not for the alertness of physicians who linked the new disease to tryptophan, the epidemiologic investigations by the state health departments and the CDC, and the prompt recall of products containing L-tryptophan by the FDA. While the epidemiological and chemical investigations indicate that the epidemic of EMS was caused by contaminated L-tryptophan, the precise contaminant(s) causing the disease is still uncertain.

Prevalence and Reasons for L-Tryptophan Usage

L-Tryptophan usage was widespread in the United States in 1989. In Oregon and Minnesota, approximately 2% of the household members surveyed had used tryptophan at sometime between 1980 and 1989. The most common reasons for tryptophan use were insomnia, premenstrual syndrome, and depression; other reasons included anxiety, headaches, behavior

disorders, obesity, and smoking cessation. Although most consumers purchased tryptophan for therapeutic use, it was marketed as a food supplement and widely available in the United States without a prescription. The manufacturers made no claims regarding its therapeutic efficacy, and the product was not regulated or approved by the FDA.

L-Tryptophan is an essential amino acid; however, sufficient quantities are present in the diet of most U.S. citizens without the need for supplements. The typical daily U.S. diet contains 1–3 g of tryptophan, which satisfies the requirement of 3 mg/kg body wt (or 210 mg/70 kg individual).

Eosinophilia–Myalgia Syndrome National Surveillance Data

As of June 1993, 1511 EMS cases had been reported to the U.S. CDC, including 37 deaths. The case definition developed by the CDC for epidemiological surveillance included (1) blood eosinophil count greater than $1000/\mu l$, (2) generalized debilitating myalgia, and (3) no evidence of infection or neoplasm that would explain the clinical findings. National surveillance data of July 1990 revealed that 84% of patients were female, 97% were non-Hispanic white, and 86% were over 34 years old (median age, 49 years). One-third of the patients required hospitalization. Ninety-seven percent of the patients with EMS had reported tryptophan use before onset of the disease, in doses ranging from 10 to 15,000 mg/day (median, 1500 mg/day). The prevalence of EMS was higher in the western United States than in other parts of the country, apparently paralleling the higher rate of tryptophan consumption in those states. The true prevalence of EMS was most likely underestimated by surveillance reports because persons with mild disease were excluded by the surveillance case definition. In addition, some cases may not have been reported to state or federal health agencies. It is likely that several thousand mild cases went unreported.

Cases of EMS were also reported in Europe, Canada, and elsewhere. In Germany, more than 100 persons became ill with EMS (as delineated by the CDC case definition). Unlike in the United States, tryptophan was available only by prescription in Germany, and thus case histories on German patients were well documented. As in the epidemiological investigations in the

United States (see below), all of the tryptophan associated with EMS in Germany was traced back to a single company.

Epidemiologic Studies

After initial studies implicated the consumption of tryptophan as a major risk factor for EMS, U.S. state and federal health agencies began investigations to further examine this association. Consumers of tryptophan were classified as either case (EMS patients) or control (non-EMS tryptophan users), and the lots of tryptophan consumed by each group were traced back to determine the tryptophan source. Before the epidemic, L-tryptophan had been manufactured by six companies, all in Japan. Analysis of the tryptophan sources for case patients and controls demonstrated a strong association between EMS and consumption of tryptophan manufactured by a single company, Showa Denko K. K. (Tokyo), a large petrochemical company.

In the Oregon study, 98% of case patients had consumed tryptophan manufactured by Showa Denko compared to 44% of controls. In the Minnesota study, 29 (97%) of the 30 case patients consumed tryptophan that was traced back to Showa Denko compared to 21 (60%) of the 35 controls [odds ratio (the ratio of the odds of the disease occurring in exposed individuals relative to the odds of the disease occurring in unexposed individuals), 19.3; 95% confidence interval, 2.5–844.9). High-performance liquid chromatography (HPLC) analysis (see below) of the tryptophan ingested by the one case of EMS in Minnesota that was not traced back to Showa Denko showed a chromatogram that was characteristic of the company's product, revealing that the tryptophan was, in fact, produced by Showa Denko. All later trace-back studies support the association between tryptophan manufactured by Showa Denko K. K. and the occurrence of EMS.

Data from a cohort of tryptophan users in a South Carolina psychiatric practice provide an estimate of the rate of occurrence of EMS (attack rate) in persons exposed to the etiologic agent. Of 157 people who consumed a single brand of tryptophan (comprising only three lots of tryptophan manufactured by Showa Denko), 29% were diagnosed as definite cases of EMS, and an additional 23% were classified as "possible cases" because they had some clinical findings of EMS (such as eosinophilia without myalgia) but did not meet

the strict CDC surveillance case definition. Thus, the pooled attack rate was 52% among persons exposed to the etiologic agent. Among those taking more than 4 g of this brand of tryptophan per day, the definite EMS attack rate was 59% and the pooled (definite and possible EMS) attack rate was 84%. These data suggest that most, or all, individuals are susceptible to EMS if exposed to sufficient quantities of the etiologic agent.

Risk Factors

Two risk factors for EMS, other than consumption of implicated tryptophan lots, have been identified: the amount of tryptophan consumed and the age of the individual. The risk of developing EMS increased with larger dosages of tryptophan and with increasing age. The tryptophan dosage most likely reflects the degree of exposure to the etiologic agent: Persons who consumed larger doses of tryptophan probably ingested greater quantities of the etiologic agent or had a greater probability of encountering a contaminated tablet that causes illness. The reason for the increased risk of EMS with age is unclear; it may be due to age-dependent physiologic changes in renal or hepatic function that delay the metabolism or clearance of a toxic substance or to age-dependent changes in the immune system. No other host factors were found to alter significantly the risk of developing EMS.

Clinical Features

EMS is a syndrome with multiple clinical presentations and variable severity. The clinical course of EMS consists of an early (acute) phase and a late, long-lasting (chronic) phase. During the early phase, most patients developed profound eosinophilia and severe myalgias. In different groups of EMS patients, the median eosinophil count has been reported to be 4000–6000 cells/μl. In addition to myalgia, the most commonly reported early symptoms included arthralgias (joint pains), weakness or fatigue, dyspnea (difficulty in breathing) or cough, rash, headache, peripheral edema (swelling), fever, and paresthesia (abnormal tingling sensations).

The majority of patients also had an elevated leukocyte count with modestly elevated levels of aldolase, a marker of muscle injury; however, creatine phosphokinase, another indicator of muscle injury, was normal in most patients. This inconsistency between the levels of these two muscle-associated enzymes, previously described in some patients with systemic sclerosis and the toxic oil syndrome (see below), is helpful in differentiating EMS from other myopathies (muscle diseases) and from eosinophilic fasciitis (see below). Approximately one-half of patients had abnormal liver function tests. The erythrocyte sedimentation rate, rheumatoid factor, and levels of IgE, complement, and cryoglobulin (all markers of immune dysfunction) were normal in most patients tested. Abnormalities in tryptophan metabolism have been reported in EMS patients.

For some patients, cessation of tryptophan ingestion led to resolution of the symptoms; however, for others, the disease evolved into a chronic phase. In this phase, EMS is a multisystem disease, with cutaneous, neuromuscular, and/or pulmonary involvement as well as immunological abnormalities. The most common features of chronic EMS are fatigue, muscle cramping, myalgia, paresthesias with objectively demonstrated hypesthesias (lessened sensitivity to touch), articular (joint) symptoms, scleroderma-like skin changes, and proximal muscle weakness. In one study, 88% of EMS patients continued to manifest more than three of these clinical symptoms after 3 years. Neurocognitive dysfunction has also been reported as a late manifestation of EMS.

Pathologic studies have demonstrated a perivascular, lymphocytic infiltrate with eosinophils in the dermis, fascia, and skeletal muscle, with variable numbers of eosinophils. The perivascular infiltrate was accompanied by thickening of the capillary and arteriolar endothelium in dermal, fascial, and muscle vessels. The frequent occurrence of microangiopathy (disease of the small blood vessels) in biopsy specimens suggests that ischemia (deficiency of blood supply) may contribute to tissue injury. Deposition of major basic protein (an eosinophil-specific protein) in affected tissue of some patients suggests that cytotoxic eosinophil degranulation products may also play a role in the pathogenesis of EMS.

The histopathologic examination of affected skin showed thickening of the fascia, deep dermal fibrosis, and accumulation of mononuclear cells and eosinophils. *In situ* hydridization and immunohistochemical studies have demonstrated increased production of type I and type VI collagen in the extracellular matrix of the affected fascia. Thus, the dermal and fascial fibrosis

of patients with EMS is likely due to stimulation of collagen synthesis by fibroblasts.

Most patients with EMS reported paresthesias, and in some patients peripheral neuropathy was the most prominent clinical feature. In some patients, persistent paresthesias have been accompanied by axonal and demyelinating abnormalities on electrophysiologic testing. Muscle biopsies showed a characteristic histopathologic picture, with extensive inflammation (fasciitis) and fibrosis in the connective tissue surrounding the muscle, but little evidence of muscle fiber damage. Perineural inflammation and type II muscle fiber atrophy with denervation features have been observed, but muscle fiber necrosis was uncommon. The severe myalgias may be related to inflammation of nerves in the fascia or muscle, peripheral nerve injury caused by granule proteins, possibly eosinophil-derived neurotoxin, or ischemia of nerves caused by occlusive microangiopathy.

Lung biopsies performed in a small number of patients revealed a vasculitis and perivasculitis with a chronic interstitial pneumonitis. Disturbances of cardiac rhythm and conduction have also been documented. Examination of cardiac autopsy specimens has demonstrated neural lesions throughout the conduction system, similar to the neuropathology seen in skeletal muscle. Inflammatory lesions of the small coronary arteries were also present. The prevalence of cardiac abnormalities among all patients with EMS is unknown, although life-threatening rhythm disturbances appear to be uncommon.

The most commonly observed disease process leading to death of patients with EMS was progressive polyneuropathy (disease involving the peripheral nerves) and myopathy (disease of muscles) that produced complications of pneumonia and sepsis or respiratory failure due to weakness. Two-thirds of EMS patients died of these complications. Other causes of mortality were cardiomyopathy (disorder affecting the muscles of the heart), primary pulmonary disease, arrhythmia (deviation from the normal rhythm of the heart), and stroke.

The response to therapy has been disappointing. Multiple therapeutic interventions have been suggested, but no clearly effective treatment has been identified. In the early phase, glucocorticoid treatment (usually prednisone) was generally helpful in treating pneumonitis, myalgias, and edema and in reducing the eosinophil count. However, some patients have not responded to high doses of prednisone, and others have had an exacerbation of symptoms when the dose was tapered. There is no evidence that prednisone therapy alters the natural history of the disease or the risk of neuropathy. Other treatments that have been included nonsteroidal antiinflammatory drugs, cyclophosphamide, hydroxychloroquine, D-penicillamine, methotrexate, octreotide (a somatostatin analog), and plasmapheresis. Many of these therapies have been tried in patients with severe illness, and insufficient information is available to assess efficacy.

The clinical and histopathologic findings of EMS overlap those of eosinophilic fasciitis (EF), a scleroderma-like syndrome characterized by tender swelling and induration (hardening) of the subcutaneous tissue, primarily in the arms and legs. Some cases of EF, in retrospect, were associated with tryptophan ingestion. However, EF is probably triggered by additional factors because few, if any, cases of EF occurring before 1986 can be attributed to the ingestion of tryptophan. Several clinical and laboratory features distinguish EMS from EF. Patients with EMS had greater frequency and severity of myalgias, fever, peripheral neuropathy, and other visceral organ involvement than patients with EF. Moreover, positive antinuclear antibody and a dichotomy between elevated serum aldolase and nonelevated creatine phosphokinase were features of EMS, but not of EF. EMS appears to be a more severe disease than EF in terms of hospitalization rate, duration of symptoms, and mortality.

Taken together, the epidemiological and clinical findings in patients with EMS could be explained by changes in the manufacturing process of L-tryptophan from 1985 to 1988 which resulted in sporadic contamination of the product, with an increase in the quantities of contaminants in 1989.

Manufacture of L-Tryptophan

The L-tryptophan produced by Showa Denko K. K. was manufactured by fermentation using the bacterium *Bacillus amyloliquefaciens*. The biosynthetic pathway of L-tryptophan is shown in Fig. E-3. Several new strains (I–V), each modified slightly to increase the biosynthesis of tryptophan, were introduced sequentially during the years preceding the outbreak of EMS (Table E-2).

In December 1988, the company introduced a new strain of *B. amyloliquefaciens* (strain V), which had been genetically modified to increase the synthesis of 5-phosphoribosyl-1-pyrophosphate, an intermediate in the biosynthesis of tryptophan (see Fig. E-3). After fermentation, the tryptophan was extracted from the broth and purified using a series of filtration, crystallization, and separation processes. The purification procedures included contact with powdered activated carbon and then granulated activated carbon. The amount of powdered activated carbon in each batch was usually ≥20 kg through 1988. In 1989, the amount of powdered activated carbon used to purify some batches of tryptophan was reduced to 10 kg. From October 1988 to June 1989, a portion of some fermentation batches also bypassed a filtration step that employed a reverse-osmosis membrane (ROM) filter to remove chemicals with a molecular weight of more than 1000 Da. According to the company, these changes did not significantly alter the purity of the tryptophan powder, which was maintained at 99.6% or greater.

Univariate analysis of retail lots of tryptophan consumed by case patients and controls demonstrated an association between development of EMS and the ingestion of tryptophan processed with 10 kg of powdered carbon per batch (odds ratio, 9.0; 95% confidence interval, 1.1–84.6; $p = 0.014$) and the use of *B. amyloliquefaciens* strain V (odds ratio, 6.0; 95% confidence interval, 0.8–51.8; $p = 0.04$). Thus, both a reduction in the amount of powdered activated carbon and use of *B. amyloliquefaciens* (strain V) were significant manufacturing changes, but the independent contribution of each manufacturing change could not be assessed because of the high correlation between them. Bypass of the ROM filter was not significantly associated with the case lots. Studies carried out by Showa Denko suggested that the "biochemical and physiological characteristics" of *B. amyloliquefaciens* (strain V) did not differ from those of earlier strains. Currently, it is unknown which particular changes in the production process contributed to the formation of the etiologic agent(s).

Contaminants Associated with EMS

Once the link between EMS and manufactured L-tryptophan had been established, chemical analyses of bulk tryptophan lots were performed by researchers at

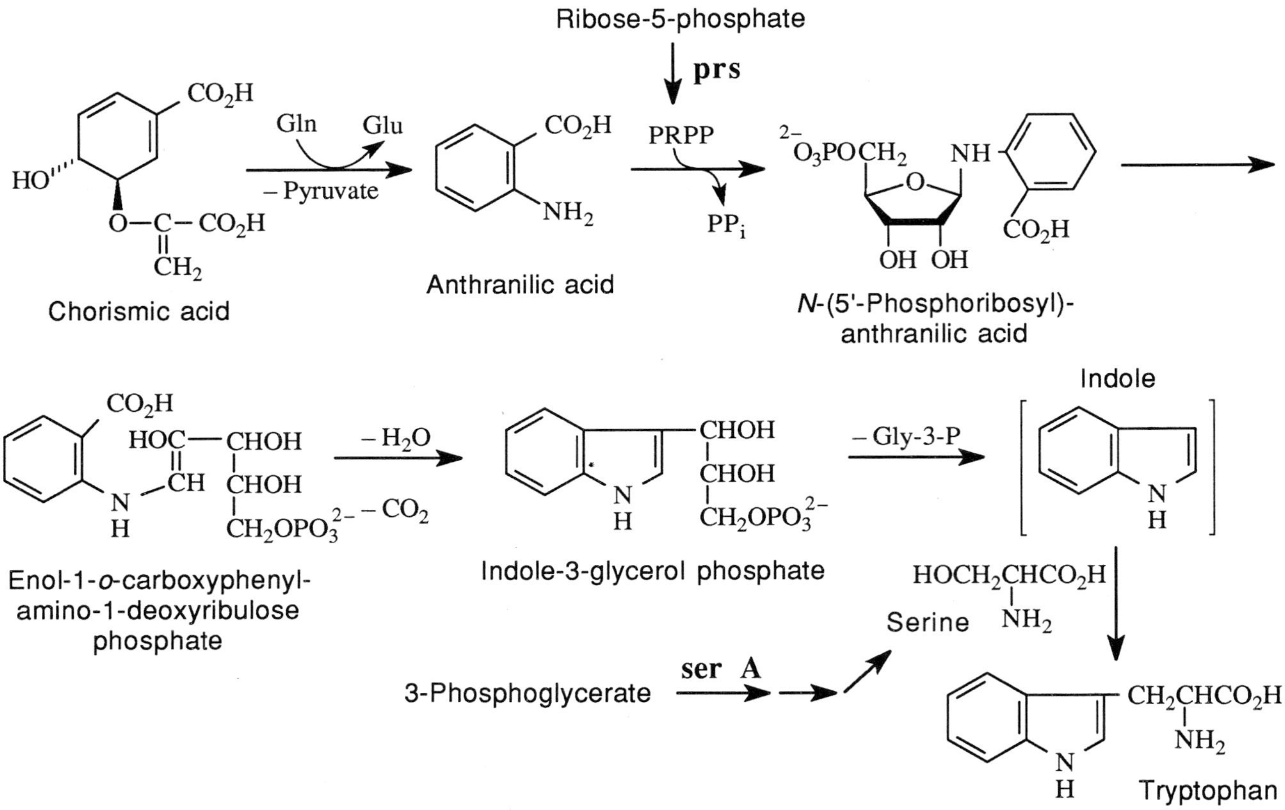

Ribose-5-phosphate

prs

Chorismic acid

Gln   Glu
— Pyruvate

Anthranilic acid

PRPP

PP_i

N-(5'-Phosphoribosyl)-anthranilic acid

Enol-1-o-carboxyphenyl-amino-1-deoxyribulose phosphate

$-H_2O$
$-CO_2$

Indole-3-glycerol phosphate

$-$ Gly-3-P

Indole

Serine $HOCH_2CHCO_2H$ NH_2

3-Phosphoglycerate   **ser A**

Tryptophan CH_2CHCO_2H NH_2

FIGURE E-3. *Biosynthesis of L-tryptophan (adapted from A. N. Mayeno and G. J. Gleich, Eosinophilia–myalgia syndrome and tryptophan production: A cautionary tale,* Trends Biotechnol. **12,** *346–352, 1994).*

the Mayo Clinic (Rochester, MN), FDA (Washington, DC), CDC (Atlanta, GA), and the Japanese National Institute of Hygienic Sciences (Tokyo) to determine if any contaminants were associated with EMS. HPLC was used to separate the contaminants in the tryptophan and revealed that each manufacturer's tryptophan produced a unique chromatographic pattern, or "fingerprint," that was distinctive for the product from each company, as shown in Fig. E-4. The chromatographic pattern consisted of multiple peaks, each of which represented a trace chemical constituent other than tryptophan, which eluted as a large, broad peak between 11 and 15 min. L-Tryptophan manufactured by each of the six companies contained impurities. The chromatogram for Showa Denko tryptophan included five "signature" peaks that were present in all tryptophan manufactured by this company (see Figs. E-4 and E-5). Initial comparison of individual peaks in case and

control lots of Showa Denko tryptophan revealed a single peak (called "peak E" or "peak 97") that was significantly associated with case lots (Fig. E-5). The chemical structure of peak E was subsequently determined to be 1,1'-ethylidenebis[L-tryptophan], or EBT (Fig. E-6). Two other contaminants were subsequently reported to be associated with case lots of tryptophan manufactured by Showa Denko. One of the peaks, labeled UV-5, eluted before tryptophan (Fig. E-5) and was determined to be 3-(phenylamino)-L-alanine (PAA) (Fig. E-6). The other peak (UV-28) eluted much later than EBT and is as yet uncharacterized. Recent HPLC studies revealed more than 60 trace contaminants in Showa Denko tryptophan, 6 of which are associated with EMS. The structures of three are known [EBT, PAA, and "peak 200" (2[3-indolylmethyl]-L-tryptophan)], but the other three have not yet been characterized. One of the uncharacterized contami-

TABLE E-2

Genetic Modifications of the Different Strains of *Bacillus amyloliquefaciens* Used to Manufacture L-Tryptophan

Strain[a]	Modification
I	Original strain of *B. amyloliquefaciens* IAM 1521
II	The tryptophan operon [coding for all enzymes catalyzing reactions from chorismate to L-tryptophan as well as for those involved in the biosynthesis of serine and 5-phosphoribosyl-1-pyrophosphate (PRPP)] of strain I was duplicated through chromosomal integration
III	The isolated tryptophan operon was attached to a more efficient promoter prior to integration into chromosomal DNA of strain II
IV	The *serA* gene (coding for phosphoglycerate dehydrogenase[b]) was amplified using a plasmid vector with strain III
V	The *prs* gene (coding for ribose phosphate pyrophosphokinase[c]) was isolated and integrated into the chromosome of strain IV

Note. Adapted with permission from A. N. Mayeno and G. J. Gleich, Eosinophilia–myalgia syndrome and tryptophan production: A cautionary tale, *Trends Biotechnol.* **12**, 346–352, 1994.

[a] Strains II–V were derived by successive modifications of strain I.

[b] Phosphoglycerate dehydrogenase catalyzes the conversion of 3-phosphoglycerate to 3-phosphohydroxypyruvate, an intermediate in the biosynthesis of serine.

[c] Ribose phosphate pyrophosphokinase catalyzes the phosphorylation of ribose-5-phosphate to give PRPP.

nants, called "peak AAA," was the contaminant most significantly associated with EMS and was recommended for characterization.

The amount of EBT present in Showa Denko tryptophan varied markedly in the period 1987–1989 (Fig. E-7), presumably reflecting alterations in the manufacturing conditions. It is likely that levels of all of the contaminants varied with time. These data are consistent with the hypothesis that a contaminant(s) in tryptophan is responsible for EMS and for the sporadic cases of EF between 1986 and 1988. Recent statistical analyses of EBT, adjusted for serial autocorrelation (to take into account that sequential lots of tryptophan may be related), revealed that higher levels of EBT are still associated with EMS, but the association ($p = 0.120$) did not achieve statistical significance. Nonetheless, the results do not vindicate EBT as a cause of EMS because misclassification of lots as case or control could weaken the association and the methods used to account for the lack of independence of observations over

time probably reduce the power of the statistical analysis.

Investigation into the origin of contaminant PAA reveals that it can be formed from aniline and serine by heating at 80°C for 6 hr under alkaline conditions (pH 11). Although aniline was not used in the biosynthesis of tryptophan, small amounts of aniline are formed from anthranilic acid, a biosynthetic precursor of tryptophan (see Fig. E-3), after heating at 80°C for 6 hr under acidic conditions (pH 2). The industrial process used by Showa Denko to purify tryptophan from the fermenation broth consisted of several steps, including anion exchange at pH 10.5, cation exchange at pH 11, and heat treatment at 80–90°C. Thus, the fermentation and purification processes used to produce tryptophan may have led to the formation of PAA as a by-product.

Connection with Toxic Oil Syndrome

The clinical and pathologic findings of EMS bear a striking resemblance to those of the toxic oil syndrome (TOS), which occurred as an epidemic in Spain during the spring and summer of 1981. Over 20,000 persons were affected, and several hundred deaths have been attributed to TOS. The similarities between EMS and TOS are summarized in Tables E-3 and E-4. Unlike EMS, respiratory symptoms (cough or dyspnea) were prominent and severe in TOS during the first week of illness (acute phase). Other early symptoms included fever, malaise, headache, nausea, and pruritic (itchy) rash. In some patients, the disease progressed to an intermediate and chronic phase that resembled EMS more closely. The intermediate phase (2–8 weeks after onset) was characterized by eosinophilia and leukocytosis (raised numbers of leukocytes). Patients whose illness progressed to the late phase developed muscle cramps and severe myalgias, peripheral edema, scleroderma-like skin changes, and polyneuropathy. The histopathological changes of skin, nerve, and skeletal muscle are remarkably similar between EMS and TOS.

The pathophysiology of both TOS and EMS involves an immunological component. Generally, early skin biopsies in both TOS and EMS showed edema and inflammatory infiltrates. Inflammatory lesions of arteries and cardiac neural structures in both EMS and TOS patients were primarily composed of lymphocytes. Per-

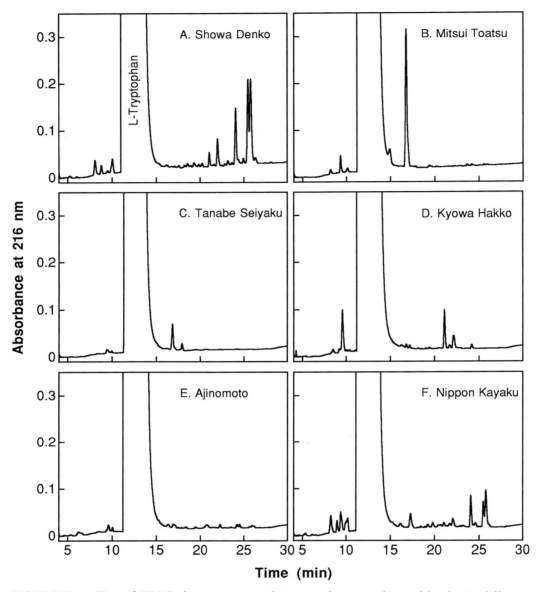

FIGURE E-4.   Typical HPLC chromatograms of L-tryptophan manufactured by the six different companies.

sistent elevated levels in the serum level of the soluble fraction of IL-2 receptor were noted in both EMS and TOS patients, suggesting chronic immune activation.

Epidemiologic investigations implicated ingestion of adulterated rapeseed oil that had been imported from France. At the time, rapeseed oil could not be legally imported into Spain as a food substance, only as an industrial lubricant after denaturation with aniline, a toxic chemical. The oil had been denatured with aniline

(to give a concentration of 2% aniline by weight) as required by law. However, the oil was then illegally de-denatured in Spain by a refining process that removed almost all of the aniline and was subsequently mixed with 10–30% of other seed oils, about 30% of animal fats, and up to 5% of a poor quality olive oil or, alternatively, chlorophyll to produce the desired color. The resulting adulterated oil was sold as pure olive oil, typically in unlabeled 5-liter containers by street vendors and itinerant salesmen.

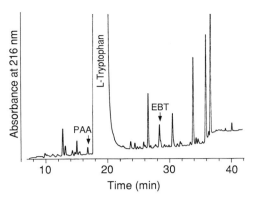

FIGURE E-5. *HPLC chromatogram of EMS-associated L-tryptophan. HPLC conditions differ from those used in Fig. E-4 (reproduced with permission from A. N. Mayeno and G. J. Gleich, Eosinophilia–myalgia syndrome and tryptophan production: A cautionary tale, Trends Biotechnol.* **12,** *346–352, 1994).*

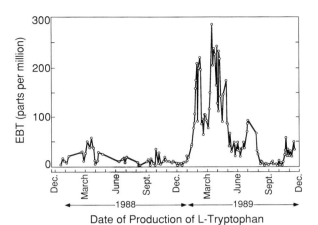

FIGURE E-7. *Levels of 1,1'-ethylidenebis[L-tryptophan] (EBT) in lots of L-tryptophan produced by Showa Denko K. K. during 1988 and 1989 (reproduced with permission from A. N. Mayeno and G. J. Gleich, Eosinophilia–myalgia syndrome and tryptophan production: A cautionary tale, Trends Biotechnol.* **12,** *346–352, 1994).*

Chemical analyses of implicated oil samples and "control" oil samples demonstrated that free aniline and aniline derivatives were significantly associated with case-related samples. Fatty acid anilides, in partic-

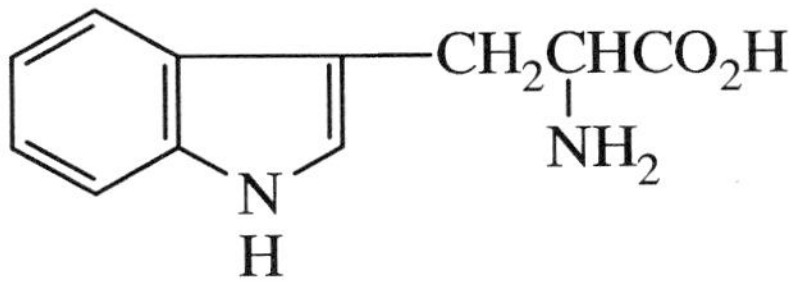

L-Tryptophan

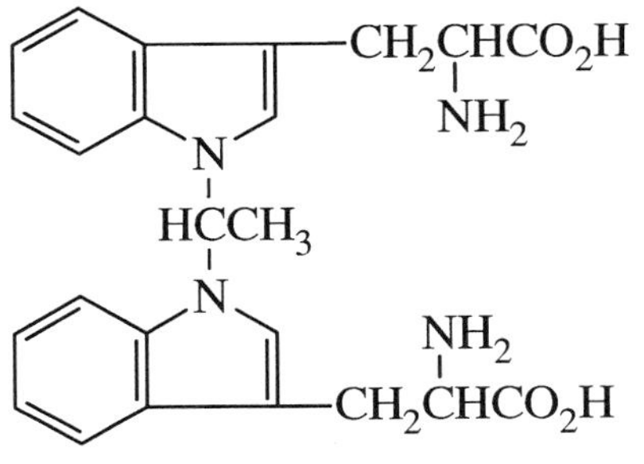

1,1'-ethylidenebis[tryptophan] (EBT)

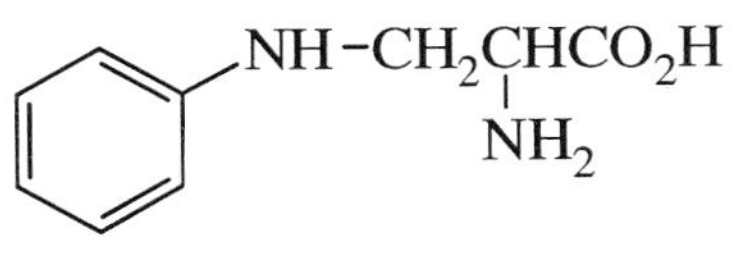

3-(Phenylamino)alanine (PAA)

FIGURE E-6. *Chemical structures of tryptophan and of contaminants 1,1'-ethylidenebis[tryptophan] (EBT) and 3-(phenylamino)alanine (PAA) associated with EMS.*

TABLE E-3

Comparison of the Clinical Features of Eosinophilia–Myalgia Syndrome (EMS), Toxic Oil Syndrome (TOS), and Eosinophilic Fasciitis (EF)

	EMS	*TOS*	*EF*
Female (%)	80	90 (late)	50
Myalgia	+++	++	±
Dyspnea/cough	+	+++ (early)	−
Pruritus	++	+	−
Rash	+	+	−
Swelling, edema	++	++	++
Muscle weakness	+	+	−
Scleroderma-like lesions/ fasciitis	++	++	+++
Heart involvement	±	+	−
Axonal polyneuropathy	++	++	−
Arthritis	+	+	+

Note. Reproduced with permission from J. Varga, L-Tryptophan-associated eosinophilia–myalgia syndrome: Clinical and pathological features of an evolving new disease and current concepts of etiology, *J. Intensive Care Med.* **8,** 229–242, 1993. +, occasional; ++, common; +++, very common.

TABLE E-4
Comparison of the Laboratory Features of EMS,
TOS, and EF

	EMS	TOS	EF
Eosinophilia	+++	++	+
Elevated IgE	−	±	−
Elevated aldolase	++	+	−
Antinuclear antibody	++	+	−
Lymphocytes in lesion	+	+	−
Eosinophils in lesion	±	±	±
Vasculitis	+	+	±

Note. Reproduced with permission from J. Varga, L-Tryptophan-associated eosinophilia–myalgia syndrome: Clinical and pathological features of an evolving new disease and current concepts of etiology, *J. Intensive Care Med.* **8**, 229–242, 1993.
+, occasional; ++, common; +++, very common.

ular oleylanilide (Fig. E-8), have been reported to be markers of TOS-causing oil. Another contaminant, 3-phenylamino-1,2-propanediol (PAP) (Fig. E-8), has been isolated from implicated oil and is chemically similar to the tryptophan contaminant PAA. Efforts to evaluate the biologic activity of the aniline contaminants have been limited by the absence of an animal model for TOS.

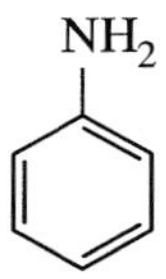

Aniline

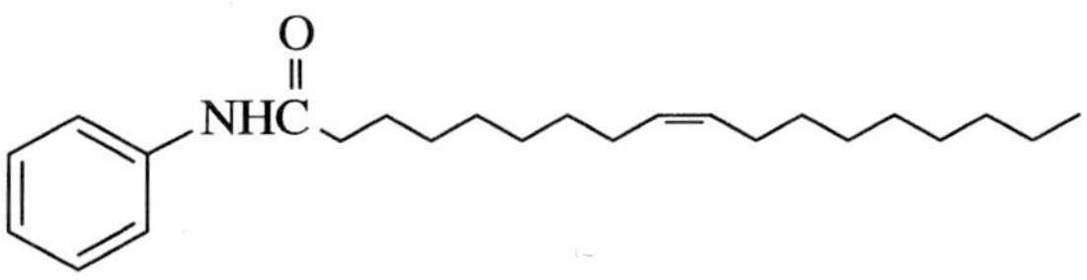

Oleylanilide

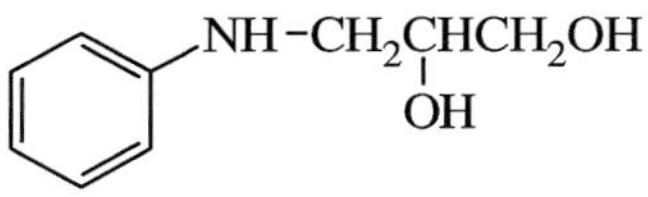

3-Phenylamino-1,2-propanediol (PAP)

FIGURE E-8. Chemical structures of oleylanide and 3-phenylamino-1,2-propanediol (PAP) associated with TOS.

The striking similarities between EMS and TOS suggest that they may share the same final pathway that leads to neuromuscular damage. The recent discovery of a chemically related aniline derivative in tryptophan preparations implicated in causing EMS suggests of a related etiology. Recently, PAP has been demonstrated to undergo biotransformation to PAA by both rat hepatocytes and human liver tissue *in vitro*, linking the two diseases to a common chemical, namely PAA (see below). This finding is the first reported chemical link between TOS and EMS.

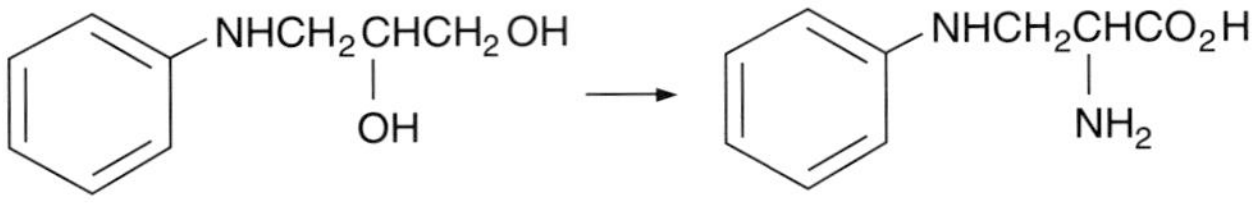

PAP   PAA

EMS Not Associated with L-Tryptophan

Some patients with EMS reported no history of tryptophan ingestion. An EMS-like syndrome has been associated with use of L-5-hydroxytryptophan (5-HTP). HPLC analysis of the 5-HTP that caused the symptoms revealed the presence of an impurity not present in 5-HTP preparations that did not cause symptoms. The structure of the impurity has not been reported. In addition, a recent pharmacoepidemiological study in Canada identified several EMS patients with no apparent history of tryptophan ingestion. These reports suggest that factors other than tryptophan ingestion can lead to the induction of EMS or EMS-like diseases.

Investigations of the Etiology and Pathogenesis

Animal Models

At this stage, without an ongoing epidemic, an appropriate animal or *in vitro* model replicating EMS may be essential in understanding the pathogenesis and etiology of the illness. Several studies using animals have been performed; however, the results of studies have been inconclusive, with no animal hitherto tested replicating all of the clinical features of the disease. Initially, the Lewis rat showed promise as a model for EMS. Muscle biopsies of Lewis rats given either implicated tryptophan (containing EBT) or U.S. Pharmacopoeia (USP) grade tryptophan (without EBT) demonstrated perimysial inflammation in 7 of 9 animals receiving implicated tryptophan compared to 0 of 10 receiving USP-grade tryptophan. A significant increase in fascial thickening was also observed in the rats receiving implicated tryptophan. However, leukocyte counts and eosinophil counts remained normal in both groups. Gastrointestinal changes were also noted with an increased number of degranulating inflammatory cells in the lamina propria of the rats that received case-implicated tryptophan. Recently, however, control L-trytophan alone was observed to cause mild myofascial thickening, alterations in peripheral blood mononuclear cell (PBMC) phenotypes, and pancreatic pathology in Lewis rats, suggesting that tryptophan itself may play a role in EMS and other fibrosing diseases. Another recent study using C57BL/6 mice found that EBT caused inflammation and fibrosis in the dermis and subcutis, including fascia and perimysial tissues, mimicking some of the clinical features of EMS. Eosinophilia was not observed.

A report published recently raises prima facie doubts about the reproducibility of the animal findings described previously. F-344 and Lewis rats, as well as BALB/c mice, were treated with one of the following substances: feed or food-grade L-tryptophan, tablets containing L-tryptophan, isopropanol extracts from L-tryptophan, synthetic EBT or PAA, and 1-methyl-1,2,3,4-tetrahydro-β-carboline-3-carboxylic acid (a breakdown product of EBT). No EMS-like symptoms were observed in any of the animals tested. The tryptophan preparations were manufactured by Showa Denko; however, neither the specific lots numbers nor their association with EMS were indicated. Thus, it is difficult to interpret the results of this study in light of the fact that only certain lots of L-tryptophan manufactured by Showa Denko were linked to EMS. A recent toxicologic study using PAA supports the negative findings; PAA (1, 10, and 100 mg/kg/day) was administered by gavage to Sprague–Dawley rats for up to 13 consecutive weeks. No EMS-like symptoms were observed. Overall, the animal studies suggest that Lewis rats and C57BL/6 mice may be useful in replicating certain aspects of EMS.

In Vitro *Models*

In vitro investigations have attempted to clarify the mechanism of immune activation but so far have pro-

vided limited data. Studies testing the hypothesis that implicated tryptophan or EBT can trigger PBMCs to release cytokines have been equivocal, although one study found that EBT activates eosinophils and induces IL-5 production from T cells. Another recent study found that certain lots of L-tryptophan could stimulate PBMCs to release granulocyte-macrophage colony-stimulating factor (GM-CSF); this response, however, was caused by endotoxin contamination and not associated with case lots of tryptophan. The mechanism of immune activation is clearly complex and may be difficult to reproduce with an *in vitro* assay. Similar difficulties have been encountered in the study of immune system activation in TOS.

Despite the negative findings of *in vitro* studies, there is evidence that cytokines may play a role in the pathogenesis of EMS. It is known that IL-3, IL-5, and GM-CSF can each induce eosinophil production and enhance *in vitro* survival. In one study, EMS patients had significantly elevated serum levels of IL-5 and a higher proportion of hypodense eosinophils compared to normal controls. Elevated levels of IL-3 and GM-CSF were not observed. Their results suggest that IL-5 is the cytokine that triggers the eosinophilia and converts peripheral blood eosinophils to the hypodense phenotype. The mechanism responsible for the elevation of IL-5 levels in the blood is, as yet, undefined.

Possible Pathogenetic Mechanisms

Although the sequence of events leading to the pathologic changes of EMS is undoubtedly complicated, a framework for possible mechanisms can be advanced. One hypothesis involves a direct effect of the etiologic agent on mononuclear cells, leading to production of cytokines, including IL-5. This cytokine could then activate tissue eosinophils and convert them to a hypodense phenotype. Effector functions would be augmented with release of cytotoxic molecules from eosinophils. Once activated, eosinophils can release additional cytokines such as IL-3, GM-CSF, IL-5, and transforming growth factor-α. A cascade of interacting cytokines could then lead to recruitment of additional inflammatory cells and increased collagen synthesis by fibroblasts.

The predominance of inflammatory changes in the fascia suggests that mediators produced by mesenchymal cells (fibroblasts and endothelial cells) may also play a role in the pathogenesis. For example, fibroblasts have been shown to augment IL-5-dependent eosinophil survival and stimulate conversion to the hypodense phenotype. Fibroblasts can also produce IL-8, which recruits neutrophils and lymphocytes when injected *in vivo*. Thus, one can speculate that the etiologic agent interacts with these cells to stimulate release of inflammatory mediators and increase collagen synthesis.

Another general hypothesis involves incorporation of the etiologic agent into metabolic or biosynthetic pathways that utilize chemically related compounds. EBT and PAA are amino acids with structural similarities to tryptophan and phenylalanine, respectively, and might function as an analog with adverse immunologic effects. If EBT, PAA, or one of their metabolites is recognized by an analogous transfer RNA, it might be incorporated into a nascent protein molecule, stimulating an autoimmune response.

The biotransformation of the toxic oil contaminant PAP to the L-tryptophan contaminant PAA links the two related diseases, TOS and EMS, to a common chemical agent, namely PAA. Both PAP and PAA are metabolized further to the *para*-hydroxylated forms, HPAP and HPAA (Fig. E-9) (unpublished data). Such compounds readily autoxidize to the benzoquinoneimine, which is reactive toward nucleophiles such as the sulfhydryl and amino moieties present on many biological molecules. Thus, upon oxidization, HPAA and HPAP may react with macromolecules as a hapten to form immunogenic targets. HPAA possesses some chemical properties similar to that of homogentisic acid (HGA), a hydroquinone derivative implicated in the causation of alkaptonuria, a connective tissue disorder resulting from an inherited abnormality in phenylalanine and tyrosine metabolism. HGA interacts with connective tissue reversibly or is oxidized enzymatically by an enzyme (polyphenol oxidase) present in connective tissue to benzoquinoneacetic acid, which covalently bonds to macromolecules.

One may hypothesize that HPAA reacts similarly, as shown in Fig. E-9. PAP is initially metabolized to PAA or HPAP. PAA is then converted to HPAA. Both HPAP and HPAA can undergo oxidation to the benzoquinoneimine. Benzoquinoneimines can hydrolyze to *p*-benzoquinone, and both benzoquinoneimines and benzoquinone readily react with nucleophilic molecules. It is also possible that PAA is metabolized to another undetermined molecule, shown as Factor X (Fig. E-9), that reacts with biological targets. Thus, HPAA may haptenize proteins, and T cell activation could result from hapten recognition.

The oxidation of HPAA and HPAP to benzoquinones may require an enzyme with properties similar to polyphenol oxidase. The inflammatory pattern and a lack of an animal model for EMS may be explained if the enzyme is localized to the connective tissue (fascia) and is specific to humans. In addition, the observation of greater respiratory symptoms during the early phase of TOS in comparison to EMS may result from accumulation and metabolism of PAP or HPAP in the

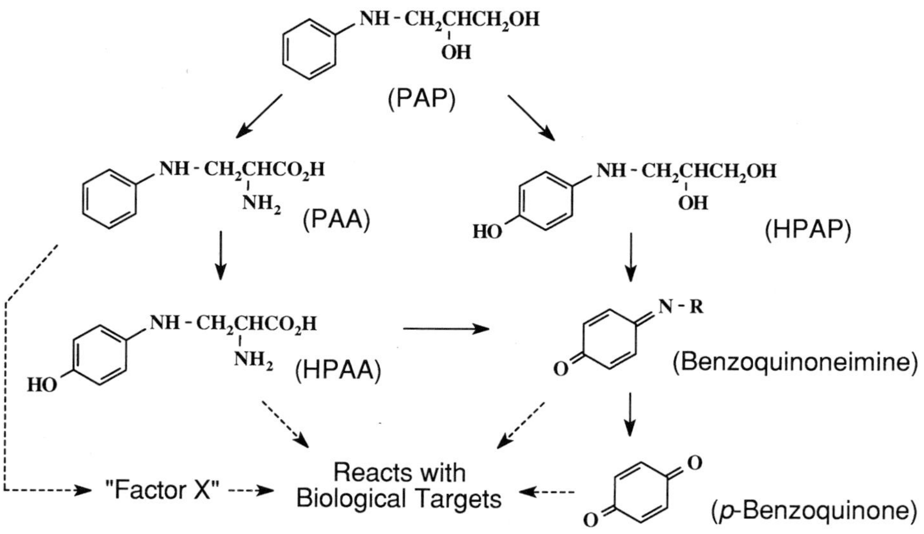

FIGURE E-9. *Hypothetical scheme for the bioactivation of PAP and PAA.*

lung. Certain drugs, such a propranolol (which has some structural similarities to PAP), are known to accumulate in the lungs.

The biotransformation of PAP to PAA must proceed through various intermediates (Fig. E-10). The first steps most likely involve stepwise oxidation of the diol to an α-keto acid or some other keto intermediate, which then undergoes transamination to give PAA. It is likely that any one of these intermediates, in the presence of hepatocytes, can be metabolized to PAA. Thus, many molecules related to PAP (e.g., phenylamino compounds) may be channeled down this pathway to give PAA. This model suggests that numerous molecules with similar chemical structures to PAP will give rise to PAA and, if PAA is indeed responsible for EMS and TOS, the model predicts that an entire class of molecules, as shown in Fig. E-10, can cause EMS/TOS-like diseases, consistent with the reports of EMS cases not associated with tryptophan ingestion.

Summary

The EMS occurred as an epidemic in the United States during 1989. It is a multisystemic disease that resulted from the ingestion of L-tryptophan manufactured by one company. The illness is clinically and pathologically similar to EF and the TOS. The syndrome is triggered by one or more contaminants in tryptophan. Contaminants currently studied include EBT and PAA, although other uncharacterized contaminants have recently been discovered. One or more of these chemicals may trigger EMS by an undefined mechanism, or they may be surrogate markers for another unidentified substance that triggers the syndrome. Consumption of high tryptophan doses and increased age have been identified as risk factors. Patients who ingested tryptophan and were diagnosed with EF during 1986–1988 had most likely EMS. The recent demonstration of the biotransformation of PAP to PAA suggests that both EMS and TOS share a common etiology, namely PAA. Ongoing research is focused on identification of contaminants in implicated tryptophan and on establishing an animal model of the diseases. Success in these endeavors would greatly increase our understanding of eosinophilic diseases and prevent the outbreak of future epidemics.

Acknowledgments

The authors are indebted to the numerous scientists, physicians, and researchers who have contributed to the investigation of EMS and TOS. We thank C. Adolphson for her critical review of the entry.

Further Reading

Eosinophilia–Myalgia Syndrome

Belongia, E. A., Mayeno, A. N., Gleich, G. J., and Kita, H. (1993). Eosinophilia–myalgia syndrome. In *Eosinophils: Biological and Clinical Aspects* (S. Makino and T. Fukuda, Eds.), pp. 421–438. CRC Press, Boca Raton, FL

Duffy, J. (1992). The lessons of eosinophilia–myalgia syndrome. *Hosp. Pract.* **27,** 65–69, 73–80, 83–90.

Kaufman, L. D. (1994). The evolving spectrum of eosinophilia myalgia syndrome. *Rheum. Dis. Clin. North Am.* 20, 973–994.

Kilbourne, E. M. (1992). Eosinophilia–myalgia syndrome: Coming to grips with a new illness. *Epidemiol. Rev.* **14,** 16–36.

Mayeno, A. N., and Gleich, G. J. (1994). Eosinophilia–myalgia syndrome and tryptophan production: A cautionary tale. *Trends Biotechnol.* 12, 346–352.

Philen, R. M., and Posada, M. (1993). Toxic oil syndrome and eosinophilia–myalgia syndrome: May 8–10, 1991, World Health Organization meeting report. *Sem. Arthritis Rheum.* **23,** 104–124.

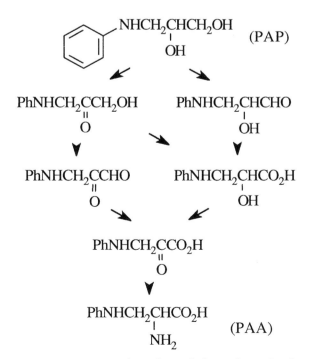

FIGURE E-10. Hypothetical metabolic pathway for the biotransformation of PAP to PAA.

Silver, R. M. (1994). The eosinophilia–myalgia syndrome. *Clin. Dermatol.* 12, 457–465.

Taylor, R., and McNeil, J. J. (1993). Eosinophilia–myalgia syndrome: Lessons for public health researchers. *Med. J. Aust.* 158, 51–55.

Varga, J. (1993). l-Tryptophan-associated eosinophilia–myalgia syndrome: Clinical and pathological features of an evolving new disease and current concepts of etiology. *J. Intensive Care Med. 8,* 229–242.

Toxic Oil Syndrome

Aldridge, W. N. (1992). The toxic oil syndrome (TOS, 1981): From the disease towards a toxicological understanding of its chemical aetiology and mechanism. *Toxicol. Lett.* **64/65,** 59–70.

Grandjean, P., and Tarkowski, S. (Ed.) (1984). *Toxic Oil Syndrome: Mass Food Poisoning in Spain,* Report on a WHO meeting, Madrid, March 21–25, 1983. World Health Organization, Copenhagen.

Kaufman, L. D., Martinez, M. I., Serrano, J. M., and Gomezreino, J. J. (1995). 12-year followup study of epidemic Spanish toxic oil syndrome. *J. Rheumatol.* 22, 282–288.

World Health Organization (1992). *Toxic Oil Syndrome: Current Knowledge and Future Perspectives,* European Series No. 42. WHO, London

Yoshida, S. H., German, J. B., Fletcher, M. P., Gershwin, M. E. (1994). The toxic oil syndrome: a perspective on immunotoxicological mechanisms. *Regul. Toxicol. Pharmacol.* **19,** 60–79.

—Arthur N. Mayeno, Edward A. Belongia, and Gerald J. Gleich

Related Topics

Blood
Epidemiology
Immune System
Neurotoxicology: Central and Peripheral

Epichlorohydrin

- ◆ CAS: 106-89-8
- ◆ SYNONYMS: 1-Chloro-2,3-epoxypropane; 3-chloro-1,2-epoxypropane; epi-chlorohydrin; chloromethyloxirane; chloropropylene oxide; glycerol epichlorhydrin; glycidyl chloride; NCI-C07001; propane, 1-chloro-2,3-epoxy-; SKEKhG

- ◆ CHEMICAL STRUCTURE:

$$CH_2\overset{\displaystyle O}{-\!\!\triangle\!\!-}CHCH_2Cl$$

Uses

Epichlorohydrin is a major raw material used in the manufacture of epoxy resin, glycerol, insecticides, adhesives, agricultural chemicals, and coatings. It is also used as a solvent in the manufacture of rubber, paper, and paint.

Exposure Pathways

Inhalation is the most important route of exposure.

Toxicokinetics

Epichlorhydrin is absorbed orally, percutaneously, subcutaneously, and by inhalation. No reports are available regarding the metabolism of these compounds.

Mechanism of Toxicity

Epichlorhydrin is an alkylating agent which causes a delayed reaction.

Human Toxicity

Toxic levels have not been established in humans. The kidneys and respiratory tract are the target organs after acute epichlorhydrin exposure. Symptoms include nausea, vomiting, and abdominal distress. It can also cause facial swelling, eye and nasal mucosal irritation, respiratory tract irritation, bronchitis, dyspnea, central nervous system depression, hepatomegaly, and kidney lesions.

Clinical Management

Emesis is not recommended. Activated charcoal slurry with or without saline cathartic and sorbitol can be used after oral ingestion. In case of inhalation exposure, good ventilation should be maintained. Skin decontamination should be performed with repeated washing with soap. Exposed eyes should be irrigated with copious amounts of room-temperature water for at least 15 min. Liver and kidney function should be monitored.

Animal Toxicity

Respiratory depression, muscle paralysis, and apnea were observed in mice exposed to 8300 ppm for 30 min. No effects were seen in animals exposed to 9 ppm. The oral LD_{50} is 90 mg/kg in rats and 236 mg/kg in mice.

—Sushmita M. Chanda

Related Topics

Pollution, Water
Respiratory Tract

Epidemiology

Epidemiology looks at the association between adverse effects seen in humans and a selected potential "cause" of interest, such as use of or exposure to a chemical, disease agent, radiation, drug, or medical device.

Epidemiology is sometimes simply defined as the study of patterns of health in groups of people. Behind this deceptively simple definition lies a surprisingly diverse science, rich in concepts and methodology. For instance, the group of people might consist of only two people—such as the case of a father suffering from rheumatoid arthritis and his daughter with vertigo. In both father and daughter the pattern of affected areas was remarkably similar, which might suggest that the distribution of joint lesions in rheumatoid arthritis is genetically determined. At the opposite extreme, studies of the geographic distribution of diseases using national mortality and cancer incidence rates have provided clues about the etiology of several diseases such as cardiovascular disease and stomach cancer. The patterns of health studied are also wide-ranging and may include the distribution, course, and spread of disease. The term "disease" also has a loose definition in the context of epidemiology and might include ill-defined conditions, such as organic solvent syndrome and sick building syndrome, or consist of an indirect measure of impairment such as biochemical and hematological parameters or lung function measurements.

Epidemiology and toxicology differ in many other ways but principally in that epidemiology is essentially an observational science, in contrast to the experimental nature of toxicology. The epidemiologist often has to make do with historical data that have been collected for reasons that have nothing to do with epidemiology. Nevertheless, the availability of personnel records such as lists of new employees and former employees, payrolls and work rosters, and exposure monitoring data collected for compliance purposes has enabled many epidemiological studies to be conducted in the occupational setting. Thus, the epidemiologist has no control over who is exposed to an agent, the levels at which they are exposed to the agent of interest, or the other agents to which they may be exposed. The epidemiologist has great difficulty in ascertaining what exposure has taken place and certainly has no control over lifestyle variables such as diet and smoking.

Despite the lack of precise data, the epidemiologist has one major advantage over the toxicologist: An epidemiology study documents the actual health experiences of human beings subjected to real-life exposures in an occupational or environmental setting. The view has been expressed that uncertainty in epidemiology studies resulting from exposure estimation may be equal to or less than the uncertainty associated with extrapolation from animals to humans. Regulatory bodies such as the U.S. EPA are starting to change their attitudes toward epidemiology and recognize that it has a role to play in the process of risk assessment. However, there is also a complementary need for epidemiologists to introduce more rigor into the conduct of their studies and to introduce standards akin to the Good Laboratory Practice standards under which animal experiments are performed.

Measurement of Exposure

Epidemiologists have placed much greater emphasis on the measure of response than on the measure of exposure. They claim that this is because most epidemiologists have been trained as physicians and are consequently more oriented toward measuring health outcomes. It is certainly true that a modern textbook of epidemiology says very little about what the epidemiologist should do with exposure assessments. However, this is probably as much a reflection of the historical paucity of quantitative exposure information as a reflection on the background of epidemiologists. Nevertheless, it is surprising how many epidemiological studies do not contain even a basic qualitative assessment of exposure. The contrast between epidemiology and toxicology is never more marked than in the area of estimation of dose response. The toxicologist can carefully control the conditions of the exposure to the agent of interest; moreover, the toxicologist can be sure that the test animals have not come into contact with any other toxic agents. An industrial epidemiologist conducting a study of workers exposed to a hepatotoxin will certainly have to control for alcohol intake and possibly for exposure to other hepatotoxins in the work and home environments. Nevertheless, it can be argued that epidemiology studies more accurately measure the effect on human health of "real-life" exposures.

If an exposure matrix has been constructed with quantitative estimates of the exposure in each job and time period, then it is a simple matter to estimate cumulative exposure. It is a more difficult process when, as is common, only a qualitative measure of exposure is available (e.g., high, medium, and low). Even when exposure measurements are available, it may not be sensible to make an assumption that an exposure that occurred 20 years ago is equivalent to the same exposure yesterday. The use of average exposures may also be questionable, and peak exposures may be more relevant in the case of outcomes such as asthma and chronic bronchitis. Noise is a good example of an exposure that must be carefully characterized and where the simple calculation of a cumulative exposure may be misleading.

Study Designs

This section provides a brief introduction to the most important types of studies conducted by epidemiologists. It is an attempt to briefly describe the principles of the major types of epidemiological studies in order to provide insight into the reporting of epidemiological studies and the assumptions made by epidemiologists. The next section will discuss the similarities and differences between the methodologies of toxicology and epidemiology.

Cohort Studies

Historical Cohort Study

When the need arises to study the health status of a group of individuals, there is often a large body of historical data that can be utilized. If sufficient information exists on individuals exposed in the past to a potential workplace hazard, then it may be possible to undertake a retrospective cohort study. The historical data will have been collected for reasons that have nothing to do with epidemiology. Nevertheless, the availability of personnel records, such as registers of new and former employees, payrolls, work rosters, and individuals' career records, has enabled many epidemiological studies to be conducted—in particular, mortality studies.

The principles of a historical cohort study can also be applied to follow a cohort of workers prospectively. This approach will be discussed further in the next section, although it should be emphasized that many historical data studies have a prospective element in so far as they are updated after a further period of follow-up. The discussion of historical cohort studies in this section will concentrate on mortality and cancer incidence studies. However, there is no reason why hearing loss, lung function, or almost any measure of the health status of an individual should not be studied retrospectively if sufficient information is available.

Mortality and cancer incidence studies are unique among retrospective cohort studies in that they can be conducted using national cancer and mortality registers even if there has been no medical surveillance of the work force. A historical cohort study also has the advantages of being cheaper and providing estimates of the potential hazard much earlier than a prospective study. However, historical cohort studies are beset by a variety of problems. Principal among these is the problem of determining which workers have been exposed and, if so, to what degree. In addition, it may be difficult to decide what is an appropriate comparison group. It should also be borne in mind that in epidemiology, unlike animal experimentation, random allocation is not possible and there is no control over the factors which may distort the effects of the exposure of interest, such as smoking and standard of living.

The principles of historical cohort studies are described in the following subsections.

Cohort Definition and Follow-Up Period

A variety of sources of information are used to identify workers exposed to a particular workplace hazard, to construct an occupational history, and to complete the collection of information necessary for tracing (see below). It is essential that the cohort be well defined and that criteria for eligibility be strictly followed. This requires that a clear statement be made about membership of the cohort so that it is easy to decide whether an employee is a member or not. It is also important that the follow-up period be carefully defined. For instance, it is readily apparent that the follow-up period should not start before exposure has occurred. Furthermore, it is uncommon for the health effect of interest to manifest itself immediately after exposure, and allowance for an appropriate biological induction (or latency) period may need to be made when interpreting the data.

Comparison Subjects

The usual comparison group for many studies is the national population. However, it is known that there are marked regional differences in the mortality rates for many causes of death. Regional mortality rates exist in most industrialized countries but have to be used with caution because they are based on small numbers of deaths and estimated population sizes. In some situations the local rates for certain causes may be highly influenced by the mortality of the patients being studied. Furthermore, it is not always easy to decide what the most appropriate regional rate for comparison purposes is, as many employees may reside in a different region from that in which the plant is situated.

An alternative or additional approach is to establish a cohort of unexposed workers for comparison purposes. However, workers with very low exposures to the workplace hazard will often provide similar information.

Analysis and Interpretation

In a cohort study the first stage in the analysis consists of calculating the number of deaths expected during the follow-up period. In order to calculate the expected deaths for the cohort, the survival experience of the cohort is broken down into individual years of survival known as "person years." Each person year is characterized by the age of the cohort member and the time period when survival occurred and the sex of the cohort member. The person years are then multiplied by age-, sex-, and time periodspecific mortality rates to obtain the expected number of deaths. The ratio between observed and expected deaths is expressed as a standardized mortality ratio (SMR) as follows:

$$\text{SMR} = 100 \times \frac{\text{observed deaths}}{\text{expected deaths}}$$

Thus, an SMR of 125 represents an excess mortality of 25%. An SMR can be calculated for different causes of death and for subdivision of the person years by factors such as level of exposure and time since first exposure.

Interpretation of cohort studies is not always straightforward; there are a number of selection effects and biases that must be considered. Cohort studies routinely report that the mortality of active workers is less than that of the population as a whole. It is not an unexpected finding since workers usually have to undergo some sort of selection process to become or remain workers. Nevertheless, this selection effect, known as the "healthy worker" effect, can lead to considerable arguments over the interpretation of study results, particularly if the cancer mortality is as expected but the all-cause mortality is much lower than expected. However, even an experimental science such as toxicology is not without a similar problem of interpretation, viz., the problem of distinguishing between the effects of age and treatment on tumor incidence.

Proportional Mortality Study

There are often situations in which one has no accurate data on the composition of a cohort but does possess a set of death records (or cancer registrations). In these circumstances a proportional mortality study may sometimes be substituted for a cohort study. In such a mortality study the proportions of deaths from a specific cause among the study deaths is compared with the proportion of deaths from that cause in a comparison population. The results of a proportional mortality study are expressed in an analogous way to those of the cohort study with follow-up. Corresponding to the observed deaths from a particular cause, it is possible to calculate an expected number of deaths based on mortality rates for that cause and all causes of death in a comparison group and the total number of deaths in the study. The ratio between observed and expected deaths from a certain cause is expressed as a proportional mortality ratio (PMR) as follows:

$$PMR = 100x \frac{observed\ deaths}{expected\ deaths}$$

Thus, a PMR of 125 for a particular cause of death represents a 25% increase in the proportion of deaths due to that cause. A proportional mortality study has the advantage of avoiding the expensive and time-consuming establishment and tracing of a cohort but the disadvantage of little or no exposure information.

Prospective Cohort Study

Prospective cohort studies are no different in principle from historical cohort studies in terms of scientific logic, the major differences being timing and methodology. The study starts with a group of apparently healthy individuals whose health and exposure are studied over a period of time. As it is possible to define in advance the information that is to be collected, prospective studies are theoretically more reliable than retrospective studies. However, long periods of observation may be required to obtain results.

Prospective cohort studies or longitudinal studies of continually changing health parameters, such as lung function, hearing loss, blood biochemistry and hematological measurements, pose different problems from those encountered in mortality and cancer incidence studies. The relationships between changes in the parameters of interest and exposure measurements have to be estimated and, if necessary, a comparison made of changes in the parameters between groups. These relationships may be extremely complicated, compounded by factors such as aging, and difficult to estimate because there may be relatively few measurement points. Furthermore, large errors of measurement in the variables may be present because of factors such as within-laboratory variation and temporal variation within individuals. Missing observations and withdrawals may also cause problems, particularly if they are dependent on the level and change of the parameter of interest. These problems may make it difficult to interpret and judge the validity of analytical conclusions. Nevertheless, prospective cohort studies provide the best means of measuring changes in health parameters and relating them to exposure.

Case–Control Study

In a case–control study (also known as a case-referent study) two groups of individuals are selected for study, of which one has the disease whose causation is to be studied (the cases) and the other does not (the controls). In the context of the chemical industry, the aim of a case–control study is to evaluate the relevance of past exposure to the development of a disease. This is done by obtaining an indirect estimate of the rate of occurrence of the disease in an exposed and unexposed group by comparing the frequency of exposure among cases and controls.

Principal Features

Case–control and cohort studies complement each other as types of epidemiological study. In a case–control study the groups are defined on the basis of the presence or absence of a given disease and, hence, only one disease can be studied at a time. The case–control study compensates for this by providing information on a wide range of exposures which may play a role in the development of the disease. In contrast, a cohort study generally focuses on a single exposure but can be analyzed for multiple disease outcomes. A case–control study is a better way of studying rare diseases because a very large cohort would be required to demonstrate an excess of a rare disease. In contrast, a case–control study is an inefficient way of assessing the effect of an uncommon exposure, when it might be possible to conduct a cohort study of all those exposed.

The complementary strengths and weaknesses of case–control and cohort studies can be used to advantage. Increasingly, mortality studies are being reported which utilize "nested" case–control studies to investigate the association between the exposures of interest and a cause of death for which an excess has been discovered. However, case–control studies have traditionally been held in low regard, largely because they are often poorly conducted and interpreted. There is also a tendency to overinterpret the data and misuse statistical procedures. In addition, there is still considerable debate among leading epidemiologists themselves as to how controls should be selected.

Analysis and Interpretation

In a case–control study it is possible to compare the frequencies of exposures in the cases and controls. However, what one is really interested in is a comparison of the frequencies of the disease in the exposed and the unexposed. The latter comparison is usually expressed as a relative risk (RR), which is defined as

$$RR = \frac{\text{rate of disease } \epsilon \text{ exposed group}}{\text{rate of disease } \epsilon \text{ unexposed group}}$$

It is clearly not possible to calculate the RR directly in a case–control study since exposed and unexposed groups have not been followed in order to determine the rates of occurrence of the disease in the two groups. Nevertheless, it is possible to calculate another statistic, the odds ratio (OR), which, if certain assumptions hold, is a good estimate of the RR. For cases and controls the exposure odds are simply the odds of being exposed, and the OR is defined as

$$v = \frac{\text{cases with exposure}}{\text{controls with exposure}} \Big/ \frac{\text{cases without exposure}}{\text{controls without exposure}}$$

An OR of 1 indicates that the rate of disease is unaffected by exposure of workers to the agent of interest. An OR greater than 1 indicates an increase in the rate of disease in exposed workers.

Matching

Matching is the selection of a comparison group that is, within stated limits, identical with the study group with respect to one or more factors (e.g., age, years of service, and smoking history), which may distort the effect of the exposure of interest. The matching may be done on an individual or group basis. Although matching may be used in all types of study, including follow-up and cross-sectional studies, it is more widely used in case–control studies. It is common to see case–control studies in which each case is matched to as many as three or four controls.

Nested Case–Control Study

In a cohort study, the assessment of exposure for all cohort members may be extremely time-consuming and demanding of resources. If an excess of death of incidence has been discovered for a small number of conditions, it may be much more efficient to conduct a case–control study to investigate the effect of exposure. Thus, instead of all members being studied, only the cases and a sample of noncases would be compared with regard to exposure history. Thus, there is no need to investigate the exposure histories of all those who

are neither cases nor controls. However, the nesting is only effective if there are a reasonable number of cases and sufficient variation in the exposure of the cohort members.

Other Study Designs

Descriptive Studies

There are large numbers of records in existence which document the health of various groups of people. Mortality statistics are available for many countries and even for certain companies. Similarly, there is a wide range of routine morbidity statistics; in particular, those based on cancer registrations. These health statistics can be used to study differences between geographic regions (e.g., maps of cancer mortality and incidence presented at a recent symposium), occupational groups, and time periods. Investigations based on existing records of the distribution of disease and of possible causes are known as descriptive studies. It is sometimes possible to identify hazards associated with the development of rare conditions from observation of clustering in occupational or geographical areas.

Cross-Sectional Study

Cross-sectional studies measure the cause (exposure) and the effect (disease) at the same point in time. They compare the rates of diseases or symptoms of an exposed group with an unexposed group. Strictly speaking, the exposure information is ascertained simultaneously with the disease information. In practice, such studies are usually more meaningful from an etiological or causal point of view if the exposure assessment reflects past exposures. Current information is often all that is available but may still be meaningful because of the correlation between current exposure and relevant past exposure.

Cross-sectional studies are widely used to study the health of groups of workers who are exposed to possible hazards but do not undergo regular surveillance. They are particularly suited to the study of subclinical parameters such as blood biochemistry and hematological values. Cross-sectional studies are also relatively straightforward to conduct in comparison with prospective cohort studies and are generally simpler to interpret.

Intervention Study

Not all epidemiology is observational, and experimental studies have a role to play in evaluating the efficiency of an intervention program to prevent disease (e.g., fluoridation of water). An intervention study at one extreme may closely resemble a clinical trial with individuals randomly selected to receive some form of intervention (e.g., advice on reducing cholesterol levels). However, in some instances it may be a whole community that is selected to form the intervention group. The selection may or may not be random.

Conclusion

Epidemiology studies can be the most powerful and persuasive tools for establishing hazards associated with chemical exposures or personal actions (such as was the case with cigarette smoking). However, due to all the factors discussed previously, such studies also tend to be somewhat insensitive. Unless one can clearly establish the symptoms and signs of disease for which there is a causal connection, such studies lose the desired specificity.

Further Reading

Lileinfeld, A. M., and Lileinfeld, D. E. (1980). *Foundations of Epidemiology.* Oxford Univ. Press, New York.
Rothman, K. J. (1986). *Modern Epidemiology.* Little, Brown, Boston.

—*Shayne C. Gad*

Related Topics

Agency for Toxic Substances and Disease Registry
Analytical Toxicology
Carcinogen Classification Schemes
Carcinogenesis
Exposure
International Agency for Research on Cancer
Medical Surveillance
National Institutes of Health
Occupational Toxicology

Ergot

♦ SYNONYMS: Ergotamine; Ergot of Rye
♦ DESCRIPTION: Ergot is the result of infection of heads of grains by *Claviceps purpurea*, producing a dark purplish fruiting body called the sclerotium, which is consumed along with the grain. Toxicity is attributed to a mixture of ergot alkaloids, principally ergotamine, but that also contains ergocristine, ergocornine, α-ergocryptine, and β-ergocryptine.
♦ CHEMICAL CLASS: Alkaloid
♦ CHEMICAL STRUCTURE:

Alkaloid §	R(2')	R(5')
Ergotamine	—CH$_3$	—CH$_2$— phenyl
Ergosine	—CH$_3$	—CH$_2$CH(CH$_3$)$_2$
Ergostine	—CH$_2$CH$_3$	—CH$_2$— phenyl
Ergotoxine group:		
Ergocornine	—CH(CH$_3$)$_2$	—CH(CH$_3$)$_2$
Ergocristine	—CH(CH$_3$)$_2$	—CH$_2$— phenyl
α-Ergocristine	—CH(CH$_3$)$_2$	—CH$_2$CH(CH$_3$)$_2$
β-Ergocristine	—CH(CH$_3$)$_2$	—CHCH$_2$CH$_3$ / CH$_3$
Bromocriptine	—CH(CH$_3$)$_2$	—CH$_2$CH(CH$_3$)$_2$

Uses

Ergot was used as early as the sixteenth century to strengthen uterine contractions. Currently, ergotamine tartrate is combined with caffeine and administered to relieve migraine headaches. Ergonovine has been used to treat postpartum hemorrhage. Derivatives of ergotamine are used to manage amenorrhea and as an adjunct in the treatment of Parkinson's disease. Hydrogenated ergot alkaloids have been used for symptoms of idiopathic mental decline in elderly patients.

Exposure Pathways

Historically, exposure occurred by consumption of contaminated grain, especially rye flour. Acute poisonings in humans are rare and are generally associated with overdosage with ergotamine tartrate medication. Poisoning by ergot-containing mixtures has been associated with attempts to induce abortion. Animal poisonings result from consumption of contaminated pasture grasses and grains. The last diagnosed human fatalities associated with consumption of ergot-containing grains occurred in 1951.

Toxicokinetics

Ergotamine is poorly absorbed orally and a considerable amount is eliminated by first-pass metabolism in the liver. Symptoms begin to appear about 4 hr after intake. The volume of distribution is estimated to be about 2 liters and the half-life of elimination ranges from 1.4 to 6.2 hr.

Mechanism of Toxicity

Ergotamine may interact with tryptaminergic, dopaminergic, and α-adrenergic receptors. Pharmacological effects are attributed to both agonist and antagonist interference with α-adrenergic receptors, producing pressor, uterotonic, and emetic effects. Action at dopaminergic and serotonin receptors may induce hypothermia. Most effects of ergotamine poisoning may be attributed to localized vascular insufficiency.

Human Toxicity

Acute symptoms are nausea, vomiting, diarrhea, severe thirst, tingling of skin, skin hypoperfusion, and chest pain. Headache, fixed miosis, hallucinations, delirium, hemiplegia, and convulsions may occur. Historically, "ergotism" symptoms are a burning sensation of the extremities and hallucinations. St. Anthony's Fire was one descriptive name given to ergotism in Europe during the Middle Ages. Peripheral ischemia of lower extremeties is present, although ischemia of cerebral, mesenteric, coronary, and renal vascular beds may also occur. Legs may become pulseless, pale, and cyanotic

with burning pain and numbness. Hemorrhagic vesiculations, pruritus, formications, and gangrene can occur. Hallucinogenic episodes may occur.

Clinical Management

Acute poisoning is treated by cathartics such as syrup of ipecac and sodium sulfate, unless the patient has already had episodes of vomiting. Gastric gavage and multidose activated charcoal may speed removal. Sorbitol may be administered. Arterial spasms may be relieved by inhalation of amyl nitrite. In advanced cases, relief by administration of vasodilators, such as nitroprusside, nitroglycerin, prazosin, and captopril, may be indicated. Diazepam may by used to prevent seizures. Anticoagulants such as heparin may be administered in cases of extreme vasoconstriction. Reaction to the ergot alkaloids is highly idiosyncratic and clinical progress should be monitored carefully.

Animal Toxicity

Peripheral vasoconstriction, particularly of the hindlimbs and forelimbs, will produce hemorrhagic vesiculations that may progress to gangrene. Embryotoxicity, malformations, growth retardation, and miscarriages may occur in pregnant females.

—*Michael J. Brabec*

Erythromycin

- ◆ CAS: 114-07-8
- ◆ SYNONYM: EES
- ◆ PHARMACEUTICAL CLASS: A macrolide antibiotic produced by the organism *Streptomyces erythreus*
- ◆ CHEMICAL STRUCTURE:

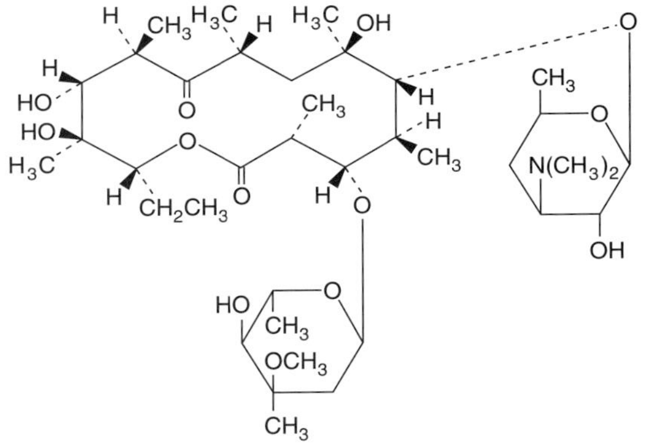

Uses

Erythromycin is used as an antibiotic.

Exposure Pathways

Erythromycin preparations are available in tablet, capsule, or liquid preparations for oral administration. Other available forms include intravenous, topical, and ophthalmic preparations. Ingestion is the most common route of accidental or intentional exposure.

Toxicokinetics

The esters and ester salts of erythromycin are absorbed more completely in the fasting state. Erythromycin estolate is absorbed with food. Erythromycin is absorbed mostly in the duodenum. Bioavailability is dependent on the specific erythromycin derivative, the presence of food in the gastrointestinal tract, and the gastric emptying time. Peak serum levels of erythromycin occur 4 hr following oral administration of erythromycin base or stearate. Erythromycin is partially deactivated by demethylation in the liver. The drug has been shown to induce microsomal enzymes and promote its own metabolism.

Erythromycin is 73.5% protein bound. The volume of distribution is 0.72 liters/kg. It is distributed in the total body water to all tissues. All tissues except the brain, middle ear, prostate gland, and eyes have higher erythromycin concentrations than the serum. Erythromycin is eliminated by renal excretion as unchanged drug in the amount of 4.5% of an oral dose. It is concentrated and excreted in unchanged form in the bile. In healthy subjects, the serum half-life is approximately 1–1.5 hr.

Mechanism of Toxicity

Erythromycin inhibits protein synthesis in susceptible organisms by binding to 50 S ribosomal subunits, thereby inhibiting translocation of aminoacryl transfer-RNA and inhibiting polypeptide synthesis. Erythromycin penetrates the wall of gram-positive bacteria more readily than that of gram-negative bacteria. A hypersensitivity reaction is probably responsible for the rare occurrence of cholestatic hepatitis following therapeutic doses of the estolate preparation of erythromycin. Erythromycin is a gastrointestinal irritant.

Human Toxicity

Toxic signs and symptoms following acute overdose of erythromycin are rare. Sensorineural hearing loss has been reported in patients taking large doses and suffering from concomitant liver or kidney disease. Local dose-related epigastric distress, nausea, vomiting, and diarrhea are common. Hepatotoxicity has been reported following therapeutic doses of erythromycin in the estolate form. Skin rash, fever, and eosinophilia may occur as a result of an allergic reaction.

Clinical Management

Basic and advanced life-support measures should be instituted as necessary. Gastric decontamination may be performed dependent on the symptomatology of the patient and the history of the exposure. Activated charcoal may be used to adsorb erythromycin. Gastrointestinal discomfort may be treated symptomatically or by reducing the dosage if appropriate. Liver function tests should be monitored if hepatotoxicity is suspected. Erythromycin blood levels are not clinically useful.

—*Carole Wezorek*

Ethane

- CAS: 74-84-0

- SYNONYMS: Bimethyl; dimethyl; methylmethane; ethyl hydride; methyl hydride; ethane, compressed (UN1035, DOT); ethane, refrigerated liquid (UN1961, DOT)

- CHEMICAL CLASS: Aliphatic hydrocarbon (C$_2$)
- MOLECULAR FORMULA: C$_2$H$_6$
- DESCRIPTION: Ethane is colorless, highly flammable gas that is lighter than air; it has a very slight odor, with a threshold odor concentration of approximately 1500 ppm.

Uses

Ethane is used as a fuel and as a raw material in the manufacture of synthetic organic chemicals (e.g., pharmaceutical and chemical industry). It occurs in natural gas at concentrations ranging from 5 to 10% and is practically inert. It is also found in the exhaust of diesel (approximately 1.8%) and gasoline (1.3–2.0%) engines. Small amounts of ethane, along with other C$_1$–C$_4$ alkanes and alkenes, have been detected in mined coal samples.

Exposure Pathways

Because ethane exists as a gas at normal temperature and pressure, exposure would occur by inhalation. Typical background concentrations detected in major U.S. cities range from 0.05 to 0.5 ppm. It is possible to spill liquid ethane from a refrigerated tank, causing frostbite upon contact with the skin due to rapid evaporation and loss of heat.

Mechanism of Toxicity

Ethane is not toxic to humans. It will, however, act as an asphyxiant at concentrations that are high enough to displace oxygen.

Human Toxicity

Ethane is not toxic to humans; studies have shown no adverse effects at concentrations of up to 50,000 ppm. Ethane is, however, classified as a simple asphyxiant. Concentrations that are high enough to displace oxygen would be expected to cause lightheadedness, loss of consciousness, and possibly death from asphyxiation.

Ethane is also highly flammable and is therefore an explosion and/or fire hazard (lower explosion limit is 3–12.5% by volume). Industrially, ethane is handled similarly to methane, and a threshold limit of 1000 ppm is commonly assumed.

Clinical Management

Persons who are exposed to high concentrations should vacate or be removed from the source of the gas and seek fresh air. Extreme care must be taken to keep areas of high concentration free from ignition sources, such as sparks from static electricity. Explosion-proof apparatus should also be used in these areas.

Animal Toxicity

Ethane has been shown to be a catabolic by-product of lipid peroxidation. Guinea pigs exposed to 2.2–5.5% of the gas for 2 hr have shown slight signs of irregular respiration that were readily reversible on cessation of exposure. As in humans, ethane acts as a simple asphyxiant at higher concentrations. Some studies have shown that certain microorganisms are able to use ethane as a nutrient, while other types of bacteria are inhibited by its presence.

—*Stephen Clough*

Related Topic

Lipid Peroxidation

Ethanol

- CAS: 67-17-4
- SYNONYMS: Ethyl alcohol; grain alcohol; methyl carbinol; ethyl hydrate
- CHEMICAL CLASS: Alcohol
- MOLECULAR FORMULA: C$_2$H$_6$O
- CHEMICAL STRUCTURE:

$$CH_3—CH_2—OH$$

Uses

Ethanol is one of the largest volume organic chemicals used in industrial and consumer products. Its primary use in industry is as an intermediate in the production of other chemicals and as a solvent. Ethanol is used in the manufacture of drugs, plastics, lacquers, polishes, plasticizers, and cosmetics. Commercial products containing ethyl alcohol include beverages, solvents for perfumes, aftershaves and colognes, medicinal liquids, mouthwashes, liniments, and some rubbing alcohols.

Exposure Pathways
Exposure to ethanol occurs principally via ingestion or inhalation.

Toxicokinetics
Ethanol is readily absorbed upon inhalation or ingestion. Absorption from the gastrointestinal tract is by simple diffusion with approximately 80% of an oral dose being absorbed in the small intestine. About 80–90% of ethanol is absorbed within 30–60 min, although food may delay complete absorption for 4–6 hr. Inhalation of ethanol vapors in the range of 5000–10,000 ppm by human volunteers indicates absorption from lungs to be approximately 62%.

Ethanol is both water and lipid soluble and therefore distributes into total body water and easily penetrates the blood–brain barriers and placenta. Ethanol has been found in the amniotic fluid of animals after a single oral dose.

The metabolism of ethanol occurs predominantly in the liver. Ethanol is oxidized by alcohol dehydrogenase to acetaldehyde and further oxidized via aldehyde dehydrogenase to acetic acid. Acetic acid is released into the blood where it is further oxidized through normal intermediary metabolism in peripheral tissues to CO_2 and water. Microsomal cytochrome P450 has also been found capable of oxidizing ethanol to acetaldehyde and may be an important pathway at high levels of ethanol exposure.

Normally, 90–98% of the ethanol that enters the body is completely oxidized, predominantly in the liver, eventually entering the citric acid cycle or utilized in anabolic synthetic pathways. The kidney and lungs excrete only 5–10% of an absorbed dose unchanged. The rate of ethanol metabolism varies between individuals and may be under genetic control.

Mechanism of Toxicity
Ethanol is a central nervous system (CNS) depressant that initially and selectively depresses some of the most active portions of the brain (reticular activity system and cortex). The mechanism of action most likely involves interference with ion transport at the axonal cell membrane rather than at the synapse, similar to the action of other anesthetic agents.

Human Toxicity
Ethanol is an irritant of the eyes and mucous membranes and causes CNS depression at very high levels of exposure. Exposure to humans of 5000–10,000 ppm has caused transient eye and nose irritation as well as cough. Exposures at 15,000 ppm produce continuous lacramation and cough, and levels of 25,000 ppm and above were judged as intolerable. Chronic exposures to ethanol vapors can result in irritation of mucous membranes, headache, and symptoms of CNS depression, such as lack of concentration and drowsiness.

Mild ethanol intoxication is observed at blood alcohol levels in the range of 0.05–0.15%. Symptoms include impairment of visual acuity, muscular incoordination, decreased reaction time, and changes in mood, personality, or behavior. At blood alcohol levels of 0.15–0.3%, visual impairment, sensory loss, muscle incoordination, slowed reaction time, and slurred speech are observed. At levels of 0.3–0.5% blood alcohol, there is severe intoxication characterized by muscular incoordination, blurred or double vision, and sometimes stupor, hypothermia, vomiting, nausea, and, occasionally, hypoglycemia and convulsions. At 0.4% and above, symptoms include coma, depressed reflexes, respiratory depression, hypertension, hypothermia, and death from respiratory or circulatory failure, often as a result of aspiration of stomach contents in the absence of a gag reflex.

Alcohol ingestion during pregnancy has been found to lead to congenital malformations that have been collectively termed fetal alcohol syndrome. Fetal alcohol syndrome is characterized by mental deficiency, microcephaly, and irritability. Affected infants typically are small and demonstrate poor muscle coordination. The severity of the effects is related to the extent of alcohol consumption by the mother during pregnancy. This syndrome has been associated with alcoholic women who drink heavily and chronically during pregnancy. There have been no reports of fetal alcohol syndrome resulting from industrial exposure.

Chronic ethanol ingestion has also been shown to produce liver damage which can eventually lead to cirrhosis of the liver and possibly death. Signs include enlarged liver, elevated serum transaminace, and jaundice.

Clinical Management
Acute and severe overexposure to ethanol via ingestion requires special attention in order to prevent aspiration

of stomach contents and ensure replacement of fluids and electrolytes.

Animal Toxicity

The acute oral, inhalation, and dermal toxicity of ethanol in animals is low. Oral LD_{50}s in rats, mice, guinea pigs, rabbits, and dogs range from approximately 6 to 18 g/kg. Inhalation LD_{50}s range from 12,000 to approximately 50,000 ppm in studies of mice, guinea pigs, and rats. Ethanol is not significantly irritating to the skin of rabbits, although it did produce eye irritation in rabbits. Subchronic and chronic toxicity testing in animals have indicated the liver as the primary site of action. Effects upon the liver observed in animals parallel those observed in humans and include fatty degeneration, focal necrosis, inflammation, and fibrosis leading to cirrhosis.

Ethanol has been studied in rats, mice, and hamsters for carcinogenicity. Due to limitations in study design and study protocol these results have to date been inconclusive. There are data from animals indicating that ethanol consumption can enhance the carcinogenic activity of other known carcinogenic agents.

Ethanol has been investigated for reproductive toxicity in male mice and rats, and while producing effects upon testes and other reproductive tissues, has generally not been shown to affect reproductive outcome or performance.

Rats, mice, and rabbits have been tested for developmental toxicity upon ethanol exposure. Inhalation of ethanol by pregnant rats at up to 20,000 ppm for 7 hr/day on Gestational Days 1–19 produced no treatment-related effects on uterine implantation or embryonic development. Similarly, 15% ethanol in drinking water of rats, mice, and rabbits, while eliciting maternal toxicity and reducing fetal weights, failed to elicit teratogenic effects. Effects were noted in the offspring of female mice maintained on liquid diets containing 15–35% ethanol dry calories for at least 30 days before and during gestation until Day 18.

—*Bradford Strohm*

Related Topics

Developmental Toxicology
Neurotoxicology: Central and Peripheral
Poisoning Emergencies in Humans

Ethanolamine

- ◆ CAS: 141-43-5
- ◆ SYNONYMS: Monoethanolamine; 2-aminoethanol
- ◆ CHEMICAL CLASS: Amine
- ◆ MOLECULAR FORMULA: C_2H_7NO
- ◆ CHEMICAL STRUCTURE:

$$NH_2—CH_2—CH_2—OH$$

Uses

Ethanolamine is used to remove carbon dioxide and hydrogen sulfide from natural gases and other gas streams. It is also used in the synthesis of surfactants, in polishes, as a dispersing agent for agricultural chemicals, and in the manufacture of pharmaceuticals.

Exposure Pathways

The low vapor pressure of ethanolamine (0.4 torr at 20°C) suggests that dermal exposure may be the most probable route of exposure to ethanolamine in the occupational setting. However, despite its wide use in industry, the toxicity of ethanolamine in humans has not been reported.

Toxicokinetics

Data on the absorption of ethanolamine via different routes are unavailable. However, based on the difference in toxicity between orally and dermally administered ethanolamine, absorption through the skin may exceed absorption through the gastrointestinal tract. Ethanolamine is thought to be metabolized primarily in the liver. No distribution data are available for ethanolamine. Ethanolamine appears to be eliminated primarily via the lungs (as carbon dioxide) and urine (as urea, glycine, serine, choline, and uric acid). Ethanolamine is a normal constituent of human urine.

Mechanism of Toxicity

The mechanism of toxicity for ethanolamine is unknown.

Human Toxicity

Toxicity in humans has not been reported and is likely to be limited to local irritation at the site of application (e.g., skin and eyes).

Clinical Management

Clinical management involves removal from exposure and treatment of symptoms.

Animal Toxicity

Although toxicological studies in animals are sparse, available data indicate that ethanolamine exposures do not generally result in systemic toxicity at sublethal doses. It appears that ethanolamine toxicity is largely limited to irritant effects, and their sequelae, at the site of application. Ethanolamine is more toxic when applied to the skin (LD_{50} of 1 mg/kg) than when administered orally (LD_{50} of 3320 mg/kg).

—Ralph Parod

Ethchlorvynol

- ♦ CAS: 113-18-8
- ♦ SYNONYMS: β-Chlorovinyl ethyl ethinyl carbinol; 1-chloro-3-ethyl-pent-1-en-4-yn-3-ol; Placidyl; Arvynol; Serenesil
- ♦ PHARMACEUTICAL CLASS: Tertiary acetylenic alcohol
- ♦ CHEMICAL STRUCTURE:

$$CH_3CH_2-\overset{\overset{\displaystyle C\equiv CH}{|}}{\underset{\underset{\displaystyle OH}{|}}{C}}-CH=CHCl$$

Uses

Ethchlorvynol is used as a sedative/hypnotic. It also possesses some anticonvulsant and muscle relaxant properties. As with most sedative/hypnotics, ethchlorvynol has abuse potential.

Exposure Pathways

Ethchlorvynol is marketed as a liquid-filled capsule. Toxicity has resulted from oral overdose and from intravenous injection of the liquid contents of a capsule.

Toxicokinetics

Ethchlorvynol is rapidly absorbed with peak plasma levels occurring within 1–1.5 hr. Approximately 90% of a dose undergoes hepatic hydroxylation and glucuronidation, and several metabolites have been identified, including hydroxyethchlorvynol. Both the parent compound and metabolites undergo enterohepatic recirculation. Binding to plasma proteins is about 35–50%. The volume of distribution is 2.5–4 liters/kg, with significant amounts distributed to, and slowly released from, adipose tissue. The elimination of ethchlorvynol appears to be biphasic, with a distribution half-life of 1–5 hr and an elimination half-life of 10–25 hr. The elimination phase is prolonged (up to 100 hr) following large overdoses. Only traces of unchanged drug appear in the urine.

Mechanism of Toxicity

The pharmacology of ethchlorvynol is much like that of the barbiturates; thus, an interaction that results in γ-aminobutyric acid-like activity is likely involved. Toxicity results in dose-dependent depression of the central nervous system (CNS).

Human Toxicity: Acute

Overdosage results in dose-dependent depression of the CNS, ranging from fatigue and lethargy to respiratory depression and coma. Coma may be profound, with a flat EEG, and has been reported to last as long as 17 days. CNS depression may be potentiated by the presence of ethanol or other CNS depressants. Hypothermia is a frequent finding, and hypotension with either tachycardia or bradycardia is common with large doses. Ataxia, nystagmus, and headache may occur. Delayed-onset (24–48 hr) noncardiogenic pulmonary edema has occurred following large overdoses and is usually associated with deep coma; however, onset may be rapid following intravenous exposure. Overdoses have also been reported to cause paradoxical excitement and pancytopenia and hemolysis. As with other sedative/hypnotic drugs, bullous lesions and pressure necrosis have been found on comatose patients, and seizures have occurred during withdrawal. Death has occurred following ingestion of 2.5 g ethchlorvynol plus alcohol, but the usual fatal dosage range is ≥10 g. Postmortem blood concentrations ranged from 14 to 400 mg/liter in one study and from 22 to 213 mg/liter in another. Blood levels, however, are not used clinically to guide treatment.

Clinical Management

The basis of treatment should be provision of supportive care, with particular attention to the level of consciousness and airway protection. All patients should have intravenous access and cardiac monitoring and should be observed for hypothermia and hypotension. Gastric decontamination should be achieved with activated charcoal. The stomach may be lavaged if the airway is secure in patients who present within 1 hr of ingestion. Emesis should not be induced with syrup of ipecac because of possible rapid onset of CNS depression. A complete blood count should be obtained to assess for anemia or thrombocytopenia. Hypotension should initially be treated by elevating the feet and administering an intravenous fluid bolus, followed by administration of vasopressors such as norepinephrine or dopamine if necessary. Pulmonary edema should be managed with positive end expiratory pressure (e.g., not diuretics or inotropic agents) if needed. There are no antidotes. Drug clearance may be enhanced by resin (15–50% removal) or charcoal (5–10% removal) hemoperfusion, but an affect on morbidity has not been demonstrated. Hemodialysis has not proven to be of benefit.

Animal Toxicity

No specific data on animal toxicity are available.

—*S. Rutherfoord Rose*

Ethene

- CAS: 74-85-1
- SYNONYMS: Ethylene; acetene; bicarburetted hydrogen; elayl; olefiant gas
- CHEMICAL CLASS: Aliphatic alkene
- MOLECULAR FORMULA: C_2H_4
- CHEMICAL STRUCTURE:

$$H_2C = CH_2$$

Uses

Ethene is used primarily as a feedstock in the production of polymers and industrial chemicals. Approximately 80% is used for production of polyethylene, ethylene oxide/ethylene glycols, and ethylene dichloride/vinyl chloride. Additionally, ethene is used for the controlled ripening of citrus fruits, tomatoes, bananas, other fruits, vegetables, and flowers.

Exposure Pathways

Because ethene is a gas, inhalation exposure is the primary route of entry.

Toxicokinetics

The inhalation toxicokinetics of ethene have been investigated in human volunteers at atmospheric concentrations of up to 50 ppm (57.5 mg/m³). The majority (94.4%) of ethene inhaled into the lungs is exhaled unchanged without becoming systemically available via the bloodstream. The remaining ethene is metabolized to ethylene oxide, which then reacts to form complexes with hemoglobin, *N*-(2-hydroxyethyl)histidine, and *N*-(2-hydroxyethyl)valine. The biological half-life is approximately 0.65 hr, being excreted in urine and feces and exhaled as CO_2. The toxicokinetics of ethene in humans and experimental animals appears to be similar.

Mechanism of Toxicity

Ethene is classified as a simple asphyxiant. In sufficient concentrations, ethene causes central nervous system depression and unconsciousness by displacing oxygen in air, which reduces the oxygen available to support cell function.

Human Toxicity

Ethene is a relatively nontoxic gas. No adverse effects are observed at concentrations of less than 2.5%. At higher concentrations, ethene exhibits the anesthetic properties associated with oxygen deprivation. Humans exposed to ethene may experience subtle signs of intoxication, resulting in prolonged reaction time. Exposure to 37.5% ethene for 15 min resulted in memory disturbances, and concentrations at 50% resulted in unconsciousness. If oxygen is deprived for a sufficient amount of time, death can occur. Ethene has been used as an anesthetic and has some advantages over those more typically used in that its effects are rapid in onset and recovery with minimal effect on other organ systems. The primary hazard associated with use of ethene, however, is its flammability and explosivity.

Clinical Management

Overexposure to ethene is treated by simply moving the victim to fresh air. Recovery is usually rapid and complete.

Animal Toxicity

Ethene has been tested in both rats and dogs in short-term inhalation exposure studies. Exposure to skin and eyes does not cause irritation. The anesthetic properties reported in humans have also been observed in experimental animals. Ethene is not a cardiac sensitizer in dogs. The toxicity and carcinogenicity of inhaled ethene was studied in Fischer 344 rats. The animals were exposed to 300, 1000, or 3000 ppm for 6 hr/day, 5 days/week for 24 months. These exposures resulted in no toxicity or carcinogenicity.

—*Patricia J. Beattie*

Related Topic

Polymers

Ethyl Acetate

- ♦ CAS: 141-78-6
- ♦ SYNONYMS: Acetic acid ethyl ester; aceticether acetidin; acetoxyethane; ethyl acetic ester; ethyl thanoate

- ♦ CHEMICAL CLASS: Aliphatic ester
- ♦ CHEMICAL STRUCTURE:

Uses

Ethyl acetate is used as a solvent in the production of pharmaceuticals and food. In food it is utilized for artificial fruit essences. Industrially it serves as a solvent for nitrocellulose, varnishes, lacquer, fingernail polishes, coatings, plastics, and inks.

Exposure Pathways

Ethyl acetate can enter the system by inhalation, ingestion, and absorption through the skin.

Toxicokinetics

Hydrolysis of ethyl acetate produces acetic acid and ethanol. The ethanol metabolic product can undergo oxidation or be excreted through expired air and urine. Metabolic products can accumulate in the vascular system.

Mechanism of Toxicity

Ethyl acetate can form peroxides; however, toxicity will be similar to ethanol toxicity.

Human Toxicity

Acute exposure produces irritation to the eyes, respiratory tract, digestive tract, and skin. Inhalation will produce central nervous system depression and can result in headache, nausea, vomiting, dizziness, narcosis, suffocation, and lowered blood pressure. Blood effects may manifest as leukocytosis and anemia. At high exposures death can result with respiratory or circulatory failure. It is not caustic but will cause a drying and cracking of skin due to its ability to solubilize fats. It is a mild membrane irritant. Prolonged inhalation can cause lung, liver, kidney, and heart damage.

Clinical Management

There is no specific antidote for ethyl acetate. Treatment should be symptomatic and supportive noting similarity to ethanol intoxication. If ingested two to four cupfuls of milk or water should be given. For inhalation exposure, the victim should be moved to fresh air and given artificial respiration if not breathing. If ethyl acetate comes in contact with skin and eyes, the affected areas should be flushed with water for at least 15 min. Ethyl acetate's defatting properties should be noted; treatment should be symptomatic.

Animal Toxicity

In rats, the oral LD_{50} was 11.3 g/kg and the inhalation LC_{50} was 1600 ppm/8 hr. The interperitoneal LD_{50} in mice was 709 mg/kg.

—*Kathryn Kehoe*

Ethyl Acrylate

- CAS: 140-88-5
- SYNONYMS: Ethyl 2-propenoate; 2-propenoic acid ethyl ester; acrylic acid ethyl ester
- CHEMICAL CLASS: Ester
- MOLECULAR FORMULA: $C_5H_8O_2$
- CHEMICAL STRUCTURE:

$$H_2C=CH-\underset{\underset{O}{\|}}{C}-O-CH_2-CH_3$$

Uses

Ethyl acrylate monomer is used to make emulsion (water-based) and solution (solvent-based) polymers. Water-based ethyl acrylate polymers are used in latex paints, caulks, leather-treating base coats, floor polishes, and textile finishes. Solvent-based ethyl acrylate polymers are used in lacquers, enamels, and lubricating oils. Ethyl acrylate has also received limited use as a fragrance and flavoring agent.

Exposure Pathways

Exposures to ethyl acrylate monomer are most likely to occur in an occupational environment via skin con-

tact and inhalation. However, the closed systems used during manufacture and transportation will limit worker exposures to those which may occur during routine process maintenance, periodic plumbing leaks, and the collection of quality control samples.

The general population does not receive a significant exposure to ethyl acrylate due to low concentrations of residual monomer in consumer products. Perhaps the greatest potential exposure of consumers to ethyl acrylate monomer is during the application of latex-based paint to the interior of a home where the large surface-to-paint ratio favors the release of any residual monomer into room air. Tests conducted in a poorly ventilated room with commercially available latex paint indicate that the range of maximum ethyl acrylate monomer concentrations in air is 2.5–8.0 ppm. In a well-ventilated room, as well as 24 hr after paint application in a poorly ventilated room, ethyl acrylate monomer concentrations in air are below the level of detection (0.2 ppm).

Toxicokinetics

Data in experimental animals indicate that ethyl acrylate is readily absorbed from the gastrointestinal tract and the respiratory tract. Ethyl acrylate is also absorbed through the skin but less readily because a large portion of the dermally applied dose appears to be retained in the dermis. Dermal absorption may also be impeded by the evaporation of ethyl acrylate from the skin if the applied dose is unoccluded.

The primary route of ethyl acrylate metabolism is its rapid hydrolysis by tissue and circulating carboxylesterases to acrylic acid and ethanol which undergo further metabolism. In the rat, it has been estimated that approximately 50% of the ethyl acrylate that passes through the upper respiratory tract is hydrolyzed by carboxylesterase in the nasal mucosa before entering the blood. Another route of ethyl acrylate metabolism is conjugation with the sulfhydryl group of glutathione.

Ethyl acrylate is rapidly distributed throughout the body. Ethyl acrylate and/or its metabolites can be detected in all organ systems, with the highest concentrations being present in the urine, expired air, and organ of entry (i.e., stomach, upper respiratory tract, and skin). Metabolites of ethyl acrylate are excreted primarily via the lungs (as carbon dioxide) and the kidneys. Approximately 60% of an orally administered dose of ethyl acrylate is eliminated from the body within 8 hr.

Mechanism of Toxicity

Pretreatment of rats with a carboxylesterase inhibitor potentiates the respiratory irritation and lethality produced by the inhalation of ethyl acrylate. This and other observations suggest that the toxicity of ethyl acrylate becomes manifest when local detoxification/defense mechanisms become overwhelmed.

Human Toxicity

Ethyl acrylate can be highly irritating to the skin, eyes, gastrointestinal tract, and the respiratory tract. Chronic exposures to high vapor concentrations may damage the nasal mucosa leading to decrements in the sense of smell and taste; these decrements may be reversible upon cessation of exposure. Ethyl acrylate can cause an allergic contact dermatitis which may cross-react with other acrylate esters.

Clinical Management

Clinical management involves removal from exposure and treatment of symptoms.

Animal Toxicity

Toxicological studies in animals indicate that ethyl acrylate exposures do not generally result in systemic toxicity at sublethal doses. Although ethyl acrylate concentrations approaching lethal doses may cause histopathological changes in the liver and kidneys, ethyl acrylate toxicity is largely limited to irritant effects, and their sequelae, at the site of application. Ethyl acrylate can produce an allergic contact dermatitis which may cross-react with other acrylic esters.

The acute LD_{50} varies between 770 mg/kg (rat, oral) and 3000 mg/kg (rat, dermal) depending on the species and exposure route being evaluated.

—*Ralph Parod*

Ethylamine

- ♦ CAS: 75-04-7
- ♦ SYNONYMS: Ethanamine; monoethylamine; aminoethane; 1-aminoethane

- ♦ CHEMICAL CLASS: Amine
- ♦ CHEMICAL STRUCTURE:

$$CH_3CH_2NH_2$$

Uses

Ethylamine is used in resin chemistry, oil refining, and in solvent extraction; as a stabilizer for rubber latex; as an intermediate for dyestuffs, medicinals, and triazine herbicides; as a corrosion inhibitor; as a urethane foam catalyst; and as a detergent.

Exposure Pathways

Exposure may occur through oral, dermal, or inhalation routes.

Toxicokinetics

Ethylamine is rapidly absorbed. Some diethylamine is excreted from the lung unchanged, and nearly one-third has been reported to be excreted unchanged in the urine following oral dosing. The metabolic pathway is similar to that of other lower aliphatic amines, although it is metabolized to a lesser extent. Initially, the amino group is dehydrogenated to the intermediate imine, which then reacts with water, forming acetaldehyde and ammonia, and then to acetic acid and urea. Ethylamine is a normal constituent of mammalian and human urine.

Mechanism of Toxicity

The toxic effects of ethylamine are due primarily to its corrosive action on tissues.

Human Toxicity

Eye exposure to ethylamine can cause corneal edema and temporary blue, hazy vision from subtle disturbances of the corneal epithelium. Ethylamine may be expected to cause skin and respiratory tract irritation and decreased olfactory sensitivity.

Clinical Management

Exposed skin and eyes should be irrigated with copious amounts of water. After inhalation exposures, the victim should be moved to fresh air and monitored for respiratory distress. Humidified supplemental oxygen (100%) should be administered with assisted ventilation as required. If coughing or breathing difficulties are noted, the patient should be evaluated for irritation, bronchitis, or pneumonitis, including chest X-rays and determination of blood gasses. If pulmonary edema is present, positive end expiratory pressure ventilation and steroids should be considered. For ingestion exposures, the stomach contents should be diluted with copious amounts of water. Emesis or lavage should be avoided.

Animal Toxicity

Severe skin irritation with extensive necrosis and deep scarring has been reported after dermal contact in the guinea pig. The LD_{50} by dermal exposure in rabbits was reported to be 390 mg/kg. Ethylamine has been shown to cause severe eye damage in rabbits. Exposure to 100 ppm by inhalation resulted in irritation of the cornea, lung and liver, and kidney damage in rabbits after 6 weeks of exposure. A 50 ppm exposure also resulted in lung irritation and corneal injury. The inhalation LC_{33} in rats was 8000 ppm for 4 hr. Orally, the LD_{50} in rats is 400 mg/kg. Ethylamine has been reported to cause adrenal cortical gland necrosis, with the adrenal gland, testis, and thyroid gland being the most sensitive to this effect.

—Janice M. McKee

Ethylbenzene

♦ CAS: 100-41-4
♦ Chemical Class: Alkyl aromatic
♦ Chemical Structure:

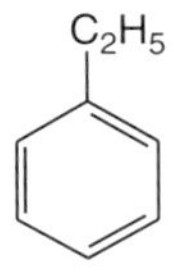

C₂H₅

Ethylbenzene

Uses

Ethylbenzene is used as an industrial solvent and as a component in automotive and aviation fuels. The majority of ethylbenzene is used in the production of styrene.

Exposure Pathways

The primary exposure route for ethylbenzene is via inhalation and the skin. Ethylbenzene is known to cross the placental barrier but has not been established as a reproductive hazard.

Toxicokinetics

Ethylbenzene distributes to the adipose tissues. It is metabolized to mandelic acid (64%) and phenylglyoxylic acid (25%). The percentage of metabolites may vary according to the route of exposure with mandelic acid formation being favored with inhalation. The primary route of excretion is via the urine. Experimental evidence indicates that the percutaneous absorption rate of ethylbenzene is 37 μg/cm^2.

Mechanism of Toxicity

Ethylbenzene is an inducer of the cytochrome P450 and cytochrome c reductase enzyme systems. Ethylbenzene acts as a mitochondrial uncoupling agent. It is believed that ethylbenzene metabolites are capable of interfering with dopamine catabolism in the brain. The tuberoinfundibular dopaminergic system may be a target for these metabolites.

Human Toxicity

Ethylbenzene is irritating to the eyes and skin. Concentrations of 875 ppm ethylbenzene are known to be irritating to the eyes of humans. Dermal application has led to erythema and inflammation of the skin. Ethylbenzene is the most irritating of the benzene series of compounds tested. Inhalation of high concentrations may cause central nervous system (CNS) excitation followed depression.

Chronic

Prolonged exposure may lead to functional pulmonary changes. These may be expressed as an increase in deep reflexes and irritation to the upper respiratory tract.

Hematological

There have been complaints of leukopenia and lymphocytosis. However, unlike benzene, ethylbenzene does not appear to cause bone marrow problems.

The OSHA PEL for ethylbenzene is 100 ppm. The odor threshold is 8.7 ppm. The U.S. EPA acceptable daily intake is 1.6 mg/day. The U.S. EPA oral reference dose (Rfd) for ethylbenzene is 0.1 mg/kg/day. The oral Rfd is based on liver and kidney toxicity observed in a subchronic rodent experiment.

Clinical Management

General life support should be maintained. Symptoms should be treated and the victim decontaminated if necessary. Persons at special risk are those with impaired pulmonary functions, particularly obstructive airway diseases. The irritant properties of ethylbenzene may exacerbate these preexisting respiratory conditions. Additionally predisposed groups include persons with liver, nervous system disorders, blood and hemopoietic disorders, and women with ovulation and menstrual cycle disorders.

Animal Toxicity

Ethylbenzene is extremely irritating in animal studies. Repeated skin application has caused blistering. Inhalation or ingestion of high concentrations has led to CNS depression with death attributed to depression of the respiratory center. Pathological observations include pulmonary edema and generalized visceral hyperemia.

Chronic

Exposure to high concentrations has led to increased liver and kidney weights in experimental animals.

—*William S. Utley*

Ethyl Bromide

- CAS: 74-96-4
- SYNONYMS: Bromoethane; bromic ether; halon 2001; hydrobromic ether; monobromoethane

- CHEMICAL CLASS: Halogenated aliphatic hydrocarbons
- MOLECULAR FORMULA: C_2H_5Br

Uses

Ethyl bromide is used as a refrigerant, an ethylating agent in organic synthesis, and in ethylation of gasoline. It was formerly used as medication (anesthetic—topical and inhalation).

Exposure Pathways

Exposure to ethyl bromide may occur through oral, dermal, and/or inhalation exposure.

Toxicokinetics

Ethyl bromide may be hydrolyzed to a significant degree resulting in formation of inorganic bromides. Glutathione *S*-alkyl transferase catalyzes conjugation in which the halogen group is replaced. Further metabolism then occurs to alkyl mercapturic acid sulfoxides. Ethyl bromide crosses the placenta.

Mechanism of Toxicity

Ethyl bromide causes irritation and has a tendency to cause fatty degeneration of the liver, renal tissue, and the heart.

Human Toxicity

Ethyl bromide is markedly irritating to the respiratory tract and the eyes. It is moderately toxic by ingestion but bromide poisoning following acute ingestion is rare. Acute effects of ethyl bromide include central nervous system (CNS) depression, coma, hypotension, tachycardia, respiratory distress, nausea, and vomiting. Ethyl bromide can produce acute congestion, edema, and liver and kidney damage. Fever may also occur. Serum bromide concentrations $\geq 50–100$ mg/dl are usually associated with signs of toxicity. Aftereffects from severe exposure may occur up to 30 hr after the exposure has ceased.

Acute or chronic exposure may result in redness of face, dilation of pupils, and a rapid pulse. The former use of ethyl bromide as a human anesthetic produced respiratory irritation and caused some fatalities, either immediately due to respiratory or cardiac arrest or delayed from effects on the liver, kidneys, or heart.

Chronic ingestion of excessive amounts of ethyl bromide may produce a toxic syndrome known as "bromism." The symptoms are behavioral changes, irritability, headache, confusion, anorexia, weight loss, lethargy, muscular weakness, and slurred speech. Chronic intoxication usually develops over 2–4 weeks or longer.

Dermal exposure to ethyl bromide can result in bromoderma, which is an erythematous, nodular, or acneform rash over the face and possibly the entire body.

Bromides cross the placenta and may be detected in the milk of nursing mothers. Case reports suggest that prenatal exposure may cause growth retardation, craniofacial abnormalities, and developmental delay.

Clinical Management

Acute oral exposure is generally treated by emesis. Emesis is most effective when initiated within 30 min of ingestion.

Chronic overexposure is treated by rehydration and administration of sodium chloride (NaCl; salt) intravenously until symptoms are alleviated. Bromide clearance may be increased by rehydrating in conjunction with administration of diuretics. Severe cases may require hemodialysis.

Animal Toxicity

Ethyl bromide is moderately toxic by ingestion and intraperitoneal routes and mildly toxic by inhalation. It is an eye and skin irritant.

Guinea pigs exposed to 2.4% ethyl bromide in air for 30 min experienced some delayed deaths and pathological changes in lungs, liver, spleen, and kidneys. Ethyl bromide is a CNS depressant causing pulmonary congestion, centrilobular necrosis of the liver, and diffuse nephritis.

A 2-year study indicated some evidence of carcinogenicity in male rats (adrenal gland and neoplasms of brain and lung) and equivocal results in female rats and male mice. Female mice experienced an increase in uterine cancer.

—*Kathryn A. Wurzel*

Ethylene Glycol

- CAS: 107-21-1
- SYNONYMS: 1,2-Dihydroxyethane; 1,2-ethanediol

- CHEMICAL CLASS: Glycols
- CHEMICAL STRUCTURE:

$$H-\overset{\overset{\displaystyle OH}{|}}{\underset{\underset{\displaystyle H}{|}}{C}}-\overset{\overset{\displaystyle OH}{|}}{\underset{\underset{\displaystyle H}{|}}{C}}-H$$

Uses

Ethylene glycol has numerous industrial and commercial applications. A major use is in antifreeze–coolant mixtures for motor vehicles. It is also used in heat-transfer fluids, airport runway and aircraft deicing fluids, hydraulic brake fluids, printers' inks, wood stains, adhesives, and pesticides. In addition, it is a solvent in various other chemicals.

Exposure Pathways

Industrial exposures to ethylene glycol liquid are most likely to occur by the dermal or ocular routes, with potential for inhalation to mists. There are also numerous cases of accidental liquid ingestion of antifreeze.

Toxicokinetics

Following exposure by various routes, ethylene glycol is rapidly absorbed, distributed, and excreted. Plasma half-lives in dogs given ethylene glycol intravenously ranged from 3.0 to 4.4 hr. When rats were exposed to ethylene glycol vapors by nose-only inhalation, plasma half-lives ranged from 34 to 39 hr and 75–80% of the initial body burden was widely distributed in animals examined immediately after exposure. Rates of metabolism and excretion vary with species, dose, and route of administration. Generally, metabolism begins immediately after administration, and excretion of most of the parent compound and metabolite is complete 12–48 hr after dosing. The major excreloy endproducts in animals are carbon dioxide in exhaled air and glycolate and unchanged ethylene glycol (25%) in the urine. Depending on the species, variable quantities of other metabolites, such as glyoxylate, hippurate, and oxalate, may also be excreted in the urine.

Metabolism occurs primarily in the liver and kidneys. The initial step is alcohol dehydrogenase-catalyzed conversion of the parent compound to glycolaldehyde. Subsequent metabolic steps are not as well characterized. Systemic toxic effects of ethylene glycol, such as metabolic acidosis and renal tubule necrosis, are generally attributed to the action of metabolites rather than the parent compound.

Excretion is mainly by way of metabolites of ethylene glycol in the urine. Some unchanged inhaled vapors could be exhaled as well.

Mechanism of Toxicity

Inhalation of ethylene glycol vapor and mist causes upper respiratory tract irritation due to its solubility in the mucus and irritation of the respiratory nerve endings.

Ingestion of large amounts of ethylene glycol leads to (1) central nervous system (CNS) depression (0.5–12 hr), (2) cardiopulmonary disturbances (12–36 hr), and (3) renal distress (36–48 hr). The CNS depression is the result of the action of unmetabolized ethylene glycol and glycoaldehyde on the CNS. Cardiopulmonary disturbances, such as tachycardia, hypertention, congestive heart failure, pulmonary edema, and cardiomegaly, have resulted from the oxalates cytotoxic effects. Renal distress is the outcome of the direct toxic effect of oxalic acid, glycolic acid, glomerulus, and tubules. Other effects on the kidney are the result of the deposition of calcium oxalate crystals in the tubules.

Human Toxicity

NIOSH has estimated that more than 1.5 million workers are potentially exposed to ethylene glycol. Ocular and dermal exposures would most likely occur in industrial settings. Occupational inhalation exposures can also occur if ethylene glycol vapors or mists are generated by heating or agitation.

The low vapor pressure of ethylene glycol virtually precludes excessive exposure to the vapors at room temperature. In a study with human volunteers exposed to about 22 hr/day at mean daily concentrations of ethylene glycol ranging from 1.4 to 27 ppm for about 4 weeks, there were some complaints of throat irritation, mild headache, and low backache. On the whole, the exposures were very well tolerated, but the upper respiratory complaints became marked as the ethylene glycol concentration was raised above 56 mg/m^3. Concentrations of 60–80 ppm resulted in a burning sensation along the trachea that became intolerable.

For worker exposures, ACGIH has established a STEL-Ceiling of 39.4 ppm, for mist and vapor combined, to minimize irritation of the respiratory passages.

Acute human exposure to ethylene glycol usually occurs when single large doses are ingested either accidentally or in intentional suicide attempts. Antifreeze-coolant fluid is the most common source of ethylene glycol, but toxicity has also resulted from contamination of potable water systems by ethylene glycol-containing heat-transfer fluids. Ingestion of large amounts of ethylene glycol leads to CNS depression, cardiopulmonary disturbances, and renal distress.

Clinical Management

Inhalation exposures to ethylene glycol mist should be monitored for respiratory tract irritation. Humidified supplemental 100% oxygen should be administered. Exposed skin and eyes should be treated for irritation from direct contact with ethylene glycol liquid, vapors, or mists.

Following liquid ingestion of ethylene glycol, vomiting should be induced within 30 min. Activated charcoal should be administered to bind the ethylene glycol and prevent further absorption. Sodium bicarbonate is used to prevent acidosis. Ethylene glycol poisoning from ingestion can be treated by administering ethanol, which competes with ethylene glycol metabolism by alcohol dehydrogenase. This enzyme has a greater affinity for ethanol (30–40 times greater than for ethylene glycol) and prevents the formation of toxic metabolites. Therefore, ethanol administration increases the urinary excretion of unchanged glycol and reduces the oxalate production. Hemodialysis may also have to be performed in severe cases.

Animal Toxicity

Following oral administration, the lethal dose in rats is about 5.5–13 ml/kg compared to about 1.4 ml/kg in humans. Therefore, animals are less susceptible to ethylene glycol toxicity from ingestion than humans.

Other available inhalation data for ethylene glycol described exposure of rats, guinea pigs, rabbits, dogs, and monkeys at 10 and 57 mg/m^3 for 8 hr/day, 5 days/week for 30 days. There were no adverse effects from these exposures. Animals exposed for 24 hr/day for 90 days at a concentration of 12 mg/m^3 exhibited moderate to severe eye irritation in rabbits and rats. On the other hand, another study exposed animals to 350–400 mg/m^3 for 8 hr/day for 16 weeks without producing any effect. Three animal teratogenicity studies were conducted and the findings were positive teratogenicity in rats and mice following oral administration. Ethylene glycol was negative in the *Salmonella* and mouse lymphoma assays and for the induction of chromosomal aberration and sister chromatid exchanges in cultured Chinese hamster ovary cells. NTP has completed the experimental phase of a long-term toxicology and carcinogenesis study, but the pathology results are incomplete.

—*Edward Kerfoot*

Related Topics

Poisoning Emergencies in Humans
Veterinary Toxicology

Ethylene Glycol Monoethyl Ether

- ◆ CAS: 110-80-5
- ◆ SYNONYMS: 2-Ethoxyethanol; Cellosolve; Dowanol; Ektasolve; ethyl glycol; hydroxy ether; Polysolv; RTECS/NIOSH KK8050000
- ◆ CHEMICAL CLASS: Glycol
- ◆ CHEMICAL STRUCTURE:

$$HOCH_2CH_2OC_2H_5$$

Uses

Major uses include a component of natural and synthetic resins; metal solvent for formulation of soluble oils; solvent for lacquers and lacquer thinners, dyeing and printing textiles and varnish removers; and anti-icing additive for aviation fuels.

Exposure Pathways

Exposure may occur by inhalation of the vapor, ingestion, and dermal contact.

Toxicokinetics

Glycol ethers of low molecular weight undergo a limited amount of destruction in the body and do not appear to be distributed in the extracellular fluids of the body. They are cleared from the plasma at a rate identical to that of creatinine. Ethylene glycol monoethyl ether was excreted via the lung at a rate slightly more than 8 mg/kg/min. Exposed workers showed measurable levels of 2-ethoxyacetic acid in urine. In another study, a good correlation was found between the average exposure and the ethoxyacetic acid excretion. Maximal urinary excretion occurred in 3 or 4 hr and the urine biological half-life was 21–24 hr.

Mechanism of Toxicity

The toxicity of ethylene glycol monoethyl ether is believed to be due to its metabolites, methoxyacetic acid and ethoxyacetic acid. The primary targets appear to be tissues with rapidly dividing cell systems and high rates of respiration and energy metabolism. It appears to inhibit respiration and the respiratory control ratio in hepatic mitochondria. Cytochrome oxidase activity is also inhibited.

Human Toxicity

Ethylene glycol monoethyl ether is low in oral toxicity and not significantly irritating to the skin. It is lightly irritating to the eyes and mucous membranes. Its vapors are irritating in acutely toxic concentrations but are not objectionable at levels considered safe. Effects on the central nervous system include headache, drowsiness, weakness, staggering gait, tremor, and blurred vision. Changes in personality are often observed. It has been recently shown to have the potential to induce reproductive toxicity. The lethal oral dose of ethylene glycol in humans is approximately 1.4 ml/kg.

Clinical Management

Victims should be moved immediately to fresh air. If not breathing, artificial respiration should be provided; if breathing is difficult, oxygen should be administered. Exposed eyes and skin should be rapidly and copiously flushed. Contaminated clothing should be removed.

Animal Toxicity

Acute exposure showed dyspnea, weakness, and slight paralysis. The kidneys showed acute nephrosis; the lung showed congestion and edema. Dogs exposed to vapor showed a slight decrease in hemoglobin and blood cells. Guinea pigs survived exposure intensities of 6000 ppm for 1 hr, 3000 ppm for 4 hr, and 500 ppm for 24 hr without apparent harm. More intense exposures caused injury of the lung, hemorrhage in stomach and intestines, and congestion of kidneys. High doses cause 100% intrauterine death. The teratogenic potential of ethylene glycol monoethyl ether has also been reported, predominantly affecting the heart and skeletal system.

The oral LD_{50} was 3 g/kg in rats, 3100 mg/kg in rabbits, and 1400 mg/kg in guinea pigs. The dermal LD_{50} was 3.6 mg/kg in rabbits.

—*Rhonda S. Berger*

Ethylene Glycol Mono-n-Butyl Ether

◆ CAS: 111-76-2

◆ PREFERRED NAME: Ethylene glycol monobutyl ether

◆ SYNONYMS: Butoxyethanol; butyl cellosolve; EGBE

◆ CHEMICAL CLASS: Glycol ether

◆ MOLECULAR FORMULA: $C_6H_{14}O_2$

◆ CHEMICAL STRUCTURE:

$$C_4H_9 - O - CH_2CH_2 - OH$$

Uses

Ethylene glycol mono-*n*-butyl ether is a widely used solvent present in many lacquers, enamels, varnishes, varnish removers, and latex paints. It has also found use in metal cleaning formulas and commercially available household cleaners.

Exposure Pathways

Exposure to ethylene glycol mono-*n*-butyl ether can occur via inhalation, ingestion, or skin absorption.

Toxicokinetics

Ethylene glycol mono-*n*-butyl ether is rapidly absorbed, distributed, metabolized, and eliminated. The urine is the primary route of excretion followed by expiration in the form of the metabolite CO_2. The carboxylic acid, 2-butoxy acetic acid (BAA), is the major urinary metabolite of ethylene glycol mono-*n*-butyl ether. BAA is formed by oxidation of the alcohol moiety of ethylene glycol mono-*n*-butyl ether through alcohol dehydrogenase and aldehyde dehydrogenase sequentially. Lesser quantities of glucuronide and sulfate conjugates have also been detected in urine.

Rats administered ethylene glycol mono-*n*-butyl ether in drinking water excrete 50–60% of the consumed dose in the urine as BAA and exhale 8–10% as CO_2. As dose increases, the proportion of ethylene glycol mono-*n*-butyl ether conjugated with glucuronic acid also increases. Similarly, oxidative dealkalation of ethylene glycol mono-*n*-butyl ether to form ethylene glycol also becomes a more prevalent route of metabolism as dose increases.

Mechanism of Toxicity

The principal toxicological effect observed upon overexposure to ethylene glycol mono-*n*-butyl ether is the destruction of red blood cells (i.e., hemolysis). BAA, the predominant oxidative metabolite of ethylene glycol mono-*n*-butyl ether, appears responsible for this hemolytic activity. It has been speculated that BAA may interact with red blood cell membranes disrupting erythrocyte osmotic balance, leading to cellular swelling and, eventually, hemolysis. *In vitro* studies have indicated that the red blood cells of rats, mice, rabbits, and baboons were susceptible to hemolysis by BAA, whereas blood from pigs, dogs, cats, guinea pigs, and humans were resistant. A number of other studies have confirmed these results *in vitro*, with red blood cells from a large cross section of the population, including those with hereditary red cell disease (sickle cell and spherocytosis) and the aged. These studies indicate that human cells are not as susceptible to hemolysis as those of rat cells tested under similar conditions. This finding suggests that humans exposed to equivalent doses of ethylene glycol mono-*n*-butyl ether would not be expected to exhibit the same spectrum or severity of hematotoxic-related effects produced in rats.

Human Toxicity

Ethylene glycol mono-*n*-butyl ether is of low to moderate toxicity upon acute exposure in humans. Metabolic acidosis has been reported in individuals poisoned through ingestion of materials containing ethylene glycol mono-*n*-butyl ether. Human volunteers exposed to 98–200 ppm ethylene glycol mono-*n*-butyl ether for 4–8 hr reported nasal and ocular irritation and disturbed taste. No abnormalities were detected in blood pressure, pulse rate, erythrocyte fragility, urinary glucose, or albumen.

Clinical Management

Management of individuals overexposed to ethylene glycol mono-*n*-butyl ether begins with removing those individuals from the source of exposure, flushing eyes and skin with water, and removing contaminated clothing. The treatment of choice for acute and severe hemolytic anemia, which may result from overexposure to ethylene glycol mono-*n*-butyl ether, is exchange transfusion. If renal failure develops as a consequence of red blood cell hemolysis, hemodialysis is the treatment of choice.

Animal Toxicity

Ethylene glycol mono-*n*-butyl ether acute toxicity has been studied in numerous species via all routes of exposure. In general, animals exhibited inactivity, weakness, and dypsnea, while autopsies revealed congested lungs and kidneys. The principal effect observed leading to death in these acute toxicity studies was damage to the kidneys. Oral LD_{50}s range from 900 mg/kg in rabbits to 250 mg/kg in rats; dermal LD_{50}s are approximately 1500 mg/kg in rabbits; and inhalation LC_{50}s (4- to 8-hr exposures) are in the vicinity of 500 ppm for cats, guinea pigs, and rats.

The subchronic toxicity of ethylene glycol mono-*n*-butyl ether has been examined in animals via oral inhalation and dermal routes of exposure. The lowest no-observed-effect level in an oral subchronic study was 80 mg/kg for rats administered ethylene glycol mono-*n*-butyl ether in feed over a 90-day period. Inhalation exposure of rats for 13 weeks, 6 hr/day, 5 days/week to ethylene glycol mono-*n*-butyl ether vapors at 25–77 ppm indicated a no-observed-effect level of 25 ppm. In a 90-day dermal study of rabbits, ethylene glycol mono-*n*-butyl ether was applied 6 hr/day, 5 days/week at doses up to 150 mg/kg. There was no evidence of systemic toxicity or skin irritation at the site of application at any of the dose levels tested.

The reproductive and developmental toxicity of ethylene glycol mono-*n*-butyl ether in both male and female animals has been the subject of numerous investigations. Ethylene glycol mono-*n*-butyl ether has been found to have no adverse effects upon the male reproductive system in mice or rats exposed orally at doses ranging from 222 to 2000 mg/kg, 5 days/week for 5 or 6 weeks. Similarly, inhalation exposure of rats for 3 hr produced no observable effects upon gross macroscopic postmortem examination.

Ethylene glycol mono-*n*-butyl ether has also been tested for effects upon the female reproductive system and the developing embryo. Mice, rats, and rabbits have been exposed during gestation at doses of 4000 mg/kg/day (oral), 424 mg/kg/day (dermal), and 25–200 ppm 6 hr/day (inhalation). No teratogenic effects were observed in the litters of dams exposed to ethylene glycol mono-*n*-butyl ether. Signs of maternal toxicity, including decreased body weight and body weight gain, were observed. At the maternal LD_{20}, ethylene glycol mono-*n*-butyl ether did induce fetal deaths in rats. BAA, the metabolite of ethylene glycol mono-*n*-butyl ether, was also studied and found to have no adverse effect upon the developing embryo *in vitro*.

—*Bradford Strohm*

Ethylene Imine

- CAS: 151-56-4
- PREFERRED NAME: Ethyleneimine
- SYNONYMS: Ethylenimine; azacyclopropane; dimethylenimine
- CHEMICAL CLASS: Imine
- CHEMICAL STRUCTURE:

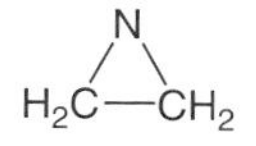

Uses

Ethylene imine is used in the manufacture of triethylenemelamine (i.e., precursor in plastics synthesis).

Exposure Pathways

Contact with skin or mucous membranes (eyes and nasal) and inhalation are the routes of exposure.

Mechanisms of Toxicity

Ethylene imine is a cross-linking agent and irritant.

Human Toxicity

Ethylene imine can be very irritating to the skin, eyes, or mucous membranes. It is mildly corrosive to skin and mucous membranes. Exposure may cause cancer. OSHA has categorized ethylene imine as a carcinogen.

Clinical Management

In acute situations, the skin should be washed thoroughly with soap and water. If ingested, an emetic should be administered or gastric lavage performed. Oxygen should be provided if breathing is difficult. If severe blood poisoning occurs, 1% methylene blue solution should be given at 1 ml/kg intravenously.

Animal Toxicity

The inhalation LC_{50} (1-hr exposure) was 185 ppm in rats, 150 ppm in mice, and 170 ppm in guinea pigs. The oral LD_{50} was 15 mg/kg in rats and mice.

—Shayne C. Gad and
Jayne E. Ash

Ethylene Oxide

- ◆ CAS: 75-21-8
- ◆ SYNONYMS: 1,2-Epoxyethane; oxirane
- ◆ CHEMICAL CLASS: Epoxides
- ◆ MOLECULAR FORMULA: C_2H_4O
- ◆ CHEMICAL STRUCTURE:

$$H_2C - CH_2 \ (O)$$

Uses

Most ethylene oxide is produced and consumed captively in the production of ethylene glycol. It is also used in production of nonionic surfactants, polyester resins, and specialty solvents. A small percentage of ethylene oxide consumption is attributed to its use as a sterilant for medical devices and pharmaceuticals.

Exposure Pathways

Ethylene oxide is a gas at room temperature; therefore, inhalation is the primary route of exposure. Handling of the condensed liquid may result in eye or dermal exposure, though rapid evaporation minimizes the opportunity for absorption.

Toxicokinetics

Absorption of ethylene oxide is rapid through the respiratory and gastrointestinal tracts. Because of its high solubility in blood, uptake depends on the ventilation rate and the airborne concentration of ethylene oxide. Studies in mice indicate almost 100% absorption following inhalation of concentrations close to the OSHA PEL of 1 ppm. At higher concentrations, the percentage absorbed, although still appreciable, decreased.

Animal data suggest two possible metabolic pathways for ethylene oxide. A hydrolysis product, ethylene glycol, has been reported as the major metabolite in the urine of dogs after intravenous injection as well as in other species. The second possible pathway is glutathione conjugation to mercapturic acid and merthio compounds. Ethylene glycol, 2-hydroxymercapturic acid, 2-mercaptoethanol, and 2-methylthioethanol were identified in the urine of rats. Substituted L-cysteine metabolites have also been identified in the urine of rodents.

Absorbed ethylene oxide is rapidly distributed throughout the body. In mice exposed by inhalation, distribution was immediate, with the highest concentrations in the lungs, liver, and kidneys. After 4 hr, it was detected in the liver, kidneys, lungs, testes, spleen, and brain. DNA adducts have been detected in the kidneys of mice exposed by inhalation and adducts have been seen in the liver and testes of rats following intraperitoneal injection.

Little absorbed ethylene oxide is excreted via exhalation. Mice exposed by inhalation excreted 74% of the radiolabel in the urine as undetermined metabolites in 24 hr and 78% within 48 hr. The half-life of ethylene oxide has been estimated to be about 0.15 hr in mice.

Mechanism of Toxicity

The mechanisms of toxicity are not yet understood; however, it is likely that, in general, the toxic effects of ethylene oxide are due to its ability to react with cellular chemicals and molecules, altering function. Ongoing dosimetry and mechanistic studies are expected to elucidate how ethylene oxide exerts its toxicity, especially with respect to its potential carcinogenicity.

Human Toxicity

Contact with liquid ethylene oxide or its solutions may result in irritation and burns. At high concentrations, ethylene oxide acts as an eye and respiratory irritant and a central nervous system (CNS) depressant. Symptoms of overexposure include nausea, vomiting, and neurological effects. Pulmonary edema may result. Cases of human poisonings involving lymphocytosis have been reported. Studies of chronically exposed populations suggest that ethylene oxide may cause allergic contact dermatitis. There is limited evidence from epidemiological studies that ethylene oxide is a carcinogen.

Clinical Management

If contact with the liquid or its solutions occurs, affected areas should be flushed thoroughly with water for at least 15 min. The areas should be observed for burns or resulting irritation. In case of inhalation of ethylene oxide, the victim should be moved to fresh air, an airway should be established, and respiration should be maintained as necessary. The victim should be monitored for irritation, bronchitis, and pneumonitis. If excessive exposure occurs, hospitalization and monitoring for delayed pulmonary edema is recommended.

Animal Toxicity

Acute inhalation LC_{50} values range from 960 ppm in dogs to 1460 ppm in rats. Effects of acute overexposure are CNS depression, lacrimation, salivation, incoordination, and convulsions. Survivors of acute toxicity studies also exhibit bronchitis and pneumonia. Histopathology has shown liver, kidney, lung, spleen, and brain injury. In lifetime studies, rats exposed to airborne concentrations of 10, 33, or 100 ppm for 6 hr/day, 5 days/week exhibited several treatment-related tumors, including brain tumors, mononuclear cell leukemia, and peritoneal mesothelioma. Ethylene oxide was also carcinogenic to mice in another study. Several groups and governmental agencies have classified ethylene oxide as a possible or probable carcinogen.

—*Daniel Steinmetz*

Related Topics

Respiratory Tract
Sensory Organs

European Commission

Environmental Research Programs of the European Commission

Research is one of the more important fields of action of the European Communities. Environmental research is recognized as being critical in supporting Community Environmental Policy, which, since the ratification of the Single European Act and the Maastricht Treaty, is integrated into all of the sectoral policies of the union (e.g., industry, agriculture, transport, energy, and development).

The research is implemented at three levels:

- Shared-cost action, also known as indirect action, where contractual research is performed in laboratories in member states of the European economic area, the costs being shared between the European Commission (EC) and the contractants

- Concerted actions, where the commission supports the coordination costs of research activities financed from other sources at national level

- In-house research (direct action), where research is performed by the staff of the commission at EC Joint Research Centers (Geel, Belgium; Ispra, Italy; Petten, The Netherlands; Karlsruhe,

Germany; and Seville, Spain) and other institutes (e.g., Joint European Torus, Culham, UK)

Community research activities date back to its inception in 1957 but, since 1984, all research has been implemented within Framework Programs (FP). The first FP ran from 1984 to 1988, the second from 1987 to 1991, and the third from 1990 to 1994. The structure and content of the fourth FP (1994–1998) is currently being discussed by the commission, the council, and the European Parliament. The FPs provide a coherent basis and direction for organizing all European Union (EU)-level research. Their overall objectives are to strengthen the scientific basis of EU policymaking, to increase the competitiveness of European industry, and to reinforce social and economic cohesion in the union.

Besides shared-cost contracts research, the programs include bursaries and training fellowships, coordination activities and support of conferences, workshops, symposia, etc.

STEP and ENVIRONMENT are two research programs concerned with the environment at large. Research pertinent to occupational medicine, food additive, etc. was therefore excluded from the outset and pilot or demonstration projects were not eligible for support because other sources of funding were available for such work.

STEP Research Program

The objectives of the STEP program are as follows:

- The provision of scientific and technical support for the environmental policy of the EC, and for other relevant EC policies such as energy, agriculture, industry, aid to developing countries, both for the solution of short-term policy questions and for the medium and long-term formulation of preventive and anticipatory policies

- The further improvement of the productivity of the overall research effort in the EC, the reduction of overlaps and the identification of gaps, through the coordination of the national research and development programs in the field of environmental research

- The promotion of overall scientific and technical quality in the field of environmental research, as

a contribution to the strengthening of the economic and social cohesion of the EC, R&D capabilities at the highest level in all parts of the EC being one of the prerequisites for its harmonious development

STEP covers nine broad research areas:

- Environment and human health
- Assessment of risks associated with chemicals
- Atmospheric processes and air quality
- Water quality
- Soil and groundwater protection
- Ecosystem research
- Protection and conservation of the European cultural heritage
- Technologies for environmental protection
- Major technological hazards and fire safety

ENVIRONMENT Research Program

The general objectives of the ENVIRONMENT program are as follows:

- To provide the scientific knowledge and technical know-how needed by the EC to carry out its new role relating to the environment, according to title VII of the EEC treaty

- To further improve the productivity of the overall research effort in the EC, to reduce overlaps and increase effectiveness through the coordination of national RTD programs in the field of environmental research

- To provide a basis for the European contribution to major international programs related to global environmental changes while focusing on topics of more specific European interest

- To reinforce the role of the EC within international conventions on the protection of the atmosphere through the improvement of fundamental knowledge on global processes related to the "greenhouse problem"

- To provide the technical basis for, and encourage the development of, environmental quality norms, safety and technical norms, and methodologies for environmental impact assessment to support the activities of the European Environmental Agency

- To contribute to the strengthening of the economic and social cohesion of the EC by promoting overall scientific and technical quality and incorporating socioeconomic aspects into the field of environmental research

The specific objectives of the ENVIRONMENT program are as follows:

- To contribute to the understanding of the processes governing environmental change and to assess the impacts of human activities

- To promote better environmental quality standards by encouraging technical innovation at the precompetitive stage and to protect and rehabilitate the environment

- To improve the understanding of the legal, economic, ethical, and health aspects of environmental policy and management

- To help solve broad problems of transnational interest through a systems approach and interdisciplinary research

The ENVIRONMENT program covers four major themes, which are discussed in the following sections.

Area I: Participation in Global Change Programs

This area provides a basis for contribution to global programs (e.g., WCRP and IGBP) with a focus on topics of more specific European interest. It includes research on (a) natural and anthropogenic climate change and climate change impacts and (b) global changes in stratospheric and tropospheric chemistry and biogeochemical cycles and their consequences for life on earth. Shared-cost action contracts within this area on UV-B effects on health are described in this document.

Area II: Technologies and Engineering for the Environment

This area includes research on technologies and methodologies related to (a) assessment of environmental

quality and monitoring, (b) rehabilitation of the environment, (c) major industrial hazards, and (d) environmental protection and conservation of Europe's cultural heritage.

Area III: Research on Economic and Social Aspects of Environmental Issues

The aim of this research area is to increase understanding of the societal and economic causes and impacts of environmental change and to provide a basis for the formulation of environmental strategies which will help achieve sustainable development. Particular attention is devoted to the study of the EC's and its member states' policies, activities, R&D, and other programs in this field in order to provide effective coordination, wherever appropriate. The key elements of this research area are indicated under the following 6 sectors: (a) methodological and strategic aspects of sustainable development, (b) integrating the objectives of economic growth with environmental quality, (c) incorporating environmental factors into sectoral policies, (d) societal aspects of environmental change, (e) socioeconomic aspects of global and regional environmental issues and actions, and (f) socioeconomic research to improve the scientific basis of environmental policies.

Area IV: Technological and Natural Risks

Research in this area is implemented under three broad headings: (a) natural risks (volcanic, seismic, and wildfires), (b) technological risks, and (c) desertification in the Mediterranean area. Within topic b on technological risks, research covers four themes: (i) risks from agricultural technologies and land-use practices to soil, surface, and groundwater quality, (ii) regional aspects of ecosystems protection, (iii) environment and human health, and (iv) risks to health and the environment from chemical substances. Shared-cost action contracts within themes iii and iv are described in this document.

Research on environmental health and chemical safety is supported predominantly under the topic Natural and Technological Hazards in this area. (Support for research on the health impacts of stratospheric ozone depletion is given under the theme of Global Change.)

The International Dimension

It is a prerequisite for participation in the program that projects must be transnational with one partner from a member state of the European Union and at least one other participant from one or more other member states of the European Economic Area.

Following earlier agreements with the EFTA countries (and the later agreement on the European Economic Area with all of them except Switzerland), the program is open to participation by laboratories in EFTA member states. Their ideas, experience, competencies, and sometimes different perspectives considerably strengthen the program and add new dimensions to it. Laboratories from outside the European Economic Area are also permitted to participate in the program but without EC funding and on a project-by-project basis.

In 1992, a program of scientific cooperation with countries of Central and Eastern Europe (PECO) was launched. Among other options, institutes from PECO countries were invited to apply to become participants in existing projects in the ENVIRONMENT (but not STEP) and other selected research programs in FP3. The program of cooperation was repeated in 1993 and a call for proposals was also issued in 1994. To date, in the fields of health and chemical safety research, 17 PECO partners have successfully joined the 46 existing projects. Among the various research areas of the program, this is one of the highest PECO participation rates, reflecting the strong interest in environmental health and chemical safety in those countries. It is expected that more will join in the future.

With the impending enlargement of the union and the increasing globalization of science, particularly in the domain of environmental research, the environmental research programs of the commission will have a key role in providing the sound scientific basis on which the union can act in developing policies which can respond to the needs and concerns of society.

Environment and Climate

Although the STEP and ENVIRONMENT research programs include projects that will run until as late as 1997, they have been succeeded by a new research program entitled Environment and Climate (1994–1998).

Environmental Health and Chemical Safety

The overall objectives of the environmental policy of the European Union, following the Maastricht Treaty, can be summarized as:

- Preserving, protecting, and improving the quality of the environment
- Protecting human health
- Prudent and rational use of natural resources
- Promoting measures at international level to deal with regional or worldwide environmental problems

Although the environmental research programs of the commission could be seen as partly addressing the fourth objective above, it is the second objective—protecting human health—which is the focus of the research areas on environmental health and chemical safety in the STEP and ENVIRONMENT research programs.

The long-acknowledged importance of the environment health link was reiterated at Rio and was again evidences recently by two major intergovernmental meetings. The first of these, a direct follow-up of UNCED, was the Intergovernmental Conference on Chemical Safety (Stockholm, April 25–29, 1994). The second is the 2nd Ministerial Conference of Environmental Health in Europe (Helsinki, June 20–22, 1994), with the associated declaration on environmental health and publication by the World Health Organization of the Environmental Health Action Plan for Europe.

In both meetings, the need to strengthen international research efforts in environmental health and chemical safety was stressed.

Environment and Health

The general objective of this research area is to provide a scientific basis for the continuing development of preventive environmental health status of populations through (a) the development of methods for exposure assessment, (b) identifying early indicators of health impairment, and (c) the development of environmental epidemiology. There was a broad degree of continuity between the STEP and ENVIRONMENT programs, although in the latter a new theme was introduced: effects of UV-B radiation on health.

Development of Methods for Exposure Assessment to Genotoxic Chemicals

Work in this area is focused on the establishment of quantitative relationships between internal (target) dose and genetic risks. It builds on an earlier initiative from the 4th environmental research program in 1988 to improve population biomonitoring capability in the EC. Work includes both the improvement of existing techniques and the development of new methods.

Identification of Early Indicators of Health Impairment from Exposure to Environmental Pollutants

The operational goal is to develop test batteries which are sensitive, specific, predictive of adverse health effects, and which are of potential applicability to population groups at risk at a stage where the effects are still reversible.

Work concentrates on early indicators for nephrotoxic, neurotoxic, respiratory, and immunotoxic effects.

Development of Environmental Epidemiology within the European Community

The operational goal is to support the development of an EC capability in epidemiological surveillance of environmental health and to provide a coordinated framework for advancing methodology development in environmental epidemiology.

Work in this area included the comparative evaluation of data sets in member states concerning human exposure to pollutants and health statistics (e.g., morbidity and mortality) with investigation of the scope for data linkage, the development of methods of analyze data in order to identify adverse health effects, and the coordination of environmental epidemiology within the EC. Cooperation with international activities (e.g., of WHO and IPCS) was assured.

Research aimed at the establishment of baseline data for environmental epidemiology studies at EC level was also permitted in the ENVIRONMENT program as were feasibility studies in environmental epidemiology, including the following:

- Air pollution and health
- Urban environmental quality and health
- Possible health impacts of electromagnetic fields in the general environment

Environment UV-B Effects on Health

This research topic was implemented in Area I of the ENVIRONMENT program as part of a broader research effort on stratospheric ozone depletion and its consequences. The research area is more fully described in a sister publication to this one titled *"The Effects of Environmental UV-B Radiation on Health and Ecosystems."* The topic did not appear in the STEP program.

The research area covering environmental UV effects addressed two specific research tasks: identification of groups at risk and estimation/prediction of exposure by direct/indirect methods and understanding mechanisms of UV radiation effects, primarily on human skin and on the immune system.

The selected projects focus on both cancer induction and immune effects and address not just the mechanisms by which UV-B exerts its effects but also the identification of biological markers of susceptibility to these effects. Dosimetry is an important aspect of the projects and a cross-project working group has been established to ensure a consistent high quality of the UV-B measurements made.

It is important to note that although skin cancer incidence in Europe is rising and that there is a link between UV-B radiation and skin cancer, this does not imply that ozone depletion is responsible for observed increase. Most probably, the major factor responsible for the increased exposure of the European population to UV-B radiation is changing lifestyles including increased voluntary exposure to the sun. In taking action to reduce skin cancer incidence, it can be argued that education and self-discipline are more important than research.

Assessment of Risks Associated with Chemicals

This topic in both the STEP and ENVIRONMENT research programs includes research on risks both to human health and to the environment proper (ecotoxicology).

The research undertaken in this topic is intended to provide support for EC policy and regulatory activities on chemicals. Within the European Union, chemicals policy is now decided predominantly at international rather that national level and the commission has major responsibilities for chemicals control in the union. These include the identification of methods for the appropriate testing of newly marketed chemicals, the and drafting of regulations on the assessment of existing chemicals and on the reduction of the use of experimental animals in product testing and development. Other responsibilities include the setting of safe exposure levels. A notification scheme was introduced in September 1981 for new chemicals. Since then, about 1500 new chemicals have been introduced to the market, adding to the more than 100,000 already in use at that time.

The research undertaken is planned and implemented in close cooperation with other services of the commission, including DG V, DG XI, and both the European Chemicals Bureau and the European Center for the Validation of Alternative Testing Methods (both at the EC Joint Research Center in Ispra).

The ENVIRONMENT program was also the lead agency (on behalf of the commission) in drafting new guidelines for genetic toxicity testing for the OECD.

The objective of this research area is to advance methods for the early identification and evaluation of the health and environmental risks of chemical substances, including structure–activity relationships, in line with relevant EC regulations on chemicals.

Specific research tasks included:

Developing genetic toxicity testing with emphasis on reductions in the use of animals; test refinements for nongenotoxic carcinogenesis, aneuploidy, germ cell mutagenesis; and development of novel approaches for increased coverage of relevant endpoints, e.g., DNA technologies, multipurpose assays, and *in vitro* simulation of metabolic processes

Relating genetic toxicity to cancer and heritable mutations; evaluation of *in vivo* assays in relation to carcinogenic effects and structure–activity relationship; and improving relevance of testing methods to risk assessment.

Assessing the potential of quantitative structure–activity relationships) for predicting the fate and adverse health effects of chemicals in view of their use in regulatory schemes

—*Canice Nolan*

Acknowledgment

This entry is extracted from Report No. 15 in the *Ecosystems Research Report Series of the Environmental Research Program of the European Commission, Directorate-General for Science, Research and Development*, published by the European Commission Directorate-General XII Science, Research and Development Environment Research Program L-2920, Luxembourg.

Related Topics

Food and Agricultural Organization
International Agency for Research on Cancer
International Program on Chemical Safety
International Register of Potentially Toxic Chemicals
International Union of Toxicology
Organisation for Economic Cooperation and Development

European Society of Toxicology

The European Society of Toxicology (EUROTOX), founded in 1962, was formed by merger of the European Society for the Study of Drug Toxicity and the Federation of European Societies of Toxicology.

EUROTOX is a multinational society with industrial, university, and government toxicology researchers in 50 countries. The purpose of the society is to encourage and advance research in the field of drug toxicity and in other areas of toxicology. The society also fosters an exchange of information concerning problems in toxicology. The European Society of Toxicology is affiliated with the Society of Toxicology (see Society of Toxicology).

The society sponsors working groups and training courses in toxicology; conducts symposia, and bestows an annual Young Scientist's Award to a scientist not older than 35 years of age for an outstanding paper on the results of his or her investigations in experimental clinical toxicology.

Topics of Interest

- Effects of drugs on the human fetus
- Toxicological methods and their reliability
- Toxicity problems of organs such as the liver and the nervous system
- Carcinogenesis
- Sensitization

The society publishes a directory (every 2 or 3 years), a quarterly newsletter, and Proceedings of the Annual Meeting. EuroTox holds an annual congress (with exhibits).

For additional information, contact the European Society of Toxicology, P.O. Box 230, NL-6700-AE WAGENINGEN, The Netherlands.

—David M. Krentz and
Harihara M. Mehendale

(Adapted from information supplied by EUROTOX.)

Related Topics

Academy of Toxicological Sciences
International Life Sciences Institute–North America
National Center for Toxicological Research
Society of Environmental Toxicology and Chemistry
Society of Toxicology

Excretion

Excretion is the process by which chemicals are definitely eliminated from the body. When chemicals gain access to the body, they usually do so as lipid-soluble molecules. In order to be eliminated, most of them must first undergo biotransformation to become more water soluble and, consequently, more easily excreted. Biotransformation and excretion are therefore the main processes involved in the elimination of chemicals.

The most important routes for excretion are urine, feces, and exhaled air; others include milk, sweat, saliva, tears, and hair.

The mechanism responsible for excretion of chemicals is usually (but not exclusively) passive diffusion; lipid-soluble, electrically neutral molecules find a passage through cellular membranes by solubilizing within the lipids of the membrane. The concentration gradient of the chemicals between each side of the membrane acts as a driving force for this process, directing the movement of the molecules from the side of the membrane with a high concentration to the side with a low concentration. Other mechanisms for excretion include filtration through pores in cell membranes and active transport provided by specialized carrier proteins (for more on mechanisms by which chemicals cross membranes, see Absorption).

Urinary Excretion

This is the most important route for excretion. The kidney is made of several functional units called nephrons. The initial segment of the nephron is a tuft-like structure called the glomerulus, which acts as a filter for plasma; once filtered under the driving force of circulating blood pressure, plasma fluid becomes diluted urine. During the process of filtration, some of the substances that are dissolved in the plasma can also filter freely altogether with plasma water. The next segment of the nephron is a long tubular structure that allows exchange of water and solutes between the newly formed urine and the blood circulating in the kidney.

Chemicals behave as solutes; they may filter through the glomerulus and then undergo exchange along the tubular segment to be partially reabsorbed into the blood or definitely excreted. Electrically neutral molecules are subject to reabsorption from urine into blood by simple passive diffusion, moving along a concentration gradient. For drugs that are weak electrolytes, as it is the case for many therapeutic agents, urinary pH has a considerable influence on excretion. At a moderately alkaline pH, weak organic acids are present mostly as ionized or electrically charged molecules; this prevents their diffusion from urine into blood and facilitates their elimination with the voided urine. The same occurs with weak organic bases at a moderately acidic pH. Altering the pH of urine is currently used to enhance elimination of chemicals in certain drug poisonings (e.g., salicylates and phenobarbital).

Passive diffusion is not the sole mechanism by which chemicals are exchanged between urine and blood in the tubular segment of the nephron. For some agents, active transport, even against a concentration gradient, is another means by which molecules may be transferred from urine to blood, but much more often from blood to urine. There are two types of specialized transport systems by which chemicals can be actively secreted in the urine. One is for the organic acids like penicillin and certain diuretics; another, less well understood, is for organic bases like quinine. Like most active transport systems, the latter can be saturated at high concentrations of the transported chemicals and can also be blocked by other chemicals sharing the same transport system; the secretion of penicillin into urine can be prevented by concomitant administration of a drug known as probenecid.

In patients with severe renal impairment, kidney function can be effectively substituted using an artificial kidney. The latter exploits the properties of semipermeable membranes to allow elimination of endogenous waste materials. The same principle can be used to help remove certain freely diffusible chemicals in severe cases of poisoning (e.g., bromides, ethylene glycol, isopropyl alcohol, and lithium). This procedure is called hemodialysis.

Fecal Excretion

This is the second most important route for excretion. Chemicals present in feces are mainly those excreted in the bile but also, to a much smaller extent, those diffusing passively through the intestinal wall. Of course, chemicals that are not completely absorbed during their passage in the gastrointestinal tract are also found in the feces.

All chemicals absorbed in the gastrointestinal tract first reach the liver, where they normally undergo biotransformation to new, more water-soluble molecules (metabolites). Some of these will eventually be excreted in the bile. Thus, in addition to playing an important role in the digestion and intestinal absorption of fats, bile is also involved in the elimination of chemicals from the body.

Bile is formed by liver cells and is collected and transported in the biliary system, which comprises a series of ducts, from extremely small ones to larger ones, branching like a tree throughout the liver. In contrast to what happens in the kidney, the driving force for bile secretion is not the pressure of the circulating blood. It is rather a drawing pressure that is generated within the system of ducts by the presence of various solutes in the bile, creating a passive movement of fluids from liver cells and intracellular spaces (osmotic pressure). Solutes that contribute to create such pressure are bile acids but also smaller molecules like sodium, chloride, and bicarbonates. Bile collected at the very smallest ducts, next to each single liver cell, is later modified in larger ducts by processes of reabsorption or secretion of electrolytes and water. Ultimately, bile empties into the first segment of the small intestine, the duodenum.

Some endogenous and foreign chemicals, usually molecules with a molecular weight larger than 325 Da, will appear in bile at concentrations exceeding that in plasma by a factor of 10–1000. Biliary excretion is thus an important route of elimination for such chemicals. Bilirubin is an example of an important endproduct of endogenous metabolism of red blood cells that is normally excreted in bile. The excretion of foreign chemicals is supported selectively by either one of at least three active, carrier-mediated and saturable transport systems: one for organic acids (dyes like sulfobromophtalein and indocyanine green and various glucuronide conjugates of chemicals), one for organic bases, and one for neutral substances (ouabain, a cardiac stimulant). Lead (the metal) is also actively transported.

An important portion of the more water-soluble metabolites secreted in bile is ultimately excreted in the feces. Some metabolites, however, may undergo further enzymatic modification by the intestinal bacterial flora to a state of greater lipid solubility. This metabolic step facilitates reabsorption of such chemicals and extends their life in the body. The process in known as the enterohepatic cycle.

Exhaled Air

Chemicals present in blood that are gases or possess a high degree of volatility diffuse passively into the alveolar air of the lung until they reach equilibrium. The concentration of these chemicals in the air phase is directly proportional to their concentration in blood, and the latter in turn is in equilibrium with the concentration of the chemicals in the tissues. This phenomenon can be applied to monitor noninvasively the presence and the concentration of gases and volatile substances in blood. A practical example of such application is the indirect measurement of alcohol present in blood by analyzing for ethanol in exhaled air with an instrument known as the Breathalyzer.

In industrial settings, exhaled air is used to monitor exposure to volatile organic solvents. A major drawback of this approach is the very high sensitivity of the analysis to rapid changes in exposure concentrations; such changes are rapidly reflected by parallel fluctuations in the concentrations of exhaled air. Under these conditions, point measurements represent exposure poorly over an entire workshift. When exposure concentrations fluctuate, it is recommended to analyze exhaled air the morning after exposure. At this time, blood concentrations of solvents are in equilibrium with concentrations in fatty tissues. The latter present the advantage of slowly and progressively taking up and releasing solvents, thus integrating the previous day exposure quite independently of the pattern of exposure. Point measurements of solvents in exhaled air the morning after exposure are therefore proportional to exposure during the entire previous workshift.

Milk

Human milk is essentially a solution of sugars and minerals forming a suspension medium for other important nutrients like fat globules and proteins. Normal milk components are derived from maternal blood. Any extraneous chemical that enters blood circulation may also eventually appear in milk.

The transport of foreign chemicals from maternal blood into breast milk can proceed by a number of different mechanisms. Uncharged lipid-soluble molecules may diffuse passively through membranes, whereas small water-soluble and small charged molecules may cross membranes through minute pores or water channels. In addition, lactating cells may secrete nutrients like proteins and fat droplets; both can carry foreign chemicals, either bound to proteins or dissolved into fat droplets.

Nursing mothers taking medications can expect to transfer minute amounts of drugs to their child. However, at maternally therapeutic doses, the amounts transferred are too low to produce pharmacological effects. Over-the-counter analgesics like aspirin and acetaminophen, at usually recommended doses for the mother, should not represent a risk for the nursing infant. The same holds true for nonsteroidal antiinflammatory agents (like ibuprofen) frequently used in self-medication for common aliments like arthritic conditions and musculoskeletal pain. For all prescription drugs, it is strongly recommended that nursing mothers ask a physician or pharmacist about the compatibility of medication with breast feeding.

Caffeine, a central stimulant found in commonly consumed beverages, like coffee, tea, and certain soft drinks, is excreted in breast milk. Although newborn infants eliminate caffeine very slowly, normal consumption of caffeine is not contraindicated during nursing. Heavy coffee drinking, of course, is not recommended.

Ethanol diffuses readily in the water fraction of breast milk. Nursing mothers should refrain from chronic consumption of alcoholic beverages since such action is conducive to adverse effects on the intellectual and psychological development of the infant.

Drugs of abuse, like cocaine and heroin, are excreted in breast milk in amounts that may be clinically effective; such exposure is formally contraindicated during breast-feeding. Although no adverse effects in the infant have been reported in the case of mothers using marijuana, caution should be exercised.

Finally, lipid-soluble chemicals like the insecticide DDT, polychlorinated biphenyls, and methylmercury are excreted readily as dissolved chemicals into milk fat droplets. Lead is secreted into milk using the same transport system as calcium. Nursing mothers may therefore transfer environmental contaminants to their infants—not to the point, however, of negating the well-established benefits of breast-feeding, provided the milk is not too heavily contaminated with these chemicals.

Saliva

Saliva is not an important route of excretion since most of the chemicals present in saliva will eventually reach the gastrointestinal tract to be reabsorbed or eliminated in the feces. The unbound fraction of several therapeutic drugs may diffuse passively from plasma into saliva. This provides a noninvasive means of indirectly monitoring plasma concentrations of drugs like lithium, phenytoin, and theophylline. Metals like lead, cadmium, and mercury are also present in saliva.

Hair

Hair is an unexpected and only minor route of elimination for certain chemicals, especially metals. However, the presence of metals in hair has been used as a practical means of monitoring exposure to such chemicals.

Hair is formed from matrix cells present in a bulb-shaped follicle located in the dermis. During growth, hair is exposed to circulating blood and extracellular fluids; certain chemicals can then diffuse into cells producing the hair root and eventually the hair strand, where they will be fixed. The interesting aspect about monitoring exposure to metals using hair is the fact that metals will distribute along the hair strand exactly in the sequence of deposition while hair is growing. Knowing that human hair grows at a rate of approximately 1 cm per month makes it possible to monitor retrospectively exposure to certain toxic elements, like arsenic, cadmium, mercury, and lead, during the past several months and to establish the duration of exposure.

Part of the evidence that Napoleon Bonaparte suffered poisoning during his exile at St. Helen Island rests upon finding increased concentrations of arsenic in hair samples taken from the emperor's scalp.

In 1971, consumption of homemade bread prepared with flour containing a mercurial fungicide in Iraq led to severe intoxication. The degree of exposure to mercury was monitored using hair as the biological sample.

A threshold for neurotoxic effects in children born to exposed pregnant mothers was established at values slightly above 10 μg/g of maternal hair.

Currently, hair is used routinely to monitor exposure to methylmercury in fish-eating native populations of northern Canada. The objective is to adjust consumption of contaminated fish so as not to exceed recognized as safe concentrations of 30 μg/g of hair in the general adult population and 15 μg in fertile women.

Analysis of metals in hair is of limited practical value for monitoring exposure to metals in occupational settings due to the very distinct possibility of hair contamination with exogenous metals present in the ambient air.

Further Reading

Rozman, K., and Klaassen, C. D. (1996). Absorption, distribution, and excretion of toxicants. In *Casarett and Doull's Toxicology. The Basic Science of Poisons* (C. D. Klaassen, Ed.), 5th ed., pp. 91–112. McGraw-Hill, New York.

Roebuck, B. D. (1992). Absorption, distribution, and excretion of chemicals. In *A Primer of Environmental Toxicology* (R. P. Smith, Ed.), pp. 59–76. Lea & Febiger, Philadelphia.

—Jules Brodeur and Robert Tardif

Related Topics

Absorption
Biotransformation
Distribution
Kidney
Liver
Metals
Pharmacokinetics/Toxicokinetics

Exposure

Introduction

The biosphere is composed of elements contained in the periodic table. The elements are present throughout the biosphere and constitute the air, soil, and water. The elements also serve as the foundation of all chemical compounds found in nature. In the context of plants, fish, mammals, etc. the process of birth results in exposure to the elements and/or compounds in the biosphere. In the broadest sense, the term exposure is defined as the condition of being in contact with or exposed to foreign matter. It should be clearly understood that exposure or mere contact with a chemical, bacteria, etc. does not necessarily mean that a harmful or toxic outcome will happen. In certain situations exposure does not produce an effect (no observed effect), while in other conditions an adverse or toxic effect can occur. There are conditions in which exposure produces a beneficial effect; for example, the administration of insulin to treat diabetes mellitus. With respect to toxicology, exposure becomes a concern when contact between an individual and foreign material results in a harmful consequence in the host. It should be noted that exposure refers not only to (1) gases (e.g., ozone and nitrogen dioxide), inorganic chemicals (containing, for example, aluminum, fluoride, cadmium, and lead), microbial organisms (e.g., protozoa, virus, and bacteria), but also to (2) "organic chemicals" that contain carbon and some of which may be vital to life (e.g., vitamins and hormones). This latter class of compounds also comprises the majority of chemicals to which a human may be exposed. However, the damage to mammals may be equivalent whether there is exposure to organic or inorganic material. For the purposes of toxicology, the mere contact of an individual, plant, or fish with an element, chemical compound, or bacteria constitutes exposure but not necessarily a harmful outcome.

The definition of exposure may be oversimplified because it is dependent on a variety of factors. It is well established that, in children, ingestion of paint chips contaminated with lead resulted in behavioral and learning disabilities. In contrast, the presence of lead in the atmosphere and in human blood need not produce central nervous system (CNS) disturbances in adults. Using our definition one could state that children were exposed to lead where an adverse effect was seen but this did not occur in the adults, implying exposure with no observed effect in this latter group. Clearly, both children and adults were exposed to lead, but only one segment of the population is more sensitive and displays a response. In order to appreciate the concept of exposure, various factors involved must be taken into consideration.

Routes of Exposure

In order to survive, mammals require air to breathe, water to drink, and nutrients derived from soil to eat. As indicated previously, the biosphere comprises air, water, and soil; it contains elements vital to life as well as chemicals and microorganisms that could upon exposure be harmful. In the case of humans, it can be seen that a route of exposure to a chemical or element by air is termed "inhalation" or by swallowing water or food is termed "ingestion." It should be noted that mammals exist in a biosphere where chemical contact can occur with the skin and, hence, entry is via the skin and is termed "dermal." The treatment of certain diseases with drugs requires the penetration of a needle through the skin into the body. This route of administration is termed "injection" and is intended to be beneficial; however, it can result in harmful effects as in the case of heroin addicts who inject the compound to produce analgesia and a "euphoric-like" state. It should also be noted that the physiological conditions of pregnancy and lactation introduce two other routes of exposure. During pregnancy there is an exchange of constituents and nutrition from mother to fetus and removal of wastes from the fetus through the placenta. The presence of chemicals in the mother would thus cross the placenta to the fetus resulting in exposure of the fetus. Finally, the suckling infant is dependent on maternal milk for nutrition. During lactation chemicals also enter and may remain in mammary tissue. This latter type of exposure is termed "lactational" or "mammary." Although the mother in pregnancy and lactation provides an environment to protect the fetus or infant, the placenta or mammary tissue does not represent a protective barrier. Indeed, placental or lactational exposure to chemicals is easier and greater than in the adult and consequently puts the fetus or infant at far greater risk to toxic reactions. Unlike the fetus, where origin of exposure can only be attributed to the mother, the lactating infant can be exposed via the mother but also directly to the constituents of the biosphere.

Although exposure can be initiated through various routes, it is the ability of the offending substance to reach the blood that results in the most prominent effect. Thus, the greatest reaction to exposure occurs with injection into a vein termed "intravenous." In descending order of exposure to produce an effect there is inhalation, subcutaneous, intramuscular, oral, and topical. Subcutaneous and intramuscular involve the injection of a substance below the skin layer and into muscle, respectively, while topical is direct application to skin. In the case of placental and lactational, the chemical is already in the mother's blood and is thus similar to intravenous for the fetus or infant.

Exposure to chemicals is not limited to one route. It is well established that spraying of crops with insecticides can result in inhalation of the chemical. Furthermore, by ingestion of these crops, the individual is thus exposed orally, by inhalation and potentially dermally if there is skin contact. The route of exposure is important but one should take into account that there may be more than one source of the substance. This is especially true in the case of cadmium exposure. While it is obvious that workers in a smelter developed kidney and bone problems due to cadmium inhalation, exposure was more evident in cigarette smokers because cadmium is a by-product of cigarette smoke.

Dose

In an industrialized society, individuals are continuously exposed to numerous chemicals by several routes. If one took a sample of the exposed population, the adverse reaction may not be generalized but rather specific or select. One of the explanations for this random reaction to exposure involves dose or concentration of offending material. Regardless of the route of exposure, the toxic material must reach a target site at sufficient amounts to produce an effect. The concept of target site presumes the presence at the tissue of a receptor to which the chemical can attach or interact and subsequently cause a reaction. The inability of chemical to initiate a reaction despite attachment leads to no effect. In this latter scenario, there is exposure but there is a no-observed-adverse-effect level (NOAEL). If, on the other hand, there is exposure that results in a harmful reaction, the lowest level or amount of chemical required to produce an effect is termed lowest-observed-adverse-effect level (LOAEL). Clearly, the difference between NOAEL and LOAEL is the amount of chemical to which individuals are exposed. In addition, the concept of LOAEL presumes that the chemical will

reach a sufficient amount at the target site. Hence, it is possible for two individuals to be equally exposed but the reaction is selective. As indicated previously, the concentration of chemical at the tissue receptor site is one of the factors to account for the difference in the response observed.

Although it is not the focus of this entry, it should be borne in mind that various factors affect the concentration of toxicant at the target site. The ability of a substance to reach the blood or tissue is dependent on the phenomenon of pharmacokinetics, which includes absorption, distribution, biotransformation, and excretion. Exposure to a chemical may be considered the first step in a chain for the substance to reach its target receptor. With oral exposure, the chemical is absorbed from the gastrointestinal system, whereupon it is distributed to the blood or tissues, one of which is the liver. The substance can be degraded (biotransformation) or converted to a more toxic entity (metabolite) in the liver. The removal of the substance is primarily through the kidneys into urine. Thus, with equivalent oral exposure to a chemical, a difference in biotransformation or excretion between two individuals results in differences in concentration of toxicant at the target sites. This may explain the basis for a difference in response to exposure.

Population

It is evident that exposure is not a generalized phenomenon because all individuals in a population are not exposed to the same environment. Although mammals all share the biosphere, the concentration of chemicals differs throughout the atmosphere. Consequently, the type of exposure can be dependent on occupation, lifestyle, and state of health. In general, exposure associated with the workplace is termed occupational or industrial. Easily identifiable in this type of exposure is the individual who sprays crops with pesticides or the worker in a tanning factory. However, occupational exposure also refers to individuals working in an office where there are supplies and equipment and to school children exposed to indoor air pollutants. Occupational exposure is thus deemed as the contact of humans with chemicals outside the home but associated with a job. Exposure to toxicants outside the home and not associated with a job can occur and be considered recreational or lifestyle. This latter category should not exclude exposure inside the home; for example, an individual can be exposed to secondhand cigarette smoke inside or outside the home but away from the job; and this can be termed as recreational exposure. The health status reflects a different type of exposure. An individual suffering from a chronic disorder may be prescribed a specific medication. As a consequence of poor health, this individual is exposed bearing in mind that the drug is taken for beneficial reasons yet adverse, unwanted effects occur. While categorization of exposure based on occupation or lifestyle is distinctive, the fact remains that the majority of the population is subjected to all of these types of exposure. In attempting to elucidate the consequences of exposure, the origin or source of toxicant can be determined. However, one must also consider that the offending toxicant per se may not by itself produce damage but that other exposure comes into play. An example of this is an individual who works in a flour mill and is exposed to grain dust. This worker does not display any clinical signs initially but subsequently moves to a home in an area where there is a great deal of smog. As a result, respiratory difficulties are seen in this individual. It is thus possible that the smog augmented this worker's response to grain dust. This example points out an important phenomenon in our industrialized society. The location of the home can play a crucial role in exposure. It is clearly established that the closer the proximity to a waste site, the greater is the chance that the incidence of adverse reactions will occur in the population in this area. Similarly, individuals living in a high smog area are more likely to develop respiratory difficulties due to greater exposure to ozone and nitrogen dioxide. In essence, the human population is exposed to chemicals, regardless of origin, but efforts should involve minimization to these exposures.

Duration

The consequences of exposure are dependent not only on the amount of chemical but also on time. In general, exposure to a low level of toxicant for 1 hr would presumably result in less damage than that seen after 1 day, a month, or a year. The duration of exposure is thus a critical factor in the observed outcome on the health of the ecosystem. It should be borne in mind that a spillage of chemical into a river may result in immediate human exposure, and that cleanup eliminates this type of toxicant contact. However, chemicals tend to accumulate in sediment, may travel in drinking water, or may be ingested by fish. Clearly, the ecosystem provides a further source of exposure to this same chemical, and the time factor is increased. In an ideal setting, one could speculate an immediate, singular exposure; however, in practical environmental conditions, the chemical tends to linger leading to a longer duration for potential toxicant contamination.

The duration of exposure can be subdivided into three categories as follows: acute, subchronic, and chronic. Acute exposure is defined as exposure to a chemical for less than 24 hr, where the route is generally oral but can be intravenous, subcutaneous, inhalation, or dermal. In a laboratory setting, one can readily control an acute exposure and observe the adverse consequences. When considering the human situation, the term acute exposure can result in a life-threatening crisis such as an industrial accident as in Bhopal, a drug overdose as in the treatment of asthma with a methylxanthine, or a suicide attempt. As indicated previously, an industrial accident can have immediate, acute consequences but the persistence of a chemical also leads to chronic exposure.

The use of mammalian cells to study scientific phenomena rather than whole animals has increased over the years in an effort to understand the basis of disease and mechanisms of chemical actions. The incubation of cells with chemicals also denotes an acute exposure. Although the use of cells plays an important role in exposure studies, whole animal experimentation is still a necessity prior to the registration of chemicals and drugs for general public usage. A major and important advantage for cellular acute exposure is the diminished need for whole animal studies.

Extension of the exposure period to 3 months for animals with a short life span (rodents) is termed "subchronic." In mammals with a longer life span (dogs, monkeys, and humans), subchronic is defined as exposure to the chemical for 1 year. Beyond 3 months in rodents and 1 year in nonrodent animals, this type of exposure is defined as "chronic." The routes of exposure in the subchronic and chronic simulate human conditions with the toxicant either applied to the diet or administered by gavage, inhalation, or skin application. Clearly, injection on a chronic basis is not normally experienced, with an exception being a diabetic who receives daily drug injections for several years.

When exposure occurs over a prolonged period different scenarios can arise. It is possible to be exposed daily to a single amount of toxicant, which can be termed "repeated daily exposure." In another situation, there can be exposure for a specific duration followed by no exposure and then another period of contact with the toxicant, which is termed "intermittent." In certain circumstances, it was found that intermittent exposure to photochemical smog produced greater damage to the lung compared to chronic low-level toxicant exposure. With chronic exposure, the lungs appear to adapt to continuous exposure. In the case of intermittent exposure, repair processes are initiated and the lungs appear more susceptible to chemical-induced damage.

In laboratory settings, the scientist can control the exposure parameters of amount and type of chemical, route of administration, duration, etc. The data generated are then used to simulate human conditions to predict risk. However, humans are subjected to a multitude of factors that impinge on exposure. Thus, in the interpolation of exposure from the experimental situation to the human setting, caution should be exercised, but at the same time misinformation should not be disseminated. By necessity, exposure to chemicals will occur in our biosphere but the consequences for the general population are not always adverse.

Further Reading

Hayes, A. W. (1989). *Principles and Methods of Toxicology*, 2nd ed. Raven Press, New York.
Lu, F. W. (1996). *Basic Toxicology: Fundamentals, Target Organs, and Risk Assessment*, 3rd ed. Taylor and Francis, Washington, DC.

—*Sam Kacew*

Related Topics

Exposure Assessment
Exposure Criteria
Levels of Effect in Toxicological Assessment
Medical Surveillance
Modifying Factors of Toxicology
Permissible Exposure Limit
Pharmacokinetics/Toxicokinetics
Resistance to Toxicants
Short-Term Exposure Limit
Threshold Limit Values

Exposure Assessment

Exposure assessment is one of the four major steps in the risk assessment paradigm, as defined by the National Academy of Sciences. Exposure assessment is defined as the qualitative or quantitative determination or estimation of the magnitude, frequency, duration, and rate of exposure. The quantitative amount of exposure usually refers to the dose or intake (e.g., mg/kg body weight/day) or amount of contact with a chemical or physical agent at the biological exchange boundaries (e.g., lungs, skin, and gastrointestinal tract). Exposure can be viewed as an external dose or an absorbed dose.

Exposure may be evaluated using measurement or modeling approaches. The determination of which approach is taken, and what level of sophistication is appropriate, depends on a number of factors, such as the purpose of the specific assessment, the quality and quantity of available data, and the experience of the exposure assessor. There are three basic types of exposure assessment methods: direct exposure assessment, reconstructive exposure assessment, and predictive exposure assessment.

Direct exposure assessment consists of monitoring approaches to quantitate the amount of a chemical in an environmental media (e.g., air). Examples of direct monitoring approaches include the radiation dosimetry badge, personal air monitoring approaches to measure breathing-zone concentrations of contaminants in air, and passive dermal dosimeters such as those used to collect pesticide residues to measure occupational exposures during agricultural use of pesticides.

Reconstructive exposure assessment uses biological monitoring data in conjunction with pharmacokinetic modeling to estimate the levels of exposure to a chemical that resulted in the measured levels in biological tissues and/or fluids. Biological monitoring consists of the measurement of the concentration of a chemical or its biotransformation products in biological tissues or fluids (e.g., adipose tissue, blood, and urine) or the measurement of the amount of chemical bound to a target molecule (e.g., DNA-bound chemical).

Predictive exposure assessment, involving modeling of exposures, is perhaps the most widely used approach to exposure assessment, and for many exposure assessors is synonymous with the term "exposure assessment." Predictive exposure assessments can be of varying sophistication, ranging from simplistic screening-level approaches using worst-case assumptions to more sophisticated modeling efforts involving random number-based simulations to define the likely distribution of exposure in a human population of concern. Predictive exposure assessment usually consists of a number of steps, including determination of exposure pathways, construction of likely exposure scenarios, estimation of environmental concentrations (sometimes through sophisticated environmental fate modeling), and calculation of human exposures.

Important to the concept of exposure assessment is the profile of exposure that a person experiences. This includes:

- Exposure duration, which is the length of time over which each discrete exposure event occurs (e.g., minutes and hours);

- Exposure frequency, which is a measure of how often an exposure occurs (e.g., 1 day/year, 5 out of 7 days as in the case of some occupational exposures, or continuous exposure over a number of years or a lifetime, as in the case of ambient air pollutants);

- Exposure chronology, which may provide a measure of timing of the exposure relative to the "windows of opportunity" for key toxicological effects. The exposure metric should be consistent with the time frame of the dose–response term relevant to the toxicological endpoint of concern. For example, exposure of women to a teratogen (i.e., a birth defect-causing agent) is of most concern during pregnancy, and exposure to a reproductive toxicant (such as one inducing spontaneous abortions) is of concern primarily during reproductive years, whereas exposure to a carcinogen is of concern over an entire lifetime. Thus, exposure chronology, in combination with knowledge about the toxicological endpoint, can help determine the most appropriate averaging time (e.g., 1 day versus lifetime) for calculating exposure to a chemical.; and

- Exposure patterns, which usually reflect the time and location relationship between sources of exposure (e.g., motor vehicles, home appliances,

consumer products, and volatile organic compounds in drinking water) and human activity patterns (i.e., where people spend their time in a given day). This brings to bear the concept of different microenvironments in which exposures occur, such as pollutant exposures inside residences, pollutant exposures in outdoor air, pollutant exposures in buildings, and pollutant exposures in vehicles (e.g., cars, buses, trains, and airplanes).

Exposure assessment is complicated by the fact that pollutants can move along various pathways between the source of contamination (e.g., landfill or smokestack) and human receptors. Historically, these exposure pathways have been treated separately. However, because many human exposures can occur through a variety of environmental pathways and by different routes (e.g., inhalation, ingestion, and dermal contact), exposure assessors have recently begun to take a total human exposure approach that takes into account all exposure pathways and routes.

Because every person in a given population is likely to experience a different exposure from a given source due to different factors, such as different inhalation rates, skin surface area contacted, and different frequency and duration of exposure due to different time–location–activity patterns, the recent trend has been to integrate exposure assessment with uncertainty analysis. Estimates of exposure are bounded by a wide range of possible values, with the variability in exposures being driven by the variability in exposure parameters. Thus, one can obtain an exposure distribution curve, where each given exposure level has a specific probability of occurring in an exposed population. Thus, measures of central tendency for exposures in a population (e.g., 50th percentile) may be widely separated from the theoretical upper bound estimate of exposure obtained by assuming worst-case values for some exposure parameters (e.g., concentration in air, frequency of exposure, and duration of exposure).

—Gary Whitmyre

Related Topics

Dose Response
Hazard Identification
Medical Surveillance

Risk Assessment
Uncertainty Analysis

Exposure Criteria

All chemicals have the potential to cause harmful effects at high enough dosage. Exposure criteria (exposure limits, toxicity values) define a dose or exposure level that is likely to be below the level at which adverse effects are expected to occur under particular conditions. Most commonly, criteria are expressed either as a dose (e.g., milligram of chemical per kilogram of body weight per day) or a chemical concentration in a media (e.g., milligrams of chemical per cubic meter of air, per kilogram of soil, or per liter of water) (Table E-5). Criteria, however, may be expressed in other ways, including blood levels or excreted concentration of the chemical or a metabolite.

Human exposure in the United States is primarily regulated by four federal agencies (U.S. FDA, OSHA, U.S. EPA, and the Consumer Product Safety Commission) as well as by numerous state agencies. The laws that empower these agencies were written at different times and regulate different environments, activities, and commodities. Criteria developed by each are often based on different assumptions and different procedures and may be designed to protect different populations. What is deemed acceptable for chemical exposure under one law or regulation may not be acceptable under another.

Certain criteria serve as guidelines and others as legally binding standards. For example, OSHA permissible exposure limits are workplace standards and define acceptable concentrations in the air in a work environment, but NIOSH recommended exposure limits, AIHA workplace environmental exposure limits , and ACGIH Threshold Limit Values provide guidance for limiting exposures in the workplace.

In general, criteria are developed based on toxicity studies and assumptions about exposure. For criteria that are expressed as media concentration, assumptions about the length of time over which exposure is ex-

pected to occur and the way the body is likely to contact, absorb, and react to the chemical are factored into the final concentration that is determined to be acceptable. Typically, compliance feasibility is considered when developing criteria that are applied as legally binding standard.

Commonly, a dose or dose range that has been shown to not cause adverse health effects in animals and/or humans is identified. This dose is typically adjusted downward by some factor to account for the uncertainty involved in extrapolating from the available studies to the population and situation in which it is to be applied. It is common for the reduction to be made by dividing by specific factors that are meant to account for uncertainty. The approach and methodology used to adjust the dose varies with the agency or group that is developing the criteria.

Exposure criteria may be based on human or animal data. When human studies are available and appropriate, they are used. In the absence of human data, animal studies are used. Route specific criteria (e.g., inhalation, ingestion, dermal) may be useful in evaluating potential hazards and risks. Chemicals may have local effects that either are different from one route to another or occur at different dose levels. For example, the respiratory tract may be irritated at a concentration that would not affect the skin. In addition, a chemical may be absorbed to a different degree or metabolized differently when entering the body through different routes. Chemicals that are absorbed through the gastrointestinal tract and enter the blood are transported directly to the liver. Significant metabolism in the liver, which typically decreases toxicity, may increase toxicity. Effects may then differ in character or degree according to whether the chemical had been inhaled or absorbed through the skin.

The length of time over which exposure is expected to occur should be considered in evaluating the potential for adverse effects under particular conditions. Multiple criteria for the same chemical may be relevant to evaluating the potential for adverse effects to occur under particular conditions. Criteria for exposures that occur over different time periods (e.g., acute, chronic, or subchronic) may be defined. Typically, a higher exposure can be tolerated for over a shorter time period, and criteria protective of acute exposures (e.g., OSHA Ceiling values, ACGIH EPRGs) are often higher than those protective of longer periods (e.g., averaged over a work life or a lifetime).

TABLE E-5
Examples of Exposure Criteria

Agency	Criteria	Effect	Expressed as dose or media concentration	Route	Media	Target population
USEPA	RfC	Noncancer (subchronic, chronic)	Media (mg/m³)	Inhalation	Air	Public
USEPA	oral RfD	Noncancer (subchronic, chronic)	Dose (mg/kg/day)	Ingestion	Soil, water, food	Public
USEPA	MCL	All	Media (mg/liter)	Ingestion	Drinking water	Public
OSHA	PEL	All	Media (mg/m³)	Inhalation	Air	Workers
NIOSH	REL	All	Media (mg/m³)	Inhalation	Air	Workers
ACGIH	TLV	All	Media (mg/m³)	Inhalation	Air	Workers
AIHA	WEEL	All	Media (mg/m³)	Inhalation	Air	Workers
USEPA	NAAQS	All	Media (mg/m³)	Inhalation	Air	Public
ACGIH	EPRG	Acute (1 hr)	Media (mg/m³)	Inhalation	Typically max. airborn concentration	Workers, but are referred to for guidance for the public
ATSDR	MRLS	Noncancer, (acute, intermediate, chronic)	Dose	Inhalation and oral		Public
USEPA	HA	10 day to lifetime	Media (mg/liter)	Ingestion	Drinking water	Public
ACGIH	BEI	All	Biological monitoring	Any	Any	Workers
USEPA	DWEL	Noncancer, chronic	Media (mg/liter)	Oral	Drinking water	Public
USEPA	MCLG	Any, chronic	Media (mg/liter)	Oral	Drinking water	Public

Note. Abbreviations used: USEPA, United States Environmental Protection Agency; RfC, reference concentration; RfD, reference dose; MCL, maximum contaminant level; OSHA, Occupational Safety and Health Administration; PEL, permissible exposure limit; ACGIH, American Conference of Government Industrial Hygienists; AIHA, American Industrial Hygiene Association; NIOSH, National Institute for Occupational Safety and Health; REL, recommended exposure limit; TLV, threshold limit value; WEEL, Workplace exposure limit; NAAOS, National Ambient Air Quality Standards; EPRG, emergency response planning guidelines; ATSDR, Agency for Toxic Substance and Disease Registry; HA, health advisory; BEI, biological exposure index; DWEL, drinking water equivalent level; MCLG, maximum contaminant level, goat.

For regulatory purposes, it is common to separate chemicals into two groups, carcinogens and noncarcinogens. Noncarcinogenic effects include all effects (e.g., nervous system damage, liver damage, kidney damage, and reproductive effects) except cancer. In general, it is assumed that there is a threshold under which noncancer effects will not occur. Criteria define a dose or exposure level that is likely below the threshold. Chemicals that are regulated as potential human carcinogens are assumed to carry some risk of the effect occurring at even the lowest dose; therefore, no threshold can be defined as "safe" or having no risk. Certain groups and agencies account for this by making a larger reduction in the dose or dose range that has been identified as the basis for the criteria. Others develop criteria by defining a dose that is associated with an acceptable excess lifetime cancer risk. U.S. EPA typically uses the

threshold versus nonthreshold assumption when developing criteria in the absense of data that suggests it is not appropriate.

Cancer as a Nonthreshold Effect

Cancer refers to a collection of diseases that are characterized by cellular growth that does not respond to normal homeostatic controls. Cells may grow faster than normal, grow where there should be no growth, and/or may spread to other sites and tissues in the body. Cancer has been linked to a number of causes, including viral infection, radiation exposure, chemical exposure, and genetic predisposition.

Only a limited number of chemicals are known human carcinogens. Often there is a complex array of data that have to be evaluated and used to predict the likelihood that a chemical has the potential to cause cancer in humans. The body of evidence may include epidemiological, animal, and *in vitro* studies.

A number of agencies and groups evaluate the likelihood that chemicals are human carcinogens. Usually the likelihood is expressed within the context of a carcinogen classification scheme. The purpose of these schemes is to summarize the results from the evaluation and identify and communicate conclusions concerning the potential for the chemical to cause cancer in humans. The following agencies classify chemical carcinogenicity:

- International Agency for Research on Cancer (IARC)
- U.S. EPA
- ACGIH
- OSHA
- National Toxicology Program (NTP)
- NIOSH

These groups and agencies use their own systems and may not reach the same conclusion for a given chemical. If a chemical is identified as a potential human carcinogen, then certain agencies will limit exposure based on the assumption that any dose may be associated with some risk.

U.S. EPA develops factors that can be used to estimate the risk associated with exposure to potential human carcinogens that are known (class A), probable (class B), and sometimes possible (class C) human carcinogens. The classification scheme used by U.S. EPA is currently being revised and may change to a narrative form in the near future.

The risk associated with a particular dose is estimated from the best available data. Human studies will be used when available; however, in most cases, critria are based on animal studies. Animal studies are typically conducted on relatively small groups of animals exposed to relatively high doses of the chemical of interest. A dose–response curve is then generated from the data and extrapolated to predict the potential impact of exposure on large groups (e.g., a human population) exposed to low doses. This relationship is used to identify exposure levels associated with acceptable risk (e.g., one additional cancer over a lifetime per million people exposed).

A number of models can be used to generate dose–response curves and extrapolate data to low doses. Models include the one-hit, multistage, gamma multihit, probit, and Weibull models. Each is based on a series of assumptions, and when sufficient data are available to identify the model that is most appropriate, that model can be used. Currently, most chemically induced cancers are thought to occur after a normal cell is transformed and then altered by a series of mutagenic and nonmutagenic steps in a process often described as multistep/multistaged. In the absence of data suggesting that a particular model is appropriate, U.S. EPA uses the linearized multistaged model, as certain assumptions built into the model are consistent with this theory. The multistage model incorporates the following assumption: there are a number of biological stages in the carcinogenic process, cancer is initiated by mutation, there is no threshold for the effect, and humans are as sensitive as the most sensitive animal species.

The slope of the dose–response curve in the low-dose region is used as an indicator of the potency of a carcinogen. A factor called the slope factor is based on the data and used to relate dose with risk. When the slope of the dose–response curve is not clearly straight, the slope of the straight line from 0 (0 = excess risk) to the dose at 1% excess risk is used as the basis for the slope factor. An estimate of the likely upper limit of the slope, rather than the slope itself, is typically used. The slope factor is usually expressed a 1/(mg/kg-day) and can be used to estimate risk as shown below.

$$\text{Cancer risk} = \text{dose (mg/kg/day)} \times \text{slope factor (mg/kg/day)}^{-1}.$$

The unit risk is defined as the upper bound additional lifetime cancer risk associated with either 1 μg/L in water or 1 μg/m^3 in air. What risk is acceptable varies and may be defined in law, regulation, and/or policy. In general, only very low risks are acceptable when dealing with environmental contamination (e.g., ranging from a 1 in 10,000 to 1 in 1 million additional lifetime risk of cancer).

Chemicals that can cause cancer may also cause other harmful effects. As a rule, criteria based on cancer risk are significantly lower than those developed to prevent noncancer effects. From a regulatory standpoint, the criteria developed to protect against the development of cander are typically the lowest.

Noncarcinogenic Threshold Effects

In general, criteria developed to protect against noncancer effects are based on the assumption that there is a threshold under which no adverse health effects will occur. This means that exposure to the chemical up to a certain amount should be well tolerated. This may occur for a number of reasons. For example, a subthreshold dose may change a small percentage of cells in an organ that performs a particular function. The number of cells affected may not be high enough to impact the ability of an organ function. Over time, cells may return to normal or even be replaced, and the organism will continue to function normally.

Criteria are often based on an experimentally defined dose; that is, a dose at which no adverse health effect (no-observed-adverse-effect level, NOAEL) is identified and used as the basis of the criteria. If not available, then the lowest dose at which effects have been observed (lowest-observed-adverse-effect level, LOAEL) can be used. Commonly, the dose is reduced by dividing it by factors that are intended to account for uncertainty.

For example, U.S. EPA develops oral reference doses (RfDs). RfDs are defined as an estimate (with uncertainty spanning perhaps an order of magnitude) of a daily exposure to the human population (including sensitive subgroups) that is likely to be without appreciable risk of deleterious effects during a lifetime. It is defined in terms of a dose of the chemical in milligrams of the chemical per kilogram of body weight per day (mg/kg-day). RfDs are based on a thorough review of the toxicological literature. All studies (animal and human) are evaluated, and the noncancer effect that occurs at the lowest dose (the critical effect) is identified. The lowest NOAEL or LOAEL is identified and the RfD developed based on that dose. U.S. EPA commonly divides the NOAEL or LOAEL by several factors (usually 10) to account for the uncertainty associated with defining a human "threshold" that will be protective of large numbers of people. Standard U.S. EPA factors include:

- Animal to human
- Human to sensitive human
- Subchronic to chronic
- LOAEL to NOAEL

Modifying factors that account for the completeness of the database may also be applied and can range from 0 to 10.

$$\frac{NOAEL}{MF \times UF} = RfD$$

RfDs are commonly used to evaluate whether exposures are acceptable. Exposure may occur via drinking water or contact with soil. Dose has to be estimated in terms of mg/kg-day and then compared to the RfD. If the estimated dose exceeds the RfD, the exposure may not be acceptable.

Clearly, using a single data point as the basis of the criteria is not ideal. Other approaches (e.g., benchmark dose) may also be used when the appropriate data are available.

New methods and procedures for developing criteria are being evaluated and applied by various agencies. Efforts are focused on using as much of the available data as possible, including data from multiple studies and mechanistic and pharmacokinetic data. For example, the data on lead are sufficient to support regulating exposure based on an approach that is not consistent with either the threshold or nonthreshold approaches described above. The most sensitive effect appears to be neurotoxicity, a noncarcinogenic effect (even though it is classified as a B2 carcinogen by U.S. EPA). Although neurotoxicity is a noncancer effect, a threshold has not been identified. Exposure limits for lead, created to be protective of the general public, are often developed using the U.S. EPA's Integrated Exposure Uptake Biokinetic (IEUBK) model (1994). Young children appear to be most sensitive to the effects of lead. The IEUBK model generates a probability distribution of blood-lead levels and in young children provides an estimate of blood-lead levels associated with exposure to common environmental sources (e.g., soil, dust, water).

—*Betty J. Locey*

Related Topics

Carcinogen Classification Schemes
Carcinogenesis

Dose–Response Relationship
Exposure
Exposure Assessment
Hazard Identification
Levels of Effect in Toxicological Assessment
Permissible Exposure Limit
Risk Characterization
Risk Assessment, Ecological
Risk Assessment, Human Health
Risk Communication
Risk Management
Sensitivity Analysis
Threshold Limit Value
Uncertainty Analysis

Eye Irritancy Testing

Introduction

Virtually all man-made chemicals have the potential to end up in the eyes of people. In fact, many (cosmetics and shampoos, for example) are intended to be used in such a manner that ocular exposure is inevitable in a large number of cases. The term "Draize test" is a misnomer for the entire class of predictive eye irritation tests performed in rabbits and is sometimes generalized to all eye irritation testing performed in animals.

In the early 1930s, an untested eyelash dye containing *p*-phenylenediamine ("Lash Lure") was brought onto the market in the United States. This product (as well as a number of similar products) rapidly demonstrated that it could sensitize the external ocular structures, leading to corneal ulceration with loss of vision and at least one fatality. This occurrence led to the revision of the Food, Drug and Cosmetic Act of 1938. To meet the provisions of this act, a number of test methods were proposed. Latven and Molitor and Mann and Pullinger were among those to first report on the use of rabbits as a test model to predict eye irritation in humans. No specific scoring system was presented

to grade or summarize the results in these tests, however, and the use of animal with pigmented eyes (as opposed to albinos) was advocated. Early in 1944, Friedenwald *et al.* published a method using albino rabbits in a manner very similar to that of the original (1944) Draize publication but still prescribing the description of the individual animal responses as the means of evaluating and reporting the results. Although a scoring method was provided, no overall score was generated for the test group. Draize (head of the Dermal and Ocular Toxicity Branch at U.S. FDA) modified Friedenwald's procedure and made the significant addition of a summary scoring system.

During the 40 years since the publication of the Draize scoring system, it has become common practice to call all acute eye irritation tests performed in rabbits "the Draize eye test." However, since 1944, ocular irritation testing in rabbits has significantly changed. Clearly, there is no longer a single test design that is used, and there are different objectives that are pursued by different groups using the same test. This lack of standardization has been recognized for some time and attempts have been made to address standardization of at least the methodological aspects of the test (such as how test materials are applied and scoring performed), if not the design aspects (such as numbers and sources of test animals). For the purposes of the remainder of this entry, the term Draize test has been replaced with eye irritancy testing.

Ocular irritation tests are significantly different from the other local tissue irritation tests on a number of grounds. For the pharmaceutical industry, eye irritation testing is performed when the material is intended to be put into the eye as a means or route of application for ocular therapy. There are a number of special tests applicable to pharmaceuticals or medical devices that are beyond the scope of this discussion since they are not intended to assess potential acute effects or irritation. In general, however, it is desired that an eye irritation test that is utilized by this group be both sensitive and accurate in predicting the potential to cause irritation in humans. Failing to identify human ocular irritants (lack of sensitivity) is to be avoided, but of equal concern is the occurrence of false positives.

The primary eye irritation test was originally intended to predict the potential for a single splash of chemical into the eye of a human being to cause reversible and/or permanent damage. Since the introduction of the original Draize test 40 years ago, ocular irritation testing in rabbits has both developed and diverged. Indeed, clearly there is no longer a single test design that is used and different objectives are pursued by different groups using the same test. This lack of standardization has been recognized for some time, and attempts have been made to address standardization of at least the methodological aspects of the test, if not the design aspects.

The common core design of the test consists of instilling either 0.1 ml of a liquid or 0.1 g of a powder (or other solid) onto one eye of each of six rabbits. The material is not washed out, and both eyes of each animal (the nontreated eye acting as a control) are graded according to the Draize scale (Table E-6) at 24, 48 and 72 hr after test material instillation. The resulting scores are summed for each animal. The major variations involve the use of three additional rabbits which have their eyes irrigated shortly after instillation of test material. There are, however, many variations of these two major design subsets (i.e., with and without irrigation groups).

Even though the major objective of the Draize scale was to standardize scoring, it was recognized early that this was not happening; instead, different people were "reading" the same response differently. To address this, two sets of standards (also called training guides to provide guidance by comparison) have been published by regulatory agencies through the years. In 1965, the U.S. FDA published an illustrated guide with color pictures as standards. In 1974, the Consumer Product Safety Commission published a second illustrated guide which provided 20 color photographic slides as standards. The U.S. EPA (1979) also supported the development of a guide with color plates/slides, which is still available from NTIS.

A second source of methodological variability has been in the procedure utilized to instill test materials into the eyes. The general consensus is that the substance should be dropped into the cul-de-sac of the conjunctiva formed by gently pulling the lower eyelid away from the eye, then allowing the animal to blink and spread the material across the entire corneal surface. In the past, however, there were other application procedures (such as placing the material directly onto the surface of the cornea).

There are also variations in the design of the "standard" test. Most laboratories observe animals until at least 7 days after instillation and may extend the test to 21 days after instillation if any irritation persists (in

fact, U.S. EPA labeling requires such an extension). These prolonged postexposure observation periods are designed to allow for evaluation of the true severity of damage and for assessing the ability of the ocular damage to be repaired. The results of these tests are evaluated by a descriptive classification scale (Table E-7)

TABLE E-6

Scale of Weighted Scores for Grading the Severity of Ocular Lesions (Draize, 1944)

Cornea
  A. Opacity–degree of density (area which is most dense is taken for reading)
    Scattered or diffuse area, details of iris clearly visible — 1
    Easily discernible translucent areas, details of iris slightly obscured — 2
    Opalescent areas, no details of iris visible, size of pupil barely discernible — 3
  B. Area of cornea involved
    One-quarter (or less) but not zero — 1
    Greater than one-quarter, less than one-half — 2
    Greater than one-half, less than whole area — 3
    Greater than three-quarters up to whole area — 4

Iris
  A. Values
    Folds above normal, congestion, swelling, circumcorneal ingestion (any one or all of these or combination of any thereof), iris still reacting to light (sluggish reaction is possible) — 1
    No reaction to light, hemorrhage; gross destruction (any of these) — 2
    Scoring equals A × B
    Total possible maximum = 10

Conjunctivae
  Redness (refers to palpebral conjunctival only)
    Vessels definitely injected above normal — 1
    More diffuse, deeper crimson red, individual vessels not easily discernible — 2
    Diffuse beefy red — 3
  B. Chemosis
    Any swelling above normal (includes nictating membrane) — 1
    Obvious swelling with partial eversion of the lids — 2
    Swelling with lids about half closed — 3
    Swelling with lids about half closed to completely closed — 4
  C. Discharge
    Any amount different from normal (does not include small amount observed in inner canthus of normal animals) — 1
    Discharge with moistening of the lids and hair just adjacent to the lids — 2
    Discharge with moistening of the lids and considerable area around the eye — 3
    Scoring (A + B + C) × 2
    Total maximum = 20

Note. The maximum total score is the sum of all scores obtained for the cornea, iris, and conjunctivae.

TABLE E-7
Severity and Persistence of Irritation

Inconsequential or complete lack of irritation: Exposure of the eyes to a material under the specified conditions caused no significant ocular changes. No staining with fluorescein can be observed. Any changes that do occur clear within 24 hr and are no greater than those caused by normal saline under the same conditions.

Moderate irritation: Exposure of the eye of the material under the specified conditions causes minor, superficial, and transient changes of the cornea, iris, or conjunctivae as determined by external or slit-lamp examination with fluorescein staining. The appearance at the 24-hr or subsequent grading of any of the following changes is sufficient to characterize a response as moderate irritation: opacity of the cornea (other than a slight dulling of the normal luster), hyperemia of the iris, or swelling of the conjunctivae. Any changes that are seen clear within 7 days.

Substantial irritation: Exposure of the eye to the material under the specified conditions causes significant injury to the eye, such as loss of the corneal epithelium, corneal opacity, iritis (other than a slight injection), conjunctivitis, pannus, or bullae. The effects clear within 21 days.

Severe irritation or corrosion: Exposure of the eye to the material under the specified conditions results in the same types of injury as in the previous category and in significant necrosis or other injuries that adversely affect the visual process. Injuries persist for 21 days or more.

such as that described in NAS publication No. 1138, which is a variation of that reported by Green *et al.* This classification is based on the most severe response observed in a group of six nonirrigated eyes, and data from all observation periods are used for this evaluation.

Different regulatory agencies within the United States have prescribed slightly different procedures for different perceived regulatory needs. There have also been a number of additional grading schemes, but these will not be reviewed here.

Current *in Vivo* Test Protocols

Any discussion of current test protocols (or of any proposed *in vitro* alternatives) must start with a review of why tests are performed. What are the objectives of eye irritation testing, and how are these different objectives reflected not just in test design and interpretation but also in the regulations requiring testing and in the ways that test results are utilized?

There are four major groups of organizations (in terms of their products) which require eye irritation studies to be performed. These can be generally classified as the pharmaceutical, cosmetic and toiletries, consumer product, and industrial chemical groups. There are also minor categories of use (which we will not consider here) such as for military agents.

For the pharmaceutical industry, eye irritation testing is performed when the material is intended to be put into the eye as a means or route of application or for ocular therapy. There are a number of special tests applicable to pharmaceuticals or medical devices which are beyond the scope of this discussion because they are not intended to assess potential acute effects or irritation. In general, however, it is desired that an eye irritation test that is utilized by this group be both sensitive and accurate in predicting the potential to cause irritation in humans. Failing to identify human ocular irritants (lack of sensitivity) is to be avoided, but of equal concern is the occurrence of false positives.

The cosmetics and toiletries industry is similar to the pharmaceutical industry in that the materials of interest are frequently intended for repeated application in the area of the eye. In such uses, contact with the eye is common, though not intended or desirable. In this case, the objective is a test that is as sensitive (as that in the preceding paragraph), even if this results in a low incidence of false positives. Even a moderate irritant would not be desired but might be acceptable in certain cases (such as deodorants and depilatories) in which the potential for eye contact is minimal.

Consumer products which are not used for personal care (such as soaps, detergents, and drain cleaners) are approached from yet a different perspective. These products are not intended to be used in a manner that either causes them to get into the eyes or makes that occurrence likely. However, because of the very large population that uses them and the fact that their modes of use do not include active measures to prevent eye contact (such as the use of goggles and face shields), the desire is to accurately identify severe eye irritants. Agricultural chemicals generally fit in this category, though many of them are covered by specific testing requirements under the Federal Insecticide Fungicide and Rodenticide Act.

Finally, there are industrial chemicals. These are handled by a smaller population (relative to consumer products). Eye contact is never intended and, in fact, active measures are taken to prevent it. The use of eye irritation data in these cases is to fulfill labeling requirements for shipping and to provide hazard assessment information for accidental exposures and their treatment. The results of such tests do not directly affect the economic future of a material. It is desired to accurately identify moderate and severe irritants (particularly those with irreversible effects) and to know if rinsing of the eyes after exposure will make the consequences of exposure better or worse. False negatives for mild reversible irritation are acceptable.

To fulfill these objectives, a number of basic test protocols have been developed and mandated by different regulatory groups. Table E-8 gives an overview of these protocols. Historically, the philosophy underlying these test designs made maximization of the biological response equivalent with having the most sensitive test.

One widely used study design, which begins with a screening procedure as an attempt to avoid testing severe irritants or corrosives in animals, is described in the following section.

Rabbit Eye Irritancy Testing: Widely Used Study Design

Test Article Screening Procedure

1. Each test substance will be screened in order to eliminate potentially corrosive or severely irritating materials from being studied for eye irritation in the rabbit.

2. If possible, the pH of the test substance will be measured.

3. A primary dermal irritation test will be performed prior to the study.

4. The test substance will not be studied for eye irritation if it is a strong acid (pH of 2.0 or less) or strong alkali (pH of 11.0 or greater) and/or if the test substance is a severe dermal irritant [with a primary dermal irritation index (PDII)] of 5–8) or causes corrosion of the skin.

5. If it is predicted that the test substance does not have the potential to be severely irritating or corrosive to the eye, continue to, Rabbit Screening Procedure.

Rabbit Screening Procedure

1. A group of at least 12 New Zealand White rabbits of either sex are screened for the study. The animals are removed from their cages and placed in rabbit restraints. Care should be taken to prevent mechanical damage to the eye during this procedure.

2. All rabbits selected for the study must be in good health; any rabbit exhibiting sniffles, hair loss, loose stools, or apparent weight loss is rejected and replaced.

3. One hour prior to instillation of the test substance, both eyes of each rabbit are examined for signs of irritation and corneal defects with a hand-held slit lamp. All eyes are stained with 2.0% sodium fluorescein and examined to confirm the absence of corneal lesions.

Fluorescein staining: Cup the lower lid of the eye to be tested and instill one drop of a 2% (in water) sodium fluorescein solution onto the surface of the cornea. After 15 sec, thoroughly rinse the eye with physiological saline. Examine the eye, employing a hand-held long-wave ultraviolet illuminator in a darkened room. Corneal lesions, if present, appear as bright yellowish-green florescent areas.

4. Only 9 of the 12 animals are selected for the study. The 9 rabbits must not show any signs of eye irritation and must show either a negative or minimum fluorescein reaction (due to normal epithelial desquamation).

Study Procedure

1. At least 1 hr after fluorescein staining, the test substance is placed in one eye of each animal by gently pulling the lower lid away from the eyeball to form a cup (conjunctival cul-de-sac) into which the test material is dropped. The upper and lower lids are then gently held together for 1 sec to prevent immediate loss of material.

2. The other eye remains untreated and serves as a control.

TABLE E-8
Regulatory Guidelines for Irritation Test Methods

Agency	Draize	FHSA	NAS	OECD	IRLG	CPSC	TOSCA	FIFRA
Test species	Albino rabbit	Same	Same[a]	Same	Same	Same	Same	Same
Age/weight	NS[b]	NS	Sexually mature/less than 2 yrs. old	NS	Young adult/2.0	NS	NS	NS
Sex	NS	NS	Either	NS	Either	NS	NS	NS
No. of animals/group	6	6–18	4 (minimum)	3 (minimum)	3 (preliminary test);[c] 6	6–18	6	6
Test agent volume and method of instillation liquids	0.1 ml on the eye	Same as Draize	Liquids and solids; two or more different doses within the probable range of human exposure	Same as Draize	Same as Draize	Same as Draize	Same as FHSA	Same as FHSA
Solids	NS	100 mg or 0.1 ml equivalent when this volume, Weighs less than 100 mg; direct instillation into conjunctival sac	Manner of application should reflect probably route of accidental exposure	Same as FHSA	Same as FHSA	Same as FHSA	Same as FHSA	Same as FHSA
Aerosols[e]	NS	NS	Short burst of distance approximating self-induced eye exposure	1-sec burst sprayed at 10 cm	1-sec burst sprayed at approx 4 in.	NS	Same as OECD	Same as OECD
Irrigation schedule	At 2 sec (3 animals) and at 4 sec (3 animals) following instillation of test agent (3 animals remain nonirrigated)	Eyes may be washed after 24-hr reading	May be conducted with separate experimental groups	Same as FHSA; in addition for substances found to be irritating; wash at 4 sec (3 animals) and at 30 sec (3 animals)	Same as FHSA	Same as FHSA	Same as FHSA	Same as FHSA
Irrigation treatment	20-ml tap water (body temp.)	Sodium chloride solution (USP or equivalent)	NS	Wash with water for 5 min using volume and velocity of flow which will not cause injury	Tap water or sodium chloride solution (USP or equivalent)	Same as FHSA	NS	NS
Examination times (postinstillation)	24 hr 48 hr 72 hr 4 days 7 days	24 hr 48 hr 72 hr	1 day 3 days 7 days 14 days 21 days	1 hr 24 hr 48 hr 72 hr	24 hr 48 hr 72 hr	24 hr 48 hr 72 hr	Same as OECD	Same as OECD
Use of fluorescein	NS	May be applied after the 24-hr reading (optional)	May be used	Same as FHSA	Same as FHSA	Same as FHSA	Same as FHSA	Same as FHSA
Use of anesthetics	NS	NS	NS	May be used	May be used	NS	May be used	May be used
Scoring and evaluation	Draize et al.	Modified Draize et al. (1944) or a slit-lamp scoring system	CPSC (1976)	CPSC (1976)	CPSC (1976)	CPSC (1976)	CPSC (1976)	CPSC (1976)

Note. Abbreviations used: FHSA, Federal Hazard Substance Act; NAS, National Academy of Sciences; OECD, Organization for Economic Cooperation and Development; IRLG, Interagency Regulatory Liaison Group; CFR, Code of Federal Regulations.

[a] Tests should be conducted on monkeys when confirmatory data are required.

[b] NS, not specified.

[c] If the substance produces corrosion, severe irritation or no irritation in a preliminary test with three animals, no further testing is necessary. If equivocal responses occur, testing on at least three additional animals should be performed.

[d] Suggested doses are 0.1 and 0.05 ml for liquids.

[e] Currently, no testing guidelines exist for gases or vapors.

[f] Eyes may also be examined at 1 hr and 7, 14, and 21 days (at the option of the investigator).

[g] Office Pesticide Assessment.

3. For testing liquids, 0.01 ml of the test substance is used.

4. For solid or pastes, 100 mg of the test substance is used.

5. When the test substance is in flake, granular, powder, or other particulate form, the amount that has a volume of 0.01 ml (after gently compacting the particles by tapping the measuring container in a way that will not alter their individual form) is used whenever this volume weighs less than 10 mg.

6. For aerosol products, the eye should be held open and the substance administered in a single, 1-sec burst at a distance of about 4 in. directly in front of the eye. The velocity of the ejected material should not traumatize the eye. The dose should be approximated by weighing the aerosol can before and after each treatment. For other liquids propelled under pressure, such as substances delivered by pump sprays, an aliquot of 0.01 ml should be collected and instilled in the eye as for liquids.

7. The treated eyes of six of the rabbits are not washed following installation of the test substance.

8. The treated eyes of the remaining three rabbits are irrigated for 1 min with room-temperature tap water, starting 20 sec after instillation.

9. To prevent self-inflicted trauma by the animals immediately after instillation of the test substance, the animals are not immediately returned to their cages. After the test and control eyes are examined and graded at 1 hr postexposure, the animals are returned carefully to their respective cages.

Observations

1. The eyes are observed for any immediate signs of discomfort after instilling the test substance. Blepharospasm and/or excessive tearing are indicative of irritating sensations caused by the test substance, and their duration should be noted.

Blepharospasm does not necessarily indicate that the eye will show signs of ocular irritation.

2. Grading and scoring of ocular irritation are performed in accordance with Table E-6. The eyes are examined, and grades of ocular reactions are recorded.

3. If signs of irritation persist at Day 7, readings are continued on Days 10 and 14 after exposure or until all signs of reversible toxicity are resolved.

4. In addition to the required observation of the cornea, iris, and conjunctiva, serious effects (such as pannus, rupture of the globe, or blistering of the conjunctivae) indicative of a corrosive action are reported.

5. Whether or not toxic effects are reversible depends on the nature, extent, and intensity of damage. Most lesions, if reversible, will heal or clear within 21 days. Therefore, if ocular irritation is present at the 14-day reading, a 21-day reading is required to determine whether the ocular damage is reversible or nonreversible.

Limitations

Commonly used methodological variations to improve the sensitivity and accuracy of describing damage in these tests are inspection of the eyes with a slit lamp and instillation of the eyes with a vital dye (very commonly, fluorescein) as an indicator of increases in permeability of the corneal barrier.

To assess the adequacy of the currently employed eye irritation tests to fulfill the objectives behind their use, we must evaluate them in terms of (a) their accuracy (how well do they predict the hazard to humans) and (b) whether comparable results can be obtained by different technicians and laboratories, and (c) what methods and designs have been developed and are being employed as alternatives to rabbit eye irritation tests.

Assessing the accuracy of rabbit eye irritation tests—or indeed, of any predictive test of eye irritation—requires that the results of such tests be compared to what happens in humans. Unfortunately, the human database for making comparisons is not large. The concerns, however, have been present almost as long as the tests have been performed.

Rabbit Eye Irritancy Testing: Alternative Methods

The alternatives that have been proposed and adapted for the performance of rabbit eye irritation tests themselves should be reviewed. These alternatives have been directed at the twin objectives of making the tests more accurate in predicting human responses and at reducing both the use of animals and the degree of discomfort or suffering experienced by those that are used.

Alternative Species

Dogs, monkeys, and mice have all been suggested as alternatives to rabbits that would be more representative of humans. Each of these, however, also has shown differences in responses compared to those seen in humans and pose additional problems in terms of cost, handling, lack of database, and so on.

Use of Anesthetics

Over the years, a number of authors have proposed that topical anesthetics be administered to the eyes of rabbits prior to their use in the test. Both OECD and IRLG regulations provide for such usage. However, numerous published and unpublished studies have shown that such use of anesthetics interfere with test results usually by increasing the severity of eye irritation findings.

Decreased Volume of Test Material

An alternative proposal (one which a survey showed has been adopted by a number of labs) is to use a reduced volume/weight of test material.

In 1984, Freeberg *et al.* reported a study in which they evaluated 21 different chemicals at volumes of 0.1, 0.03, 0.01, and 0.003 ml. These are materials on which the human data were already available. It was found that the volume reduction did not change the rank order of responses, and that 0.01 ml (10 μl) gave results which best mirrored those seen in humans.

In 1985, Walker reported an evaluation of the low-volume (0.01 ml) test which assessed its results for the correlation with those in humans based on the number of days until clearing of injury and reported that 0.01 ml gave a better correlation than did 0.1 ml.

While it must be pointed out that there may be some classes of chemicals for which low-volume tests may give results less representative of those seen in humans, it seems clear that this approach should be seriously considered by those performing such tests.

Use of Prescreens

This alternative may also be considered a tier approach. Its objective is to avoid testing severely irritating or corrosive materials in many (or in some cases, any) rabbits. This approach entails a number of steps which should be considered independently.

First is a screen based on physicochemical properties. This usually means pH, but also should be extended to materials with high oxidation or reduction potentials (hexavalent chromium salts, for example).

Though the correlation between low pHs (acids) and eye damage in the rabbit has not been found to be excellent, all alkalis (pH 11.5 or above) tested have been reported to produce opacities and ocular damage. Many laboratories now use pH cutoffs for testing of 2.0 or lower and 11.5 or 12.0 and higher. If a material falls outside of these cutoffs (or is so identified due to other physicochemical parameters), then it is (a) not tested in the rabbit eye and assumed to be corrosive, or (b) evaluated in a secondary screen such as an *in vitro* cytotoxicity test or primary dermal irritation test, or (c) evaluated in a single rabbit before a full-scale eye irritation test is performed. It should be kept in mind that the correlation of all the physicochemical screen parameters with acute eye test results is very concentration dependent, being good at high concentrations and marginal at lower concentrations (where various buffering systems present in the eye are meaningful).

The second commonly used type of level of prescreen is the use of PDI test results. In this approach the PDI study is performed before the eye irritation study, and if the score from that study (PDII ranging from 0 to 8) is above a certain level (usually 5.0 or greater), the same options already outlined for physicochemical parameters can be exercised.

In Vitro Tests

The area of ocular irritancy has been the most active grounds in toxicology for the development of true alternative (*in vitro*) tests for the past 6 years. A complete review of this effort is beyond the scope of this entry and continues to be in a high state of flux. However,

TABLE E-9
In Vitro Alternatives for Eye Irritation Tests

Morphology
    Enucleated superfused rabbit eye system
    BALB/c 3T3 cells/morphological assays (HTD)
Cell toxicity
    Adhesion/cell proliferation
        BHK cells/growth inhibition
        BHK cells/colony-formation efficiency
        BHK cells/cell detachment
        SIRC cells/colony forming assay
        BALB/c 3T3 cells/total protein
        BCL/D1 cells/total protein
        Primary rabbit corneal cells/colony-forming assay
    Membrane integrity
        LS cells/dual dye staining
        Thymocytes/dual fluorescent dye staining
        LS cells/dual dye staining
        RCE-SIRC-P815-YAC-1/Cr release
        L929 cells/cell viability
        Bovine red blood cell/hemolysis
        Mouse L929 fibroblasts-erythrocin C staining
        Rabbit corneal cell cultures/plasminogen activator
        Agarose diffusion
    Cell metabolism
        Rabbit corneal cell cultures/plasminogen activator
        LS cells/ATP assay
        BALB/c 3T3 cells/neutral red uptake
        BALB/c 3T3 cells/uridine uptake inhibition assay
        HeLa cells/metabolic inhibition test (MIT-24)
        MDCK cells/dye diffusion
Cell and tissue physiology
    Epidermal slice/electrical conductivity
    Rabbit ileum/contraction inhibition
    Bovine cornea/corneal opacity
    Proptoses mouse eye/permeability test
Inflammation/immunity
    Chorioallantoic membrane (CAM)
        CAM
        HET-CAM
        Bovine corneal cup model/leukocyte chemotactic factors
        Rat peritoneal cells/histamine release
        Rat peritoneal mast cells/serotonin release
        Rat vaginal explant/prostaglandin release
        Bovine eye cup/histamine (Hm) and leukotriene C4 (LTC4)
          release
Recovery/repair
    Chorioallantoic membrane
Other
    EYTEX assay
    Computer-based structure–activity (SAR)
    Tetrahymena/motility

at least a brief outline or summary of approaches being pursued is called for.

As summarized in Table E-9, there are six major categories of approaches to alternative eye irritation tests. The first five of these aim at assessing portions of the irritation response (alterations in tissue morphology, toxicity to individual component cells, alterations in cell or tissue physiology, inflammation or immune modulation, and alterations in repair and/or recovery processes). These methods have the limitation that they assume that one of these component parts can or will predict effects in the complete organ system. A more likely case is that, while each may serve well to predict the effects of a set of chemical structures which have that component as a determining part of the ocular irritation response, a valid assessment across a broad range of structures will require the use of a collection or battery of such tests.

The sixth category contains tests which have little or no empirical basis, such as computer-assisted structure–activity relationship models. These approaches can only be assessed in terms of how well (or poorly) they perform.

—Shayne C. Gad

Related Topics

Analytical Toxicology
Consumer Product Safety Commission
Federal Insecticide, Fungicide, and
    Rodenticide Act
Food and Drug Administration
Good Laboratory Practices
In Vitro Test
In Vivo Test
Ocular and Dermal Studies
Sensory Organs
Toxicity Testing, Alternatives

ISBN 0-12-227221-8

90038